HANDBOOK OF PLASTIC PROCESSES

HANDBOOK OF PLASTIC PROCESSES

CHARLES A. HARPER
Timonium, Maryland

WILEY-INTERSCIENCE
A JOHN WILEY & SONS, INC., PUBLICATION

Published by John Wiley & Sons, Inc., Hoboken, New Jersey
Published simultaneously in Canada

For general information on our other products and services or for technical support, please contact our Customer Care Department within the United States at 877-762-2974, outside the United States at 317-572-3993 or fax 317-572-4002.

Wiley also publishes its books in a variety of electronic formats. Some content that appears in print may not be available in electronic formats. For more information about Wiley products, visit our web site at www.wiley.com.

Library of Congress Cataloging-in-Publication Data:

Handbook of plastic processes / [edited by] Charles A. Harper.
p. cm.
Includes index.
ISBN-13: 978-0-471-66255-6 (cloth)
ISBN-10: 0-471-66255-0 (cloth)
1. Plastics—Handbooks, manuals, etc. 2. Plastics—Molding—Handbooks, manuals, etc.
I. Harper, Charles A.

TP1130.H355 2006
668.4—dc22

2005025148

Printed in the United States of America
10 9 8 7 6 5 4

CONTENTS

CONTRIBUTORS

Steven J. Adamson, Asymtek, 2762 Loker Avenue West, Carlsbad, CA 92008
Institute of Electrical and Electronics Engineers, IEEE CPMT Chapter, International Microelectronics and Packaging Society

Nandika A. D'Souza, Department of Materials Science and Engineering, University of North Texas, Denton, TX 76203
Society of Plastics Engineers, Polymer Analysis Division

Peter F. Grelle, Dow Automotive, 6679 Maple Lakes Drive, West Bloomfield, MI 48322
Society of Plastics Engineers, Injection Molding Division

Dale A. Grove, Owens Corning Corporation, Granville, OH 43023
Society of Plastics Engineers, Composites Division

Steven Ham, Technical Consultant, 537 Hickory Street, Highlands, NC 28741
Society of Plastics Engineers, Product Designs and Development Division

Dana R. Hanson, Processing Technologies, Inc., 2655 White Oak Circle, Aurora, IL 60504
Society of Plastics Engineers, Senior Member

Alejandro Hernandez-Luna, World Wide Make Packaging, Texas Instruments, Inc., 13020 TI Boulward, MS 3621, Dallas, TX 75243
Packaging Engineer

John L. Hull, Hull Industries, Inc., 7 Britain Dri ve, New Britain, PA 18901
Society of Plastics Engineers, Platinum Level Member

Norman C. Lee, Consultant, 2705 New Garden Road East, Greensboro, NC 27455-2815
Society of Plastics Engineers, Blow Molding Division

Scott Macdonald, Maryland Thermoform, 2717 Wilmarco Avenue, Baltimore, MD 21223
Society of Plastics Engineers, Advisor

Paul Nugent, Consultant, 16 Golfview Lane, Reading, PA 19606
Society of Plastics Engineers, Rotational Molding Division

Edward M. Petrie, EMP Solutions, 407 Whisperwood Drive, Cary, NC 27511
Society of Plastics Engineers, Electrical and Electronic Division

Ajit Ranade, GE Advanced Materials, 1 Lexan Lane, Bldg. 4, Mt. Vernon, IN 47620
Society of Plastics Engineers, Sheet and Coating Technologist

Laxmi K. Sahu, Department of Materials Science and Engineering, University of North Texas, Denton, TX 76207
Society of Plastics Engineers

Will Strauss, Raytheon Company, 2501 West University Drive, MS 8019, McKinney, TX 75071
Society of Plastics Engineers

PREFACE

With the myriad of plastics, plastic compounds, and plastic types and forms, the list of end product applications is as limitless as the list of possible plastic parts is endless. We see plastic parts and assemblies in a never-ending stream of domestic and commercial or industrial applications, across every category of interior and exterior domestic application, and across every industry, from mechanical to electrical to heavy chemical to structures to art. Yet without proper processing, none of these plastic products would be possible. It suffices to say that with the breadth of plastic materials and products indicated above, processing is a major challenge. Fortunately, the strength, intelligence, and ingenuity of the army of specialists involved in all types of plastic processing has been equal to the task. To them we owe our gratitude, and to them we dedicate this book. The authors of the chapters in this book rank high among this group; and fortunately, they have achieved much through their cooperative efforts in the leading professional society in this field, *the Society of Plastics Engineers* (SPE), about which more will be said shortly. I am personally grateful to SPE for the great assistance of many of its staff and professional leaders, without whose advice and assistance I would not have been able to put together such an outstanding team of authors.

As can be seen from perusal of the subjects covered in this book, the book has been organized to fully cover each of the plastic processes that are used to convert plastic raw materials into finished product forms. The myriad of thermoplastic processes are each covered in an individual chapter, as are the thermosetting processes. The authors of each chapter detail its subject process and process variations and the equipment used in the process, discuss the plastic materials which can be utilized in that process, and review the advantages and limitations of that process. Also, since raw, molded, or fabricated parts often do not yet provide the desired end product, chapters are included on plastics joining, assembly, finishing, and decorating. Finally, and importantly, with the increasing impact of nanotechnology on plastics properties and processing, a chapter on nanotechnology is included.

As was mentioned above, success in achieving a book of this caliber can only result from having such an outstanding group of chapter authors as it has been my good fortune to obtain. Their willingness to impart their knowledge to the industry is indeed most commendable. Added to this is the fact that most of them are banded together for the advancement of the industry through their roles in the Society of Plastics Engineers. SPE has unselfishly advised me on the selection of many of the authors of this book. In addition to all of the chapter authors who are strong SPE representatives, I would like to offer special thanks to Roger M. Ferris, editor of the SPE *Plastics Engineering Journal;* Donna S. Davis, 2003–2004 SPE President; and Glenn L. Beall and John L. Hull, Distinguished Members of SPE.

Charles A. Harper

Technology Seminars, Inc.
Lutherville, Maryland

CHAPTER 1

Injection Molding

PETER F. GRELLE

Dow Automotive, Auburn Hills, Michigan

1.1 INTRODUCTION

Injection molding is one of the most widely used processes for manufacturing plastics parts. It is a major processing technique for converting thermoplastics and thermoset materials into all types of products for different end uses: from automotive to electronics, medical to sports and recreation, and building and construction to consumer products. Injection molding is a relatively new method of producing parts. The first injection molding machines were manufactured and made available in the early 1930s, whereas other manufacturing methods that may be familiar date back more that 100 years.

According to the Injection Molding Division of the Society of Plastics Engineers, *injection molding* is defined as a method of producing parts with a heat-meltable plastics material [1]. This is done by the use of an injection molding machine. The shape that is produced is controlled by a confined chamber called a *mold*. The injection molding machine has two basic parts, the injection unit, the clamping unit. The *injection unit* melts the plastic and conveys or moves the material to the confined chamber or mold. The purpose of the *clamping unit* is to hold the mold in a closed position during injection to resist the pressures of the conveying or injection and forming of the material into a specific shape, and then opens after cooling to eject the part from the mold.

Rosato [2] describes the three basic operations that exist in injection molding. The first is raising the temperature of the plastic to a point where it will flow under pressure. This is done both by heating and by grinding down the granular solid until it forms a melt at an elevated temperature and uniform viscosity, a measurement of the resistance to flow. In most injection molding machines available today, this is done in the barrel of the machine, which is equipped with a reciprocating screw. The

Handbook of Plastic Processes Edited by Charles A. Harper

screw provides the vigorous working of the material along with the heating of the material. This part of the process is referred to as the *plasticating* of the material.

The second operation is to allow the molten plastic material to cool and solidify in the mold, which the machine keeps closed. The liquid, molten plastic from the injection molding machine barrel is transferred through various flow channels into the cavities of a mold, where it is formed into the desired object. What makes this apparently simple operation so complex is the limitations of the hydraulic circuitry used in the actuation of the injection plunger and the complex flow paths involved in filling the mold and the cooling action in the mold.

The third and last operation is the opening of the mold to eject the plastic after keeping the material confined under pressure as the heat, which is added to the material to liquefy it, is removed to solidify the plastic and freeze it permanently into the shaped desired for thermoplastics.

A variety of materials can be injection molded. Table 1.1 lists the thermoplastic materials that can be processed using injection molding.

The purpose of this chapter is to break down the basic parts of the injection molding process as if you were actually taking a walking tour down the entire process. This tour is divided into four phases. The first phase is the *material feed phase* (Section 1.2). Here the focus is on material handling: how the material is dried and the preparation of the material to be injection molded. The second phase is the *melt-conveying phase* (Section 1.3). Our discussion is concentrated on the important aspects of how material goes from a solid pellet to a molten polymer. The emphasis here is on the screw, the barrel, and the nozzle. *The melt-directing phase* (Section 1.4) entails how the melt gets to its final destination, the mold cavity. In this section the sprue, runners, gates, and gate lands are reviewed as to what they do and how they

TABLE 1.1 Injection-Moldable Thermoplastic Materials

Acrylonitrile–Butadiene–Styrene (ABS)	Linear low-density polyethylene (LLDPE)
ABS/nylon blends	Polypropylene (PP)
ABS/TPU	Polyphenyl oxide (PPO)
Polyoxymethylene (POM) acetal	Polystyrene
Polymethyl methacrylate (PMMA) acrylic	Syndiotactic polystryene (SPS)
Ethylene vinyl acetate (EVA)	Polysulfone
Nylon 6	Polyether sulfone (PES)
Nylon 6,6	Thermoplastic polyurethane (TPU)
Nylon 12	Polybutylene terephthalate (PBT)
Nylon 6,12	Polyethylene terephthalate (PET)
Polyetherimide (PEI)	Liquid-crystal polymer (LCP)
Polycarbonate	Polyvinyl chloride (PVC)
Polycarbonate–ABS blends	Styrene–maleic anhydride (SMA)
Polycarbonate–PET blends	Styrene–acrylonitrile (SAN)
Polycarbonate–PBT blends	Thermoplastic elastomer (TPE)
High-density polyethylene (HDPE)	Thermoplastic polyolefin (TPO)
Low-density polyethylene (LDPE)	

affect the molding process. The last stop is the *melt-forming phase* (section 1.5). Here we discuss how to design a tool or part for the injection molding process. Section 1.6 provides an overview on how to resolve injection molding issues and gives examples of troubleshooting commonly used plastic materials.

1.2 MATERIAL FEED PHASE

When a plastic material begins its journey through the injection molding process, the first thing that is considered is how the material is delivered and stored until it is used. The next step is to determine how the material will flow to the individual machines for molding, and finally, what process is needed to prepare the material so that it can be molded. Other side processes, such as color and additive feeding, also need to be considered if these apply. However, in this section, concentration is placed on the basic factors in getting the material to the hopper.

In this section we focus on the following issues for material feed. The first is that of drying the material, a process used in preparing most thermoplastic materials for injection molding. We then explain why materials need to be dried and what needs to be considered. Then the hopper and the concept of bulk density are reviewed, how this relates to sizing storage space for materials, the elements of material mass flow, and the time and conditions involved in drying the material.

1.2.1 Drying Material

One question that is asked by many molders in the injection molding industry has been: Why do some polymer materials need to be dried? This is best explained as follows.

The chemical structure of a particular polymer determines whether it will absorb moisture. Due to their nonpolar chemical structures, a number of polymers (e.g., polystyrene, polyethylene, and polypropylene) are nonhygroscopic and do not absorb moisture. However, due to their more complex chemistry, materials such as polycarbonate, polycarbonate blends, acrylonitrile–butadiene–styrene (ABS) terpolymers, polyesters, thermoplastic polyurethanes, and nylon are hygroscopic and absorb moisture. As shown in Figure 1.1, the moisture can either be external (surface of the pellet) or internal (inside the pellet). A problem arises when the polymer processing temperatures, which can exceed 400°F (204°C), boil off the water [at 212°F (100°C)] in the polymer.

The effect that water has on a molded part is that imperfections will appear on the surface because the bubbles generated from the boiling of the moisture get trapped in the polymer, cool, and solidify in the mold. This creates *splay marks* or *silver streaks.* In some cases, as in polycarbonate and nylon-based materials, polymer degradation can occur as the water reacts with the polymer to reduce its physical and mechanical properties. Another effect results in reversing the polymer-forming reaction in the polymer, leading to *chain scission* or *depolymerization.* These types of conditions can make a polymer difficult, if not impossible, to process.

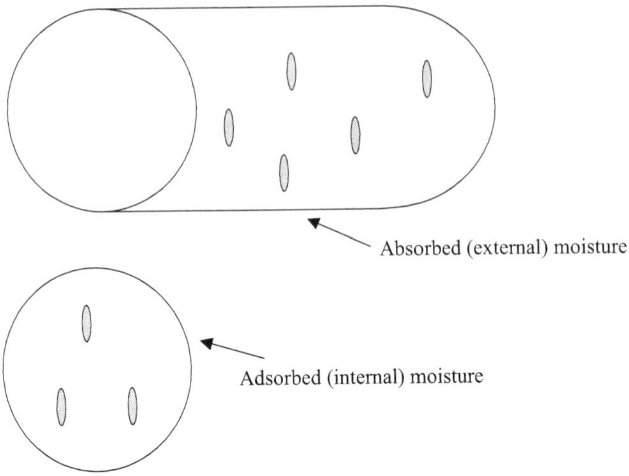

FIGURE 1.1 External and internal moisture.

The critical factor in drying plastics materials is to remove moisture not only from the pellet's exterior surface but from the pellet's interior as well. Pellets reach a moisture balance point with the surrounding environment. This is determined by the resin type, the ambient relative humidity, and time. For some resins (e.g., ABS) this is usually 0.3%, whereas for nylons this is typically 0.15%. Moisture can be driven out of the pellets under four essential conditions: (1) heat, (2) airflow, (3) dry air, and (4) time for drying effects to take place. Heat drives the moisture to the surface of the pellet. The dry air acts as a recipient or "sponge" to receive the moisture from the pellet surface. The dry airflow supplies the transportation to remove the moist air, which goes to the desiccant dryer for collection and reconditioning. All of these steps are important in drying plastic materials properly.

The delivery of air to the hopper must be such that it can absorb water from the moist pellets. The drier the air, the more effective it will be in extracting moisture from the resin. The term *dew point* is used to describe the actual amount of water in the air. The dew point temperature is defined as the temperature at which moisture will just begin to condense at a given temperature and pressure. It is a measure of the actual water in the air: the higher the dew point, the more saturated the air.

The delivery air to the hopper must be dry. Only a dew point meter can determine this. Some drying units have an onboard dew point meter, which quickly becomes unreliable due to vibration, oxidation on sensor plates, and contamination from plant air (oils, dust, etc.). After some time, an onboard unit may read −40°F (−40°C) continuously even though the actual dew point is much higher. Handheld dew point meters are suggested as an alternative because they are not exposed to continuous use and are typically stored in a dry, clean environment. Sensor plates, which are critical to the function of a dew point meter, remain clean, allowing for accurate and reliable results.

When using a handheld unit, some precautions must be taken because the unit draws a sample from the delivery air (which should be hot and dry). The air filter must be in place to avoid plugging or contaminating the sensor plates. The handheld unit draws in a sample at a very slow rate. Operation of the dryer needs to be considered because desiccant beds do swing or index at predetermined times; one bed may be acceptable while the other is faulty. Enough time should be given to measure dew point temperature to monitor all beds inside the system, which normally consists of two or three beds.

The typical life expectancy for replacement of desiccant beds is two to three years. Also, the desiccant beds must be inspected for contamination by fines, dust, and the chemical by-products of dried resins, such as lubricants and plasticizers. Desiccant beds must be properly sealed, and clean filters must always be in place to avoid the loss of drying capacity.

An insufficient dew point does not always point to bad desiccant beds. The rate of moisture pickup from the air intake may simply overwhelm the capacity of the dryer unit. This can occur for several reasons, such as inaccurate sizing of the dryer or an air leak in the return system. For air leaks it is strongly recommended that hoppers operate with the secured hopper lids and that hoses be checked for pinhole leaks because these problems can draw moist plant air into the dryer and create inefficient drying.

Hygroscopic materials can absorb more moisture from the air than can other plastic resins. This puts some demands on the molder to keep the material dry before and during molding. High-dew-point temperatures above 15°F (−9°C) are not adequate to dry most hygroscopic materials properly because the air is already saturated with moisture before contacting the resin to be dried. It is recommended that dew point temperatures of −20° to −40°F be used to dry hygroscopic materials such as nylons, polyesters, polycarbonate, and polycarbonate blends.

Table 1.2 lists recommended drying temperatures for a number of thermoplastic materials. Table 1.3 is a checklist for determining the efficiency of the dryer system and areas in the drying equipment that should be monitored.

1.2.2 The Hopper

The hopper is the section of the injection molding machine that stores material just before it enters the barrel of an injection molding machine. The hopper also has a holding area for the material as it is fed from its bulk storage (gaylords, railcars, etc.) and awaits any preconditioning of the material that may be needed, such as drying. Hopper size is a critical element in determining how to make the injection molding process efficient. The two concepts discussed here, material mass flow and bulk density, provide information on how to choose the correct-size hopper and what requirements are needed to store material prior its being sent to the hopper.

1.2.2.1 Bulk Density

Bulk density is an important material property as it relates to the injection molding process. According to Rosato [2], *bulk density* is defined as the weight per unit volume of a bulk material, including the air voids. *Material density* is defined as the weight of the unit volume of the plastic, excluding air voids.

TABLE 1.2 Typical Drying Conditions for Thermoplastic Materials

Material	Drying Conditions		
	Time (hr)	Temperature [°F (°C)]	
ABS	2–4	180–200	(82–93)
ABS/nylon	1–3	175–190	(79–88)
ABS/TPU	3–4	170	(77)
Acetal	1–4	185	(85)
Acrylic	2–3	180	(185)
Nylon 6	2–4	180–185	(82–85)
Nylon 6,6	2–4	175–185	(79–85)
PEI	4–6	270–300	(132–149)
Polycarbonate	4	250	(121)
PC–ABS	3–4	175–200	(79–93)
PC–PBT	3–4	240	(116)
PC–PET	3–4	240	(116)
Polyethylene	1–2[a]	120–140	(49–60)
PPS	2–3	300–350	(149–177)
Polypropylene	1–2	120–140	(49–60)
PPO	2–4	200–250	(93–121)
Polystyrene	1–2[a]	150–175	(66–79)
Polysulfone	4	250–275	(121)
PBT	2–4	250–280	(121–138)
PET	2–4	275	(135)
Liquid crystal Polymer	2–4	140–150	(60–66)
PVC	2[a]	170–180	(77–82)
SMA	2	180–200	(82–93)
TPE	1–2	212	(100)
TPO	1–2[a]	120–140	(49–60)

[a]Drying typically not needed.

A rough estimate of bulk density, measured in pounds per cubic foot, can be made using the following equation:

$$\mathrm{BD} = (42)\,\frac{\rho}{1.13} \tag{1.1}$$

where BD is the bulk density (lb/ft^3) and ρ is the specific gravity (g/cm^3). Table 1.4 lists the bulk densities for a number of thermoplastics based on Eq. (1.1).

Rosato [2] provides some guidelines in the interpretation of bulk density data. If the bulk density is greater than 50% of the actual density of the material, the bulk material will be reasonably easy to convey through the injection molding screw. However, if the bulk density of the material is less than 50% of the actual density , solids-conveying

TABLE 1.3 Dryer Operation Checklist

Issue	Area to Check
Drying temperature	Check operating temperature of dryer using a temperature probe at the hopper inlet.
	Check length of delivery hose. Set hose length so that there is minimal or no change in inlet temperature from set temperature.
Air drying	Use a handheld dew point meter to assure that the dew point is between −20° and −40 °F (−29° to −40 °C) range. Do not depend on dew point monitors that come with drying units.
	Check for plugged air filters that will prohibit air from entering the system.
	Inspect operation of desiccant beds to assure that they regenerate properly.
	Visually inspect desiccant beds for any contamination, such as fines, dust particles, and certain chemical additives that are by-products of some materials.
	Check for proper material mass flow.
	Inspect hose for pinhole leaks that can cause moist air to enter the system.
	Cover all hoppers with lids and make sure that the hopper system is sealed from plant air.
	If needed, apply a nitrogen blanket to keep hygroscopic materials dry in the hopper and seal the hopper.
Air delivery	Check airflow of the drying unit.
	Inspect for dirty or blocked filters due to fines and pellets.
	Inspect delivery lines for twists or kinks.
	Check material mass flow.
Mechanical/ electrical problems	Check for faulty timers for swinging desiccant beds.
	Inspect for possible disconnections of internal hoses.
	Check for faulty limit switches at the top of the hopper.
	Assure that material mass flow still matches part and production requirements.
	Insulate hoppers and hoses to improve drying efficiency.

problems can occur. When bulk density is less than 30%, a conventional plasticator usually will not handle the bulk material. Separate devices, such as crammers and force feeders, would be needed to feed the material.

1.2.2.2 Hopper Sizing for Drying and Material Mass Flow

Proper sizing of the hopper is critical and depends on the mass flow of the material. Inside the hopper, plastic material pellets move downward due to gravity, while drying air moves upward, assuming plug flow conditions. Mass flow is determined by

TABLE 1.4 Bulk Density Data for Thermoplastics

Material	Bulk Density (lb/ft^3)
ABS	42
Acrylic	42
Acetal	40
Ionomer	44
Nylon 6	41
Nylon 6,6	41
25% glass-filled nylon 6,6	49
35% glass-filled nylon 6,7	52
45% glass-filled nylon 6,8	56
PEI	52
Polycarbonate	41
PC–ABS	41
PC–PBT	42
PC–PET	42
Polyethylene	34
PPS	50
Polypropylene	34
20% talc-filled PP	40
PPO	49
Polystyrene	40
Polysulfone	50
PBT	48
PET	52
Liquid crystal polymer	50
PVC (rigid)	52
PVC (flexible)	48
SAN	40
SMA	38
TPE	48
TPO	34

three factors: (1) the shot size of the part, (2) the cycle time to manufacture the part, and (3) the number of machines supplied by the drying equipment. Figure 1.2 illustrates how to calculate material mass flow for a given material, in this case for the material ABS. The variables used are as follows:

w_p = part weight (lb)

t_c = cycle time for manufacturing the part (min)

Q_t = machine throughput (lb/hr)

M_t = mass flow (lb)

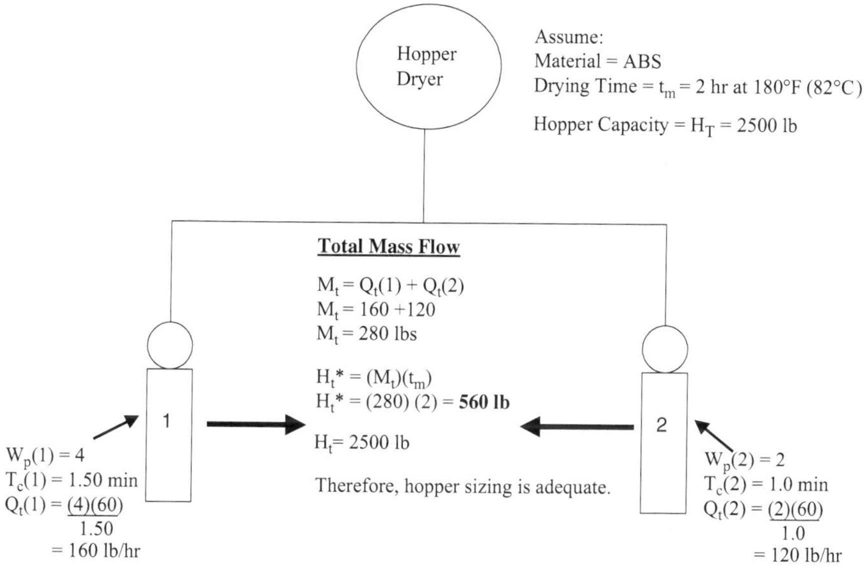

FIGURE 1.2 Determination of material mass flow.

t_m = drying time (hr)

H_T = hopper dryer capacity (lb)

H_T^* = hopper dryer capacity needed

To determine material mass flow for an individual injection molding machine, the following equation is used:

$$Q_t(x) = \frac{W_p}{T}\,(60) \tag{1.2}$$

where $x = 1, 2, 3, \ldots$. When several machines are used, the total material mass flow, M_t, is determined by adding the material mass flo w for machine 1 [Q_t (1)] together with that for machine 2 [Q_t (2)] as follows:

$$M_t = Q_t\,(1) + Q_t(2) + Q_t(3) + \cdots \tag{1.3}$$

The hopper dryer capacity needed, H_T^*, is determined with the following equation:

$$H_T^* = M_T t_m \tag{1.4}$$

The following condition determines if the hopper sizing is suf ficient. If $H_T > H_T^*$, the hopper sizing is adequate to pro vide enough drying in the system for the material.

1.3 MELT-CONVEYING PHASE

When the word *conveying* comes to mind, a number of different methods are visualized. For example, a moving belt moving articles from one location to another is an example of conveying. Another example is the use of an auger, a screwlike device that moves grain, powder, or even objects like rocks through a cylinder opened on both ends. The auger acts as the conveyor where the material is transferred within the flights of the auger in channels that have the same depth throughout the length of the auger. The open-ended cylinder acts as a guide to the conveying of material by keeping the material moving linearly. The auger–cylinder example can be used to explain melt conveying in the injection molding process.

In injection molding, an open-ended cylinder, referred to as a *barrel*, acts as a guide for the pellets and moves the pellets and melt from the hopper to the mold where the part is made. The auger, referred to as a *screw*, conveys material down through the barrel from the barrel to the mold. However, what is different in the screw and barrel from the auger and cylinder example discussed earlier is that the channels of the screw do not have a constant depth. The screw at the hopper end of the barrel will be deep, and moving forward toward the mold end of the screw, the depth of the channel becomes shallow. As all this is taking place, the inside opening of the barrel stays at a constant diameter. So, in terms of conveying, material is fed at the deep channels and conveyed into shallower channels, which cause the material to compress and pack together. This compression process increases the friction of the material against the inside wall of the barrel, providing frictional heat. In addition to this, heaters are spaced on the outside diameter of the entire length of the barrel, providing additional heat. Therefore, the frictional heat of the material in the screw plus the heat applied on the outside of the barrel together provide enough heat to convert material in pellet form at the hopper end of the screw and barrel to material in a melt form midway down the length of the barrel to the end of the barrel and screw. This simplified example provides background on the melt-conveying section.

Next, we go into more detail regarding this process by examining the barrel, screw, external heating mechanisms, venting, and nozzle sections of the melt-conveying phase.

1.3.1 The Barrel

The barrel is defined here as an open-ended cylinder that controls the linear direction of the melt-conveying process, from the hopper to the mold. This also provides a frictional surface for the plastic material, to assist in the melting of the plastic from pellet form to molten form and results in moving the material in a basically linear direction from the hopper to the mold.

One of the most important properties of the barrel is the material from which the barrel is made. The typical material is steel with a bimetallic liner. This liner is made from a steel alloy, typically a 4140 alloy. Most injection molding machine barrels are made to withstand burst strengths of approximately 22,000 lb/in^2. In special applications, barrels are made to withstand between 45,000 and 50,000 lb/in^2, especially for thin-wall injection molding [<0.0625 in. (1.6 mm)].

There are several types of barrel liners that are used for various types of materials. An abrasion-resistant liner is used for most unfilled materials or materials that contain low levels of reinforcing fillers, such as glass, talc, and mineral fillers. Another type of barrel liner is a corrosion-resistant barrel, used for materials where volatiles that can evolve from certain plastic materials will not corrode or pit the surface of the barrel. Two examples of these plastic materials are polyvinyl chloride (PVC) and polyoxymethylene, or acetal. Finally, highly abrasion resistant liners are used when a plastic material has very high loadings, percentages, or combinations of reinforcing fillers, such as glass fiber, talc, mineral filler, mineral fiber, mica, or carbon fiber.

Barrel wear is one of the problems that can be encountered in the injection molding process. There are several signs of barrel wear. During the injection phase of the molding cycle, a shot size setting used for a period of time may all suddenly provide incompletely filled parts, or *short shots*. In this case, material is back-flowing inside the barrel through a worn area of the barrel and goes back down the screw in the direction of the hopper, away from the mold. To resolve this, a repair can be made to the barrel by adding a metallic sleeve in that section of the liner to "fill in" the worn section of the barrel. Another sign of a worn barrel occurs when the screw is retracting back after injecting material into the mold. The screw should retract smoothly and evenly until it retracts to its set location. However, with a worn barrel, the screw will hesitate once or a number of times, slowing down screw retraction time and eventually slowing down the overall injection cycle. In this case, the material is flowing over the check ring and as a result, does not develop enough pressure to retract the screw. In this particular situation, complete replacement of the barrel may be required.

1.3.2 Heater Bands

Several types of heater bands are used for heating a barrel. These include tubular heaters, cartridge heaters, band heaters, and natural gas heaters.

Tubular heaters are made by suspending a coiled resistance heating element made of nichrome in a metal tube or sheath. Tubular heaters are placed on the barrel by bending and forcing them into machined grooves on the barrel surface. These heaters are held in place by peening the grooves into the grooves. Economically, use of the heaters can be expensive, due to the machining of the grooves into the hardened outside diameter of the barrel. However, the tubular heater has been known to last as long as the life of the barrel. Tubular heaters are cast in aluminum shaped to the exterior diameter of the barrel. They are effective heaters and do not require much maintenance.

Another type of heater, the *cartridge heater*, is made with nichrome wire wound on forms with a magnesium oxide type of cement. Iron–nickel chromium metal has allowed for increased heating capability. These heaters, in the shape of a pencil barrel, are placed inside a hole and supply heat to the area surrounding the hole. They are used in barrel heating but are used extensively in controlling mold temperature. Cartridge heaters require low maintenance to other types of barrel heaters. The only disadvantage with cartridge heaters is that in some applications, these heaters can cause heat to concentrate in a small area.

Band heaters are the heaters most widely used in heating barrels in the injection molding process. These heaters are made of nichrome wire wound on a form and are insulated. Mica and ceramic are used as the insulating materials in band heaters. These heaters produce a high amount of heat capacity, between 30 and 40 W/in^2, in comparison to tubular heaters (20 to 40 W/in^2) and cartridge heaters (40 W/in^2). Special heat-resistant metal alloys are used instead of copper wires for band heaters since these resist oxidation as occurs with copper wiring. The key to maximum efficiency of the band heater is the contact surface of the heater. If a band heater is not in full contact with the barrel surface, air caught inside the band heater will act as an insulator and prevent the drawing off of heat from the metal of the heater. Another problem known to cause band heater failure is plastic material coming in contact with the band heater. This can get inside the heater, shorting out the nichrome heater.

A novel method of barrel heating was developed in 2003 by the University of Duisberg–Essen in Germany using natural gas as a means to heat the barrel. *Natural gas heaters* provide heating capacities similar to those of electrical heating but with reduced energy costs. Heating of the barrel takes place by using a radial burner placed around the barrel, producing heat by convection and radiation. Work is still under way to further prove the feasibility of this novel method of barrel heating.

1.3.3 Measuring Barrel Heat: The Thermocouple

A thermocouple is used to measure and control the amount of heat being applied to the barrel by the heaters. The basic concept behind the thermocouple is that electrical energy is converted from heat energy when metals that are dissimilar are bonded or welded together. The amount of energy converted is dependent on the metals selected and the temperature. Iron and constantin, an alloy of copper and nickel, are most widely used for thermocouple materials. Two types of thermocouple are used, the J and K types. The J type is most widely used in the injection molding industry.

1.3.4 The Screw

Earlier, the analogy of the auger and cylinder was used to describe how the melt is conveyed in the injection molding process. The cylinder was just reviewed; now, it is time to discuss the auger part of the process.

The screw can be considered to be the "heart and soul" of the injection molding process, and can also be considered as the most complicated and complex section to understand. The screw is what forces the pellet, then the melt material, forward out of the nozzle into the mold. The key factor is that the material must adhere to the inside wall of the barrel. Otherwise, the screw will rotate in one spot without any forward movement.

Traditionally, the screw is divided into three parts: (1) the feed section, (2) the transition section, and (3) the metering section. In the feed section, the material in pellet form moves from the hopper section of the injection molding barrel toward the nozzle and mold section. The pellets here are still in solid form, but there has been some initial softening. The channels of the screw are deep in this area to allow the pellets to convey down the barrel. Temperature settings of the barrel are the lowest in this section, to avoid premature melting of the pellets, which can cause degradation or interfere with material feed into the barrel.

In the transition section the pellet material begins to melt and mix with unmelted pellets. In this section the channel depth of the screw becomes shallow, and this degree of shallowness increasing down the transition section. This increasing shallowness causes the melt–pellet mix to compress against the inside of the barrel wall. Frictional heat builds up, and in combination with the heat generated by the barrel heater, creates more melt to be formed within the screw flight channels. The melt pool formed as you go down the transition section increases. As the pellets reach the section where compression takes place, the volume of material inside the screw flight channel decreases until the metering section is reached.

The metering section of the screw of the standard injection molding screw acts as the pumping mechanism for the melt, forcing molten material forward accurately and completing the melting process. As the material goes forward to the front of the screw, force is generated to push the screw back in the direction of the hopper to the original, set position of the shot size. As the screw rotates and pumps the molten material through the nonreturn valve, the molten material that is accumulating in front of the valve is pushing and reciprocating the screw.

1.3.4.1 Screw Types

Over the past 50 years, a number of screw types have been developed for injection molding. In this chapter the focus is placed on three screws commonly used in the industry today: (1) the conventional screw, (2) the barrier screw, and (3) the ET screw.[1] The *conventional screw* is the screw most commonly used in injection molding machines, due to wide availability and low cost. As shown in Figure 1.3*a*, a conventional screw is recognized by its deep channels in the feed section and gradually decreasing channel depth going toward the transition and metering section. This screw design works well for most thermoplastics. However, the conventional screw is limited in performance and does not provide good melt quality or mix, in particular for color mixing. Improvements in color mixing can be achieved with the addition of a mixing head or "motionless mixer" placed at the front of the barrel beyond the metering section of the screw.

More modern screw designs utilize a *barrier flight* (Figure 1.3*b*). As the melt film is wiped off the barrel surface by the main flight, the melt is deposited into a separate melt channel. A barrier flight divides the solid and melt channels such that the clearance over the barrier flight will only allow melt to enter this channel. The main

[1]Barr ET is a registered trademark of Barr, Inc.

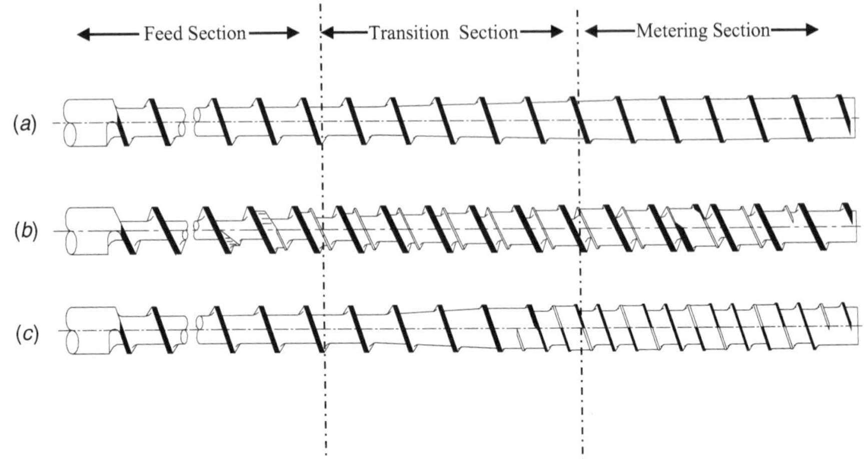

FIGURE 1.3 (*a*) Conventional, (*b*) barrier, and (*c*) ET screw designs. (Courtesy of Barr, Inc.)

function of a barrier flight is to separate the melted polymer from the solid bed and keep the solid bed from becoming unstable and breaking up prematurely. By removing the melt film continuously over the barrier flight, the solid bed surface remains intact. This allows for a greater solid bed surface area on the barrel wall to keep the viscous energy dissipation via shearing as high as possible. In addition, since the melt film thickness over the barrier flight is small, the shear energy is also high. It is believed that this type of phase separation will increase the melting rates over those of nonbarrier screws. However, since approximately 90% of the polymer is melted by the high shear in the barrier section, the melt temperatures are correspondingly higher, which is undesirable in many applications. The limitations of the barrier screw are that it is prone to high shear and higher melt temperatures than those of the conventional screw design, and is susceptible to solid pellet wedging at the start of the barrier section.

Recognizing the inherent problems and limitations of barrier screws, a solid–melt mixing screw known as the *ET screw* was developed. This principle differs from that of barrier designs in that the metering section is divided into two equal subchannels by a secondary flight. The solid bed is broken up intentionally at the end of the melting section to allow some solids to enter the mixing section. The clearance of the secondary flight is much greater than the clearance of the barrier flight on a barrier screw, allowing unmelted pellets to pass through. The depth of one subchannel decreases, while the depth of the other increases, forcing the melt to flow over the secondary flight at relatively low shear rates. Solid bed fragments mixed in the melt are broken into individual pellets by passing over the secondary flight. The pellets are mixed with the melt continually, promoting heat transfer by conduction from the melt to the pellets. Since the viscous energy dissipation via shearing in solid–melt mixing screws is low and the primary melting mechanism is by conduction, the melt temperature is reduced.

The barrier screw (Figure 1.3*b)* is characterized by the double channels found primarily in the transition section of the screw. As the pellets are conveyed in this screw design, the material is separated into the two channels: one for solid, unmelted pellets and the other for molten material. The barrier flight is undercut below the primary flight, allowing melt to flow over it. This screw design provides a higher melting rate than that of conventional screw design and gives slightly better mixing. No mixing head is needed with a barrier screw.

The ET screw (Figure 1.3*c*) has a configuration in the feed section similar to that of a conventional screw, but as the pellets enter the transition zone, the channel depth is less than that of a conventional screw. In addition, the metering section of the ET screw takes on a double-channel design. This design provides increased melting efficiency and utilizes less energy for melting the material. In addition, improved mixing and melt uniformity, as well as increased output rate and lower melt temperatures, provide the flexibility to injection mold a wide range of polymers. Its limitation is that this design is higher in cost that either the conventional or barrier screws because it is more difficult to manufacture.

Table 1.5 provides a comparison of the conventional, barrier, and ET screw designs.

TABLE 1.5 Conventional, Barrier and ET Screw Designs: Advantages and Disadvantages

Screw Type	Advantages	Disadvantages
Conventional	Cost	Performance, melt quality, poor mixing especially with colors Mixing head may be needed to improve color mixing
Barrier	Increased melting rate Slightly better mixing in comparison to the conventional screw	High shear Higher melt temperatures Prone to solid wedging Not as forgiving as a conventional screw
ET	Increased melting efficiency Increased energy utilization Increased mixing and melt uniformity Increased output rate Lower melt temperatures needed to melt material Works well for a wide range of polymers	Higher cost, due to increased difficulty in manufacturing and design of the screw

Source: Barr, Inc.

1.3.4.2 L/D Ratio

The *L*/*D* (length/diameter) ratio is an important concept for determining the sizing of an injection unit. The *L*/*D* ratio is determined by the following equation:

$$\frac{L}{D}=\frac{L_s}{D_s} \tag{1.5}$$

where L_s is the overall flight length of the screw and D_s is the nominal diameter of the screw. In the injection molding process, screws with *L*/*D* ratios of 18 : 1 and 20 : 1 are typically used. However, *L*/*D* ratios of 16 : 1, 22 : 1, 24 : 1, and 26 : 1 are also used.

1.3.4.3 Compression Ratio

Compression ratio is a term used to give an idea of how much the screw compresses and squeezes the melt–molten material mix in the screw. The intent of this concept is to divide the volume of a flight in the feed section by that of the flights in the metering section. However, the depth of the screw channel is used to calculate the compression ratio.

The equation to determine the compression ratio is

$$\mathrm{CR}=\frac{D_f}{D_m} \tag{1.6}$$

where D_f is the depth of the channels in the feed section and D_m is the depth of the channels in the metering section. Table 1.6 shows typical values of the compression ratio for various thermoplastic materials. Typical compression ration values range from 1.2 to 4.0 for most thermoplastics.

1.3.4.4 Screw and Barrel Wear

Screw and barrel wear is an area that can affect the performance and processing of thermoplastic materials. A number of factors affect how a barrel and screw wear:

- Misalignment of the screw, barrel, and drive alignment.
- Straightness of the screw and barrel. This can be especially true for a high-*L*/*D*=ratio screw (22 : 1).
- Screw design. High-compression-ratio screws wear faster.
- Barrel heating uniformity. Sections of the barrel run at low process temperatures, especially at the rear near the hopper, will show higher wear.
- Material being processed. The higher the material hardness, the greater the wear.
- Abrasive fillers, such as glass fiber, talc, mineral fillers, mineral fibers, mica, and carbon-fiber-filled materials.
- Screw surface and liner materials. Different alloy combinations will wear more.
- Insufficient support of the barrel can cause sections of the barrel to contact the screw, and as the screw rotates, wear to the barrel can occur.
- Corrosion from volatiles given off by some plastics materials.
- Excessive back pressure can cause increased wear.

TABLE 1.6 Compression Ratios for Various Thermoplastic Materials

Compression Ratio Range	Low Compression Ratio 1.2–1.9	Medium Compression Ratio 2.0–2.8	High Compression Ratio 3.0–4.5
Thermoplastic material	Polymethylmethacrylate (PMMA)	Polyoxymethylene (POM)-acetal	Polytetrafluoroethylene
	Acrylonitrile butadiene styrene (ABS) terpolymers	Cellulose acetate	Polyamides (nylon)-high melt index
	Styrene acrylonitrile (SAN) copolymers	Cellulose propionate	Polyethylene (high melt index)
	Polyvinyl chloride (PVC), rigid	Polyamides (nylon)-low melt index	Polypropylene (high melt index)
		Polyphenylene oxide (PPO)	
		Polyethylene (low–medium melt index)	
		Polypropylene (low–medium melt index) polystryene	
		Polyvinyl chloride (PVC), plasticized	
		Thermoplastic polyolefin (TPO)	
		Thermoplastic polyurethane (TPU)	
		Polyethylene terephthalate (PET)	
		Polybutylene terephthalate (PBT)	
		Polycarbonate	
		Polycarbonate/ABS blends (PC/ABS)	
		Polycarbonate/PBT blends (PC/PBT)	
		Polycarbonate/PET blends (PC/PET)	
		Acrylic styrene acrylonitrile (ASA) blends	
		Polyetherimide (PEI)	

Source: Multiple industry sources.

All of the factors mentioned can lead to wear issues for screws and barrels. Wear can be classified into three types: (1) abrasive, (2) adhesive, and (3) corrosive. *Abrasive wear* is caused by damage from fillers such as glass fibers, talc, mineral fillers, mineral fibers, mica, and carbon fibers. These materials will scrape metal off the screw and barrel over a period of time. Glass fibers are one of the worst offenders since these can abrade the root of the screw at the leading edge, usually in the transition and compression sections of the screw, where the fibers have been exposed by some melting and unmelted pellets that are squeezed against the screw and barrel. Hard surfacing materials can be applied to both the screw and barrel to reduce this wear.

Adhesive wear is gauging caused by metal-to-metal contact. In this case, certain types of metals can adhere to each other when high levels of frictional heating take place, pulling apart on further rotation of the screw. Proper clearance and alignment of the screw and barrel, compatible metallic materials, and proper hardness can reduce this type of wear.

Corrosive wear is caused by chemical attack on the barrel and screw when the plastic material is overheated and a corrosive chemical is released. The origin of this chemical can be from the material itself, such as PVC and acetal, or from additives in a particular material. The best way to prevent wear in this case is simply not to overheat the material or let material sit in the barrel for long periods of time. Corrosion-resistant coatings can be used to reduce the chance of this occurring.

1.3.5 Barrel-to-Shot Ratio

When molding a plastic part, selecting a machine with the correct barrel size is critical. Using too much of the material in the barrel can cause unmelted material to show up on a plastic part, which can affect part structural integrity. Also, using too small an amount of material in the barrel can cause material to be exposed for too long at an elevated process temperature, referred to as *residence time*, causing material to degrade and produce parts of poor structural quality. One method used in determining how much material is in the barrel related to the size of the part being molded is called the *barrel-to-shot ratio* (BSR). This ratio is calculated as

$$\text{BSR} = \frac{W_{\text{r+p}}}{\text{SC}_m}\left(\frac{\rho_m}{\rho_{\text{ps}}}\right) \tag{1.7}$$

where $W_{\text{r+p}}$ is the weight of the part plus runner system, SC_m the shot capacity of the machine (ounces), ρ_m the density of the plastic material to be molded, and ρ_{ps} the density of polystyrene. Typically, the optimum BSR values for most thermoplastic materials range from 30 to 65%.

1.3.6 Check Rings

When the screw comes forward to inject the material forward into the mold, molten material can flow back over the screw flights. To prevent this backflow from occurring, a check ring is used. A number of different types of check rings are used in injection molding. In this section, two different types are discussed; the sliding ring check valve and the ball check valve.

The *sliding ring valve* is the check valve used most commonly in the injection molding industry. When the screw moves forward during injection, the sliding ring moves back toward the rear of the screw to provide a flow path for molten material to move forward into a hollow section at the tip of the screw. After the screw moves forward and the motion of the screw has stopped, upon backward motion of the screw, the sliding ring moves forward to act as a positive shutoff, preventing molten material from back-flowing over into the screw. The flow for the material must be smooth and without interruption. The joint between the valve and screw must also be smooth, to provide a streamlined flow path for the molten plastic material and to reduce the free volume in front of the screw after injection. Check ring wear can be a problem, depending on production length and material used: in particular, materials reinforced with glass fiber, glass beads, mineral filler, and carbon fiber. Excessive wear can cause shot-to-shot variations, leading to inconsistent part weight and part size.

The *ball check valve* uses a device, referred to as a *poppet*, to block the flow of molten from back-flowing toward the screw. As the screw turns, pressure is built up behind the ball check assembly, which forces the poppet off its seat, allowing molten plastic material to flow through the valve. When the motion of the screw stops, the pressure gradient equalizes in the system and the poppet retracts and creates a positive shutoff. A spring, located at the poppet, provides the force to close the poppet. Typically, these types of valves show less wear than do sliding check ring valves.

1.3.7 The Nozzle

The nozzle, the last section of the melt-conveying phase, guides the melt of the material into the sprue bushing and into the mold. The purpose of the nozzle is to maintain the temperature of the molten material after it has been plasticated by the screw and barrel and before it enters the mold to be formed into the final part.

The nozzle typically is kept short to avoid overheating the material by increasing the residence time in plastication. These short nozzles can vary in length between 2 and 4 in. (50.8 to 101.6 mm). However, some molds are built with sprue bushings that have a deep recess into the mold, which will not allow the use of shorter nozzles. In this case, nozzle lengths as large as 8 to 12 in. (203.2 to 304.8 mm) are used. When using nozzles with deep sprue bushings, the contact seal between the nozzle and sprue bushing becomes more critical and should be inspected. Another consideration when using a long nozzle is that more heat is required to heat the nozzle. A longer band type or other heater is needed since the contact area for the heater is now longer. Increased residence time is another result of using longer nozzles.

Another part of the nozzle that is critical is the nozzle tip. It is important that the orifice or opening of the nozzle tip match the opening of the sprue bushing. If these areas match up, material will get caught and hang up in this area, leading to material degradation due to excessive shear. A rule of thumb is to have the orifice diameter be smaller in diameter by 20% than the diameter of the sprue bushing.

1.3.7.1 Nozzle Types

There are several types of nozzle designs that are commonly used in injection molding: (1) the standard nozzle, (2) the reverse taper nozzle, (3) the spring-operated valve type, and (4) the mechanical shutoff. A *standard nozzle* uses an open channel with no mechanical or spring-loaded valve to convey material through the nozzle. As the name implies, this is the standard type of nozzle found on many injection molding machines. A *reverse taper nozzle* utilizes a tapered section inside the nozzle at the tip of the nozzle. One major use of this nozzle is to prevent materials such as nylon from drooling. After the screw retracts back after injecting material, the screw moves back into its final set position, and by suction, the reverse taper pulls material back into the nozzle to prevent drooling. A *spring-operated valve-type nozzle* uses an internal check valve held closed by an internal or external spring. Injection pressure opens the valve, allowing material to flow through the nozzle to the sprue bushing. When injection pressure is decreased, the spring closes off flow through the nozzle. The only disadvantages of this valve is that the flow is more restricted than in other nozzle types, and due to the complexity of the nozzle, material can hang up inside the valve. A *mechanical shutoff valve* uses a device such as a sliding ring to control flow through the nozzle. When the screw moves forward, the sliding ring moves out of position to allow material to flow through the nozzle. When the screw retracts, the sliding ring prevents the material from flowing. As a result, material builds up behind the ring to prepare for the next shot of material. This is the most commonly used and simplest nozzle design used. However, after long periods of operation, these rings wear excessively, leading to problems with material hangup and inconsistent filling of the cavity, due to material leaking back over the valve into the barrel.

1.4 MELT-DIRECTING PHASE

When molten polymer leaves the barrel through the nozzle, its flow path begins to be guided toward its final destination, the mold cavity. The path the molten material takes goes through a series of turns, twists, and restrictions as it approaches the mold cavity. This section of the process is referred to here as the melt-directing phase. As simple as this may sound to describe, there are many things along this pathway that can affect the final performance of both the process and ultimately, the part performance. In this section we review the critical issues that affect how a material is directed to the mold cavity.

1.4.1 The Sprue

The sprue is the first channel that molten material is exposed as it goes from the barrel and nozzle into the mold and begins directing the molten material toward the mold cavity. The main interface between an injection molding machine's nozzle and the runner system that starts the melt-directing process is referred to as the *sprue bushing*. The design of this bushing is by no means standard to every molding process. The sprue bushing is designed as a "shelf item," available through a number of manufacturers. In general, the design of a typical sprue starts with a large-diameter opening at the nozzle–sprue bushing interface, tapering to a smaller-diameter opening as it enters the runner system. Two factors to consider in selecting the proper sprue bushing are the depth and radius of the sprue bushing where the nozzle makes contact, and the overall sprue length needed for the type of runner system to be used. The depth and radius of the sprue are critical so that proper sealing can take place between the nozzle and the sprue bushing to avoid leakage of molten material. The type of mold, cold versus hot runner, also is critical in the design of a sprue bushing. For cold runner, the overall diameter is kept small to reduce the cooling needed in the sprue, thus keeping the overall molding cycle time minimal. For hot runner systems, larger-diameter tapered sprues can be used, since in this case the material remains molten inside the tool until it enters the cavity. Later in this section, both cold and hot runners are examined. Figure 1.4 shows the design of a typical sprue bushing.

The opening of the sprue at the nozzle interface typically comes in a variety of diameters. Standard sizes range from 0.125 in. (3.2 mm) to approximately 0.500 in. (12.7 mm). A rule of thumb for the opening at the sprue–runner interface is that the diameter of the sprue match the diameter of the runner. It is recommended that the entry of the sprue bushing not be less than 0.125 in. (3.2 mm) or larger that 0.500 in. (12.7 mm). The diameter of the nozzle should be 0.125 in. (3.2 mm) smaller than the opening of the sprue bushing diameter.

Another part of the sprue, which is sometimes neglected in its design, is the cold slug well. The well is located at the end of the sprue at the interface of the sprue and runner and is in the direct line of the sprue. The cold slug well plays an important

FIGURE 1.4 Sprue.

part in material directing and has two functions. First, the cold slug provides a location for cold material that is entering the mold to be directed to, which allows the hotter material to flow straight to the cavity. Otherwise, introduction of cold material into the mold cavity can cause surface defects, such as a blemish or cold flow mark, or create a weakness in the part, causing the part to fail prematurely. The other function of the cold slug is to provide an anchor to cause the sprue to be pulled away from the sprue bushing and ultimately be ejected from the mold after the part is cooled. Figure 1.5 shows several designs of a cold slug well: the annular ring, the reverse taper, and the Z-pin.

1.4.2 The Runner System: Cold Versus Hot Runners

The runner system is defined here as a series of channels that direct the molten polymer from the sprue bushing to the gate, the entry port to the mold cavity. There are two key factors to consider when designing runner systems for directing material to the cavity. The first consideration is to design the runner channel system so that it gets material to the cavity in the shortest, most direct route and avoids a lot of bends, twists, and turns in getting material to the gate. Another factor to consider is that the runner system is balanced. In this case, balancing refers to directing the molten polymer flow to the cavity so that the sections of the part in the cavity fill completely at the same time without sections of the part either filling too fast or too slow.

Two basic types of runners are used in injection molding, the cold runner and the hot runner. In *cold runner* design, the runner starts to cool as it enters the sprue, and then enters into the runner. The skin of the runner solidifies first, and during the mold

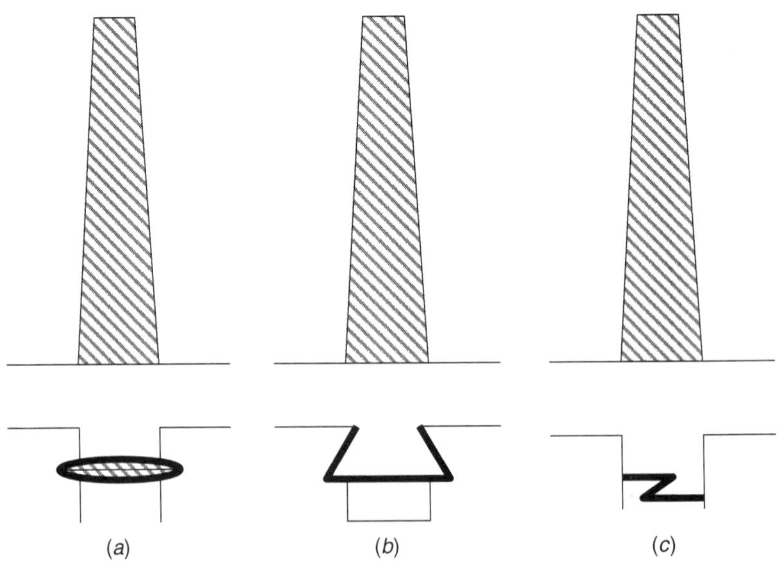

FIGURE 1.5 (*a*) Annular ring, (*b*) taper, and (*c*) Z-pull cold slug wells.

cooling process, cools the rest of the runner cross section. Upon ejection of the part, the cold runner is ejected from the tool. The *hot runner* tool operates with a system of heater bands located inside the tool, and heaters, called *manifold heaters*, which are located inside the runner system. Molten materials flow around the heaters, and along with heaters in the tool this material stays molten until it reaches the cavity. In a sense, hot runner systems act as an extension of the barrel by keeping the material molten until it enters the cavity. When the part is ejected from the mold after cooling, no runner is ejected as in cold runner systems. The material stays molten in the hot runner and then is injected into the cavity.

1.4.2.1 Cold Runner Systems

Cold runners should be designed to have a high volume-to-surface ratio. Such a section will minimize heat loss, premature solidification of the molten resin in the runner system, and pressure drop. Several cold runner cross-sectional geometries are used in injection molding, including the full round, half round, trapezoidal, and modified trapezoidal. Figure 1.6 illustrates these types of designs.

The *full-round runner design* is the most efficient type of runner system and is widely used. Full-round runners are easy to eject and are easily machined using a standard end mill. However, this type of runner needs to be machined into both halves of the mold and can be more expensive to machine. Also, matching both halves of the runner is critical to the functioning of this design.

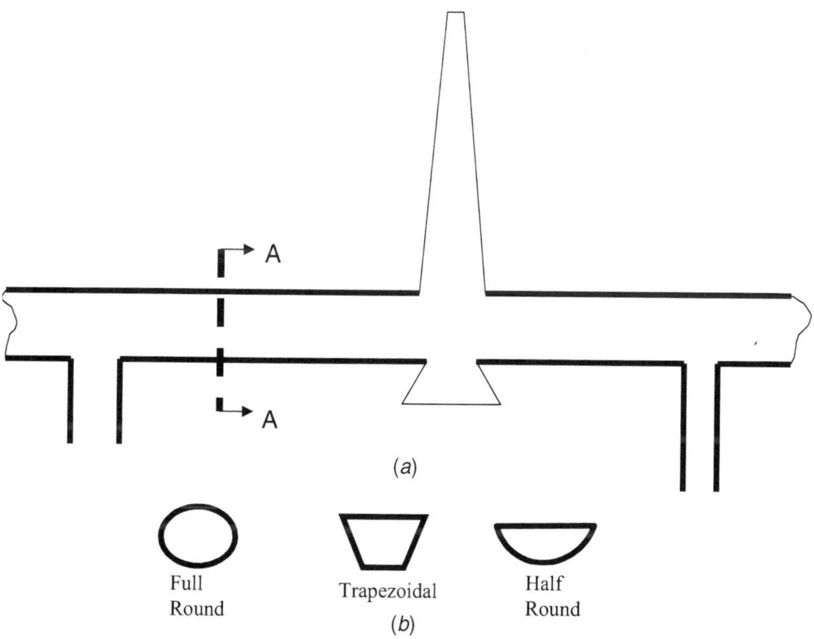

FIGURE 1.6 (*a*) Runner; (*b*) section A-A runner types (full round, trapezoidal, and half round).

The *half-round runner design* allows for machining on one side of the mold with a circular end mill. However, a low volume-to-surface area is present in this runner design.

The *trapezoidal runner design* is less expensive to machine than a full-round design since, again, machining takes place on one-half of the tool. The trapezoidal runner should be designed with a taper between 2 and 5° per side, with the depth of the trapezoid equal to its base width. This configuration will provide excellent volume-to-surface area. However, if sharp corners are used at the base of the trapezoidal runner, part ejection may be hindered.

The *modified trapezoidal runner system* is another variation of the standard trapezoidal runner system. It offers the same features as those of the trapezoidal runner system but includes a radiused base. This provides ease of part ejection and is also easy to machine. Modified trapezoidal runners have been used with a great deal of success with semicrystalline materials, such as polyethylene, polypropylene, and both nylon 6 and nylon 6,6.

Cold runner diameter size takes into account several factors, such as part volume, part flow length, machine capacity gate size, and cycle time. A general rule of thumb is that a cold runner should have a diameter equal to the maximum part thickness but within the range 0.175 to 0.400 in. (4 to 10 mm) diameter to avoid premature freeze-off or excessively long cycle times. The cold runner should be large enough to minimize pressure loss, yet small enough to maintain satisfactory cycle time. Smaller runner diameters have been used successfully as a result of computer flow analysis, where smaller diameter increases the shear of the material, increasing melt temperature. This helps tremendously in maintaining melt temperature and increasing the flow of the molten polymer. Larger runners are not as economical as smaller cold runners because of the amount of energy that goes into forming the runner system and the cost of regrinding the materials that solidifies within them. Table 1.7 lists a number of thermoplastic materials with recommended runner diameter sizes.

When considering runner diameter, another factor to consider is runner length. One of the objectives of either hot or cold runners is to get the material to the cavity by the shortest, most direct route. The longer the length of the runner system, the more "system" pressure is found in filling the cavity. This can lead to narrowing the injection molding window for molding a given part and causing other difficulties in molding a part.

The layout of the runner is critical for properly filling out the part or parts in a mold. An H-configured runner system is recommended in multicavity molds that have similar parts, as shown in Figure 1.7. Balancing the runner system assures that all mold cavities will fill at the same rate and pressure. A more popular method of assuring runner balancing is utilizing computer-aided mold-filling analysis. This methodology allows the designing of molds with minimum-size runners that deliver melt at the temperatures recommended by the material supplier, reduce regrind, reduce barrel temperatures, reduce injection pressures, and conserve energy while minimizing the possibility of degrading the material. Another advantage that computer-aided simulation provides is an artificially balanced runner system, which fills multicavity tools at the same time and pressure, eliminating overpacking of more easily filled cavities.

TABLE 1.7 Cold Runner Size Recommendations for Injection Molding

Material	Minimum Diameter [in. (mm)]	Maximum Diameter [in. (mm)]
ABS	0.187 (4.750)	0.375 (9.525)
Acetal	0.125 (3.175)	0.375 (9.525)
Acrylic	0.312 (7.925)	0.375 (9.525)
Cellulosics	0.187 (4.750)	0.375 (9.525)
Nylon 6	0.062 (1.575)	0.375 (9.525)
Nylon 6,6	0.062 (1.575)	0.375 (9.525)
Polycarbonate	0.187 (4.750)	0.375 (9.525)
Polycarbonate–ABS blends	0.187 (4.750)	0.375 (9.525)
Polyester	0.187 (4.750)	0.375 (9.525)
Polycarbonate–PET	0.187 (4.750)	0.375 (9.525)
Polycarbonate–PBT	0.187 (4.750)	0.375 (9.525)
Polyethylene	0.062 (1.575)	0.375 (9.525)
Polypropylene	0.187 (4.750)	0.375 (9.525)
PPO	0.250 (6.350)	0.375 (9.525)
Polysulfone	0.250 (6.350)	0.375 (9.525)
Polystyrene	0.125 (3.175)	0.375 (9.525)
PVC	0.125 (3.175)	0.375 (9.525)
SAN	0.187 (4.750)	0.375 (9.525)

Source: Multiple industry sources.

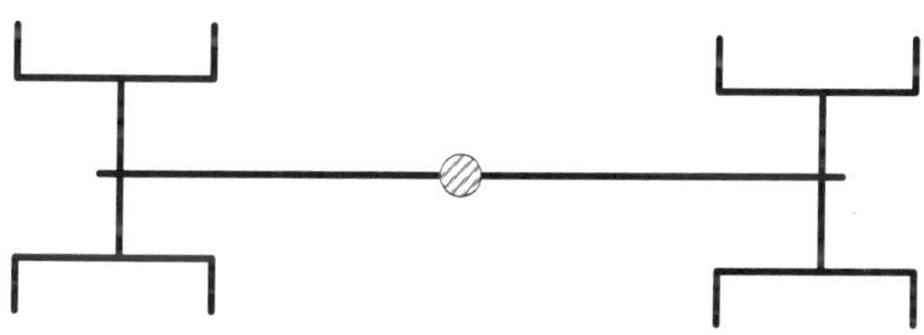

FIGURE 1.7 H-configured runner layout.

1.4.2.2 Hot Runner Systems

Cold runner systems have a runner that is ejected from the tool after the part is cooled. However, hot runner systems, referred to by some as *runnerless systems*, do not have a runner that is ejected with the part. As mentioned earlier hot runner systems differ from cold runner systems by extending the barrel of the injection molding machine and acting as an extension of the machine nozzle. A hot runner system maintains a major portion of the molten polymer material at approximately the same temperature and viscosity as the polymer in the barrel. There are two general types of hot runner systems, insulated and conventional.

An *insulated hot runner system* allows the molten polymer to flow into the runner and then cool to form an insulating skin of solid, cooled material along the walls of the runner. This insulating layer decreases the diameter of the runner and helps maintain the temperature of the molten polymer as it awaits injection into the mold cavity.

The insulated runner is designed so that while the runner volume does not exceed the cavity volume, the entire molten polymer in the runners is injected into the mold during each shot. This full consumption of molten material is necessary to prevent excess buildup of the insulating skin and to minimize any drop in melt temperature.

A *conventional hot runner system* allows increased control over melt temperatures and other processing conditions, as well as increased freedom in mold design, especially in molding large multicavity tools. Hot runner systems retain some of the advantages of the insulated runner system over the cold runner system, and eliminate some of the disadvantages. For example, startup procedures are easier than with insulated runner systems. However, increased costs are present with hot runner systems, due to more complex mold design, manufacturing, and operation. This is due to the need to install a heated manifold located inside the tool in the path of the molten polymer flow, balanced heating needed by these manifolds, and minimized molten polymer hang-up areas.

The heated manifold acts as an extension of the barrel of the injection molding machine by maintaining a totally molten material from the nozzle to the gate. To accomplish this, the manifold includes heating elements and controls for keeping the melt at the desired temperature. Installation and control of the heating elements are difficult. Insulating the rest of the mold from the heat of the manifold is also difficult. The challenge is to keep the heat of the manifold away from the cyclic cooling of the cavity without affecting the overall cycle time.

Using heated manifolds causes another concern with thermal expansion of mold components, such as movable slides and inserts. This is a major consideration to ensure the maintenance of proper alignment between the manifold and the cavity gates. Currently, there are many suppliers of hot runner systems, and the main considerations in selecting a hot runner system are cost and design limitations.

Table 1.8 lists the advantages of cold, insulated, and hot runner systems.

1.4.3 The Gate

The gate is the last major passageway for material to flow from the injection molding machine barrel to the mold cavity. The gate directs the flow of molten material from the runner channel system into the mold cavity.

The location of the gate on the molded part plays a major role in how the part will perform, as well as the quality, properties, and performance of the part. A number of items need to be considered when selecting for or locating a gate on a part.

- The gate needs to be located so that gases builtup during processing can escape through a parting line, ejector pin, porous insert, or vented area without leaving a burn mark or poor surface finish.
- The gate should be located where the material can flow into a wall, core pin, or other part feature rather than an empty space, to prevent jetting or "worming" of the polymer into the part surface.
- The location and size of the gate vestige or scar on the part should be in a location where part functionality is not compromised.

TABLE 1.8 Cold Runner, Hot Runner, and Insulated Runner Systems: Advantages and Disadvantages

Runner Type	Advantages	Disadvantages
Cold	Lower tooling costs vs. hot and insulated runners Low maintenance costs vs. hot and insulated runners. Easier mold start-up Ability to perform color changes faster	Higher sensitivity to runner balancing Higher scrap due to sprue and runner Higher tool wear since runner, gate, and cavity are all exposed to higher pressures Inconsistent part weight due to gate freeze-off Higher material shear Slower cycle times vs. insulated and hot runners
Insulated	Reduced material shear Lower sensitivity to runner balancing More consistent volume of polymer to part; more shot-to-shot consistency Faster molding cycles Improved part finish Decreased tool wear Elimination of scrap from sprue and runners	In general, more complex tool design and higher tooling costs Higher tool maintenance costs More complex startup procedures vs. cold runner systems Increased chance of thermal degradation to the material Color changes slower
Hot	Increased cycle times Improved part surface quality Elimination of scrap from sprue and runners No degating of part necessary	More complex tooling, resulting in higher tooling costs Temperature control more critical since material residence time is increased, leading to the potential of thermal degradation Color changes more complex and difficult Potential hang-up or dead spot areas, which can cause thermally degraded material to flow into the melt stream

Source: Multiple industry sources.

- Gating is recommended at the thick est section of the part, to allow flow to go from a thick to a thin section, which can cause part defects such as v oids.
- The gate is an area of high stress and should be located in a part that is not exposed to high forces or stresses in its end use.
- Gates should be located so that flo w does not occur around core pins, depressions, and holes, leading to the formation of weld or knit lines.

A number of types of gates are used for injection molded parts. In this section, we concentrate on some of the more popular gate designs used.

1.4.3.1 Edge Gates

Edge gates are used most often in large part designs and also where thin walls are used in a part. Examples of where these types of gates have been used are in business machine and electronics enclosures and in automotive glove box doors. One of the advantages of edge gates is that it provides the widest molding window since, due to its design, low shear rates are found. This gate is placed along the side or width of a part, and the width can range anywhere from 0.500 to beyond 12 in. (12.7 to 305 mm). The recommended thickness of an edge gate is approximately 0.40 to 0.50 times the nominal wall thickness where the edge gate is located.

Figure 1.8 illustrates an edge gate design. Gate vestiges for this type gate can be quite large however. To reduce this problem, a straight land can be used between the edge gate and the edge of the part that can facilitate part trimming.

1.4.3.2 Fan Gates

Fan gates are similar to edge gates, but in this case, the gate fans out at an angle from the runner system to the part where in the edge gate, the gate comes out in a straight line from the runner to the part. Fan gates, used in applications such as automotive trim parts and electronics covers and enclosures, provide reduced pressures and clamp tonnage over other conventional gate designs and are excellent when flow lengths are short. As in edge gates, fan gates allow for a wide process window and reduce overpacking issues since the pressure is lower than that found in tunnel gates. The disadvantages of using fan gates include the inability to trim off the gate since a larger area must be trimmed through, leaving a large gate vestige. Increased scrap may also be found, due to the difficulty of trimming off fan gates. One solution to reducing the vestige and alleviating the trim issue is to use a "chisel" cross section for the fan gate. This allows the fan gate to break off from the part cleanly and

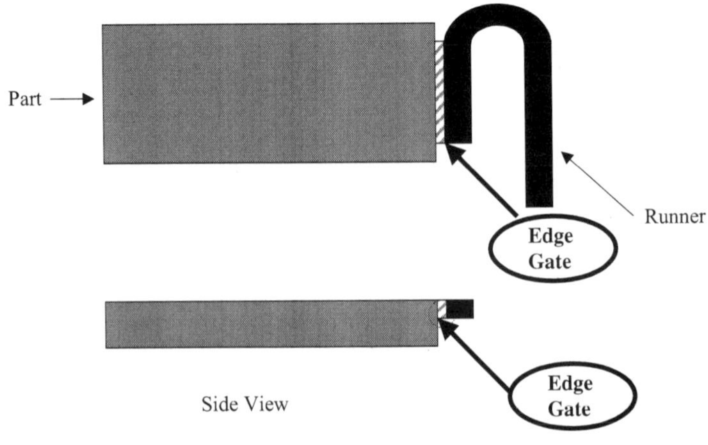

FIGURE 1.8 Edge gate design.

evenly. Figure 1.9 shows a fan gate and a chisel-type design to reduce gate vestige issues.

1.4.3.3 Pinpoint Gates

Pinpoint gates are used mostly in single- or multicavity three-plate tools, or where multiple gates are used in a part. Pinpoint gates are also used with a thin nominal wall thickness, but for a part with a large surface area. One big advantage of pin point gates is their ease of de gating from a part without the use of special de gating tools or fixtures to remove the gate. However, pinpoint gates have the potential to creat high shear on the molten material. It is suggested that the recommendations given by a material supplier for a given material to be molded using a pinpoint gate be followed. These types of gates are used in applications such as electrical switches and consumer applications such as children's toys. Figure 1.10 illustrates a pinpoint gate design.

1.4.3.4 Tunnel Gates

As the name implies, tunnel gates, convey material below a parting line of the mold and "tunnel" into the cavity. This type of gate is used on smaller parts and thin-walled parts due to its small size. A big advantage in using tunnel gates is its ease in separating the part from the runner and gate system upon the part ejecting from the mold cavity. Similar to pinpoint gates, its big disadvantage is that due to the high shear rates, which can cause the material to degrade, the injection molding window is narrowed. Once again, It is suggested the material supplier's recommendations as to the optimum gate design to use for a selected material. Tunnel gates are used in applications such as electronic connectors and small parts for medical applications. Figure 1.11 shows a typical tunnel gate.

1.4.3.5 Ring Gates

Ring gates are used primarily around cylindrical cores. In this case the material flows around a core pin and flows down to fill the pin evenly. A major advantage that

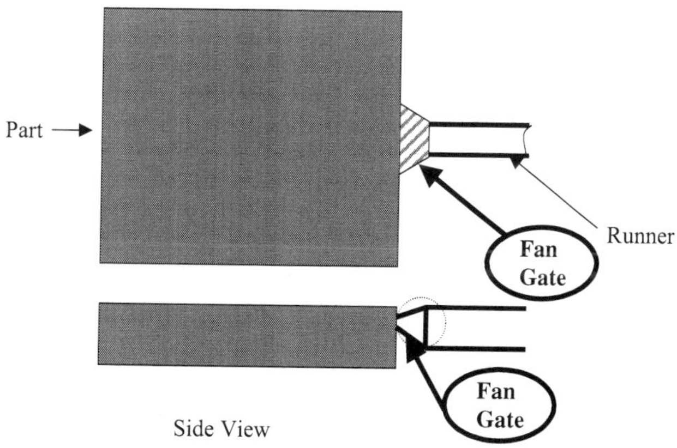

FIGURE 1.9 Fan gate design.

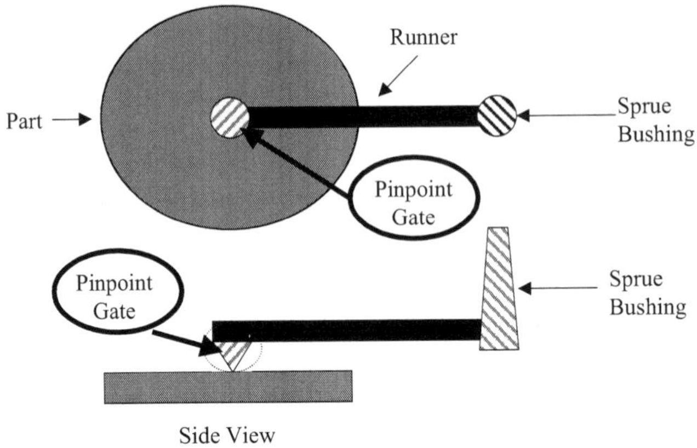

FIGURE 1.10 Pinpoint gate design.

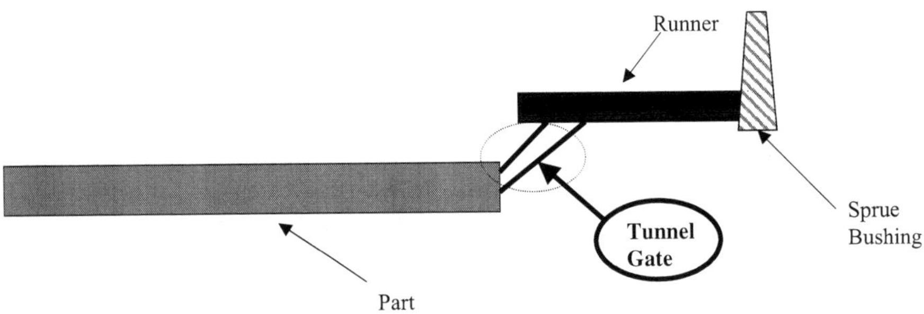

FIGURE 1.11 Tunnel gate design.

ring gates have is that these gates can prevent the formation of trapped air, which can cause weld or knit lines. Another advantage that ring gates have is to prevent a phenomenon, core pin shift that occurs when molding around circular cores. Core pin shift is caused when the pressure and flow around a core pin are unbalanced around a core pin, causing the pin to deflect or move. This shift creates a nonuniform wall thickness around the circumference of the core pin. Ring gates provide even flow around the core pin to minimize core pin shift and create a more uniform wall around the core. A major disadvantage is that ring gates require a fixture to remove the gate from the part before the part is sent to the part end user. Applications for which ring gates include medical syringes and in ammunition applications such as shotgun shells. Figure 1.12 illustrates a ring gate design.

1.4.3.6 Diaphragm (Disk) Gates

Diaphragm, or disk gates, operate opposite to a ring gate, in that the material flows from the cylindrical core to its perimeter. These gates are used mostly for single-cavity tools in fabricating single-shaped parts, such as cylindrical parts, that have small or medium-sized internal diameters. As found with ring gate designs, one advantage of diaphragm gates is the reduction in core pin shift when molding tube-shaped parts. Applications where diaphragm gates have been used are in toner bottles for business machines and industrial applications such as hydraulic fluid reservoirs. Figure 1.13 shows a diaphragm (disk) gate.

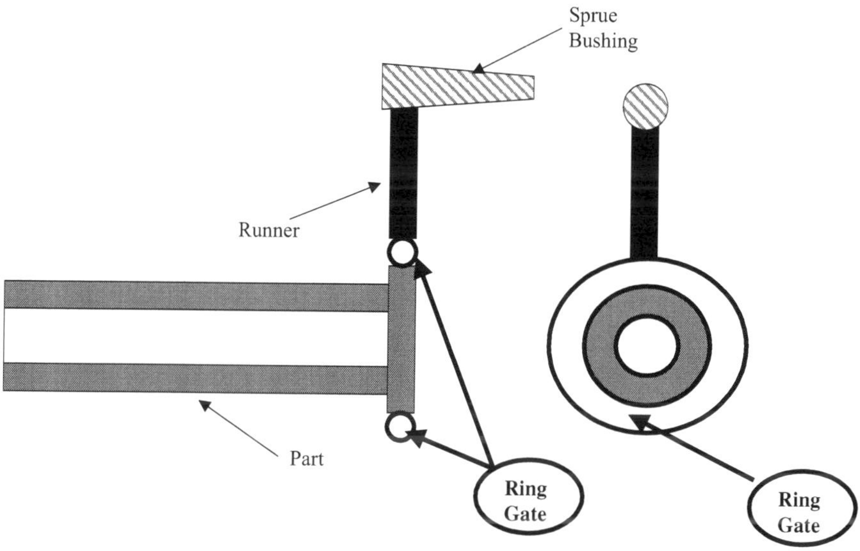

FIGURE 1.12 Ring gate design.

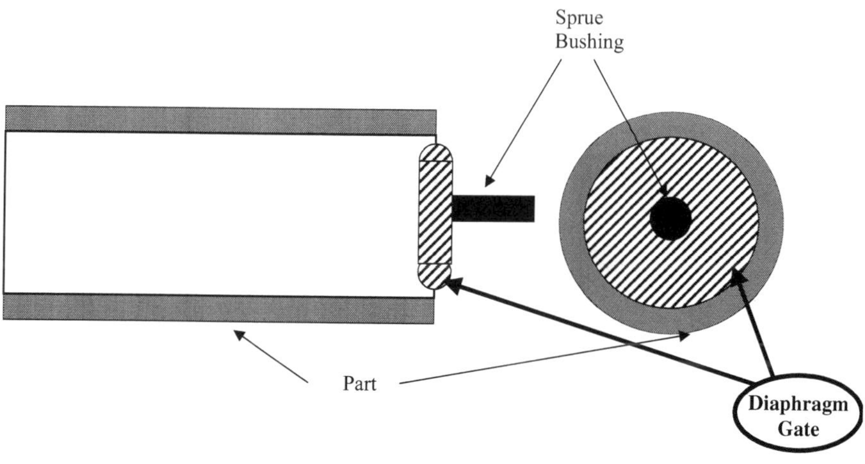

FIGURE 1.13 Diaphragm (disk) gate design.

1.4.3.7 Sprue Gates

Sprue gates are used in large parts made in a single-cavity tool. Typically, the gate is located at the center of the part to allow for even flow through the part. In this gate design, a reverse taper placed at the end of a sprue gate to remove the part from the mold upon ejection, acts as a cold slug well to capture material coming from the end of the nozzle of the barrel and prevent it from entering the runner system. There are other variations of the reverse taper, including ring pullers and Z-pullers, which were discussed earlier in this section. A big advantage of sprue gate designs is the simplicity in designing this type of gate versus tunnel and pinpoint gates. However, these types of gates require special degating tools and fixtures to remove the gate from the surface. Another problem with the use of sprue gates in standard injection molding parts is the formation of a sink mark on the surface behind where the gate meets the part surface. These gates work well with structural foam injection molding since the foaming agent expands this area where the sink marks occur and eliminates the sink mark. End-use applications where sprue gates are used are in business machine applications such as business machine enclosures, and in enclosures for electronics applications such as switch boxes. Figure 1.14 shows a sprue gate design.

1.4.3.8 Tab Gates

Tab gates usually extend from the runner into the part and can have the same thickness as the part. However, to degate the part without the use of special degating tools and fixtures, the gate thickness should not exceed 50% of the part's nominal wall. These types of gates are found in consumer applications such as storage drawers and in small enclosure for electronics. Figure 1.15 illustrates a common tab gate design

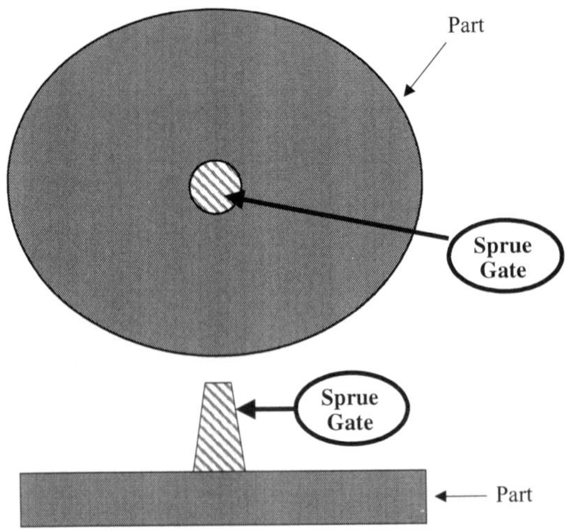

FIGURE 1.14 Sprue gate design.

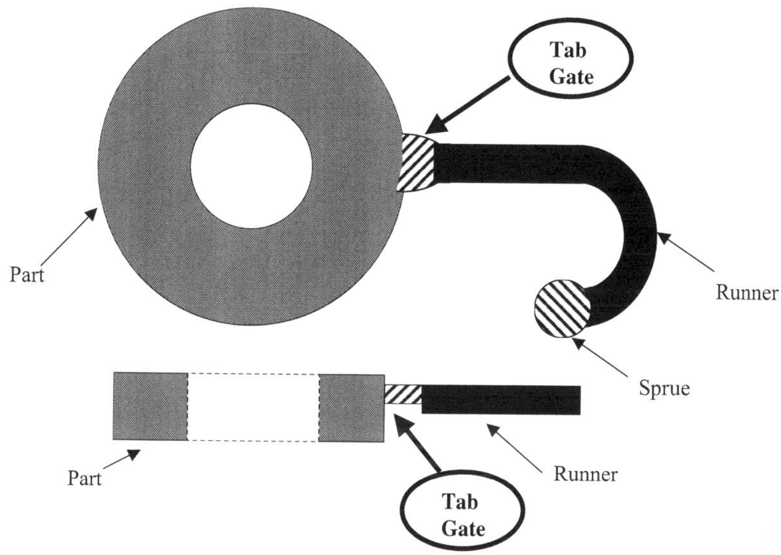

FIGURE 1.15 Tab gate design.

Table 1.9 provides a summary of all gate types discussed in this section, and lists advantages and disadvantages of using each of these types of gates.

1.4.4 Gate Land

A gate land (Figure 1.16) is a section of the melt-directing system that is used in some but not all gate designs. It acts as an interface between the gate and the molded part, is used to facilitate degating of the gate and the runner from the part, and provides a smooth edge or v estige to the gated area. The gate land is used mostly in, such tab, edge, and fan-type gates. For most thermoplastic materials, the gate land should not exceed 0.025 in. (0.64 mm) in length. Land lengths greater than 0.025 in. increase the shear rate of the material exponentially and can lead to conditions such as thermal degradation or defects such as splay marks.

1.5 MELT-FORMING PHASE

When the molten polymer travels through the channels that guide its flow path, such as the sprue, runner, and gate, it arrives at its destination, the mold cavity. In this location, the shape, size, and design of the cavity take the configuration of the final part to be molded. The molten material flows into this cavity, takes the shape of this chamber, and is cooled from a molten material to a solid mass of polymer . As simple as this sounds, there are many aspects to forming the molten material into its final solid mass that need to be considered to fabricate a part for its final end use.

TABLE 1.9 Gate Designs: Advantages and Disadvantages

Gate Type	Advantages	Disadvantages
Edge	Low-shear gate, therefore providing a wide molding window. Restricts flow to prevent jetting into the cavity. Good for multicavity tools. Optimum for medium and thick part sections. Can be placed outside the aesthetic surface.	Degating is difficult. A mechanical degating device, such as cutters, grinders, and a holding device, may be needed to degate part.
Fan	Low-shear gate, therefore providing a wide molding window. Can degate easily, pending material used and if using a chisel (angled) taper into the part. Can be placed outside the aesthetic surface. Provide reduced injection pressures. Optimum in shorter flow lengths.	Can be difficult to gate if flexible materials are molded. Cutters or degating fixtures nay be needed. Increased scrap may be realized, due to trimming off of gate.
Pinpoint	Optimum in single- or multicavity tools. Good in thin-walled parts and larger surface areas. Gates are self-degating and do not need any mechanical degating equipment.	Narrower processing window since this is a high-shear gate.
Tunnel	Good for small parts [less than 2 lb (0.91 kg)]. Gates are self-degating and do not need any mecahnical degating equipment. Part and runner separate with ease.	Narrower processing window since this is a high-shear gate. High pressure losses can be encountered. Potential of jetting, causing poor surface aesthetics. Can cause weak weld or knit lines.
Ring	Optimum for molding around cores in cylindrical parts. Eliminates trapped air that can cause weld lines. Reduces core pin shift, which can cause variations in wall thickness in a finished part.	Special degating tool or fixtures needed to degate part.
Diaphragm (disk)	Optimum for molding single-shaped parts with small or medium-sized internal diameters. Improved flow around core. Reduces core pin shift, which can cause variations in wall thickness in the finished part.	Special degating tool or fixtures needed to degate part.

TABLE 1.9 (*Continued*)

Gate Type	Advantages	Disadvantages
Sprue	Optimum for large parts. Low tooling cost since it uses only one gate. Gate design simple since it uses a reverse taper.	Degating tool needed due to large diameter of gate. Leaves a huge vestige on a part.
Tab	Optimum for larger parts. Prevents jetting into the part. Enables area of greatest strain in part to be removed from the part.	Design must be no more than 40 to 50% of the nominal wall. Otherwise, mechanical degating tools may be needed.

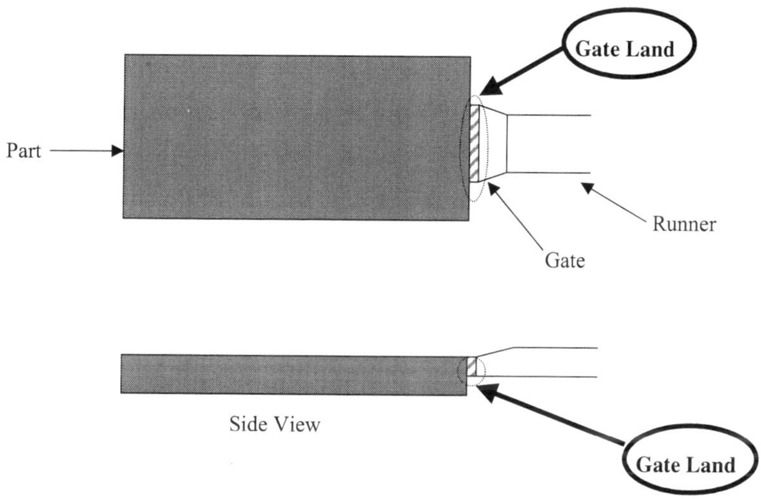

FIGURE 1.16 Gate land design.

In this section we arrive at the final destination of this trip through the injection molding process, the melt-forming phase. We discuss the aspects of forming a part into its final end use design and the f actors that control the forming of the melt. The section begins with a discussion of the clamp of the molding machine, which acts as the force to resist the pressures needed to inject the part through and form the part. Ne xt, we review the process of mold cooling, which involves how to use a cooling medium properly to transfer this molten polymer into a solid polymer mass. A detailed discussion of how form the parts will examine how to design a part for the injection molding process follo wed by a discussion on v enting the injection mold. Finally , we review mold shrinkage ef fects on f inal part dimensions.

1.5.1 The Clamp

The clamp of an injection molding machine is what closes the mold, holds it closed under pressure during injection of the molten material and during cooling of the material to form a solid part, and opens the mold so that the part can be ejected and removed from the mold. In injection molding, four different types of clamping are used: hydraulic, mechanical, hydromechanical, and electric.

1.5.1.1 Hydraulic Clamp System

A hydraulic clamp operates using hydraulic fluid and pressure to close, hold, and open the clamp. This system consists of a large main cylinder which is located at the rear of the machine. There are three platens: cylinder, movable, and stationary. The central cylinder and two transverse cylinders, moving at a higher speed than the central cylinder, control the motion of the mold tool without mechanical linkages or connections. The clamp also utilizes a larger clamp cylinder that is used to build the required clamp tonnage for the injection molding machine. The basic concept of the hydraulic clamp is to direct hydraulic fluid used to move the clamp forward. Hydraulic oil is fed from a storage tank, which fills the main area at the large cylinder. The oil is fed by gravity if the oil storage reservoir is located overhead, and is fed by suction if the reservoir is located below the clamp cylinder. As the cylinder moves forward, the oil flows from the reservoir to the area behind the cylinder. Once the clamp is closed, the prefill valve is closed, trapping the oil behind the cylinder. High-pressure fluid is used in this area, and the volume of oil in this area is compressed, increasing the pressure in this area to a level used for holding the mold closed during injection and cooling of the molten material. Upon opening of the mold, the hydraulic fluid is directed away from behind the cylinder and is returned to the storage reservoir after the prefill valve opens. Figure 1.17 shows a hydraulic clamping system.

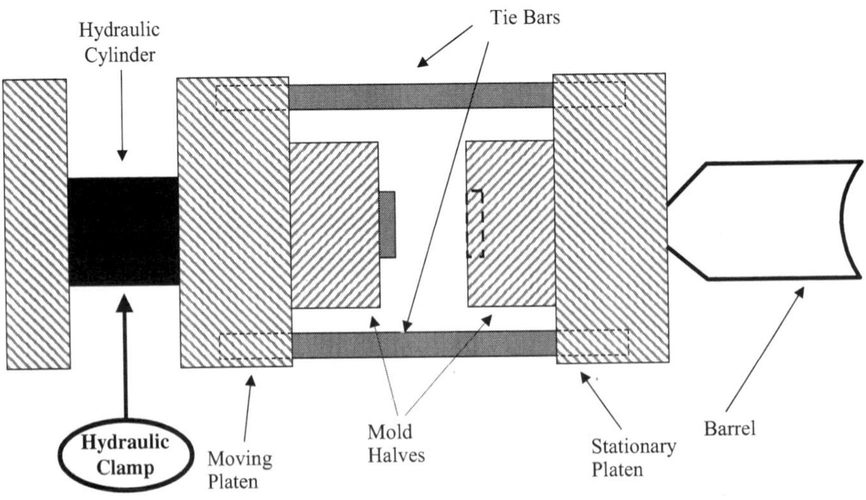

FIGURE 1.17 Hydraulic clamp design.

1.5.1.2 Mechanical Clamp System

The mechanical clamp system has been used for many years in injection molding and was in most common use until development of the hydraulic clamp. The concept of a mechanical clamping system utilizes a mechanical linkage, referred to as a *toggle*, which develops the clamping force needed to hold the mold closed during the injecting and cooling portion of the injection molding cycle.

A mechanical clamp consists of three platens, four tie bars, and a toggle system activated by a hydraulic cylinder. Two types of mechanical toggle are used: the single-toggle lever and the double-toggle lever. The single-toggle lever is used for machines built for lower clamping forces (25 to 50) tons but has been used with machines built for as much as 200 tons of force. The double-toggle lever is used for clamping forces as high as 1000 tons. In the single-toggle lever system, a small hydraulic cylinder is used with the single-toggle lever for closing the clamp. The cylinder travels at a constant speed with a slowdown built in as the two mold halves close. This system allows for only short opening strokes. For the double-toggle lever system, a center hydraulic driving system is used, and larger opening strokes are realized, depending on the length of the driving system.

In a mechanical toggle system, the hydraulic cylinder causes the toggle to stretch or collapse the toggle mechanism similar to the way that the human elbow extends or contracts the human arm. The mold is fully closed once the toggle has stretched and is locked in place. To assure that the toggle is not overstretched or not stretched enough, after the mold is installed between the platens, the clamp is moved forward until the mold actually "snaps" the mold halves to a closed position. This is adjusted at the rear of the machine using the die height adjustment. If the toggle is overstretched, the mold may open slightly, causing material to flash excessively between the mold halves. To prevent overstretching, machines with toggle systems are equipped with a limit switch that will switch off the hydraulic valves operating the linkages. Figure 1.18 shows a mechanical toggle clamping system.

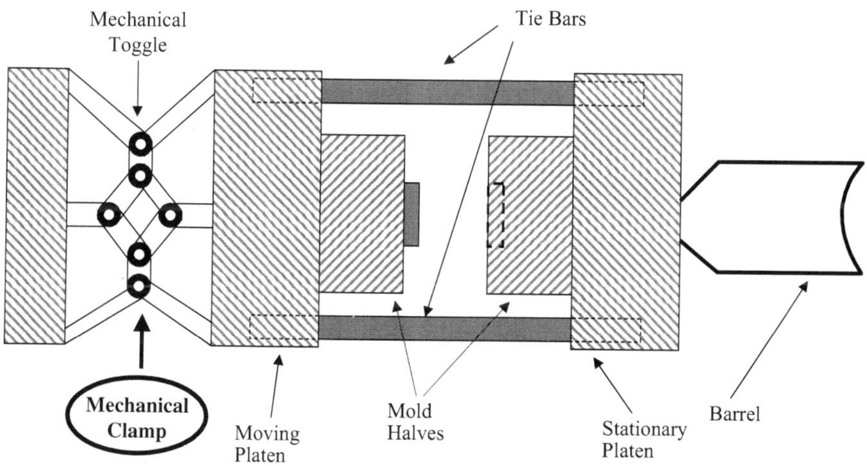

FIGURE 1.18 Mechanical toggle clamp design.

1.5.1.3 Hydromechanical Clamp

The hydromechanical clamp system uses a mechanical means for closing and opening a mold under high speeds. It consists of tw o small fast-travel cylinders and one large central clamp cylinder. However, in this system, a locking pad is used behind the large central cylinder, built for clamp tonnages similar to those of a hydraulic clamping system. The purpose of using a locking pad is to reduce the size of the large cylinder. Otherwise, the cylinder needed would be long, and the amount of oil needed would be excessive. The locking pad also keeps the clamp shortened.

A hydromechanical clamping system operates as follo ws. First, the mold closes using the two small, fast cylinders until it is fully extend. At this full extension point, a switch indicates the correct pressure of the pressure column and then mo ves the locking pad into position behind the large central clamp cylinder. The clamp system moves forward and locks the mold platens in place and holds the mold tool in place through the injection and cooling cycle. Upon opening of the mold, the clamp cylinder pressure is reduced, the travel cylinders move forward, and the locking pad moves out of position. At this point the small f ast travel cylinders open the mold. Figure 1.19 illustrates a hydromechanical clamping system.

1.5.1.4 Electric Clamp System

Developed in the late 1980s, the electric clamping system operates similar to the mechanical clamping system, but in this case, no main hydraulic c ylinder is used to move the toggle. An electrically driven motor is used to mo ve the toggle forward and backward into position using a ball scre w mechanism. The electric clamping system provides an ener gy-efficient mechanism to accomplish all machine functions and is totally mechanical. Velocity and pressure are adjusted independently . Injection rates are high and provide a rapid response to profile changes. Energy savings been realized at 60 to 70% of hydraulic clamping systems.

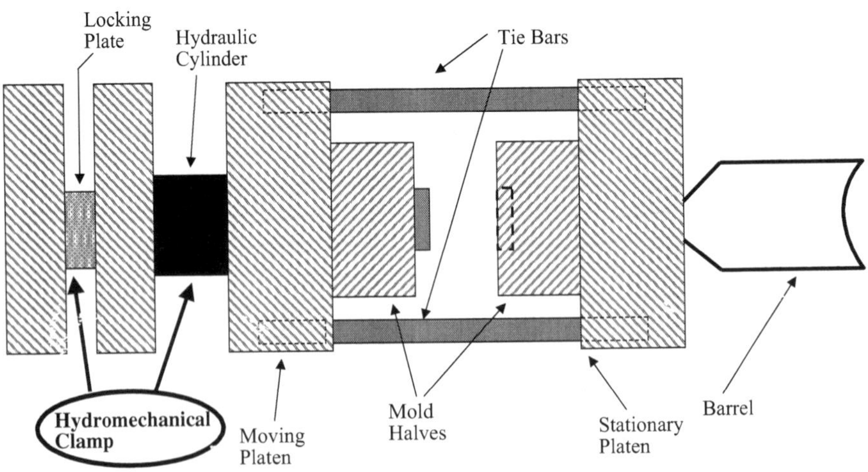

FIGURE 1.19 Hydromechanical toggle clamp design.

1.5.1.5 How Much Tonnage Is Needed? The Projected Area

One of the main concerns with molding an injection molded part is selecting the correct clamp size needed to mold the part. To do this, the projected area of the part is used. The *projected area* is defined as the area of the part, including the runner system, which is present in the parting line of the tool. The required clamp for a part can be determined using the equation

$$C = A_p T \tag{1.8}$$

where C is the clamp required (tons), A_p the projected area (in^2), and T the tons/in^2.

The T value used depends on the material from which the part is being made. Table 1.10 lists data for tons/in^2 for a number of plastic materials.

1.5.2 Cooling the Injection Mold

Once the molten material is injected into the mold cavity, the next objective in the injection molding process is to turn the molten polymer into a solid mass. This is done by decreasing the temperature of the molten material by cooling the injection mold.

In thinking in terms of cooling the molten material, the basic principles of heat flow are used. Heat flows from a higher-temperature source to a lower-temperature source. The big factor in how fast heat flows is the difference in temperature between the higher-and lower-temperature source. The larger the difference in temperatures between these two source, the more rapid the rate of heat flow between them. There are three methods of heat transfer: conduction, radiation, and convection. Of these, conduction is the most important in the injection molding process.

TABLE 1.10 Tonnage Values for Various Thermoplastic Materials

Material	Tons/in.^{2}
ABS	2.4
PC	4.0
PC–ABS blends	2.7
PS	1.0–1.5
PP	2.2
HDPE	2.0–2.2
SAN	1.0–1.5
Nylon 6,6	3.0–5.0
TPU	3.0–5.0
TPO	2.0–2.3
PMMA	2.5–3.0
Nylon 6	3.0–5.0
Acetal (POM)	5.0–7.0
PBT	3.0–4.0
PET	3.0–4.0

Source: Multiple industry sources.

In an injection mold, the mold material is responsible for the conduction process. The mold material moves the heat from the molten polymer material to the mold and the cooling lines. Heat is taken away from the molten material by the mold, and this heat is taken away by the water or cooling medium flowing through the cooling lines. Once the process of heat flow is completed and the equilibrium of the temperature is achieved, the heat flow works backward to hold the temperature of the part at the desired temperature.

The following equation is used to determine the rate of heat transfer:

$$Q = HA\ \Delta T \tag{1.9}$$

where Q is the heat transfer rate, H the heat transfer coefficient, A the area of the mold cavity, and $\Delta T = T_m - T_c$ (T_m is the temperature of the mold cavity and T_c is the temperature of the coolant).

Table 1.11 lists values for the heat transfer coefficient for various materials used in molds for injection molding.

Another important variable when reviewing mold cooling requirements is the flow of the coolant medium, especially when coolant flows through cooling channels and passages. A measure for determining the conditions for flow is that described by the Reynolds number. The Reynolds number is determined by the following equation:

$$N_{\mathrm{Re}} = \frac{4Q\rho}{\pi D \mu} \tag{1.10}$$

where N_{Re} is the Reynolds number, Q the heat transfer rate, ρ the specific gravity, D the diameter of the channel, and μ the dynamic viscosity of the transfer medium. If $N_{\mathrm{Re}} > 2100$, flow is determined to be turbulent; if $N_{\mathrm{Re}} < 2100$, flow is considered to be laminar. To cool the tool efficiently, turbulent flow is found to provide the most optimum heat conductivity. Turbulent flow can provide as much as two to three times the heat conductivity as laminar flow.

TABLE 1.11 Heat Transfer Coefficients: Mold Materials

Mold Material	Heat Transfer Coefficient [Btu/ft-°F-hr (J/s·m·°C·m)]
Aluminum	100 (173)
Beryllium copper	62 (107)
Brass (60 : 40 alloy)	70 (121)
Copper	222 (384)
Kirksite	62 (107)
Nickel	52 (91)
Stainless steel	10 (17)
Tool steel	21 (36)
P-20 H-13	12 (21)

Source: Multiple industry sources.

The temperature of the cooling medium can affect the kinematic viscosity of the cooling medium. Figure 1.20 shows a plot of kinematic viscosity versus water temperature. This dynamic viscosity at a given temperature can be substituted into the Reynolds number equation to determine the type of flow to be experienced.

1.5.2.1 Cooling Line Positioning

Location of the cooling lines is critical to achieving efficient cooling of the part and improving part productivity. Figure 1.21 illustrates guidelines for an optimum layout of cooling channels. One point to keep in mind is that the cooling channel diameter should be large enough to have reasonably low pressure drop but not so large that excessive flow rate is needed to obtain maximum cooling efficiency via turbulent flow. The guidelines to follow for determining a proper cooling line layout are:

- Keep the cooling lines at a uniform distance from the part walls. Otherwise, cooling will not be consistent across the part.
- Distribute cooling to match the amount of heat in various sections of a part. For example, a nonuniform thick area needs more intensive cooling than a thin area.
- Avoid a metal-to-metal interface between cooling channels and the mold surface. Install channels directly in cores and cavities, not into the back plates alone.
- Limit hoses for cooling lines and fitting restrictions. The inside diameter should be at least the size of the cooling channel inside diameters.
- To avoid losing heat to ambient temperature surroundings, insulation should be used on all pipes and lines.
- Avoid more than 10 bends in any cooling line circuit, to minimize the pressure drop.

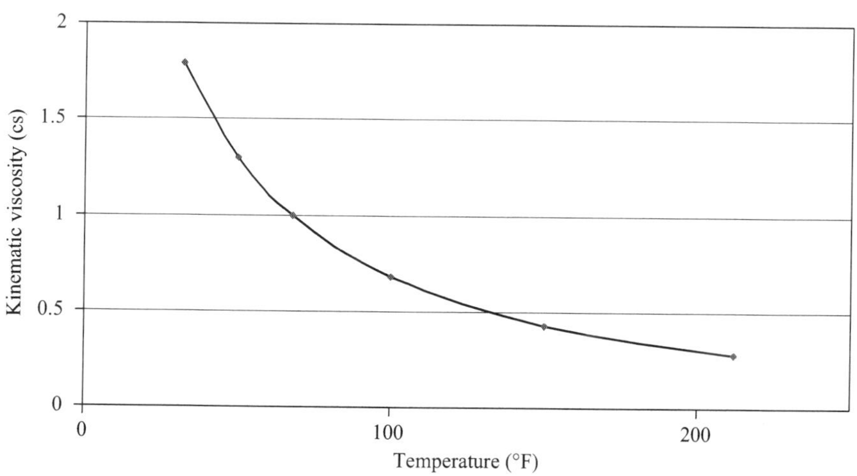

FIGURE 1.20 Kinematic viscosity versus water temperature. (From Ref. 2.)

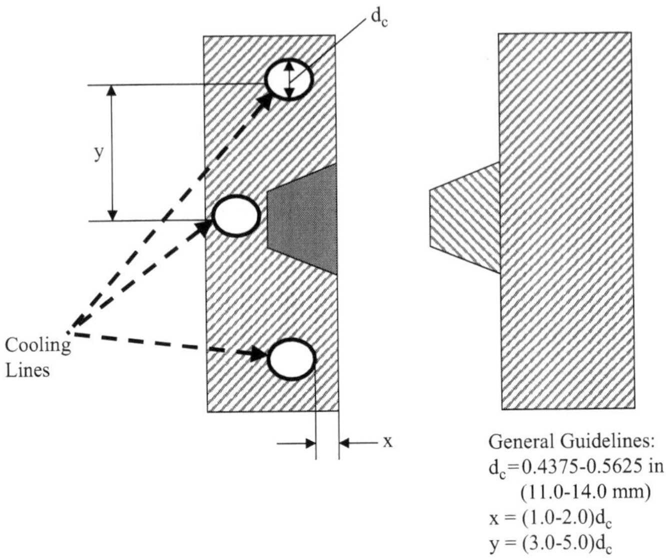

FIGURE 1.21 Guidelines for layout of cooling line channels.

1.5.2.2 Cooling Channel Schemes

A number of various cooling channel schemes or layouts are commonly used in injection molding. For cooling a core-shaped part, a V channel (Figure 1.22), basic fountain (Figure 1.23), and fountain baffle channel (Figure 1.24) are used to provide optimum cooling. For irregular-shaped parts, the baffle design concept is used, which circulates flow up close to the mold surface and provides turbulent flow inside the mold. (Figure 1.25).

Another method used, which does not utilize a liquid medium such as water, is the conductive insert. This method provides spot cooling to a specific area, such as a thick section, where placing a cooling line can be difficult. These inserts are made from materials such as aluminum or beryllium copper. Figure 1.26 illustrates this conductive insert method.

1.5.3 Part Design for Injection Molding

In this section of the melt-forming phase we review general guidelines in the design of parts for the injection molding process. Both product design and processing of the material are two important steps in designing a plastic part. Both of these factors must be considered in determining the feasibility of manufacturing a plastic part. Other factors that also need to be considered are the economics, quality, appearance, and performance of the plastic part.

As seen thus far, the injection molding process involves feeding pellet material, melting this pellet material from pellets to molten plastic material, and conveying this melt through a barrel using a screw, directing this melt from the barrel through sprues,

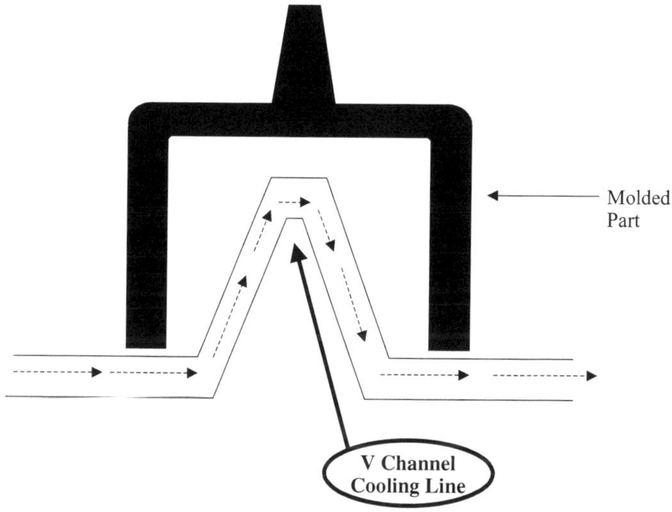

FIGURE 1.22 V-channel cooling lines.

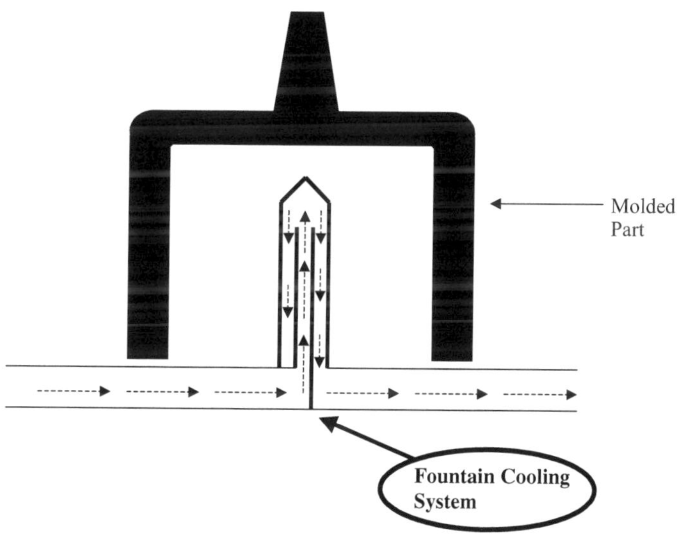

FIGURE 1.23 Fountain channel cooling lines.

runners, and gates, and then forming this melt inside the mold cavity. The mold cavity may also have such features as cores, inserts, and slides that form the part.

When designing a plastic part, there are several considerations in selecting the right material:

- *Service temperature*. What will be the short- and long-term environments to which the material and part will be exposed? Properties of materials change

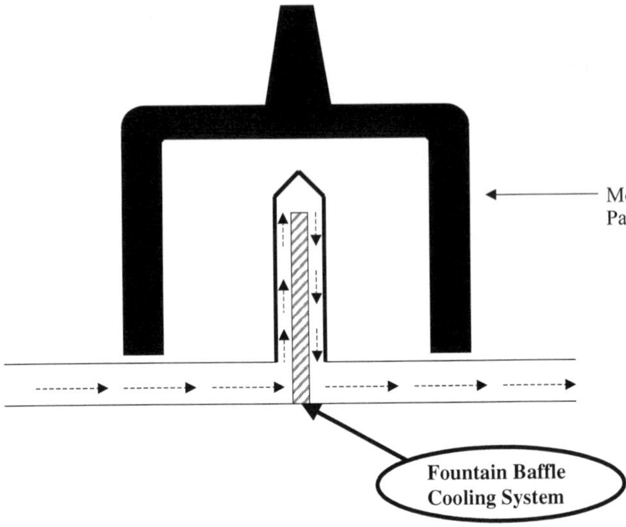

FIGURE 1.24 Fountain baffle cooling line system.

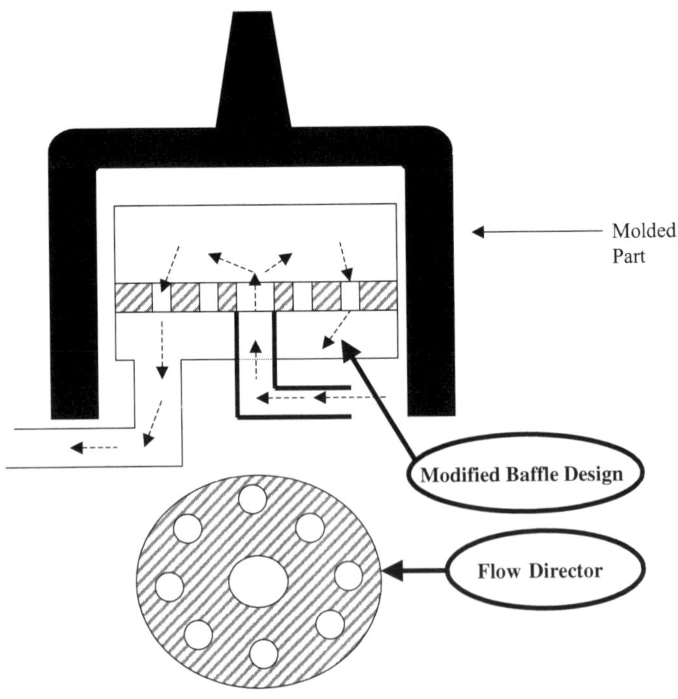

FIGURE 1.25 Modified baffle design with flow director.

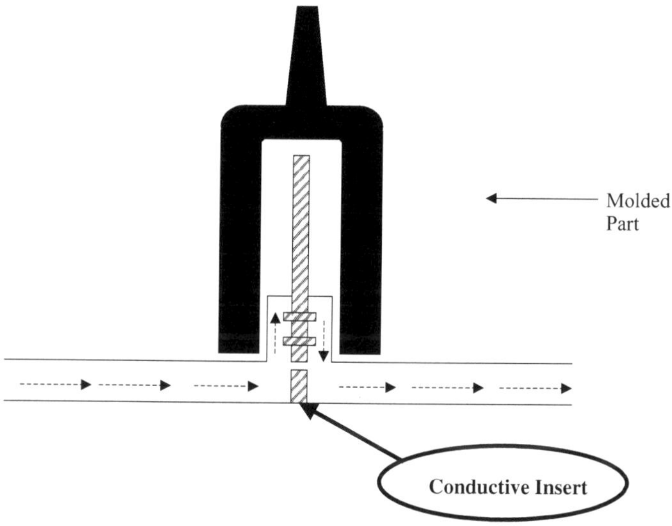

FIGURE 1.26 Conductive inserts.

with temperature; high temperatures make plastic material more ductile, low temperatures make it more brittle. In real-time manufacturing, a part may be exposed to these changes with manufacturing, such as assembly, finishing, and shipping, and in end use.

- *Chemical environment.* To which chemicals is the part exposed (solvents, oils) and to what environments (e.g., ultraviolet light)?
- *Part economics.* What is the part cost per cubic inch? This can be determined as follows:

$$\text{cost/in}^3 = \text{cost/lb} \times \text{specific gravity (g/cm}^3) \times 0.0361 \tag{1.11}$$

- *Product specifications.* Are there any agency or OEM (original equipment manufacturer) specifications that the material or part needs to meet? Examples are U.S. Food and Drug Administration, FMVSS, Underwriters' Laboratories, and National Science Foundation.

The following aspects of injection molding part design will provide general guidelines on how to form a melt into a part with maximum integrity and aesthetics.

1.5.3.1 Nominal Wall Thickness

A great starting point for reviewing part design is at the ground floor. The nominal wall is the ground floor for part design since this has the strongest influence on many of the key part characteristics, such as performance, surface aesthetics, appearance, moldability, and economics. The best part thickness is often a trade-off between strength versus part weight, durability versus part cost. Nominal wall thickness needs to be the first item on the part design checklist since this can be the most expensive item to change, due to tool modifications and production problems that can sneak up quietly. It is important that the thickness of a part be as constant, or nearly as constant, to provide optimum plastic part design, part economics, and moldability. Constant wall thickness also assures that uniform mold shrinkage will prevent part warpage problems.

If transitions in wall thickness are needed, the key consideration is to make wall transitions less drastic and abrupt. Figure 1.27 illustrates both poor and optimum part nominal wall thickness design.

The flow of the plastic material is another consideration in determining nominal wall thickness. The flow length, defined as the distance from the gate to the last section of the part to be filled, must be within acceptable limits for the plastic material selected. Factors such as thin walls can resist melt flow, leading to high part stresses, aesthetic part surface problems, and part-filling problems, limiting the processability and process window of the part. Parts with very thick nominal walls can also cause problems with excessive cycle times, affecting part and molding economics, and overpacking the parts, leading to high levels of molded-in stresses, causing parts to fail prematurely. One concept that is used is the *flow ratio* which is defined as follows:

$$\text{FR} = \frac{FL_{\text{max}}}{t_{\text{avg}}} \tag{1.12}$$

where FR is the flow ratio, FL_{max} the distance in inches (or millimeters) from the gate to the farthest point to be filled, and t_{avg} the average nominal wall thickness across that same distance [in. (mm)].

For parts made from most thermoplastics, a rule of thumb is to avoid designing with nominal wall thicknesses above 0.150 in. (4.0 mm). Nominal wall thicknesses above 0.150 in. cause excessively long cycle times, due to the long cooling times needed to remove heat from a thick wall, and increase the risk of voids in a part, affecting part performance negatively. If wall thicknesses higher than 0.150 in. (4.0 mm) are required, other injection molding process technologies, such as structural foam or gas-assist injection molding, should be considered. These are explained in a later chapter.

Some variation in nominal wall thickness is allowed in designing parts. As a general rule, for thermoplastic materials with a shrinkage of less than 0.010 in./in.,

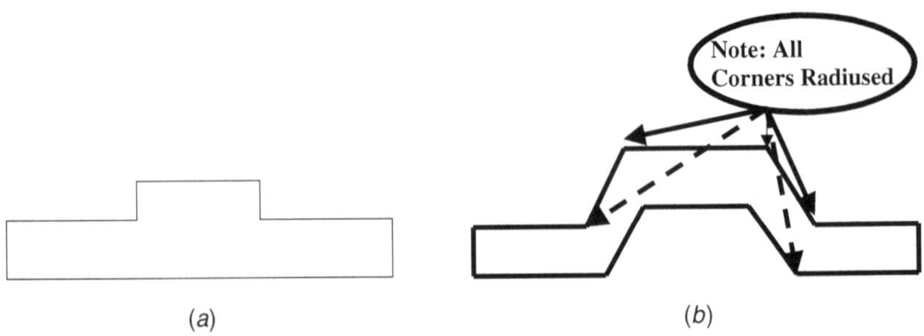

FIGURE 1.27 (*a*) Poor versus (*b*) optimum wall thickness design.

a wall thickness variation of ±25% is acceptable. If shrinkage is greater than 0.010 in./in., a thickness variation of ±15% is acceptable.

A problem encountered in nominal wall design is a part designed with thin areas surrounded by thick perimeter sections, creating *racetracking*. In this case, molten plastic material flows rapidly around the thick section of the part faster than the material in the thinner section. The flow from filling the thick section arrives before the slower front, creating a pocket in the form of a gas trap. Short shots, or incompletely filled parts, result. Figure 1.28 illustrates the racetracking phenomenon. Keeping the nominal wall as uniform as possible avoids racetracking.

Another way to achieve uniform walls in thick parts is to core out thick sections (Figure 1.29). This allows improved cooling of the parts, leading to reducing long cycles times found with thick nominal walls without sacrificing part structural integrity.

1.5.3.2 Removing Sharp Corners: Radiusing the Part

One of the most common errors made in plastics part design is the use of sharp corners. Designing with sharp corners can cause the potential of premature part failure in final use since sharp corners are areas of high stress development and drastically reduce the impact performance of a part. In designing with plastics, radiused corners or transitions are highly recommended to eliminate areas of high stress concentration. This will increase the strength of the corner and transition and facilitate flow of the polymer material. Figure 1.30 illustrates a recommended design for using radii in corners.

1.5.3.3 Ribs, Bosses, and Gusset Design

Ribs are longitudinal protrusions on a plastic part that may be used for a number of purposes. They may be used to add strength to a part or to add aesthetic features to a part. Also, a rib may be used to minimize warpage in a part. The use of ribs also

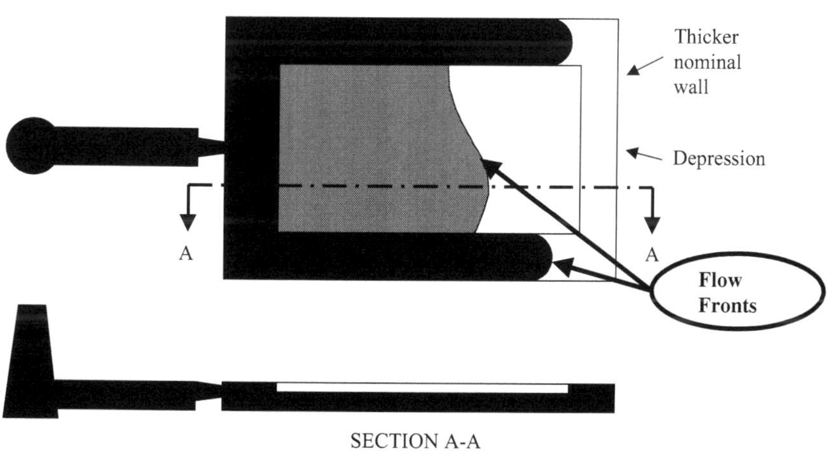

FIGURE 1.28 Racetracking phenomenon.

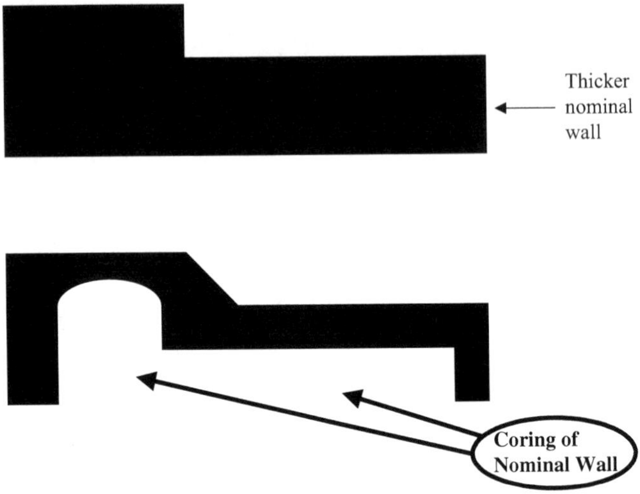

FIGURE 1.29 Coring of plastic parts.

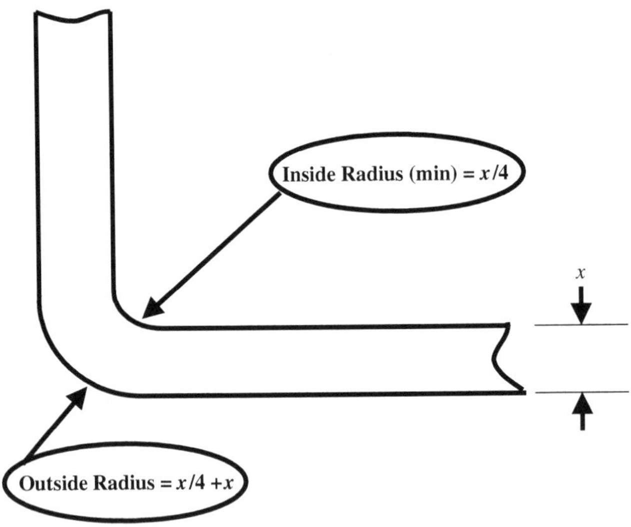

FIGURE 1.30 Radii design for corners.

results in parts that are lighter in weight; use less material, and have shorter molding cycles. Ribs are also used as an alternative to increasing the nominal wall thickness of a part, which leads to increased molding cycles and sink marks on the surface. For rib design, a thickness of 0.5 times the nominal wall thickness and a maximum height of 3.0 times the nominal wall thickness add a minor amount to the total part weight but can provide the same rigidity as designing a thick nominal wall section.

As a result, a few tall, narrow ribs may have more influence on part integrity and rigidity than that of only one tall rib. The advantages of using ribs are that, first, narrow ribs improve part cooling, reducing cycle time, and second, ribs can reduce the chance of sinks marks provided that the rib thickness does not exceed 0.50 times the nominal wall thickness. Figure 1.31 illustrates proper rib design.

Bosses can be defined as protruding, hollowed-out posts on a part that assist in the assembling of one plastic part with another piece, plastic, metallic, or other type of material. Because bosses are frequently used to fasten members, by using methods such as inserts, self-tapping screws, drive pins, expansion inserts, cut threads, and plug and force fits, they are exposed to different forces, strains, and stresses that are not typically found in other sections of the plastic part. As a general guideline, the outside diameter should remain between 2.0 and 2.4 times the outside diameters of the insert, screw, or drive pin. To reduce or eliminate sink, the guidelines provided for boss wall thickness are the same as those used for ribs, 0.50 times the nominal wall thickness. Figure 1.32 shows recommended boss design.

Gussets are reinforcing plates used with both ribs and bosses to add further structural support and improve structural integrity. In designing gussets, there are two parts, the gusset rib thickness and the length of the reinforcing rib to the gusset. The gusset rib thickness should be designed between 0.050 and no greater that 0.75 times the nominal wall thickness. The length of the reinforcing rib of the gusset should not exceed 2.0 times the nominal wall to avoid sinks from appearing on the part. Figure 1.33 illustrates gusset design.

The location of ribs and gussets can affect the mold design of a part. Gate location can work in concert with the position of ribs and gussets to improve part filling.

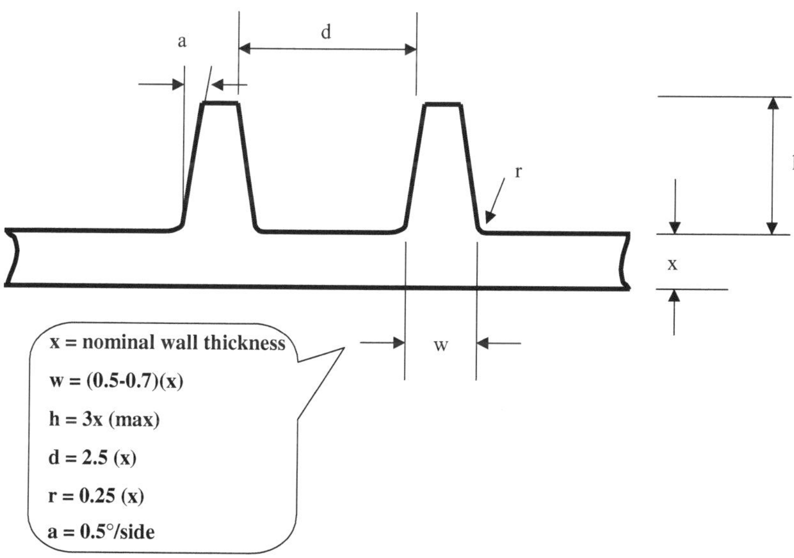

FIGURE 1.31 Rib design for injection-molded parts.

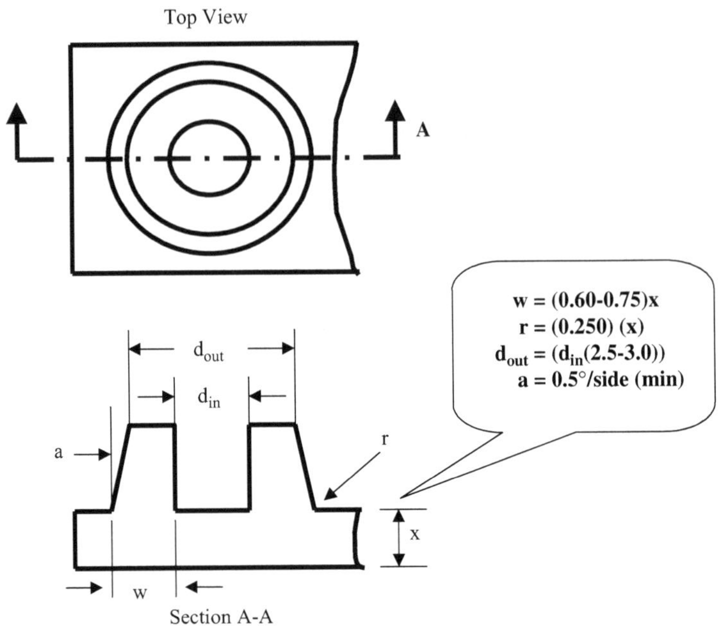

FIGURE 1.32 Boss design for injection-molded parts.

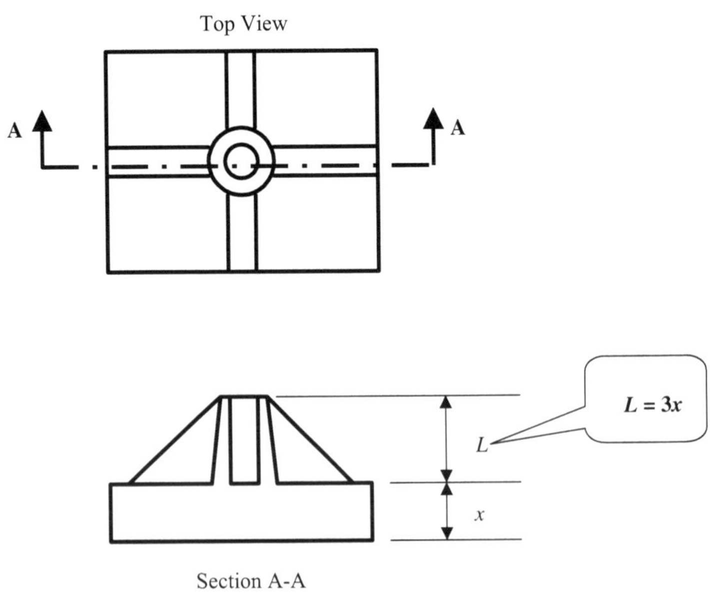

FIGURE 1.33 Gusset design for injection-molded parts.

In this case, the ribs and gusset act as an internal runner or flow director, directing flow in a specific direction. However, poorly located or poorly designed ribs and gussets can hinder flow and can also create burn marks in the part as a result of trapping gas in the tool, otherwise known as *dieseling*. This occurs when ribs and gusset are isolated where gas can be trapped.

1.5.3.4 Draft Angle

Draft angle is critical in part design since this directly affects how easily a part can be ejected from a tool without causing damage to the part or tool. A slight angle is recommended for all thermoplastics to facilitate part ejection. This draft angle is determined by a number of factors, such as plastic material used, part geometry, and mold ejection system used, whether mechanical or air assist. A draft angle of 1° draft per side (Figure 1.34) is used for most thermoplastics, with a minimum of $\frac{1}{2}^\circ$ draft per side. Some materials, such as styrene–acyilonitrile (SAN), can require up to as much as 2° per side. The guideline above is recommended for untextured part surfaces.

For textured part surfaces such as grained surfaces, an additional 1° of draft is added for every 0.025 mm depth of texture, to eject the part and avoid drag marks from forming on the part.

1.5.3.5 Hole Design

Holes are added to a part to provide assistance in assembly, disassembly, or other functions. Design of holes affect the molding cycle time since mechanical action is needed to design some holes, causing the overall cycle to slow down. Hole design can also affect part structural, integrity due to the formation of weld or knit lines.

Hole design can be approached in the same manner as explained in the section on boss design. A general rule is that a wall located near a hole should be a minimum of one hole diameter away to obtain maximum part strength and facilitate melt flow. Figure 1.35 shows this recommended hole design. What this rule takes into account is the formation of weld or knit lines on the backside of the hole near the adjacent wall. In a plastic part, a through hole is preferred to a "blind hole." The reason for

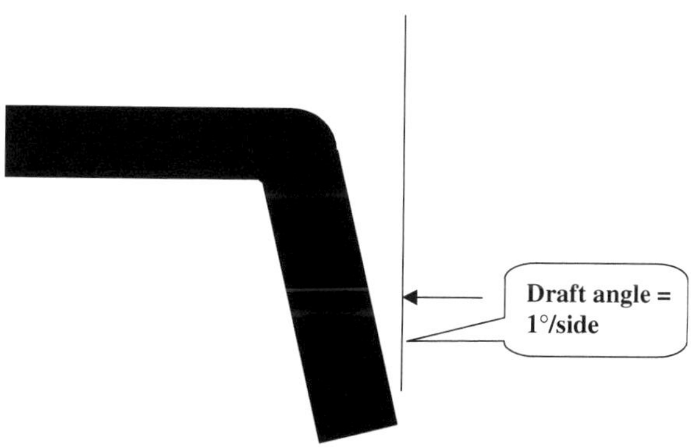

FIGURE 1.34 Draft angle.

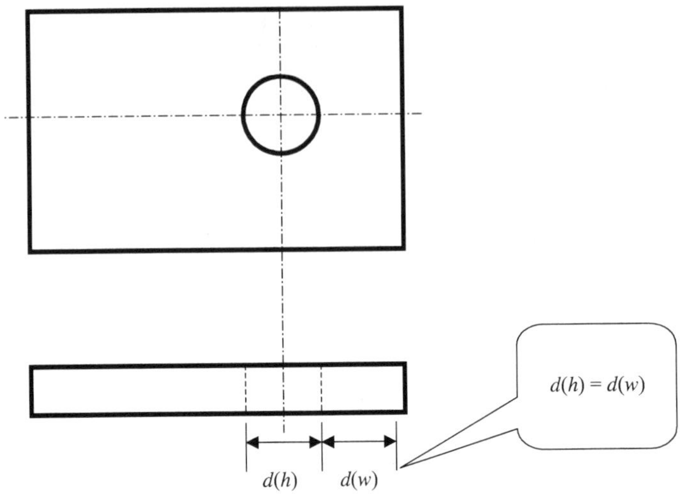

FIGURE 1.35 Optimum hole design.

this is that for a through hole, core pins that are used to form the holes are supported by both mold halves and can withstand the high injection pressures of the plastic melt. Blind holes use an unsupported core pin, which when exposed to the high injection pressures of the plastic melt can bend or deflect, commonly referred to as *core pin shift*. Due to this limitation, if a blind hole is required in a part, its depth hole should not exceed 2.0 times its diameter. In addition, hole roundness or concentricity can be affected by using core pins to form blind holes, since these core pins can become overheated and cause the plastic material formed by the core pin not to cool sufficiently and become distorted during ejection. Alternative mold materials other than steel, such as beryllium copper, can improve core pin heating.

1.5.3.6 Undercuts

Because of the way they are oriented, some features designed into injection molded parts, place sections of the mold in the way of ejection of a plastic part. These protrusions, referred to as *undercuts*, are typically used in articles such as enclosures to provide a snap fit to assemble parts.

The biggest problem of using undercuts is in removing the part from the mold. Sometimes the part can bend enough to eject the part from the mold without damaging the part or the mold, depending on the depth and shape of the undercut together with the plastic material's flexibility, or flexural modulus. Undercuts can be ejected from a mold only if they are located away from stiffening members such as corners or ribs. The part must also have enough room to flex and deform.

The maximum allowance for using an undercut can be measured using the equation shown in Figure 1.36, $U(T)$ is the undercut size measured as a percentage. Recommendations for $U(T)$ vary according to the plastic material used.

For rigid high-flexural-modulus polycarbonates, polycarbonate blends, ABS, filled nylon 6 and nylon 6,6, and polyesters and polyester blends, it is recommended that

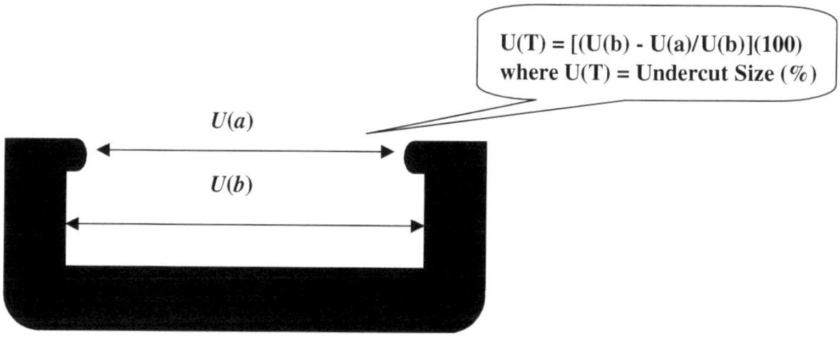

FIGURE 1.36 Undercut design.

undercuts be avoided. However, if undercuts need to be utilized in a part design from these products, a $U(T)_x$ of no more than 2% should be used, provided that the part walls are flexible and beveled to facilitate part ejection. For more flexible or lower-modulus materials such as unfilled nylon 6, polypropylene, polyethylene, and thermoplastic polyurethanes, a *U*(*T*) value of 5 to 10% can be tolerated. However, use of a *U*(*T*) value above 5% risks permanent damage to a part, particularly in polypropylene parts.

For some filled or reinforced plastic materials, such as nylon 6 and nylon 6,6, collapsible cores or split-cavity undercuts are used and are recommended to reduce high stresses in the part. Mold temperature can have an affect on the amount of undercut allowed. For molding at mold conditions below 100°F (38°C), the maximum allowable undercut, *U*(*T*) is restricted to 1%. For mold temperature as high as 200°F (93°C), a *U*(*T*) value of 2% can be used.

1.5.3.7 Thread Design

Molded-in threads can be designed into parts made of thermoplastic materials. Two types of threads are used in plastics part design: internal, found internally in a protrusion such as a boss, and external, located on the outside surface of a boss. When threads are designed, the thread root should be radiused and should not have "feathered" edges. Threads can also form undercuts, and special care should be taken to eject the part from the tool as is taken with undercuts. Usually, mechanical actions such as unscrewing and collapsible cores should be used.

Most standard thread designs can be used in thermoplastics, but keep in mind some of the limitations for the types of threads used. Extra fine threads should be avoided since these are very difficult to fill completely, and if overpacking the mold is needed to fill the cavity, high molded-in stresses can bring potential failure to the part and to the fastening of the part. Also, extra fine threads can strip and may not withstand the high torque forces needed in tightening and fastening the parts sufficiently. No sharp corners should be used, as these can act as a stress concentration in a part, causing the part to fail structurally. To avoid stress concentrators, a radius at the root of the thread is recommended. A gap of 0.31 in. (0.8 mm) should be left at the base of the screw to avoid bottoming out the screw and causing stress concentrations. The minimum active thread length should be 1.5 times the pitch diameter of the thread. The minimum wall thickness around an internal thread should be 0.50 times the major diameter of the thread. Finally, the use of tapered pipe threads can cause excessive stresses in internal threads, and as the threaded part is tightened, more hoop stress can develop. Using threaded pipe threads should be avoided. However, if a mating plastic part is fastened to a metal tapered thread, the external threads on the plastic part can be designed to avoid excessive hoop stresses. Also, a straight thread and rubber O-ring can be used to produce a seal at the part joint.

1.5.3.8 Designing with Weld Lines

Weld lines, known also as *knit lines*, are formed when two melt fronts converge and join, forming a thin, fine line in a part. In a plastic part, weld lines can be formed as a result of flow interruptions that occur as the plastic melt fills the cavity. These restrictions include thin sections, holes, slots, and other types of cored-out areas. Weld lines that form as a result of these interruptions affect the part strength and part aesthetics. There are several factors to keep in mind with coping with weld line development.

Gating number and location affects the number of weld lines that can be found in a molded part. For example, a single gated part usually produces no weld lines if the plastic melt has not been interrupted. However, two gates yield one weld line, and three gates will produce two weld lines. For a part with flow interruptions, the number of weld lines that will be found in the part is the number of gates minus one, so the fewer the gates, the fewer the weld lines. If gates are located in a part so that the plastic melt fronts flow toward each other in a part, a weld line will form once the two melt fronts converge. Here again, if one gate is used, one melt front is produced and no weld line is found. The rules above are simple, but part designs are complex, and using more than one gate or gating where two melt fronts flow toward each other cannot be avoided.

The magnitude of mechanical and aesthetic effects on weld lines depends on a number of variables. Specific resins which flow easily or have lower viscosity and high melt flow rates will produce a weld line that can be less noticeable and produce higher strengths at the weld line. On the other hand, using filled and reinforced grades of thermoplastics can produce weak weld line strength from both having stiffer flow and due to the orientation of the fillers and reinforcements, which is perpendicular to flow at the weld line interface. Part geometry and flow length play a part in weld line strength. For example, if a plastic material needs to flow a long distance around a metal core, as in molding a television housing or automotive bezel, a weld line is created that may not be acceptable aesthetically. Finally, process conditions will cause the formation of weld lines. If a mold is too cool, the plastic melt will set up before the melt fronts join together, giving weak weld line strength and noticeable weld lines.

Several design solutions can improve weld line strength and aesthetics. If a weld line is found in an aesthetic area or in a location where part structural integrity is compromised, increasing the wall thickness about 0.005 in. either side of the weld line may be enough to shift the weld line to a different location. Also, moving the gate location apart also can shift a weld line into another location. Flow directors, usually ribs designed in a part to guide flow to a certain area of the part, is another method to shift the weld line location. Using a finish on the part, such as texturing, can hide the weld line, thus give the surface better part aesthetics. An overflow located at the converging melt fronts is another technique. This permits the material at the weld line to flow into a pocket, allowing the two melt fronts to bond together with higher injection pressure. However, this overflow may need to be removed manually since this may interfere with the part's functionality. Finally, a new technology introduced in the late 1990s is using a porous metal insert to remove trapped gases at the converging melt fronts and providing a higher weld line strength. This is recommended mostly for textured surfaces since on smooth surfaces a slight gloss difference can be found on the part. Figure 1.37 shows both the overflow and porous insert methods for resolving weld line issues.

1.5.3.9 Hinge Design

Merriam-Webster's *Collegiate Dictionary* defines a *hinge* as "a jointed or flexible device on which a door, lid, or other swinging part turns." Hinges can be used for some, but not all, thermoplastics. Where hinges are needed materials that have excellent toughness and ductility, such as polyethylene, polypropylene, nylon 6, and nylon 6,6, are used primarily. Two basic types of hinges are used in plastics part design; molded-in hinges and formed hinges.

The *molded-in hinge* is used more widely since it can be molded by the standard injection molding process. An important polymer property that affects the durability and long-term performance of a hinge is molecular orientation. Molecular orientation transverse to the center axis of a hinge, will produce a hinge that has greater durability and is bettter able to withstand many flexing cycles. This orientation can be

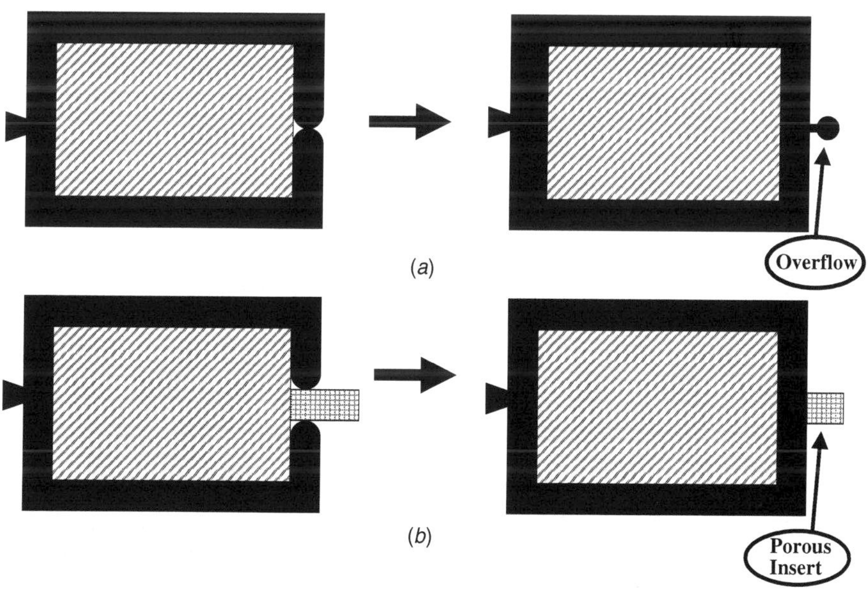

FIGURE 1.37 Weld line solutions: (*a*) overflow design; (*b*) porous insert design.

achieved when the polymer melt flows across the hinge at a high speed at high melt temperatures. If flow moves along the hinge, the flow is similar to converging melt fronts found in weld line formation, causing the hinge to rupture, reducing its durability and shortening its use. Using slower injection speeds may cause the hinge to delaminate and also cause the hinge to fail prematurely. Figure 1.38 shows the critical parameters for designing a hinge. Gate location is another critical factor in hinge design. As mentioned previously, flow across a hinge is found to provide optimum performance. The key for gate location is that gates should be placed so that flow is directed across the hinge and the polymer melt flows evenly across the hinge with minimal hesitation. Figure 1.39 illustrates optimum gate locations to maximize hinge performance, as well as other designs that are not recommended. A number of products, such as video cassette cases, automotive glove box doors, and packaging and container enclosures, utilize molded-in hinges.

Formed hinges are created outside the injection mold tooling as part of an auxiliary process. Normally, two matched dies are brought together and penetrate the wall, pinching the nominal wall thickness of the part. A die is used to compress a slot into a flat surface. This die is heated and allows the material to move away from the die upon contact. The cross section of the part is compressed to form the thin section of the hinge.

In designing hinges, the elasticity of the material is also important in determining hinge performance. Analysis of the elasticity of a hinge can be categorized according to whether the strain in the hinge is purely elastic or whether plastic strain is also present. Plastic strain can be classified as either pure bending strain or bending plus tensile strain. The design of the hinge durability depends on the type of stress it is exposed to in the end-use application.

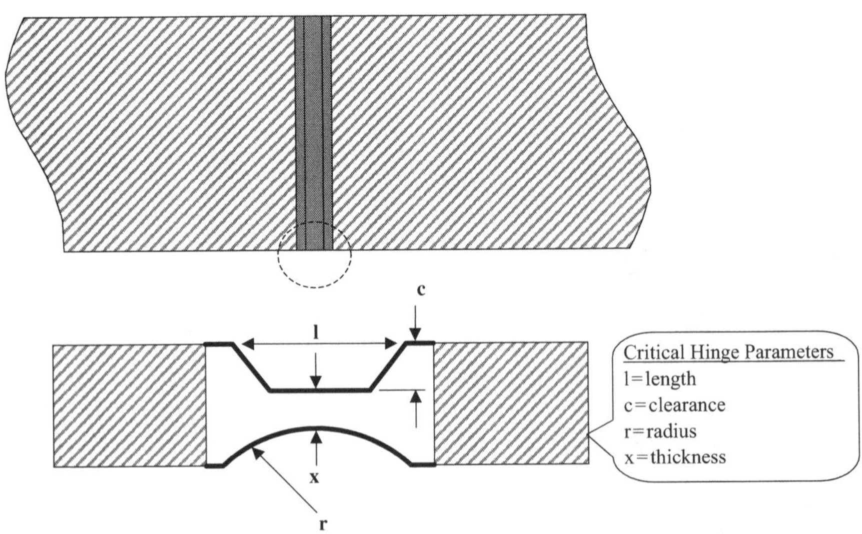

FIGURE 1.38 Critical parameters for a living hinge.

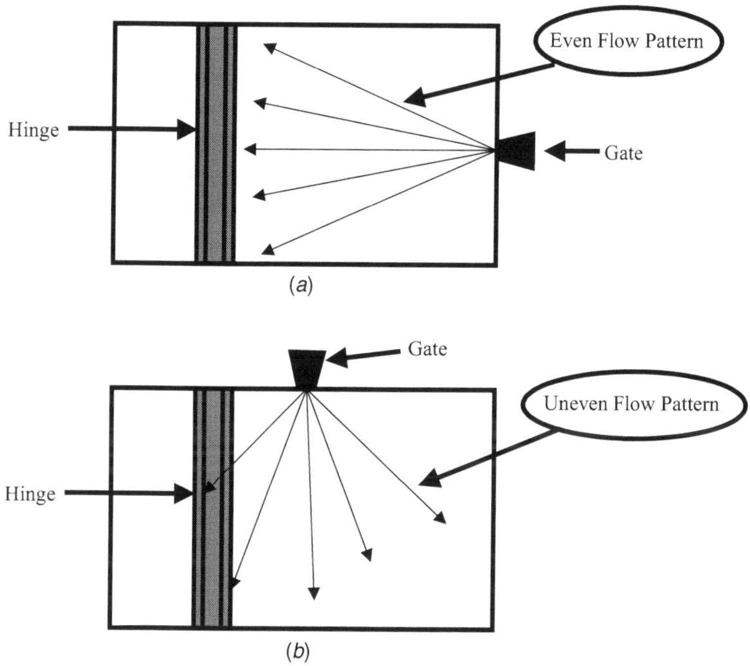

FIGURE 1.39 Optimum gate locations for a living hinge.

A critical parameter in designing hinges is the ratio of hinge thickness to hinge width. This ratio is referred to here as the *hinge ratio* (HR).This can be determined by the following equations:

$$\mathrm{HR} = \frac{H_x}{H_w} \tag{1.13}$$

where H_x is the hinge thickness, in. (mm) and H_w is the hinge width, in. (mm).

If

$$H_x < \frac{L_c}{\pi} - \frac{L_c}{H_w} = 1 - (1 - P_r E)\ 0.5$$

where L_c is the closure limit, P_r is the Poisson's ratio, E is the elongation at break (%). This equation provides just a quick check to determine upfront if hinge f ailure will occur. If elongation at break is less than 100%, the hinge ratio will be relatively insensitive to Poisson's ratio.

1.5.4 Venting the Injection Mold

During the melt-filling phase of the injection molding process,air can be trapped and caught inside the mold ca vity. This trapped air in ca vity is compressed, increasing pressure and temperature at the melt front. This air can rise in temperature similar to superheated steam and can cause defects such as the following:

- Burn marks, also referred to as dieseling
- Dulled or poor surface finish
- Weak weld lines, causing the part to fail in service prematurely
- Short shots or incompletely filled parts
- Variations in part dimensions
- Corrosion inside the mold cavity, damaging the cavity surface

This highly compressed air needs a location to escape entrapment in the mold cavity. The purpose of venting is to allow this highly compressed air to evacuate the cavity and escape out into the atmosphere. There are a number of guidelines in using venting in injection molding tooling. Table 1.12 provides guidelines for a properly venting injection molding tool.

1.5.5 Mold Shrinkage

Mold shrinkage is one of the properties of a material that is needed in determining the final part dimensions required. This value indicates the amount of contraction that a part exhibits after it has been removed from the mold and is then cooled and conditioned at 73°F (23°C) for a period of 48 hours.

TABLE 1.12 Guidelines for Mold Venting

Vent Location Recommendations	Vent Sizing
1. Space vents approximately 2 in. (50 mm) along parting lines	1. Land depth 0.0015–0.002 in (0.04–0.05 mm) recommended
2. Space vents approximately 2 in. (50 mm) along parting lines	2. Land length 0.0015–0.0025 in. (0.04–0.06 mm) recommended
3. Locate venting in end of fill areas	3. Land and relief widths 0.25 in. (6.4 mm) minimum recommended
4. Place vents in areas where gas may be trapped, such as corners and deep ribs	4. Relief depth 0.010–0.020 in. (0.25–0.50 mm)
5. Place vents in the runner system	
6. Locate vents in ejector pins by machining a flat section along the length of the ejector pin	
7. Place vents in sleeve ejectors and moving cores, if these are used in tool	
8. Use porous inserts in areas where melt fronts converge (see Figure 4.21)	
9. Use overfills at areas where melt fronts converge (see Figure 4.21)	

Source: Multiple industry sources.

Shrinkage is determined by the following equation:

$$S = \frac{L_f - L_o}{L_o} \tag{1.14}$$

where S is the mold shrinkage (in./in. or mm/mm), L_f the final part dimensions (in. mm), and L_o the mold cavity dimensions (in. mm). Mold shrinkage can be affected by a number of variables. Slower part cooling, injection pressures, and part thickness can change the mold shrinkage for a given material. The addition of fillers and reinforcements, such as glass fiber or mineral filler, can reduce mold shrinkage from what is found with an unreinforced or neat material. However, the type and classification of plastic material has the most influence, as shown in the following paragraphs.

Plastic materials can be classified into two categories: amorphous and semicrystalline. *Amorphous materials* typically have no defined melting point but have a broader melting point range. When amorphous materials cool during the cooling phase of the injection molding process, they return to a rigid polymer. The polymer chains that make up the amorphous material have no specific orientation. As a result, these amorphous materials have lower mold shrinkage. Examples of amorphous materials are polycarbonate, ABS, and polystyrene.

Semicrystalline materials do have a defined crystalline melting point. The polymer chains arrange themselves in ordered molecular configurations. These ordered areas are crystals that form when the polymer is cooled from its molten state. As a result of the cooling and formation and the increased packing of the molecular chains in these crystalline areas, the mold shrinkage for semicrystalline materials is higher than for amorphous materials. The degree of crystallinity in a semicrystalline material is affected not only by the chemical structure of the polymer, but by the process conditions that are used to manufacture the part. In particular, the rate of cooling during the injection molding process has the greatest effect on the degree of crystallinity. Slight variations in the molding process, such as mold temperature, melt temperature, and part thickness, play a major role in determining the final dimensions in semicrystalline materials. Examples of semicrystalline materials are nylon, polypropylene, and polyethylene.

Table 1.13 lists a number of plastic materials, both amorphous and semicrystalline, and their mold shrinkage.

1.6 TROUBLESHOOTING THE INJECTION MOLDING PROCESS

1.6.1 Typical Injection Molding Problems and Solutions

The injection molding process is basically a straightforward process involving injecting molten material into a defined shape and letting it cool until it solidifies. However, resolving a problem in injection molding can make this simple process more complex to understand.

In this section we describe a number of problems that are commonly encountered with injection molding. If there were a basic rule that could be followed, it could be described by the injection process triangle (Figure 1.40). The injection molding process is influenced by three factors: (1) pressure, temperature, time. Any changes

TABLE 1.13 Mold Shrinkage for Thermoplastics

Material	Mold Shrinkage (in./in.)
ABS	0.004–0.007
Acrylic	0.002–0.010
ABS–nylon	0.010–0.012
ABS–TPU	0.005–0.007
Acetal	0.020–0.035
Nylon 6	0.007–0.015
Nylon 6,6	0.010–0.025
PEI	0.005–0.007
Polycarbonate	0.005–0.007
PC–ABS	0.005–0.007
PC–PBT	0.008–0.010
PC–PET	0.008–0.010
Polyethylene	0.010–0.030
PPS	0.001–0.002
Polypropylene	0.008–0.030
PPO	0.005–0.007
Polystyrene	0.004–0.008
Polysulfone	0.001–0.003
PBT	0.017–0.023
PET	0.017–0.023
SMA	0.005–0.007
TPO	0.012–0.016

Source: Multiple industry sources.

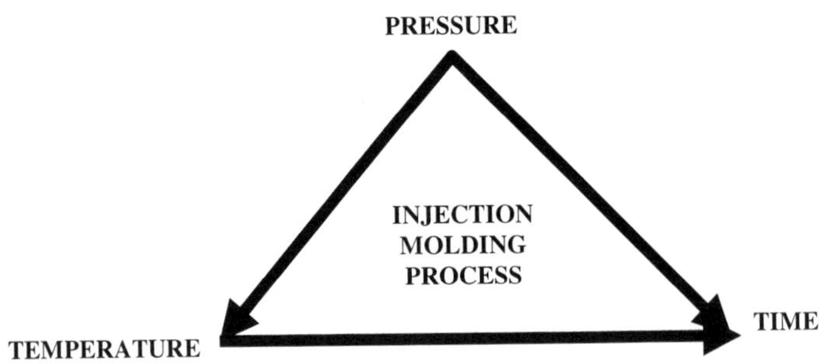

FIGURE 1.40 Injection molding triangle.

in one leg of the triangle affects the others. For example, an increase in cooling time can affect the temperature of the part or the temperature of material as it sits in the barrel of the machine. Also, a change in the melt temperature of the material can affect the pressure needed to fill the part cavity. The combinations can become endless when you think of this triangle.

In this section we review common problems found in injection molding as it pertains to the cause and to ho w to resolve these problems. The solutions provided for resolving these common molding problems will be covered in the same way that this chapter has been covered by breaking these down into the phases of the injection molding process. In addition, troubleshooting guides for a number of commonly used thermoplastic materials are provided in Section 1.6.2 to address problem solving for individual materials.

1.6.1.1 Blush at the Gate

Blush is a hazy surface imperfection that is found at the gate of the part. It appears as a visual difference in color and/or gloss surrounding the gate and is usually in the direction of flow. Blush occurs as the melt flows through the gate. Due to the high pressures and velocity going through a gate, a phenomenon called *melt fracture* occurs. The material is squeezed through the small gate opening, and as this takes place, the polymer chain is extended. The material expands rapidly as it hits the cavity wall, the polymer material will cool, and the polymer chains will contact. Due to this cooling and contraction, a haze or discoloration will appear at the gate area, referred to as gate blush. The higher the injection speed, the more pronounced the gate blush.

Melt-Conveying Phase A number of solutions can be applied to resolve gate blush. The first is to reduce the injection speed so that the polymer chains do not extend as much as with higher speeds. This will lay down a clean skin of material without gate blush. Velocity profiling of the process can also be used to start with a slow injection speed, building to a higher injection speed as the cavity is filled. One way is to increase the melt temperature to provide easier flow to the material. An increase in nozzle temperature can also produce a higher melt temperature as it enters through the gate.

Melt-Directing Phase One solution, which is often overlooked, is the use of a cold slug well. The purpose of a cold slug well is to receive the first slug of material from the nozzle and keep it from interfering with the flow path of the hotter material behind. Lack of a cold slug well can force this colder material into the gate and into the part showing up as gate blush. It is highly recommended to check if a cold slug well is present at the large-diameter end of the sprue.

Melt-Forming Phase Another solution to eliminating gate blush is to increase the mold surface temperature so that the cooling rate at the gate will be slowed and will be closer to the cooling rate of the rest of the part. Gate size can be increased to provide less resistance from the material in entering the cavity.

1.6.1.2 Brittleness of Parts

Part brittleness is caused most often by a loss of molecular weight in the polymer material. A reduction in molecular weight results in a reduction of the mechanical properties of the material, such as tensile strength, elongation, and impact strength.

Material Feed Phase Brittleness from material degradation can be caused if materials are not dried sufficiently. This applies specifically to hygroscopic materials. In this case, a chemical reaction called hydrolysis takes place when materials are melted with moisture present. When hydrolysis occurs, a loss in molecular weight results in a reduction in physical properties. Excessive use of regrind can cause part brittleness since regrind has already been exposed to heat and as a result suffers a loss in molecular weight. Keeping regrind at lower levels can avoid the effects of molecular weight reduction.

Melt-Conveying Phase This loss of brittleness and molecular weight can be the result of several issues. Melt temperatures may be higher than recommended process temperatures. In this case, melt temperatures at the middle of the recommended melt temperature range should be used. Another cause of this molecular weight reduction is a result of material sitting in the barrel for long periods of time, referred to as residence time. The barrel-to-shot ratio (BSR) is a measurement, in terms of percentage, used in determining residence time. Equation (1.7) gives a formula for determining the BSR.

For most materials, the BSR should fall between 30 and 65%. If this falls well below 30%, this will indicate a residence time problem that can lead to brittleness. There are other solutions that can be applied to eliminate part brittleness. A reduction in nozzle temperature will reduce melt temperature avoiding brittleness from occurring. High screw speeds and back pressure can "overwork" the material, resulting in an increase in melt temperature. Reducing screw speed and lowering back pressure can prevent resin melt temperature from exceeding recommendations.

Melt-Directing Phase The gate location affects the orientation of the polymer molecules. Polymer materials exhibit two types of orientation behavior: isotropic and anisotropic. In *isotropic behavior*, mechanical strength varies little in the machine and cross-machine direction. This condition provides flexibility in gate location. In *anisotropic behavior*, mechanical strength is higher in the machine direction than in the cross-machine direction. In this case, gate location may becomes more critical.

Melt-Forming Phase Part design can play a major role in part brittleness behavior. Increasing wall thickness, robust rib designs, avoidance of sharp corners, and addition of radii can reduce the chance of brittle parts.

1.6.1.3 Burn Marks

Burn marks are normally black discolorations found on a part surface. However, depending on the material, this discoloration can range in color from black to tan to white. There are two basic causes of burn marks.

Material Feed Phase Contamination is one possible cause of burn marks. This can occur when the melt temperature of the contaminant is lower than the plastic material being molded. The burned contaminant can take the form of burn marks. It is recommended that any contaminated material be eliminated from the material feed system.

Melt-Conveying Phase Excessive temperature is caused primarily by high melt temperature, and as discussed previously under brittleness, excessive residence time.

Again, a BSR value below 30% can cause the potential of overheating a material, leading to gas evolution from the material. Other solutions include reducing melt temperatures to the midrange of the melt temperature recommended. Reductions in injection speed can reduce the chance of overheating the material. Decreasing back pressure and screw speed will lower melt temperature. Material can also hang up in areas such as check rings or in hot manifolds and become exposed to excessive heating. Check rings should be checked for hang-up areas and areas of wear, and hot runner manifolds should be designed to eliminate hang-up locations. Finally, heater bands can malfunction, causing a nearby zone to overheat and overcompensate by overheating. Calibrating and checking heater band function need to be checked to avoid potential overheating and burn marks on parts.

Melt-Directing Phase Gate diameter is critical to prevent part burn marks. Too small a gate diameter leads to high shear conditions, which produce burn marks in the part. Gate size should be designed according to recommendations from the material supplier. Another source of burn marks in the gate is burred or rough gate orifices resulting from rough machining or gate wear. Gate areas should be polished and checked for wear, and repaired. One option is to use a gate in an insert and replace the worn gate with an insert instead of remachining a new gate.

Melt-Forming Phase Dieseling occurs when gas is trapped in a mold cavity, typically in the last section of the part to be filled. The pressure of the melt compresses this trapped air, and the combination of pressure and heat burns the material at the melt front, leading to dieseling. Dieseling on the part takes the form of a burn mark and poor surface finish. There are several options to resolve dieseling. Reducing injection fill time will allow more time for gases to escape into the vents without being trapped, assuming that the vents are all clear of clogged material. Cleaning and maintaining vent areas will also help in eliminating the dieseling effect. An insufficient number of vents can also create dieseling since there are not enough passages for the trapped gas to escape. It is recommended to use as many vents as possible without causing flashing of the tool. Fine adjusting of clamp pressure can be used to eliminate dieseling since a slight reduction in clamp pressure can act as a large-scale vent to remove trapped gas. However, there is a fine line in using this remedy, as too low a clamp pressure leads to potential part flashing. Gate locations can cause dieseling, especially if flowing around a hole or at a weld line. Vents can be added into near a hole or where the two melt fronts converge to form the weld line. Also, porous metal inserts can be placed at weld lines and around holes to remove trapped gas.

1.6.1.4 Delamination

Delamination is a condition that exists when the surface skin of a molded part can be physically separated from the part, which can lead to a reduction in mechanical strength.

Material-Feed Phase One of the main causes of delamination is incompatibility of mixed materials due to material contamination. For example, when a material such as nylon is contaminated or mixed with another material, such as polyethylene, delamination will take place, due to the difference in the rheological properties of the two materials. Contamination can also occur from moisture in material that has not been dried sufficiently. Bubbles or blisters are formed and trapped on the part surface in a fully packed cavity, and the skin formed by the trapped moisture can be separated. Recommended drying temperatures and times should be used based on material supplier recommendations.

Melt-Conveying Phase Nonuniform melt temperatures are yet another cause of delamination. Colder material will flow much slower than the hotter material, causing the cold skin to separate from the hotter forming skin. Better mixing and better melt uniformity as well as increasing melt temperature will is needed to prevent delamination in this case. Better screw design will also help in providing improved melt uniformity.

Melt-Directing Phase High packing pressures can cause delamination at the gate. When a part shrinks away from the gate or sprue bushing, high packing pressure can force excess material into the gap between the part and mold cavity. A thin layer of material is formed on the surface of the part. Upon part ejection, this thin layer can separate and delaminate. Reducing pack pressures will prevent this buildup of excess material from forming.

Melt-Forming Phase Too hot a mold temperature can cause delamination, causing the skin to stick on the hot mold surface and separate. Reducing mold temperature will eliminate this condition.

1.6.1.5 Dimensional Stability

Dimensional stability problems occur when a part is inconsistent in part size or in part weight from shot to shot. Parts that are underweight can suffer in structural integrity; overweight parts can also suffer from structural integrity due to molded-in stresses and increased part costs.

Melt-Conveying Phase Shot inconsistency can occur when a number of conditions occur. When injection pressure transfers on time or hydraulic pressure, part weight and size can vary since any change in hydraulic pressure can cause too little material to fill one shot and too much material the next. Transfer due to screw position is found to provide improved part weight and dimensional stability. The screw position can be set to fill the cavity about 95%, and then the remaining 5% can be filled on hold pressure.

The pack pressure is increased to fill the part with sufficient material, the injection forward time is increased, and the fill time to force more material into the cavity faster is also increased. The cushion is examined visually to verify if it is consistent. If variations can be seen, adjustments to shot size and back pressure can be made to improve cushion consistency. A worn check ring can also cause cushion variation. This can be noticed by observing the screw position after moving forward.

If the screw shows some rotation after the screw has moved fully forward at the end of its stroke, this indicates a worn check ring. Replace the check ring if this condition is seen. Increasing melt temperature within the recommended temperature range will allow the melt to pack out the cavity more than at lower melt temperatures, resulting in increased part size.

Melt-Directing Phase Undersized nozzle or sprue size can restrict the amount of material that enters through the gate and into the cavity. Enlarging the nozzle and sprue will permit more material to flow into the gate and cavity.

Melt-Forming Phase Too short a closed mold time will cause the part to underpack and shrink more. Increasing mold closed time will keep more pressure on the melt and cavity and increase the part size. High mold temperatures can result in too small parts. This can be noticed even more in semicrystalline materials such as polyethylene and polypropylene. Lowering mold temperatures will allow the material to cure faster in the mold and reduce shrinkage. Fixturing parts are another method that can be used to prevent parts from being undersized. This method involves holding the part after it has been ejected from the mold and is still warm in a fixture or frame that allows the part to cool in a fixed position. Conversely, parts can be made that are oversized, indicating that not enough shrinkage is occurring. The following are solutions for resolving oversized parts.

- In the material-conveying phase, excessive packing pressure can push too much material into the cavity. Reducing packing pressure permits the material to shrink away from the cavity.
- In the melt-forming phase, the cooling time can be reduced, allowing the material to be warmer when it is ejected, thus allowing the part to shrink more once it is removed from the mold. A cold mold surface will cure the part faster, resulting in a cooler ejected part that will not shrink much more. A hotter mold will slowly cure the part, and upon ejection, will be warm and continue to shrink.

In materials such as polypropylene (PP), the rate of crystallinity plays a role in determining mold shrinkage. A slow rate of crystallinity will cause a material to shrink more. To prevent undersized parts, nucleating agents are used in PP to increase the rate of crystallinity, causing the shrinkage to be increased. As shown in a study by Gipson, Grelle, and Salamon [41], using nucleated grades of PP that are commercially available can be used to increase shrinkage of the final part.

1.6.1.6 Ejector Pin Push Marks

The role of the ejector pin is to push the part from the mold after cooling. However, localized deformation can occur on the part as a result of the forces placed on the part by the ejector pins. This deformation can take the form of stress whitening on the part surface. Ejector pin push marks can be caused by several factors.

An insufficient number of ejector pins for ejecting a part can cause high ejection forces, forming concentrated areas of stress on the part. Adding more ejector pins to the mold can alleviate this and facilitate part ejection. Another method to reduce the possibility of pin push marks is to add added draft angles to eject the part easier and to polish any imperfections on the part surface resisting part ejection.

Placement of ejector pins is critical to preventing pin push marks. Areas that can resist ejection should be avoided in placement of ejector pins. Uneven ejector pin lengths can force a part to eject unevenly, forcing ejector pins to distort and deform the part. Ejector pins should all be machined to the same length. In addition, placement of ejector pins should be distributed so that ejector forces are spread evenly over the ejector plate.

Ejecting a part that has not cooled sufficiently can cause an ejector pin to distort or even puncture and push through the part. Improved cooling of the part either by increasing cool time or decreasing mold temperature will allow the part to cure enough to resist pin push. Ejecting from cores provides a potential for pin push, especially with high-shrinkage materials such as polypropylene and polyethylene. If ejection takes place after the part has shrunk onto the core, the amount of force needed to eject the part is high, giving way to formation of pin push marks. To reduce ejection forces, it is recommended that the part be ejected before the part shrinks fully onto the core.

Another problem in molding around cores is that a vacuum can be formed beneath the core, causing the part to resist being ejected. Vents have been used to try to force this vacuum out, but this method has had limited success. An air poppet assist located in the core to reduce ejector forces has been used with a lot of success.

1.6.1.7 Flashing

Flashing is defined as excess material that has exceeded the boundary of a mold beyond the mold cavity. Flash can be found in a number of areas. The parting line is one of the more frequent locations where flash is found. When a mold does not clamp fully, this provides a space for material to flow. Vent areas are another location of flash, especially if vents are cut too deep and/or wide. If not fully closed, movable mold sections such as slides can move back, due to the pressure of the melt front permitting material to flash into the space resulting from the relocation of the slide.

Material Feed Phase One cause of flash on a part is inadequate drying of the material, especially hygroscopic materials that can absorb moisture. The addition of water to many hygroscopic materials, such as nylon and polycarbonate, can reduce the viscosity of the material by breaking down the molecular weight of the polymer. Moisture in the form of bubbles creates an easy-flowing polymer. Drying the material according to recommendations needs to be followed to avoid this phenomenon.

Melt-Conveying Phase One of the most common remedies for eliminating flash is to reduce injection pressure. The best way to look at flash is to separate injection pressure into injection booster (first stage) and holding pressure (second stage). Injection booster pressure can cause flash in several ways:

- If the feed setting or shot size is too high, excessive material can flash around the mold. Reducing the feed setting should reduce the amount of material being jammed into the cavity.
- Excessive melt temperatures allow the material to flow easily, especially if a mold parting line is not fully clamped. Using recommended melt temperatures, specifically in the midrange recommended, can reduce the chance of flashing.
- Injecting the material at a high speed can cause excessive heat due to shear, causing the material to flow easier. Decreasing injection speed will reduce shear heating, decreasing the melt temperature.
- Flow properties of the material, such as viscosity and melt flow rate (MFR), affect the amount of flashing found. Low-viscosity, high-MFR materials will have a tendency to flow easier, raising the potential of flashing. High-viscosity, low-MFR materials can resist flashing since they display stiffer flow properties.

Holding pressure can cause flashing, especially when packing pressures are too high, forcing more material into the cavity. Reducing hold pressure will reduce the chance of flashing. Reducing the amount of screw feed or reducing the cushion will reduce the risk of flashing, due to less material being forced into the cavity.

Decreasing melt temperature will also prevent flashing since the viscosity of the material increases resisting flow. Too long a melt residence time in the barrel will overheat the material and reduce viscosity, allowing the material to flow easily. Once again, using no less 30% of the barrel capacity and using the right-size barrel for the part being molded helps prevent overheating and ultimately flashing the mold.

Melt-Directing Phase An insufficient number of gates can create high injection pressures and thus increase the chance of flashing in a part. This can be resolved by either increasing the number of gates to reduce injection pressure by distributing the pressure over more gates, or by using other gating technologies, such as sequential valve gating, which uses hydraulically control pins that control the flow in each gate. This will make it possible to control how and when the material enters each gate.

Melt-Forming Phase If mold temperatures are too high, the material will flow more easily increasing the chances of part flashing. Too deep a flow path allows the material to follow right outside the mold cavity area. Vent depths depend on the type of material used. Consulting with the material supplier as to vent depths suggested is a good option. If an injection machine clamp is too low for the part being molded, the mold platen can actually be blown open, causing material to flow between the open halves of the mold. The projected area of the part is one method used to determine the size of the machine and clamp tonnage. Equation (1.1) can be used to determine the clamp tonnage required.

Obstructions between the mold halves created from surface imperfections on the mold surface can prevent the mold from closing, causing a location for material to flow and flash to occur. Polishing these surface imperfections will allow the mold to close completely. Conversely, a worn-out area in a mold can cause localized flashing to be found on a part. Machining the tool to improve the flatness of the tool can eliminate flash. Another costly option is to build another tool.

Support inside the tool to resist the high pressures applied is important in preventing flash. When a tool is designed with not enough support to the mold surface, flexing of the mold surface may occur, which will flash the tool. In the mold design phase, sufficient support of the mold surface and cavity needs to be designed to prevent flexing of the surface.

1.6.1.8 Gloss

Gloss is defined as the amount of light that is reflected off a surface. The rating used for gloss is based on how much light is reflected at different angles and the degree of light that is scattered. Parts with high gloss reflect the majority of light with very low scatter. Low gloss is low reflection at differing angles with large scatter. The correct amount of gloss is determined by the requirements of the end-use application. The following solutions will provide ways to control gloss and assume an untextured part surface.

Melt-Conveying Phase Cold melt temperatures provide low gloss on untextured surfaces. If higher gloss is needed, processing at higher temperatures is needed. In some cases, too hot a melt temperature can also cause low gloss. In this case, additives that are mixed in with the base resin, such as plasticizers or some flame retardants, can bloom on the surface. Reducing the melt temperature may help in this situation. Low packing pressure is another cause of low gloss, since the plastic material is not fully packed onto the cavity surface and does not replicate the surface. Increasing pack pressure will allow the melt to pick up all the details of the mold cavity. Injection speeds that are too slow will reduce the gloss on the surface since these also may cause the melt not to replicate the mold surface. Increasing injection speed will raise the melt temperature and pack out the part.

Melt-Forming Phase Cold mold temperatures will give low gloss surfaces since this causes the melt front to cool too soon and not replicate the cavity. Higher mold temperatures will keep the melt warmer for a longer time, enough to fill the cavity. The type of mold surface, such as a dull surface, gives a low gloss, and polishing and cleaning the cavity will increase the gloss. Insufficient venting in the mold can cause gases to be trapped, creating the dieseling effect. This causes the surface of the part to show low gloss. Following recommendations for venting a specific material can prevent this. Part sticking in the cavity or on a core upon ejection results in drag marks on the part, resulting in a low gloss surface finish. To prevent this, the cavity or core should be checked for machine marks that may be found on the surface that are not in the direction of part ejection. These machine marks can act as an undercut and resist part ejection. Machining of the core and cavity in the direction of ejection will prevent this from occurring.

Our discussion of gloss has been in terms of untextured surfaces. The effects on gloss are different for textured surfaces.

Textured parts are made up, on a microscopic scale, of a group of peaks and valleys. As a result, light scattering is different from that in untextured parts where light is reflected at a very low scatter, giving high gloss. However, textured surfaces cause more light scatter, producing low gloss. So in this case, the texture provides low gloss. If a part is not fully packed out, or if the mold and melt temperatures are too cold, the peaks and valleys of the texture will not fill completely, creating a microscopic short shot. This will result in high gloss on the textured part. High mold and melt temperatures, high packing pressures, and high injection speeds are needed to achieve low gloss since they will all help to fill all the peaks and valleys completely and give the textured part low-gloss surface.

1.6.1.9 Jetting

Jetting is a phenomenon that occurs in injection molding when molten plastic material flows through the gate and forms a wormlike formation without laying down a skinned surface. This formation becomes cold, and as molten material is injected around it and bonds poorly with the cold material, a weakness develops in the part. This cold, solidified formation can also trap air and create a void in the part, also reducing the part's structural integrity. The surface appearance can vary as a result of differences in color or gloss.

Melt-Conveying Phase One of the primary contributors to jetting is excessive injection speed. At high injection speeds, the molten material will shoot rapidly into and across the cavity, forming a "worm." Reducing injection speed and using injection speed profiling will eliminate jetting. In using injection speed profiling, it is recommended that filling be slow at the start and increase as the gate is formed. Conversely, slower injection speeds can cause the melt to cool too quickly, leading to jetting. In this case, increasing melt temperatures will reduce the chance of jetting. Low melt temperatures can also cause jetting because at lower melt temperatures, more pressure is needed to force the material to flow. This higher pressure will again force the material to shoot across the cavity. Raising the melt temperature will allow the material to flow easily and at reduced pressures.

Melt-Directing Phase Gate design is very critical in resolving jetting issues. Gate diameters that are too small form a restriction for the melt that causes the injection pressure in the runner to increase. This higher pressure will force the melt to shoot across the cavity. Increasing the gate diameter will reduce the runner injection pressure. Gates located in an area of a part where the melt is hitting a hollow cavity or core opening will cause the melt to worm. Relocating the gate into a barrier, such as a cavity wall, will prevent worm formation and eliminate jetting.

Melt-Forming Phase Low mold temperatures contribute to the jetting phenomenon by causing the melt too cool too quickly as it hits the cavity, forming poor bonding between this colder material and the hotter material flowing behind. Increasing mold temperature will keep the melt warmer and provide better bonding to the hotter flowing material.

1.6.1.10 Nozzle Drool

At the nozzle of the barrel, molten material can flow continuously, especially during part ejection. This can be noticed inside between two open mold platens. Once in contact with the cooler mold surface, this continuously flowing material cools and solidifies. As the mold closes on this cooled material, this solidified material can be caught in the closed mold and cause mold damage. This process of continuously flowing material out of nozzle is referred to as *drool*. Drooling can be found not only at the nozzle but can also occur at the sprue bushing, hot tip, or at the gate.

Material Feed Phase The presence of moisture causes a reduction in the molecular weight of the material and promotes drooling as well. Drying the material according to recommended times and temperatures is needed to prevent drooling in this case.

Melt-Conveying Phase The excessive melt temperature of the material is another cause of drool. Barrel temperatures set too high will increase the melt temperature of the material. Lowering the barrel temperatures, starting with the middle zones, will lower the material viscosity and reduce drool. Lowering the setting of the nozzle temperature will also reduce drool. Location of the nozzle heater too close to the nozzle orifice can cause drool. Relocating the nozzle heater farther back on the nozzle will reduce the temperature at the tip and orifice of the screw. Nozzle shutoff valves have been used as another option to prevent material from drooling between mold halves. High back pressures can create drooling due to excessive working of the material, increasing the melt temperature. Lowering back pressures can eliminate or reduce drool. The use of decompression or "suck back" forces the melt back into the nozzle and can be used to prevent nozzle drool.

Melt-Directing Phase In the case of hot runner molds, high manifold, gate, or hot tip settings can cause drooling to occur away from the nozzle in the tool. Lowering the temperatures in one or all of these areas will decrease the chance of drool.

Melt-Forming Phase Drooling is noticeable when two mold halves are in the open position when a part is ejected from the cavity. Too long a mold open time will give more time for molten material to drool. Reducing mold open time can help by not allowing much time for drool to occur between the mold halves.

1.6.1.11 Nozzle or Hot Tip Freeze-off

In comparison to nozzle drool, material at the nozzle can solidify and restrict and block the flow of material. This is found mostly during the cooling cycle of the process or if the injection molding process is interrupted. This condition is called *nozzle freeze-off*.

Melt-Conveying Phase Melt temperature that is set too low can cause the material to cool rapidly and freeze off. Running melt temperatures higher will slow the time required for the melt to cool and keep the melt molten longer. Back pressures that are too low do not provide enough mixing to the material to raise the melt temperature. Increasing back pressure will also help raise the melt temperature. At the nozzle, if the nozzle tip temperature is too low or the nozzle orifice is too small, nozzle freeze-off can happen. Poor control of the nozzle heater can create cold spots at the nozzle, causing freeze-off. Positioning nozzle heaters on the nozzle and maintaining heat control in this area can reduce the risk of freeze-off. Raising the nozzle tip can maintain the higher melt temperature from the barrel, preventing freeze-off. Using a larger nozzle orifice size will slow cooling of the material in the nozzle and prevent freeze-off. Insufficient decompression or suck back can rapidly cool material at the nozzle tip and freeze off the nozzle. Using more decompression will force material back into the hotter nozzle.

Melt-Directing Phase In hot runner molds, controlling manifold temperatures is critical in preventing freeze-off. Manifold temperatures that are too cold can cause freeze-off. When hot tips are installed, it a check with the manifold designer is recommended so that after the hot tips expand, a gap is provided to prevent surface contact of the hot tips.

Melt-Forming Phase Mold temperatures that are too cold, especially for hot tips in hot runner tools, can freeze-off the melt from entering the cavity. Higher mold temperatures will help in keeping the melt front at the tip molten.

1.6.1.12 Screw Slippage

After the screw injects material forward into the cavity, the screw retracts back into position to prepare to inject material into the cavity for the next shot. As the screw retracts, hesitation of the screw can sometimes be noticed. This hesitation can prevent the screw from retracting back in time before the mold opens. The common resolution to this is to extend the screw retract time, but this lengthens the overall process cycle. This can negatively affect the economics of the process. This hesitation is referred to as *screw slippage*.

Material Feed Phase In the hopper of the injection molding machine, material in pellet form can prematurely melt or soften in the hopper and conglomerate into large clumps of material. This phenomenon is called *bridging*. When this occurs, material is blocked from entering the rear flights of the screw, causing the screw to slip and spin and slowly retract. Cooling water is run at the feed throat of the hopper to keep this area cool and prevent premature melting or softening of the material. Another problem that can be found in a hopper that contributes to screw slippage is feed problems in the hopper feeding systems. Excessive fines, long strands, and angel hair from the pellet material can clog a material handling system and block material feed.

Melt-Conveying Phase Controlling the rear zone of the barrel is one item critical to controlling screw slippage, depending on the material. For amorphous materials such as ABS, polycarbonate, and polystyrene, lowering the rear zones will reduce screw slippage since these materials do not require high input for melting. However, for semicrystalline materials such as nylon, polypropylene, and polyethylene, more heat input is required to alter the crystalline polymer to act as an amorphous polymer for processing. Increasing the rear zone is needed in this case. Another concept tried in many molding facilities is to run a reverse temperature profile. This involves running the barrel profile from the rear to the nozzle from high to low temperature as opposed to the traditional low to high. Melt temperatures that are too low will cause excessive slippage in the screw; raising the melt temperature will help. Heater band consistency is also important since hot spots can occur on the barrel and slow down screw recovery. Calibrating heater bands and replacing malfunctioning heater bands can reduce screw slippage. High back pressure promotes screw slippage, and reducing back pressure can prevent this. Excessive wear of the barrel and screw causes screw slippage since the material will not melt sufficiently.

1.6.1.13 Short Shots

A short shot is a molded part that is underfilled or not filled completely. Visually, this can be seen as sections, such as the last areas of the part to fill, or projections of the part that are missing.

Melt-Conveying Phase If the melt temperature is too low, material cannot flow easily to fill out a part completely. Raising the melt temperature will permit easier flow to fill out more sections of the part. Insufficient packing pressure also prevents the material from filling out the cavity, creating short shots. Increasing the packing pressure exerts more pressure on the melt to flow farther and pack out the part. Injecting material at a slow speed causes the viscosity of the material to increase, restricting flow. Increasing the injection speed provides more shear heat to the material, reducing its viscosity and increasing flow. A worn check ring in the nozzle creates inconsistent cushion, causing shot-to-shot variation of material entering the cavity; some parts are filled completely and others are short. If shot size or feed setting is also set to low, the part will not fill out completely. Enough shot size needs to be provided to assure that enough material enters the cavity and fills it completely. Another source of insufficient material filling the cavity can be traced to a blockage in the nozzle from cold material or other foreign matter.

Melt-Directing Phase As a blocked nozzle can cause a blockage inhibiting material from flowing into the cavity, a blocked gate can also restrict flow into the cavity, producing short shots. One of the more common blockages is cold material breaking off from the runner or stuck in the gate from a previous shot. Also, contamination such as metal can block a gate as well. It is recommended to remove these blockages in the gate and also to locate the source or sources of these blockages. Too small a gate size can freeze off before the mold is full, resulting in incomplete fills. A larger gate size will cool faster and allow more material into the cavity.

Melt Forming Phase Low mold temperatures are a common cause of short shots since this causes the melt front to cool too soon before the part is full. Increasing the mold temperature can keep the melt front from cooling prematurely, before the cavity is full. If not enough vents are used in a mold cavity or if no vents are used, air can be trapped and acts like a cushion, halting material flow. The addition of vents will force this trapped air out of the mold, facilitating material flow into the cavity. Long part lengths can cause the molten material to cool fast, before filling out the cavity. A number of options to filling long-length parts is to use flow leaders, which are concentrated areas of increased thickness in the part that guide flow to a specific area of the part. Additional gates also can be added to supply more entry areas for the molten material to enter the part. Finally, using a material with a higher viscosity, or a high melt flow rate, will fill the part easier and faster and reduce the chance of short shots.

1.6.1.14 Sink Marks

Sink marks on a plastic part appear as a depression or dip on the part surface. These are created when the shrinkage of the molten core of a part wall is greater that that of the skin. This larger shrinkage of the molten core "sucks" in the surface, causing a depression to occur.

Melt-Conveying Phase High melt temperatures cause sink marks because they prolong the cooling time of the part. This slows down the cooling of the molten core. Reducing the melt temperature will reduce the cooling time of the part and reduce the sink marks. If the pack pressure or holding pressure is too low, not enough material will be pushed against the cavity wall. This lack of pressure will not supply enough resistance to the molten core shrinking away from the cooler part wall. Higher packing pressure or hold pressure will provide more pressure on the cavity wall to cool the skin and core to prevent sink marks. In the same way, lower injection forward time can create sink marks since both pressure and time are needed to assure that the part is cooled and packed sufficiently. High injection speeds increase melt temperature due to high shear, and lowering injection speed will reduce melt temperature and sink marks. Worn check rings creating shot-to-shot inconsistencies in packing the part can cause sinks. Parts need enough cooling time in the cavity to solidify the skin and core; short cooling times will prevent the molten core of the part from solidifying creating sink marks. As with short shots, obstructions in the nozzle restrict flow, underpack the part, and also cause sink marks.

Melt-Directing Phase If the gate freezes off too soon, the part will not pack out completely, allowing sink marks to form. Gate seal time needs to be examined to determine if the gate is freezing off before the cavity is full. Runner size can also play a role in causing sink marks. Using smaller cross-section runners, such as half-rounds, will cool and solidify faster and as a result restrict flow to the cavity. Moving to larger-diameter runners, such as full round, trapezoidal, and modified trapezoidal runners, will keep the runner more molten and hotter, allowing the material to flow more easily into the cavity.

Melt-Forming Phase Mold temperatures that are too cold will set up the skin of the part much faster than the molten core. Increasing mold temperature will slow down the skin from cooling. However, sinks can be seen on both sides of the part, and increasing mold temperature may not improve the situation. It is recommended to decrease the mold temperature on the side where the most visible sink is found. Flow of the material into a part can play a part in preventing sink marks by directing the flow from thick to thin wall thicknesses instead of from thin to thick. Localized cooling with the addition of water bubblers, cooling pins, thermal pins, or steel or metal alloy inserts will cool the part. If there is freedom in the part design,coring out thick sections will reduce the chance of sink marks by improving part cooling. Also, care needs to be taken designing projections such as ribs, bosses, and gussets, since intersections of these projections and the nominal wall are prime locations for sink marks. It is recommended to keep these intersections at no more than 50% of the nominal wall thickness.

1.6.1.15 Black Specks

Black specks appear in a part as random or numerous small black spots. They can give the appearance of dirt particles that can cause poor surface aesthetics on the part. These can be noticeable in transparent, translucent, and opaque parts.

Material Feed Phase Contamination of the material is one of the prime causes of black specks. This contamination can take the form of dirt, dust, or other types of foreign matter. Dust from fines will fall to the base of the hopper, sit at the back end of the screw at the barrel, and over time will break into the melt-conveying stream, showing up as black specks in the part. Materials that are notorious for fines are polycarbonate, acrylic, and some styrenics. Pellets feeding through a material handling system will strike sharp corners in hose lines and conveying systems. As a result, chips will break off the pellets of these types of materials and create dust particles. Dedusting equipment, which uses air to separate dust and pellets, can be used to reduce fines that can be delivered to the hopper. Contamination from regrind can also cause black specks to be found in a part. Color concentrates that are not compatible are another source of black specks.

Melt-Conveying Phase High melt temperatures can promote the formation of black specks from polymer degradation. High injection speeds cause excessive shear to the material, also degrading the polymer. Long residence times in the barrel will also overheat material and cause degradation and black speck formation. Screw and barrel wear can cause molten material to hang up in worn areas and break off sporadically during the process cycle in the form of black specks. In looking forward in the barrel at the nozzle, two sources of black specks can be found. Material can hang up in the check ring as it can in a worn barrel and fall into the melt stream as a black speck. Inspecting the check ring for wear or using a check ring design with no dead spots can help. If high decompression is used, the potential can occur to trap air in the nozzle and cause black specks. Reducing decompression will prevent air trapping and black speck formation.

Melt-Directing Phase Hot runner manifolds are a major source of black specks since inside the manifold, sharp corners for directing the melt can be found, as well as dead spots areas where the melt can stagnate. Manifolds should be designed with few or no dead spots for material to hang up and cause black specks.

1.6.1.16 Splay Marks

Splay marks, also referred to as silver streaks, appear as a blister that radiates over the surface of a part from the gate. If you were to take a small, sharp metal blade and scrape one of these splay mark blisters, the blister will appear to delaminate away from the surface. These types of defect cannot only look poor on the part surface, but if the part requires finishing, such as painting or electroplating, this defect can be enhanced.

For this discussion, splay is categorized into three types. The first is moisture splay, caused by moisture entering the material melt stream; thermal splay, caused by overheating the material; and shear splay, which results from exposing the polymer material to high shear during processing.

Material Feed Phase Moisture from material that has not been dried enough or at all time is one of the prime causes of splay and should be one of the first causes checked. Moisture in material is a source of gas, which can get into the melt stream and show up later in the cavity as splay. To avoid the occurrence of splay, plastic material, especially hygroscopic material must be dried according to the material supplier's recommendations. Fines are another cause of splay, but in this case it is heat splay. Fines can overheat and burn faster than larger pellets, and this burned material can take the form of thermal splay marks.

Melt-Conveying Phase Excessive melt temperature can cause outgassing of the material, resulting in thermal shear. High melt-temperature settings, high back pressures, and high screw rotation are three causes of thermal splay. Reducing either melt temperature, reducing back pressure, or decreasing the screw rotation will eliminate thermal splay. Low back pressures can also be a source of splay since this is yet another source of trapping air in the melt. High injection speeds lead to shear splay since the material is exposed to high shear conditions. Using slower injection speeds reduces the risk of shear splay. High nozzle temperatures can promote the formation of thermal splay marks from keeping the melt too hot going into the sprue bushing. Using decompression may force moisture into the melt, showing up as moisture splay in the part.

Melt-Directing Phase Sharp transitions in the runner systems of both hot and cold runner systems can lead to shear splay of the material. Smooth transitions should be designed into these areas.

Melt-Forming Phase Even in part design, sharp corners should be avoided, as these cause shear splay. Radiused corners are recommended not only for reducing shear splay, but also to avoid part stresses. With materials that require running mold temperatures below 75°F (24°C), condensation can form on the mold surface especially on humid days, forming moisture splay on the part surface. Wiping the tool on a regular basis will help reduce this in addition to reducing the mold open time.

1.6.1.17 *Voids*

Voids are formed inside the wall of a part and can adversely affect the structural performance of the part. Except in transparent parts, voids can be invisible to the naked eye. They can be noticed by either dissecting the part and visually examining the core of the wall, or by using sophisticated x-ray machines to locate voids. Voids are formed as a result of shrinkage of the molten core after the wall has solidified on the cooler mold surface. The shrinkage of the molten core causes the layers of material to pull away, forming an opening in the part, or void.

Melt-Conveying Phase Higher melt temperatures slow down the cooling rate of the polymer in the mold, causing higher shrinkage in the hotter core, leading to voids. Lower melt temperatures can help reduce the formation of voids. Injecting the material at high speeds raising melt temperatures due to shear heating can also cause overheating. Low packing pressures will allow voids to form since there is not enough resistance applied on the core to prevent shrinkage. Also, short packing times will not provide enough time to hold pressure on the core to prevent shrinkage. Nozzle orifice diameters that are too small are another source of shear heating increasing melt temperature.

Melt-Directing Phase A gate that freezes-off too soon will prevent material from filling out the cavity completely, causing the molten core to shrink. Measuring gate seal time can indicate whether the cavity is full before the gate closes. Small runner diameters will cool faster and hinder cavity filling, creating the potential for voids to form in the part. Larger-diameter runners, such as full round, trapezoidal, and modified trapezoidal are recommended.

Melt-Forming Phase Mold temperature is critical in preventing voids. A cold mold will only enhance the formation of the skin surface long before the core cools sufficiently. As a result, the core forms more slowly and shrinks more. Increasing mold temperature slows down the cooling rate, and the formation of the skin prevents voids. Improvements can be made in mold cooling, such as the addition of bubblers, cooling pins, thermal pins, or steel inserts to drive out more heat from the part. Designing the part to flow from a thick to a thin wall eliminates the formation of voids. Coring out thick sections allows for improved cooling of the part, reducing voids.

1.6.1.18 *Warpage*

When plastic material is injected into a mold cavity, it is cooled and solidifies to form a part that is made to final part dimensions. However, if the cooling of this part in the mold is not uniform, differences in part shrinkage can occur, causing the part to distort. This condition is referred to in the injection molding industry as *warpage*. Warpage is also the result of excessive stresses that are built into a part during the molding process.

Melt-Conveying Phase Injection speed can affect part warpage in two ways. An injection speed that is too fast will increase the melt temperature, forcing the melt to overpack the cavity. This overpacking creates molded-in stresses in the part, which warp the part. Conversely, if the injection speed is too slow, the viscosity of the material increases the amount of packing pressure needed to fill the cavity. This higher packing pressure adds molded-in stresses in the part, causing warpage. Long injection forward times contribute to part warpage by allowing more time for more material to overpack the cavity. Check ring wear can cause shot-to-shot consistency, creating overpacked parts.

Melt-Directing Phase Part gating is a contributor to part warpage. If the gate has not sealed off before the cavity is full, overpacking occurs, causing warpage. The gate seal needs to be measured to assure that the cavity is not full before gate seal takes place. If not enough gates are used to fill a part, higher packing pressures occur to fill out the part, producing molded-in stresses. Increasing the number of gates reduces the packing pressure by distributing the pressure over a number of gates. Smaller gate sizes cause a high degree of molecular orientation in the polymer that causes different shrinkage in different directions of the part (machine vs. transverse). Larger gate sizes reduce orientation.

Melt-Forming Phase Insufficient cooling times do not provide enough time for the part to cool, increasing shrinkage. Increasing cooling time or fixturing the part are options to prevent the part from distorting. Warpage can be corrected by changing the mold temperature in one mold half. Running the mold halves at different temperatures can cause shrinkage in one direction to help overall warpage. Parts have a tendency to warp toward hotter mold halves since this causes higher shrinkage. For semicrystalline materials such as polypropylene, nylon, and polyethylene the mold temperature is decreased on the mold half that the warpage is toward. For amorphous materials such as polycarbonate, ABS, and styrenics, the mold temperature is decreased on the mold half that the warpage is toward. However, parts molded using different mold half temperatures may warp again later in use if the part are exposed to elevated temperature conditions. Location and placement of cooling lines should be reviewed to ensure that even efficient cooling is taking place to the part. Warpage may be caused by mechanically induced stresses caused in part ejection. If ejection forces are uneven due to ejector pin placement or length, stresses can be induced in the part causing warpage. Improving or modifying part design can reduce part warpage. Uniform wall thickness helps avoid warpage issues since major differences in wall thickness promote warpage; thin sections cool faster, thicker sections cool slower. Removing sharp corners can reduce molded-in stresses and reduce stresses that can be caused during part ejection.

1.6.1.19 Weld or Knit Lines

Weld lines, also called knit lines, are formed when two melt fronts converge and join forming a thin fine line in the part. This can be seen in untextured parts and in highly polished surfaces. The two converging melt fronts cool rapidly and bond poorly when they join. At this melt junction, gas is also trapped by the converging melt fronts. These fine lines cause a major weakness in the part that can caused part failure in its end use. Differences in gloss can also be noticed in weld line areas.

Material Feed Phase Moisture in the material due to insufficient drying can form at the melt front, forming an interface between the two melt fronts. This creates a weakness in the parts structure. Using recommended drying conditions for the material is advised.

Melt-Conveying Phase Low melt temperatures cause the melt fronts to cool faster, causing converging melt fronts to bond poorly. Increasing melt temperatures increase flow and improve the melt front bonding. High packing and hold pressures are needed to adequately force the two melt fronts to join and bond. Increasing hold time or packing time will also improve weld strength. Slow injection speeds will allow the melt front to cool early in the part fill, and by the time the two melt fronts join, poor and weak bonding of the melt front occurs. Packing pressures need be increased to fill out the part completely to assure strong weld line strength or even to eliminate weld lines. Variations in cushion due to worn check rings can cause inconsistent part quality, due to inconsistent weld line formation and strength.

Melt-Directing Phase Gate location plays a key role in weld line formation. Gates can be located to shift the position of the weld lines away from areas that can affect part performance. Also, gating technologies such as sequential valve gating can be used. This technology involves using hydraulics that open gates one at a time in order to fill apart.

Melt-Forming Phase Low mold temperatures cause the melt front to cool fast and set up before the converging melt fronts join. This creates weak weld lines and more pronounced weld lines. Raising the mold temperature will keep the material molten longer, resulting in stronger weld lines and even elimination of weld lines. Poor venting at the location where melt fronts come together results in gas trapping at the melt front, preventing the melt fronts from coming together. Adding vents, using overflows, or using porous metal inserts can force air out, improving the weld or knit line. Finally, increasing the wall thickness no more than 0.005 in. along either side of the weld lines can shift the location of the weld away from critical areas.

1.6.2 Troubleshooting Guidelines For Thermoplastic Materials

The next pages list a number of troubleshooting guidelines for a number of thermoplastic materials. Table 1.14 lists general process conditions for various thermoplastic materials. Tables 1.15 through 1.25 provide troubleshooting guides for a number of thermoplastic materials.

TABLE 1.14 General Process Conditions for Thermoplastic Materials

Material	Melt Temperature [°F (°C)]	Mold Temperature [°F(°C)]
ABS	425–520 (218–271)	100–150 (38–66)
ABS–Nylon	485–530 (252–277)	100–200 (38-93)
ABS–TPU	410–425 (210–218)	80–100 (27–38)
Acetal	400–440 (204–227)	180–200 (82–93)
Acrylic	420–485 (216–252)	140–220 (60–104)
Nylon 6	450–530 (232–277)	100–200 (38–93)
Nylon 6,6	520–575 (271–302)	100–200 (38–93)
PEI	640–780 (338–416)	150–350 (66–177)
Polycarbonate	530–600 (277–316)	170–210 (77–99)
PC–ABS	500–570 (260–299)	150–200 (66–93)
PC–PBT	490–530 (254–277)	150–200 (66–93)
PC–PET	490–530 (254–277)	150–200 (66–93)
Polyethylene	360–585 (182–302)	55–130 (131–154)
PPS	590–620 (310–327)	275–325 (135–163)
Polypropylene	375–510 (191–266)	60–140 (16–54)
PPO	540–590 (282–310)	160–200 (71–93)
Polystyrene	450–500 (232–260)	60–140 (16–54)
PBT	480–500 (249–260)	95–150 (35–66)
PET	540–570 (282–299)	200–220 (93–104)
Liquid crystal polymer	640–700 (338–371)	180–250 (82–121)
PVC	375–400 (191–204)	70–100 (21–38)
SMA	490–510 (254–266)	120–180 (49–82)
TPO	430–475 (221–246)	80–120 (27–49)

Source: Multiple industry sources.

TABLE 1.15 Injection Troubleshooting Guide: ABS (Acrylonitrile Butadiene Styrene)

Problem	Cause	Solution
Short shots or insufficient part fill	Injection pressure too low	Increase injection pressure.
	Melt temperature too low	Raise melt temperature.
	Nozzle orifice size too small	Use a larger nozzle size.
	Hopper empty or obstructed	Remove any obstructions from hopper or fill hopper with more material.
	Nozzle flow path blocked	Remove any blockage from nozzle and use caution.
	Insufficient venting in mold	Check vent sizes and vent according to material supplier recommendations. Also, check for clogged or obstructed vents.
	Gate and runner sizing too small	Increase size of gates and runners.
	Mold temperature too low	Increase mold temperature to at least the midrange recommended by material supplier.

TABLE 1.15 (*Continued*)

Problem	Cause	Solution
Bubbles, gas entrapment, burn marks	Injection pressure too low	Increase injection pressure.
	Injection speed too high	Reduce injection speed.
	Residence time in barrel too long	Use a machine that will allow using between 30 and 65% of barrel capacity per shot.
	Gate location in part going from thin section to thicker section	Relocate gate to allow flow to go from thick to thin sections.
	Runner size too small	Increase runner size.
	Insufficient venting in mold	Check vent sizes and vent according to material supplier recommendations. Also, check for clogged or obstructed vents.
Splay marks, gate blemish, and other surface defects	Nozzle orifice size too small	Use a larger nozzle size.
	Injection speed too high	Reduce injection speed.
	Nozzle flow path blocked	Remove any blockage from nozzle and use caution.
	Barrel temperature profile too high	Decrease barrel temperature profile and, in particular reduce middle barrel zones.
	Mold temperature too low	Increase mold temperature to midrange by material supplier.
	Gate size too small	Increase gate size without sacrificing part degating.
	Insufficient venting in mold	Check vent sizes and vent according to material supplier recommendations. Also, check for clogged or obstructed vents.
	Excessive level of fines in regrind	Remove excessive fines by using auxiliary equipment, such as dedusters.
	Resin not dried sufficiently	Review recommendations by material supplier for drying and check drying equipment operation and dryer hose conditions and connections. If using regrind, check for larger pellet size.
	Contamination in material	Remove any contamination, review material handling procedures or contact material supplier.
Weak weld or knit lines	Nozzle temperature too low	Increase nozzle temperature.
	Air trapped at joining melt fronts	Add vents at melt junction or use porous metal inserts.
	Injection pressure too low	Increase injection pressure.
	Injection speed too low	Increase injection speed.

TABLE 1.15 (*Continued*)

Problem	Cause	Solution
	Mold temperature too low	Increase mold temperature to at least the midrange recommended by material supplier.
	Runner size too small	Increase runner size.
	Resin not dried sufficiently	Review recommendations by material supplier for drying and check drying equipment operation and dryer hose conditions and connections. If using regrind, check for larger pellet size.
Sink marks	Injection pressure too low	Increase injection pressure.
	Melt temperature too high	Lower melt temperature.
	Hopper empty or obstructed	Remove any obstructions from hopper or fill hopper with more material.
	Cycle time too short	
	Injection speed too low	Increase injection speed.
	Insufficient shot capacity of machine	Utilize a machine with enough shot size for the part.
	Nozzle flow path blocked	Remove any blockage from nozzle and use caution.
	Runner size too small	Increase runner size.
	Gate size too small and freezing off before part is filled	Increase gate size and measure gate freeze-off time.
Excessive flashing in mold	Injection pressure too high and "blowing" the mold open	Reduce injection pressure.
	Clamp pressure too low	Increase clamp pressure.
	Packing pressure too high	Reduce packing pressure.
	Melt flow rate of material too high	Select a material that has a lower melt flow rate which will not compensate part performance.
Part or sprue sticking	Packing pressure too high	Reduce packing pressure
	Too much material feeding into cavity	Reduce shot size.
	Injection pressure too high	Reduce injection pressure.
	Barrel temperature too low, causing sprue to stick	Increase barrel temperature.
	Melt temperature too high, causing part in cavity to stick	Reduce melt temperature.
	Nozzle not seated properly into sprue bushing	Check alignment of nozzle to sprue bushing.
	Nozzle orifice not matched to orifice of sprue bushing	Check alignment of nozzle and sprue bushing orifice interface.

TABLE 1.15 (*Continued*)

Problem	Cause	Solution
	Sprue bushing has burrs or undercuts	Machine all burrs from sprue bushing and remove undercuts.
	Mold temperature too high	Reduce mold temperature.
	Burrs, undercuts, or other surface defects on mold	Machine all burrs, undercuts, and other surface defects.
	Material does not have sufficient levels of internal lubricants	Use a material grade with an increased level of lubricants.
Dark streaks in part	Machine not completely purged	Follow proper purge procedures and clean out barrel or nozzle manually if needed.
	Melt temperature too high	Reduce melt temperature.
	Material hung up in barrel and nozzle	Remove any sharp areas or other areas where material can get hung up and break off after material change.
	Contamination such as oils, greases, and mold cleaners getting into material	Clean any excessive oils, greases, and mold cleaners off the tool surface before startup.
	Insufficient venting in mold	Vent tool according to recommended guidelines for material.
Gloss too high or low[a]	Melt temperature too low	Increase melt temperature.
	Mold temperature too low	Increase mold temperature.
	Additive in material blooming to surface	Reduce melt temperature, improve venting, or review material with supplier.
	Insufficient venting in mold	Vent tool according to recommended guidelines for material.
	Injection pressure too low	Increase injection pressure.
	Injection speed too low	Increase injection speed.
	Improper mold finish	Review part finish requirements.
	Weld/knit lines present in part	Review solutions in weld line section.
	Burn marks in part	Reduce melt temperature and check for proper venting.
	Blush or splay at gate	Increase gate size.
	Grade of material has too high or too low gloss	Review requirements and availability for high- or low- gloss grades.

Source: Multiple industry sources.

[a]Textured surfaces have a reverse effect on gloss from that on untextured surfaces. Packing pressure, higher melt, higher melt, and higher injection speeds will reduce gloss. This results in packing out the texture peaks and valleys, resulting in the low gloss provided by the texture. For smooth surfaces, the above will give a shinier, higher-gloss surface.

TABLE 1.16 Injection Troubleshooting Guide: PMMA (Polymethyl Methacrylate)

Problem	Cause	Solution
Burning and dieseling	Insufficient venting in mold	Check vent sizes and vent according to material supplier recommendations. Also, check for clogged or obstructed vents.
	Injection speed too high	Reduce injection speed.
	Mold temperature too high	Reduce mold temperature.
	Melt temperature too high	Reduce melt temperature.
	Clamp pressure too high	Reduce clamp pressure.
	Gate located into projection such as rib or boss	Relocate gate or improve venting of projection.
Cold slugs	Nozzle temperature too low	Increase nozzle temperature
	No cold slug well located at end of sprue	Install a cold slug at end of sprue to receive cold slug.
Crazing in part	Residue depositing in part	Remove any residue deposited in mold cavity.
	Injection speed too slow	Increase injection speed.
Delamination of part surface	Melt temperature too low	Increase melt temperature.
	Mold temperature too low.	Increase mold temperature.
	Material contaminated	Remove any contamination, review material handling procedures, or contact material supplier.
Excessive cycle time	Mold open time too long	Reduce mold open time.
	Mold temperature too high	Reduce mold temperature.
	Melt temperature too high	Reduce melt temperature.
	Injection speed too slow	Increase injection speed, but not high enough to cause shearing of material.
	Screw recovery time too long	Reduce screw recovery time
	Screw forward time too long	Reduce screw forward time.
	Parts too hot after ejection	Quench part in cool bath, but be aware of possibly creating molded in stresses in part.
Flashing of part	Clamp pressure too low	Increase clamp pressure.
	Injection pressure too high	Reduce injection pressure.
	Injection speed too high	Reduce injection speed.
	Melt temperature too high	Reduce melt temperature.
	Mold temperature too high	Reduce mold temperature.
	Mold surface has locations of excessive wear	Repair mold surface and tool flatness.
	Damage in tool preventing tool from completely closing	Repair any tool damage by grinding, filling, or other machining techniques.
	Clamp tonnage of machine insufficient for part to be molded	Move tool to a machine with a higher clamp tonnage.

TABLE 1.16 (*Continued*)

Problem	Cause	Solution
Gate blemish	Injection speed too high, causing shear at gate	Reduce injection speed initially until gate area is filled.
	Injection hold pressure too high	Reduce injection hold pressure.
	Gate design too small, causing jetting effect	Increase or modify gate size to remove jetting effect.
Internal bubbles and voids	Back pressure too low	Increase back pressure.
	Barrel temperatures too low	Increase barrel temperatures starting with the middle zones.
	Screw speed too high	Reduce screw speed.
	Inconsistant feed of material from hopper	Increase hopper (feed) temperature, but avoid caking and bridging in hopper.
Internal bubbles and voids (thick parts)	Screw forward time too short	Increase screw forward time.
	Injection pressure too low	Increae injection pressure.
	Injection speed too low	Increase injection speed.
	Gate and runner size too small and freezing off before part is completely filled	Increase gate and runner size, and measure gate freeze-off time.
	Resin not dried sufficiently	Review recommendations by material supplier for drying, and check drying equipment operation and dryer hose conditions and connections. If using regrind, check for larger pellet size.
	Parts too hot after ejection	Quench part in cool bath, but be aware of possibly creating molded-in stresses in part.
Part weight variation	Screw recovery inconsistent	Monitor screw recovery time for shot-to-shot change.
	Check ring showing signs of wear	Replace check ring in nozzle.
	Back pressure too low	Increase back pressure.
	Screw forward time too short	Increase screw forward time.
	Injection pressure too low	Increase injection pressure.
	Injection speed too low	Increase injection speed.
	Gate and runner size too small and freezing off before part is completely filled	Increase gate and runner size and measure gate freeze-off time.
Sink marks	Cushion size insufficient	Increase cushion.
	Injection pressure too low	Increase injection pressure.
	Screw forward time too short	Increase screw forward time.
	Melt temperature too high	Reduce melt temperature.
	Mold temperature too high	Reduce mold temperature.
	Injection speed too high	Reduce injection speed.

TABLE 1.16 *(Continued)*

Problem	Cause	Solution
	Gate and runner size too small and freezing off before part is completely filled	Increase gate and runner size and measure gate freeze-off time.
	Insufficient venting in mold	Check vent sizes and vent according to material supplier recommendations. Also, check for clogged or obstructed vents.
	Gate size too small and freezing off before part is filled	Increase gate size, and measure gate freeze-off time.
	Check ring showing signs of wear	Replace check ring in nozzle.
	Parts too hot after ejection	Quench part in cool bath, but be aware of possibly creating molded-in stresses in part.
Short shots: incomplete filled parts	Cushion size insufficient	Increase cushion.
	Injection pressure too low	Increase injection pressure.
	Injection speed too low	Increase injection speed.
	Screw forward time too short	Increase screw forward time.
	Melt temperature too low	Increase melt temperature.
	Mold temperature too low	Increase mold temperature.
	Nozzle orifice size too small	Use a larger nozzle size.
	Gate size too small	Increase gate size without sacrificing part degating.
	Check ring showing signs of wear	Replace check ring in nozzle.
	Screw recovery inconsistent	Monitor screw recovery time for shot-to-shot change.
	Melt capacity of barrel too small	Move tool to a machine with a larger barrel capacity.
Splay marks	Resin not dried sufficiently	Review recommendations by material supplier for drying and check drying equipment operation and dryer hose conditions and connections. If using regrind, check for larger pellet size.
	Back pressure too low	Increase back pressure.
	Melt temperature too high	Increase melt temperature.
	Barrel temperatures too high	Reduce barrel temperatures starting with the middle zones.
	Injection pressure too low	Increase injection pressure.
Sprue breakage	Injection speed too high, causing shear of material	Reduce injection speed.
Sprue sticking	Nozzle not seated properly into sprue bushing	Check alignment of nozzle to sprue bushing.
	Nozzle orifice not matched to orifice of sprue bushing	Check alignment of nozzle and sprue bushing orifice interface.

TABLE 1.16 (*Continued*)

Problem	Cause	Solution
	Sprue bushing has burn marks and rough surface	Polish out burn marks and rough areas.
	Injection pressure too high	Reduce injection pressure.
	Screw forward time too long	Reduce screw forward time.
	Burrs, undercuts, or other surface defects on mold	Machine all burrs, undercuts, and other surface defects.
	Taper of sprue has insufficient draft and may have a rough surface	Increase draft of sprue and polish sprue bushing.
	Sprue puller not grabbing part out of cavity for ejection	Review sprue puller design, and increase cycle time if needed.
	Material does not have sufficient levels of internal lubricants	Use a material grade with increased level of lubricants or add external lubricant such as mold release.
Warpage	Mold-closed time too short	Lengthen mold-closed time.
	Mold temperature too high	Reduce mold temperature.
	Melt temperature too high	Reduce melt temperature.
	Injection speed too low	Increase injection speed.
	Injection pressure too high	Reduce injection pressure.
	Parts too hot after ejection	Quench part in cool bath, but be aware of possibly creating molded-in stresses in part.
Weak weld or knit lines	Injection speed too low	Increase injection speed.
	Injection pressure too low	Increase injection pressure.
	Melt temperature too low	Increase melt temperature.
	Mold temperature too low	Increase mold temperature.
	Screw forward time too short	Lengthen screw forward time.
	Air trapped at joining melt fronts	Add vents at melt junction or use porous metal inserts.

Source: Multiple industry sources.

TABLE 1.17 Injection Troubleshooting Guide: POM (Acetal)

Problem	Cause	Solution
Black or brown streaks	Cycle time too long	Reduce overall cycle time.
	Material building up in corners or dead spots in barrel, screw flights, check ring, or nozzle	Remove sharp corners or dead spots in barrel, screw, and nozzle. Review check ring design.
Blush, surface haziness	Injection speed too high	Reduce injection speed.
	Mold temperature too low	Increase mold temperature.
Gate smear	Mold temperature too low	Increase mold temperature.
	Injection speed too high	Reduce injection speed.

TABLE 1.17 *(Continued)*

Problem	Cause	Solution
	Gate size too low and restricting flow and producing high shear	Enlarge gate size.
	Gate location in highly aesthetic area	Move gate to nonaesthetic location.
Jetting	Melt temperature too low	Reduce melt temperature.
	Injection rate too high	Reduce injection speed.
	Mold temperature too low	Increase mold temperature.
	Gate size too small restricting flow and producing high pressures for material entering the cavity	Enlarge gate size.
	Gate location in highly aesthetic area	Move gate to nonaesthetic location.
Odor issues	Gassing occurring during processing	Measure melt temperature using an air shot.
	Melt temperature too high	Reduce melt temperature.
	Overall cycle time too long	Reduce overall cycle time.
	Material building up in corners or dead spots in barrel, screw flights, check ring, or nozzle	Remove sharp corners or dead spots in barrell, screw, and nozzle. Review check ring design.
	Material contaminated	Remove any contamination, review material handling procedures,or contact material supplier.
	Barrel-to-shot ratio too low; barrel size too large for part	Use a machine with a smaller barrel and lower barrel-to-shot ratio.
Part sticking in mold	Injection pressure too high	Reduce injection speed.
	Injection speed too high	Reduce injection pressure.
	Screw forward time too long	Reduce screw forward time to avoid overfilling cavity.
	Undercuts or machining marks in tool holding part from ejecting	Remove and machine out undercuts and machining marks in cavity.
	Overall cycle time too short	Increase cycle time.
Pits, orange peel, and surface wrinkling	Injection pressure too low	Increase injection pressure.
	Injection speed too low	Increase injection speed.
	Screw forward time too short	Increase screw forward time.
	Mold temperature too low	Increase mold temperature.
	Melt temperature too low	Increase melt temperature.
	Insufficient venting in mold	Check vent sizes and vent according to material supplier recommendations. Also, check for clogged or obstructed vents.
	Gate size too small, restricting flow and producing high pressures for material entering the cavity	Enlarge gate size.

TABLE 1.17 (*Continued*)

Problem	Cause	Solution
Residue buildup in mold cavity	Injection speed too high	Reduce injection speed.
	Melt temperature too high, causing additives in the material to juice	Reduce melt temperature.
	Material building up in corners or dead spots in barrel, screw flights, check ring, or nozzle	Remove sharp corners or dead spots in barrell, screw, and nozzle. Review check ring design.
	Material contaminated	Remove any contamination, review material handling procedures, or contact material supplier.
	Gate size to small, restricting flow and producing high shear	Enlarge gate size.
	Insufficient venting in mold	Check vent sizes and vent according to material supplier recommendations. Also, check for clogged or obstructed vents.
Short shots or incomplete filling of cavity	Cushion not holding consistent from shot to shot	Maintain consistent cushion.
	Check ring leaking back into barrel	Replace worn check ring.
	Injection pressure too low	Increase injection pressure.
	Injection speed too low	Increase injection speed.
	Melt temperature too low	Increase melt temperature.
	Mold temperature too low	Increase mold temperature.
	Insufficient venting in mold	Check vent sizes and vent according to material supplier recommendations. Also, check for clogged or obstructed vents.
	Overall cycle time too short	Increase cycle time.
	Press size too small for part; insufficient clamp	Use a larger machine with increased clamp tonnage.
Sink marks	Check ring leaking back into barrel	Replace worn check ring.
	Injection pressure too low	Increase injection pressure.
	Screw forward time too short	Increase screw forward time.
	Injection speed too low	Increase injection speed.
	Mold temperature too high	Decrease mold temperature.
	Gate size small, restricting flow and producing high shear	Measure gate freeze-off time. Enlarge gate size.
Splay marks	Melt temperature too high	Reduce melt temperature.
	Overall cycle too long	Reduce cycle time.
	Injection speed too high, causing shear	Reduce injection speed.
	Injection rate too low, causing material viscosity to decrease, leading to stiff flow and shear through a gate	Increase injection speed.

TABLE 1.17 (*Continued*)

Problem	Cause	Solution
	Gate size too small, restricting flow and producing high shear	Measure gate freeze-off time. Enlarge gate size.
Sprue sticking	Nozzle not seated properly into sprue bushing	Check alignment of nozzle to sprue bushing.
	Nozzle orifice not matched to orifice of sprue bushing	Check alignment of nozzle and sprue bushing orifice interface.
	Sprue bushing has burrs or undercuts	Machine all burrs from sprue bushing and remove undercuts.
	Nozzle temperature too low	Increase nozzle temperature.
	Burrs, undercuts, or other surface defects on mold	Machine all burrs, undercuts, and other surface defects.
	Taper of sprue has insufficient draft	Increase draft of sprue.
	Sprue puller not grabbing part out of cavity for ejection	Review sprue puller design.
	Material does not have sufficient levels of internal lubricants	Use a material grade with increased level of lubricants.
Variation in dimesions from shot to shot	Injection pressure too low	Increase injection pressure.
	Cushion not holding consistent from shot to shot	Maintain consistent cushion.
	Check ring leaking back into barrel	Replace worn check ring.
	Screw forward time too short	Increase screw forward time.
	Umelted particles in part	Remove unmelted particles.
	Press size too small for part; insufficient clamp	Use a larger machine with increased clamp tonnage.
Warping of parts	Mold temperature not uniform across mold cavity and core	Use separate mold cooling units to cool the mold, or relocate cooling lines.
	Mold temperature too high	Reduce mold temperature.
	Insufficient cooling time	Increase cooling part of cycle.
	Screw forward time too short, creating insufficient part fill in cavity	Increase screw forward time.
	Screw forward time too long, creating excessive stresses in part	Decrease screw forward time.
	Injection pressure too low, creating insufficient part fill in cavity	Increase injection pressure.
	Injection pressure too high, creating excessive stress in part	Reduce injection pressure.
	Water channels coated with scale and residue, creating poor heat transfer	Clean out scale and residue buildup in cooling lines.
Weak weld lines	Injection pressure too low	Increase injection pressure.
	Injection speed too low	Increase injection speed.
	Mold temperature too low	Increase mold temperature to at least the midrange recommended by material supplier.

TABLE 1.17 *(Continued)*

Problem	Cause	Solution
	Melt temperature too low	
	Excessive mold release used	Avoid use of mold release.
	Insufficient venting in mold	Check vent sizes and vent according to material supplier recommendations. Also, check for clogged or obstructed vents.
	Press size too small for part; insufficient clamp	Use a larger machine with increased clamp tonnage
Voids in parts	Injection pressure too low	Increase injection pressure.
	Screw forward time too short	Increase screw forward time.
	Injection speed too high	Reduce injection speed.
	Melt temperature too high	Reduce melt temperature.
	Check ring leaking back into barrel	Replace worn check ring.
	Gate size too small, restricting flow and producing high shear	Enlarge gate size.
Unmelted particles	Barrel temperature too low	Increase middle-zone barrel temperatures.
	Back pressure too low	Increase back pressure.
	Screw speed too high	Reduce screw speed.
	Overall cycle too short	Increase overall cycle time.
	Barrel-to-shot ratio too high	Use a larger machine with lower barrel-to-shot ratio.

Source: Multiple industry sources.

TABLE 1.18 Injection Troubleshooting Guide: Nylon (Polyamide)

Problem	Cause	Solution
Black streaking	Barrel and screw not completely purged	Follow purging procedures as recommended by material supplier.
	Area at hopper overheating, causing material to burn or degrade	Lower rear-zone temperatrure.
	Hopper feed throat water turned off	Check water circulation at hopper for any blockages, or turn on feed throat water.
	Material contaminated	Remove any contamination, review material handling procedures, or contact material supplier.
	Oil or grease coming from mold	Remove any oil from mold surface.
	Material drying time too long	Follow recommended drying conditions from supplier, or check for malfunctioning dryer heater.
	Barrel-to-shot ratio too low; residence time in the barrel too long	Move mold to a machine with a smaller injection cylinder to reduce residence time.

TABLE 1.18 *(Continued)*

Problem	Cause	Solution
Brittleness	Melt temperature too high, causing material to degrade	Reduce melt temperature.
	Back pressure too high, causing material to overheat	Reduce back pressure.
	Screw speed too fast	Decrease screw speed.
	Injection speed too fast	Decrease injection speed.
	Barrel-to-shot ratio too low; residence time in the barrel too long	Move mold to a machine with a smaller injection cylinder to reduce residence time.
	Material contaminated	Remove any contamination, review material handling procedures, or contact material supplier.
	Mold temperature too high	Reduce mold temperature.
Burn marks	Melt temperature too high, causing material to degrade	Reduce melt temperature.
	Back pressure too high causing material to overheat	Reduce back pressure.
	Screw speed too fast	Decrease screw speed.
	Injection speed too fast	Decrease injection speed.
	Barrel-to-shot ratio too low; residence time in the barrel too long	Move mold to a machine with a smaller injection cylinder to reduce residence time.
	Overall cycle time too long	Reduce overall cycle time.
	Insufficient venting in mold	Check vent sizes and vent according to material supplier recommendations in the area of the joining melt fronts.
	Gate located around hole or other depression, causing weld line to form	Move gate to different location.
	Gate size too small, causing material to shear	Increase gate size.
	Nozzle, sprue bushing, runners, or gates have sharp corners or burrs in flow path	Remove and polish out all sharp corners or burrs.
	Melt flow rate of material too low	Convert to higher melt flow grade.
	Material drying time too long	Follow recommended drying conditions from supplier, or check for malfunctioning dryer heater.
Color fading	Melt temperature too high, causing material to degrade and discolor	Reduce melt temperature.
	Barrel-to-shot ratio too low; residence time in the barrel too long	Move mold to a machine with a smaller injection cylinder to reduce residence time.
	Overall cycle time too long	Reduce overall cycle time.

TABLE 1.18 (*Continued*)

Problem	Cause	Solution
	Color concentrate made for low-temperature materials	Use a carrier material for the color that has a higher process temperature.
	Excessive amount of fines in regrind	Reduce level of fines by using fines separation equipment, such as a deduster.
	Grade of nylon not heat stabilized	Convert to heat-stabilized grade of nylon.
Delamination	Melt temperature too low	Increase melt temperature.
	Mold temperature too low	Increase mold temperature.
	Barrel and screw not completely purged	Follow purging procedures as recommended by material supplier.
	Back pressure too low	Increase back pressure.
	Screw speed too slow	Increase screw speed.
	Material contaminated	Remove any contamination, review material handling procedures, or contact material supplier.
	Color concentrate carrier not compatible with base resin	Use a carrier material that is compatible with nylon.
Discoloration of material	Barrel and screw not completely purged	Follow purging procedures as recommended by material supplier.
	Melt temperature too high	Reduce melt temperature.
	Barrel-to-shot ratio too low; residence time in the barrel too long	Move mold to a machine with a smaller injection cylinder to reduce residence time.
	Rear heat zones set too high	Lower rear-zone temperatrure.
	Back pressure too high, causing material to discolor	Reduce back pressure.
	Injection speed too fast	Decrease injection speed.
	Material contaminated	Remove any contamination, review material handling procedures, or contact material supplier.
Excessive shrinkage (small parts)	Overall cycle time too short	Increase overall cycle time.
	Back pressure too low	Increase back pressure.
	Insufficient cushion	Increase cushion according to recommendations of material supplier.
	Injection time too short	Increase injection time.
	Injection speed too slow	Increase injection speed.
	Melt temperature too low, causing melt to cool quickly	Increase melt temperature.
	Melt temperature too high, causing material to overpack in cavity	Lower melt temperature.

TABLE 1.18 (*Continued*)

Problem	Cause	Solution
	Check ring worn out, causing shot-to-shot variation	Replace worn-out check ring.
	Mold temperature too high	Reduce mold temperature.
	Gate size too small	Increase gate size.
	Runner size too small	Increase runner size.
	Sprue size too small	Increase sprue size.
	Nylon material not nucleated	Switch to a nucleated grade of nylon.
	Parts ejecting too hot from mold	Quench parts, but be aware of molded-in stresses that may result.
	Inefficient mold cooling creating uneven cooling of mold and part	Review water channel locations and circuit and modify.
Flashing of parts	Injection hold time too long	Reduce injection hold time.
	Clamp pressure too low	Move tool to press with a higher clamp pressure.
	Injection pressure too high	Reduce injection pressure.
	Injection speed too fast	Reduce injection speed.
	Melt temperature too high	Decrease melt temperature.
	Mold temperature too high	Decrease mold temperature.
	Mold part line mismatched	Rematch mold parting line or realign mold in press.
	Temperature controller malfunctioning	Replace temperature controller.
	Insufficient venting in mold	Check vent sizes and vent according to material supplier recommendations in the area of the joining melt fronts.
	Resin not dried sufficiently	Review recommendations by material supplier for drying, and check drying equipment operation and dryer hose conditions and connections. If using regrind, check for larger pellet size.
	Contamination on mold mating surfaces preventing mold from fully closing	Clean and polish out any contamination in mold.
	Melt flow rate of material too high	Switch to lower melt rate material or nucleated grade of nylon.
Nozzle drool	Resin not dried sufficiently	Review recommendations by material supplier for drying and check drying equipment operation and dryer hose conditions and connections. If using regrind, check for larger pellet size.

TABLE 1.18 (*Continued*)

Problem	Cause	Solution
	Nozzle temperature too high	Lower nozzle temperature.
	Injection speed too fast	Reduce injection speed.
	Melt temperature too high	Decrease melt temperature.
	Decompression ("suck back") set too low	Increase decompression.
	Nozzle orifice too large	Switch to a nozzle with a small orifice.
	Standard taper nozzle used	Switch to a reverse taper nozzle.
	Nozzle not flush against sprue bushing	Check interface contact area between nozzle and sprue bushing.
	Nozzle or sprue bushing is cracked or has burrs	Replace nozzle or sprue bushing.
	Overall cycle time too long	Reduce overall cycle time.
Nozzle freeze-off	Nozzle temperature too low	Increase nozzle temperature.
	Heater band malfunctioning	Replace heater band.
	Heater band located too far back on nozzle	Relocate nozzle heater band to front of nozzle.
	Watt capacity of nozzle heater insufficient	Switch to high watt heater.
	Overall cycle time too long	Reduce overall cycle time.
	Nozzle orifice size too small	Switch to a nozzle with a larger orifice.
	Standard taper nozzle used	Switch to a reverse taper nozzle.
	Mold temperature too low	Increase mold temperature.
Screw slippage	Screw worn out	Check screw flights and replace screw.
	Barrel worn out	Measure clearance between barrel and screw and replace or repair barrel lining.
	Feed throat at hopper blocked	Remove blockage from feed throat.
	Hopper empty	Refill material in hopper or check for malfunctioning material handling system.
	Back pressure too high	Reduce back pressure.
	Rear zone of barrel either too high or too low	Follow material supplier recommendations for rear-zone settings.
	Resin not dried sufficiently	Review recommendations by material supplier for drying and check drying equipment operation and dryer hose conditions and connections. If using regrind, check for larger pellet size.

TABLE 1.18 *(Continued)*

Problem	Cause	Solution
Short shots	Insufficient cushion	Increase cushion according to recommendations of material supplier.
	Insufficient screw feed setting, reducing shot size	Increase screw feed setting to increase shot size.
	Check ring worn out causing shot-to-shot variation	Replace worn-out check ring.
	Injection speed too slow	Increase injection speed.
	Injection pressure too low	Increase injection pressure.
	Melt temperature too low	Increase melt temperature.
	Heater band malfunctioning	Replace heater band.
	Hopper feed throat water turned off, causing material to bridge in hopper	Check water circulation at hopper for any blockages or turn on feed throat water.
	Hydraulic system on machine malfunctioning	Consult with machine supplier and perform diagnostic check of machine hydraulic system.
	Barrel capacity for part too small	Move part to machine with a larger barrel capacity.
	Mold temperature too low	Increase mold temperature.
	Gate size too small	Increase gate size.
	Runner size too small	Increase runner size.
	Mold not properly balanced	Check a series of short shots to check flow pattern of part, or run 2D or 3D mold-filling analysis.
	Melt flow rate of material too low to fill part	Switch to a material with a higher melt flow rate.
	Insufficient venting in mold	Check vent sizes and vent according to material supplier recommendations in the area of the joining melt fronts.
	Nozzle orifice size too small	Switch to a nozzle with a larger orifice.
	Material is overdried	Follow material drying conditions as recommended by the material supplier.
Sink marks	Injection hold time too short	Increase injection hold time.
	Injection pressure too low	Increase injection pressure.
	Melt temperature too high	Lower melt temperature.
	Overall cycle time too short	Reduce overall cycle time.
	Injection speed too slow	Increase injection speed.
	Insufficient cushion	Increase cushion according to recommendations of material supplier.

TABLE 1.18 (*Continued*)

Problem	Cause	Solution
	Check ring worn out, causing shot-to-shot variation	Replace worn-out check ring.
	Part gated at heavy section of part	Relocate to thinner nominal wall location.
	Gate size too small	Increase gate size.
	Mold temperature too high	Reduce mold temperature.
	Runner size too small	Increase runner size.
	Sprue size too small	Increase sprue size.
	Melt flow rate of material too low to fill part	Switch to a material with a higher melt flow rate.
Splay or silver streaking	Resin not dried sufficiently	Review recommendations by material supplier for drying and check drying equipment operation and dryer hose conditions and connections. If using regrind, check for larger pellet size.
	Mold temperature too high	Reduce mold temperature.
	Nozzle temperature too high	Lower nozzle temperature.
	Nozzle length too long	Switch to a shorter nozzle.
	Nozzle, sprue bushing, runners, or gates have sharp corners or burrs in flow path	Remove and polish out all sharp corners or burrs.
	Injection speed too fast	Reduce injection speed.
	Gate size too small	Increase gate size.
	Insufficient number of gates causing high shear condition	Add additional gates to part.
	Barrel-to-shot ratio too low; residence time in the barrel too long	Move mold to a machine with a smaller injection cylinder to reduce residence time.
	Back pressure too high	Reduce back pressure.
	Water contamination in material	Find source of water contamination.
	Decompression (“suck back”) set too high, pulling trapped air into melt stream	Reduce decompression.
	Nozzle orifice too small	Switch to nozzle with a larger orifice.
	Gate size too small	Increase gate size.
	Runner size too small	Increase runner size.
	Sprue size too small	Increase sprue size.
	Material contaminated	Remove any contamination, review material handling procedures, or contact material supplier.
Sticking in mold cavity	Injection pressure too high	Reduce injection pressure.
	Injection hold time too long	Reduce injection hold time.
	Overall cycle time too short	Reduce overall cycle time, in particular the cooling cycle.

TABLE 1.18 (*Continued*)

Problem	Cause	Solution
	Melt temperature too high	Reduce melt temperature.
	Insufficient draft angle in part	Increase draft angles according to recommendations from material supplier.
	Undercuts in part preventing part from ejecting from mold	Remove undercuts from part.
	Mold temperature too low	Increase mold temperature.
	Insufficient lubricant in material	Convert to material with higher level of internal lubricant.
Sticking in sprue	Injection pressure too high	Reduce injection pressure.
	Injection hold time too long	Reduce injection hold time.
	Nozzle temperature too low	Increase nozzle temperature.
	Standard taper nozzle used	Switch to a reverse taper nozzle.
	Interface of nozzle to sprue bushing not aligned	Realign nozzle–sprue bushing interface.
	Burrs and other surface imperfections in sprue bushing	Polish out sprue bushing.
	Insufficient taper in sprue bushing	Increase taper in sprue bushing.
	Screw puller not working properly or has poor design	Switch to a different design sprue puller.
	Nozzle orifice smaller than sprue bushing opening	Switch to a nozzle with a larger orifice which is the same diameter as the sprue bushing.
	Insufficient lubricant in material	Convert to material with higher level of internal lubricant.
Voids	Insufficient cushion	Increase cushion according to recommendations of material supplier.
	Check ring worn out, causing air to be trapped in melt	Replace worn-out check ring.
	Injection pressure too low	Increase injection pressure.
	Injection time too short	Increase injection time.
	Injection hold pressure too low	Increase Injection hold time.
	Injection speed too fast	Reduce injection speed.
	Melt temperature too low	Increase melt temperature.
	Melt temperature too high, causing overheating of material	Reduce mold temperature to recommended conditions.
	Gate size too small	Increase gate size.
	Runner size too small	Increase runner size.
	Insufficient venting in mold	Check vent sizes and vent according to material supplier recommendations in the area of the joining melt fronts.

TABLE 1.18 (*Continued*)

Problem	Cause	Solution
	Mold temperature too high, causing material to cool slower	Lower mold temperature.
	Mold temperature too low	Increase mold temperature.
Warpage	Overall cycle time too long	Reduce overall cycle time.
	Melt temperature too low, hindering material flow and creating high stresses in part	Increase melt temperature to within the range recommended.
	Melt temperature too high, causing part to overpack in mold	Decrease melt temperture to within the range recommended.
	Injection hold time too short, insufficiently packing the part	Increase injection hold time.
	Injection hold time too long, causing part to overpack	Reduce injection hold time.
	Injection speed too slow, creating high stresses in part	Increase injection speed.
	Injection speed too fast, overpacking part in cavity	Decrease injection speed.
	Insufficient number of ejector pins	Modify or improve ejector pin setup or system.
	Ejector pin damaged or too short	Replace damaged ejector pin or replace short ejector pin.
	Mold opening speed too fast	Reduce speed to open mold.
	Gate size too small	Increase gate size.
	Runner size too small	Increase runner size.
	Sprue size too small	Increase sprue size.
	Part not cooling evenly in mold	Review cooling line circiut and either add cooling lines or redirect cooling water flow.
Weld or knit lines	Injection pressure too low	Increase injection pressure.
	Injection speed too slow	Increase injection speed.
	Melt temperature too low	Increase melt temperature.
	Mold temperature too low	Increase mold temperature.
	Air trapped at joining melt fronts	Add vents at melt junction, overflow channels, or use porous metal inserts.
	Gates located at pins or around holes and openings, causing melt front to form	Relocate gate.
	Nominal wall thickness too thin	Increase nominal wall or switch to a higher melt flow rate material.
	Core pin shifting causing weld line to form on one side of core	Improve support structure of core pin to prevent shift.

Source: Multiple industry sources.

TABLE 1.19 Injection Troubleshooting Guide: Polycarbonate

Problem	Cause	Solution
Black specks	Machine not purged properly	Follow guideline for purging material and, if necessary, manually clean out check rings, barrel, and screw.
	Excessive fines in material	Use equipment such as a deduster to separate fines from pellets. Also, check material handling system for any sharp corners in delivery lines.
Burn marks	Nozzle temperature too high	Reduce nozzle temperature.
	Injection speed too fast	Reduce injection speed.
	Injection time too long	Reduce injection time.
	Insufficient venting in mold	Check vent sizes and vent according to material supplier recommendations. Also, check for clogged or obstructed vents.
	Gate size too small	Increase gate size without sacrificing part degating.
	Barrel heater malfunctioning	Check and calibrate heaters or replace malfunctioning heaters.
Discoloration	Machine not purged properly	Follow guideline for purging material and, if necessary, manually clean out check rings, barrel, and screw.
	Melt temperature too high	Reduce melt temperature.
	Nozzle temperature too high	Reduce nozzle temperature.
	Overall cycle time too long	Reduce cycle time.
	Material contaminated	Remove any contamination, review material handling procedures, or contact material supplier.
	Insufficient venting in mold	Check vent sizes and vent according to material supplier recommendations. Also, check for clogged or obstructed vents.
	Machine size too large for the part causing a long residence time of the material in the barrel; barrel-to-shot ratio too low	Move mold to a machine with a shorter barrel and higher barrel-to-shot ratio.
Jetting	Injection speed too fast	Reduce injection speed.
	Melt temperature too low	Increase melt temperature.
	Mold temperature too low	Increase mold temperature.
	Gate size too small, restricting flow and producing high pressures for material entering the cavity	Enlarge gate size.

TABLE 1.19 (*Continued*)

Problem	Cause	Solution
	Gate located in an area where the material injects into a large space, such as a thick wall or deep area	Relocate gate to direct flow path into a shallow wall or tab.
Part brittleness	Resin not dried sufficiently	Review recommendations by material supplier for drying and check drying equipment operation and dryer hose conditions and connections. If using regrind, check for larger pellet size.
	Material contaminated	Remove any contamination, review material handling procedures, or contact material supplier.
	Regrind levels too high	Reduce regrind levels.
	Melt temperature too high	Reduce melt temperature.
	Barrel heater malfunctioning	Check and calibrate heaters or replace malfunctioning heaters.
Sinks and/or voids	Injection pressure too low	Increase injection pressure.
	Injection hold time too short	Increase injection hold time.
Sinks	Injection speed too slow	Increase injection speed.
Voids	Injection speed too fast	Reduce injection speed.
Sinks	Mold temperature too low	Increase mold temperature.
Voids	Mold temperature too high	Reduce mold temperature.
Sinks	Melt temperature too low	Increase melt temperature.
Voids	Melt temperature too high	Reduce melt temperature.
	Gate size and/or runner size too small, freezing off before part is completely filled	Enlarge gate size.
Sticking in cavity	Injection pressure too high	Reduce injection pressure.
	Injection hold time too long	Reduce injection hold time.
	Injection time too long	Reduce injection time.
	Mold-closed time too short	Increase mold-closed time.
	Mold temperature too high	Reduce mold temperature.
	Melt temperatures too high	Reduce melt temperature.
	Insufficient part draft used in part design	Increase draft angles on part to facilitate part ejection.
	Undercuts found in mold	Machine and polish out undercuts.
Sticking on mold core	Injection pressure too high	Reduce injection pressure.
	Injection hold time too long	Reduce injection hold time.
	Injection time too long	Reduce injection time.
	Mold-closed time too long	Reduce mold-closed time.
	Melt temperature too low	Increase melt temperature.
	Melt temperatures too high	Reduce melt temperature.
	Insufficient part draft used in part design	Increase draft angles on part to facilitate part ejection.
	Undercuts found in mold	Machine and polish out undercuts.

TABLE 1.19 (*Continued*)

Problem	Cause	Solution
Splay marks, silver streaks, and splash marks	Resin not dried sufficiently	Review recommendations by material supplier for drying and check drying equipment operation and dryer hose conditions and connections. If using regrind, check for larger pellet size.
	Material contaminated	Remove any contamination, review material handling procedures, or contact material supplier.
	Melt temperatures too high	Reduce melt temperature.
	Melt temperatures too high	Reduce melt temperature.
	Overall cycle time too long	
	Injection speed too fast	Reduce injection speed.
	Mold temperature too low	Increase mold temperature.
	Gate size too small, restricting flow and producing high shear for material entering the cavity	Enlarge gate size.
	Back pressure too low	Increase back pressure.
	Decompression ("suck back") set too high, pulling trapped air into melt stream	Reduce decompression.
Warpage, part distortion	Cooling rate of both mold halves not consistent with each other	Balance out cooling rate of both mold halves. General rule of thumb: Keep both mold halves within ±10°F of each other.
	Parts warping after part ejection	Check handling of part after ejection.
	Injection hold time too short	Increase injection hold time.
	Mold closed time too short	Increase mold-closed time.
	Parts too hot after ejection	Quench part in cool bath, but be aware of possibly creating molded-in stresses in part.
Weld lines or knit lines	Injection pressure too low	Increase injection pressure.
	Injection speed too slow	Increase injection speed.
	Injection hold time too short	Increase injection hold time.
	Mold temperature too low	Increase mold temperature.
	Melt temperature too low	Increase melt temperature.
	Vents in weld line area clogged with foreign matter	Clean out all vents.
	Insufficient venting in mold	Check vent sizes and vent according to material supplier recommendations in the area of the joining melt fronts.
	Gate location promoting formation of weld/knit lines	Relocate gates to move weld lines.

Source: Multiple industry sources.

TABLE 1.20 Injection Troubleshooting Guide: Polycarbonate Blends

Problem	Cause	Solution
Black specks	Machine not purged properly	Follow guideline for purging material and, if necessary, manually clean out check rings, barrel, and screw.
	Barrel heater malfunctioning and overheating the material	Check and calibrate heaters or replace malfunctioning heaters.
	Excessive fines in material that are degrading in the barrel	Use equipment such as a deduster to separate fines from pellets. Also, check material handling system for any sharp corners in delivery lines.
Brown streaks/ burning	Resin not dried sufficiently	Review recommendations by material supplier for drying and check drying equipment operation and dryer hose conditions and connections. If using regrind, check for larger pellet size.
	Melt temperature too high	Reduce melt temperature.
	Nozzle temperature too high	Reduce nozzle temperature.
	Nozzle length too long	Install a shorter length nozzle.
	Nozzle blocked	Remove blocked area in nozzle.
	Nozzle tip too small	Increase nozzle tip diameter to 80% of sprue bushing, maximum.
	Flow building in dead spots of check ring, runner, or hot manifold system, if used	Remove dead spots in check ring, runner, or hot manifold, and clean areas on a routine basis.
	Gate and runner system too small, causing high shear condition on material	Increase gate and runner size to reduce pressure in runner system to reduce shear to material.
	Material contaminated	Remove any contamination, review material handling procedures, or contact material supplier.
	Insufficient venting in mold	Check vent sizes and vent according to material supplier recommendations in the area of the joining melt fronts.
	Vents in weld line area clogged with foreign matter	Clean out all vents.
	Check ring worn out	Inspect check rings for cracks and worn valve seats.
	Worn or improperly fitted screw and barrel configuration	Measure both barrel and screw for possible wear.
Brittle parts	Resin not dried sufficiently	Review recommendations by material supplier for drying and check drying equipment operation and dryer hose conditions and connections. If using regrind, check for larger pellet size.

TABLE 1.20 (*Continued*)

Problem	Cause	Solution
	Mold temperature too low	Increase mold temperature to reduce part stress.
	Melt temperature too low	Increase melt temperature to reduce part stress.
	Injection pressure too high	Reduce injection pressure to reduce part stress from overpacking.
Charring of material/ dieseling	Injection speed too fast	Reduce injection speed.
	Melt temperature too high	Reduce melt temperature.
	Mold temperature too high	Reduce mold temperature.
	Insufficient venting in mold	Check vent sizes and vent according to material supplier recommendations.
	Clamp pressure too high, trapping gas in part	Reduce clamp pressure.
Discoloration of material	Melt temperature too high	Reduce melt temperature.
	Back pressure too high, overworking material	Reduce back pressure.
	Drying time too long, causing material degradation	Reduce drying time and measure hopper capacity.
	Material contaminated	Remove any contamination, review material handling procedures, or contact material supplier.
	Screw clearance too large	Measure screw-to-barrel clearance.
Flashing of part	Injection pressure too high	Decrease injection pressure.
	Mold damaged or mold surface has wear areas	Repair damaged tooling.
	Resin not dried sufficiently	Review recommendations by material supplier for drying and check drying equipment operation and dryer hose conditions and connections. If using regrind, check for larger pellet size.
	Melt temperature too high	Reduce melt temperature.
	Vents machined too deep	Fill existing vents and remachine.
	Mold platen misaligned	Reset and align platens.
	Clamp tonnage too low	Move mold to a machine with a lar ger tonnage capacity.
Gate blush	Melt temperature too low	Increase melt temperature.
	Injection speed too fast	Reduce injection speed.
	Gate size too small, creating high shear condition	Increase gate diameter or size to reduce shear.
	Sharp edge in gate area	Radius all sharp corners at gate. F an the gate if posible.
	Cold slug not used	Install a cold slug well at end of sprue.

TABLE 1.20 (*Continued*)

Problem	Cause	Solution
Jetting	Injection speed too fast	Reduce injection speed.
	Gate location into open area	Move gate into location where it flows to an obstruction, like a cavity wall.
	Gate size too small, creating high pressure condition into the cavity	Increase gate diameter or size to reduce pressure entering cavity.
Splay/silver streaks	Resin not dried sufficiently	Review recommendations by material supplier for drying and check drying equipment operation and dryer hose conditions and connections. If using regrind, check for larger pellet size.
	Melt temperature too high	Reduce melt temperature.
	Back pressure too high, overworking material	Reduce back pressure.
	Screw speed too fast	Reduce screw speed.
	Barrel heater malfunctioning and overheating the material	Check and calibrate heaters or replace malfunctioning heaters.
	Back pressure too low, trapping air in melt	Increase back pressure.
	Decompression ("suck back") set too high pulling trapped air into melt stream	Reduce decompression.
	Injection speed too fast	Reduce injection speed.
	Injection pressure too high	Reduce injection pressure.
	Gate size too small, creating high shear condition	Increase gate diameter or size to reduce shear.
	Nozzle orifice too small	Change to orifice with a larger orifice size.
	Nozzle length too long	Change to a nozzle with a shorter length.
	Nozzle blocked	Remove blocked area in nozzle.
	Check ring worn out or damaged	Inspect check rings for cracks and worn valve seats.
	Insufficient venting in mold	Check vent sizes and vent according to material supplier recommendations.
	Cold slugs found at nozzle or hot runner tip	Increase nozzle temperature.
	Sharp corners or transitions at or near gate	Radius all sharp corners adjacent to gate.
	Material contaminated	Remove any contamination, review material handling procedures, or contact material supplier.

TABLE 1.20 (*Continued*)

Problem	Cause	Solution
	Sharp corners along flow path in areas such as sprue, runner, and in part design	Radius all sharp edges. Recommended: 0.015–0.305 in. (0.40–7.75 mm).
Short shots/ incomplete part fill	Insufficient screw feed setting, reducing shot size	Increase screw feed setting to increase shot size.
	Injection pressure too low	Increase injection pressure.
	Injection speed too slow	Increase injection speed.
	Melt temperature too low	Increase melt temperature.
	Inconsistent drying of material causing processing of wet to dry material	Review recommendations by material supplier for drying and check drying equipment operation and dryer hose conditions and connections. If using regrind, check for larger pellet size.
	Mold temperature too low	Increase mold temperature.
	Machine clamp tonnage insufficient	Move mold to a machine with a larger tonnage capacity.
	Nozzle blocked	Remove blocked area in nozzle.
	Gate, sprue, and/or runner undersized	Enlarge runner.
	Insufficient venting in mold	Check vent sizes and vent according to material supplier recommendations.
	Flow length too long for material	Use a material with a higher melt flow rate or redesign part design or tool design.
Sinks/voids	Insufficient screw feed setting, reducing shot size	Increase screw feed setting to increase shot size.
	Injection hold pressure too low	Increase injection hold pressure.
	Injection hold time too short	Increase injection hold time.
	Part section too thick	Core out thick sections of part or use ribs in place of thick sections.
	Gate freezing off too fast	Enlarge gate and check gate freeze-off time.
	Insufficient old cooling	Review cooling line circuit, and either add cooling lines or redirect cooling water flow.
Sticking parts	Mold temperature too high	Reduce mold temperature.
	Injection pressure too high	Reduce injection pressure.
	Undercuts found in mold	Machine and polish out undercuts.
	Rough or damaged surface on mold	Polish and repair mold surface.
	Insufficient draft in part	Increase draft on part.
	Injection hold pressure too high	

TABLE 1.20 (*Continued*)

Problem	Cause	Solution
	Resin not dried sufficiently	Review recommendations by material supplier for drying and check drying equipment operation and dryer hose conditions and connections. If using regrind, check for larger pellet size.
	Overall cycle too short, causing part to eject too hot	Increase overall cycle by increasing cooling time.
Warpage	Part not cooling evenly in mold	Review cooling line circuit and either add cooling lines or redirect cooling water flow.
	Injection hold time too long	Reduce injection hold time.
	Orientation due to glass filler orientation	Change gate location to compensate for glass filler orientation or switch to unreinforced grade.
Weld lines/ knit lines	Mold temperature too low	Increase mold temperature.
	Melt temperature too low	Increase melt temperature.
	Injection speed too slow	Increase injection speed.
	Injection pressure too low	Increase injection pressure.
	Injection hold pressure too low	Increase injection hold pressure.
	Vents in weld line area clogged with foreign matter	Clean out all vents.
	Insufficient venting in mold	Check vent sizes and vent according to material supplier recommendations in the area of the joining melt fronts.
	Gate size too small	Increase gate size to prevent premature gate freeze-off.
	Runner size too small	Increase runner diameter to force more material into cavity.

Source: Multiple industry sources.

TABLE 1.21 Injection Troubleshooting Guide: Polyester

Problem	Cause	Solution
Delamination	Material contaminated	Remove any contamination, review material handling procedures, or contact material supplier.
	Gate size too small	Increase gate diameter.
Short shots/ incomplete filled parts, surface rippling	Hopper empty or obstructed	Remove any obstructions from hopper or fill hopper with more material.
	Insufficient screw feed setting, reducing shot size	Increase screw feed setting to increase shot size.
	Injection pressure too low	Increase injection pressure.

TABLE 1.21 (*Continued*)

Problem	Cause	Solution
	Injection speed too slow	Increase injection speed.
	Cushion size too large	Decrease cushion.
	Mold temperature too low	Increase mold temperature.
	Overall cycle time too short	Increase overall cycle time.
	Melt temperature too low	Increase melt temperature.
	Gate size too small	Increase gate diameter.
Sink marks	Injection pressure too low	Increase injection pressure.
	Injection hold pressure too low	Increase injection hold pressure.
	Injection hold time too short	Increase injection hold time.
	Injection time too short	Increase injection time.
	Gate freezing off too fast	Enlarge gate and check gate freeze-off time.
	Gate located at thin section, causing flow from thin to thick walls	Reloacte gate to thick wall section.
Splay marks	Injection speed too fast	Reduce injection speed.
	Mold temperature too low	Reduce mold temperature.
	Gate size too small	Increase gate diameter.
	Gate area temperature too cold	Add electric cartridge heaters at gate or circulate hot oil in area.
	Melt temperature too high	Reduce melt temperature.
	Resin not dried sufficiently	Review recommendations by material supplier for drying and check drying equipment operation and dryer hose conditions and connections. If using regrind, check for larger pellet size.
	Nozzle temperature too high	Reduce nozzle temperature.
	Injection time too long	Reduce injection time.
	Cushion size too large	Reduce cushion size.
	Decompression ("suck back") set too high, pulling trapped air into melt stream	Reduce decompression.
Sticking in cavity	Injection pressure too high	Decrease injection pressure.
	Injection hold time too long	Reduce injection hold time.
	Cushion size inconsistent	Check condition of check ring and replace if needed.
	Mold closed time too short	Increase mold closed time.
	Mold temperature too high	Reduce mold temperature.
	Insufficient draft in part	Increase draft on part.
	Undercuts found in mold	Machine and polish out undercuts.
	Material does not have sufficient levels of internal lubricants	Use a material grade with increased level of lubricants, internal or external, such as mold release.

TABLE 1.21 (*Continued*)

Problem	Cause	Solution
Voids	Injection pressure too low	Increase injection pressure.
	Injection time too short	Lengthen injection time.
	Cushion size too large	Decrease cushion size.
	Mold temperature too low	Increase mold temperature.
	Gate freezing off too fast	Enlarge gate and check gate freeze-off time.
Warpage	Part not cooling evenly in mold	Review cooling line circuit and either add cooling lines or redirect cooling water flow.
	Injection time too short	Lengthen injection time.
	Cushion size too large	Decrease cushion size.
	Mold temperature too low	Increase mold temperature.
	Gate freezing off too fast	Enlarge gate and check gate freeze-off time.

Source: Multiple industry sources.

TABLE 1.22 Injection Troubleshooting Guide: PP/PE (Polypropylene/Polyethylene)

Problem	Cause	Solution
Brittle parts	Degradation of material	Decrease barrel temperatures starting with middle zones.
		Residence time in barrel too long; measure barrel-to-shot ratio.
	High level of molded-in stresses in part	Increase barrel temperatures starting with middle zones.
		Reduce injection speed.
		Remove sharp corners and transitions in parts and radius corners.
		Reduce injection packing and hold times.
	Material contaminated	Remove any contamination, review material handling procedures, or contact material supplier.
		Check compatibility of color concentrate with base resin.
Excessive shrinkage	Cooling time too short	Increase cooling time.
	Injection pressure too low	Increase injection pressure.
	Mold temperature too high	Reduce mold temperature.
	Melt temperature too high	Reduce melt temperature.
	Gates and runner size too small	Enlarge gates and runners.
	Excessive wall thickness variation	Design part with consistent nominal wall thickness.

TABLE 1.22 (*Continued*)

Problem	Cause	Solution
Flashing of parts	Injection pressures too high	Reduce injection pressure.
	Clamping pressures too low	Increase clamp pressure.
	Melt temperature too high	Reduce melt temperature.
	Barrel heater malfunctioning	Check and calibrate heaters or replace malfunctioning heaters.
	Back pressure too high	Reduce back pressure.
	Obstruction on mold surface preventing mold from closing completely	Check compatibility of color concentrate with base resin.
Flow marks	Melt temperature too low	Increase melt temperature.
	Injection speed too slow	Increase injection speed.
	Mold temperature too low	Increase mold temperature.
	Poor surface finish on mold	Polish cavity and core surfaces.
	Flow marks located in areas that compromise part integrity	Relocate flow marks away from areas where strength is critical.
Gas burn marks (dieseling)	Injection speed too high.	Reduce injection speed.
	Melt temperature too high	Reduce melt temperature
	Insufficient venting in mold	Check vent sizes and vent according to material supplier recommendations in the area of the joining melt fronts.
Gloss		
Too low[a]	Injection speed too slow	Increase injection speed.
	Melt temperature too low	Increase melt temperature.
	Mold temperature too low	Increase mold temperature.
	Injection pressure too low	Increase injection pressure
	Poor surface finish on mold	Polish cavity and core surfaces.
	Injection speed too high	Reduce injection speed.
Too high[a]	Melt temperature too high	Reduce melt temperature.
	Mold temperature too high	Reduce mold temperature.
	Injection pressure too high	Reduce injection pressure.
	Mold surface highly polished	Add texture to surface.
	Mold not properly balanced	Check a series of short shots to check flow pattern of part or run 2D or 3D mold-filling analysis.
	Machine undersized for part	Check barrel-to-shot ratio as well as volumetric requirements of machine and heating capacity of machine.
	Check ring worn out	Inspect check rings for cracks and worn valve seats.
Nozzle drool	Melt temperature too high	Reduce melt temperature.
	Hot runner manifold and/or barrel drop temperatures too high	Check and calibrate heaters or replace malfunctioning heaters.

TABLE 1.22 (*Continued*)

Problem	Cause	Solution
	Overall cycle time too short	Increase cycle time.
	Decompression ("suck back") shut off	Utilize decompression.
Part sticking in cavity	Injection pressure too high	Reduce injection pressure.
	Overall cycle time too short	Reduce overall cycle time.
	Surface imperfections on mold	Polish cavity and core surfaces. Utilize high lubricity mold coatings or finishes.
	Insufficient draft angles in part	Increase draft angles on part.
	Gates located near projections, such as ribs, bosses, and holes	Relocate gates away from projections.
Part sticking on core	Injection pressure too high	Reduce injection pressure.
	Overall cycle time too short	Reduce overall cycle time.
	Surface imperfections on mold	Polish cavity and core surfaces. Utilize high lubricity mold coatings or finishes. Draw machine cores (machine in direction of flow).
	Insufficient draft angles in part	Increase draft angles on part.
	Undercuts on part preventing part ejection	Remove undercuts on part.
	Insufficient number of ejector pins	Modify or improve ejector pin setup or system.
Poor color dispersion	Incompatible color concentrate	Change to a compatible color concentrate.
	Melt temperature too low	Increase melt temperature.
	Back pressure too low	Increase back pressure.
	Poor mixing in barrel	Replace standard machine screw with an improved mixing screw such as a barrier-type screw. Use static mixer or mixing nozzle.
Poor weld lines/knit lines	Melt temperature too low	Increase melt temperature.
	Mold temperature too low	Increase mold temperature.
	Injection time too short	Increase injection time.
	Weld lines located in areas that compromise part integrity	Relocate weld lines away from areas where strength is critical.
	Air trapped at joining melt fronts	Add vents at melt junction, overflow channels, or use porous metal inserts.
Rough Part Surface	Melt temperature too low	Increase melt temperature.
	Mold temperature too low	Increase mold temperature.
	Insufficient venting in mold	Check vent sizes and vent according to material supplier recommendations in the area of the joining melt fronts.
	Poor surface finish on mold	Polish cavity and core surfaces.

TABLE 1.22 (*Continued*)

Problem	Cause	Solution
Short shots	Insufficient screw feed setting, reducing shot size	Increase screw feed setting to increase shot size.
	Injection time too short	
	Melt temperature too low	Increase melt temperature.
	Mold temperature too low	Increase mold temperature.
	Gates and runner size too small	Enlarge gates and runners.
	Nozzle and gates blocked due to foreign matter	Remove foreign matter from gates and nozzle.
	Insufficient venting in mold	Check vent sizes and vent according to material supplier recommendations in the area of the joining melt fronts.
	Melt flow rate of material too low	Switch to a higher melt flow rate material.
	Mold not properly balanced	Check a series of short shots to check flow pattern of part or run 2D or 3D mold-filling analysis.
Sink marks	Injection pressure too low	Decrease injection pressure.
	Injection time too short	Lengthen injection time.
	Injection speed too fast	Decrease injection speed.
	Gates size too small	Enlarge gates.
	Gates located in thin section of part, causing flow path to go from thin to thick section	Relocate gates in thick section of part.
	Nominal wall too thick	Core out part to reduce wall thickness or thin part and use ribs.
	Wall too thick at projections (ribs, bosses, gussets)	Reduce wall thickness at projections.
Splay marks	Resin not dried sufficiently	Review recommendations by material supplier for drying, and check drying equipment operation and dryer hose conditions and connections. If using regrind, check for larger pellet size.
	Material contaminated	Remove any contamination, review material handling procedures, or contact material supplier.
	Moisture accumulating on mold	Set mold temperatures above dew point.
	Flow path disrupted due to burr, sharp corner	Polish out burrs and radius sharp corners.
	Injection speed too high	Reduce injection speed.
	Melt temperatures too high	Reduce melt temperature.
	Back pressure too high	Reduce back pressure.

TABLE 1.22 (*Continued*)

Problem	Cause	Solution
	Injection pressure too low	Increase injection pressure.
	Mold not properly balanced	Check a series of short shots to check flow pattern of part or run 2D or 3D mold-filling analysis.
Voids	Moisture in material	Dry material, if needed.
	Material contaminated	Remove any contamination, review material handling procedures, or contact material supplier.
	Insufficient venting in mold	Check vent sizes and vent according to material supplier recommendations in the area of the joining melt fronts.
	Gates located in thin section of part causing flow path to go from thin to thick section	Relocate gates in thick section of part.
	Injection speed too high	Reduce injection speed.

Source: Multiple industry sources.

[a]Textured surfaces have a reverse effect on gloss in comparison to untextured surfaces. Packing pressure, higher melt, higher melt, and higher injection speeds will reduce gloss. This results in packing out the texture peaks and valleys resulting in the low gloss provided by the texture.

TABLE 1.23 Injection Troubleshooting Guide: Polystyrene

Problem	Cause	Solution
Brittleness	Melt temperature too high	Reduce melt temperature.
	Material contaminated	Remove any contamination, review material handling procedures or contact material supplier.
	Mold temperature too low, creating molded-in stresses	Increase mold temperature.
	Screw speed too high	Reduce screw speed.
	Back pressure too high, causing material to overheat	Reduce back pressure.
	Insufficient venting in mold	Check vent sizes and vent according to material supplier recommendations in the area of the joining melt fronts.
	Gate located in area of part in area where part performance is compromised	Move gate to different location.
	Injection speed too high	Reduce injection speed.

TABLE 1.23 (*Continued*)

Problem	Cause	Solution
Blush at gate	Mold temperature too low	Increase mold temperature.
	Injection speed too high, causing material to shear at gate	Reduce injection speed.
	Melt temperature too low, causing material to shear at gate	Increase melt temperature.
	Melt temperature too high, causing material to degrade	Reduce melt temperature.
	Injection pressure too low	Increase injection pressure.
	Gate size too small	Increase gate size.
	Runner size too small	Increase runner size.
	Sprue size too small	Increase sprue size.
Burn marks at weld lines	Injection speed too high	Reduce injection speed.
	Melt temperature too low, causing material to shear at gate	Increase melt temperature.
	Melt temperature too high, causing material to degrade	Reduce melt temperature.
	Injection speed too high	Reduce injection speed.
	Back pressure too high, causing material to overheat	Reduce back pressure.
	Nozzle orifice size too small	Increase nozzle orifice size.
	Gate size too small	Increase gate size.
	Runner size too small	Increase runner size.
	Sprue size too small	Increase sprue size.
	Injection pressure too low	Increase injection pressure.
	Gate located around hole or other depression, causing weld line to form	Move gate to different location.
Delamination of surface	Material contaminated	Remove any contamination, review material handling procedures, or contact material supplier.
	Mold temperature too low	Increase mold temperature.
	Melt temperature too low	Increase melt temperature.
	Injection speed too high.	Reduce injection speed.
	Nozzle orifice size too small	Increase nozzle orifice size.
	Land at gate too long causing high-shear condition to material	Reduce length of land to no more than 0.025 in.
	Injection pressure too high	Reduce injection pressure.
Flashing of part	Injection pressure too high	Reduce injection pressure.
	Clamp pressure too low	Move tool to press with a higher clamp pressure
	Injection speed too fast	Reduce injection speed.
	Melt temperature too high	Decrease melt temperature.
	Mold temperature too high	Decrease mold temperature.
	Mold part line mismatched	Rematch mold parting line.

TABLE 1.23 (*Continued*)

Problem	Cause	Solution
Jetting	Injection speed too fast	Reduce injection speed.
	Melt temperature too low	Increase melt temperature.
	Melt temperature too high	Decrease melt temperature.
	Mold temperature too low	Increase mold temperature.
	Nozzle orifice size too small	Increase nozzle orifice size.
	Land at gate too long causing high-shear condition to material	Reduce length of land to no more than 0.025 in.
	Gate size too small	Increase gate size.
	Runner size too small	Increase runner size.
	Sprue size too small	Increase sprue size.
	Nozzle temperature too low	Increase nozzle temperature or check for malfunctioning heater band.
Poor dimensional stability	Melt temperature too low	Increase melt temperature.
	Melt temperature too high	Decrease melt temperature.
	Injection hold time too short	Decrease injection hold time.
	Injection pressure too low	Increase injection pressure.
	Back pressure too low	Increase back pressure.
	Mold temperature too low	Increase mold temperature.
	Mold temperature too high	Decrease mold temperature.
	Inefficient mold cooling creating uneven cooling of mold and part	Review water channel locations and circuit and modify.
Poor surface finish	Material contaminated	Remove any contamination, review material handling procedures, or contact material supplier.
	Injection pressure too low	Increase injection pressure.
	Melt temperature too high	Reduce melt temperature.
	Mold temperature too low	Increase mold temperature.
	Surface imperfections on mold	Polish cavity and core surfaces. Also, utilize high lubricity mold coating or finishes.
Sticking in cavity	Injection pressure too high	Decrease injection pressure.
	Injection hold time too long	Decrease injection hold time.
	Overall cycle time too short	Increase overall cycle time to allow more part cooling.
Sprue sticking and breaking	Burrs and other surface imperfections in sprue bushing	Polish out sprue bushing.
	Nozzle not seated properly into sprue bushing	Check alignment of nozzle to sprue bushing.
	Mold temperature too low	Increase mold temperature.

TABLE 1.23 *(Continued)*

Problem	Cause	Solution
	Injection hold time too long	Decrease injection hold time.
	Nozzle temperature too low	Increase nozzle temperature or check for malfunctioning heater band.
	Surface imperfections on mold	Polish cavity and core surfaces. Also, utilize high-lubricity mold coating or finishes.
	Material contaminated	Remove any contamination, review material handling procedures, or contact material supplier.
Warpage	Mold temperature too high	Decrease mold temperature.
	Melt temperature too high	Decrease melt temperature.
	Overall cycle time too long	Reduce overall cycle time.
	Injection speed too fast	Reduce injection speed.
	Injection pressure too high	Reduce injection pressure.
	Back pressure too high	Reduce back pressure.
	Inefficient mold cooling, creating uneven cooling of mold and part	Review water channel locations and circuit and modify.
Wave marks	Insufficient screw feed setting, reducing shot size	Increase screw feed setting to increase shot size.
	Mold temperature too high, causing flow to be turbulent	Decrease mold temperature.
	Injection hold time too short	Increase injection hold time.
	Injection pressure too low	Increase injection pressure.
	Melt temperature too high	Decrease melt temperature.
	Insufficient venting in mold	Check vent sizes and vent according to material supplier recommendations in the area of the joining melt fronts.
Weak weld/ knit lines	Mold temperature too low	Increase mold temperature.
	Injection speed too slow	Increase injection speed.
	Melt temperature too low	Increase melt temperature.
	Injection pressure too low	Increase injection pressure.
	Insufficient venting in mold	Check vent sizes and vent according to material supplier recommendations in the area of the joining melt fronts.
	Gate size too small	Increase gate size.
	Runner size too small	Increase runner size.
	Sprue size too small	Increase sprue size.

Source: Multiple industry sources.

TABLE 1.24 Injection Troubleshooting Guide: Polyurethane Elastomers

Problem	Cause	Solution
Brittleness or bubbles	Insufficient venting in mold	Check vent sizes and vent according to material supplier recommendations in the area of the joining melt fronts.
	Resin not dried sufficiently	Review recommendations by material supplier for drying and check drying equipment operation and dryer hose conditions and connections. If using regrind, check for larger pellet size.
	Cold slug not used	Install a cold slug well at end of sprue.
	Injection speed too fast	Reduce injection speed.
	Melt temperature too high	Reduce melt temperature
	Gate size too small	Increase gate size.
	Runner size too small	Increase runner size.
Decrease in material toughness	Melt temperature too high, causing material degradation	Reduce melt temperature.
	Melt temperature too low, causing high stresses to be molded in the part	Increase melt temperature.
	Resin not dried sufficiently	Review recommendations by material supplier for drying and check drying equipment operation and dryer hose conditions and connections. If using regrind, check for larger pellet size.
Flash on part	Melt temperature too high	Reduce melt temperature.
	Injection pressure too high	Reduce injection pressure.
	Injection speed too fast	Reduce injection speed.
	Excessive number of vents	Check vent sizes and vent according to material supplier recommendations in the area of the joining melt fronts.
	Clamp pressure too low	Move tool to press with a higher clamp pressure.
	Resin not dried sufficiently	Review recommendations by material supplier for drying and check drying equipment operation and dryer hose conditions and connections. If using regrind, check for larger pellet size.
	Mold not properly balanced	Check a series of short shots to check flow pattern of part or run 2D or 3D mold-filling analysis.

TABLE 1.24 (*Continued*)

Problem	Cause	Solution
Flash on parts with short shots	Resin not dried sufficiently	Review recommendations by material supplier for drying and check drying equipment operation and dryer hose conditions and connections. If using regrind, check for larger pellet size.
	Melt temperature too high	Reduce melt temperature.
	Clamp pressure too low	Move tool to press with a higher clamp pressure.
	Injection speed too fast	Reduce injection speed.
Nozzle drool	Melt temperature too high	Reduce melt temperature.
	Resin not dried sufficiently	Review recommendations by material supplier for drying and check drying equipment operation and dryer hose conditions and connections. If using regrind, check for larger pellet size.
	Back pressure too high	Reduce back pressure.
	Nozzle temperature too high	Reduce nozzle temperature.
Parts too flexible	Melt temperature too high, causing material degradation	Reduce melt temperature.
Short shots or incomplete filled parts	Resin not dried sufficiently	Review recommendations by material supplier for drying and check drying equipment operation and dryer hose conditions and connections. If using regrind, check for larger pellet size.
	Melt temperature too low	Increase melt temperature.
	Injection pressure too low	Increase injection pressure.
	Injection hold time too short	Increase injection hold time.
	Insufficient screw feed setting reducing shot size	Increase screw feed setting to increase shot size.
	Insufficient venting in mold	Check vent sizes and vent according to material supplier recommendations in the area of the joining melt fronts.
	Injection speed too slow	Increase injection speed.
	Gates too small	Increase size of gates so as to not interfere with part ejection or degating.
	Runners too small	Increase size of runners.
	Mold not properly balanced	Check a series of short shots to check flow pattern of part or run 2D or 3D mold-filling analysis.

TABLE 1.24 (*Continued*)

Problem	Cause	Solution
	Nozzle orifice too small	Use a nozzle with a larger orifice diameter.
	Nozzle temperature too low	Increase nozzle temperature.
	Mold temperature too low	Increase mold temperature.
	Check ring worn out	Inspect check rings for cracks and worn valve seats.
Sinks marks	Injection pressure too low	
	Insufficient screw feed setting, reducing shot size	Increase screw feed setting to increase shot size.
	Melt temperature too low, causing insufficient packing of part	Increase melt temperature.
	Injection hold time too short	Increase injection hold time.
	Melt temperature too high, causing part not to cool sufficiently	Decrease melt temperature.
	Gate size too small	Increase gate size.
	Runner size too small	Increase runner size.
Splay marks	Resin not dried sufficiently	Review recommendations by material supplier for drying and check drying equipment operation and dryer hose conditions and connections. If using regrind, check for larger pellet size.
	Melt temperature too high, causing material to degrade	Decrease melt temperature.
	Injection speed too fast	Reduce injection speed.
	Barrel-to-shot ratio too low, causing long residence time in the barrel	Move part to machine with a higher barrel-to-shot ratio between 30–65%.
	Material contaminated	Remove any contamination, review material handling procedures, or contact material supplier.
Sticking of part in cavity	Overall cycle time too short	Increase overall cycle time.
	Melt temperature too high causing overpacking of part	Reduce melt temperature.
	Mold surface too glossy	Use fine texture on surface or utilize high-lubricity mold coating or finishes.
	Mold temperature too low, causing part not to shrink enough to eject part	Increase mold temperature.
	Mold temperature too high causing part to overpack in mold	Reduce mold temperature.

TABLE 1.24 *(Continued)*

Problem	Cause	Solution
	Ejector pins not ejecting part properly	Review ejector pin location or number and design of ejector pins and modify.
	Rough or damaged surface on mold	Polish and repair mold surface.
	Resin not dried sufficiently	Review recommendations by material supplier for drying and check drying equipment operation and dryer hose conditions and connections. If using regrind, check for larger pellet size.
	Injection hold time too long	Reduce injection hold time.
	Injection pressure too high	Reduce injection pressure.

Source: Multiple industry sources.

TABLE 1.25 Injection Troubleshooting Guide: Polyvinyl Chloride—Rigid and Flexible

Problem	Cause	Solution
	Rigid PVC	
Blisters	Melt temperature too high	Decrease melt temperature.
	Injection speed too fast	Reduce injection speed.
	Insufficient venting in mold	Check vent sizes and vent according to material supplier recommendations in the area of the joining melt fronts.
Burning in part	Gate size too small	Increase gate size.
	Injection speed too fast	Reduce injection speed.
Burning in runner	Melt temperature too high	Decrease melt temperature.
	Injection speed too fast	Reduce injection speed.
	Nozzle orifice size too small	Switch to nozzle with larger orifice size.
	Screw speed too fast	Reduce screw speed.
	Barrel-to-shot ratio too low; residence time in the barrel too long	Move mold to a machine with a smaller injection cylinder to reduce residence time.
Flashing of parts	Melt temperature too high	Decrease melt temperature.
	Clamp pressure too low	Move tool to press with a higher clamp pressure.
	Injection pressure too high	Reduce injection pressure.
	Mold part line mismatched	Rematch mold parting line or realign mold in press.
Orange peels	Injection speed too slow	Increase injection speed.
	Melt temperature too low	Increase melt temperature.
	Mold temperature too low	Increase mold temperature.

TABLE 1.25 (*Continued*)

Problem	Cause	Solution
Sink marks	Feed from hopper blocked or has emptied	Remove blockage and check material-handling system for proper operation.
	Injection pressure too low	Increase injection pressure.
	Injection hold time too short	Increase injection hold time.
	Part nominal wall too thick	Core-out thick part sections or use ribs plus a thinner nominal wall thickness.
Weld/knit lines	Gates located at pins or around holes and openings causing melt front to form	Relocate gate.
	Melt temperature too low	Increase melt temperature.
	Mold temperature too low	Increase mold temperature.
	Flexible PVC	
Black specks	Barrel and screw not completely purged	Follow purging procedures as recommended by material supplier.
	Blockage of material in barrel	Remove and clean out blockages.
	Check ring worn out, creating dead spots in the check ring for material to hang up	Replace worn-out check ring or use a check ring with fewer flow restrictions.
Blisters or bubbles	Melt temperature too high	Decrease melt temperature.
	Insufficient venting in mold	Check vent sizes and vent according to material supplier recommendations in the area of the joining melt fronts.
Burned spots	Injection speed too fast	Reduce injection speed.
	Insufficient venting in mold	Check vent sizes and vent according to material supplier recommendations in the area of the joining melt fronts.
	Melt temperature too high	Decrease melt temperature.
Dull streaks in part	Melt temperature too high	Decrease melt temperature.
	Gate size too small	Increase gate size.
	Runner size too small	Increase runner size.
	Mold temperature too low	Increase mold temperature.
	Gates located at pins or inserts, causing melt front to form	Relocate gate or supply localized heating at or around inserts.
Excessive shrinkage	Melt temperature too low, causing part not to fully pack out cavity	
	Melt temperature too high, extending cooling time of part	

TABLE 1.25 *(Continued)*

Problem	Cause	Solution
	Gate size too small; gate freezing off too soon before cavity off is full	Increase gate size.
	Barrel and screw not completely purged	Follow purging procedures as recommended by material supplier.
	Screw speed too fast.	Decrease screw speed.
Flashing of parts	Melt temperature too high	Decrease melt temperature.
	Clamp pressure too low	Move tool to press with a higher clamp pressure.
Orange peels at gate	Injection time too long	Reduce injection time.
	Injection pressure too high	Decrease injection pressure.
	Melt temperature too high	Decrease melt temperature.
Short shots	Mold temperature too low	Increase mold temperature.
	Feed from hopper blocked or has emptied	Remove blockage and check material handling system for proper operation.
	Injection pressure too low	Increase injection pressure.
	Gate size too small; gate freezing off too soon before cavity is full	Increase gate size.
Sink marks	Injection pressure too low	Increase injection pressure.
	Injection hold time too short	Increase injection hold time.
	Injection hold pressure too low	
	Part nominal wall too thick	Core-out thick part sections or use ribs plus a thinner nominal wall thickness.
	Insufficient feed of material to fill cavity	Increase feed setting and shot size.
Weld lines/ knit lines	Melt temperature too low	Increase melt temperature.
	Runner size too small	Increase runner size.
	Gates located at pins, inserts, or depressions, causing melt front to form	Relocate gate away from pins, or increase wall thickness in thin depressed areas of part.
	Mold temperature too low	Increase mold temperature.

Source: Multiple industry sources.

ACKNOWLEDGMENTS

The author thanks Jef fery Myers of Barr , Inc. and Mark Spalding of The Dow Chemical Company for their contributions to this chapter.

REFERENCES AND BIBLIOGRAPHY

1. Society of Plastics Engineers, Injection Molding Division, www.injectionmolding.com.
2. D. V. Rosato, *Injection Molding Handbook*, 2nd ed., Chapman & Hall, New York, 1995, pp. 183–186, 250–251, 297–298.
3. *Vydyne Molding Guide*, Dow Chemical Company/Solutia Inc., Midland, MI, 2000, pp. 10–24, 26–33.
4. Charles A. Harper, *Modern Plastics Handbook*, McGraw-Hill, New York, 2000, pp. 5.86–5.87.
5. Irvin I. Rubin, *Injection Molding Theory and Practice*, Wiley, New York, 1972, pp. 48–52.
6. Spirex Inc., Spirex auto shut-off valve, www.spirex.com.
7. Timothy Womer, The injection unit, in *Plastics Technician's Toolbox*, Society of Plastics Engineers, Brookfield Center, CT, 2002, pp. 11–15.
8. Karen Wood, K-2001: nozzles for every occasion and more, *Injection Molding*, February 2002, www.immnet.com.
9. Tony DeLigio, The taming of the scre w, *Injection Molding*, February 2002, www.immnet.com.
10. Suzy Witzler, Purchasing basics: screws and barrels, *Injection Molding*, April 2000, www.immnet.com.
11. Nicholas Kudlik, Comparison of different screw designs for injection molding machines, *Proceedings of the Society of Plastics Engineers Annual Technical Conference (ANTEC)*, New York, 1999, www.4spe.org.
12. Timothy Womer, Basic screw geometry, *Proceedings of the Society of Plastics Engineers Annual Technical Conference (ANTEC)*, San Francisco, CA, 2002, www.4spe.org.
13. Greg Angel, Injection molding and color: a basic look at design, processing, and troubleshooting, *Proceedings of the Society of Plastics Engineers Annual Technical Conference (ANTEC)*, San Francisco, CA, 2002, www.4spe.org.
14. Johannes Wortberg, Novel barrel heating with natural gas, *Proceedings of the Society of Plastics Engineers Annual Technical Conference (ANTEC)*, Nashville, TN, 2003, www.4spe.org.
15. *Designing with Thermoplastics*, Dow Chemical Company, Midland, MI, 1992, pp. 56–69.
16. DSM, Engineering plastics cold runners systems, www.dsm.com, 2004.
17. DSM, Engineering plastics hot runner systems, www.dsm.com, 2004.
18. J. Robertson, Innovative gating techniques and new resin technology for thin w all injection molding, *Proceedings of the Society of Plastics Engineers Injection Molding RETEC Conference*, Toronto, Ontario, Canada, 1995, p. 163.
19. J. H. DuBois et al., *Plastics Mold Engineering*, Van Nostrand Reinhold, New York, 1965, pp. 429–452.
20. D. Kazmer and S. Carter, Studies of plastic boss design and methodology, *Proceedings of the Society of Plastics Engineers Annual Technical Conference (ANTEC)*, New York, 1999, www.4spe.org.
21. DSM, Engineering plastics design venting, www.dsm.com, 2004.
22. A. B. Glanvill, *The Plastics Engineer's Data Book*, Industrial Press, New York, 1974, pp. 11–13.

23. E. Bernhardt, *Processing of Thermoplastic Materials*, Reinhold Publishing Corporation, New York, 1959, pp. 308–311.
24. G. Markus, Water cooling, in *Plastics Technician's Toolbox*, Society of Plastics Engineers, Brookfield Center, CT, 2002, p. 145.
25. P. Lipp et al. Clamp end, in *Plastics Technician's Toolbox*, Society of Plastics Engineers, Brookfield Center, CT, 2002, p. 1.
26. Electric machines and new processes catch fire, *Plastics Technology*, November 2002, www.ptonline.com.
27. NPE news wrap-up: hot runners, *Plastics Technology*, July 2003, www.ptonline.com.
28. Natti Rao, *Design Formulas for Plastics Engineers*, Hanser Publishing, New York, 1991, pp. 39–41.
29. Paul Tres, *Designing Plastic Parts for Assembly*, Hanser Publishing 2000, pp. 154–158.
30. P. Grelle et al., The effects of process conditions, nominal wall thickness, and flow length on the shrinkage characteristics of injection molded polypropylene, *Proceedings of the Society of Plastics Engineers, Annual Technical Conference (ANTEC)*, New York, 1999, pp. 443.
31. Sidney Levy and J. H. DuBois, *Plastics Product Design Engineering Handbook*, Chapman & Hall, New York, 1984, pp. 126–132.
32. Robert Malloy, *Plastic Part Design for Injection Molding*, Hanser Publishing, New York, 1994, pp. 15–19, 63–67.
33. Ronald B. Beck, *Plastic Product Design*, 2nd ed., Van Nostrand Reinhold, New York, 1980, pp. 125–130.
34. *INSPIRE™ Molding Guide*, Dow Chemical Company, Midland, MI, 1999, pp. 10–14, 16–21.
35. *INSPIRE™ Design Guide*, Dow Chemical Company, Midland, MI, 1999, pp. 12–19.
36. *General Polymers Injection Molding Troubleshooting Guide*, Ashland Chemical, General Polymers Distribution, 1998.
37. J. D. Retzlaf, Injection molding troubleshooting pocket guide, in *Plastics Technician's Toolbox*, Society of Plastics Engineers, Brookfield Center, CT, 2002, pp. 11–15.
38. P. F. Grelle, Injection molding engineering thermoplastics in comparison to commodity materials, in *Plastics Technician's Toolbox*, Society of Plastics Engineers, Brookfield Center, CT, 2002, p. 1
39. *Injection Molding and Troubleshooting*, 2nd ed., Advanced Process Engineering, 1996, pp. 6–27.
40. *Injection Molding Reference Guide*, 4th ed., Advanced Process Engineering, 1996, pp. 65.
41. P. Gipson, P. Grelle, and B. Salamon, The effects of process conditions, nominal wall thickness, and flow length on the shrinkage characteristics on injection molded polypropylene, *Proceedings of the Society of Plastics Engineers, Annual Technical Conference (ANTEC),* New York, N.Y., 1999, www. 4spe.org

CHAPTER 2

Assisted Injection Molding

STEPHEN HAM
Consultant, Highlands, North Carolina

2.1 INTRODUCTION

Low-pressure injection molding has been practiced in the industry since the 1950s. It was not uncommon for molders to add a small amount of sodium bicarbonate to eliminate sink marks. During the 1960s and 1970s, injection-molded structural foam became a standard method for very large injection-molded applications and those requiring thicker wall sections. During the late 1980s the gas-assist process emerged as another way to injection-mold large parts under low pressure. Gas assist also enabled injection moldings with hollow sections. The addition of foaming agents and gas injection are methods of assisting the injection-molded process with two of its biggest problems: filling the mold with molten plastic, and compensating for plastic shrinkage as the material cools. The methods of assist have ranged from the use of baking soda (sodium bicarbonate) during the 1950s, to nitrogen gas used as a blowing agent (i.e., gas injection), and a wide range of chemical blowing additives. Water injection is the latest development for assisted injection molding. Assisted injection molding has traditionally been employed in applications with one or more of these characteristics:

- Usually involves thermoplastic resins
- Injection molding process
- Uses an alternative means to fill and/or pack plastic in a mold
- Commonly used for structural or load-bearing applications
- Commonly used for larger plastic components

Handbook of Plastic Processes Edited by Charles A. Harper

Although smaller injection-molded parts certainly qualify as structural, they are addressed in Chapter 1. These smaller parts can be produced in machines ranging from 300 to 500 tons. The industry has substantial capacity in this size range. The low-pressure techniques associated with assisted molding have a greater reward on large parts. The molding machine hour rate savings of going from 1500 tons to 1000 tons is far greater than going from 500 tons to 300 tons. Assisted injection molding is defined as a thermoplastic injection molding process producing a molded article under lower pressures, both injection pressure and clamping force, resulting in lower residual stress. The resulting article has a high strength-to-weight ratio and can be considered structural. The following injection molding processes are considered assisted injection molding and are covered in this chapter:

- Low-pressure structural foam
- High-pressure structural foam
- Expanding-mold structural foam
- Co-injection structural foam
- Counterpressure structural foam
- Gas-assisted injection molding
- Structural web molding
- External gas molding
- Water injection molding

2.1.1 Why Consider Low-Pressure Options

In today's competitive environment, manufacturers of durable products are seeking ways to improve performance and reduce cost. By investing initially in good tooling, these goals can be achieved through the assisted variants of the injection molding process. Proper application of these processes offers designers an opportunity to optimize their products and a competitive advantage. The primary application goal of these processes is to obtain stiff molded parts with functional interior features and a cosmetically acceptable exterior. Parts consolidation is a benefit of this approach, together with the associated cost and labor savings. Achieving the desired degree of stiffness is often a challenge to a designer. There are several approaches to consider, such as the specified wall thickness, fillers such as glass added to the polymer, the rib configuration, and external reinforcements.

Wall thickness is the backbone of a part. It affects stiffness by a power of 3. Doubling wall thickness achieves eight times the rigidity (2×2×2). The inherent resin stiffness is a factor. Adding fillers such as glass or minerals can increase the flexural modulus of the base resin. Part configuration is determined based on stiffness requirements and the modulus of the resin. Deeper ribs are dictated for lower-module resins. The gas-assist process in some cases will require oversized ribs, which also increase overall stiffness. External reinforcement is also a practical solution when beam stiffness is needed on a highly detailed and functional part. Simple part configurations, which require great stiffness, are often better left to inherently flat materials such as sheet metal and wood, offering greater stiffness-to-cost ratio. Plastic materials must provide additional features to offer an overall cost advantage.

Table 2.1 illustrates the wall thickness most commonly employed for each process together with its comparative machine clamping pressure requirements. Molding machines with a higher clamp tonnage carry a higher cost per hour, thereby increasing piece-part cost and energy consumption.

2.1.2 General Part Design Considerations

Product designers and piece-part designers have the ability to control the success of the ultimate product. The goal should not be merely to design a product that meets the demands of marketing but rather, to create a product that will defeat the competition. Spending a tooling budget can achieve this goal by optimizing the design to exactly meet the needs of the application at the lowest possible manufacturing cost. Selection of a process for a given application has to be closely matched with the resin specification. Hundreds of resin options, coupled with dozens of conversion technologies, result in an incomprehensible option list. In reality, a review of the market shows that certain resins team up with certain processes consistently across a variety of markets. The designer or product development engineer should evaluate the prevailing resin–process options employed in similar products as a starting point. Then consider resin–process options used in other markets, where a slight change in formulation or process could extend the option to your application. Consider new resin options and process improvements that may not have been available when the competition tooled-up. Once the option list has been narrowed, take a close look at the design. The part geometry and end-use application will sometimes clearly dictate the best process. When a composite of materials is desired, co-injection will allow single-operation fabrication. Hollow sections are easily formed by gas or water injection. Extreme variations in wall section are achieved with an expanding mold. Several factors should be considered:

1. *Wall thickness.* The processes covered in this course are best when a uniform wall thickness is employed (expanding mold will accommodate more transition).

TABLE 2.1 Wall Thickness and Clamp Tonnage

Process	Typical Wall (mm)	Typical Clamp Tonnage (tons/in^2)
Injection molding	3	3
Structural foam	6	0.25
Co-injection SF	5	0.50
Counterpressure	5	0.50
Expanding mold SF	3/6	3
Gas assist thin wall	3	1
Structural web	6	0.50
External gas molding	3	0.75–1
Water-assisted molding	5/3	1–2

The backbone of any design is the part's wall thickness. This determines the strength, overall cost, and amount of design flexibility. Generally, the thicker the wall, the greater the design flexibility but also the cost. Thicker walls allow deeper ribs and larger bosses and also provide improved weld line strength. With gas assist, thinner walls are necessary, to prevent fingering of the gas from the rib pattern. Each process has an optimum wall thickness.

2. *Ribs and bosses.* The functionality of a design dictates the use of molded-in features, which are accomplished as ribs and bosses. Ribs require draft, which causes a buildup of thickness at the rib base. Deep ribs require thicker wall sections to observe the aspect ratio. The exception is gas assist, which offers oversized ribs when connected to a gas passage.

3. *Louvers and openings.* When the flow front separates for a core pin or other obstruction to flow, the area of flow re-forming is the weld line. Strength is diminished in this area. Each of the processes has a special consideration for weld lines.

4. *Flow length.* Large-part designs need to consider the best process to use to optimize the design. This usually means finding the lowest possible manufacturing cost. Since material makes up approximately half of the unit cost, the goal is to have the thinnest material possible. The flow length table can narrow the options.

5. *Other molding conditions.* These include draft, radii, fillets, gate provisions, ejection plan, overall parting line, exposed gas injection gate, spillover gate, and appearance (cosmetic) needs.

2.1.3 Cosmetic Limitations

Because of its varying stress levels and corresponding nonuniform shrinkage, injection molding is notorious for sink marks. Much of the problem comes from the skin freezing up, which isolates the injection pressure from the flow front. Low-pressure structural foams (SFs) internal foaming pressure evens out stress and shrinkage but introduces surface swirl caused by the gas blowing agents escaping on the leading edge during the filling stage. Counterpressure structural foam overcomes the swirl problem by countering the leading-edge escape with an in-mold back pressure. However, this frozen-off leading edge on the flow front causes exaggerated weld lines. Gas-assisted injection molding and structural web use internal gas channels to convey pressure to the leading edge. Appearance problems arise when there is a different texture read in the areas of higher pressure adjacent to the gas channels. External gas molding also applies gas to the injected plastic, but externally, from one of the mold halves. Although providing excellent exterior appearance, the interior surfaces are wavy and irregular. Water-assisted molding uses water instead of gas to fill and pack the plastic. Water is less compressible than gas, giving better core out and weight reduction, but hesitation lines are visible on short-shot moldings.

The actual overlap between these processes is less than one would expect. Typically, some designs are better suited for thicker walls, dictating structural foam or structural web. Some designs have naturally thick areas, which can be cored out. When the wall transition exceeds two fold, gas assist techniques should be considered. In addition, overall size is a consideration, due to clamp tonnage requirements and molding machine size.

Each of the assisted injection molding processes offers appearance problems just by its very nature.

- *Injection molding:* sink marks
- *Structural foam:* swirl
- *Expanding mold SF:* mold witness line, sink marks
- *Co-injection SF:* sinks or swirl
- *Counterpressure SF:* knit lines
- *Gas-assisted molding:* gloss variation, hesitation line
- *Structural web molding:* flow lines, sink, gloss
- *External gas molding:* nonuniform surface on one side
- *Water-assisted molding:* hesitation, gloss variation

2.1.4 Design Sequence

When dealing with a highly functional design, a different sequence of design events can be considered. To have the function determine the design, lay out the rib pattern first and then focus on the wall section (and therefore the process):

1. Design for maximum function in the molded article.
2. Determine ribs and bosses to achieve the desired function.
3. Use rib/wall aspect ratios to determine wall thickness.
4. Remember that wall thickness will indicate process and resin.
5. Evaluate potential gating locations.
6. Consider flow lengths for part size and wall thickness.
7. Determine weld lines and corresponding loss of strength.
8. Consider the cosmetic requirements.

2.1.5 Plastic Flow and Shrinkage Considerations

The piece-part designer does not need to be as concerned about the actual processing details as about knowing the design parameters of the process. Uniform wall thickness has always been the essential ingredient in good piece-part design. Nonuniform wall thickness only adds another variable to the process. The designer has an influence on the process, and this is particularly true for assisted injection molding processes. For the designer, the filling stage of the process should be considered the starting point. It is the point where empty space becomes the molded article. This is where uniform wall thickness is very important. Variations in

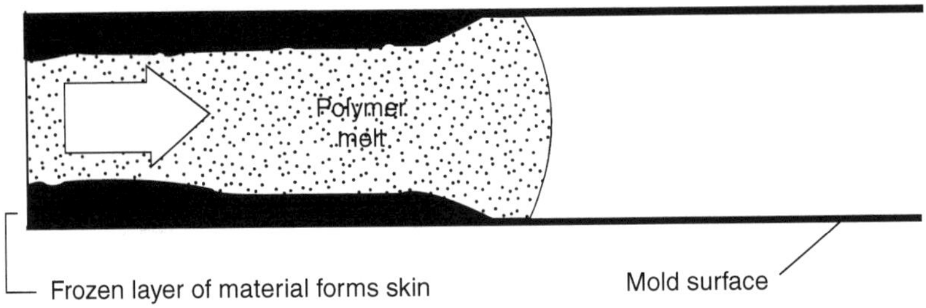

FIGURE 2.1 Injection molding flow front.

thickness cause filling rate variation and therefore an increase in stress and cosmetic problems. With this perspective, we can visualize each of the assisted processes during the critical filling stage. Since the foam and gas processes are nothing more than hybrid injection molding processes, it is logical that injection molding be the starting point for understanding the process basics.

Figure 2.1 represents a cross section through the mold at the filling stage. Molten plastic is flowing from left to right, from the injection point, which is called the *gate location*, toward the end of the flow. This leading edge forms the flow front. The essential point of this illustration is that skins are formed on the cool mold surface, thus encapsulating the melt flow. With solid injection molding it is necessary to inject all the resin to the end of flow prior to skin closure.

Another important concept that one must understand is resin shrinkage. The transformation from melt temperature to ambient causes the spacing of molecules in the polymeric chains to become closer or more tighter packed. In crystalline or highly symmetrical molecular chained polymers, this volumetric shrinkage can be 20% or higher. In more randomly oriented chains or amorphous resins,this shrinkage is in the range of 5%. Injection molding attempts to compensate for this through an extended packing phase, which pushes more resin into the mold as this shrinkage, occurs. The obvious problem is skin freeze-off, which isolates areas away from the gate. We can now understand why solid injection molded parts often show signs of high stress.

Figure 2.2 illustrates how structural foam compensates for flow and shrinkage with two major changes. There is a thicker wall section (often twice as thick) and a

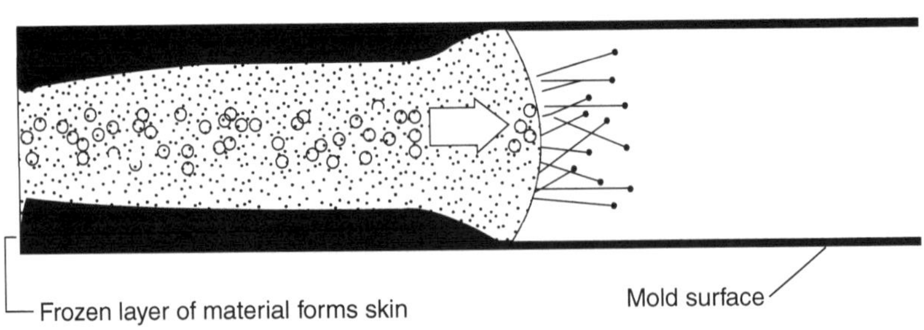

FIGURE 2.2 Structural foam flow front.

blowing agent which is added to the molten plastic in the molding machine. The thicker wall allows easier flow of a low-pressure short shot. In the later stages of filling, the blowing agent actually provides injection pressure on the flow front. The foam serves as an internal cushion for shrinkage and as a stress reliever.

The drawback to structural foam is the surface finish of the molded article. During the critical filling stage, the blowing agent is escaping through the flow front, becoming trapped between the mold surface and the skin of the part as it is formed, resulting in a condition called *swirl*. Figure 2.3 illustrates counterpressure structural foam, which was invented as a method to eliminate swirl. This goal is accomplished by applying a resistance pressure (nitrogen gas) against the flow front which is enough to keep the blowing agent in solution in the polymer melt and behind the flow front. This is accomplished by a pressure-tight mold and special gas pressure and timing sequences on the molding machine. Counterpressure structural foam succeeds in eliminating swirl and still maintains low stress through internal foaming but at the cost of the thicker wall of structural foam and therefore higher weight. This is compensated by greater physical properties.

Figure 2.4 shows how gas assist injection molding is capable of producing thin-walled parts with low stresses by incorporating a stream of gas (usually, nitrogen) within the polymer melt to enhance flow and compensate for shrinkage. With this

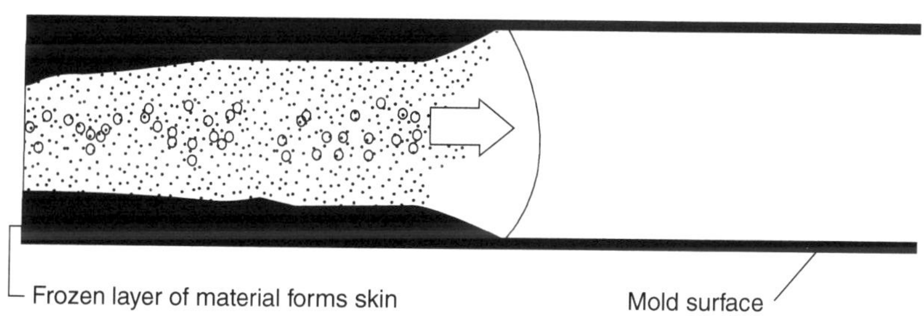

FIGURE 2.3 Counterpressure flow front.

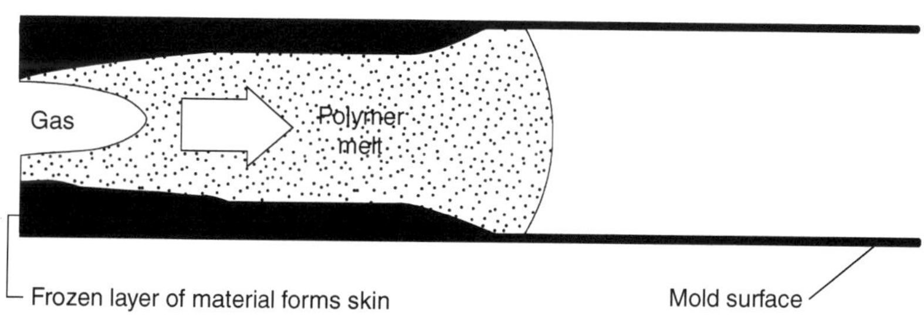

FIGURE 2.4 Gas-assisted flow front.

process, the part designer becomes a process engineer as well. The designer's placement of ribs and wall thickness determines the location of the gas passages, which are like in-part manifolds for material flow.

Properly designed rib patterns can facilitate thin-walled (half that of structural foam) gas-assisted parts in two ways. The ribs will become flow enhancers, and when cored out with gas they become shrink compensators. The rib pattern then adds to overall part stiffness, overcoming the stiffness loss of thin walls.

2.1.6 Flow Length Considerations

The data in Table 2.2 give a sense of the relative flow of each process. The flow length ratio will depend on process conditions and part geometry and is intended for comparative analysis and as a rule of thumb. The ratio shown is the flow length divided by the wall thickness. Figure 2.5 shows the calculations for predicting cavity pressure with low-pressure structural foam.

- *Example:*
  - 12.5-in. flow length from nozzle/0.250-in wall thickness = 50 : 1 flow ratio.
  - 50 : 1 = 300 lb/in^2 average cavity pressure.
- Keep *L*/*T* flow ratio as low as possible.
- Processing problems increase as *L*/*T* increases.
- 140 *L*/*T* is a practical upper limit.

This is an example of how to apply the *L*/*T* ratio to determine clamp tonnage needs. Practical limits can be determined for each type of machine. The data are based on flow from any one injection nozzle. Multiple injection points will decrease the flow length and the *L*/*T* ratio and therefore the cavity pressure. Depending on the process and equipment, there could be a need for a hot runner system in the mold.

TABLE 2.2 Comparative Flow Length-to-Wall Thickness Ratios

Process	Wall (mm)	Polypropylene	HIPS	ABS	Polycarbonate
Injection molding	3	130	100	85	75
Low-pressure SF	6	200	160	140	120
Co-injection SF	5	185	145	125	105
USM/expanding mold SF	3/6	130	100	85	75
Counterpressure SF	5	175	150	130	90
Gas assist/Thin wall	3	175	150	130	90
Structural web	6	200	160	140	120
External gas	3	195	157	138	118

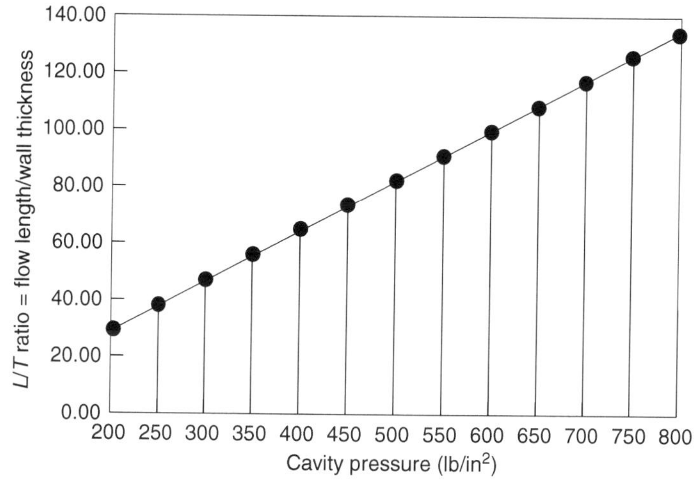

FIGURE 2.5 Cavity pressure versus flow length table. Information is approximate and to be used for estimation and example only. (Courtesy of Uniloy/Milicron.)

2.2 STRUCTURAL FOAM: INJECTION MOLDING WITH FOAMING AGENTS

The low-pressure structural foam process can be broken down into four stages of filling. These four stages apply to all the SF processes in terms of short-shot filling and in-mold pressures. Figure 2.6*a* shows a cross section through a clamped mold and nozzle. The clamped mold has an empty cavity ready for its short shot of molten plastic preblended polymer and blowing agent. The nozzle rod is in the closed or shutoff condition, and a charge of material is being held under pressure behind the nozzle. The density reduction of the part is determined by the amount of material injected versus the full volume of the mold cavity.

In Figure 2.6*b* the nozzle rod retracts and instantly the charge of plastic begins to enter the mold cavity. There is a drop in pressure as the material enters the open mold cavity, so the blowing agent begins to expand. Mold venting is liberal by injection molding standards, for several reasons. Since structural foam has a greater wall, there is naturally more air in the cavity to be displaced as resin flows. The blowing agent escaping off the leading edge of the flow front itself becomes more gas to be displaced prior to complete filling of the mold. This open nozzle phase will last from 1 second to as much as 15 to 20 seconds. Material viscosity, part wall thickness, flow restrictions, and injection pressure all have an effect on injection time. Typically, long injection time results in more swirls. The machine is in the high-pressure clamp mode at this stage.

Figure 2.6*c* illustrates the third stage, where the charge of material has been injected and the nozzle rod closes, isolating the mold cavity from any injection pressure. At this point, blowing agent pressure is filling the remaining open mold cavity. Venting is particularly critical during this blowing agent filling stage. Nondisplaced

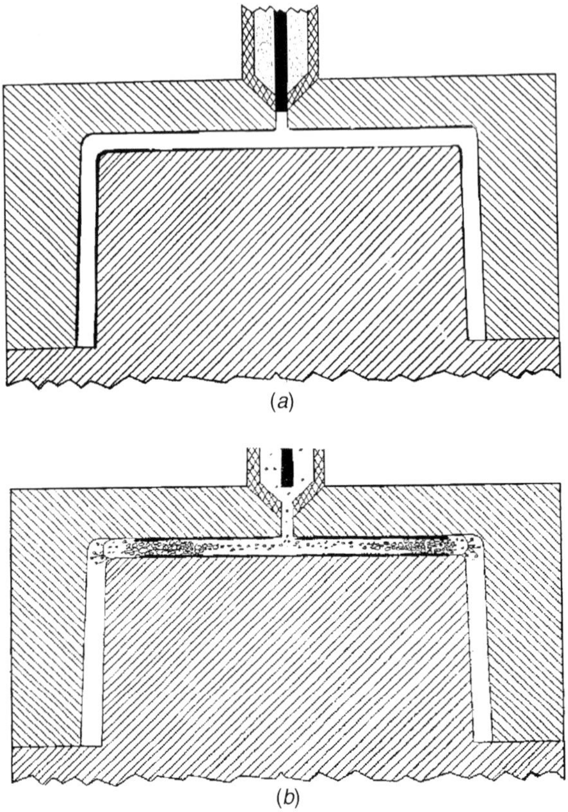

FIGURE 2.6 Structural foam filling stages.

gas in front of the flow will cause restrictions. Proper venting at the end of flow will improve weld line strength and reduce short-shot rejects. A visual reference can be observed by watching the venting time after nozzle closure. This should continue for a few seconds as the cavities become packed out under blowing agent pressure. High clamp tonnage continues through this stage.

Figure 2.6*d* shows the mold cavity after complete filling. The blowing agent continues to exert pressure, thereby creating an internal cushion. This cushion compensates for resin shrinkage and balances out the internal stresses. At this point, venting has stopped and solid skins are formed outside the cushion of foam. A cooling cycle of 30 to 300 seconds follows, depending on wall thickness and thick spots. Table 2.3 shows a general guideline for overall cycle time for low-pressure structural foam.

The major drawback of structural foam is the surface finish of a molded part. While blowing agent pressure provides for low-pressure molding and an internal cushion, it causes the surface to have a blushed appearance referred to as *swirl*, which results from blowing agent gas escaping from the leading edge of the resin flow front

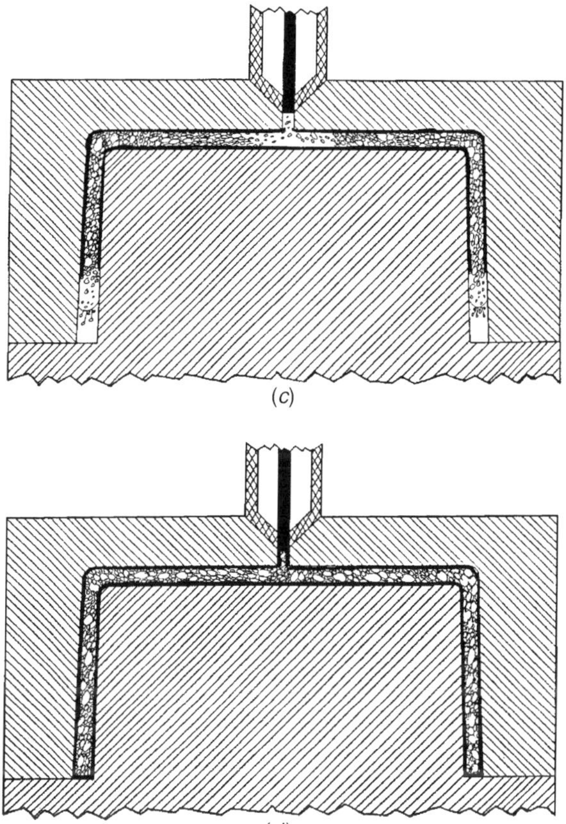

(c)

(d)

FIGURE 2.6 (*Continued*)

TABLE 2.3 Wall Thickness verses Cycle Time

Wall Thickness (in.)	Cycles/hr
0.125–0.157	30–40
0.157–0.187	25–35
0.187–0.196	20–30
0.196–0.236	15–25
0.236–0.275	12–24
0.275–0.314	10–18
0.275–0.393	5–12

Source: Data from Uniloy/Milicron.

during the filling stages. The low-pressure condition inside the mold is necessary for foaming to occur, but it allows the bubbles to escape from the plastic and become trapped between the surface of the mold and the skin of the part. This swirl appears as a sunburst pattern with trails tracing the flow of the material. Figure 2.7 illustrates the formation of swirl on the part's surface. The swirled surface finish has often restricted structural foam's use to nonvisual types of applications or has meant that the price of the final product allowed for the cost of a painting operation. Over the past several years, molding technology has been developed that produces swirl-free surface finishes on structural foam parts. This technology can be broken down into three types: (1) high-pressure expanding mold, (2) co-injection structural foam, and (3) counterpressure structural foam.

2.2.1 Low-Pressure Structural Foam Equipment

Structural foam was initially conceived as a method to produce large thermoplastic parts with the same advantages that injection molding offered to smaller parts. Much of the equipment technology is the same as injection molding. The structural foam industry had its humble beginnings using purpose-built two-stage-injection multiple-nozzle machines. These machines were characterized by oversized platens, material accumulators feeding built-in multiple-nozzle hot runner manifolds, low clamp tonnage, and low injection pressure. One of the advantages of this equipment is its ability to mold very large parts or several smaller parts simultaneously on large molding platens. Over the years, this family of equipment has evolved to follow the latest philosophies of structural foam theory. Today's refinements still include two-stage injection through multiple gating nozzles, but now the nozzles are controlled sequentially. Much higher injection pressures are available to ensure quick injection of material. Digital and programmable controllers now manage temperature injection, clamp pressures, timing, platen travel, and extruder operation. Figure 2.8 illustrates a typical layout of material accumulator, manifolds, and nozzles.

In Figure 2.9*a* the preinjection stage has the mold in the closed and clamped position, material accumulator charged, and the nozzle rod in the closed position. The mold is clamped under full clamp tonnage as quickly as possible after the preceding part removal stage. The material accumulator has been precharged during

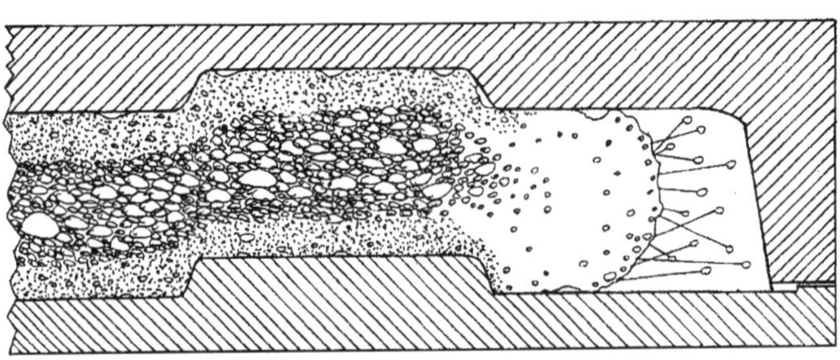

FIGURE 2.7 Swirl formation.

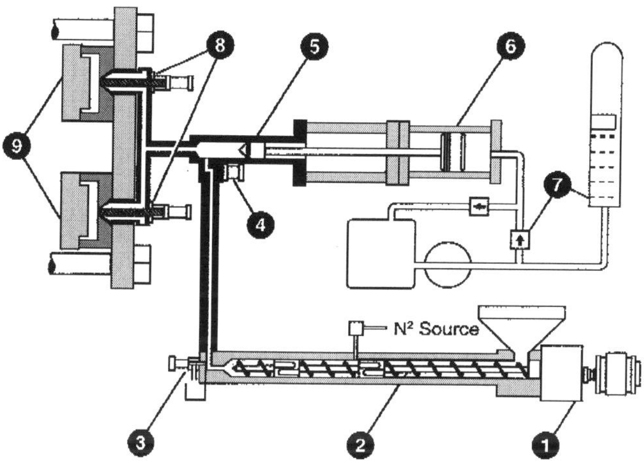

1. Variable speed electric motor and gear reducer for efficient screw rotation.
2. Two (2) stage extruder screw and barrel with N^2 gas port and dynamic mixer.
3. Purge valve for color and material change.
4. Shut off/non return valve.
5. Injection accumulator with plunger.
6. Injection hydraulic cylinder.
7. Gas over oil accumulator system with proportional control valves.
8. Multiple injection nozzles with individual controls.
9. Molds can be filled sequentially.

FIGURE 2.8 Multi-nozzle manifold. (Courtesy of Uniloy/Milicron.)

the previous cycle's cooling stage. Table 2.4 provides a general rule of thumb for clamp tonnage requirements for low-pressure structural foam. Flow length, resin type, and processing conditions all have an obvious effect on clamp tonnage requirement, so Table 2.4 is presented for comparison. The point is that reducing wall thickness has a direct effect on increasing filling pressures and therefore machine clamp requirements. Reduced wall sections cycle faster, however. Refer to Table 2.3 for estimated cycle times.

During the filling stage (Figure 2.9*b*), the mold remains clamped under full tonnage; the material accumulator has begun to discharge material because the nozzle is in the open position with the nozzle rod retracted. The action of the nozzle rod is coordinated with the travel of the material accumulator. This allows precise shot sizes and sequential nozzle opening. Sequential injection has two potential rewards. On large multiple-gated parts it is possible to have a cascade molding effect where the material injects from behind the flow front. This increases weld line strength and enables longer flow lengths. In addition, multiple cavities can be family

TABLE 2.4 Structural Foam Wall Thickness and Tonnage

Wall Thickness (in.)	In-Mold Pressure (lb/in^2)	Clamp Pressure ($tons/in^2$)
0.150	2000	1.0
0.200	1200	0.6
0.250	500	0.25
0.300	400	0.20

molded, with each cavity having its own precise shot of plastic. Family molding of as many as 10 cavities of smaller parts is common. Some specially built machines have multiple barrel–screw units, each feeding its own manifold and making possible parts of different materials from the machine cycle.

In Figure 2.9*c* the packing stage is shown beginning, the nozzle being closed after the precisely measured short shot has been injected. The blowing agent pressure pushes the flow front to the end of flow. In Figure 2.9*d* the blowing agent has completed mold filling and continues to exert pressure as it cools. The cooling stage that follows is determined largely by the wall thickness of the part or, to be more precise, by the thickest individual section of the part. There are two cycle-time considerations with structural foam. First is the need to withstand the force of ejection, just as with solid injection-molded parts. The second consideration is unique to structural foam. If the part is removed too quickly, the internal blowing agent pressure may be greater than that of the skins trying to contain it. The result, called *post blow*, resembles a bulge or knot on the surface of the part. This is one way of finding the thickest area of the part. Unfortunately, the design is captured in steel at this point and is expensive to correct. During the cooling stage there is a reduction in the clamping force from the machine, as the injection and blowing agent pressures are no longer a factor. In addition, the machine will recharge its material accumulator prior to the next cycle.

The final phase is the ejection stage, when the machine platen moves open, which in turn opens the mold and the mold's ejection system pushes the part free. Once the part is removed, the machine is closed by the platen progressing to the closed position, and the necessary clamp pressure is then returned to the preinjection requirements.

Low-pressure molding explains the most common defect of structural foam molding—short shots. While every structural foam part began as a deliberate short shot in the machine barrel, it is supposed to expand into a full shot through the action of the blowing agent. This is not a closed loop system, however, and after the shot is injected, there is generally no way to compensate for a nonfilling area. In addition, the pressures at the far end of flow are reduced to less than 25% because of total dependence on blowing agent pressure to move the flow front.

Low filling pressure on the leading edge is also an explanation for the very slow development of computerized flow simulation for structural foam. While an experienced process engineer can use injection molding programs, the best results come from a heavy use of previous experience (trial and error). It is fortunate that

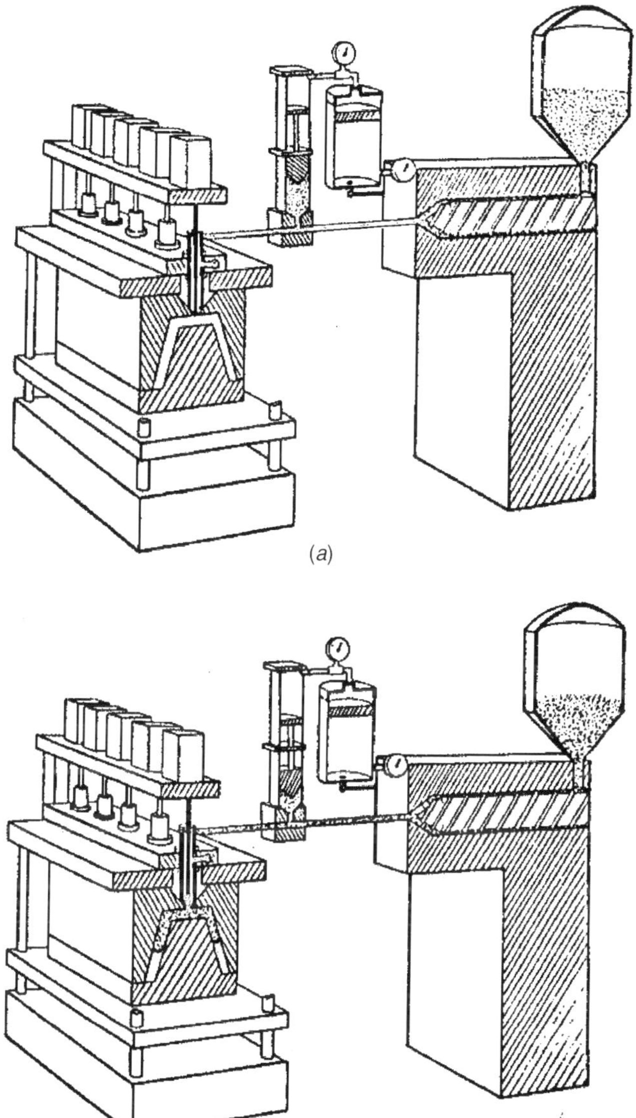

(a)

(b)

FIGURE 2.9 Multi-nozzle machine operation stages.

structural foam is such a forgiving process, due to the thicker wall sections employed on structural foam parts. Gate location and weld line integrity are important issues for the mold designer, process engineer, and part designer. Multiple gate locations become advantageous when large parts are designed with flow-restricting features.

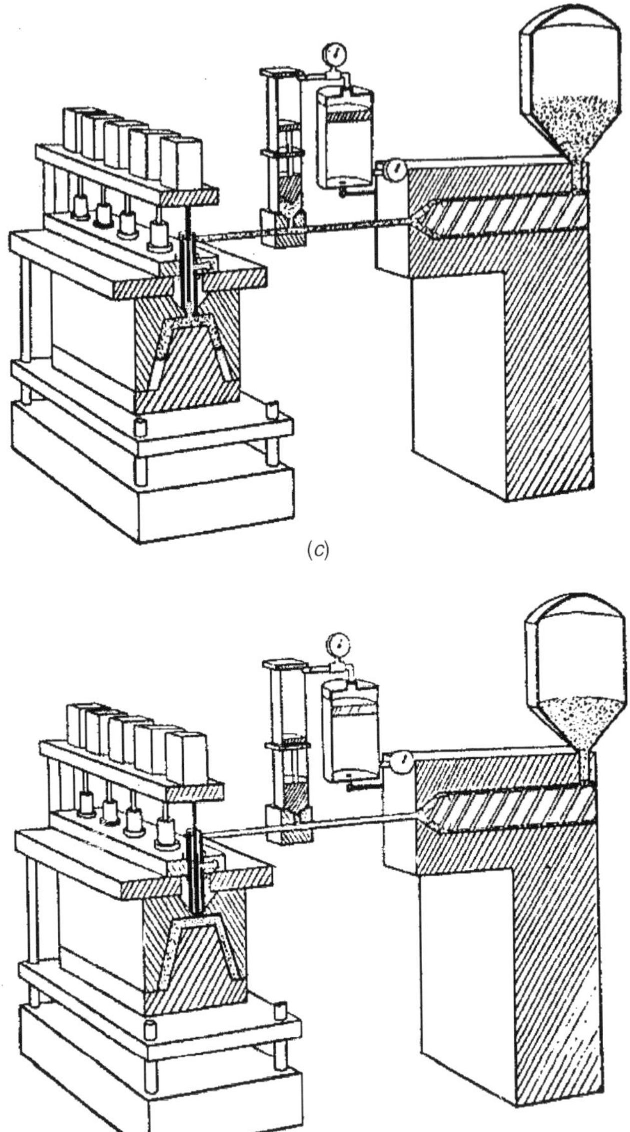

(*d*)

FIGURE 2.9 (*Continued*)

The same basic stages hold true when a single-nozzle injection machine is used. In these applications, some special modifications are beneficial. A shutoff nozzle prevents drooling caused by blowing agent prefoaming in the barrel. It also provides a more controlled pressure drop of the injected short shot.

FIGURE 2.10 Multi-nozzle Uniloy machine. (Courtesy of Uniloy/Milicron.)

Later refinements of structural foam molding equipment have included faster injection speeds, which produce improved surface finishes. Most structural foam theories hypothesize that less swirl can develop if the injection time is shortened. Increased injection speed is also helpful on parts with a reduced wall section, where more packing pressure is required. Faster injection speeds, however, tend to produce higher cavity pressures, and therefore higher molded-in stress can result (and higher clamp tonnage, to offset increased injection pressure). Figure 2.10 shows a purpose-built low-pressure structural foam machine.

2.2.2 High-Pressure Structural Foam Equipment

An alternative to a multiple-nozzle structural foam machine is a converted high-pressure injection molding machine. A standard high-pressure injecting molding machine is suitable for structural foam molding with the addition of a shutoff nozzle. The concept is to isolate the mold cavity from the machine barrel to prevent blow agent from escaping. This is also true with the gas-assisted process. In both processes it is necessary for the cavity to become pressurized by the gas or blowing agent. By adapting a built-in manifold to the machine, it is not necessary to put a hot runner system on each mold. A wise design incorporates the industry standard 6-in. spacing pattern. This works well with a single-nozzle reciprocating machine. Refer to Chapter 1 for complete details on this type of machine.

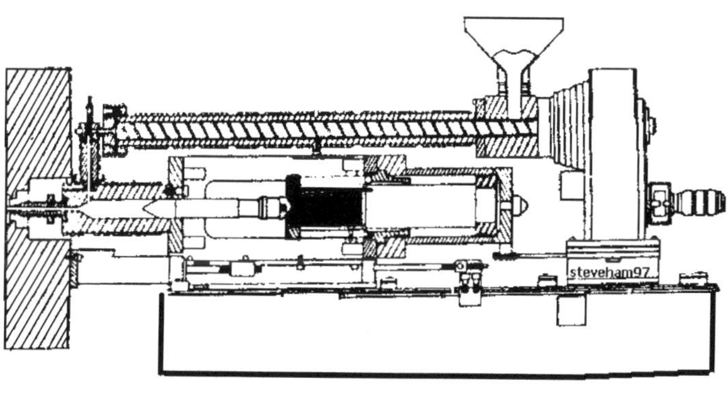

FIGURE 2.11 Two-stage injection machine.

Another major category of equipment is a hybrid of the first two. These are purpose-built single-nozzle structural foam machines. Generally, these are two-stage machines with a material accumulator feeding a single nozzle, which has shutoff capability (see Figure 2.11). Fast injection speeds are obtained and oversized platens are utilized.

2.3 CO-INJECTION STRUCTURAL FOAM

As the name implies, co-injection structural foam is the product of nearly simultaneous injection of two dissimilar but compatible materials. The goal is to produce a part with the foam core encapsulated within a solid skin. The heart of this technology lies in the molding equipment, which is a highly developed hybrid. There are three types of equipment for co-injection production. Purpose-built co-injection machines utilize two separate extruders, one that produces the skin component, which is a material without blowing agent, the other which provides the foam material with a chemical blowing agent. The blowing agent is heat activated but remains compressed under pressure. Reciprocating screws, which feed a special two-channel nozzle, provide injection pressure. The nozzle provides two concentric flow channels for the respective materials, which allows the skin to encapsulate the core. The hydraulic nozzle is controlled electronically to allow for the variable injection velocity of each flow channel. Hot runner manifolds can be designed with special co-injection features. Two barrels are usually involved with these systems. The latest equipment approach retrofits a special machine barrel with two material paths to allow separate materials.

In all of these equipment approaches, the same basic concept is followed. The first phase (Figure 2.12*a*) is injection of solid skin only. This is the formation of the leading edge and the entire outer skin. The second phase (Figure 2.12*b*) follows almost simultaneously with injection of solid skin and a short shot of foamed core. The solid-skin leading edge becomes smaller as it is stretched to the end of the flow. The final phase (Figure 2.12*c*) is shutoff of the injection nozzle at a predetermined time when the mold cavity is less than completely full. The foamed core responds to the pressure drop by expanding to fill out the mold cavity. The leading edge of solid skin prevents the foam from breaking through; thus encapsulation occurs.

The major advantage of the co-injection structural foam process is the surface finish, which is almost injection-like in appearance, when an injection-grade material is used for the skin component. The foamed core offers the advantages of low-pressure structural foam. It is noteworthy that no mixing of the materials occurs yet the core material is completely distributed. One of the unique features of the process is the capability of having different materials for skin and core. This can be of benefit when one material cannot provide all the properties required. It is also conceivable to use an inexpensive core material and prime material for the skin, but this is not advisable for applications requiring Underwriters' Laboratory approved materials.

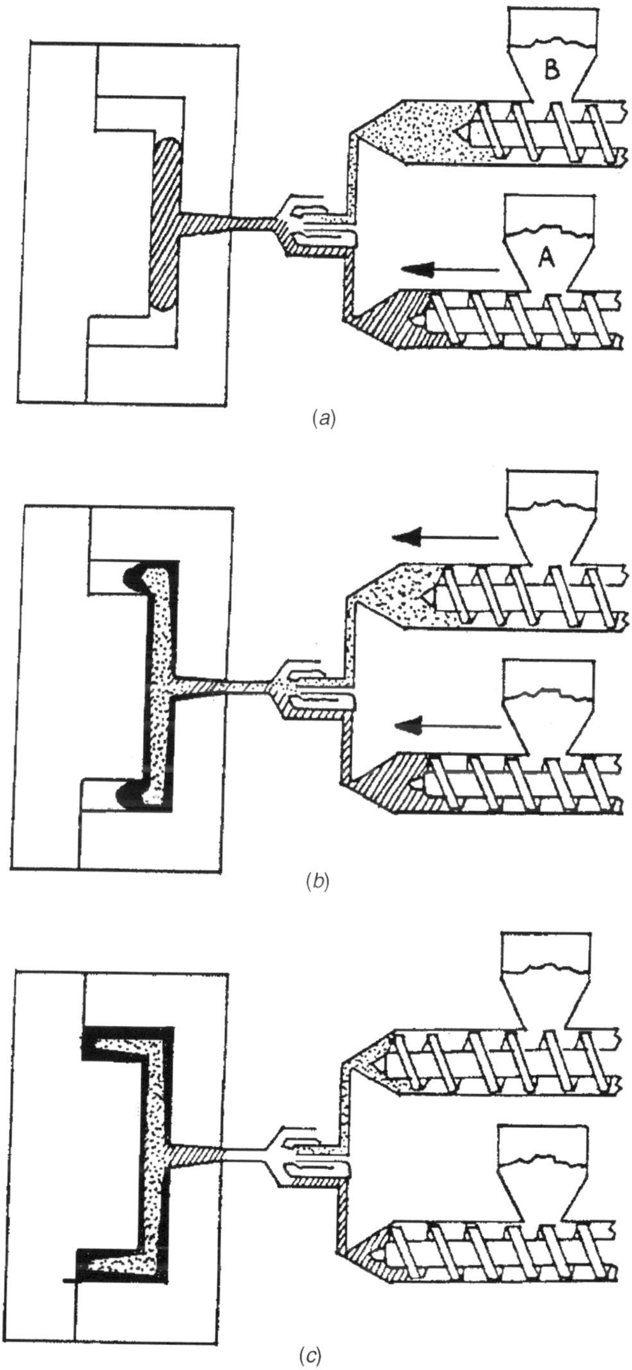

FIGURE 2.12 Co-injection structural foam molding stages.

Co-injection structural foam will have a variable filling profile depending on the pressure of the core component. Since co-injection structural foam is a short-shot process, the filling profile will vary according to the distance from the gate, among other variables. Table 2.5 shows compatibility of materials for co-injection structural foam. Since a mechanical bond occurs between the skin and core materials,there has to be physical compatibility. In addition, resin shrinkage is a factor where dissimilar shrinkage will cause delamination.

2.4 COUNTERPRESSURE STRUCTURAL FOAM

The next swirl-free molding technique is the use of counterpressure structural foam, a process previously called gas counterpressure. With the advent of gas-assisted injection molding there was too much gas in the terminology for these two very different processes. The concept behind counterpressure is to utilize a gas-pressurized mold, which through controlled venting allows the foam expansion stages of the cycle to occur after a smooth surface has been formed. The gas most commonly used is nitrogen of at least 98% purity. The purpose is to improve the molded surface finish by preventing swirl formation. Counterpressure keeps the blowing agent compressed until solid skins are formed. Numerous developmental efforts, starting as early as the mid-1970s, have produced several different counterpressure technologies in Europe, the United States, and Japan.

The process begins with a pressure-tight mold (Figure 2.13*a*), that is charged with gas counterpressure prior to the injection of plastic. To maintain pressure, the mold parting line, ejectors, moving slides, and nozzle seat must be sealed in some way. The sealing techniques vary considerably, but the sealing concept is the same. When the proper in-mold counterpressure is obtained, a predetermined short shot containing a dispersed and compressed blowing agent is injected into the mold (see Figure 2.13*b*). During the injection sequence, controlled venting should occur to allow consistent counterpressure as the unfilled mold volume is being reduced. The counterpressure prevents the blowing agent from expanding, thus eliminating swirl, which is normally caused by the gas bubbles being trapped between the molded surface and the skin of the part. Again, many techniques are practiced, but the concept is the same. The extent of the short shot is open to discussion. Good results are obtained with full-shot counterpressure, where the cavity is filled but not packed. A cavity pressure drop is caused by the volumetric shrinkage of the resin. The blowing agent, compressed up to this point, senses the pressure reduction and expands to fill out the voids and to keep the walls packed against the mold cavity during the cooling stage.

At a predetermined time (Figure 2.13*c*), the counterpressure in the mold is released through exhaust vents. This venting process causes a pressure drop inside the mold cavity, thereby allowing instantaneous expansion of the blowing agent, which is now encapsulated by solid skin. Normal mold cooling and part ejection cycles follow, and the sequence is repeated. The parts produced by gas counterpressure have 5 to 10% density reduction, compared to 15 to 20% for regular low-pressure structural

TABLE 2.5 Co-injection Materials Compatibility[a]

Material	ABS	ASA	CA	EVA	PA 6	PA 66	PC	PE-HD	PE-LD	PMMA	POM	PP	PPO mod.	PS-GP	PS-HI	PBT	TPU	PVC-W	SAN	TPR	PET	PVA	PPS	PC& PBT	PC& ABS
ABS	G	G	G				G	N	N	G	N	N	N	N	N	G	G	G	G		G	B		G	G
ASA	G	G	G	G			G	N	N	G	N	N	N	N	N	G	G	G	G			B		G	G
CA	G	G	G	B				N	N			N	N	N	N	G	G	G	G						
EVA		G	B	G				G	G			G		G											
PA 6					G	G		B	B		N	B		N	N		G								
PA 66					G	G	B	B	B		N	B		N	N		G								
PC	G	G				B	G	N	N			N		N	N	G	G	G	G		G		G	G	G
PE-HD	N	N	N	G	B	B	N	G	G	B	B	N		N	N	N	N	B						N	N
PE-LD	N	N	N	G	B	B	N	G	G	B	B	G		N	N	N	N		N					N	N
PMMA	G	G						B	B	G		B		N	N			G	G						
POM	N	N			N	N		B	B		G	B		N	N	N			N		N				
PP	N	N	N	G	B	B	N	N	G	B	B	G	B	N	N	N	N	B	N	G				N	N
PPO mod.	N	N	N									B	G	G	G	N	N	N	N					N	N
PS-GP	N	N	N	G	N	N	N	N	N	N	N	N	G	G	G	N	N	B	N					N	N
PS-HI	N	N	N		N	N	N	N	N	N	N	N	G	G	G	N	N	B	N					N	N
PBTP	G	G	G				G	N	N		N	N	N	N	N	G	G	G	G						
TPU	G	G	G		G	G	G	N	N			N	N	N	N	G	G	G	G						
PVC-W	G	G	G	N			G	B		B		B	N	B	B	G	G	G	G						G
SAN	G	G	G				G	N	N	G	N	N	N	N	N	G	G	G	G		G	B		G	G
TPR												G								G					
PETP	G						G				N								G		G			G	G
PVAC	B	B																	B			G			
PPSU							G																G		
PC&PBTP	G	G					G	N	N			N	N	N	N				G		G			G	
PC&ABS	G	G					G	N	N			N	N	N	N			G	G		G				G

[a]G, good bonding; B, bad bonding; N, no bonding.

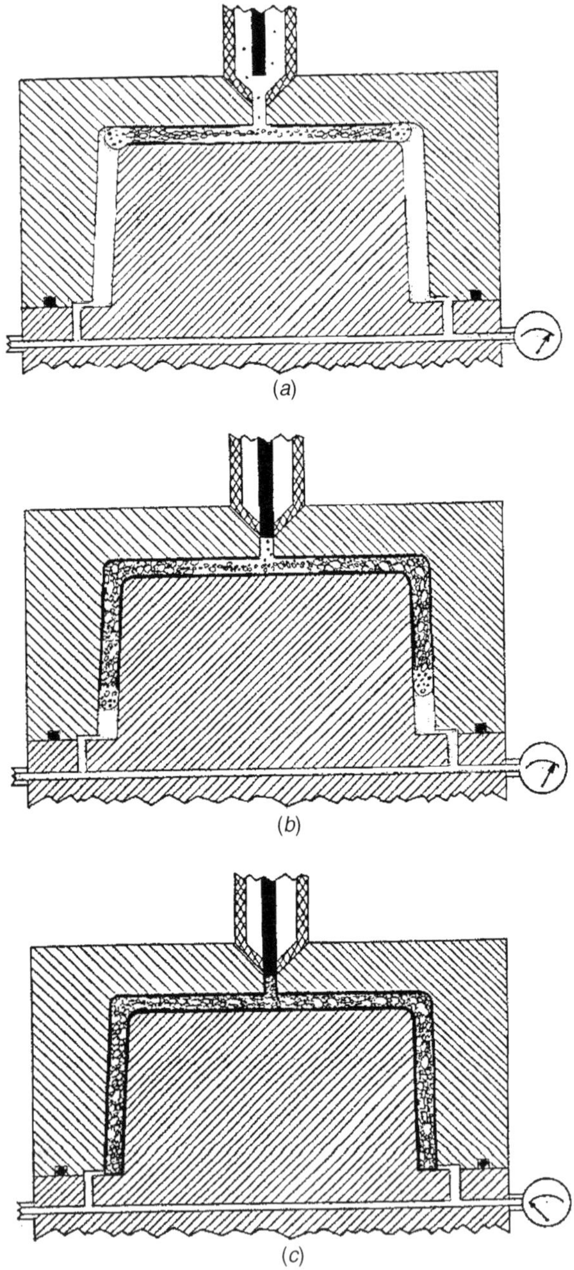

FIGURE 2.13 Counterpressure mold-filling stages.

foam. This is due to the formation of thicker skins. Gas counterpressure structural foam parts, while denser, do display more uniform cell structure throughout the part. This results in improved physical properties.

The gas counterpressure technique requires machines with greater clamp tonnage than that with low-pressure structural foam, due to the increased in-mold pressure caused by the gas counterpressure. The gas most commonly used for the counterpressure is nitrogen, but other inert gases may produce the desired results. The amount of additional clamp pressure required ranges from 25 to 100%, depending on the venting technique used during the injection of plastic into the mold. This also explains why steel molds are generally required for this process. The advantage of this technology is the ability to produce improved surface finishes on parts produced in conventional structural foam machines. All that is required is a specially designed mold and a sequentially controlled pressure source. The surface finish is improved to the point where secondary painting can be eliminated in many applications. In highly cosmetic applications, the cost of painting is reduced significantly, to 33 to 50% of that of low-pressure structural foam. The counterpressure process is capable of producing molded parts with a near injection-molded surface, which greatly reduces finishing costs.

There are two types of counterpressure structural foam as well as nonfoamed counterpressure. Full-shot counterpressure structural foam is similar to solid injection molding but without the cushion. The blowing agent serves as an internal cushion for shrinkage. In effect, the density reduction equals the volumetric shrinkage of the resin inside the closed mold cavity. Short-shot counterpressure is a true low-pressure process where the blowing agent actually aids with the filling pressure. Two more considerations are that often the packing pressures for injection molding are higher than the filling pressures, so full-shot counterpressure can still have a reduction in machine clamping force requirements. The other consideration is that adding blowing agent changes the viscosity of the molten material, often increasing flow lengths. Solid counterpressure involves no blowing agent, and the counterpressure is a tool for uniform flow front pressure and closed-loop process control. Counterpressure structural foam has been in existence since 1977 and in production since 1983. Until recently, the process has been a molder-developed technology. No industry standards have ever been established. The counterpressure adaptations involve two major areas, machine control and mold construction. The molding machine is modified to gain control of the injection profile. A counterpressure gas source is integrated into the system. This gas source will need to create 200 to 500 lb/in^2 for approximately 1 minute. The volume of gas needed will be the volume of the mold cavity plus that of the venting system and connecting lines. Either external or internal controls are added to monitor and control the sequencing of the counterpressure gas. Compressed air or nitrogen have been used successfully. Best results are achieved with nitrogen of at least 98% purity. A purpose-built mold is employed which has been sealed completely and equipped with a venting system and pressure transducers. When figuring clamp tonnage requirements it is necessary to include the area inside the O-ring. Figure 2.14 is a basic schematic of the process.

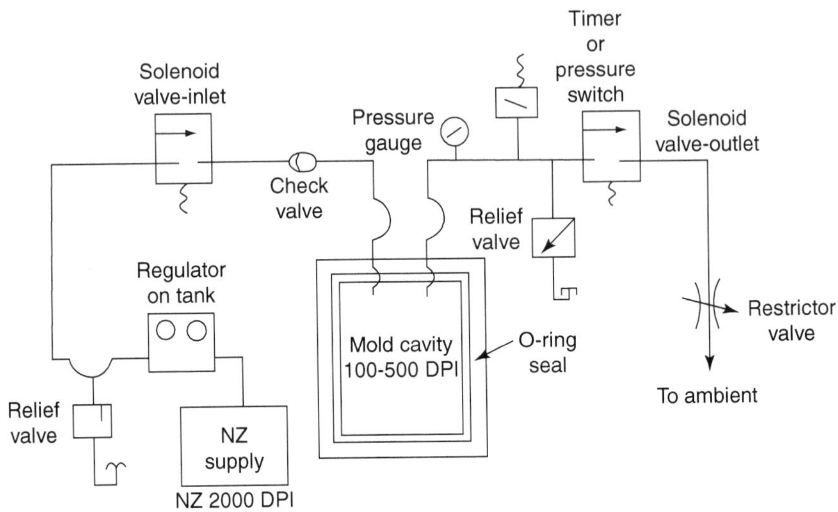

FIGURE 2.14 Counterpressure process.

2.5 EXPANDING MOLD STRUCTURAL FOAM

United Shoe Machinery (USM) patented an e xpanding mold process, which is a unique combination of structural foam and injection molding. The equipment used is a high-pressure reciprocating-screw injection molding machine. This is the closest thing to true high-pressure structural foam, but even this process has a low-pressure phase to allo w blowing agent expansion. The process was licensed to F arrell Machine in the 1970s, but very few machines were built, and these were seldom used for this unique process. Later attempts at the process ha ve been successful it the processor level. The mold design is one of the most important aspects of the process. Figure 2.15*a* shows the first step, injecting a premeasured shot of material and blowing agent into the mold under high pressure and with complete f illing. The mold is packed under a high-pressure cushion, as in injection molding. The high in-mold pressure prevents blowing agent expansion, and thus no swirl is created.

Once the skins are established at a predetermined time, a specialized action in the mold progresses to a retracted position, increasing volume in the mold cavity and thus reducing pressures and allowing blowing agent expansion. Figure 2.15*b* illustrates the mold with the insert expanded and the blowing agent acting. Mold expansion controls where foaming occurs. The skins, which were formed prior to blowing agent expansion, are not subject to swirling. Higher levels of density reduction are feasible based on the increase in cavity volume. Areas with no mold expansion remain solid. This is a feasible process for parts that require large variations in wall thickness.

The disadvantages include design limitations because of tooling constraints caused by the expansion. There is a noticeable witness line on a part at the point of mold movement. The clamp tonnage of the molding machine will be lik e that of

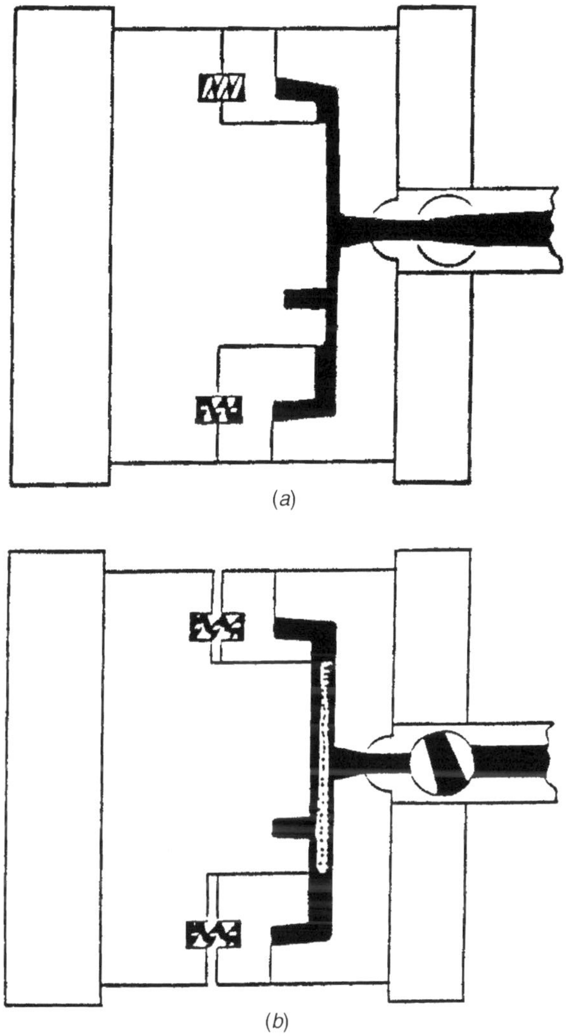

FIGURE 2.15 Expanding structural foam mold-filling stages.

injection molding. The mold cost will be substantially higher, due to the moving action and the increased pressure requirements.

2.6 GAS-ASSISTED INJECTION MOLDING

Gas-assisted injection molding is a process in which gas (nitrogen) is injected together with molten plastic to produce a part that has at least some hollow sections.

The purpose is to assist with the normal injection molding problems of plastic flow and shrinkage. There are several categories of gas-assisted applications.

It seems natural to apply this process to the production of hollow parts. Although this is a possible application area, it is not as effective for large diameters as blow molding. Typically, blow molding will produce a weight reduction of up to 75%. Gas assist will achieve a 30 to 40% weight reduction in hollow areas. Gas assist will make sense in these applications when there are some injection-molded details going beyond the scope of blow molding. The main advantage of gas assist with regard to hollow parts is the ability to combine a hollow part with an otherwise flat piece or to add details resembling those of injection molding.

Gas assist really comes into its own with thin-walled structural parts. A designer has the ability to create parts at the cost of thin walls with the strength of thick walls. This is a short-shot technique, which cores-out oversized ribs with a stream of gas. The resulting hollow tubes within the molded article give an incredible strength-to-weight ratio. Compared to a part with tall ribs as its means of stiffness, there is approximately a 25 to 40% increase. Containing the bubble inside the rib pattern is the essential issue with design and processing. An optimized design will not have the margin of error to allow the bubble to penetrate the wall section in what is called *fingering*. Thick-walled structural parts comprise a class of parts that can be thought of as resembling structural foam parts, where the foam has been replaced by an interconnected web of hollow sections. The concept is that structural foam parts obtain strength primarily from the solid skins. By eliminating the blowing agent and completing the short shot with a shot of gas, there is no swirl, and in concept the gas webs will provide an internal cushion in the way that foam does. It is difficult to achieve any greater density reduction than foam achieves, and from a structural standpoint, the wall has to be designed for the worst-case web. Structural foam will have more uniform physical properties. Since gas-assist parts have the ability to obtain stiffness from oversized ribs, increasing wall thickness takes away the low weight (lower cost) benefit inherent in thin-walled gas assist. Thick-walled gas assist makes sense when the application simply demands a thicker wall, such as in an existing mold or because of ergonomics.

Full-shot injection molding can sometimes benefit from a gas cushion instead of the usual plastic cushion. With this approach the gas is injected after the full shot of resin and will compensate for resin shrinkage. Often, the gas is injected directly into an article at a given thick spot or problem area. Gas injected into molten resin will immediately seek the path of least resistance. The gas will go to the thickest area of the part and always find the easiest way around a corner, a condition referred to as racetracking. The gas bubble will be profiled given a consistent section through which to flow. That is, the gas bubble starts large and is reduced in diameter as it approaches the end of flow.

The process can best be described in the five steps of a short-shot molding. In Figure 2.16*a* the molten plastic is injected into a closed mold under high pressure. In Figure 2.16*b* the gas injection process is activated and the gas and molten plastic flow into the cavity together. In Figure 2.16*c*, plastic injection stops while gas continues to flow into the cavity. The gas pushes the plastic in front of it to finish filling the cavity. It will move toward the areas of highest temperature and lowest pressure. In Figure 2.16*d*, once the cavity is filled, the gas continues to push the plastic against

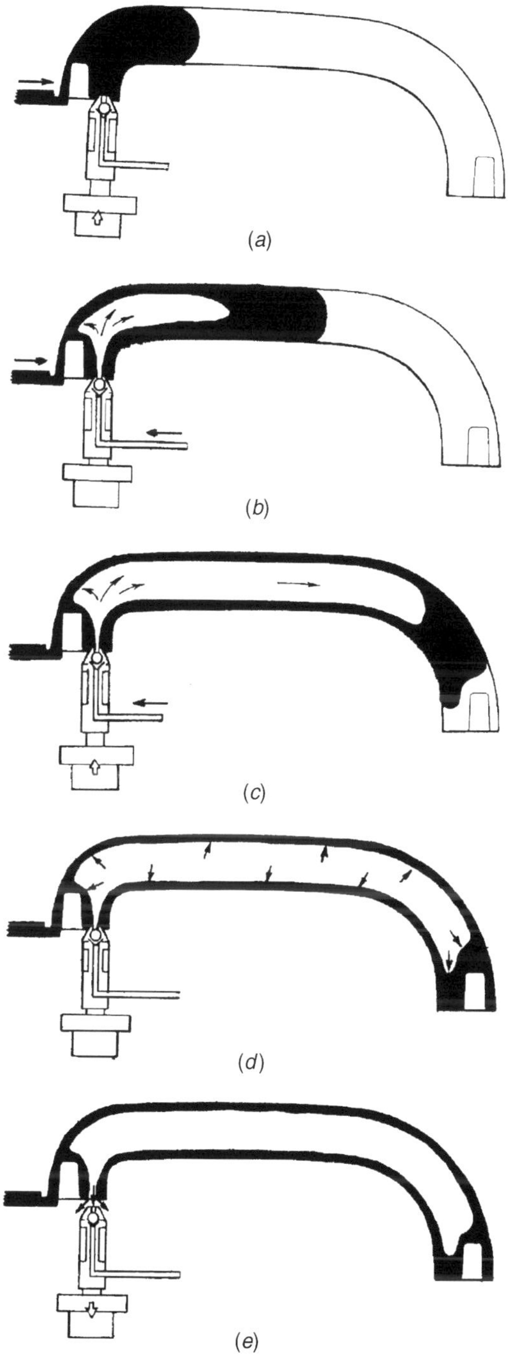

FIGURE 2.16 Gas-assisted mold-filling stages. (Courtesy of Cinpres Gas Injection Ltd.)

the cool surfaces of the cavity. This action shortens the cooling cycle, discourages sink marks, and improves dimensional reproducibility. In Figure 2.16 *e* the plastic part has cooled sufficiently to retain its shape, the gas nozzle is retracted to vent the gas, and the part can be ejected. Of all the structural plastic processes, gas assist has the most to gain from a designer's vision of the molding process. The designer becomes the mold designer and the process engineer, with the engineer's ability to control the flow of both plastic and nitrogen.

Ribs are the design element that will establish the gas passage. Gas will follow the course of least resistance. Thicker areas in the part have greater volume and therefore lower pressures, thus attracting the gas bubble. The effective way to establish these thick areas is through the aspect ratio relative to wall thickness. In effect, these thick areas will become manifolds or gas passages connecting to a common gas injection point. The gas passages should be three to six times the thickness of the wall section. Lower aspect ratios become ineffective and allow fingering; high aspect ratios become susceptible to gas breakthrough where the gas stream gets ahead of the resin flow front during filling.

These gas passages are housed in gas runner ribs. Deliberate wall thickness variations, which extend like a rib, are also considered projections. Gas passages should extend to the end of the part. Oversized stiffening ribs become the basic geometry for the gas passage. Many variations of rib design are feasible. A practical approach for a deeper rib is to stack a conventional rib on a gas passage rib with proper aspect ratios. This is a good solution to the deep rib draft problem of too thin at the top versus too thick at the bottom. Several rib variations are shown in Figure 2.17. A key to the success of a product development program is to make full use of the molded components. The piece-part design is more important than ever with gas-assisted injection molding. The rib pattern becomes the lowest resistance course and therefore a manifold for both plastic (during filling) and gas. Rib placement is enhanced through computerized mold filling simulation. The remainder of the part design closely follows established design practices, particularly holding a uniform wall section, thus allowing an accurate computer model. The part designer ultimately controls the success of any gas-assist program. By following established design rules, many unnecessary variables are eliminated.

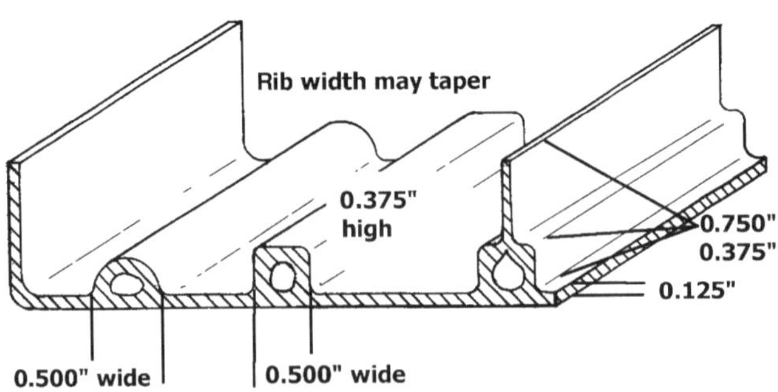

FIGURE 2.17 Rib design.

The ultimate step in control of the gas bubble is with spillovers or overflow cavities. The expulsion of plastic is a displacement of the incoming gas volume. This is considered an upgraded process to basic gas-assisted injection molding. The technology can be licensed from many of the suppliers of gas assist equipment. The advantages include exact control of the volume of gas being injected and therefore of the gas passage profile. Initial mold filling is a full shot of plastic and more easily controlled than a short shot.

2.6.1 Gas-Assisted Injection Molding Equipment

Nearly all suppliers of gas-assist equipment use the same basic process concept; follow a short shot of resin with a shot of gas (nitrogen) which displaces the plastic in the thicker, easier-filling areas of the molded article. The result is the ability to create an interconnected series of hollow channels in a part. These channels are contained in the rib pattern, which for the part designer is the basis of the functional requirements of the design. However, for the process engineer these channels are a hot runner system built into the molded article. The challenge is to determine which gas technology is right for a given application. Several considerations are necessary. A common question is: Is the gas injection technology controlling the gas by pressure or volume? Both types are used successfully. Application requirements may dictate the answer. Gas injection is controlled by monitoring the gas volume or the pressure. For volume control, a fixed volume of nitrogen is prepressured before injection. The gas is pushed from its reservoir into the mold. The pressure in the part depends on the ratio of volume versus the cylinder volume. Timing, gas pressure, and injection speed control the gas injection profile. Pressure systems utilize a reservoir of pressure supplying pressure to proportional valves. Pressure profiles are achieved by controlling the gas pressure to each valve. Cost is usually an overriding factor in process selection.

Another question is: Will the process be adapted to an existing injection molding machine or will a new machine be acquired? Several equipment manufacturers offer injection molding machines with gas assist apparatus integrated into the machine and controls. The most desirable situation for production is to specify new equipment that has the optimal features for a given application. This includes the correct gas technology for part configuration. The machine platen can be matched to the intended mold size, and the clamping unit can be matched to the needs of the particular part geometry. Using existing injection machines for gas assist (or foam, for that matter) usually results in a machine platen maxed-out with molds but the clamp unit running at 25 to 35% capacity. Although this is not necessarily detrimental to the molding, it is a detrimental to the cost. This is especially true in large parts, where the difference between 1000 tons and 3000 tons amounts to several hundred thousand dollars in the capital budget. Matching of machine barrel size to shot size is also critical. To obtain optimal throughput, the shot should be 70% of the barrel capacity. New equipment can be specified with optimal plastification capacity. Existing injection molding machines will have barrels that are too small for gas assist. The result could be a part design in gas with the wall thickness optimized for fast cycles but the machine unable to produce enough plastic melt to k eep up. The cost of ne w equipment may not be justif ied if injection machines are a vailable. Thousands of successful gas programs have been produced in existing equipment, so this is usually an acceptable trade-off.

When selecting equipment suppliers, considerations include the right to patents and licenses. Is technical service a vailable in terms of training, design assist, and troubleshooting? Is there a forum for technology update? Ho w reliable is the equipment?

2.6.2 Gas Injection Through the Nozzle

Figure 2.18 illustrates a combination shutoff nozzle and gas injection apparatus. This is the original form of gas assist, going back to a European patent in 1978. Today, several technologies emplo yee though the machine nozzle of the gas. This means that resin and gas ha ve the same entry point. This method limits gas pressure and timing. Balanced filling has to be designed into the molded part's runners, rib design, and wall thickness. It is best when used in applications with symmetrical or balanced filling. Gas is vented through a two-way nozzle prior to mold opening or with sprue break when the mold first opens. Sprues and runners are hollo w and are connected to the gas channels throughout the part. Nozzle gas injection is not used with hot runner systems because the gas pressure will displace resin in the manifold. The common practice is to save cost by eliminating the hot runner system by designing a rib pattern into the molded article.

2.6.3 Gas Injection into the Mold

Special gas injection pins are placed in the mold to inject gas into the runner system or directly into the article. Figure 2.19 sho ws the gas being injected into the molten plastic as it enters the mold cavity. The pins can be either a two-way gas flow design or a one-w ay check v alve. The two-way valve allows the gas to pass through the

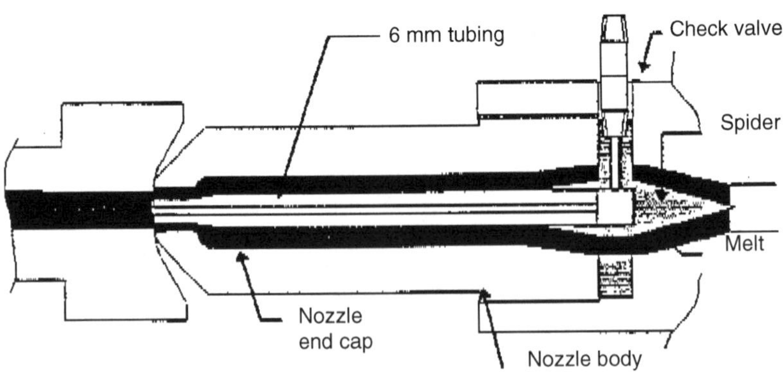

FIGURE 2.18 Gas injection through the machine nozzle.

molten resin once it has entered the mold. The pin allows the injection of the gas and also the venting of the gas prior to mold opening. The gas channels in the article remain open from the injection point to the farthest point of gas penetration. Vented gas is either collected for recycling or released to the atmosphere. The two-way valves sometimes become plugged by volatiles, glass fibers, and hot resin that enter during venting and become obstructions in the small clearances in the pin mechanism.

Development continues. Gas controllers will have a gas purge feature as well. Figure 2.20 illustrates the one-way or check valve type. This type requires that the pin be removed from the mold cavity prior to mold opening and part ejection. The one-way valve releases the vented gas when the pin is retracted by hydraulic

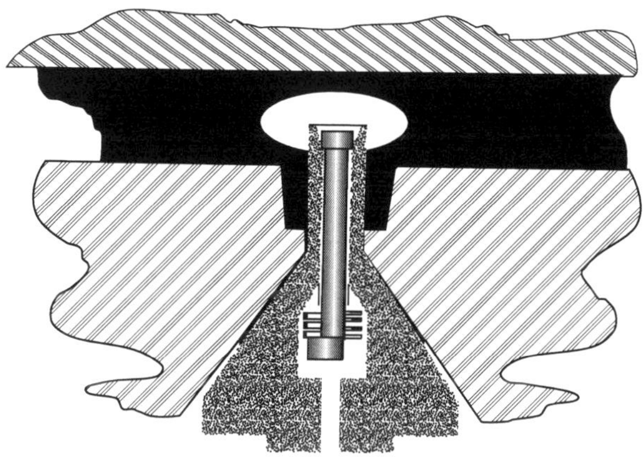

FIGURE 2.19 Gas injection pin mounted in the mold cavity.

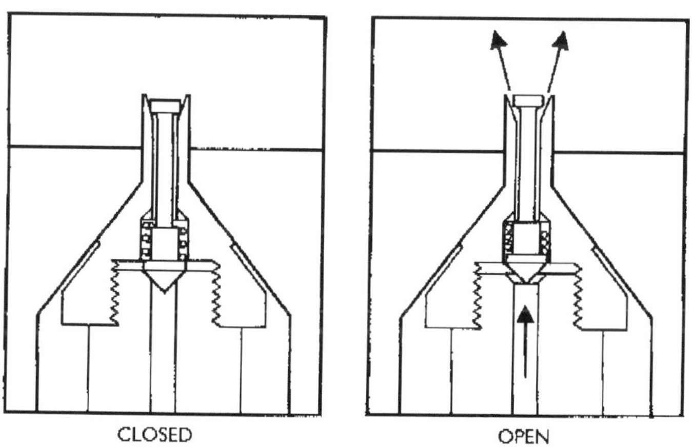

FIGURE 2.20 Gas injection pin operation. (Courtesy of Cinpres Gas Injection Ltd.)

movement or a moving core in the mold. Regardless of the type of pin, far more important is its location in the mold. Pins placed too close to the resin injection point will result in molten plastic being forced into the pin. The standard gas injection pin positions the gas in the center of the melt stream. Best results are achieved with a boss in the part design to serve as a shoulder for the gas pin's stability. High-pressure designs have a back taper in the cap and a longer tower to contain the pin while in the forward position, protecting it from the plastic pressure. A 90° pin is available to prevent the blemish of gas blasting against an appearance surface. Gas injection can be in the same direction as plastic flow. Venting occurs when the gas nozzle retracts from the article prior to mold opening.

2.7 EXTERNAL GAS MOLDING

The external gas molding (EGM) process is based on injecting gas at controlled pressures and timing at one surface of a mold after plastic has been injected into the mold completely. The gas forms a microthin layer between the plastic and the adjacent mold surface. The gas is thereby used as a medium for applying uniform pressure to the plastic, forcing it against the opposite mold surface as it cools, shrinks, and solidifies, precisely replicating the mold surface. Stress-free molding is achieved by allowing more shrinkage to occur inside the closed mold. With conventional molding much of this shrinkage occurs after the part has been removed from the mold. Imagine a five-sided box 1 foot square with 1-in. sidewalls. Even with a low-shrink resin such as polycarbonate, 0.062 in., $\frac{1}{16}$ in., or 1.5 mm of shrinkage must occur to the part. According to the laws of physics, most of this shrinkage will occur after ejection from the steel core. With EGM, much of the shrinkage still occurs outside the mold, but there is no orientation stress, so the part shrinks uniformly and flat. This accounts for EGM's typical cycle-time improvement of 15%. The molded parts have very little stress, so the cooling cycle does not have to be extended on warpage-oriented designs. EGM is a true low-pressure injection molding technology. Clamping tonnage rarely exceeds 1 ton/in^2. Since there is little or no high-pressure packing phase, there are energy savings in two ways. The machine hydraulics have to work less to achieve high-pressure clamping because the plastic is at lower pressure. Second, the lower injection pressures require less work on the injection side of the system. The same logic is true with electric molding machines because the amount of work is still reduced.

Alternative methods of sealing to prevent escape of gas are necessary as either integrally molded seals (as shown in Figure 2.21) or mechanical seals. As the incoming gas ribbon separates the molded plastic skin away from the mold surface on one side of the mold, the plastic is pushed into the opposite mold half. The ribbon of gas moves toward the parting line, and to avoid pressure loss there has to be a seal. Special provisions have been devised to seal the mold cavity, such as the use of integral seals (Figure 2.22) and O-ring seals. Patented sealing expertise is included in the technology package that comes with the purchase of the EGM license. With EGM, gas is injected at one or more positions from gas injectors. Typically, the gas injectors

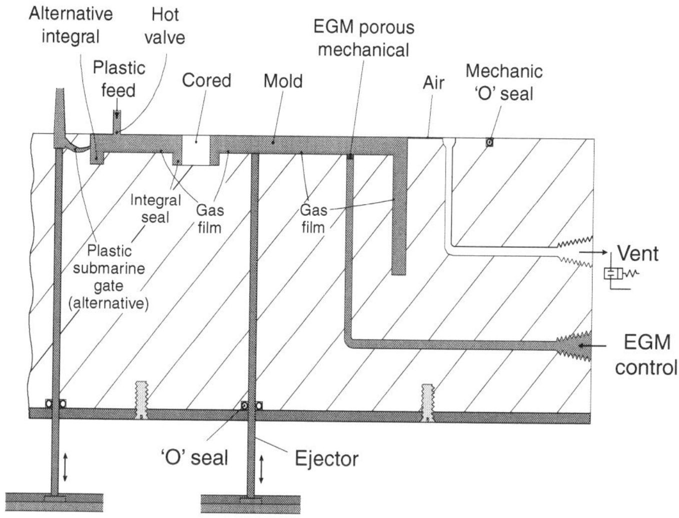

FIGURE 2.21 External gas molding tool details. (Courtesy of Cinpres Gas Injection Ltd.)

are located on the ejection side of the mold and can be incorporated without interfering with water lines or ejector pins.

The process can be used with most injection molding technologies and materials as a stand-alone accessory to the basic injection molding machine. EGM can be used as a short-shot method for low-pressure molding. It can be used for full-shot molding with or without packing. In applications with excessively thick ribs (i.e., an aspect ratio of 3 to 10), EGM can be used in conjunction with internal gas injection. It works very well with counterpressure molding, either foamed or solid. There is a further capability with EGM. At the time of mold opening, a second stream of gas can be injected to aid in the ejection process. With many of EGM's applications being for cosmetic parts, this air ejection approach helps to reduce blemishes and stress halos sometimes encountered with pin ejection. Reductions of molded-in stress are achieved due to lower packing pressures. The advantages of low-pressure molding have long been understood; lower machine clamp tonnage can allow lower hourly machine rates, softer tooling metals, lower power consumption, and reduced maintenance costs. Lower part weight results from less packing. Molding cycle time for low-pressure EGM parts are possible, as no internal stresses have to be overcome by longer cooling (in-mold fixturing).

The University of Bradford in the UK has completed extensive research into energy consumption of various plastic molding processes. The results of this work indicate that regular internal gas-assisted molding yielded a 5% energy saving over conventional injection methods. EGM yielded an 8% improvement over conventional methods. This happens on both ends of the injection molding machine. Obviously there is less clamping tonnage, but less conspicuous is the energy used by

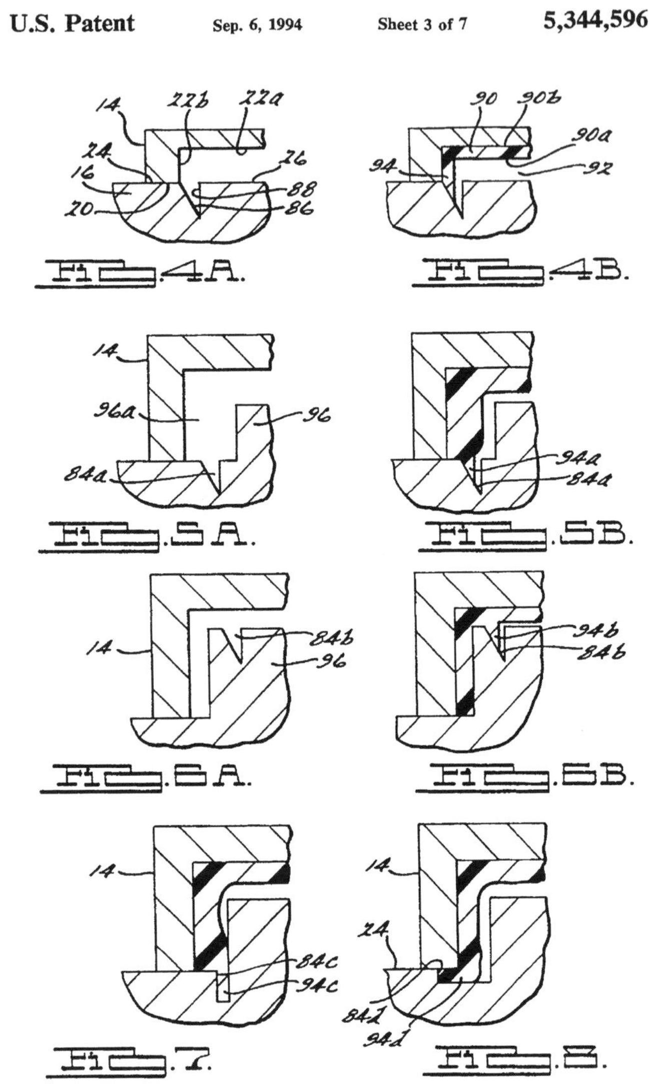

FIGURE 2.22 Integral seals for external gas molding.

the injection unit during conventional molding's packing–cushion phase. Further, the study was able to determine true production output by comparing pounds per hour of shippable product. EGM showed an amazing 18% improvement over conventional injection molding. Dr. Dawson at Branford concluded that the energy savings might seem small but that over time they are very significant. He also felt that the energy savings of EGM are outweighted by the saving achieved in both material reduction and, more significantly, cycle time. The University of Manchester in the UK has constructed test molds to evaluate the EGM process, and there is ongoing research into its benefits. Further, the school has an ongoing instruction program for EGM, including dedicated molding machines for education and research.

The EGM process is subject to patents owned by Asahi Corporation and Cinpres Gas Injection.

2.8 MATERIALS, GAS, AND BLOWINGS

Plastics, or polymers, are actually long molecular chains, based on a carbon molecule with four bonds, packed very closely together. Based on the type of polymer, these chains are either symmetrical or asymmetrical. The symmetrical chains allow very tight spacing or packing together, thus becoming more crystalline in structure. Examples of these semicrystalline materials are polyethylene, polypropylene, nylon, and thermoplastic polyester. Higher shrinkage and good chemical resistance characterize these materials. The other type of molecular formation is more random or asymmetrical, thus forming amorphous resins. Examples include polystyrene, polycarbonate, acrylic, and several alloys, such as acrylonitrile–butadiene–styrene (ABS) and styrene–acrylonitrile (SAN). These materials shrink less than do semicrystalline materials and have a lower resistance to chemical attack. In short, this is how thermoplastics are classified.

In addition to the basic polymer types, there are numerous fillers and additives to enhance the performance or appearance of the molded part. Examples include glass for stiffness or colorant for appearance. All structural foam applications have an additive by their very nature—a blowing agent. Several considerations go into the specification of material by the designer. Specifications should be called out on the drawing clearly and concisely. Almost every thermoplastic can be, and probably has been, foamed, either in daily production or laboratory testing. The blowing agents used range from inert gas, which is referred to as the *physical process*, to heat-activated concentrates, usually termed the *chemical process*. These are added to the molding resin during the melt stage. The same statement could probably be made that most types of materials have been successfully gas assisted.

High-density polyethylene (HDPE) is probably the earliest commercial resin. Its low cost, ease of processing, chemical resistance, and low-temperature impact strength make it a natural for material handling applications. Today, this market and this resin account for the largest sector of the structural plastic industry. Polypropylene is another semicrystalline material, very similar to polyethylene. It is also a low-cost resin with excellent processing characteristics and good chemical resistance due to its crystalline structure. This material has excellent elongation properties as shown by the fact that it makes an excellent living hinge. Living hinges are difficult to obtain in structural foam, due to wall thickness and weld line concerns, but are very practical with gas assist, although gas passage location and weld lines have to be considered carefully. Painting is difficult with semicrystalline materials, due to their inherent chemical resistance, which makes paint adhesion a problem. Special paint primer–barrier coats have been developed to address this situation.

High-impact polystyrene is a common structural foam resin. Often used in furniture applications, it is a stiffer but more brittle plastic. Its amorphous structure makes it easily paintable. Various amounts of rubber are compounded by the resin supplier to give more impact strength but at a sacrifice of stiffness. The material is commonly used for large panels that require both stiffness and low cost. Polystyrene accepts flame-retardant additives very well. Several resin companies supply Underwriters' Laboratories recognized polystyrene for the television and electronics industries. These materials are used for cabinets and internal parts. Polystyrene or its base resin styrene monomer is commonly used as an ingredient to produce polymer alloy resins such as ABS, SAN, or PPE [sometimes known as polyphenyl oxide (PPO)]. These materials fall into a family of polymers called *engineering materials* that have a combination of desirable properties, such as stiffness, temperature resistance, impact, and appearance, but at a higher cost.

Polycarbonate is another amorphous resin. It is a true engineering material, with both higher stiffness and improved impact strength. Commodity materials can be modified through additives to achieve one of these properties at a sacrifice of the other. Polycarbonate achieves both through molecular chain structure. Processing is critical to obtain the full potential of the resin. Moisture or heat degradation causes a loss of impact strength. Many blends have become popular where attributes from two different resins can be compounded for improved properties. Examples include polycarbonate–ABS, thermoplastic polyolefin, and GTX products.

The parts produced by counterpressure typically show 5 to 10% less density reduction than expected with conventional low-pressure structural foam. This is explained by the formation of thicker skins. Counterpressure structural foam parts, although denser, display more uniform cell structure throughout the part.This results in improved physical properties.

Structural foam offers a superior strength-to-weight ratio and reduced levels of molded-in stress. A common misconception is that structural foam has greater strength than an equivalent wall thickness of solid material. This simply is not the case, with the exception of hollow bosses, where the foam core offers shock absorption of the hoop stress exerted on the inside diameter. Generally, the loss of physical properties is directly proportional to the percentage of density reduction.

Structural foam exhibits many benefits as a processing technique for thermoplastic resins. A high stiffness-to-weight ratio is a natural benefit of a honeycomb structure. Structural foam offers the ability to produce very large parts with low clamp pressure using multiple-injection-nozzle molding equipment with oversized platens. The low-pressure nature of structural foam makes this possible. The most important benefit, however, is the low level of residual stresses found in structural foam parts. Again, this is due largely to the low-pressure nature of structural foam.

Residual stresses are those forces in the internal molecular structure of a part,that are present in the absence of external loading. These stresses develop from the resin shrinkage. Although the differences in thermal expansion that occurs for the various components in multiphase materials will result in minor amounts of stress,the major source of stress for all types of materials results from nonuniform flow during the mold-filling stage and to a lesser extent from nonuniform cooling. The effects of high levels of stress are displayed in nonuniform shrinkage, warpage, sink marks, loss of chemical resistance, and poor dimensional tolerances. The designer must be aware of the potential of residual stresses. For instance, maintaining uniform wall thickness contributes to an even flow and therefore to lower residual stresses. The designer must also recognize different processes when specifying dimensional tolerances.

Structural foam exhibits much lower levels of residual stress than that in solid injection molded parts, primarily for two reasons. First, the low filling pressure allows the part to have a more relaxed flow. Second, an inner layer of foam in the compost profile becomes a stress equalizer. The result of this low residual stress is the ability to mold dimensionally accurate parts which exhibit fewer tendencies for sink marks on the surface opposite ribs and bosses. The combination of these two factors makes structural foam a natural choice for combining several parts. This parts consolidation offers reduced cost of final assembly to end users.

2.8.1 Blowing Agents

The most common additive for the structural foam processes is the blowing agent. This component of the material mix is broken down into two categories, physical gas and chemically generated gas. Consider the effect of the size and quantity of bubbles forming the foam, as illustrated in Figure 2.23. Variations in the process and blowing agent will change the type of foam or cell structural. Generally, chemical blowing agents produce smaller cells. A faster injection speed produces smaller bubbles. Cooler mold surfaces make thicker skins.

The inert gas most commonly used for structural foam is nitrogen gas. The early purpose-built multiple-nozzle equipment utilized compressed nitrogen gas for the blowing agent as well as injection pressure. As a blowing agent, it is injected into the extruder when the resin has reached the melt stage. Nitrogen gas is a very inexpensive blowing agent but is limited by rather poor gas dispersion and requires a lengthy outgassing period of up to 14 days prior to painting.

Chemical blowing agents (CBAs) were developed to allow high-pressure injection molding machines to mold structural foam parts. These CBAs are heat-activated concentrates which are added to the molding resin prior to the plasticization step. These CBAs are available in several forms. Liquid dispersions are pumped directly into the feed throat of the barrel through a metering device. Powders and palletized CBAs can be preblended with the molding resin prior to plasticization. CBAs can also be metered in directly at the feedthroat. Flake and bar stock forms of CBA, which require special metering equipment, are also available. Generally, the CBAs, produce better surface finishes and improved physical properties, due to better gas dispersion and the resulting improved cell structure. CBAs also require less outgassing time for parts prior to painting. These blowing agents are now used extensively on purpose-built low-pressure machines as well as converted high-pressure injection molding machines.

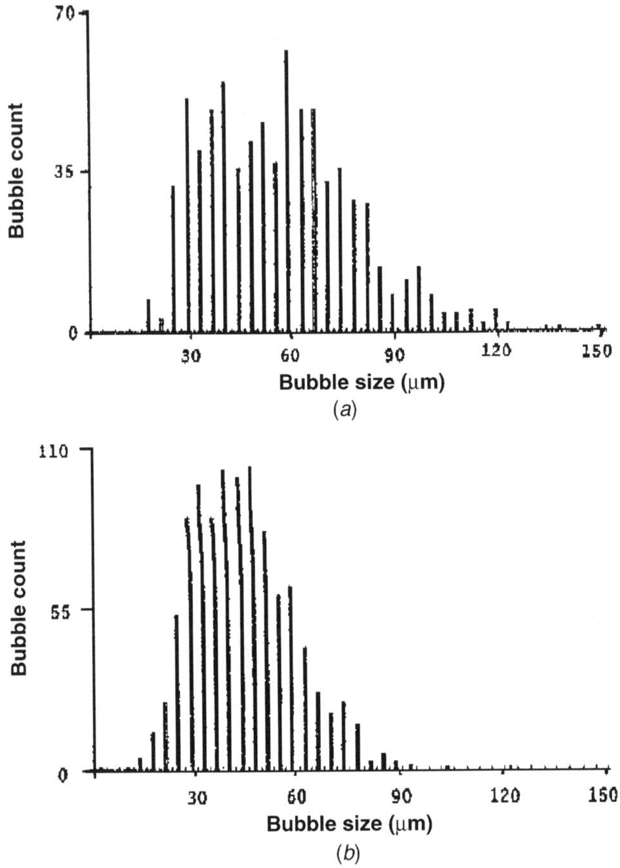

FIGURE 2.23 Bubble size and population considerations: (*a*) wide profile; (*b*) narrow profile.

Azodicarbonamide (AZO) has been the CBA in most common use since the early 1970s, but this is changing. As technology has improved for molding process control and resins have become more refined, there is a need for more sophisticated CBAs. AZO is an exothermic blowing agent, which means that it gives off heat as it reacts, along with some by-products. The gas generated from this reaction is nitrogen gas, which has several benefits, including safety. The nitrogen molecule is large and takes longer to dissipate through the solid skins. This release of trapped blowing agent is called outgassing. Water, another of the by-products, is a major problem for some engineering resins, such as polycarbonate. 5-Phenyl tetrazole is a CBA that was developed specifically to avoid this moisture problem.

There is a growing use of endothermic blowing agents, which give the immediate benefit of reduced heat and faster cycle times. Bicarbonate of soda and citric acid are the two most common products. Endothermic CBAs can act in a delayed reaction working directly behind the flow front, thereby reducing swirl. Adding a blowing agent to a resin changes its viscosity significantly and often works as a flowenhancer, together with its self-expanding force. This helps to explain structural foam's flow length ratio of up to 200 : 1. The by-products of the CBA reaction are of concern in some applications. For instance, water is a common by-product, which adversely effects materials that are sensitive to moisture. Experience has shown that some of these by-products lead to corrosion of steel molds when coupled with certain resins, such as ABS. Nickel plating has been employed on steel mold cavities to prevent surface pitting. Stainless steel mold cavities will also prevent this corrosion. Recent formulations of pH-neutralized citric acid blowing agents address this problem.

Looking at blowing agents from a scientific standpoint brings out two important concepts. First, think in terms of the gas solubility of the two most common blowing agent gases. Nitrogen gas has poor solubility, requiring 3000 lb/in^2 to go into solution, and carbon dioxide, which requires only 750 lb/in^2 for the same conditions. Lower-pressure molding systems are possible with the carbon dioxide gas systems found in the endothermic blowing agents. These differences in vapor pressures are being employed in a new family of endo/exo blowing agent systems, where there is better control of bubble size.

Second, consider melt acidity. This seems insignificant until recognizing that the polymer system is a smorgasbord of chemistry. In addition to the main resin, there are fillers for impact, heat, and processing. Ultraviolet screeners and pigments are added. Flame-retardant materials include some wild ingredients. This mixture of material then goes into the molding machine barrel at several hundred degrees under as much as 20,000 lb/in^2 of pressure for some extended period. In the laboratory this is called a *reactor*; in the plastics industry we call it a *number nine machine*. Recent studies have shown that melt pH can be as low as 1.5, that is in the range of strong acids. This has a negative affect on all metal surfaces that it contacts. In addition, this greatly affects the physical properties of the molded part.

Delivery of CBAs to a resin melt is similar to that used by injection molders for other additives. This includes blending of pellets or powder at the source of material entering the barrel. Some metering devices are used to disperse liquid CBA directly into one of the later stages of the melt stream in the barrel. A typical cost for both the exothermic and endothermic types is $2.50 to $5.00 per pound with a mixture ratio of 1 to 2%.

Physical blowing agents are in their natural form as they are introduced into the process, generally a gaseous state. Nitrogen gas is by far the most common of this type. It provides an inert, clean, low-cost option. The structural foam process has its historical roots with a nitrogen blowing agent. It became the early industry standard with low-pressure structural foam machines, which constituted the majority of industry capacity. The gas is mixed with the polymer melt in the barrel, accumulator, or through a special nozzle. Thorough mixing is always an issue with a nitrogen gas blowing agent. In recent years it has become feasible to have an in-house nitrogen generator. A market has been created for nitrogen generators through the widespread use of gas-assisted injection molding. Structural foam molders benefit from this more competitive in the market. The conventional source of nitrogen has been bottles of compressed gas or bulk liquid nitrogen. This offers the lowest-cost option, with average costs of $0.0025 per pound of processed resin.

2.8.2 Gas Types and Sources

Gas-assisted injection-molded parts obviously require some type of gas, nitrogen in most cases. It is no accident that this is the same gas type, that is used in many structural foam applications. Nitrogen is a clean, inexpensive, relatively safe, and commonly used gas. It has a high expansion rate of 696 : 1 when changing from the liquid phase to a gaseous phase (at –320°F). Traditionally, the injection molding industry has used liquid nitrogen in either cylinder containers or a bulk nitrogen storage system. With a growing need for higher flow rates it has become economical to invest in a nitrogen generating system.

Each process will have its own flow rate requirements. When serving as a blowing agent, there is a relatively small flow rate. Gas-assisted molding will have a higher flow requirement. Counterpressure structural foam and external gas molding can have the highest flow rate requirements when using nitrogen as the source of counterpressure. The entire volume of the mold cavity, all connecting lines, and sometimes the ejection system must be pressured.

Cylinders offer the lowest capital cost and highest per unit cost. There are increased safety issues with the mere handling of containers of 2600 + lb/in^2. The typical 55 in.×9 in. diameter cylinder will have a 19% waist at 500 lb/in^2 (304 standard cubic feet full and 57 empty). It is common to manifold several together to create a larger reservoir. Nitrogen purity is at least 99.99%, which is far more than is actually needed by injection molding.

Bulk storage requires investment in a reservoir vessel of 600 to13,000 gallons that is able to maintain the liquid phase of –320°F. Liquid nitrogen is converted to its gaseous state through a vaporizer on demand. Bulk nitrogen is capable of high flow rates. The level of purity is 99.99%.

A nitrogen generator is actually a separator device. Outside air is drawn through an intake filter and compressor. The compressed air passes through a series of filters and an activated carbon tower to remove residual moisture and oil. As clean air is heated to a controlled temperature and fed through the separator, the nitrogen flows into one stream and all the other gases into another stream. This separation process uses the difference in permeability rates to segregate the various components. The nitrogen stream flows through a control valve, which controls the flow rate. A buffer tank is placed in line to accommodate peak loads. The by-products are vented to atmosphere. Membrane separators require periodic replacement of the membranes. Carbon molecular sieve separators maintain efficiency by performing a constant burn-off of redundant sieves. While one sieve unit is in operation, another is being cleaned. Purity varies between 98 and 99.5%, which is generally adequate for the demands of injection molding. The generator option requires the greatest capital expenditure and the lowest cost per cubic foot of nitrogen. Regardless of the type of gas source, there will be a need for high-pressure compressors and distribution control units.

2.9 APPLICATIONS

The variety of structural plastics processes coupled with the hundreds of resins available produces an infinite number of combinations. Early applications such as those of low-pressure structural foam tended to gather around previously proven combinations. Inexplicably the wall thickness was uniformly standardized at its ideal 0.25 in. almost from the beginning. HDPE was employed across the board for materials handling. PPO–modified styrene (Noryl FN-215) found its way into the fast-growing computer and electronics industry probable, due to acceptance by Underwriters' Laboratories. Polypropylene copolymer was used in seating, speakers, and batteries. ABS was the transportation standard. Due to wide variation in performance due to processes and resin, deviation caused designers to overspecify performance. Designers had a serious problem predicting load and impact performance because of a lack of reliable physical property data and relevant test procedures in determining the data. Over the past 30 years, incredible amounts of data have become available, along with computers able to crunch the information down to an applicable level. As the structural plastics industry has matured, it has become specialized in terms of industries served, specialized processes, precise resin selection, process control, and overall optimization.

The material handling market is a natural application to examine. Initially, plastic pallets were offered in standard sizes on an off-the-shelf basis. In was difficult for end users to match their load requirements with off-the-shelf products. Best-case matching resulted in a cost disadvantage over wood of 4 to 1. More common was a formula of plastic pallets costing six to eight times that of wood. The material handling industry has benefited from better design options through relevant resin data and optimized processes and resins. An example is the comparison of the 0.25 in. wall structural foam container in Figure 2.24 versus that shown in Figure 2.25 of a

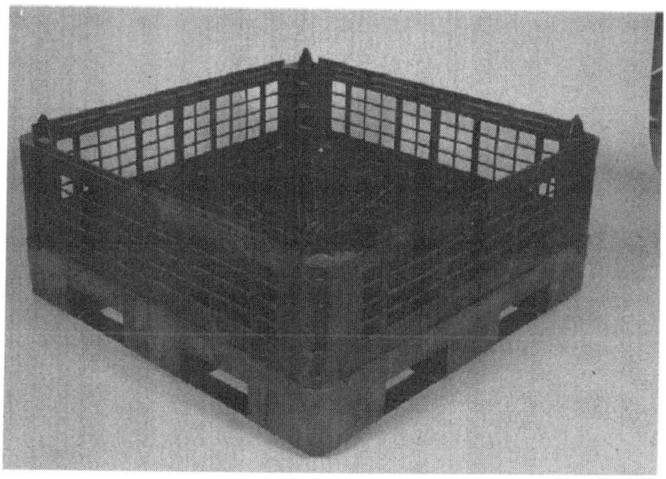

FIGURE 2.24 Structural foam collapsible container.

FIGURE 2.25 Gas-assisted shipping container.

0.12 in. wall section gas-assist container. The structural foam design had to double up in the corners to gain hoop strength. The gas design puts a thick gas channel around the collar of a one-piece design. The gas design gives more strength, thinner nominal walls, and lower costs.

Another market worthy of a design trend review is the television industry. This is a high-volume consumer arena, where a 30-year downward spiral of consumer prices has driven down manufacturing costs annually. With the high production volumes, new tooling is justified every year or two. Product designers are constantly striving to reduce costs and optimize design.

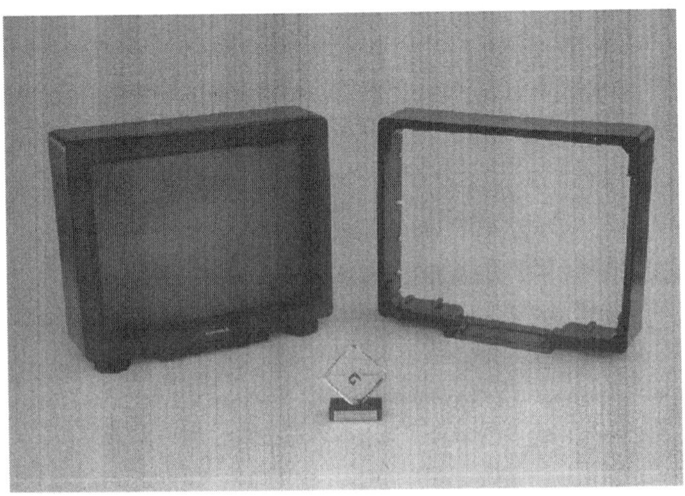

FIGURE 2.26 Counterpressure structural foam television cabinet wrap.

One of the early applications w as wood replacement cabinets. Lo w-pressure structural foam HIPS (high-impact polystyrene) w as molded from molds with cast beryllium cavities to simulate the w ood grain. The cabinets were f inished with a hand-highlighted lacquer system. Smaller cabinets went to injection molding,but the larger screen sets still utilize gas-assisted injection molding. Figure 2.26 sho ws a 35-in. TV cabinet wrap that supports the weight of a glass tube. 0.25-in. thick counterpressure structural foam in flame-retarded polystyrene. The cabinet was sprayed with high-gloss black pearl lacquer. Later versions have been designed with a much thinner 0.125-in. wall thickness and oversized structural ribs. The gas assist process injects nitrogen into the thick rib sections.

One of the early a ward winners was the unicell cabinet (Figure 2.27), molded in low-pressure structural foam polypropylene with mineral fillers. Produced on a multiple-nozzle machine with aluminum molds, it represents furniture as one of the original markets. Grumman's mass transit v ehicle of the early 1990s featured a clamshell design in 10% glass-f illed polycarbonate molded via lo w-pressure structural foam molded on multiple-nozzle machines. The seat carries the load of two passengers with no supporting legs. The seat was totally cantilevered from the sidewall of the bus for preassembly ease. Assisted injection-molded processes ha ve been widely used in

FIGURE 2.27 Unicell medical cabinet in structural foam.

FIGURE 2.28 Steelcase cachet chair: example of gas-assisted injection molding. (Courtesy of Cinpres Gas Injection.)

FIGURE 2.29 Delphi super plug: gas-assisted application.

seating applications. Internal seat frames, stadium seating, institutional seating and table, and high-end decorator chairs are examples. Figure 2.28 illustrates the Steelcase Cachet chair, molded with gas-assisted injection molding employing a unique back flow through overflow cavities of material to control polymer orientation.

General Motors specified gas-assisted injection molding for its award-winning inner door structure the Super Plug (Figure 2.29). The resin is a glass-filled polycarbonate–polybutylene terephthalate blend. The multiple functions of the Super Plug are illustrated. Thin-wall design provides for flow channels feeding into ribs. The gas then displaces the plastic to fill the part. Sixty-one individual pieces are consolidated into this design. Assembly-line labor was reduced from 5 minutes to 1 minute.

The swimming pool industry has long utilized structural plastics for their obvious advantages of corrosion resistance, light weight, and low cost. Playground equipment such as sliding boards and swing seats are accepted applications. Underground applications take advantage of plastic's corrosion and chemical resistance. Utility boxes, swimming pool panels, and storm chambers are examples.

One of the early gas-assist applications in the electronic office equipment market was perhaps the most challenging. The use of flow simulation to establish the location and sizing of the gas runner ribs was employed successfully. The application was the front panel and paper feed-in section of a new blueprint copier that was under development (see Figure 2.30). Traditionally, this had been a sheet metal application; however, Xerox engineering felt that the multiple-piece sheet metal design and associated tolerance buildup could be avoided with a single molded part. Gas assist allowed two large cutouts for the paper path, which surrounded a center section of considerable complexity. Heat also provided a challenge, as warm paper would be contacting the output section of the part, and hot air would exit the opening, thus warping any residual stress.

FIGURE 2.30 Xerox blueprint copier: example of gas-assisted injection molding.

2.10 MOLDS AND TOOLING

The assisted injection molding processes discussed are closed mold techniques, which gives process control to both sides of the part. Typically, a mold consists of a cavity half (A plate), which produces the exterior surfaces of the molded part, and a core half (B plate), which produces the interior part surfaces. The plane where these two plates come together is referred to as the *parting line*. In addition to defining the part, the mold directs the material flow and controls the cooling. The mold also has a built-in ejection system to facilitate removal of the molded part. The part designer must take into consideration such tooling aspects as draft, ejection surfaces, and gate location. Figure 2.31 shows a typical mold for the large-part molding industry. Being low-pressure, the processes covered have the capability of using lighter construction techniques for the molds. In the early days of the industry it was very common to use cast aluminum alloy molds. These molds were cast to shape from a pattern and mounted to a steel mold base. Cooling jackets were either machined or cast. However, cast molds are less desirable because the shrinkage, in the casting process combines with the molding shrinkage, making close-tolerance parts almost impossible to mold without a great deal of mold work. Controlling thermal expansion is an issue with cast alloy molds that affects a part's dimensional repeatability. In addition, the casting process has an inherent porosity, creating imperfections in the surface and limiting the marketability of the parts molded. Low-cost cast tools are often considered for prototyping in structural foam. Faster turnaround time is feasible on deeper draw parts where the machining of deep cavities is very time consuming. The shortcuts taken on cast molds are not as detrimental when prototyping if the mold is removed from service on the planned intervals. More than one program has failed due to attempting to extend the life of a cast tool beyond its predicted life cycle.

Machined aluminum molds have replaced cast aluminum molds and are used more commonly in the structural foam industry. These aluminum molds have steel bases, guided ejection, and steel inserts in wear areas. The best results are achieved from billet cuts. A large part size will sometimes dictate fabricated aluminum cavities. This compromise can be offset by proper support plates, good mold design, and styling provisions for the witness lines on the molded part. Tight tolerances can be machined directly into the mold. The part-to-part repeatability is excellent. Mold surfaces can be polished to achieve smooth ejection and better surface finish on the molded part. These tools are generally considered useful for up to 50,000 parts, at which time parting line refurbishment is generally necessary. However, the swirl marks caused by the blowing agent and low-pressure molding, limit the type of molded-in texture that can be defined on a part surface. Aluminum is superior to steel in terms of thermal transfer, so cycle times will favor aluminum molds. One of the best aspects of a machined aluminum mold is that it is machined. This gives the mold precise reproduction of the part design database. This practice can also be applied to cast molds. An example might be a deep draw part, with tight tolerances, requiring fast mold delivery. In this case, the time required for hogging

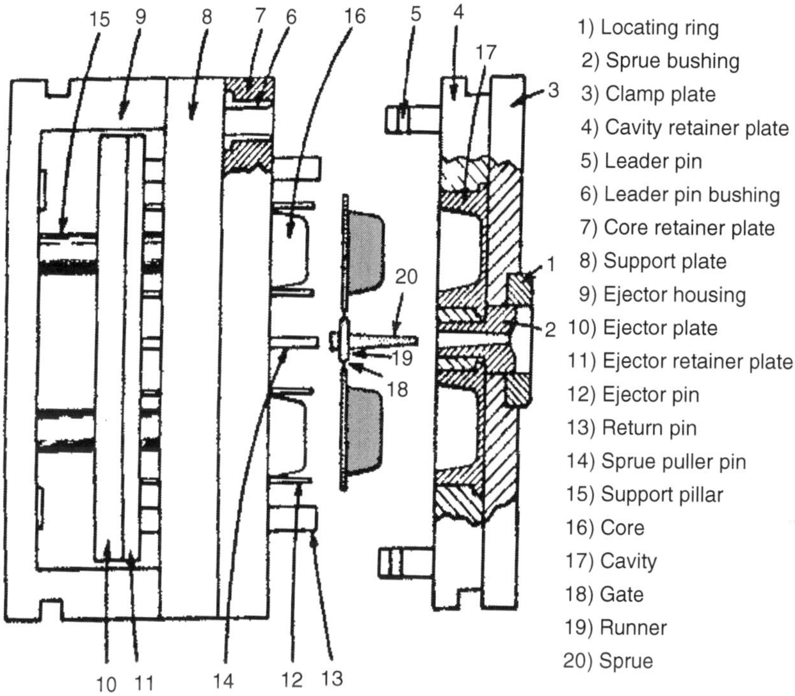

FIGURE 2.31 Basic mold.

the deep cavity is avoided by casting a deep cavity with enough extra metal to allow a final machining of the surfaces. Although this approach does resolve some of the dimensional problems, it still results in a very soft mold cavity with many of the same problems. Does it make sense to invest all that machining into a soft metal? The answer is yes in only a few cases.

High-volume applications or molded-in textures generally require more elaborate steel mold materials, such as 4140, P-20, or HR-13 materials. Preferably, these molds are also cut from billet as opposed to fabricated pieces. The new swirl-free structural foam molding systems also generally require steel molds to obtain the best possible part surface and to cope with the higher pressures that are sometimes present. Programs requiring a high level of part-to-part repeatability (e.g., 1.33 cP·K) are often better prescribed in steel molds. In recent years a new type porous of mold steel has been developed for injection molding tools. This porous steel allows for venting in areas otherwise difficult to vent. These new steels are slow to become prominent in structural foam because of theirs low pressures and greater wall thickness. With thicker wall sections there are fewer difficult-to-fill areas.

It is customary in the structural plastics industry for the custom molder to purchase molds from an outside mold maker; however, occasionally, there is in-house toolmaking capability. There are obvious pros and cons to both. In either event, a set of detailed mold drawings should be generated. The procedure actually begins at the quotation phase, when the molder will provide a layout sketch or a detailed description of details to the mold maker for purposes of an accurate quotation. When the order is released to the selected mold maker, it is accompanied by the database or drawings from the part designer and some very important information from the molding process engineer. This information includes resin shrinkage factor, gate location and method, cooling layout, and venting requirements. The mold maker will generate a preliminary mold design for customer approval. This early layout approval allows the mold steel and other materials to be ordered concurrently with mold design and Computer-assisted machine programming. Final mold drawings and a mold bill of materials should be expected by the midpoint of the mold-building program.

Evaluation samples are required prior to mold approval and production release. It has become commonplace in the injection molding industry for large mold makers to have tryout presses available for sample molding. This is uncommon for large structural foam parts, where large machines and special nozzle arrangements are often employed. Even if the mold maker has a 3000-ton injection machine, the samples produced are not from the same process as those from a multiple-nozzle machine. It is always good practice to use mold maker samples for obvious errors but to rely on final process samples for final measurements and data documentation. The SPI Mold Makers Division provides a form, which is a good example of the type of information expected from a mold maker's quotation.

2.10.1 Gating

There is an obvious need to predict the plastic filling patterns and to arrange the gate locations accordingly. Gate locations are of critical importance to the part designer and to the molding process engineer. The mold designer needs to be aware of these concerns.

Multiple-nozzle gating provides solutions to many concerns about weld line location and flow lengths. Direct sprue gates feeding into thicker, ribbed, or high-flow areas of the part are preferred. Direct sprues feeding into edge gates along the edge of the part or in an opening are ways to avoid appearance areas. Reverse ejection is another solution. This involves gating the part from the nonappearance or backside by flipping the mold over. This means that the part will be injected and ejected from the same side. This happens to be the stationary half of the mold, so there is a need to add hydraulic or mechanical ejection. This also requires heated nozzle extenders to reach through the added mold buildup.

Runners in open areas should be opened up to a full round $\frac{1}{2}$ in. or more in diameter. It is also a good idea to have sprue-puller ejector pins under the sprues. In some cases a submarine gate may be used, but a 0.12-in. minimum diameter is needed to allow adequate resin flow.

2.10.2 Venting

Large-part molding has always involved special attention to venting. With the advanced technologies, venting should no longer be viewed as the necessary displacement of air with plastic. Modern pressure transducers and controllers allow the venting process to be a way to control mold filling. The thin-wallet variants will follow traditional venting philosophy. In venting structural foam, which has a greater wall, there is naturally more air in the cavity to be displaced as resin flows. The blowing agent escaping off the leading edge of the flow front itself becomes more gas to be displaced prior to complete filling of the mold. Typically, vents are cut in the corners, in difficult-to-fill areas, and at the end of flow. These vents are cut in stages. They should measure $\frac{1}{2}$ in. wide and 0.002 to 0.003 in. deep adjacent to the cavity and then open up to 0.010 to 0.015 in. over the first $\frac{1}{4}$ in. of cut. They are usually cut into the cavity half of the mold.

2.10.3 Cooling

The cooling cycle for structural foam is critical, as this is the longest part of the overall molding cycle. Cooling line placement and design may have a great effect on overall cost. Careful consideration must be given to the cooling channel geometry, placement, and circuitry. Lack of adequate cooling is usually exhibited through *post blow,* the blowing agent pressure bulging the skin after ejection. One other consideration is that ejector pins should be of larger diameter to avoid penetrating the skin. This becomes a concern at diameters, which are less than half the wall thickness of the part.

2.10.4 Ejection

Ejection follows normal rules except that the ejector pins' diameters should be no less than one-half of the wall thickness. This is to avoid piercing through the skin and into the foam layer. There should generally be at least eight pins. Due to the size of many structural foam parts, guided ejection systems are necessary to prevent any binding of the movement, which can cause an increase in process instability and scrap. The knockout pattern must be determined in conjunction with the planned molding machine.

2.10.5 Texture

When using low-pressure structural foam, there are three reasons to reconsider the normal mold texturing of injection molding. The most obvious problem is the swirled surface, which makes fine textures impossible to detect. Only heavy textures are recommended (0.003 to 0.005 in.). The packing pressures are much lower even with the advanced processes that have good surface capability. To compensate, a deeper etching of the same texture will achieve good results. For example, with counterpressure, add 0.0005 in. to the desired 0.0015 in. The third consideration is painting, which is a common secondary operation for cosmetic structural foam parts. When trying to reduce paint steps down to only a color coat over texture, add another 0.0005 in. to the texture depth to compensating for paint buildup when the articles are painted.

2.10.6 Counterpressure Details

Counterpressure structural foam has been in existence since at the latest 1977 and in production since 1983. Until recently the process has been a molder-developed technology. No industry standards have ever been established. The counterpressure adaptations involve two major areas, machine control and mold construction. The purpose-built mold is constructed with the following additional details:

- A venting system is used that feeds into a sealed internal system of vent channels. The venting system should follow the traditional venting philosophy of placement at the end of flow and at weldlines. Generous vents work best. Screen protectors improve reliability.
- Parting line seals are required. One simple example is the addition of a 0.25 in. × 0.25 in. groove around the mold cavity and exterior to the venting track. The seal should be installed in one half of the mold, usually the cavity half. A rubber O-ring is installed. A flat parting line works best. If this is not possible, a gradual ramping of the seal surface will help reduce O-ring wear.
- The entire ejector system must be sealed. One practical solution for ejector pins is to use the same sealing techniques that hydraulic cylinders employ. The air volume inside the sealed portion of the ejection system is a concern for control and cost reasons.
- The nozzle seats must be sealed.
- Side actions and movements must be sealed.

The mold maker is a critical link in the success of counterpressure molding. With the need for a venting system and parting line seals exterior to it, an oversized mold footprint will result. With an improved surface finish, the mold cavity will be more polished than with low-pressure structural foam. Counterpressure molds will have added costs, usually \$2500 to \$5000, to cover these added requirements.

2.10.7 Tooling Checklist

Tooling are brief checklists of items to consider through the phase of developing the mold.

Part Print Review Checklist

1. Overall size and configuration
2. Resin/process

3. Moldability (draft and parting line)
4. Wall thickness
5. Press size and requirements
6. Quantity forecast for mold life cycle
7. Delivery date needed
8. Filling/packing simulation
9. Mold cooling simulation
10. Type of database and compatibility with mold shop (three-dimensional wire frame)

Mold Design Checklist

1. Molding process (solid, foam, counterpressure, gas)
2. Mold material (aluminum, steel grade, cast)
3. Type of construction (two-plate, hot runner, reverse ejection, slides, lifters, sleeve ejectors, strippers)
4. Texture
5. Class of tool (machines into solid, fabricated, standard mold base)
6. Prototypes needed
7. Slides (material, ware plated, water, locks, returns, movement source, side angle draft)
8. Lifters (movement space in mold and part, material of mold and lifter, water)
9. Mold size (depth, 4 to 5-in. steel border, 3-in. border depth, molding pressure, counterpressure)
10. Support pillars (maximum number)
11. Water lines (straight lines, parallel cavity/core, zones; avoid loops, think Btu)
12. Gates

2.11 PART DESIGN

Structural foam, gas assist, and the hybrid processes are variations of injection molding. A thorough understanding of injection molding part design is an excellent starting point and allows the discussion to focus on what is different and why. The steps for the part designer are viewed with equal importance by the process engineer who will actually be responsible for molding good parts. The focal points are flow and shrink. The mold has to be filled with material before the skins freeze-off and the material will shrink during cooling. The wall thickness determines the flow of injected resin. Any wall thickness irregularities cause uneven skin freeze-off. With structural foam the foam core is elastic, in that it fills in the thicker sections and is thus more forgiving. On the other hand, its low filling pressures may cause excessive rejects with restrictive and nonuniform wall sections.

Achieving structure is an obvious requirement of a structural plastic process. As wall thickness is the backbone of the design, it is a starting point to gain structure. The problem is that wall thickness also dictates the costs and may not be the most effective way to gain structure. Consider some elements of design that affect structure:

- Resin flexural modules
- Wall thickness, which increases stiffness by a power of 3
- Reinforcement ribs
- Part configuration
- Reinforcement attachment
- Matching performance profiles

In general, good design includes features and molded-in functions that are not practical in other materials. If the design goal is to replace a piece of steel or wood plate, injection-molded thermoplastics will not be cost competitive. On the other hand, if the sheet of steel has to go through several forming and secondary operations, a one-piece molding can have advantages in piece costs, dimensional consistency, and assembly costs. Let's consider how several of the processes should approach this structure question.

For conventional injection molding the range in wall thickness has varied from 0.009 to 4.5 in.; the practical limits are 0.035 to 0.25 in. Straight injection-molded parts basically rely on resin and part geometry to gain structure. Drastically increasing wall thickness is not a good option. Extensive ribs are limited in height and width. Commonly, thin-walled injection parts attach to other mating parts before the assembly takes on structure.

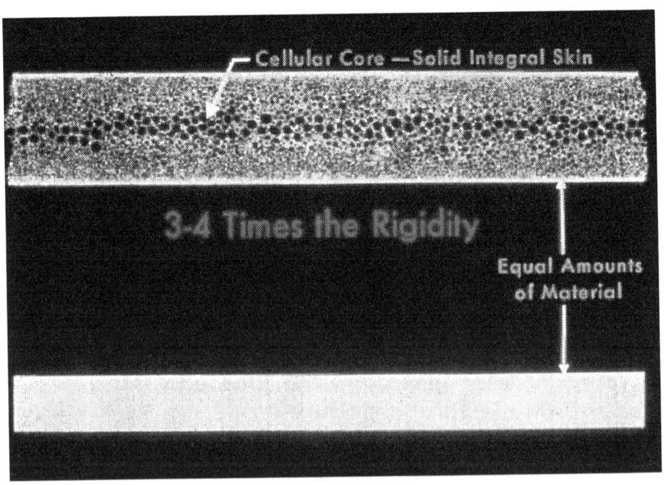

FIGURE 2.32 Structural foam vs. solid wall.

Structural foam, counterpressure, co-injection, and thick-wall gas injection have the option of thicker walls at reduced density. For these processes it is feasible to rely on the cubing ef fect of w all section increase. In theory , if the density could be reduced by 50%, yielding twice the thickness at the same weight, the calculated stiffness would be eight times greater (see Figure 2.32). In reality, this in not practical, and lower density produces reduced physical properties. The result is that overall stiffness per unit of material increases by three- or four fold. Reinforcing ribs are very practical with the thicker wall processes. The design rules allow a higher aspect ratio, the rib width at the base is greater , and less draft is required. Co-injection structural foam has two material components to consider. The I-beam loading profile says that most of the stiffness comes from the skin component of the composite. Refer to Table 2.5 for co-injection material compatibility.

Thin-wall gas assist has the obvious strength advantage of deliberately oversized ribs. The process offers the option to put the reinforcement ribs exactly where needed. To determine wall thickness, the overall gas strategy has to be considered.

- The part configuration is critical.
- Gas channels should extend to the extremities of the molded part.
- Gas channels must be well separated and should not intersect.
- Overall flow determines gas passage size.
- Resin viscosity has to be considered.
- General rule: channel diameter = 3 × wall thickness.

2.11.1 Importance of the Corner Radius

When a uniform wall has to make a bend (e.g., for the sidewalls), a radius in the corner is essential. Stress caused by resin shrinkage from two directions will build up in the corners, which can be distributed with the correct radius. Each material and process will have it own best formula for a proper radius. As a general rule, the inside radius should at 25% of the wall thickness. The outside radius should be thicker yet, by the addition of the wall thickness to the inside radius (0.060 in. inside radius + 0.125 in. inside radius wall = 0.185 in. outside radius). This is to ensure uniform flow during the filling stage. Since the wall thickness for structural foam is greater, the corner radius must conform (Figure 2.33). The draft requirements for

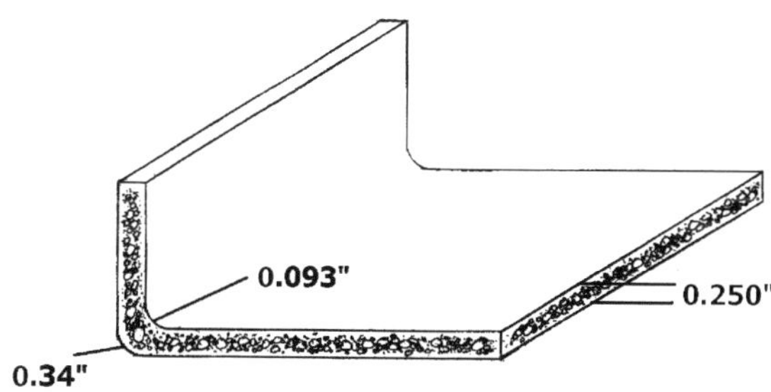

FIGURE 2.33 Low-pressure structural foam wall thickness and corner radius details.

structural foam are slightly less than those for injection molding, primarily because of the thicker walls and lower stresses. Typically, 1.5° of draft is the minimum draft allowed for sidewalls. Less draft can lead to ejection problems. Some naturally lubricated materials such as polyethylene can stand a bit less draft. In addition, isolated areas such as card guides can be designed with little or no draft. Ejection is always a consideration when making these decisions. A greater draft angle improves molded part quality, eases ejection, improves cycle time, and reduces cost.

2.11.2 Draft

Articles produced in closed molds from materials that shrink must have taper on the plane perpendicular to draw. The degree of draft has a major effect on cycle time and therefore on cost. Molding cycle time is largely the cooling of the molten plastic down to a temperature that will allow the article to withstand the force of ejection. Designs with greater draft will require less force to eject and can be removed from the mold hotter and quicker. Thicker-walled parts will cycle longer but can with-stand more force at ejection. The general rule of thumb for thicker-walled parts is that 1.5° of draft is minimum. For the thinner-walled processes, 2° is minimum. Designing parts with minimum draft will cause longer cycle time and higher molding scrap. The resin selected has a definite bearing on draft requirements. High-shrink resins will generally need more draft, especially on the core side. Glass and other fillers reduce shrink and can withstand a greater ejection force, but there is less part flexibility, so more force is sometimes required to eject. Some resins (e.g., polypropylene) are naturally more lubricated and will have less drag, requiring less ejection force.

Textured molds will require additional draft beyond nominal. Often, the draft is specified by the texturing house. Too often the problem is their late involvement—after the mold is built. The texture type and depth can be limited by this approach. A rule of thumb for texture depth is 1° for each 0.001 in. of texture depth. It is feasible to have a slightly shallower texture depth on sidewall with a texture transition in the corner.

Anticipating the effect of shrinkage on the molded article, it is feasible to design more draft on a surface that will be subject to more shrinkage drag. An example is a five-sided box, where the article will shrink to the core. Cavity walls could have much less draft because the plastic is actually shrinking away from that surface. In the case of polypropylene at 0.25-in. wall, this shrinkage could amount to 0.005 in. The tendency then would be to minimize outside draft and double up on inside draft. The resulting variation in wall thickness could be unacceptable for deep parts. Nonuniform wall sections will show differences in shrinkage rates, resulting in stress and warpage. A more acceptable technique is to design no draft standoffs that no draft plane across their top surface. The percent of sidewall affected by this design should be less than 10% of the total side wall. Added ejection may be necessary in these areas to put ejection force directly over the affected area.

The final solution to a no-draft requirement is to put side action movement into the mold. The sidewall of the mold can be retracted mechanically in the no-draft areas. This solution has several disadvantages. Obviously, mold price goes up by several thousand dollars. The mold base size increases to accommodate the footprint required by the perpendicular movement. In some situations this can mean moving to a larger machine just to find enough platen area. Unit costs are thus affected. The larger-platen machine could have a larger barrel, causing a mismatch of shot size. When there is a moving piece of steel in the mold, there will be a witness mark at the parting line. On exterior surfaces, this can be cosmetically unacceptable.

2.11.3 Rib Design: Know Your Processes

Whereas wall thickness will establish the cost of a molded article,it is rib design that determines its functionality. Frequently, the rib requirements will dictate the wall thickness. Rib width is expressed as its aspect ratio relative to the wall thickness, and rib height is also relative to a wall thickness ratio. Typically, deeper ribs require thicker walls unless the application allows the penalties paid for nonuniform flow and shrink. Injection-molded parts are limited to rib widths of 0.5 to 0.75 aspect ratio. Rib height is three times wall thickness; One way to offset short ribs is to increase their quantities. For best results, the corner radius at the rib base should be 50% of the wall thickness; 25% is the minimum.

Structural foam ribs must follow prescribed aspect ratios but allow more leeway because of the self-expanding nature of the material. This is illustrated in Figure 2.34. This compensates for volumetric shrinkage of the resin. Lower stresses result in flatter and more repeatable parts. The part designer has a major effect on the filling pattern based on rib design. The rib width at the base can be as high as the wall thickness, but a reduced ratio of 0.75 is preferred if sink marks are a concern. The rib pattern is a less expensive detail to add later if it turns out to be needed, because it involves removing metal from the mold. On the other hand, ribs are expensive to eliminate after mold completion, as metal has to be added to the mold. If later ribbing changes are anticipated, the mold designer should be advised so that cooling lines and ejection are designed accordingly.

Thick-walled gas-assist designs often find their origin as low-pressure structural foam. Frequently, gas assist is substituted to eliminate swirl or the need for blowing agent. Density reduction results in reduced physical properties. A thick gas part with improved impact strength will probably weigh more than its structural foam counterpart.

In the application illustrated in Figure 2.35, the stiffening rib also provides a bonding surface for the filler piece, which is solvent, bonded. Low-pressure structural foam allows a 1 : 1 rib profile. The resin is HIPS. Molded-in wood grain is painted with acrylic lacquer.

With thin-wall gas assist, many rib configurations are feasible (see Figure 2.17). Rib sizing is critical to success of the molded article. Gas blow-through and fingering are the primary concerns. Gas assist moldings fall into three types. The designer needs to recognize which type of part he or she is designing,as there are special rules for each type.

1. *Hollow moldings:* designs with no nominal wall other than the section around the gas passage. Generally, the gas volume is at least 30% of the overall volume. The

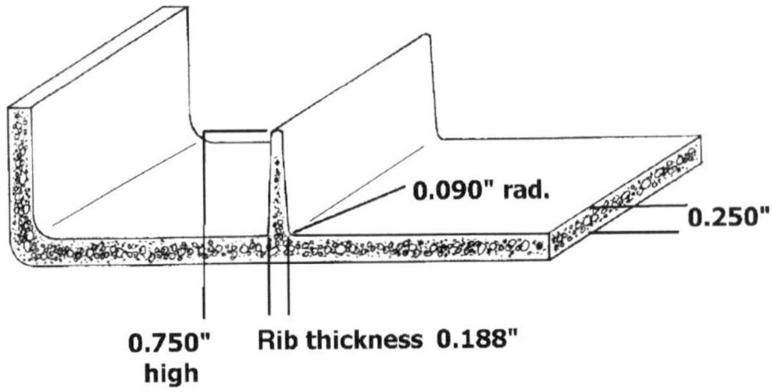

FIGURE 2.34 Low-pressure structural foam rib design details.

FIGURE 2.35 Low-pressure structural foam rib.

gas core is restricted to the lar gest circular area of the cross section, with a length-to-width ratio of at least 10. The maximum diameter is 30 mm.

2. *Solid moldings:* injection-molded designs with a problem or thick section. With this type of molding, the gas is injected after the resin, as a cushion to compensate for volumetric shrinkage.

3. *Short-shot moldings:* specifically designed to achie ve a greater core-out for added weight reduction and cycle-time improvement. These moldings are produced with a simultaneous gas and resin injection stage, with the gas boosting the resin flow. Optimization of part design, specifically a rib-to-w all thickness prof ile, is essential for success. The remainder of this design discussion will address this optimization for short-shot moldings.

Designing uniform wall thickness becomes challenging when there is a change in planes. Side walls are a continuation of the uniform wall with an intersection point at the corners. The corners will be subject to shrinkage pulling in different directions; therefore, a substantial radius is required to spread the stress. From a flow standpoint, a uniform section is required throughout to maintain a constant flow rate and filling. The final consideration for sidewalls is the draft angle, which is a matter of physics and tooling. There has to be enough taper on a part to eject it from its mold.

Ribs are the design element that will establish the gas passage. We have to remember laws of physics as we determine gas passage strategy. Gas will go the course of least resistance. Thicker areas in the part have greater volume and therefore lower pressures, thus attracting a gas bubble. The effective way to establish these thick areas is through the aspect ratio relative to wall thickness. In effect, these thick areas will become manifolds or gas passages connecting to a common gas injection point. These gas passages should be two to four times the thickness of the wall section. Lower aspect ratios become ineffective and allow fingering; high aspect ratios become susceptible to gas breakthrough where the gas stream gets ahead of the resin flow front during filling. These gas passages are housed in gas runner ribs. Deliberate wall thickness variations which extend like a rib are also considered projections. Gas passages should extend to the end of the part. Oversized stiffening ribs become the basic geometry for the gas passage. A linear area of increased wall thickness can be considered as a rib. An abrupt transition in section will help reduce fingering.

Multiple areas of linear wall increase can be designed to achieve ergonomic edges and follow the overall styling lines of the product. Placement of projections in corners reduces the risk of gloss variation. Multiple passages must be connected back to the gas injection point and should extend to the ends of the part. The natural buildup of thickness in corners is an obvious location of a gas passage. Again, gloss variations are hidden in the corner. Tighter aspect ratios apply in corners, as thickness builds up naturally. Many variations of rib design are feasible. A practical approach for a deeper rib is to stack a conventional rib on a gas passage rib with

FIGURE 2.36 Gas-assisted injection molding rib pattern.

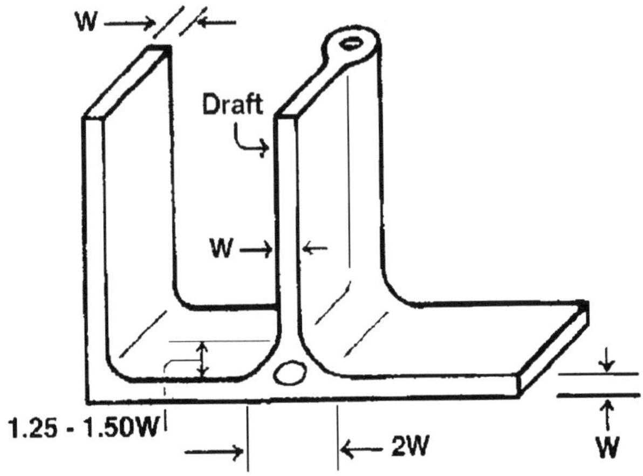

FIGURE 2.37 Gas-assisted injection molding rib and boss design guidelines.

proper aspect ratios. This is a good solution to the deep rib problem of plenty of draft versus too thin at the top versus too thick at the bottom.

Connecting the boss with a gusset or full-height rib avoids gas entrapment (Figure 2.37). Gas passages containing projections should extend to the end of flow and must have continuity back to the gas injection point. Fingering is the penetration of the gas layer into the solid wall area. This is to be avoided, as it reduces the integrity and strength of the wall section. Some designs will dictate reducing wall thickness at the end of flow, which helps maintain constant gas penetration velocity. This is important to control fingering. Fingering indicates a problem: usually with the process but sometimes with the design. A uniform gas penetration rate is an important consideration in determining the correct wall thickness. The designer can think of this as a bubble velocity issue, where maintaining a constant velocity to the end of flow is the goal. The primary factors of penetration rate are gas passage size and location relative to wall thickness.

Another important consideration of the short-shot filling puzzle is that of displaced plastic. For a gas bubble to penetrate past a certain point, there has to be a place for plastic to move. Proper design will give some wall section or an overflow provision for this plastic displacement. To determine wall thickness, the overall gas strategy has to be considered. Since the gas passage is controlled by the aspect ratio and gas penetration rate, it has less dimensional control than does a "tooled" feature. Additionally, the diameter of the core-out will be reduced continuously toward the end of flow. The designer needs to think of the gas passage as a constantly changing profile.

2.11.3.1 Ribs Are Critical

When everything else is equal, a lower-cost product will usually achieve a higher market share and greater commercial success. The injection molding process and its variants have long been considered optimum for many applications. Consider the cost elements of an injection-molded component. The greatest single cost element is that of the raw material: the plastic resin that has to be purchased by the molding company. Further analysis shows that most of the resin (weight) is in the skin or nominal wall of the component, whereas most of the strength is provided by the rib structure. This phenomenon results from sizing formulas that make the skins (a uniform wall section) serve as a platform for the rib structure as a means of net shape fabrication. To look at structure in a new way, consider the strength of a flat plate versus that of a reinforcing beam. Stiffness is a direct function (by a power of 3) of thickness or depth for both. However, the reinforcement beam, in the form of a structural rib, is a far more efficient use of material and potentially, a competitive advantage if fully exploited. A comparison of the strength-to-weight ratio for a nominal wall versus a reinforcement rib reveals an exponential advantage for ribs. If the concept of structural ribbing is to increase strength with the least amount of weight, there is no purpose for the nominal wall section other than appearance and to fill in the holes between structural ribs. By allowing a greater rib-to-wall aspect ratio, the internal and external gas-assisted processes allow designers to reduce weight and increase the strength-to-weight ratio (competitive advantage).

2.11.3.2 Reinforcement Ribs and Aspect Ratios

While each type of molding process will have its own rib design guidelines, thus influencing rib strength, it is the rib depth that ultimately establishes strength. Rib design is a compromise of limitations of draft, aspect ratio, ejection, and material filling. The result is a rather limited rib height for high-pressure injection molding. Taller ribs require thicker wall sections at their foundation.

There is actually very little difference in performance (less than 4%) between the process types. It is the rib configuration (depth) that establishes its strength. Rib depth determines strength but it is rib width that can dictate wall thickness. A reinforcing rib's width is a compromise of plastic filling, ejection force, and draft angle all wanting a wider rib versus sink mark prevention and a thinner rib. In most cases, some draft is going to be necessary (0.5 to 2° per side), and when coupled with minimum ejection or filling constraints, becomes the rib's minimum width. Useful design guidelines have been developed focused on the aspect ratio of the rib width to wall section (see Table 2.2).

2.11.3.3 Sink Marks

The aspect ratio of rib width to wall section is a useful tool in controlling sink marks by limiting the mass at the intersection point. Areas of high mass will shrink differently than thinner areas. The aspect ratio of rib thickness to wall thickness will range from 0.5 to 5.0+, depending on the molding process.

The concept of *center of mass* brings further enlightenment to resin shrinkage and the resulting sink marks (Figure 2.38). Consider a centerline running through the midsection plane of the exterior skin or nominal wall section. Any molding orientation is evenly split between the two halves, and shrinkage will be uniform. When a rib intersects the nominal wall section, there is a change in the center of mass and therefore a shrinkage concentration. The result is sink marks. In the EGM process

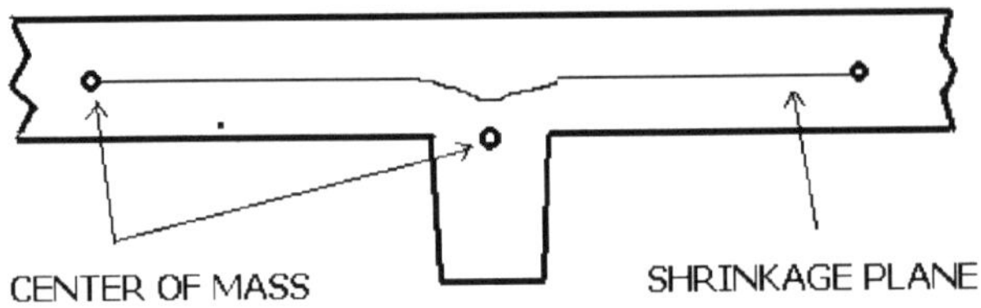

FIGURE 2.38 Center of mass. Mass = shrinkage = sinks.

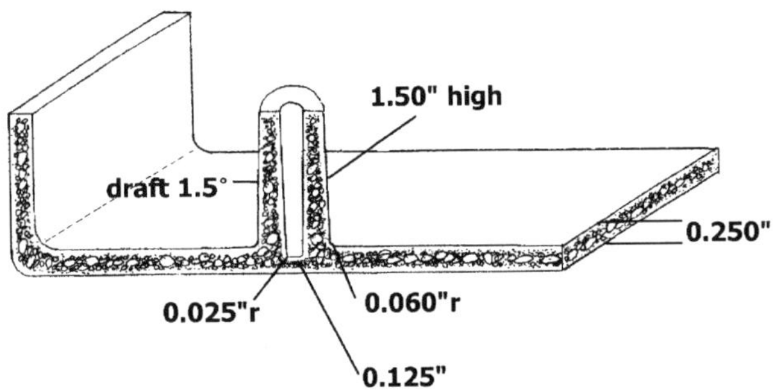

FIGURE 2.39 Low-pressure structural foam boss design.

the center of mass is moved much closer to the center plane of the wall section. Often, plastic at the base of the rib is displaced into the shrinkage vacuum. Depending on the part's geometry, cycle-time savings of 15% have been experienced by the EGM process.

2.11.4 Design Sequence

The following is the new process selection/part design sequence of events:

1. Conduct a value analysis.
2. Determine the functional requirements.
3. Design the features into the ribs and bosses.
4. Determine the wall thickness, molding process, and resin.
5. Finalize the design and details.

The procedure is best for new designs or total product redesigns where a "clean sheet" or "blank screen" is unrestrictive. This *design for function format* follows a logical procedure to collect and analyze data, which leads to products that are highly functional and designed for value. By analyzing the needs of the application, the functional details are determined. Ribs and bosses form a functional platform, which is then attached to the wall section best suited. Using aspect ratios and flow length data, the wall thickness and resin are determined. The final design details such as cosmetics, tolerances, and specifications complete the design task. This sequence differs from the industry norms but is well suited for innovative designs.

2.11.5 Bosses and Inserts

Injection molding has strict rules for boss design to avoid sink marks and other problems related to flow and shrink. Rib thickness is limited, which means that rib height is limited. This can result in a weak boss after assembly. One of the best solutions is to add a gusset or rib to the boss to maintain strength. Boss design is more flexible with structural foam. The thicker wall section gives a bigger starting point. The allowable aspect ratios are also greater. One other consideration is that hoop stress normally associated with hollow bosses is minimized by the layer of foam, which isolates the inner skin from the outer skin. Even reduced profile boss walls will have at least a microscopic layer of foam, due to volumetric shrinkage of the resin. Co-injection structural foam follows the guidelines of structural foam bosses. For thin-wall gas assist, the rules for boss design follow those of conventional injection molding. Free-standing bosses will cause problems with filling, venting, and ejection, just like injection molding. Free-standing bosses attached to a rib or at a gusset to avoid sleeve ejectors in the mold are another solution but are more expensive and require regular maintenance to avoid being clogged. If an oversized boss is necessary, it should be connected to a gas passage rib so that volumetric shrinkage can be gassed. Thick-wall gas assist will follow the guidelines of structural foam except that the core pin should not extend into the nominal wall under the boss. The concern is excessive fingering, which could result in a weak boss.

A primary application of a boss is to accept metal hardware as part of the assembly process. All of the low-pressure injection molding processes excel in providing functional details to minimize assembly steps. The thread-forming screw is a good example of this ability (Figure 2.40). Structural foam has the ability to produce exceptionally strong bosses. The hoop stress normally associated with metal hardware in plastic bosses is absorbed in the foam core. Gas assist has the ability to accept larger threaded inserts due to its ability to have oversized bosses. The sizing of the holes is crucial. For greater performance with the mating hardware, threaded inserts can be installed.

2.11.6 Louvers and Weld Lines

When the flow of plastic is separated during filling by an opening or a design feature such as a louver, a weld line results (Figure 2.41). The weld line will have less strength, due to a lack of homologation of the molecular structure. Flow and shrink become considerations here also. Louvers cause restrictions to flow, resulting in less pressure exactly where it is needed. As volumetric shrinkage occurs, the resin in the weld line area will try to pull apart. With injection molding, there is enough pressure on the leading edge to force some strength back into the joint. Properly designed gas assist parts have the benefit of shorter flow lengths and uniform shrinkage being

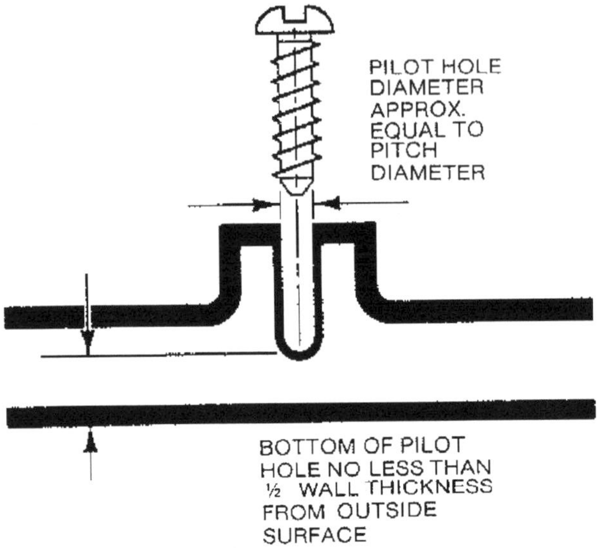

FIGURE 2.40 Boss design for threaded hardware.

FIGURE 2.41 Low-pressure structural foam louvers and weld line.

cushioned by the injected gas. Co-injection structural foam has the problem of the leading edges from the flow fronts consisting of solid resin. This is favorable in terms of weld line strength, but a substantial area will be devoid of foam and subject to sink marks. A common practice for co-injection structural foam molders is having a material overflow to offload the problem.

2.12 BASIC TROUBLESHOOTING

Structural foam and gas-assist molding technologies are essentially injection molding. Most of the regular injection molding problem solving techniques still apply. The main difference is that instead of a plastic packing cushion, gas pressure (from either direct nitrogen injection or foaming) does the packing. Short-shot molding methods will have filling problems that differ from high-pressure injection of a full shot. In this section we describe a few of these unique differences.

Structural foam is a short-shot process, and the mold filling depends on the internal gas pressure from foaming. When there is a lack of blowing agent pressure or an increase in material viscosity, the mold does not fill completely. This is referred to as a *short-shot reject*. Venting is an important consideration for structural foam. Inadequate venting can restrict filling as well. If too little blowing agent causes short-shot rejects, too much blowing agent can cause another problem, called *postblow*. This is when the strength of the molded part's skins is too weak to prevent foaming after the part is removed from the mold. Typically, postblow indicates incomplete cooling and appears as a bulge in the surface of the part, often around thick sections. Extending the cooling cycle strengthens the skins and also reduces the pressure of the blowing agent within the molded part. Residual blowing agent trapped within the molded part can cause further problems and require some outgassing. Painting a structural foam part before proper outgassing can lead to outgassing blemishes and blisters. Elevated post molding heat treatment accelerates the outgassing process.

Gas-assisted injection molding also has its own set of problems. With a short shot of plastic followed by a volume of gas, there are issues with gas penetration. If the gas penetrates too far into the plastic flow front, there can be *break through*, an open gas channel extending beyond the advancing plastic. The result is a rejected part. A less dramatic example of overpenetration is a condition called *fingering*, where the gas penetrates the thin areas that were designed to be solid. The other extreme is inadequate gas penetration, leaving solid, overly thick areas that were designed to be cored with gas. All of these conditions are controlled by the gas injection profile. The design of the molded part has a direct effect on the gas assist process and should be included in the gas plan. There is one more unique aspect of the process, a *venting ben*, which is a rough finish on the inside of the gas passage caused by extreme gas pressure variations between the injection and the venting stages. This is more commonly seen in hollow parts or high-aspect-ratio projections, where there is a high percentage of gas core-out (30% plus). The rough finish takes the form of pitting or small volcanoes. It results from nitrogen being forced into the polymer during the high-pressure injection phase and then being released during the venting stage.

2.13 WATER-ASSISTED INJECTION MOLDING

Reduced molding cycle time is the principal advantage of using water instead of nitrogen gas for the assist. The notion of using a liquid medium is not unexpected considering phase change due to the temperatures and pressures going on inside the mold during the process. Indeed, many of the gas-assisted injection molding patents refer to the pressure medium as a fluid rather than a gas. Water-assisted injection molding can be applied to a wide range of resins, including unfilled polypropylene and glass-filled polymide. There are obvious problems associated with the use of water, such as safety due to electrocution, steam, hot water, and slipping. One water-related issue is the evacuation of water from the molded article prior to being handled by human hands. Some adaptations of water assist include the use of compressed air or nitrogen gas at various stages of the process. The compressed gas is injected after water evacuation or as a means of water evacuation. Several variations of technology have been demonstrated with very promising results, including the method of initial assist and coring with nitrogen gas followed by water and then compressed air or nitrogen. It is also helpful to pierce the molded article as it cools to allow a continuous flow of water or supercooled nitrogen.

In one of the techniques, substantial core-out can be achieved by injecting a short shot of resin followed by the injection of water. Since the water flows into a dead end, a unique water evacuation method is used. Late in the cooling cycle water pressure is reduced, and compressed air is injected through the water gate. Since the water gate is at the gate end of the molding, pressure builds up at the opposite end. When the air pressure is released, the remaining water is blown out during the venting stage. The resulting interior passage is very smooth and much more uniform in thickness than if nitrogen had been used for the initial core-out. This can be explained by the difference in compressibility between nitrogen and water. Water-assisted methods are undergoing further development as of this writing. The equipment for the process requires the ability to control precisely the volume and pressure of high-temperature water as well as gas or air. Water recirculation has to be considered. A major limitation is the tendency for the shock cooling effect to freeze a web of plastic adjacent to the inner walls. Molded articles with both thick and thin sections can have premature plastic freeze-off, and a pressure drop results. Product design should involve the molding process engineer. The process works best in tubular products with external ribs and detail.

CHAPTER 3

Sheet Extrusion

DANA R. HANSON

Processing Technologies, Inc., Aurora, Illinois

3.1 INTRODUCTION

The sheet extrusion markets have seen many changes over the last decade. Sheet processors have been challenged to keep pace with the changes and refinements in resins as well as processing technologies. Since sheet extrusion has rapidly evolved into a global market, today's sheet processors must approach the marketplace from this perspective. To stay competitive, all costs must be accounted for and controlled strictly in day-to-day operations. Cost savings include those that may be derived from processing efficiencies, resin and other raw material costs, labor rates, and related overhead costs. Processing efficiencies are considered paramount to control costs associated with production effectiveness. Examples of these include (1) minimizing scrap rates and/or off-spec sheet goods, (2) producing high rates of production on a per line basis to keep the impact of production labor burden rates minimized, and (3) maximizing the production system's *uptime* (i.e., time available to run production versus *downtime* associated with process-related and/or equipment-related issues).

The focus of this chapter is to provide a thorough explanation of sheet extrusion as it relates to process techniques, advances in equipment, and commonly used resins and structures. Sheet extrusion is broken down into several basic elements. The essential main equipment components, the "tools," are each examined to review their relative role in the overall scheme of producing high-quality sheet. A detailed explanation is given as to the benefits of using various types of equipment for specific sheet extrusion applications. The sheet extrusion novice is provided with a great deal of background on many key topics relating to extruding sheet, and there are also many beneficial elements for intermediate to expert processors. The chapter contains many advanced explanations regarding techniques for producing multilayer

Handbook of Plastic Processes Edited by Charles A. Harper

structures, provides a comprehensive review of individual resin-to-resin compatibilities, and cites many examples of end uses and methods commonly practiced by some of the industry's leading processors. A detailed explanation is also given of how heat transfer analysis results are used to size various components within the sheet extrusion system. An expansive rate estimation matrix provides an accurate means of estimating the sizes of extruders that are required within the extrusion complex for a variety of mono- and multilayer structures.

The chapter contains a great deal of technical data, descriptions of various processing methods and techniques, sheet extrusion equipment configurations, and so on. The reader is cautioned to use this material as general information only and is further encouraged to research the information as it relates to a particular topic of interest such that it may be determined useful or practical for use. Actual results that are described herein may vary. The author reserves the right to make changes, revisions, and/or updates to the contents herein without prior notice.

3.2 CRITICAL COMPONENTS

Every sheet line is not created equal, just as every sheet application is not created equal. It is critical to assemble the appropriate components (the "tools") for the desired sheet process. Without the correct equipment, the end results may not meet the objectives of the desired process. Often, processors attempt to produce sheet types on equipment not originally configured for that particular process. There are some sheet types that possess a relatively wide window of processing flexibilities. Given a properly configured sheet production system that is equipped with a reasonable degree of sophistication will enable the user to process sheet in a tighter window of control. At the same time, this configuration will also be able to run less critical sheet processes.

In this section we examine many critical components of a sheet production system along with specific measures that can be taken to avoid pitfalls of applying a system configuration incorrectly. Since feed stocks and formulations vary from application to application, material handling and feeding and blending equipment are not covered in this review. It is important, however, to match the proper material handling equipment to the material types being processed. Proper care must be practiced when selecting the correct material handling system. Briefly, consider the following questions when determining the best solution for a given application.

- What types of materials need to be transported (pellets, granules, powders, regrinds, and/or liquids)?
- Will edge trim be fed back into the system and what challenges can this present?
- Are any of the ingredients hydroscopic (moisture absorbing)?
- Are any of the ingredients corrosive in nature or unstable in any fashion?
- What types of storage (i.e., bulk) devices will be used and do they exist or do they need to be acquired? (When adding storage devices, keep in mind the

respective storage volume of each and the logistics of how they will be refilled such that production does not get interrupted.)

- What is the relative distance from the storage source to the feed and blending unit that will provide materials to the extruder(s)?
- What are the rates of consumption for each material type? (These should conservatively be increased by no less than 20% as a buffer to assure proper feed rate capacities are achieved.)
- How many ingredients does the process require for each extruder? (This will establish the number of component feeders for blending into each extruder.)
- What level of precision is required with the desired process? (This will help determine the type of system that will be required, i.e., gravimetric- or volumetric-based systems.)
- How many other lines will be installed or exist in the general vicinity, and will they share features or function from common elements of the material handling system?
- What type of control system (i.e., distributed among several lines, dedicated to a single line and/or integrated into the control scheme for the balance of the sheet system) will be required?

Keep in mind that the environmental conditions that exist for a particular geographic location play a critical role in determining the need for specific features of a material handling system. More specifically, if it is a characteristically humid climate, special measures may be required for predrying materials that ordinarily do not require drying in other geographic regions. Discuss these issues with your material handling vendor to assure that proper equipment configurations are achieved. The aforementioned questions will help narrow the field with respect to formulating the proper specifications for the material handling and feed and blending equipment.

Now that the material handling and feed and blending equipment specifications have been addressed, the core equipment components need to be reviewed. Figure 3.1 highlights the major components of the sheet production system. In the following sections we examine critical elements of each of these components as well as other components and processes. The contents serve as a fundamental guideline for structuring some basic specifications for determining which elements are required and why.

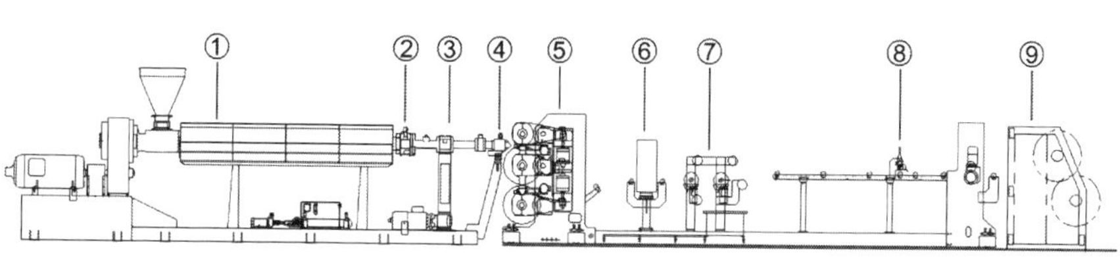

Figure 3.1 Major components of a sheet line: 1. the extrusion complex, 2. polymer filtration, 3. valve, gear pump, and static mixer, 4. die and feed block, 5. roll stand, 6. gauge scanner, 7. antistatic coating, 8. slitting station, 9. winding systems. (Courtesy of PTi, Aurora, IL.)

3.2.1 Extrusion Complex

The extrusion complex is fundamentally one of the more important elements of the sheet production system. It will determine the overall rate of the line (provided that the downstream components have been properly sized such that they collectively do not represent a bottleneck to the production rate). In addition to the overall output of the sheet system, the extrusion complex will melt and mix the molten polymer to assure that a uniform and stable melt flow is achieved. This is critical to assure that stable laminar flow is achievable throughout the balance of the extrusion process up and through the sheet die.

Monolayer structures typically feature extrusion complexs that consist of a single extruder. However, it is possible to have a dual-extruder process for a monolayer structure, as would be the case in a tandem extrusion process. The tandem process features a primary extruder for mixing, melting, and pumping and a secondary extruder for melt conditioning prior to discharge. Special applications of ultrahigh-capacity lines and/or materials that are melt and mixing sensitive may facilitate a tandem configuration. For the purposes of this review, focus will be maintained on single-screw extruders configured individually or as multiples for co-extrusion (multilayer) structures. The following discussion highlights the basic design elements for the extruder(s) within an extrusion complex for a typical sheet line.

3.2.1.1 Screw Speed Range and Torque Required

The speed range and required torque relate to the gearbox (transmission) ratio and motor size (hp or kW). A minimum service factor of 1.6 is a good practice to assure that suitable component life is maintained. The gearbox will have an integral thrust bearing which will be rated in hours of production at a given head pressure. This rating is typically reported on the basis of the B-10 or L-10 statistical failure scales at 5000 lb/in^2 gauge (340 bar) head pressure. Each gearbox has both a mechanical and a thermal rating. The mechanical rating is a measure of the gearing strength using well-known and accepted standards as with AGMA (American Gear Manufacturers Association) ratings. The thermal rating is a measure of the gearbox's ability to convectively dissipate heat resulting from the lubrication oil being sheared between the teeth of the gearing. This shear action is characteristic of an operating functional gearbox. The thermal rating value reflects the equivalent size of the motor that would place the gearbox in a state of thermal equilibrium operating at full speed while drawing the load equal to the thermal rating value. The equilibrium level is the temperature value that the gearbox will rise to and not ascend above. The thermal rating value assumes that no external cooling is applied. Most gearbox applications have a low characteristic thermal rating and will require either an internal cooling coil or external lube oil cooling system to increase the thermal rating to the level at or above the mechanical rating of the gearbox. This will prevent the gearbox from increasing in temperature to an unacceptable level. Most gear cases will operate at temperatures of 140°F (60°C) or lower in order to maintain suitable performance of petroleum-based oil lubricants. Synthetic oils permit higher operating temperatures while maintaining acceptable levels of lubrication performance.

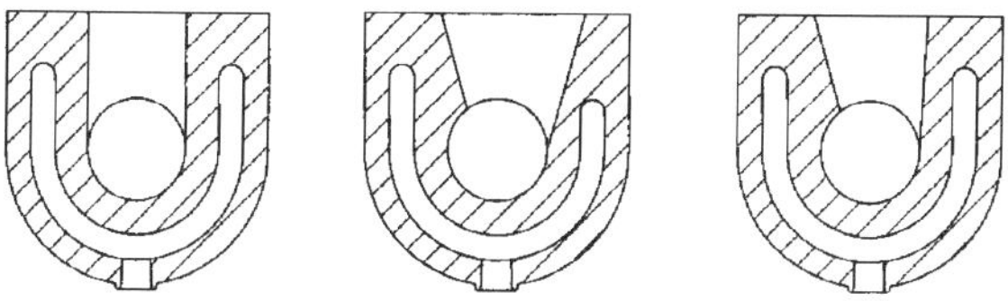

FIGURE 3.2 Three styles of feed sections: (*a*) center fed; (*b*) full tangential; (*c*) hybrid tangential. (Courtesy of PTi, Aurora, IL.)

3.2.1.2 Feed Section Configuration

The feed section configuration should be matched to the types of materials that the extruder will be processing. Several geometries are available, as depicted in Figure 3.2. The first is a center-fed configuration and features a circular, square, or rectangular opening (Figure 3.2*a*). These are typically used with pellets and granules as the major feed component. The second configuration is full tangential-fed geometry (Figure 3.2*b*). This feed section typically features a square or rectangular opening. These are typically used for applications processing mostly rigid regrinds as the major feed component. The third configuration is the hybrid tangential (Figure 3.2*c*). This feed section features a square or rectangular opening.

The hybrid tangential feed section that PTi uses is best and most practical when feeding combinations of pellets and regrinds. It is relatively universal in its ability to feed well and without surging while permitting self-cleaning action during product changeovers. The center-fed configuration was popular in the 1970s and 1980s and was essentially a by-product of the injection molding and/or blow molding extruder arrangements. It soon became out of fashion, due to the high demand of variability and consumption of various reground materials. The full tangential geometry was originally developed for use in the reclaim industries and is good for feeding up to 100% regrinds. But it can generate surging as a detrimental process characteristic and is difficult to clean during product changeovers. The full tangent is also overly aggressive when running feed stocks consisting primarily of pellets.

3.2.1.3 Barrel Length-to-Diameter (L/D) Ratio

The *L/D* ratio must be suitably long to permit the melt stabilization necessary to deliver properly conditioned extrudate downstream to the sheet die. Extrusion machinery has seen an increase in the length of the barrels over the past several decades. This can be attributable to the superior feed screw designs which are necessitated by the additive mixing that is required. This longer machine characteristic is also attributed to advances in polymer science that have evolved high-performance polymers. PTi's standard extrusion system features an *L/D* ratio of 32 : 1. Longer machines up to 36, 40, 44, and even 48 : 1 *L/D* are not uncommon.

The barrel length requirement is a derivative of the process that the machine will undertake. The process determines the need for feeding, melting, mixing, degassing, venting, and even additive side feeding, all of which affects the overall length requirements of the extruder. Co-extruders for multilayer applications have a basic length requirement since it is common for these machines to have a dedicated and simple formulation that does not require special considerations. Table 3.1 highlights typical machine barrel configurations for primary and co-extruders.

3.2.1.4 Barrel Cooling Type (Air or Water)

Barrel cooling is typically available in two basic forms: air cooling and water cooling. Both are heated electrically. Zone control is achieved using well-established proportional–integral–differential (PID) logic to hold and control each zone to the desired temperature. Many commercially available control manufacturers offer proven versions of this technology—in both individual PID instruments, cluster instruments for ganged zone control, and PLC (programmable logic controller)-based PID control.

The type of barrel cooling depends greatly on the formulation of the feed stock, screw design, head pressure, and the degree of mixing that is required by the process. PTi offers both styles, although the majority of its sheet lines are equipped with water-cooled barrel configurations. This is due to the variability of the resins that most sheet lines are required to produce. If the sheet line is dedicated to a specific process, the feed screw can be optimized for a given resin, and subsequently an air-cooled barrel configuration can be implemented.

3.2.1.5 Venting Requirements

Extruder barrels are often vented in order to degas and devolatize a polymer. Vacuum systems are added to enhance and aid in the removal of the gas, moisture, and

TABLE 3.1 Typical Machine Barrel Configurations

Resin[a]	Primary Extruder Barrel *L/D*	Co-extruder Barrel *L/D*
PP	32	30, 32
APET	32, 36	30, 32
RPET	32, 36	30, 32
ABS	36, 40	36, 40
PS	32, 36	30, 32
HIPS	32, 36	30, 32
EVOH	N/A	30, 32
Admer	N/A	30, 32
PMMA	32	30, 32
PE	32	30, 32
PC	32	30, 32

Source: PTi, Aurora, IL.

[a]ABS, acrylonitrile–butadine–styrene; APET, Amorphous polyethylene terephthalate; EVOH, ethylene vinyl alcohol; HIPS, high-impact polystyrene; PC, polycarbonate; PE, polyethylene; PMMA, polymethyl methacylate (acrylic); PP, polypropylene; PS, polysulfone; RPET, recycled polyethylene terephthalate.

vapors. Vacuum systems also increase the efficiency of the vent performance. Multiple vents can be used to assure that the polymer is properly conditioned and void of these gases, particularly if the polymer characteristically exhibits excessive amounts of gas, moisture, and vapors. Table 3.2 lists polymer and corresponding venting requirements.

Venting requirements may change if the feed stock consists of a high degree of regrind. The source of the regrind also plays an important role in determining the venting requirements. Regrinds generated as a result of edge trim scrap usually do not create a cause for venting. However, if the materials are not kept clean and dry, they will normally require venting. For applications involving thermoforming where high amounts of regrind are run back through the primary extruder, venting is normally required to degas and remove air pockets that can be tramped in the polymer melt.

Vacuum systems are commonly used to enhance the venting performance. Many styles of vacuum systems are available. The most common system is a liquid ring vacuum pump. This pump style is a cost-effective and reliable vacuum pump solution but requires either a water recirculation system for the vacuum water effluent or will have a constant discharge of water that will go to the wastewater stream and often, to drain water. Caution must be practiced when dealing with vacuum water effluent since it can contain chemicals not suitable for drain water systems. Another vacuum system type is the dry pump. This pump features a scrubbing cycle to remove contaminants from the barrel vent without using water. This pump type is relatively expensive due to the supporting scrubber system and can also be more maintenance intensive due to the propensity for contamination buildup on the internals of the system. This is dependent on the nature of the contaminants being drawn off the barrel through the vent. Many inks, monomers, and vapor by-products contain vapor solids that literally plate out on the internals of everything that they encounter. There are

TABLE 3.2 Polymer Venting Requirements

Resin[a]	Venting
PP	Not required
APET	Not required
RPET	Often required
ABS	Required, often double vented
PS	Required for general purpose, not for crystal
HIPS	Required
EVOH	Not required
Admer	Not required
PMMA	Not required
PE	Not required
PC	Not required

Source: PTi, Aurora, IL.

[a]Abbreviations are defined in Table 3.1.

several commercially available suppliers of these different vacuum systems and both types have been used prolifically for many years. PTi typically implements a water ring style since it is a cost-effective solution that delivers good performance. In addition to this vacuum system, PTi offers a recirculation system to minimize the vacuum water effluent discharge. The system is configured with a surge pot and a proprietary-designed vent stack and lid arrangement that prevents molten polymer from entering into and destroying expensive component parts. However, dry vacuum systems are becoming more and more prevalent in dryerless PET processing.

3.2.1.6 Centerline Heights

PTi engineers its sheet lines for functionality. This includes the centerline requirements of the extrusion complex. Today's sheet lines differ vastly from those of previous decades. The outputs are much higher and the co-extrusion structures are more and more common. Not every installation is the same. User's plant space allocation can differ from installation to installation. Sheet line manufacturers must be flexible with their product configurations to accommodate the end user's needs. PTi provides for this flexibility. It is common for an extrusion complex consisting of multiple extruders to have multiple centerlines. The arrangement of the extruders will dictate the requirement for the approach of the extruder. Each extruder must allow for feed screw removal. This must be accounted for in the original design of the layout of the sheet line. Often, the roll stand has limited space that it can retract, and therefore the extruders upstream must permit means of removing the feed screw other than in-line with the machine direction of the roll stand. PTi commonly applies different centerlines coupled with various approach angles to permit minimum disassembly using the space that is available. Table 3.3. highlights the primary nip centerline of the roll stands versus roll diameters for down-stack configurations.

PTi provides a minimum distance to the floor of 12 in. when the bottom roll is in the opening position. SPI/ANSI B151.7 Guidelines require roll gap openings for new machinery to be not less than 4 in. for safe operation [1]. Thus, for down-stack arrangements the basic formula becomes the sum of middle and bottom roll diameters plus 4 and 12 in. The individual extruder centerlines can vary up or down from the primary nip centerline due to the feed block geometry and each machine's

TABLE 3.3 Primary Nip Centerline for Down-Stack Configurations

Roll Diameter (in.)	Primary Centerline (in.)
16.0	44.0
18.0	52.0
24.0	64.0
30.0	76.0
36.0	88.0

Source: PTi, Aurora, IL.

approach via the feed pipes such that the feed screws may be removed along the axial direction of the barrel.

3.2.1.7 Expansion Allowance Provisions

The overall length of the extruder barrel assembly through the feed pipes can expand thermally by as much as 1 in. when going from ambient to operating temperatures. Combining this with other co-extruders coupled together via the feed block to form an extrusion complex yields a definite design challenge. Not every extruder will expand the same since the overall effective lengths vary for different machine sizes. PTi employs several methods of expansion techniques in its designs to facilitate a balanced and unencumbered path for thermal expansion. It is very important to permit each extruder the ability to expand unencumbered. Otherwise, extreme stresses can manifest in the flange connections and can lead to leaks and even component failure.

One method that PTi employs is mounting the extruder on casters and tracks to permit rearward expansion. With each extruder mounted in this fashion, the expansion is away from the connection points at the feed block. Not always is it as simple as this. With feed pipes that run at complex angles, the expansion can be a vector skewed to the tracks versus a straight path along the tracks. PTi addresses this issue with the side-to-side movement permitted on the caster axles. When the machines are installed on the tracks, the machine is originally at ambient temperature. By inserting a shim in the expansion side of the machine the caster position is established such that when the shims are removed, the machine will permit out-of-plane expansion. Another technique is to mount the feed block and die on gib and weighs. This allows the primary extruder, which is usually in-line with the feed block and die to expand through its barrel supports and on down through the gib and weighs. Combining this technique with the simple track method provides the necessary combinations required to permit unencumbered thermal expansion. Keep in mind that some sheet line configurations require the sheet die to be fixed and the extrusion complex to expand away from the sheet die. This is also easily achieved using the aforementioned method of casters and tracks.

3.2.1.8 Feed Screw Performance

This is one of the most important aspects of the extrusion complex. Without proper screw designs the end result is rarely achieved. The topic of feed screw designs has long been one of proprietary technology and even patented technology. Manufacturers of extrusion machinery rarely will share their trade secrets for proprietary feed screw designs.

Feed screw construction must be matched to the application and barrel linings. Base metals used to make screws can vary from tool steel through stainless steel. Most feed screws are made from a base material of 4140HT tool steel. Flight lands will feature inlays of Colmonoy No. 56 or 83 as well as Stellite materials [2]. When using premium high-carbide-lined barrels (Xaloy No. 800 or equivalent), Colmonoy No. 83 is recommended. Most feed screws manufactured in the United States today are equipped with Colmonoy No. 56 flight land inlays and are applied to normal carbide-lined barrels (Xaloy No. 102 or equivalent) [3].

Feed screw designs have evolved from the original standard and barrier designs from the 1970s to the advanced and barrier designs of the 2000s. Variations of these designs include a wave- or lobe-style screw, where the melt is conveyed down the screw length and is transferred back and forth in an exchange fashion. Variations of the barrier also include hybrid barriers, which differ from a true barrier in that the barrier section is compressed over a tighter region on the feed screw. PTi typically employs a hybrid barrier when applying a general-purpose feed screw to multiple resins, including polystyrene, polypropylene, and polyethylene terephthalate (PET) processing on one design. Additionally, PTi typically cores the feed screws through the feed zone for optional screw cooling or heating, depending on the application.

Mixing sections were developed to provide intense mixing performance at a specific location on the feed screw while limiting the amount of shear that is introduced into the melt stream. There are several popular designs, including mixing pins, blister rings, fluted channels, twisted fluted channels, and many others. Often, these mixer sections are introduced as removal screw tips such that they may be altered and/or changed without requiring a re-cut on the feed screw itself. Whether a mixing section is required for your application is really a function of the process objectives. Check with your machinery or feed screw supplier for the proper choice of feed screw configurations and mixing section selections.

The performance measure of a good feed screw design includes one that (1) provides a consistent melt quality without surging, (2) delivers the desired rate of performance at a controlled and reasonable melt temperature, (3) offers good mixing while permitting ease of product changeovers, and (4) does not experience rapid wear in any specific regions along the screw length. However, it is beyond the intended scope of these discussions to present details of specific screw designs and mixer styles. There are many other publications that go into greater detail regarding this subject matter.

3.2.2 Polymer Filtration Devices

Positioned immediately downstream of the extruder barrel at the discharged end of the machine, the polymer filter device plays an integral role in assuring that the melt stream is suitably clean and void of contaminants. Melt stream contaminants consist of gels (unmelted and/or congealed polymer formations), paper, ferrous and nonferrous metals, glass, and degraded and/or carbonized polymer. To assure that the melt stream is suitably void of these contaminants, a properly configured polymer filter must be used. There are several possibilities to consider. The following discussion summarizes the most common styles of polymer filtration devices, describes the relative functions, and highlights their practical use for specific applications.

3.2.2.1 Barrel Clamp Method

The most fundamental method for filtering polymer is a split-style barrel clamp and a recessed breaker plate that supports a screen pack (Figure 3.3). This method is a crude but cost-effective solution. It serves a basic purpose of filtration and works reasonably well for small-bore co-extruders ($1\frac{1}{2}$ through $2\frac{1}{2}$ in. in diameter). However, there are

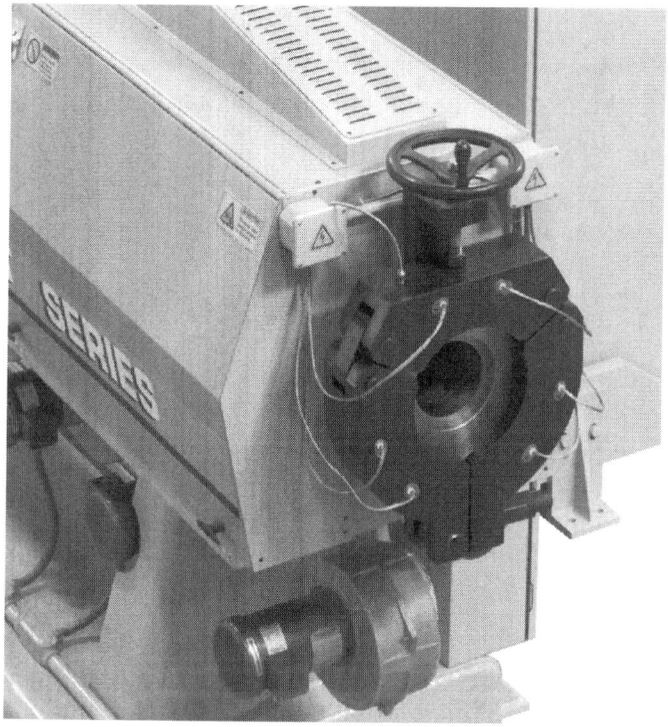

FIGURE 3.3 Barrel clamp method. (Courtesy of PTi, Aurora, IL.)

sheet applications using much larger extruders (6 through 12 in. in diameter) that use this method, due to the coarse screen pack (or no screen pack) requirement. It is best when used with virgin pellet-form feed stocks. Changing the screen packs is a cumbersome task while the line is not in operation. Keep in mind that once a clean screen pack is installed, it builds up contaminants progressively until the next time it is changed. As it builds up contaminants, the differential pressure across the screen pack is steadily increasing. This means that the extruder must work harder to maintain a consistent rate of performance. At the same time, the back pressure (head pressure) on the barrel is increasing, and so is the amount of shear being introduced into the polymer. Typically, this method is not used for applications requiring fine filtration, due to the required method of changing the screen packs. It is best when used on an application that does not require frequent screen pack changes (i.e., less than once every other day or so).

3.2.2.2 Manual Screen Changer

Another cost-effective solution is a manual-style screen changer. Figure 3.4 shows a typical manual screen changer. These are commonly used on applications similar to the barrel clamp method that do not require frequent screen changes. Typically, these are used on smaller machines, ranging from 1- through $4\frac{1}{2}$-in. bores. The line must be shut down and void of head pressure (or minimal head pressure present) to permit a

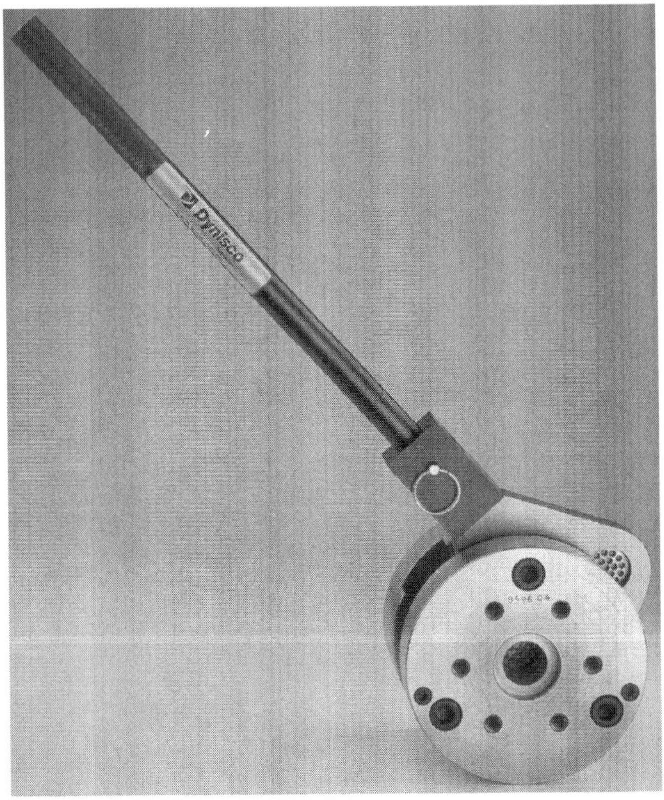

FIGURE 3.4 Manual screen changes. (Courtesy of Dynisco, Hickory, NC.)

screen change. The screen is changed by stroking the lever arm manually from position to position. The seal is pressure activated, so when in operation the head pressure maintains the seal pressure. This is why the unit cannot be changed while pressure is present due to the drag force against the seal surface. These are commonly applied to applications of co-extruders processing dedicated resins which are characteristically clean. They are used on applications requiring infrequent changes since the machinery must be shut down to facilitate a screen change. Some units permit feed screw removal through the body of the screen changer.

3.2.2.3 Ribbon-Style Screen Changer

A rather unique polymer filter is the ribbon-style screen changer. It uses a coil of screen medium of a desired mesh arrangement and advances the screen ribbon continually across the melt stream face of the filtered area. The screen advances using pressure differential and seal temperature as the driving forces. The changer was developed for film and sheet applications of long, continuous runs. The polymer usually consists of extremely clean virgin resin materials. Figure 3.5 shows an example

FIGURE 3.5 Ribbon-style screen changer. (Courtesy of HITECH, Hackensack, NJ.)

of a ribbon-style screen changer. The market application for this device is not as widespread as some of the other styles that are being presented herein, but there remains a market for these nevertheless. Increasing amounts of regrinds and recycled materials are finding their way into sheet applications. This presents a challenge for this style of filtering device. The primary reason is the limited velocity with which the screen can be advanced. It is best for relatively clean materials. If the process encounters a slug of heavily contaminated material, it can virtually blind off the screen area faster than the screen can advance across the melt stream, thus creating a high-pressure condition. And if not corrected quickly, machine shutdown will occur due to high head pressure. This type of polymer filter is in the median to upper end of the price range for the units being examined. Partial disassembly of this unit must be performed to remove the feed screw.

3.2.2.4 Hydraulic Slide Plate Screen Changer

This is by far the most popular style of polymer filter used today. It is cost-effective and offers a fundamentally practical solution to polymer filtration. Figure 3.6 shows a typical hydraulic slide plate screen changer arrangement. It is commonly used in nearly every industrial segment of the plastics industry. This unit features a pressure-activated seal and is leak-free. The guards offer interlocks for safe operation, and the screen packs are readily assessable and easy to change. The maintenance is low given the simplicity of the design. In applications involving contaminated materials, the slide plate offers a rapid transfer from position to position. The operator must open the guard and remove the dirty screen and install the clean screen during each change. This should not take more than 1 to 2 minutes to execute. It is also common to oversize the unit for applications involving recycled or contaminated materials as a means of getting more effective screen area. Typically, the feed screw can easily be removed through the body of this type of screen changer.

3.2.2.5 Bolt-Style Screen Changer

This style of polymer filter was introduced in Europe and has migrated into the domestic U.S. markets in a big way. Use of this type of filter is steadily on the rise. The initial price point for the bolt style was much higher than that of a conventional slide plate filter. However, it has become more and more cost-competitive over the recent years, such that the premium for a basic bolt changer is now relatively small. Figure 3.7 shows a bolt-style screen changer.

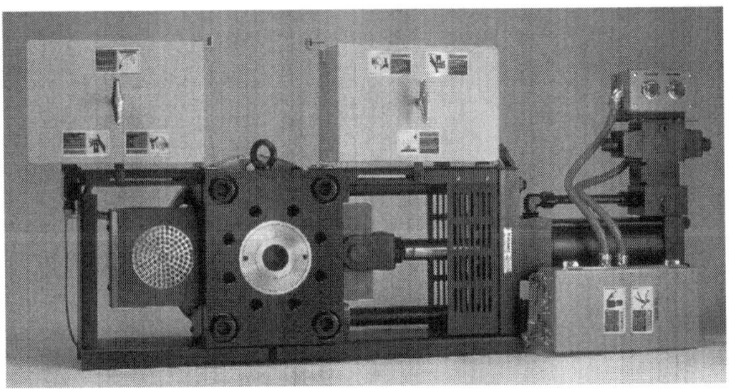

FIGURE 3.6 Hydraulic slide plate screen changer. (Courtesy of Dynisco, Hickory, NC.)

FIGURE 3.7 Bolt-style screen changer. (Courtesy of Kreyenborg Industries, Lawrenceville, GA.)

Bolt-style screen changers come in a variety of configurations. A two-cavity single-bolt unit offers several of the benefits of the bolt changer at a cost-competitive price. It is simple and very functional and provides some of the continuous filtering benefits of the dual-bolt unit for most polymers. A dual-bolt two- and four-cavity filter provides a great deal of filtering flexibility. The dual-bolt two-cavity unit offers continuous filtration since either each of the bolts can be in-line performing filtering functions, and as the screens become dirty, each bolt may be brought off line to have a new screen installed without interrupting the polymer flow and filtering performance. The dual-bolt four-cavity filter offers a large overall filtering area, provides continuous operation, and is also available as a back-flushing unit for extended screen pack life. Each bolt comes equipped with automatic bolt positioning via a PLC controller. This makes the operation relatively easy and avoids pressure interruptions due to incorrect bolt positioning. PTi commonly applies both single- and dual-bolt systems to its sheet lines. Bolt style units are good for polyethylene terephthalate (PET) applications as well. The cost for these filters ranges from low to high, depending on the features, style, and size. A user can get a simple bolt changer for a modest premium over an equivalent slide plate. Going to a dual-bolt automatic back- flushing system can cause the price to rise rather quickly, but the added cost is often justified on screen pack savings alone given some of the recycled materials that are being processed into sheet in today's markets. Feed screws cannot pass through the bodies of this style of screen change, and complete removal is required eventually.

3.2.2.6 Rotary-Style Screen Changer

The rotary changer features a round disk that contains multiple breaker plates (Figure 3.8). The indexing mechanism rotates the disk incrementally as the pressure builds as a result of contaminants collecting on the screen packs. The portion of the disk that is not in-line provides easy access to the operator for changing dirty screen packs. This unit provides a great deal of flexibility and continuous operation. It is more expensive than most screen changers given the mechanical nature of its design. The seal area for this style rarely manifests leaks and is very reliable.

There are specific processing advantages in a continuous operation for a polymer filter of this type. It is becoming increasingly popular with producers of recycled PET sheet. It offers a great-performing fine mesh filter with minimal degradation and intrinsic viscosity drop across the body of the unit. Other styles do not permit direct flow of the polymer through the body and have passageways that the polymer must travel. This can contribute to longer-term processing issues if the unit is not kept clean. The rotary-style systems are pressure controlled and relatively automatic. They offer a high-quality solution at a premium price. Some users prefer this style over any other and will pay the difference in price. The feed screw cannot be removed through the body of this filter. Complete removal of the filter from the extruder must take place in order to remove the feed screw. Various models are available, including larger units for high-throughput applications (up to and over 4000 pph). The screen packs for this style of filter are typically kidney or elliptically shaped, due to the geometric limitations of the disk into which they are recessed.

FIGURE 3.8 Rotary-style screen change. (Courtesy of Gneuss, Mattews, NC.)

Individually, the screen packs do not provide as much surface area as some units of other styles, but it is clear when summing up the total effective area of the screen packs on the disk that the areas are on a par with, if not greater than, other styles.

Table 3.4 summarizes generic attributes and characteristics of the various polymer filters presented herein. (This comparison represents the opinions of the author of this chapter, not an industry consensus.)

3.2.3 Valves, Gear Pumps, and Static Mixers

Positioned immediately downstream of the polymer filter is a series of several possible component scenarios. This all depends on the desired process objectives and

TABLE 3.4 Generic Attributes and Characteristics of Various Polymer Filters[a]

	Clamp	Manual	Ribbon	Slide Plate	Bolt	Rotary
Cost	1	2	3–4	3	3–5	4–5
Screw removal through the unit	Y	Y	N	Y	N	N
Continuous filtration	N	N	Y	N	Y	Y
Automatic control	N/A	N/A	Y	N	Y	Y
Flexibility	1	2	2–3	3	5	5
Commonly used for co-extruders	2–3	4–5	1	4–5	1–2	1
Commonly used for primary extruders	1	1	3–4	4–5	3–4	3–4

Source: PTi, Aurora, IL.

[a]1, lowest; 2, lower; 3, average; 4, higher; 5, highest; N, no; Y, yes; N/A, not applicable.

degree of equipment versatility. A monolayer sheet process will have a different equipment configuration requirement to that of a multilayer co-extruded sheet process, including but not limited to the requirement for a feed block and feed pipes running from the various co-extruders to the appropriate ports on the feed block. Feed blocks and sheet dies are presented in the following section, where we concentrate on what is located between the polymer filter and the feed block and/or sheet die.

3.2.3.1 Valve Adapters

Sheet lines are typically required to process a variety of resins and feed stock forms. When configuring a sheet line it is important to identify a thorough list of desired resins and formulations for each extruder in the process. Feed screws are designed accordingly. Consider the basics of a two-stage feed screw (feed zone, transition, metering, degassing/venting, second-stage feed zone, second-stage transition, and second-stage metering). The geometry of the feed screw is fixed. But the resin formulations can vary radically. The bulk density as measured in pounds per cubic feet is an indication of the unit weight of the premixed and preblended feed stock. When considering all of the desired formulations it is important to examine the range of feed stock bulk densities as a possible indicator of component configuration requirements.

Consider an application involving processing virgin pellets. The bulk density of typical pellets can range from 35 to 45+ lb/ft^3. Now consider the bulk density of this same application with a formulation constituent of 50% regrind flake added to the blend. Regrind bulk density can range from 10 to 30 lb/ft^3, depending on the sheet thickness that is sent to the grinder to become regrind. At 50% on the lower end of the scale this constituent can have a large impact on the extruder's ability to process the material with a performance similar to that of the heavier all-virgin pellet blends. Now consider increasing the amount of regrind to 100%. This magnifies the problem.

An old-fashioned but effective means of counteracting the adverse affects of high concentrations of regrind is a valve adapter. PTi commonly uses this technique with applications involving processing significant percentages of reground materials back into the sheet process. It is also a good practice on applications involving multiple formulations of different polymers. The basic function of the valve is to permit the operator to increase the amount of back pressure that the feed screw is pushing against (i.e., commonly referred to as the *head pressure* of the extruder). This assures that all of the flights in the last metering zone of the feed screw are properly filled. If the metering zone of the feed screw is not filled properly, the machine can experience reduced rates of performance and significant degrees of surging.

The streamlined bore of the valve adapter permits easy product changeovers and is essentially self-cleaning. The valve stem represents an indication of the relative position of the valve. The valve stem is a precision-machined part that features packing seals such that it does not leak polymer out the top of the valve body. It can be dismantled completely for maintenance purposes. The benefits of this device are clear. It provides the operator with the ability to alter the performance of the feed screw when processing material blends that characteristically are radically different than what the original screw was designed to process. A typical feed screw can accommodate variations of bulk densities ranging from ±10 to 20%. Consider an example of processing materials of 100% reground flake derived from 0.010- to 0.020-in. thick sheet. The resulting bulk density could range from 12 to 20 lb/ft^3, well beneath the 20% rule of thumb. Using a valve adapter, the operator could close down the valve stem, thus increasing the amount of head pressure on the extruder and improving the net results and performance of the machine.

The valve adapter is a very flexible device. It can be run fully open such that it is invisible to the process; or it can be fully closed such that it is contributing to increasing the amount of head pressure on the extruder. Keep in mind that extruders are pumps, and when the screw flights of the metering zone are not kept filled, the efficiency of the pump goes way down. A balance must be maintained such that the head pressure does not increase excessively. A good practice is to keep it at the relative level as it is when processing 100% virgin pellets. Since the valve is increasing back pressure, it will contribute slightly to a rise in melt temperature due to the shear that is gained from the increased pressure. However, it will be less than that of a gear pump, static mixer, or screen changer device.

3.2.3.2 Gear Pumps

As discussed previously, the extruder is a pump. It possesses a screw that is turning in the bore of the barrel. The fluid being conveyed is Newtonian and viscous in nature. Compared to other types of pumps, the extruder is not very efficient and often produces pressure variations at the discharge end. These variations can be as high as ±150 to 200 lb/in^2 gauge (10.2 to 13.6 bar). This is also known as the *screw beat*.

Sheet is formed as a result of polymer being pumped through the sheet die and into the nips of rotation chrome rolls. The sheet die establishes the initial gauge or thickness of the sheet and delivers a curtain of polymer uniformly across the die's width into the nip of the sheet takeoff unit. If the feed screw from the extruder is creating a pressure oscillation, the sheet die will experience this same adverse affect since the fluid behaves hydraulically and transmits the pulsations downstream. These pulsations become very critical when attempting to process a highly transparent and/or thin gauge sheet. A common method of isolating this phenomenon is to use a gear pump.

Figure 3.9 shows a typical gear pump. When examining a typical feed screw geometry, the metering zone (also referred to as the pumping zone) of the feed screw has a rather large volume of displacement for every screw rotation. This is a predominant cause of the pressure oscillations. The gear pump features intermeshed counterrotating gears that, when turning, displace comparatively small volumes of polymer for every rotation of the gear pump. This creates a buffer between the upstream extruder and the downstream devices of the feed block and/or sheet die, depending on whether the application calls for a monolayer or multilayer structure. Pumps are sized according to the materials being processed and the output range desired for the pump. Gear pumps are also categorized by the differential pressure ratings. The most common pumps operate at a maximum differential pressure of 3500 lb/in^2 gauge (240 bar). High-pressure pumps are available, particularly for applications involving highly viscous polymers such as those common to HMWHDPE with fractional melt indexes.

Another benefit of using a gear pump is that the pump can be used to establish the discharge pressure of the extruder. Gear pumps feature pressure controllers that use PID algorithms to control the speed of the pump. The gear pump becomes the focal point of control for an extruder/gear pump combination. The control logic slaves the extruder control to the gear pump control such that the pressure set points raise and lower the extruder speeds accordingly, to maintain the desired pressures. By setting the

FIGURE 3.9 Gear pump. (Courtesy of Maag, Charlotte, NC.)

suction pressure of the gear pump, the extruder head pressure can be increased or decreased. This permits an operator to reduce the head pressure the extruder would otherwise create for a given application. This is beneficial in that it reduces the net shear rates the feed screw is putting into the molten polymer, which will net out a lower melt temperature. The gear pump itself introduces shear into the molten polymer such that the process will realize a net gain in melt temperature across the pump device.

The startup and shutdown procedures for gear pumps are critical to the operation and continued operation of a pump. Improper procedures can result in high-pressure spikes, causing the gear shafts to flex, which can lead to galling and eventual seizure of the pump. Also, if the pump is operated outside its intended range, the pump shafts can run dry and also become galled. The gears within the pump are supported by bearings. These bearings are physically lubricated by the molten polymer that is being pumped. The shafts and bearings are sealed with specialty seals that prevent polymer from leaking out of the rotating shafts. Some gear pumps feature water-cooled seals to assure leak-free operation. This is particularly common when processing low-viscosity polymers.

PTi commonly employs gear pumps on the majority of its sheet lines. The processing benefits are essential to the production of high-quality sheet goods with tight gauge tolerances (±2 to 3% of target gauge is common). A properly maintained and operated gear pump can run for years without a failure. PTi commonly uses an integral base support for the motor and gear reducer that is required to drive the gear pump. This helps maintain a clean and compact equipment area for operators to work around. A typical machine configuration is shown in Figure 3.10.

3.2.3.3 Static Mixers

Plastic polymers are Newtonian in nature; that is, the rate of shear and viscosity are directly proportional. The lower the viscosity at a given velocity, the lower the shear rate, and so on. As polymers are pumped through the extruder, the boundary layer between the barrel wall and the screw flight land is sheared, thus creating viscosity differentials across the flow area once discharged out of the extruder. Gear pumps also impart regions of high shear into the melt as a result of the individual gear teeth wiping molten polymer against the inner lining of the gear pump body. These shear regions can manifest as defects into the newly created gear lines. Viscous fluid flow teaches us that for applications involving combined flow lower-viscosity fluids flowing in a pipe tend to migrate to the center, while higher-viscosity fluids collect along the circumference of the pipe wall. This is simply a matter of viscous fluids finding their lowest state of energy as they are conveyed down and through a pipe.

When producing a high-quality sheet, the processor wants to avoid color variations, opacity and/or transparency defects, gear lines, potential sheet die problems related to the separation of high- and low-viscosity regions within the molten polymer, other surface defects, and so on, that are caused by improperly mixed nonhomogenous extrudates. A practical and cost-effective solution to this is the static mixer shown in Figure 3.11.

The static mixer features a barrel body with flanges on each end for mounting purposes, a retaining ring for securing the mixer elements in place within the body, and

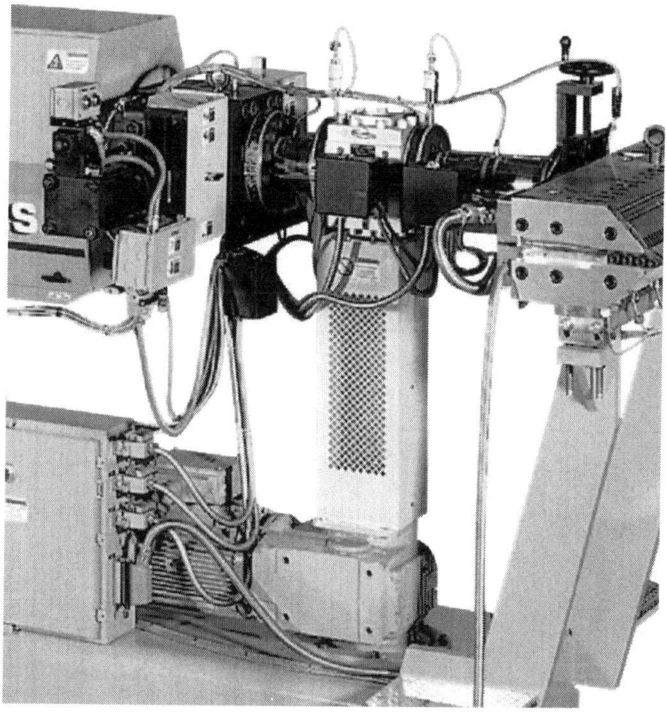

FIGURE 3.10 Gear pump configuration. (Courtesy of PTi, Aurora, IL.)

FIGURE 3.11 Static mixer. (Courtesy of Maag, Charlotte, NC.)

the mixer elements themselves. There are several styles of static mixer. The most common static mixer is the twisted ribbon style. This style of mixer element provides excellent mixing results while maintaining an easy-to-clean and often self-sweeping action during product changeovers. This style also minimizes the pressure drop, which is characteristic across all static mixers. PTi commonly uses the twisted ribbon style in conjunction with gear pumps. The mixer features stainless steel mixer elements that have been precisely twisted and welded to opposing twisted elements. The mixer element assembly is turned on a lathe to assure that a tight and precise fit is achieved in the mixer barrel. This prevents short-circuiting the flow of polymer around the mixer elements and forces flow through the element passages. This style of static mixer yields excellent results when processing a variety of polymers. By increasing the number of elements, the overall length of the mixer is increased and the subsequent number of mixing divisions is increased. Table 3.5 shows the geometric progression of the number of mixing divisions that are achieved versus the number of elements.

Static mixers can be used in nearly all sheet applications. They are not recommended, however, on polymer applications that are heat sensitive and prone to degrade, such as rigid and flexible polyvinyl chloride(PVC). The ribbon-style static mixer is a cost-effective and process-beneficial approach to maintaining a well-mixed and thermally homogeneous melt stream. They are not intended to replace the mixing performance on the extruder feed screw but rather, serve as a mixing device that is located prior to the inlet of the feed block and/or sheet die.

3.2.4 Feed Pipes and Adapter Networks

The layout of the extrusion complex for a multilayer co-extruded sheet line is an extremely important aspect of project planning. It must address plant space utilization efficiencies along with the need for operator and maintenance access. Some co-extrusion systems can consist of four, five, or even six or more extruders that feed into a common feed block(s) and sheet die system. Each extruder must be arranged to permit unencumbered thermal expansion while permitting a feed screw removal plan that includes minimal machinery dismantling.

PTi uses two basic approaches in its layout schemes for co-extrusion systems. The first is the branched or fanned pattern. This approach arranges the extruders in a geometric pattern that literally forms a branched or fanned pattern when viewed from the plan view perspective. The extruders are arranged such that the length of the adapter piping is minimized while suitable access for operation and maintenance is not compromised. All control panel doors and guard access must be functional. NEC Section 110-16 is a safety and electrical code of compliance for

TABLE 3.5 Number of Mixing Divisions per Number of Elements

Number of Elements	Number of Divisions
3	8
4	16
6	64
9	512

Source: Maag, Charlotte, NC.

assuring suitable access to the equipment and the relative space surrounding it [4]. Figure 3.12 displays a typical branched arranged co-extrusion system.

The second arrangement places smaller co-extruders onto a mezzanine structure above one or more of the other larger extruders. This is done as a means of preserving plant floor space, particularly in areas that require multiple lines to be installed adjacent to one another. The mezzanine-positioned extruders are typically air cooled and arranged with ''low boy'' centerlines. The advantage to this approach is the savings in plant floor space. The disadvantage is the additional cost for the structure and the inconvenience of accessing the mezzanine-positioned extruders for operation and maintenance activities.

The design of the feed pipes and adapter networks is somewhat of an art form. Since the extruders must be arranged to meet the objectives of access and plant floor space availability, the feed pipes and adapters are laid out last. When planning the system layout, PTi tries to maintain practical numbered angle values, that is, 10, 15, 20, ... degrees for the angles between adjacent extruders as opposed to fractional or odd-numbered angles. This allows a more sensible and practical installation of the system, along with feed pipes and adapters that feature similar -angled facing on union blocks and mounting fixtures for feed blocks.

PTi uses a very heavy duty approach to its feed pipe and adapter design. This practice yields a system that can hold up over the course of time without failure or manifestation of leaks due to the numerous thermal cycles each system will experience in the course of its life span. Additionally, the heavy wall structures and solid steel block approach to angle blocks provides a thermally stable system of control. Ambient conditions will not affect these systems negatively, due to their high thermal mass. Complex feed pipes and adapter networks are shown in Figure 3.13. PTi's feed pipe designs

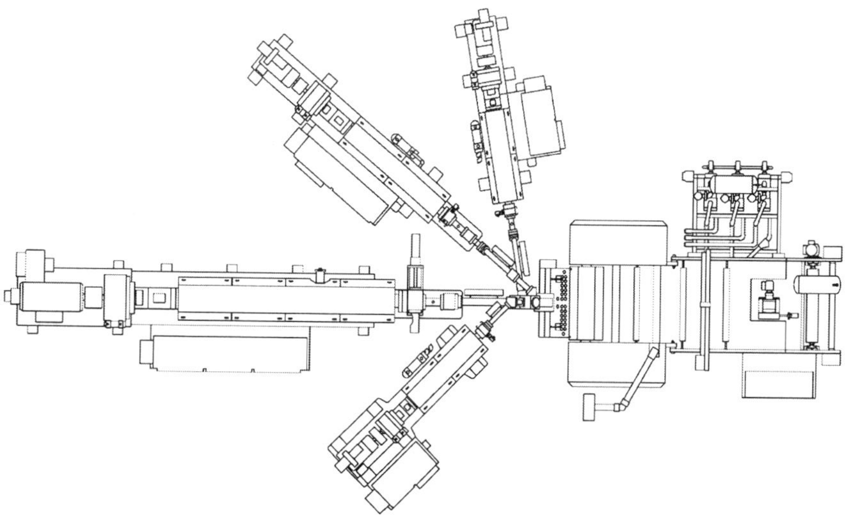

FIGURE 3.12 Typical co-extrusion configuration. (Courtesy of PTi, Aurora, IL.)

feature polished bores with thick walls and flanges at each end. Optionally, the bores are either nickel or chrome plated, depending on the application. The angle blocks are also polished on the flow surfaces and heated with cartridge-style heaters to assure that proper zone-temperature uniformity is achievable. The mating surfaces feature positive pilot and counterbored interfaces, with the angled approaches featuring mating cones to assure a positive seal even in the event of slight misalignment at installation.

Table 3.6 summarizes the extruder bore size, rated capacity, feed pipe bore size, and pressure loss per unit of length of the feed pipe for a comprehensive matrix of extruder and co-extruder sizes. These are calculated using the classical Bernoulli's formulation and drag coefficients for pipe flow. This table assembles a model for predicting pressure drops in adapter piping, as an example of using Polystyrene as a base resin. Other resins can easily be modeled using the corresponding thermal physical and melt viscosity properties. It is a good practice to keep the overall pressure drop in the adapter pipes for each extruder below 400 lb/in^2 gauge (27.7 bar).

3.2.5 Die and Feed Block Technology

Extrusion dies are used in the production of flat, continuous webs of various materials. Their most common applications are found in the extrusion of thermoplastics,

FIGURE 3.13 Co-extrusion feed pipes and adapters. (Courtesy of PTi, Aurora, IL.)

either as webs by themselves or as laminates to other materials. The main function of the die is to spread the material being processed to the desired width and thickness in a controllable and uniform manner. Figure 3.14 displays an image of a typical sheet die. In general terms, flat dies used in the polymer industry can be grouped into two major categories, based on the product that is to be produced. One is for film dies used for the production of thin-gauge materials usually less than 0.008 in. (0.203 mm), and the other is for sheet dies that are used to process thicker materials.

TABLE 3.6 Pressure Loss per Unit Length of the Feed Pipe

Extruder Size [in. (mm)]	Capacity [pph (kg/h)]	Typical Feed Pipe Bore [in. (mm)]	Pressure Loss per Foot of Length	
			psig	bar
1.5 (40)	100 (45)	.875 (22)	89.1	6.1
2.0 (50)	180 (80)	1.000 (25)	94.0	6.5
2.5 (65)	330 (150)	1.000 (25)	172.4	11.9
3.0 (75)	500 (230)	1.250 (32)	107.0	7.4
3.5 (90)	720 (330)	1.250 (32)	154.0	10.6
4.0 (100)	980 (440)	1.375 (35)	143.2	9.9
4.5 (115)	1380 (630)	1.375 (35)	201.7	13.9
5.0 (125)	1920 (870)	1.500 (38)	198.1	13.7
5.5 (140)	2160 (980)	1.500 (38)	222.9	15.4
6.0 (150)	2840 (1290)	1.750 (44)	158.2	10.9
6.5 (165)	3620 (1640)	1.750 (44)	201.6	13.9
7.0 (175)	4280 (1940)	2.000 (50)	139.7	9.6
8.0 (200)	5520 (2500)	2.000 (50)	180.2	12.4

Source: PTi, Aurora, IL.

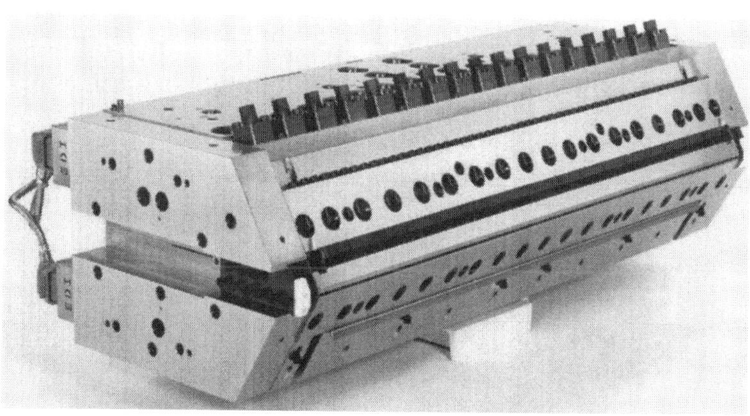

FIGURE 3.14 Sheet die. (Courtesy of EDI, Chippewa Falls, WI.)

3.2.5.1 Die Body Materials

Extrusion dies are made from a wide variety of materials, the most common being P-20, which is a medium-carbon tool steel and is typically chrome plated after construction. A few different grades of stainless steel and some exotic alloys are also used, particularly in applications where corrosion is a concern. Different plating techniques are also used. The aggressiveness of the thermoplastic material that is to be extruded and the overall process itself are the primary factors in determining the appropriate material selection. Other factors to consider include the overall mechanical properties of the die body materials. Hardness and strength at operating temperature and the availability of the material as a high-purity large billet are all considerations. Additionally, the machine-ability and polish-ability are also key factors. Flow channels within the die are usually polished to a surface finish of 2 to 4 μin. RMS and are mirrorlike in appearance. Figure 3.15 shows the interior view of the die body. This is done to enhance the appearance of the material being extruded.

3.2.5.2 Boltless Dies

Many processors require frequent and immediate splitting of the die halves for cleaning purposes. This is due to a large number of color changes or sometimes, the corrosive nature of the materials being processed (e.g., rigid PVC). This frequency drastically affects machine uptime and worker productivity. Traditional die design requires a row of body bolts to secure the upper and lower halves of the die together. The removal of these bolts can be time consuming. A new style of boltless die solves this problem. The boltless die, an example of which is depicted in Figure 3.16, utilizes a mechanized technique that allows the die body halves to be clamped and released via a series of hydromechanical clamps positioned along the backline of the die manifold.

3.2.5.3 Manifold Design

The area of flat die design that has seen the most significant improvements in the past five years is the flow channel or manifold. Aided by computational fluid design programs, state-of-the-art hybrid manifolds tuned specifically to polymer and process conditions are designed by solving three-dimensional flow equations. The basic manifold for an extrusion die is a constant cross section or t-slot design. In all cases the internal geometry should be streamlined, and care must be taken to avoid dead spots that could lead to polymer stagnation. The t-slot manifold relies on a large manifold area and a lip

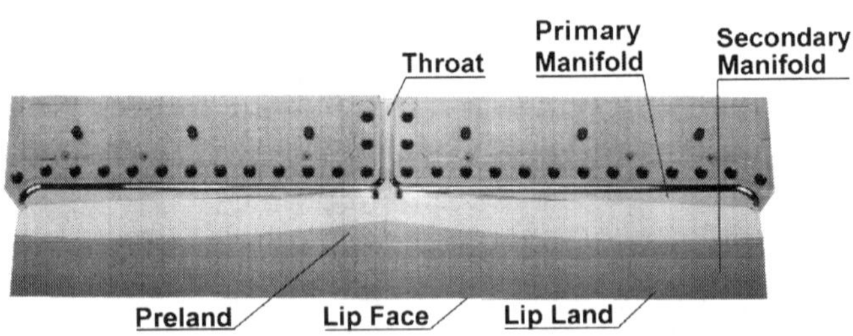

FIGURE 3.15 Interior die view with Multi-Flow V. (Courtesy of EDI, Chippewa Falls, WI.)

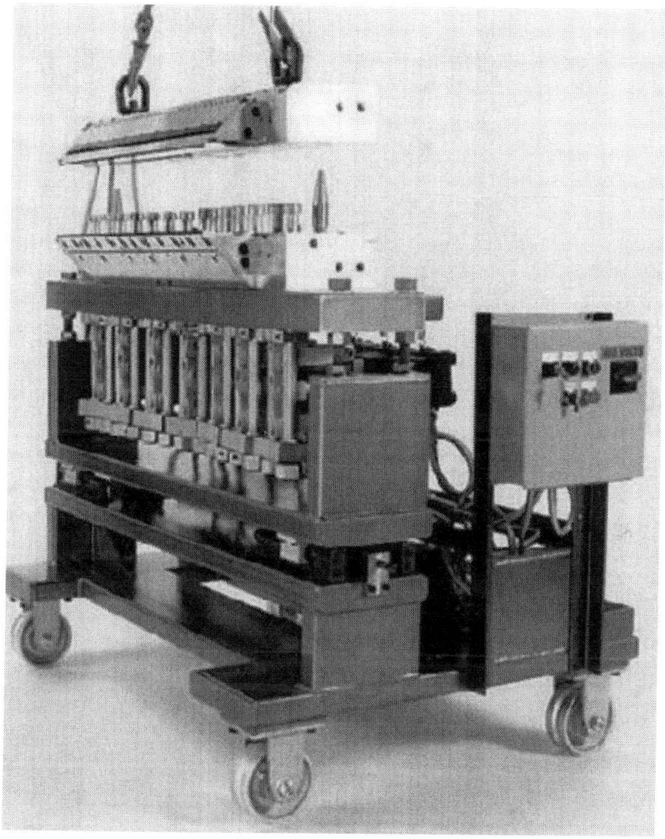

FIGURE 3.16 Boltless die. (Courtesy of EDI, Chippewa Falls, WI.)

land that is long enough to create ample pressure drops to force the material to the ends of the die. This simple manifold design is used when low-viscosity materials that are not thermally sensitive are being processed. The coat hanger manifold design shown in Figure 3.17 can be tuned for a window of materials and flow rates. This manifold is designed by establishing the same resistance to flow across the manifold and pre-land at various points across the die. The manifold shape is typically a teardrop that reduces in size from the center to the end of the die. This reduction in volume reduces residence time in the die, which is an important design consideration if thermally degradable materials are being processed. After the pre-land, the polymer flows into the secondary manifold. This is an area of the flow channel that allows the material to move laterally again if required and is also used to control the total pressure drop of the die in conjunction with the final lip land.

There are a few disadvantages with coat hanger manifolds. The backline of the manifold is farther from the exit of the die at the center than it is on the ends. This may result in a die that will deflect open more at the center, causing uneven melt distribution.

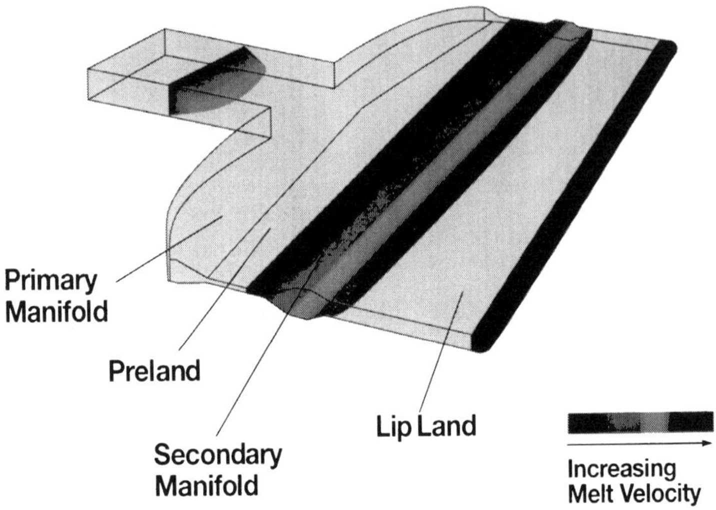

FIGURE 3.17 Coat hanger manifold. (Courtesy of EDI, Chippewa Falls, WI.)

Gauge band distribution occurrences forming "M" or "W" patterns of heavy- and light-gauge developments are also common with coat hanger manifolds. These are a result of the linear flow channels within this type of die.

The newest manifold design is a hybrid, which addresses the drawbacks of the designs mentioned earlier. The teardrop shape is elongated, with an aspect ratio that changes from the center out to the ends of the die. This results in less interface distortion in co-extrusion applications. The pre-land is a nonlinear shape that reduces the geometric gauge band patterns associated with coat hanger manifolds. This manifold also has a straight backline, which allows for straight-line body bolt placement. This results in a die that deflects uniformly, further improving the distribution of the die.

The newest and most effective design used to achieve uniform deflection is a new type of flat extrusion die with a tapered, sculpted shape machined into the external geometry that differs in shape dramatically from conventional die designs. This approach provides an unprecedented combination of the following features:

- Uniform die body deflection across the width of the die
- Streamlined melt flow distribution
- Shorter residence times
- Faster purges
- Rapid achievement of target gauge
- Overall scrap reduction
- Improved product quality

Figure 3.18 reveals the unique geometry of the contour die.

3.2.5.4 Deckling Methodologies

A deckle has two basic functions. Used primarily to change the die slot width, allowing greater flexibility for controlling the resulting overall sheet width. A deckle can also be used to control the edge bead on the web. The material to be extruded and the extrusion process together determine whether or not the die can be deckled. Degradable materials preclude themselves since deckles dam the flow, causing stagnate areas where degradation will probably occur. The most common deckle is a fixed external type that can be bolted to either half of the die, as shown in Figure 3.19. External deckles can also be designed to be adjusted on-the-fly and are typically driven by power screws or rack and pinion gears. Figure 3.20 shows a die that is equipped with a deckling mechanism of this type.

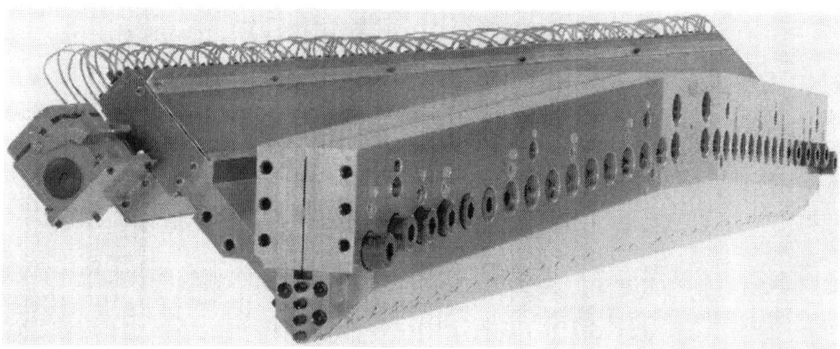

FIGURE 3.18 Contour die. (Courtesy of EDI, Chippewa Falls, WI.)

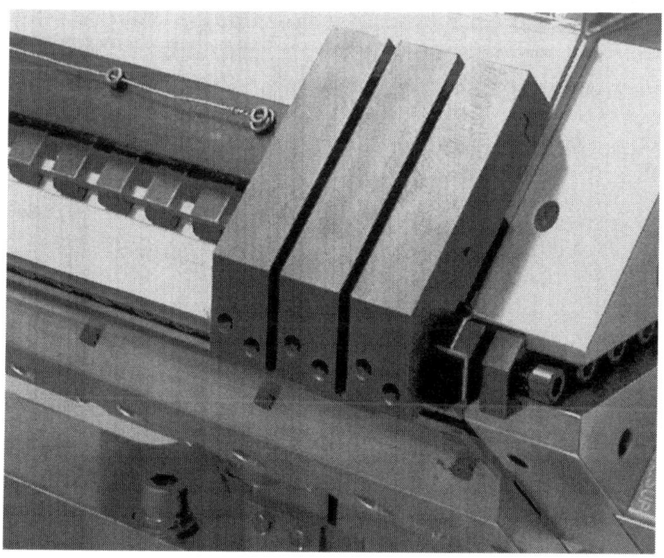

FIGURE 3.19 Fixed-side exterior deckle assembly. (Courtesy of EDI, Chippewa Falls, WI.)

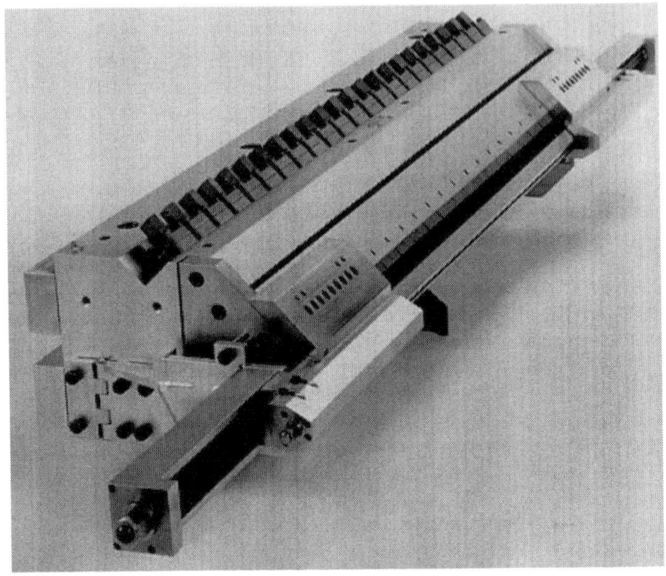

FIGURE 3.20 Fast-deckle die. (Courtesy of EDI, Chippewa Falls, WI.)

3.2.5.5 Temperature Control

Temperature control of a die is critical for good performance. A value of ±1°F is typical using today's control systems. If there is a cold area in the die, the material flow in that area will slow and the result will be a thin-gauge region. Hot areas result in greater flow and the potential to burn the material. Type J thermocouples are most often used to measure the die body temperature. The die body is usually cored and heated with electric cartridge heaters. External plate heaters, liquids, and vapors are also used to heat the die body. The wattage used to heat a die is determined by the operating temperature and the mass of the die itself. Thermocouple placement and die zone layout are important for proper control of the die and to prevent hot and cold spots. Insulation of the die should be used whenever practical to protect against operator injury, reduce cold areas, and to conserve energy. Die body cooling, when required, can be done with air or liquid. Heat tubes are also used occasionally to equalize the temperature of the die.

3.2.5.6 Adjustment Techniques

Extrusion dies that are designed for extrusion coating and the processing of film and sheet are typically adjustable. Dies designed to produce thinner -gauge sheet materials [i.e., lip openings typically under 0.080 in. (2.032 mm)] and extrusion coating dies are each provided with a simple flex lip adjustment. The die can have adjustments for flow tuning early in the flow stream after the manifold or late in the flow stream using a lip adjustment system. The intent of the adjustment in either case is to control the final shape of the extruded materials. The selection of a style of die depends on the end product that is to be produced. Sheet dies that are operated with larger lip openings will most often have a restrictor bar. They will also have a removable lip such that several lip geometries may be substituted to compensate for a broader range of adjustability. Figure 3.21 shows a restrictor bar. Since the restrictor bar is located upstream of the flexible lip, it is an aid in establishing good polymer distribution. Restrictor bars can be externally adjustable or fixed. In either case the principle of operation is the same. The gap in a specific area of the flow channel is either increased or decreased by profiling the bar. The change in resistance to flow by the change in channel height will encourage or discourage the flow of polymer in that area.

The restrictor bar's relationship to the polymer flow can be at 45, 70, or 90°. The angle usually depends on the final sheet thickness. Externally adjustable restrictor bars are most common in thick sheet extrusion and in applications where the polymer varies by type or by rheological properties. Sheet dies designed for thicker sheet gauge applications [i.e., typically over 0.500 in. (12.700 mm) thick] normally include restrictor bars at 90° to the flow channel and sliding upper lips for increased gap adjustment.

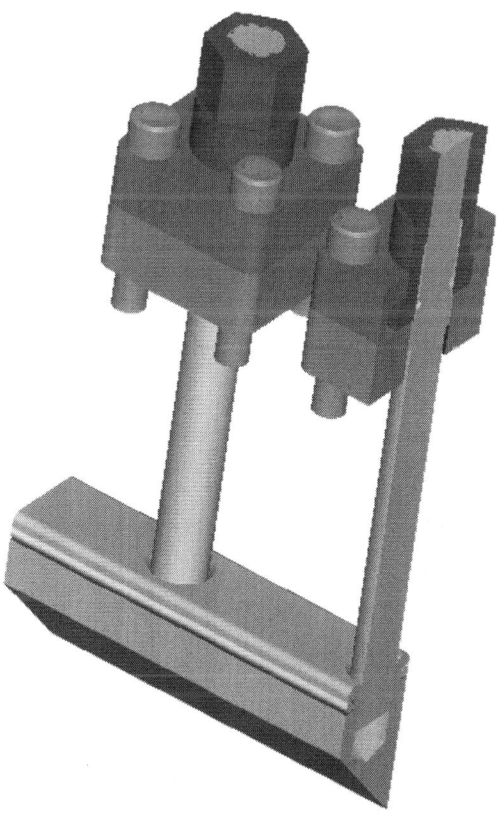

FIGURE 3.21 Cross section of a restrictor bar. (Courtesy of EDI, Chippewa Falls, WI.)

The simplest lip adjustment system is the push-only adjustment. Screws are used to push the die lip closed. Dies are usually operated by closing down on the high-gauge areas. The die lip will also respond in most cases if the adjusting screw is loosened, which allows low-gauge areas to be corrected. However, a more positive means of controlling the lip (particularly when trying to eliminate low-gauge areas) is the push–pull adjustment. By attaching the adjustment bolt mechanically to the flex lip movement can be made in both directions. The push–pull system can be provided with a differential screw thread system. This system has finer linear resolution per revolution of the adjustment nut, which makes it easier for an operator to make small changes to the die lip. Differential adjustments are also available in push-only dies. Fast-Gap is a device that has been in use for several years. It allows the lip of a die to be moved open or closed with a single adjustment screw located on the end of the die. Typically, Fast-Gap is used on the lower lip and has been a real time-saver in many applications. Traditionally, the die had to be disassembled to change the opening range and a different lip installed. With Fast-Gap, the die gap can be changed from a single point adjustment ±0.100 in. (2.540 mm) while the line is running [5]. The die lip can also be opened wide to clean or flush the die if something gets hungup in the lip land. The Fast-Gap technology is shown in Figure 3.22.

Autoflex systems (Figure 3.23) work in conjunction with a measurement device located downstream from the die. The material thickness is measured online and the results are fed back into a microprocessor that sends an output signal to either open or close the flexible lip of the die. The motion is typically produced by heating or cooling a series of closely spaced adjustment blocks which expand or contract to cause an adjustment to occur. Heat is most commonly provided by electric cartridge heaters and cooling through a steady flow of air. Motors, piezoelectric elements, and liquid heating and cooling have also been used in automatic die applications. But the most simple and reliable technique is the electric cartridge heater combined with air cooling. This approach remains the industry standard.

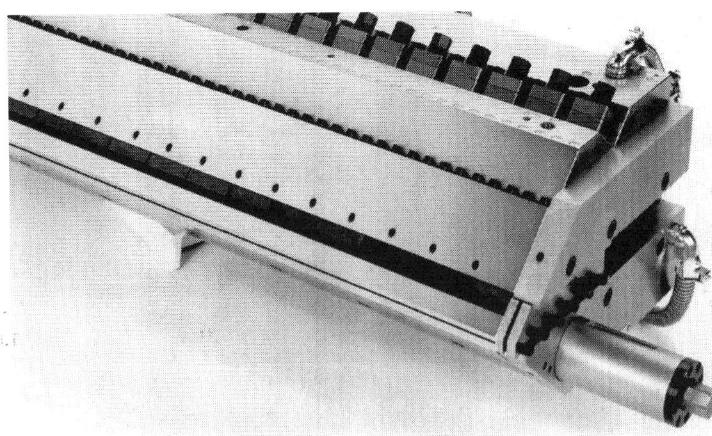

FIGURE 3.22 Die with Fast Gap. (Courtesy of EDI, Chippewa Falls, WI.)

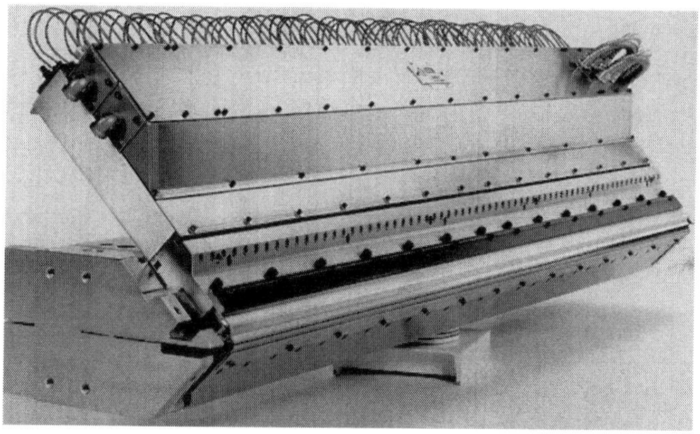

FIGURE 3.23 Autoflex die. (Courtesy of EDI, Chippewa Falls, WI.)

The current Autoflex system addresses the heat and range issues that limited some earlier Autoflex designs. This will allow auto-dies to be used in more applications, primarily where wider lip openings are required and more temperature-sensitive materials are to be processed. The heated translator has been moved out of the die body. This allows it to run at a mean temperature that is lower than that of the die body. The available stroke is now ±0.015 in. (0.381 mm) and the actuators run at maximum temperatures that exceed the die body by approximately 100°F (38°C) [5]. This new system does require more room on one side of the die.

The newest innovation to Autoflex makes it more applicable to a greater range of product thicknesses. The utilization of a lever arm at the end of the translator allows for an effective adjustment of ±0.030 in. (0.762 mm) [5]. Benefits include:

- Larger stroke range for faster gauge changes with less operator input
- Greater lip action with the same amount of thermal expansion
- Substantial improvement in overall system response time
- Faster convergence to the desired product gauge after invoking a dimensional change

3.2.5.7 Lip Scrapers

Some materials tend to build up as a residue on the lip exit area during processing. Eventually, these residues are released onto the sheet's surface as it is exiting the die lip, causing appearance problems. It is best to scrape this buildup off during roll changes so that no saved sheet is affected. However, it is sometimes difficult and dangerous for the operator to reach across the die lip area with a cleaning tool. The lip scraper is a semiautomatic device that allows the operator to clean the lip surface effectively without risk of personal injury or subsequent equipment damage due to the practice of improper manual cleaning techniques. Figure 3.24 shows an example of this device.

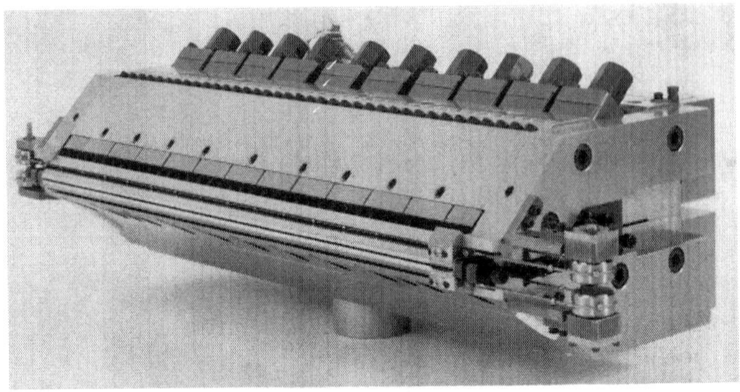

FIGURE 3.24 Die with lip scrapers. (Courtesy of EDI, Chippewa Falls, WI.)

3.2.5.8 Co-extruded Structures

Multilayer thermoplastic films, sheets, and coatings provide properties that cannot be achieved feasibly or economically using a single material. This versatility puts multilayer structures in great demand. Multilayer structures provide the ability to achieve each material's desired properties while minimizing unwanted characteristics. There are two basic methods commonly practiced to create co-extrusion structures: the co-extrusion feedblock with a single manifold extrusion die, and utilizing a multimanifold extrusion die often coupled to an appropriately configured co-extrusion feed block. To decide which type of approach is best for a particular application, consider the following factors:

- Layer tolerance requirements
- Layer viscosity ratio
- Layer thickness ratios
- Material thermal and chemical interactive stabilities
- Number of layer combinations and thickness range combinations
- Space– line layout considerations
- Capital investment constraints
- Ease of cleaning and handling of die system considerations

3.2.5.9 Co-extrusion Feed Block with a Single Manifold Extrusion Die

The feedblock approach combines layers through a rectangular entrance area that forms a 4.00-in. (100.00 mm) -wide rectangular sandwich that is fed into the die. The layers are then spread to the width of the die slot. The flow conditions within the feed block and die must remain laminar at all times such that the layers remain discrete without mixing or developing turbulence. Typically, the layer-to-layer distribution can be held at ±15% through a feed block and then ±5% through a well-designed die manifold. The feed block/single manifold die approach works well for individual layers that range in thickness from 10% of the overall structure and up. The layer-to-layer performance becomes more critical for layers that fall below 10% [5]. This is particularly important when the overall thickness of the final structure is such that the achievable tolerance for an individual layer compromises the performance contribution to the structure itself.

Individual skin layers (i.e., layers positioned on the outside surfaces of a structure) that range from 5 to10% of the overall structure compromise the ability to achieve the desired layer-to-layer thickness tolerances, due to the high shear stresses that are characteristically formed in the melt flow between the skin layers and the die's inner flow surfaces. Individual skin layers that fall below 5% of the overall structure do not typically yield favorable results using the feed block/single manifold die approach. To process thin skin layers with good results, a multimanifold die is recommended.

Conformance to the aforementioned feed block guidelines for layer distribution and individual layer sizing are not a unilateral guarantee for success. Conditions can be introduced that create process difficulties using the feed block approach. Layer distortion can be caused fundamentally by viscosity differences between the layers, and regions of high shear stress can develop along the flow geometry. This creates interfacial instabilities to occur, making layer control difficult to achieve and maintain. There are flow channel design techniques that can minimize this distortion, and feedblock profiling is commonly used to compensate for this phenomenon. Figure 3.25 displays a fixed-geometry feed block.

A rational approach to feed block design is to utilize exchangeable combining flow inserts. These flow inserts can be profiled to compensate for interface distortion. For example, if the skin layers tend to be heavy on the ends of the die when using the standard flow inserts, new inserts can be profiled to put less skin material on the ends of the feed block sandwich. These flow inserts create a geometry that will combine the layers at or near the downstream equilibrium position. Also, they are shaped such that all of the layers combine in a very streamlined, parallel-path manner. For quick changes to layer sequences, models are available that include a selector spool as displayed in Figure 3.26.

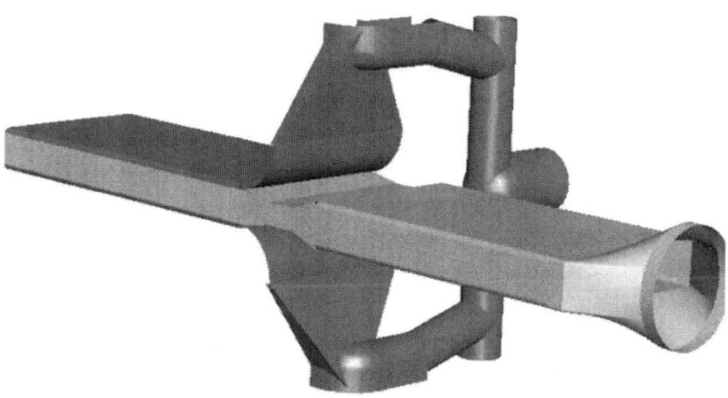

FIGURE 3.25 Polymer flow in a fixed-geometry feedblock. (Courtesy of EDI, Chippewa Falls, WI.)

FIGURE 3.26 Selector spool in feedblock. (Courtesy of EDI, Chippewa Falls, WI.)

For greater flexibility some feed block designs permit adjustments during production. These feed blocks can be "tuned" for viscous encapsulation, secondary flows, and changes in layer thickness ratios. Rotation of the combining spools provides precise and repeatable control of the layer distribution. Each spool has a plasma-coated exterior that enables it to rotate without binding. Due to the tapered and long land exit geometry in the spool, rotation has a pronounced effect on the final layer distribution. This adjustable feed block is shown in Figure 3.27.

The type of distribution manifold design used can also greatly influence layer uniformity. Current accepted manifold designs all incorporate an elongated teardrop cross-sectional shape. This shape tends to minimize shear stress levels at the layer interfaces such that interfacial deformation is also minimized. The most sophisticated manifold design available today is an elongated teardrop shape with a nonlinear pre-land. It is accurately designed utilizing finite element modeling techniques that solve full three-dimensional flow equations. The manifold backline is straight and parallel to the lip exit, which allows for a straight row of body bolts. Since all of the body bolts are equidistant from the lip exit, die body deflection will be quite uniform. This feature provides distribution that is not sensitive to changes in rate, so less die adjustment is required after a rate change.

The primary manifold has an elongated teardrop cross-sectional shape that has been shown to resist a phenomenon known as *viscous encapsulation*. This occurs when lower-viscosity layers are driven toward the ends of the die at a higher rate than are the adjacent higher-viscosity layers. The driving forces for viscous encapsulation are high

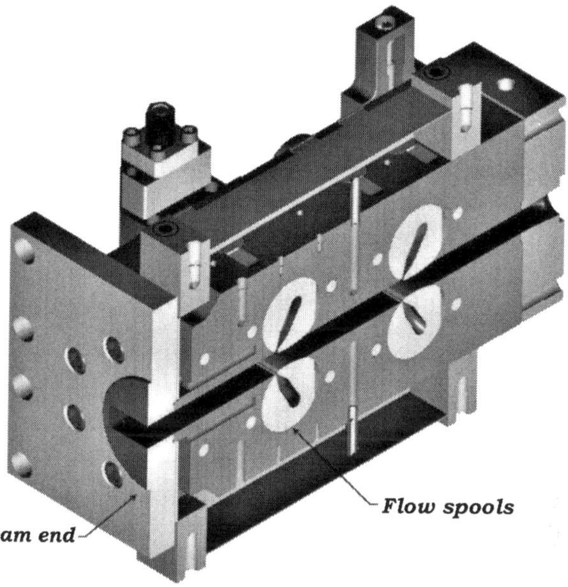

FIGURE 3.27 Cross section of an adjustable feedblock. (Courtesy of EDI, Chippewa Falls, WI.)

viscosity ratios between layers and high shear stress levels at the layer interfaces. As material flows down a teardrop-shaped manifold, high shear stress levels will develop along the manifold back wall. The layer interfaces are found along the manifold back wall, so a different manifold shape that reduces shear stress levels in this area will dramatically improve layer uniformity. The elongated teardrop manifold shape accomplishes this. As the aspect ratio of manifold length to the manifold height increases,the shear stress levels at the manifold back wall decreases. A key feature of these manifold designs is that the manifold aspect ratio increases along the width of the die such that the flow path offers excellent resistance to the viscous encapsulation effect, even on very wide dies. Figure 3.15 also displays a proprietary manifold designed by Extrusion Dies Incorporated, referred to as Multi-Flow V.

3.2.5.10 Multimanifold Extrusion Dies

The multimanifold approach has a separate distribution manifold, for each layer or partitioned combinations of layers dedicated to each manifold, as in the case when combining this technique with a feed block(s) upstream. Each layer is spread to the full die slot width before they are combined. The combination of full-width layers occurs just before the lip exit, so there is not sufficient time for layer-to-layer interfacial distortion to occur. This approach will always provide the best layer uniformity particularly for structures involving large melt viscosity ratios (i.e., typically greater than 1.5 : 1) and/or very thin skin layers [i.e., typically from 10% to less than 5% of the overall structure, but not less than 0.001 in. (0.025 mm) as a practical lower technological threshold for co-extruded sheet applications]. Each manifold is able to achieve ±5% layer-to-layer uniformity as opposed to ±15%, which is typical for feed block performance [5]. Additionally, structures that require melt temperature differentials of adjacent layers are easily processed in a multimanifold die, due to the limited amount of time that the polymers are allowed to flow together.

This is especially important when processing materials of vastly different coefficients of thermal expansion and/or core layers that are foamed. With a multimanifold die, a processor can run materials at different melt temperatures to keep the net structure dimensionally stable. Processing foamed core structures can be challenging using a feed block/single manifold die approach, due to the extended heat history that is characteristic of that combination. For example, consider a three-manifold die. By dedicating the center manifold to a foamed core, the processor can keep the layer separated and maintain a lower melt temperature profile, which is critical to cell size formation, uniformity, and net density reductions. Prior to exiting the die, the three manifolds combine at what is referred to as the *combing point* and exit the die quickly, greatly reducing the amount of time that the three layers are in contact together within the die.

Common multimanifold dies incorporate two, three, four, and even five separate manifolds. The basic restriction is the physical size of the resulting die. For this reason they are often used in conjunction with one or more feed blocks. Often, the structure can be configured to allow compatible resins to be processed in the same manifold so that a fewer manifolds are required. In addition to being used with feed blocks, multimanifold dies are used in conjunction with combining or distribution blocks. These allow the various layers to be rerouted to different manifolds, to achieve a variety of structure configurations. A multimanifold die is displayed in Figure 3.28.

FIGURE 3.28 Multimanifold die. (Courtesy of EDI, Chippewa Falls, WI.)

The multimanifold approach offers the best method for creating "naked edge" structures (i.e., a full-width core layer with narrower skins). Since there is no spreading after combination, the naked edge remains clearly defined, as depicted in Figure 3.29. This is particularly important when processing barrier sheet materials such as those that contain EVOH (ethylenevinylalcohol). The processor's objective is to optimize use of the expensive barrier resins. It is also essential to minimize the amount of EVOH that is present in the regrind materials due to processing difficulties that can manifest. One process phenomenon is commonly referred to as *blooming*. This occurs as a result of processing high concentrations of EVOH particulate contained in the regrind layer(s) that are adjacent to the adhesive–EVOH–adhesive barrier sandwich layer on a time–temperature basis. The use of a multimanifold die addresses this issue by fundamentally shortening the amount of residence time the regrind–adhesive–EVOH–adhesive–regrind sandwich is exposed to the skin layers in a die.

3.2.5.11 Optimizing Co-extrusion Tooling Selections

Multimanifold die designs can be constructed to meet most application requirements. They provide a greater range of application-specific flexibilities. Particularly for co-extrusion applications that require combining materials that yield high melt viscosity ratios and/or characteristically possess very thin skin layers and/or require combinations of materials that are adversely affected by prolonged exposure to one another in the same manifold (i.e. barrier structures for blooming avoidance or foaming applications for preserving cell size formation,uniformity, and net density reductions), the multimanifold approach is the best choice. If the materials being processed are of similar viscosities and the layers desired are of similar thicknesses,

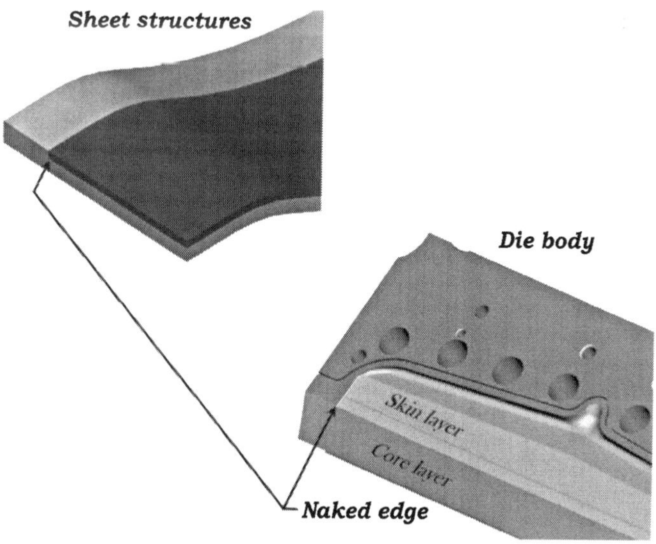

FIGURE 3.29 Naked edge technology. (Courtesy of EDI, Chippewa Falls, WI.)

the feed block approach is the best choice. The overall objective should be to keep the design as simple as possible while achieving the desired process capabilities.

3.2.5.12 Die Carts

To support and align a die and other tooling (feed blocks, distribution blocks, etc.) properly, a die cart is sometimes necessary. The cart can function purely as a support device or can be configured to aid in on-line die splitting to facilitate cleaning. Figure 3.30 shows a die cleaning cart with an in-line splitting feature. Due to their large size or complexities (i.e., multimanifold dies) some sheet dies are cumbersome to handle while cleaning. Specifically designed die cleaning carts are a great benefit to the maintenance staff. They allow a die to be split and cleaned while being held by the cart at all times, furthering safety and productivity. Figure 3.31 shows a cleaning cart designed to handle multimanifold dies.

3.2.6 Roll Stacks

Roll stands vary in configuration depending on the sheet type that is going to be produced. Consider the variety of resins being used to produce sheets. Some exhibit high melt strengths; others are low in comparison. Furthermore, consider the number of thicknesses and widths being produced. Some applications require thin gauges at narrow widths versus thick gauges at wide widths.

When determining the proper configuration for a roll stand, many additional factors need to be assessed. These include surface texture, laminating or embossing effects, mono- or co-extrusion requirements, and post-sheet processing requirements such as in-line thermoforming, winding roll stock, or shearing and stacking for cut sheet applications.

Each roll stand arrangement consists of a variety of high-quality features and design elements. PTi has developed a complete and comprehensive portfolio of these

FIGURE 3.30 Die cart with in-line splitting. (Courtesy of EDI, Chippewa Falls, WI.)

FIGURE 3.31 Multimanifold die cleaning cart. (Courtesy of EDI, Chippewa Falls, WI.)

in order to provide superior Tools to those who run sheet on a daily basis. Included in this portfolio are the following tools.

3.2.6.1 Heavy Duty Precision Frames

The frame superstructure of a roll stack is critically important (1) to assure precise machine operation throughout the life cycle of the equipment, (2) to provide the ability to produce high-quality sheet that e xhibits superior surface appearances as well as excellent clarity for transparent applications, and (3) to assure vibration-free operation. An integral part of the frame is the manner in which the rolls are interconnected to the roll stand itself.

PTi employs three different high-quality approaches for interfacing the rolls with the frame. The first is the pivot arm design. This approach requires a heavy-duty roll actuation arm made from a casting. It comes either as a split-arm or a cantile vered arm design. Each approach uses a large-diameter precision pin and bushing arrangement to secure the arms to the side frames. The split-arm style features a special geometry that permits the cylinder actuation location to actually straddle the frame. This is commonly used for individually driven rolls with large diameters and high roll load-bearing applications [i.e., roll diameters of 30 in. (760 mm) and larger, and roll loads of 800 lb per linear inch (140 N/mm)]. The second approach, the cantilever support, is used universally for chain-driven scenarios using a wide range of roll diameters from 12 to 24 in.(300 to 610 mm). The last approach uses linear bearing-mounted roll bearing blocks,as used for direct-driven rolls in PTi's Revolution roll stand design. It features linear bearings that are mounted onto precision rails that are partially captured into the side frames,with the linear bearings themselves partially captured into the roll bearing blocks. This provides a very strong and integral design and is used for a wide range of roll diameters [i.e., roll diameters ranging from 12 to 36 in. (300 to 914 mm)].

3.2.6.2 Precision Casters and Tracks

This is an area that is typically overlooked by many. The importance of precision casters and tracks contribution cannot be taken lightly. Many manufacturers of sheet production machinery still use V-groove casters mounted onto simple tracks that use inverted angle iron as the rail. This was the standard years ago b ut is long outdated and does not lend itself well to a precision system. PTi designed a precision solid steel wheel that is bearing mounted and supported by a v ery heavy duty wheel hub. The hub is adjustable up and do wn to provide a means of le veling and aligning the roll stand. The wheel face features either flat or groo ved geometry. The grooved side mates with a machined groo ve on the track plates. PT i requires that the track plates be laser leveled and grouted into place. It is similar to ho w any piece of precision equipment would be installed. Otherwise, the frame will be subjected to twisting as it traverses up and down the tracks. PTi users all attest to the benefit of precision equipment, even for the casters and tracks.

3.2.6.3 Air-Boosted Pneumatic Roll Actuation

PTi offers both pneumatic and boosted pneumatic roll actuation systems. The boosted pneumatic method permits the plant air supply to be increased to nearly three fold the value using a proprietary boosting system. This is great for applications that require additional roll loads using pneumatic c ylinders. Some users are opposed to using hydraulics for roll actuation. Using an air booster system, users are able to achie ve higher roll loads [i.e., loads as high as 600 lb/linear inch (105 N/mm) can be achieved, depending on roll width, product width, and air cylinder diameter].

3.2.6.4 Auto-GAP and Hydraulic Roll Actuation

PTi developed a precision roll positioning system using serv o hydraulics called Auto-GAP. This is a single- stage servo valve that is driven using a small electric servomotor to position the spool of the v alve. It is extremely accurate and repeatable. The Auto-GAP system offers an operating range in 0.001-in. (0.025-mm) increments with the same incremental repeatability . This system is a vailable fully integrated with PTi's TITAN PLC-based control system that is interfaced with the recipe management module of the control system. This is a very convenient method of controlling the line. The servo technology is a modulation-based system that will correct to the proper setting if momentarily forced open due to e xcessive roll loads.

The hydraulic roll actuation is also a good means of roll actuation. It is particularly good for heavy rolls characterized by large diameters and wider widths. Some resins require higher roll loads to run good sheet. Hydraulics is a great w ay to achieve high roll loads while offering a practical and cost-effective method of opening and closing the rolls. PT i uses an accumulator with all of its roll actuation systems as a safety measure in the e vent of a power failure.

3.2.6.5 Heavy-Duty Roll Journals and Precision Bearings

Roll journal designs structurally determine how strong and rigid the roll shafts actually are when exposed to high loads. If the sheet application calls for high roll loads, the roll shafts must be suitably strong to avoid undesired flexing and thus creating processing difficulties. Applications involving specialty rolls that permit a controlled and predictable flexure are in a different category. They are designed to flex in a contour that matches the mating roll. These are good when used on applications where the aspect ratio of the roll is high. The aspect ratio is defined by the width divided by the diameter, either in inches or millimeters. An aspect ratio of less than 3.5 would not be a good application for a controlled flexure roll while an aspect ratio greater than 3.5 would begin to realize the benefits intended with this roll design. Keep in mind that all rolls flex when placed under load. They are essentially giant springs and therefore have an associated stiffness.

Another vital element of maintaining precision capability is the roll bearings themselves. This is often the source of a tolerance problem. The ability to run precision sheet can be determined partly by a roll's ability to turn in the bearings without excessive running out. All bearings have an associated run-out. They are manufactured for specific applications. Some are tighter than others. The manufacturers of bearings offer data as to the relative performance of bearings. PTi employs a highly precise bearing in all of its roll stand designs. The style of bearing is a spherical roller bearing. The grade of bearing is commonly used in the printing press industry and is characteristically accurate. Once the rolls are assembled with the bearings and seal and are installed in the arms or bearing blocks, it is easy to test them for their relative run-outs. An easy method is using a dial indicator and plastic tip to assure that the roll face does not get scored. Slowly rotating the roll through one complete rotation will reveal its run-out. PTi rolls typically run-out below 0.0005 in. (0.01 mm).

3.2.6.6 Individual Roll Drives

The majority of all sheet lines manufactured by PTi today have individually driven rolls. The overall trend in the industry is moving in this direction, although several manufacturers produce chain-driven systems. The design for the roll drive has evolved from servo to direct current (dc) to alternating current (ac) drives.

Most individually driven roll drives today are ac based. This is due to the increased torque performance and digital control formats for the ac motor and drive controller technology. Each drive comes equipped with communications for interfacing with computer and PLC-based control systems and 0.01% speed resolution control platforms. Additionally, the ac motors offer 1000 : 1 speed-to-full torque ratios, which will deliver the full torque rating for the motor at 1/1000 times the base motor speed. This performance is greatly improved over ac motors of the past. More and more these systems are becoming the standard for today's roll drives.

However, PTi is developing a new technology for driving rolls, the torque drive– based system. It consists of a high-pole-count dc servomotor. It is available in a hollow shaft arrangement such that it slips over the roll shaft and uses a taper lock interface to secure it to the shaft. The encoder that is integral to the motor has 2 million pulses per revolution. The motor delivers full torque at zero revolutions. It features a water-cooled motor jacket so that it won't overheat at low speeds. One major benefit of this approach is that it does not use a gearbox to transfer and amplify its torque. This provides a roll drive scenario that is highly precise, provides full torque at even the lowest speeds, and delivers truly chatter-free drive power to the roll shaft. The motor is sized based on the torque required to accomplish the job. Its unique geometry and performance make this technology worth exploring. This emerging technology could possibly be the new trend in driving roll systems.

3.2.6.7 High-Flow Water Circuits

Each PTi sheet takeoff system features multiple water circuits to provide PID-controlled zones to each roll. The oversized roll journals permit using large-bore water passages. These oversized water passages provide a means of transporting high volumes of water at relatively low head losses. In other words, PTi's water circuits can deliver high water volumes through the various components, thus minimizing the temperature differential that is normally experienced across the roll face. The general guideline for designing a sensible water circuit limits the velocity that water can flow through pipes to 12 to 18 ft/sec. This is a limiting factor that serves as a sizing guideline for selecting other components, including pipe diameters, valves, and other fittings.

By implementing a system that displaces high volumes of water, PTi roll stands have the lowest-temperature differentials in the industry. This is a basic principle of convective heat transport using water as the transporting vehicle. The roll circuit operation is similar to that of a heat exchanger. Energy is being exchanged or transported from one location to another. In the case of the roll, heat is either being removed from the roll or supplied to the roll in the case where heat is required. For the case where heat is being removed, the outer shell of the chrome roll is the conducting medium for transmitting energy from the plastic to the water passing through the roll on the inside of the shell.

In a basic heat exchanger application, the higher the water volume available, the lower the temperature difference required to transport the thermal energy that is being transmitted. This is the basis for the reduced temperature diferential across the roll face. PTi designs its water circuits to limit this to 1 to 2°F. The supply side and return side are alternated from roll to roll to eliminate a stackup of temperature differentials, which could lead to a colder side of the side when examining it across the face.

3.2.6.8 Heavy-Duty Conveyor Units

The frame of the conveyor is often used to support such ancillary devices as coaters, masking stations, gauge scanners, and slitters. It is important that the idler rollers be supported and arranged such that the frame support will maintain the alignment over the years of service. PTi implements a unique method of mounting its idler rolls as well as other devices onto the conveyor frame. A common mounting arrangement is used such that the idler rollers can be positioned at different locations as required. Additionally, other devices may be added or removed as needs dictate. It is important to keep the conveyor unit aligned, particularly when processing sheet that may be prone to scratches or marring. A misaligned idler roller can be a potential source of adverse cosmetic defects in a sheet's surface.

3.2.6.9 Safety Element

PTi's equipment products feature strategically placed safety devices to protect operators from possible injury as well as to minimize the possibility of equipment damage. This is achieved by a combination of emergency pushbuttons and pull chords (the type that must be reset manually) located at hazardous locations, including roll nip points, inbound roll locations and other pinch points, control panel interface locations, and general areas up and down both sides of a system such that the operator could trigger a shutdown from nearly any point in the vicinity of the equipment. Safety-color-coded rigid guarding (usually, safety yellow is used) is applied to areas that require impediments to prevent access to specific areas of hazardous conditions. These include outbound nip points, rotating shafts, pinch points in general, casters, and other moving parts. Additionally, it is a good practice to implement a battery of safety labeling on the equipment. The labels should clearly define the hazard as well as portray a pictogram displaying a clear picture of the hazard.There are several well-published guidelines that need to be followed and serve as excellent references. The Society of Plastics Industries (SPI) and the American National Standard's Institute (ANSI) have collaborated on several important aspects of equipment safety as it relates to sheet lines: the *American National Standard for Plastics Machinery* [1], the *American National Standard for Plastic Sheet Production Machinery* [6], and the *American National Standard For Plastic Film Sheet Winding Machinery* [7].

PTi has implemented many of the aforementioned design elements and features into a portfolio of high-quality sheet production systems, each designed for a particular use and range of processes [8]. These include the V-series (vertical), J-series (J approach), I-series (inclined), H-series (horizontal), and revolution series (vertical). PTi provides all of its roll stands as full-length systems for producing either wound roll stock or cut sheet and as compact systems for feeding a thermoformer directly for in-line processing.

V-Series Down-Stack The V-Series Down-Stack is designed for commodity sheet with thin to mid gauges, narrow to medium sheet widths, and low to medium melt strengths. An optional mezzanine positioned unwind is available for top-side lamination. A smaller top roll permits a close-approach sheet die-to-nip point operation. Figure 3.32 is an example of this style of roll stand. They are in common use in food packaging, cups and lids, agricultural products, consumer products, electronic components, medical equipment, and lawn and garden products.

V-Series Up-Stack The V-Series Up-Stack is ideal for both specialty and custom sheet in mid to thick gauges, medium to wide sheet widths, and medium to high melt strengths. An optional mezzanine-positioned unwind can be used for bottom-side laminating or a conveyor-positioned unwind for top-side laminating. Typically all rolls feature similar diameters. They are in common use in automotive products, construction products, consumer products, industrial products, and lawn and garden products.

Compact V-Series Down-Stack The Compact V-Series Down-Stack is designed specifically for in-line thermoforming commodity sheet ranging from thin to mid gauges, narrow to medium widths, and low to medium melt strengths. A smaller top roll permits a close-approach sheet die-to-nip point configuration. Figure 3.33 provides an image of the compact system. They are in common use in food packaging, agricultural products, medical equipments, and cups and lids.

FIGURE 3.32 V-Series down-stack. (Courtesy of PTi, Aurora, IL.)

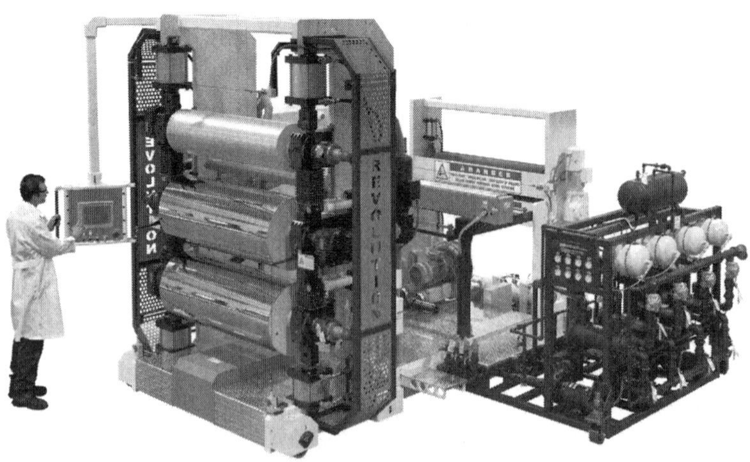

FIGURE 3.33 Compact V-Series down-stack. (Courtesy of PTi, Aurora, IL.)

J-Series Up-Stack. The J-Series Up-Stack is designed for commodity, specialty, and custom sheet ranging from thin to thick gauges,medium to wide widths, and low to high melt strengths. Both mezzanine- and conveyor-positioned unwinds are optionally available for top-and bottom-side laminating, respectively. A smaller bottom roll permits a close-approach sheet die-to-nip point configuration. The unique features of this roll stand can be seen in Figure 3.34. They are in common use in automotive products, food packaging, consumer products, industrial products, and lawn and garden products.

I-Series Up-Stack The I-Series roll stand implies “inclined”. The web path for an inclined roll stand is up versus down. So this stand is available only as an up-stack. It is designed for commodity and specialty sheet ranging from thin to medium gauges and narrow to medium widths. Both mezzanine- and conveyor-positioned unwinds are optionally available for top- and bottom-side laminating. A smaller bottom roll permits a close-approach sheet die-to-nip point configuration. This roll stand is shown in Figure 3.35. They are in common use in food packaging,electronic components, medical equipments, and office products.

H-Series Roll Stack The H-Series Roll Stack is ideal for specialty and custom sheet applications ranging from medium to thick gauges, narrow to medium widths, and low to medium melt strengths. Rolls can be textured and vary in diameter for timed-pattern embossing. A floor-positioned unwind station permits extrusion coating for scrim and other substrates. The unique features of this roll stack are shown in Figure 3.36. They are in common use in automotive products, construction products, and office products.

FIGURE 3.34 J-Series up-stack. (Courtesy of PTi, Aurora, IL.)

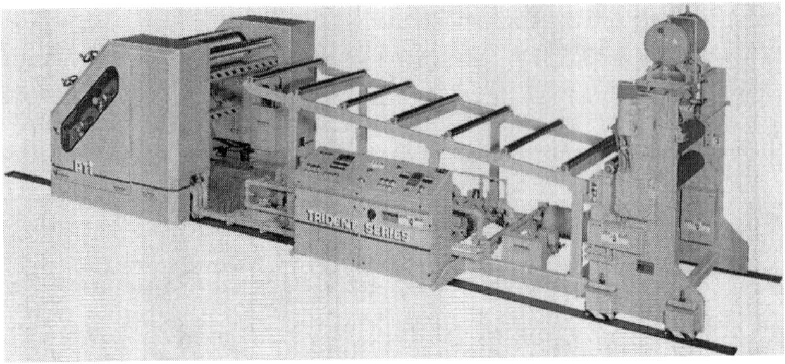

FIGURE 3.35 I-Series up-stack. (Courtesy of PTi, Aurora, IL.)

FIGURE 3.36 H-Series roll stack. (Courtesy of PTi, Aurora, IL.)

Revolution Roll Stack PTi's newest addition to its roll stand designs is the Revolution roll stack. It was developed to provide users with the added flexibility needed to address issues relating to custom sheet applications. This is especially important when production runs are short and changeovers to various pattern rolls are frequent. This design is also beneficial to producers of various sheet goods, due to its unique blend of operational features and conveniences. These include direct-acting cylinder placement for roll actuation; linear bearing-guided roll bearing blocks; easy operational access in front, behind, and at the side of the unit; direct-driven rolls; swing pendant controls; and fully enclosed decking to keep material from collecting beneath the unit. The enclosed lower platform also provides a surface for the operator to stand on during web-up and other activities. Helper arms features tool-less connections and literally install in seconds to assist with roll removal exercise during roll change-outs. The Revolution comes equipped with a unique vertical roll stand positioning system that permits motorized height adjustment of the roll stand such that the attitude of the primary nip can be changed relative to the sheet die approach. Figure 3.37 shows this new and exciting roll stand design. They are in common use in food packaging,cups and lids, agricultural products, consumer products, electronic components, medical equipment, and lawn and garden products.

3.2.7 Hot Nip Laminating and Masking

PTi offers a couple of unique approaches to modifying the surfaces of sheet during production. There are specific differences regarding which method is best to use to achieve a desired affect. The first method is that of hot nip laminating. This is the process of applying a film substrate to the surface of the sheet using the primary nip point on the roll stand to ''weld" the two materials together. This assures that a bond is achieved, and the likelihood of delamination is minimized. The other method is that of masking. This method is commonly used for applying a protective film to sheet after it is formed, as a means of preventing marring and/or the occurrence of other surface defects caused during handling.

3.2.7.1 Hot Nip Laminating

PTi offers a unique method of welding film to sheet as it is being extruded. The method of hot nip laminating provides processors with a cost-effective technique for modifying the surface of the sheet for various desired effects. This includes laminating film materials that (1) are heat sealable, such as plastic-coated foils and/or paper products that are to be applied to thermoformed packaging goods; (2) provide oxygen and moisture barrier properties for extended-shelf-life packaging for applications that involve altering the performance of the sheet goods beyond what is capable using the processors existing sheet extrusion equipment; (3) are desired to create a unique cosmetic

FIGURE 3.37 Revolution roll stack. (Courtesy of PTi, Aurora, IL.)

and/or decorative affect on the sheet surface, including multicolored preprinted materials; and (4) increase the coefficient of friction of the sheet surface.

Caution must be practiced to assure that suitable compatibilities exist between the sheet process and the film that is being laminated, so that the film does not melt, distort, or tear and that it is properly bonded to the sheet at the primary nip point. Additionally, proper equipment arrangements are necessary to deliver the film material to the nip point in order to remove all wrinkles.

3.2.7.2 Masking

This is the process of laying a protective film on a sheet surface on one or both sides of the sheet to prevent marring, scratching, and other undesirable surface defects during the handling of sheet products. The masking station is typically located downstream of the primary roll stack. It is desirable to mask the sheet when it is still warm. A good application temperature ranges from 110 to 140°F (43 to 60°C). This helps the masking film bond to the sheet surface while permitting it to release when it is desirable to remove the protective film. Films used for masking have been specially formulated to achieve static cling and adherence to a sheet's surface. The film is loaded into unwind positions that feature tension control and alert the operator when the position is low on film and a changeover is required. The unwind stations are adjustable side to side such that the film can be aligned properly with the location for masking. The masking film can be edge trimmed in line or can be ordered to a pretrimmed width which is typical and more cost-effective than slitting in line. Edge trim is usually taken before the application of the masking, or it can be accounted for in determining the width of the masking film such that the edge trim knives do not cut into the masking film, causing cross-contamination.

The position where the masking stations are located along the conveyor are dictated by the ancillary components that are to be adjacent either upstream and/or downstream from these positions. A gauge scanner should be located upstream from the masking stations such that the scanner is not recording the thickness of the film as part of a sheet's scanned profile. Additionally, the edge trim slitting positions must be coordinated with the masking station location, as mentioned above. It is good practice to use a pneumatically actuated rubber roller to apply contact pressure on the masking film and sheet at the point of application. Immediately upstream from this position the film must pass over a specialty roller that spreads out the wrinkles as it is rotating to assure a high quality result. This applies to both top- and bottom-position masking stations.

3.2.8 Slitting

PTi offers two basic categories for in-line slitting. In-line slitting is the operation of removing unwanted edges as well as subdividing the sheet width into multiple smaller widths. Edge trim is removed to reduce the sheet to the desired net width as well as removing the raw edges that do not typically meet with the quality standards desired. The sheet edges experience a thickening phenomenon as a result of the transformation that occurs from molten polymer at the primary nip point to cooled and polished sheet. Sheet typically draws down from the exit point at the sheet die lips. The drawdown that occurs at the sheet die, also referred to as *neck-in*, contributes to this undesirable edge condition. Typically, the edges are trimmed by a minimum of 1.0 to 2.0 in. (25 to 50 mm) per side. This assures that the nontrue and thickened portion of the edge is removed and that only within-specification sheet remains. Wider bands of edge trim can be taken to obtain the desired net width if necessary. Caution needs to be practiced in doing so since the overall efficiency and yield of the production line is compromised by this practice. Sheet die deckling is the preferred method of reducing the untrimmed width of the sheet such that minimal edge trim can be removed. This is a more effective and efficient means of controlling and producing a desired net sheet width.

The number of subdivisions that the sheet can be slit into depends on the practical ability to handle the multiple widths. When winding, how many ribbons the winder and corresponding core rods can accommodate is often a limiting factor. Differential shaft winding is a good means of handling several slit widths of sheet on the same winder shaft. For this application it is best to locate the slitter knives immediately upstream of the winder. This will keep the individual sheet widths organized prior to winding. Off-line slitting is also an option if the process objective calls for many slit widths. Typically, in-line slitting using conventional winding technology is limited to two or three separate widths. However, it is possible to handle more, but it does represent some logistical considerations for the winding system.

3.2.8.1 Razor Knife Slitting

A common and cost-effective solution to slitting is the razor knife arrangement. It is by far the most prevalent means of achieving slit-to-width objectives. Figure 3.38 shows a typical razor knife slitting station. It features a knife holder for each slitting position desired. The knife assemblies are secured to the slitter bar by means of a quick-release clamp that allows the assembly to be positioned back and forth across the width of the sheet to the desired slitting location. The bar features a scale such that the operator has a reference point to track the location of the knife assemblyThe bar geometry and knife assembly are keyed together to assure that the knife is held in place and does not rotate or twist on the bar mechanism. The slitter bar is secured at either end and is simply rotated into and out of the slit position manually. The knives may be operated with or without heat. Some applications may yield improved results with heated knife blades. This is accomplished using a special heater to trickle heat into the knife blade location. The razor blades feature a beveled leading edge that cuts into the sheet. For most applications this is an acceptable means of slitting sheet. However, there are some applications that require a square-cut edge. Depending on the geometry of the leading edge (i.e., double or single bevel), one or both sides of the slit sheet can have a corresponding beveled edge. This will not affect the quality of the wound rolls during winding, but it does have an impact on the specific geometry of the slit edge. If the application calls for a square edge, an alternative slitter needs to be used. Keep in mind that some plastics slit better using a rotary style knife system in lieu of a razor slitter, particularly those that have possess a highly crystalline characteristic and are relatively brittle to slit. However, most

FIGURE 3.38 Dual-razor-knife slitting station. (Courtesy of PTi, Aurora, IL.)

polymers can be slit successfully using a razor-style slitter, from thick to thin sheet applications.

3.2.8.2 Rotary Knife Slitting

An alternative approach to razor slitting is the application of rotary slitters. This approach offers a greater degree of precision and an overall cleaner cutting performance. Razor knives are sometimes prone to flexing, while the rotary assembly will not move in the event of a side load. Rotary knives come in a variety of sizes and blade arrangements, depending on what the application requires. Rotary slitters have been used in many different markets, including coated and plated paper and paperboard, polycoated and uncoated foils, and coated printed materials, along with a myriad of plastic sheet goods. Figure 3.39 shows a typical rotary slitter arrangement. The slitter assembly features upstream and downstream guide idlers rollers to assure that a sheet enters and exits the slitter properly; a driven anvil roller, which supports the mating knife rings; the support rail to which the slitter knife assemblies are mounted; the slitter knife assembly; side frames; controls; and guarding.

The knives are available in various service levels, depending on the severity of the application. Blade diameters are available in 3.5-, 5.9-, and 7.8-inch (90-, 150-, and 200-mm diameters). Multiple-knife assemblies can easily fit onto the supporting rail to achieve the desired number of slits. The knives retract out of the slitting position

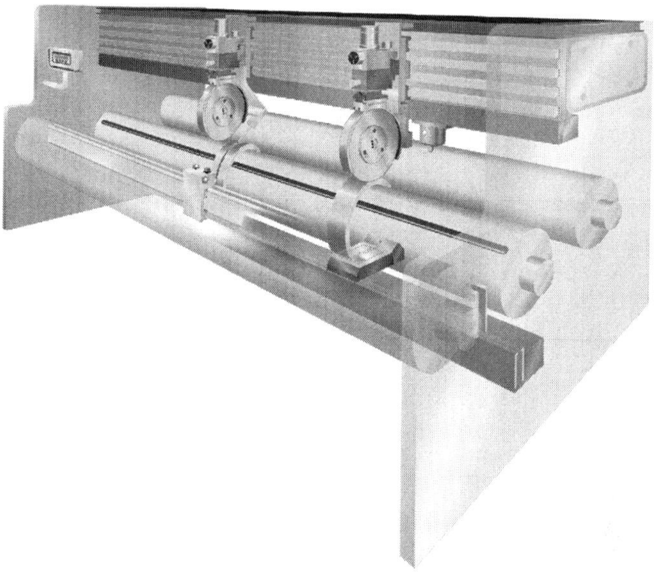

FIGURE 3.39 Rotary slitter arrangement. (Courtesy of Maxcess International, Camas, WA.)

away from contact with the knife rings. This arrangement permits shear cutting and crush scoring depending on the objective for a given process.

3.2.9 Antistatic Coating

During the production phase of producing sheet the web has a tendency to gain a static charge or static electricity. The surface of the sheet may also have a characteristically high coefficient of friction (COF) when placed in contact with the same sheet materials that it contacts during the thermoforming process. Once formed, the parts are trimmed, counted, stacked, bagged, boxed, and eventually shipped to the end user. The end user then places the desired contents into the container. It is usually at this stage that problems can manifest with the removal of formed parts from the bagged sleeves. As one part is separated from the other, the parts have a tendency to stick together. This is referred to as *nesting*.

A common method of minimizing the negative affects of static and nesting parts is the application of a thin film of fluid onto the surface of the sheet prior to the thermoforming stage of the process. There are many common solutions that are used by processors to improve the antistatic and denesting performance of the sheet. PTi recommends the use of an approved solution that is certified for its end use, particularly for food-grade applications. Some plastics have a greater tendency to experience static buildup and nesting. The geometry of the thermoformed part can also play a role in its nesting characteristics: The deeper or taller the part, the greater the likelihood of nesting. This is due to the two surface areas that come into contact. Keep in mind that friction or drag is directly correlated to a surface area. The COF is a coefficient of drag as it relates to an area of contact.

PTi offers two different configurations for addressing the issue of antistatic coating of a sheet during its production. The first such system is a simple submersion tank that places the sheet web fully into a solution, then squeegees the excess fluid off using rubber-coated nip rollers, and collects the excess back into the tank. This style is the most practical and cost-effective, and is simple and easy to use. The tank can have a recirculation pump circuit incorporated in the design to regulate the temperature of the solution. If the sheet is hot going into the tank, it will exchange energy with the tank and the solution will eventually rise in temperature. The cooling circuit keeps the solution from getting overly hot. Figure 3.40 shows a submersible-style antistatic coating system.

The other style of treatment is the contact roller method of coating. This uses a series of rollers to guide a web of sheet through a desired path to control the wrap contact area on the application roller. The sheet does not enter the solution directly. It comes into contact with a specialty roller that through surface tension collects the fluid on its surface and applies it to the sheet as the web passes over it. The application roller is driven and can operate in reverse or in the same direction as the sheet. The solution tank also features a pumping circuit such that its temperature is regulated. The desired temperature of the solution typically ranges from 90 to 110° F (32 to 43° C). The sheet surface temperature should be some what higher than the solution temperature such that the excess fluid wets out onto the surface of the sheet and then evaporates, leaving the surface dry to the touch while the antistatic chemical remains on the sheet's surface. Figure 3.41 show a contact roller antistatic coating system.

FIGURE 3.40 Submersible-style antistatic coating system. (Courtesy of PTi, Aurora, IL.)

FIGURE 3.41 Contact roller antistatic coating system. (Courtesy of PTi, Aurora, IL.)

Note that the contact roller style permits the operator to coat one side only rather than both sides as is the case with the submersible approach. This can also be arranged such that each sides may be coated independently. Processors must have the flexibility of coating one side and not the other for applications involving printed inks that are not compatible with coated surfaces. Both of the antistatic coaters presented herein are easy to use. The contact roller method offers greater flexibility but is more expensive than the submersible approach.

3.2.10 Gauge Scanning and Automatic Profile Control

Despite modern process control technology, including microprocessor-based temperature and drive controls and sophisticated melt pumps to regulate polymer flow, the final determination of thickness is still under the control of the machine operator. Without some sort of automated device to measure the thickness of the product and to trim the process automatically to optimize sheet thickness, final product quality is a function of operator expertise and dedication. The machine operator's tendency is to be protective of product specifications, and if an error is to be made, he or she will run the product heavy to be safe. This factor, combined with the limited visibility as to cross-web thickness when using manual techniques such as a hand micrometer, results in a product with excess variance as well as raw material wastage due to overweight product. In addition, limited visibility creates additional unnecessary scrap at startup and during product changeovers.

To eliminate these problems, online gauging systems have come into common use. The first of these devices were developed over 50 years ago. Over time, these systems have been improved and modernized, particularly with the addition of computer controls. Sheet extrusion and thermoforming processors have every incentive to encourage the use of a gauging system so as to maximize sheet thickness within tolerances. The gauging system is a valuable tool that maximizes the yield of raw material, minimizes startup scrap, and ensures that the product that arrives at a customer's doorstep meets critical specifications for thickness with minimal variation. It can be shown that a good gauging system will return its cost in a year—even less when one considers the use of ever-more expensive polymers.

3.2.10.1 Gauge Selection Factors

There are a number of different measurement technologies available to make continuous online product measurement across the width of the web. The technologies employed include nucleonic devices (beta transmission, gamma backscatter), infrared absorption sensors, x-ray transmission sensors, and air and laser caliper sensors, among others. Each of these sensor technologies has unique characteristics,and it is important to choose the sensor with the best balance of operating properties for the application and end use. For example, a sensor that might work very well on wide, clear films, may work poorly on thick sheet, and vice versa.

Each installation, then, is considered on its merits. Some of the parameters important to determine the best gauge to use on an extrusion line are as follows:

- Product thickness and thickness range
- Whether the products are clear, translucent, or opaque
- Whether the product is of uniform density, foam, rough surface, or high regrind content
- Whether or not the sheet is co-extruded
- Whether or not the materials being processed are commodity grade (PE, PP, PVC, PS, PET) or high value (EVOH, nylon, engineering resins).

We examine each of these issues to understand why they are important to gauge selection.

Product Thickness and Thickness Range Each sensor type has limits on its operating range. For example, a krypton beta gauge (which measures mass and infers thickness through assumption of density) has an operating range of approximately 10 to 1000 g/m^2 [9]. At a density of 1.0, this equates to a product from 0.4 to 40 mils thick—thinner than 0.4 mil and the signal-to-noise ratio of the gauge becomes unacceptable. Above 40 mils the krypton gauge cannot measure the product; the beta radiation will be completely absorbed by the product, rendering no measurable signal. So it is important to consider your current and future product thickness operating range when selecting a measurement sensor.

Product Opacity For most gauging technologies, opacity is not an issue. However, for sensors that work on the principle of light measuring the sheet (specifically, infrared sensors and laser caliper sensors), the optical nature of the sheet is of critical importance. To make a measurement, infrared sensors need to be able to "see" through the sheet (with infrared light, not visible light). Laser caliper sensors, on the other hand, locate the top and bottom surfaces of a sheet, and need to be able to do this correctly to provide a stable thickness measurement. This is why this optical factor is important in gauge selection.

Uniform Density of the Product As thickness is usually the parameter the sheet processor is trying to control, it is important that gauge selected be able to measure thickness correctly. Nuclear, infrared and x-ray gauges actually measure the mass of a sheet. When the sheet density is constant, the thickness can be inferred directly. When sheet density varies, the sheet processor needs to consider carefully whether mass or thickness is the variable that needs to be controlled. This will help determine the correct sensor for the application. Factors that affect product density include the, regrind content, the additive content (colors, pigments, fillers), co-extrusion of materials with different densities, and Changes in surface roughness.

Co-extrusion Factors In addition to its effect on product density, co-extruded products usually raise a second measurement challenge: Can the individual layers be measured independently? Unfortunately, when it comes to sheet extrusion, the options are pretty limited. The only established technology that can approach this need is infrared absorption, and there are some limitations on this approach. First, infrared absorption measures components, not layers; so in an A–B–C–B–A co-extrusion, the infrared (IR) sensor may be able to determine how much of materials A, B, and C are present, but not how much of B is in the second versus the fourth layer. Second, to make this component measurement, the various materials in the co-extrusion must be chemically different. Polyolefins, for example, share almost identical IR spectral signatures, so they cannot be differentiated from each another. Third, for an IR sensor to work, the material must not be opaque to IR light. These three factors limit the use of IR on sheet extrusion to clear barrier-grade sheet materials; although IR can be used on most clear products, it does not find much use in sheet extrusion, due to its inability to measure opaque materials. To date, the best approach for co-extrusion control on sheet extrusion lines is to use gravimetric weigh hopper controls to make sure that the average weight of each component is correct, and then use a mass or caliper gauge to measure total thickness.

Economic Considerations Although the quality of the sheet is important to the processor, one of the primary drivers for purchasing a gauging system is the economic return that the improved thickness control and subsequent raw material and scrap savings bring. It is easy to understand that if two processors are running machinery with similar output rates and one is processing a material twice as expensive as the other, the former processor has twice the economic incentive to utilize a gauging system as the latter. Also, the first processor may be interested in greater accuracy, higher scanning speeds, and so on, and as a result, may select more advanced sensors and controls than the second processor would.

3.2.10.2 Type of Gauges

As mentioned earlier, online gauges fall into two major categories: mass-measuring devices and thickness-measuring devices. In general, mass-measuring devices tend to be more accurate, reliable, and simpler in design than thickness-measuring devices, so when they can be employed (e.g., if density is not an issue), they are the preferred approach.

Thickness-Measuring Devices Two types of thickness-measuring devices have emerged commercially: the air caliper sensor and the laser caliper sensor. The air caliper sensor uses one or two pneumatic servo-driven air pistons that create a constant air pressure between the piston and the sheet surface. In a dual-sided system, these two pistons sit directly opposed to each other, scanning back and forth across the sheet. As the sheet thickness varies, the pressure between the piston and the sheet changes, and the pneumatic servo responds to bring the piston back into balance. One of the pistons contains a magnetic reluctance sensor (which generates an eddy current), and the other piston contains a metal target. The reluctance sensor is used to determine the distance between the two pistons, from which a thickness measurement is derived.

One optional configuration of the air caliper uses one movable piston on top of the sheet with a backing roll. The piston rides on the sheet surface as in a dual-piston system; the magnetic reluctance sensor monitors the distance between the piston and the backing roll to determine sheet thickness. For this to work, the sheet must make intimate contact with the backing roll across its entire face.

Air caliper sensors have several limitations. They can be subject to errors created by changes in air temperature, which directly affect the air pressure at the head(s). Changes in the sheet surface can also create measurement errors. The use of a magnetic reluctance sensor requires that no metal foils or additives be present in the product. In addition, any dirt, moisture, or oil in the air can create operational and/or reliability problems with the measurement sensor.

Laser caliper gauges, like air caliper sensors, are available in both single-sided (measuring against a backing roll) and dual-sided designs. The principle of operation is fairly straightforward (Figure 3.42). A laser light beam is reflected off the sheet surface, back to a detector array, at an angle. Depending on the location of the sheet surface, the light strikes a different point on the array, which indicates the position of that surface of the web. By subtracting the two positions, one can derive the product thickness. To gain acceptable accuracy, one must also use a magnetic reluctance sensor to detect any up or down movement of the two sensor heads on the scanning rails, which can be on the order of several mils.

Laser sensors have to be used carefully, as the way light interacts with the surface of a product can affect accuracy. Since the sensor uses a magnetic reluctance compensation scheme, no metal foils or additives can be present in the product. Laser based systems are also fairly expensive.

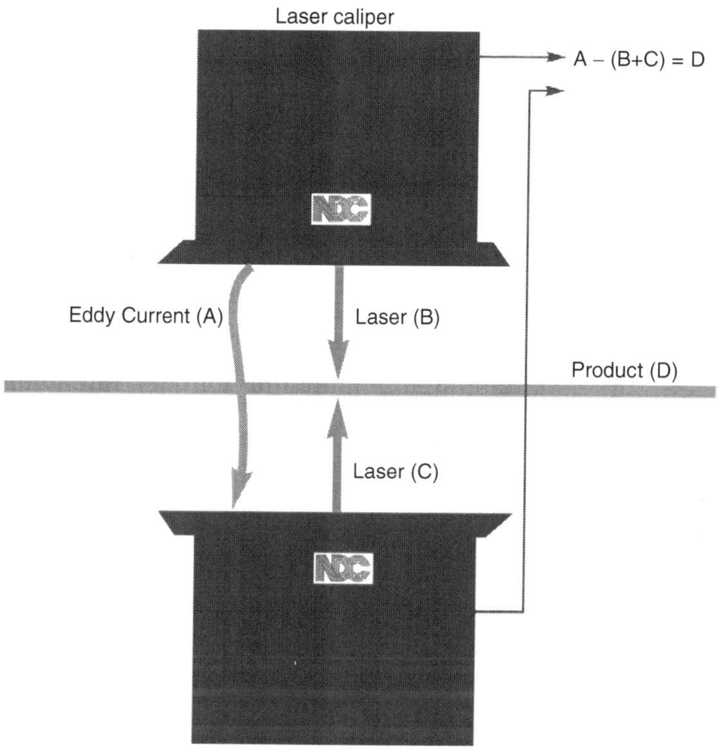

FIGURE 3.42 Laser caliper gauge. (Courtesy of NDC, Irwindale, CA.)

Mass-Measuring Devices Mass-measuring devices fall into four types: beta transmission, gamma backscatter (GBS), x-ray transmission (XRT), and infrared transmission. We have already discussed IR to some de gree. It has limited use in sheet extrusion, as the product cannot be opaque. Except for specialized use on clear-barrier sheet products, it is not normally considered for this application.

X-ray transmission sensors offer the advantage that they typically do not require government licensing for operation. XRT sensors operate on the principle of photon transmission. A high-voltage power supply excites an x-ray tube, which in turn outputs a stream of high-energy photons that impinge on the sheet. Some of these photons are reflected, some are absorbed, and the rest pass through the sheet. On the other side of the sheet, an ionization chamber collects the remaining photon and converts this energy into an electrical signal. The mass of the sheet is inversely proportional to the output of the ion chamber.

There are two challenges to the use of XRT sensors. First, the high-voltage power supply must provide a rock-solid output or the gauge will drift. Today's digitally controlled power supplies can manage this issue, at some expense. The second problem, fundamentally larger, is that the gauge is extremely sensitive to the chemical composition of the material. This effect is particularly prominent with mineral additives and fillers, such as calcium and titanium, commonly found in colorants and pigments. With XRT, this is a *fourth-order effect*, meaning that the gauge is *very* sensitive to additives. If a processor plans to run different materials on its sheet process line, the x-ray gauge will not be attractive.

Beta transmission gauges work similar to XRT gauges; the main difference is that a nuclear isotope is used to generate the signal used for measurement. Isotopes are selected that emit beta particles in the ideal range to make a thickness measurement. For sheet extrusion, two common isotopes are used: krypton (^{85}Kr) for lighter-gauge products (0.4- to 40-mil sheet) and strontium (^{90}Sr) for heavier-gauge products (5.0- to 250-mil sheet) [9]. As beta transmission gauges exhibit less composition sensitivity (beta transmission exhibits a second-order sensitivity to material composition), they are preferred over XRT gauges. However, beta gauges do require government licensing, although they are very safe to use when safety procedures are adhered to.

One other negative of the beta transmission sensor is that the isotopes used have relatively short half-lives (10 years for ^{85}Kr, 28 years for ^{90}Sr). This means that their output signal changes from day to day, and the gauges must be standardized on a frequent basis to minimize errors. Also, since the sensor consists of two components (source and detector, mounted on either side of the sheet), and these components must be scanned together very precisely to prevent measurement errors due to source–detector misalignment, a relatively expensive scanning frame called an O-frame must be employed. Figure 3.43 shows a beta sensor's operation.

The gamma backscatter sensor sees more widespread use on sheet extrusion processes than any other sensor. The GBS sensor emits photons via its isotope source

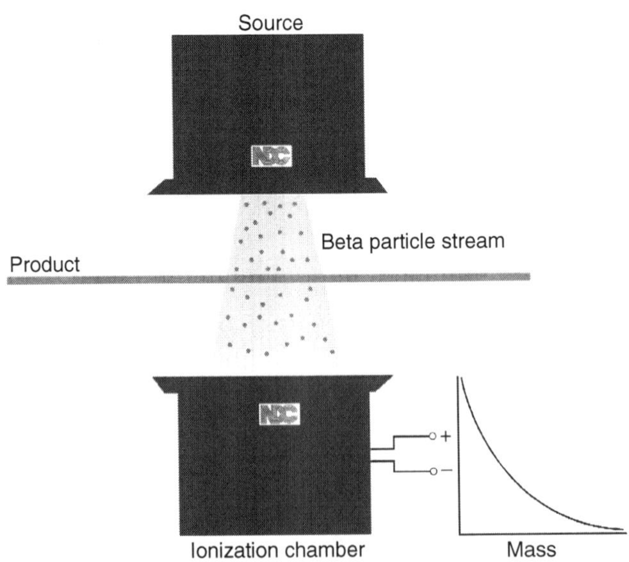

FIGURE 3.43 Beta transmission sensor. (Courtesy of NDC, Irwindale, CA.)

(americium 241). Some of these photons are absorbed by the product, or pass through. The photons of interest are the ones that reflect back to the detector, also known as Compton effect photons. It is easy to discriminate these primary reflected photons, as they are at a specific energy level. The photons are converted to light pulses and counted, and the energy count is linear in relation to thickness. Thus, the sensor is easy to calibrate. Since it is single-sided (both the source and detector electronics are housed in a single, flashlight-sized tube), the scanner mechanism is simple and relatively inexpensive. The isotope has a long half-life and therefore does not require standardization. Compton photons show the least sensitivity to composition of any of the mass-measuring devices, making calibration simple. For most processors, a single calibration will work on most, if not all, of their products. Figure 3.44 shows a GBS sensor's operation. Figure 3.45 shows a gamma backscatter sensor in scanning operation on a sheet extrusion process. The combination of simplicity in design, easy calibration, inherent accuracy, and trouble-free, long-term operation are the reasons why the gamma backscatter sensor sees such widespread acceptance by sheet extrusion processors. Once the sensor selection is made, the scanner design falls in quickly by default. The final decision rests with the operator interface, and what, if any, automatic thickness controls will be implemented.

Systems fall into two types: basic systems with PLC-like user interfaces, and higher-level polycarbonate-based systems, which still include proprietary electronics to manage sensor–scanner control and data processing. An important point for consideration is who will be using the system most of the time (typically, a machine operator) and making sure that the system has displays and operating functions that the level of person using the system will be able to grasp easily. A typical system for sheet will display the following data:

- Thickness profiles across the width
- Trends in average thickness in the machine direction
- A pictorial summary data (histograms, etc.)
- Numerical reports of quality throughout a job, batch, pallet, or reel

Many other measurement options are available to assist in diagnosing and fine tuning the sheet line. Displays are typically on a touch screen for ease of operator use and are simple to understand and operate with minimal training.

Many sheet producers use only the display and report capabilities of their systems. However, control, by closing the loop on haul-off speed (for average thickness) or thermal bolt heaters (for cross-direction profile control), are also in use at many plants. These options further enhance the payback on investment by always ensuring that the product is as flat as possible and always on target thickness.

A final consideration in gauging system selection is high-quality local support. The plant personnel become quite dependent on the system to optimize their process; hence it can seem like a crisis the day that a system is not available to control the product. Therefore, equipment and vendor selection are key to ensuring that the gauging system will probably not develop faults and that good local service is available at fair rates in the event it is ever needed.

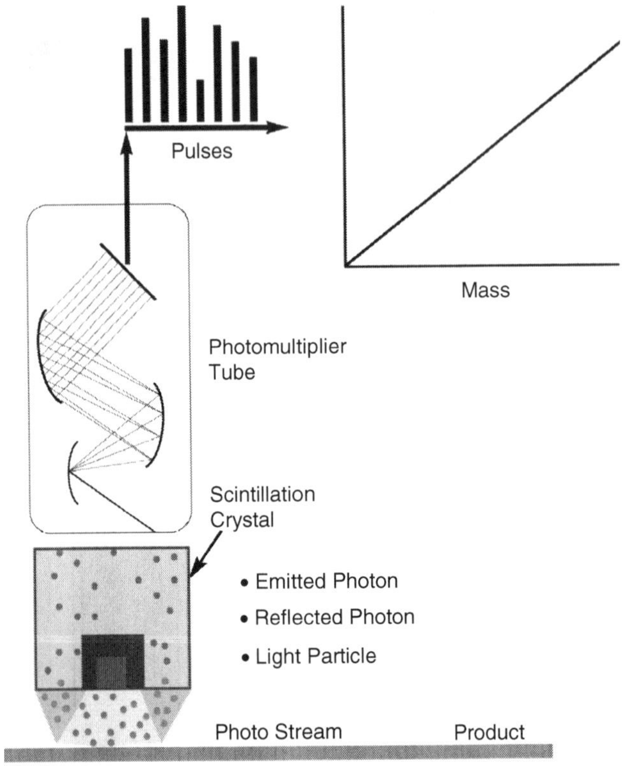

FIGURE 3.44 Gamma backscatter sensor. (Courtesy of NDC, Irwindale, CA.)

FIGURE 3.45 Gamma backscatter in scanning operation. (Courtesy of NDC, Irwindale, CA.)

3.2.11 Approaches to Winding Systems

Sheet is often supplied in roll form to various converting operations, such as roll-fed thermoforming, die cutting, and some printing applications. Rolls are sometimes shipped to end users for seamless installations as in roofing membranes or pond and landfill liners. Whatever the end use of the rolls may be, they must be supplied with straight edges at desired thickness that are free of gauge bands and be of uniform tension throughout.

The winding system to produce the rolls is as important in a sheet system as any other component. If the rolls are loosely wound, they will not unwind properly and will also cause problems in shipping. Inconsistent roll diameters will cause problems in shipping (i.e., fitting the rolls onto a shipping skid), and roll placement in converting equipment can be problematic. If the wound roll exhibits uneven edges, it will wander when unwinding into a secondary piece of machinery. For these reasons and others, roll geometry plays as significant a role as gauge uniformity and surface quality do in the resale of extruded sheet.

There are several designs of winders available. Each represents a different approach and subsequently, varying degrees of sophistication with respect to the technology that they represent. These can range from simple manual winders to complex designs with automatic functions and roll unloading features. The winders being presented herein include the manual, A-frame, and turret-style winders. Additionally, there are auxiliary devices that accompany the winder systems. An accumulator is such a device. It slows the web down or in some cases stops it altogether, such that the web may be handled manually in a safe manner. This is used most commonly with the higher line speeds often associated with thinner gauge materials. As an alternative to the accumulator, winders can also be equipped with automatic web transfer devices.

3.2.11.1 Manual Winders

The simplest and most common type of winder is a manual fixed-shaft winder. As the name implies, it consists of two or more winding spindles installed within steel side frames as depicted in Figure 3.46. The winding shafts are mounted into core chucks that are fixed into their respective positions relative to each other. The core chucks are typically driven individually with independent speed or torque regulation for each shaft. The core rod shafts that are placed into the chucks can be a pneumatic type of shaft with expanding lugs to secure the cardboard cores or simple solid shafts designed to support and hold the cores in place during winding activities. The pneumatic shafts have a rubber bladder inside which, once inflated, expands a set of lugs, which grip the inside of a winding core to keep it securely in place. In either case the core rod shafts are removed with the finished roll and fitted with another core to wind the next roll.

The word *manual* applies to the thread-up of the machine and cutover of the web. The sheet is manually threaded through support idlers and affixed to the winding core by means of two-sided tape or spray adhesive to get the roll started. The speed is increased on the winding shaft or arbor to take up the slack in the web as it enters

FIGURE 3.46 Manual fixed-shaft winder. (Courtesy of PTi, Aurora, IL.)

the winder. The winder will then simply continue to wind a roll to a predetermined diameter which is usually based on length. A counter is used to measure the length of sheet on the roll and provides an indication when the preset level is reached. When the full roll buildup has been reached, it is time for the operator to cut and transfer the web. The operator cuts the web with a utility knife and places the tail from the sheet onto an empty core which is on standby and once again wrapped in adhesive. The full roll is removed and the cycle starts over. There must always be a spare core rod shaft ready to insert into the position being emptied in order to keep pace with the winding production since there is not time to re-prepare the removed core rod for reuse. Manual winders typically require one more winding position than the number of webs being wound. This serves as the rotation winding position when transferring webs during a changeover. For instance, a single web line will have two winding shafts, a two-web line will have three shafts, and so on.

Manual fixed-shaft winders have a limitation to the maximum roll diameters that can be wound sensibly due to the spacing between the shafts. Typically, this type of machine will wind rolls to a maximum diameter of 40 to 44 in. (1000 to 1100 mm). A winder with more than two shafts may be impractical at these large diameters because the top shaft will be required to be very high, and roll removal becomes a logistics problem. An alternative to the fixed-shaft machine is an A-frame or portable winder.

3.2.11.2 A-Frame Winders

The A-frame winder consists of a very simple tubular steel frame arranged in an "A" pattern as displayed in Figure 3.47. It .features a single winding position consisting of a permanently configured winding arbor or removable pneumatic core rod shaft similar to the manual-style winder. The maximum diameter of the wound roll will be determined by the centerline distance that the core rod shaft is above the cross-bar support for the frame. These machines are often used to wind jumbo rolls of diameter 60 to 72 in. (1525 to 1830 mm). Two or more of these A-frame winders are used in-line, depending on the number of webs being wound simultaneously. The drive system can be with a standard ac or dc drive or portable hydraulic motor. If the drive is electric, it is mounted on a pedestal connected to the winding shaft by a flexible drive shaft. The drive system is designed specifically to be independent from the frame of the winder because the A-frames are caster mounted for mobility and are often used for storage of the large-diameter rolls. The ability to store wound rolls on the A-frames is an important feature of this style of winder. It negates the necessity to remove and handle the very large jumbo rolls. Multiple A-frames are commonly used for the winding and storage.

3.2.11.3 Turret Winders

Another type of winder is a turret-style machine. An example of which is displayed in Figure 3.48. The turret winder features a rotating turret assembly that supports each of its winding positions in a common chassis. The chassis is supported via side frames and other structural elements. The turret winder features a winding active position to the inside of the machine with a standby position always outboard of the machine. When the roll has reached its full diameter, the turret is indexed forward and the full roll is then rotated to the outside of the machine frames for convenient unloading and transfer of the web to the empty winding shaft. When the index is completed, the empty shaft drive is energized and increases in speed to match the speed of the sheet web coming off the line. This eliminates a slackening of the sheet, which can cause a poor start on the new roll. The advantage of this style of machine

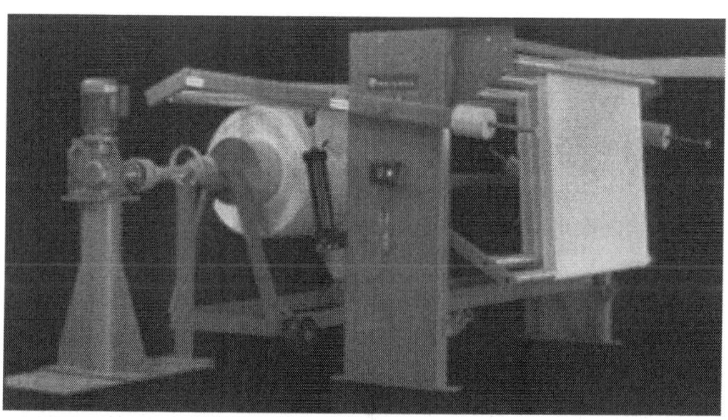

FIGURE 3.47 A-frame winder. (Courtesy of Menzel, Spartanburg, SC.)

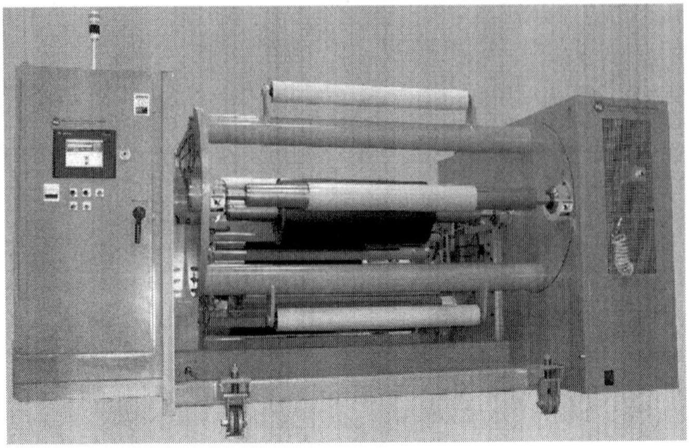

FIGURE 3.48 Turret-style winder. (Courtesy of Parkinson, Woonsocket, RI.)

is the ability to wind fairly large diameters at a relatively low centerline. Cutover to the empty core rod shaft is very convenient, as the roll is moved to a comfortable working height. Roll removal is simplified, as a cart can be used to support the roll from underneath.

A lay-on roll is often used on a turret machine to improve the quality of the wound rolls. *A lay-on roll* is an idler roll mounted within arms designed to ride on the winding roll as it builds up in diameter. The lay-on roll helps to push out air trapped between the layers of sheet and also aids in web tracking to provide straighter edges.

3.2.11.4 Accumulator

When winding relatively thin gauges of sheet, the line speed can become excessively high, making manual transfers of the web challenging and often unsafe. The transfer is accomplished when the operator physically cuts the web across the width of the sheet with a razor knife. A web traveling at high speeds becomes difficult to handle and safety becomes a serious issue. To alleviate this concern, an in-line accumulator is used as displayed in Figure 3.49. The accumulator is a sheet storage device which has a carriage containing idler rolls that are designed to rise and fall with web tension. When the operator begins to transfer from full roll to a prepared core rod, the web slackens and the accumulator rolls automatically take up the slack by spreading apart. By taking up the slack the web at the active winding position slows to a point where the operator can transfer the web safely and easily.

3.2.11.5 Automatic Web Cutoff and Transfer

Another solution to high line speeds with thin-gauge sheet is an automatic cutoff mechanism built into the machine. The automatic cut and transfer takes the place of an operator performing a manual cut with a handheld knife. The empty core still needs to be covered in an adhesive material for the start of the new roll, but this is

FIGURE 3.49 In-line accumulator. (Courtesy of Menzel, Spartanburg, SC.)

the only participation of the human operator. The automatic web cutoff and transfer method works in a sequence similar to that of manual or A-frame winders. The difference is that this method occurs automatically without the operator's assistance. Once the length counter has measured the preset length, the full roll will be indexed out of the winding active position and the new (empty) winding core brought into the active position. At this point the cut sequence is initiated to complete the transfer of the web from the full winding position to the empty ready position.

The cut is typically done in one of two ways. One cut method is a plunge knife. This consists of a fixed blade mounted on the cut arm which is driven down into the web path pneumatically or mechanically. When it contacts the web, the blade cuts through the sheet that is running across and just above the empty core. An idler roll positioned beneath the cut blade applies the web to the empty core,where it is caught in the adhesive on the surface and the new roll begins. A second method of cutting the web and transferring it automatically is with a flying knife. The sequence is the same as in the plunge knife above, but the knife is a traveling knife which is installed in a rodless pneumatic cylinder. Once the cut is activated, the blade is driven across the web, slicing the sheet as it travels, forming a triangular shape to the web at the cut location. The triangular cut leader of the web is contacted by an idler roll and applied to the surface of the tacky empty core as above. Additionally, winders are available with many automatic features, such as roll unloading, automatic spray systems to apply adhesive to the cores, shaft extraction (from the full roll), and programmable tension controls though a programmable logic controller. The automatic functions are designed for efficiency and personnel safety.

Edge trim on all winding operations must be taken to eliminate rough edges and/or edge beads which can occur at the sheet die during the extrusion process. The trim knives can be simple utility razor slitters or rotary anvil slitters, depending on the thickness and type of material. In any case it is advisable to place the slitting equipment as close as possible to the winding shaft. This placement will reduce uneven roll edges that will result from the tendency of the sheet to wander on the idler rolls that handle it entering and within the winder. Sheet wander is further reduced by precise alignment of the machine to the upstream components feeding it. The closer to perfectly parallel that the machine is installed, the more the sheet will follow a straight path to the winding shaft.

The simple fixed-shaft winders are provided only with basic controls for speed and tension control. Sophisticated turret winders require more complex control systems since the machine has critical sequencing operations for roll indexing and lay-on roll positioning. Tension control is the most important control function on a winder. As the roll gets larger, the surface speed of the roll increases which will create more tension on the web as it is drawn through the machine. The winder must be programmed to taper this tension so that the tension of the wind throughout the roll remains fairly even. A roll buildup calculator is one method used to sense and control this tension. The calculator compares the speed of the shaft of the drive motor with the line speed to determine where the diameter is, and the tension is adjusted electronically. Typical winding tension in sheet applications is 1 to 5 lb per linear inch across the web. Dancers and load sensing rolls are also used to control tension by measuring the tension in the web as it moves through the winder and making minute adjustments to the drive controls to compensate for changes in web tension and to maintain the winding tension desired.

3.2.12 Shearing, Sawing, and Stacking

Extruded sheet is commonly supplied in pieces which have been cut to a desired size with a corresponding cut-to-dimension tolerance. The sheet is transported on skids for converting operations such as heavy-gauge thermoforming, die cutting, slip sheets, and offset printing presses, to name just a few. The width of the sheet is trimmed to the desired size with in-line edge-trimming stations. The length is typically cut in-line with a shear, crosscut saw, or a specialty device commonly referred to as a *sheeter*.

3.2.12.1 Shear Cutting

The most common method of cutting sheet to length is the in-line power shear as depicted in Figure 3.50. A power shear for cutting plastic is very similar to those

FIGURE 3.50 Inline power shear. (Courtesy of Famco, Kenosha, WI.)

used in the sheet metal industry. They typically consist of a heavy-duty cutting table with an integral cutting head located above it. The opening between the table surface and the cutting head is minimal, usually less than 1.00 in. (25 mm). The table has a very smooth surface over which the sheet slides as it being pushed forward by the extrusion process. The shear is located immediately after the pull rolls so that a positive feed is ensured. There is a short section of conveyor between the pull roll unit and the shear to convey and support the sheet.

The sheet is fed on to the surface of the shear and continues moving through it. When a predetermined length has been measured by digital counter, a hold-down clamp is energized. The clamp is part of the head on the machine and driven by cylinders, typically pneumatic. This mechanism clamps the sheet to the bed or tabletop, holding it stationery prior to the cut, keeping the sheet square relative to the cutting blade, which is located on the edge of the head. The cutting point is where the moving blade on the head meets the stationery steel table edge. As the sheet is clamped into place, the cutting head is activated. The head is driven down and the blade mounted on the head's surface penetrates the sheet on the edge of the shear cutting table creating a guillotine effect between the moving blade on the head and the fixed table surface. The cutting head motion is controlled either by a clutch/brake assembly, hydraulic cylinders, or pneumatic cylinders.

The cutting table surface is covered in a nonmarring surface such as Micarta mounted onto steel. The hold-down clamp is covered in neoprene or a similar material again to prevent scratching and other damage to the sheet surface. The opening between the cutting table surface and the moving head are meticulously guarded to prevent hands and tools from being placed in danger from the cutting blades.

The sheet is continuously being pushed forward by the pull roll unit since sheet extrusion is a continuous operation. The slack will need to be taken up, and sheet is relatively stiff in nature. Therefore, a specialty conveyor section is placed immediately upstream of the shear device. The shape of the conveyor features a bowed affect so the sheet will naturally follow this contour instead of binding against the blade surface during cutting. When the hold-down clamp contacts the sheet and impinges it to the cutting table, the sheet web is stopped momentarily. This causes the sheet upstream to have a tendency to fold back on itself. The bow in the conveyor allows the sheet to form a small curvature above the conveyor for the period that the web is stopped. The inertia from the web pushes the sheet forward again when the cut is finished and the hold-down clamp has been released.

The squareness of the cut can be very important, depending on the secondary operation that the sheet will go through. Offset printing requirements, for example, have very stringent specifications for sheet squareness. An in-line power shear can cut sheet to a tolerance of ±0.125 in. (3.175 mm) of square for each sheet. This can be tightened to ±0.0625 in. (1.5875 mm) with machine modifications. If the requirements are tighter than this, a squaring shear will be required to trim the cut sheet to tighter specifications.

3.2.12.2 Saw Cutting

Some plastic sheet materials are brittle due to the crystalline nature of the polymer of which they are comprised. General-purpose polystyrene (GPPS) and acrylic (PMMA) are two that fall into this category. These are examples of sheet that would shatter if clamped and cut through by force from a guillotine type of shear as described above. Additionally, thick sheet applications and/or sheet that contain fillers may not be suitable for shear cutting.

For this type of sheet a powered circular saw is used for both the edge trim and crosscut, as shown in Figure 3.51. The in-line saw is located along the processing line much like the shear directly after the pull roll unit. The pull rolls provide a positive feed to the saw table. The saw consists of two sections. One section is for edge trim and the second is for crosscut. The edge trim saws are mounted in the machine direction and positioned to trim the sheet to a precise width, taking into account the width of the blades. They are adjustable for width.

FIGURE 3.51 Powered circular saw for cross-cutting. (Courtesy of Nutter, Herbon, OH.)

The crosscut section consists of a saw mounted in a traveling carriage to make a transverse cut when the preset length has been measured. The saw carriage includes a clamp that grips the still-continuous sheet web and travels along with it. The saw also begins its cycle at this time. The saw motor starts and the saw also travels in a machine direction with the sheet on the movable carriage. The clamp and saw travel with the sheet to give a cut that is perpendicular to the sheet. If it simply cut across the moving web, the cut would be on a diagonal. The saw is driven across the sheet by a rodless pneumatic cylinder. The blade is set for exact penetration into the sheet. It cuts through the sheet in the same fashion that a carpenter's saw cuts through plywood. The blades are very similar to those used for wood cutting and are specified by the blade manufacturer according to the type and thickness of the plastic to be cut. The edge trim and crosscut saws themselves are also similar to a construction contractor's saw. The motor typically operates above 3000 rpm. As with any rotating or moving blade, safety is a major concern. The blades are carefully guarded to make access while in operation virtually impossible.

Dust collection is also a critical part of the saw cutting and trimming system. Collection hoses and shrouds with a vacuum drawn on them are mounted just above the cutting blades to remove the significant amount of dust that is created from the saw cuts. The dust creates both a nuisance and a hazard. In some applications the dust can be a health concern if not suitably contained. It will also collect on exposed surfaces, including the finished product and plant floor. Dust particles can even be flammable given the right number of particles suspended in a specific volume of air. For these reasons alone, dust collection is a critical to power saw cutting operations.

3.2.12.3 Stacking the Newly Cut Sheet

After the cutting operation via a power shear or cutoff saw, the sheet is conveyed away by a powered conveyor. The conveyor is comprised of belts that are running in the machine direction with the sheet. They are powered by a variable-speed drive such that the speed of the conveyor can be adjusted to exceed the line speed of the sheet. The conveyor table then deposits the sheet into a sheet stacker. Stackers create a stack of sheet ready to be palletized and transported either to an outside customer or to another section of the plant for converting.

The simplest and most common type of stacker is a fork stacker (Figure 3.52), which has rods or forks that extend and retract in the machine direction. The sheet is conveyed onto the (extended) forks by the powered conveyor. The machine can be programmed to accept just one sheet or a small stack of sheets. Once the desired quantity of cut-to-length sheet pieces has been reached, the forks retract, depositing the sheets into a stack. This eliminates the need for operators to handle the hot sheet and possibly not building a straight stack, due to difficulties in handling the product manually.

There are many designs for conveyors and stackers. Some designs will remove a full pallet and replace the skid automatically. There are other types that use robotic arms with vacuum cups to pick the sheet off the conveyor and stack the sheet in neat piles where they can be accessed by fork truck or carts. There are other types that feature a stacking carriage. The carriage is supported by pneumatic cylinders, enabling the stacker to handle one sheet at a time while also controlling the distance

FIGURE 3.52 Fork type sheet stacker. (Courtesy of Vulcan Machinery Group, Akron, OH.)

that the sheet is dropped onto the stacked pile. This style of stacker minimizes surface scratches and marring due to stacking. The degree of automation and operational features often dictates the cost of the equipment. This is also the case in cutting sheet and handling the parts.

3.3 PROPER SIZING OF THE EXTRUSION COMPLEX

Estimating the sizes for the extruders that are required to create a desired co-extruded sheet structure must be approached carefully. Not sizing an extruder properly can compromise the system's ability to process the co-extruded structure at the desired rate of capacity, thus compromising production efficiencies. A fundamental approach to sizing the extrusion complex begins with a definition of the end product. This includes layer structure sequencing and designated resins, layer thicknesses, combined sheet width objectives, and overall desired rates of performance. The layer-to-layer resin compatibilities must be checked to assure that adjacent layers can adhere to one another chemically. If there is a question of resin compatibility, a special adhesive resin may be necessary. In Section 3.4 we present a comprehensive guide for examining layer-to-layer compatibility issues and give many examples of commonly used adhesive resin materials that promote good layer-to-layer adhesion. Once the issue of resin compatibility is addressed the extrusion system complex may be assembled. By examining the layer-to-layer sequencing of the desired co-extruded structure, an estimation of the percentage of each unique resin is derived and the overall contribution for an extrusion device may now be calculated by applying the layer percentage range to the overall rate of capacity of the production line.

The melt density of the desired resins is a fundamental variable that contributes to the calculation for determining specific rates of performance for various extruder sizes. Table 3.7 provides a summary of several commonly used resins and the corresponding melt densities for each. These are typically obtained from resin suppliers and are easily measured using common laboratory measurement techniques. By applying the melt density figures to a proprietary analysis technique, the specific rates

TABLE 3.7 Thermoplastics and Typical Average Melt Densities

Thermoplastic[a]	Density [lb/ft³ (g/cm³)]	Specific Volume [in³/lb (cm³/g)]
LDPE	57.43 (0.92)	30.09 (1.09)
LLDPE	57.43 (0.92)	30.09 (1.09)
HDPE	59.93 (0.96)	28.83 (1.04)
PP	56.81 (0.91)	30.42 (1.10)
HIPS	64.93 (1.04)	26.62 (0.96)
GPS	65.55 (1.05)	26.36 (0.95)
ABS	63.68 (1.02)	27.14 (0.98)
PET	86.15 (1.38)	20.06 (0.72)
PETG	79.28 (1.27)	21.80 (0.79)
RPVC	83.65 (1.34)	20.66 (0.75)
FPVC	81.16 (1.30)	21.29 (0.77)
PC	74.91 (1.20)	23.07 (0.83)
PMMA	74.29 (1.19)	23.26 (0.84)
EVOH	74.29 (1.19)	23.26 (0.84)
PVDC	116.74 (1.87)	14.80 (0.53)
PA	71.17 (1.14)	24.28 (0.88)
EVA	70.54 (1.13)	24.50 (0.88)
E/MMA	58.68 (0.94)	29.45 (1.06)
SBS	64.30 (1.03)	26.87 (0.97)

Source: PTi, Aurora, IL.

[a]ABS, acrylonitrile–butadine–styrene; E/MMA, ethylene methacrylic acid (ionomer); EV A, ethylene vinyl acetate; EVOH, ethylene vinyl alcohol; FPVC, flexible polyvinyl chloride; GPS, general-purpose polystyrene; HDPE, high-density polyethylene; HIPS, high-impact polystyrene; LDPE, low-density polyethylene; LLDPE, linear low-density polyethylene; PA, polymide (nylon); PC, polycarbonate; PET, polyester; PETG, glycol-modified polyester; PMMA, polymethyl methacylate (acrylic); PP , Polypropylene; PVDC, polyvinylidiene chloride (Saran); RPVC, rigid polyvinyl chloride, SBS, styrene–butadine–styrene.

of performances can be estimated for v arious extruder bore sizes. These can be seen in Table 3.8 for a v ariety of resins, corresponding to a wide range of e xtruder bore sizes. The figures are presented in rates per rpm of the feed scre w for each extruder bore size listed. The rate figures are presented in pounds per scre w rpm per hour of production. To estimate the overall performance of a particular machine size, the rate figure must be multiplied by the tar geted screw speed in rpm. F or example, high-impact polystyrene (HIPS) has a density of 1.04 g/cm³. Looking across the matrix at the column for a 5-in. e xtruder yields a rate of 14.75 lb per rpm per hour. HIPS can operate at relatively high screw speeds. Take 150 rpm for instance, and multiply it by the 14.75 rate figure to yield 2212.5 lb/hr at a screw speed of 150 rpm. Reapplying this procedure at the same machine size and resin, but increasing the scre w speed to 180 rpm, yields an estimated rate of 2655 lb/hr at this higher screw speed.

As a further example of sizing the extrusion complex consider an application that requires a three-layer structure consisting of crystal polystyrene (GPS) skins with a HIPS core with a layer sequencing of 10–80–10% at an o verall production rate of

TABLE 3.8 Rate Estimation Matrix for Common Thermoplastic Resins[a]

		Extruder Size												
	Millimeters →	40	50	65	75	90	100	115	125	140	150	165	175	200
	Inches →	1.5	2.0	2.5	3.0	3.5	4.0	4.5	5.0	5.5	6.0	6.5	7.0	8.0
Thermoplastic[b]	Density (g/cm³)	Pounds per Revolution per Hour[c]												
LDPE	0.92	0.35	0.83	1.63	2.82	4.47	6.68	9.51	13.05	17.36	22.54	28.66	35.80	53.44
LLDPE	0.92	0.35	0.83	1.63	2.82	4.47	6.68	9.51	13.05	17.36	22.54	28.66	35.80	53.44
HDPE	0.96	0.37	0.87	1.70	2.94	4.67	6.97	9.92	13.61	18.12	23.52	29.91	37.36	55.76
PP	0.90	0.34	0.82	1.60	2.76	4.38	6.53	9.30	12.76	15.99	18.47	23.84	29.36	41.73
HIPS	1.04	0.40	0.94	1.84	3.19	5.06	7.55	10.75	14.75	19.63	25.48	32.40	40.47	60.41
GPS	1.05	0.40	0.95	1.86	3.22	5.11	7.62	10.85	14.89	19.82	25.73	32.71	40.86	60.99
ABS	1.02	0.39	0.93	1.81	3.12	4.96	7.41	10.54	14.46	19.25	24.99	31.78	39.69	59.25
PET	1.40	0.54	1.27	2.48	4.29	6.81	10.16	14.47	19.85	26.42	34.31	43.62	54.48	81.32
PETG	1.27	0.49	1.15	2.25	3.89	6.18	9.22	13.13	18.01	23.97	31.12	39.57	49.42	73.77
RPVC	1.34	0.51	1.22	2.38	4.10	6.52	9.73	13.85	19.00	25.29	32.84	41.75	52.14	77.83
FPVC	1.30	0.50	1.18	2.30	3.98	6.32	9.44	13.44	18.44	24.54	31.86	40.50	50.59	75.51
PC	1.20	0.46	1.09	2.13	3.68	5.84	8.71	12.41	17.02	22.65	29.41	37.39	46.69	69.70
PMMA	1.19	0.46	1.08	2.11	3.65	5.79	8.64	12.30	16.88	22.46	29.16	37.07	46.31	69.12
EVOH	1.19	0.46	1.08	2.11	3.65	5.79	8.64	12.30	16.88	22.46	29.16	37.07	46.31	69.12
PVDC	1.87	0.72	1.70	3.31	5.73	9.10	13.58	19.33	26.52	35.30	45.82	58.26	72.77	108.62
PA	1.14	0.44	1.03	2.02	3.49	5.54	8.28	11.79	16.17	21.52	27.94	35.52	44.36	66.22
EVA	1.13	0.43	1.03	2.00	3.46	5.50	8.20	11.68	16.02	21.33	27.69	35.21	43.97	65.64
E/MAA	0.94	0.36	0.85	1.67	2.88	4.57	6.82	9.72	13.33	17.74	23.03	29.29	36.58	54.60
SBS	1.03	0.35	0.85	1.63	2.82	4.47	6.68	9.51	13.05	17.36	22.54	28.66	35.80	53.44

Source: PTi, Aurora, IL.

[a]This matrix is for reference information only. Actual rates may vary

[b]Abbreviations are defined in Table 3.7.

[c]To calculate kg/rpm·h, multiply the corresponding rate data by 0.4531.

4000 lb/hr. The skin layers require an extruder sized for an overall contribution of 20% of the structure (i.e., two skins, each 10%), and the core layer requires 80%. Therefore, the skin extruder is sized for 800 lb/hr and the core is sized for 3200 lb/hr Assume that each extruder will operate at 150 rpm. Divide both overall capacity requirements by this figure to yield 5.33 lb per rpm per hour for the skins and 21.33 lb per rpm per hour for the core. For GPS at 5.33 lb per rpm per hour, Table 3.8 yields an estimated extruder size of 3.5 in., correlating to 5.11 lb per rpm per hour. Since this estimated rate figure is lower than the required figure, the extruder would have to operate at a rate higher than the targeted screw speed of 150 rpm to keep the layer ratios consistent. Furthermore, HIPS at 21.33 lb per rpm per hour yields an estimated extruder size of 6 in. correlating to 25.48 lb per rpm per hour. Thus, the skin extruder would be a 3.50 in. bore size and the core extruder would be a 6.00-in. bore size. At 150 rpm the combined capacity estimation would be 4588.50 pounds lb/hr. Our example requires 4000 lb/hr, which is lower than the estimated overall capacity by approximately 15%. Therefore, the selections for the corresponding extruder sizes are well suited for a three-layer application.

It is generally a good practice to oversize the extrusion complex by a margin of 10 to 20% to prevent limiting the overall capacity of the line. It is also clear from the example cited that in each case the extruder can be increased in size to reduce the speed at which it must operate to yield the desired rate of performance. This can be beneficial to yield a controllable lower melt temperature given a complex co-extruded sheet structure where melt temperature–sensitive materials are involved.

3.4 CO-EXTRUSION: COMPATIBILITY OF RESINS AND END PRODUCTS

Selecting the appropriate co-extrusion structure for an application can be challenging. There are several importance aspects of this evaluation that require careful consideration to avoid costly mistakes. Fundamentals of resin compatibility coupled with appropriate hardware choices are particularly important. The science of co-extruded sheet and improvements and developments in resin technologies challenges even the most experienced processors. There are several fundamental aspects that serve as steadfast rules and guidelines to which both new and experienced processors must adhere. These are defined in detail together with the corresponding effects that each plays relative to meeting the objective of producing a high-quality co-extruded structure.

3.4.1 Measure of Compatibility

In evaluating the compatibility of various resins there are three fundamental factors that need to be examined. The first is the flow and viscosity characteristics of the materials being combined. This is an important factor to assure that the targeted viscosities are kept within reasonable limits in order to be processed through a feed block or multimanifold die. The second factor is the continuity of melt temperatures of adjacent layers. This is important since heat can be transferred between layers if one is considerably different than the other. Some materials are heat sensitive and can be affected negatively by being coupled to a higher melt temperature–based resin. A significant temperature differential between adjacent layers can also cause flow disturbance, and in some severe cases, degradation of the polymer. The third factor that needs to be addressed is that of the adhesive forces which must be present between adjacent layers of a structure. Naturally, if neighboring polymers do not exhibit mutual adhesive bonding, the net result will be a structural failure in the end product desired.

3.4.1.1 Melt Viscosity

One of the critical factors to be considered in the co-extrusion process is the flow compatibility of the materials making up the structure. Whether a process utilizes a feed block or multimanifold die, or a combination of a feed block and a multimanifold die, the flow characteristics of the adjacent layers is of paramount importance if the structure is to have proportional and layer-to-layer uniformity and consistency. Viscosity is defined as the melt viscosity of each constituent at its operating melt temperature. In this context co-extrusion is usually feasible if the melt viscosities can be kept within the range 2 to 3:1. The closer the viscosity match of the combining resins, the better both the layer and proportional distribution will be within the structure. When material viscosities deviate beyond the desired range, the lower-viscosity components tend to give heavy end flows, and a pronounced encapsulation can occur, particularly when thin skin layers (i.e., less than 5%) are desired. One of the reasons that this occurs is that the skin layer is exposed to the highest shear rates in the die. The skin layers are exposed to the die's inner flow surfaces as the structure migrates through and eventually exits the die. This effect can be extremely significant when using a feed block and a single-manifold die. A classic example of this behavior is exhibited using a feed block and single-manifold die with HIPS and GPS as depicted in Figure 3.53. The lower-viscosity GPS exhibits a strong tendency to run to the ends of the die and encapsulate the main component of HIPS. Fortunately, designers of feed blocks have been able to contour feed block inserts to overcome the GPS's tendency, and the HIPS–GPS structure is widely used throughout the industry in drinking cup applications.

Multimanifold dies, although not entirely immune to the encapsulation phenomenon, have a distinct advantage because distribution is established in individual manifolds before combining. This is accomplished just prior to exiting the die (Figure 3.54). The viscosity limits stated earlier apply to both hardware approaches (i.e., the feed

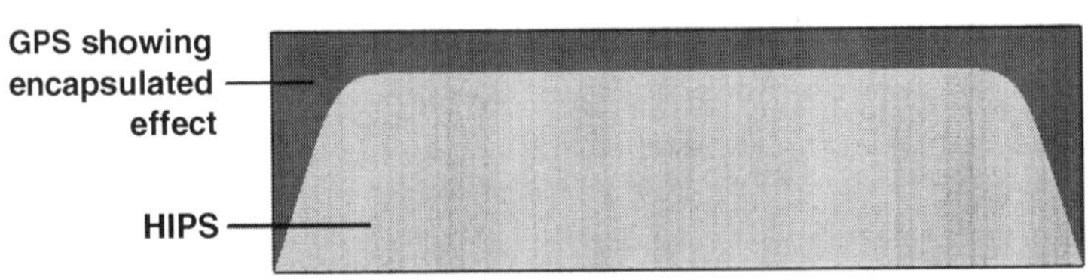

FIGURE 3.53 Encapsulation of a co-extruded structure. (Courtesy of PTi, Aurora, IL.)

block/single-manifold die as well as the multimanifold die). Exceeding the ranges will result in irregularities and distribution problems in the final product.

Flow anomalies with feed blocks can be overcome by contouring the inserts as noted in the GPS–HIPS example. Distribution exiting the feed block is altered so as to compensate for the behavior of the cap layer and adjacent interior layers within the die. In some cases the ability to groom or contour a feed block is limited, due to viscosity melt temperature differences in the adjacent layers. In these cases, use of a multimanifold die with manifolds specially designed for individual components is a better approach. Each layer has been fully distributed prior to being combined as a structure just prior to exiting the die as presented in Figure 3.54. Manifolds can be designed for a dedicated layer or a combination of layers, as is the case where a feed block is used in conjunction with a multimanifold die. The viscosity limitations must be adhered to in order that severe flow instabilities are avoided.

3.4.1.2 Melt Temperature

As in all extrusion processes, melt temperature is extremely important. In co-extrusion the importance is magnified, due to the possibility of using materials with different flow characteristics in adjacent layers. Melt temperature variations up to approximately 50° F (27.7°C) can be tolerated in a feed block. If the melt temperatures of resins in adjacent layers exceed this differential value, controlling the flow in the die manifold becomes more difficult. Shear rates throughout the die vary widely. The skin or cap layers are exposed to the highest shear rates. The core layer experiences shear

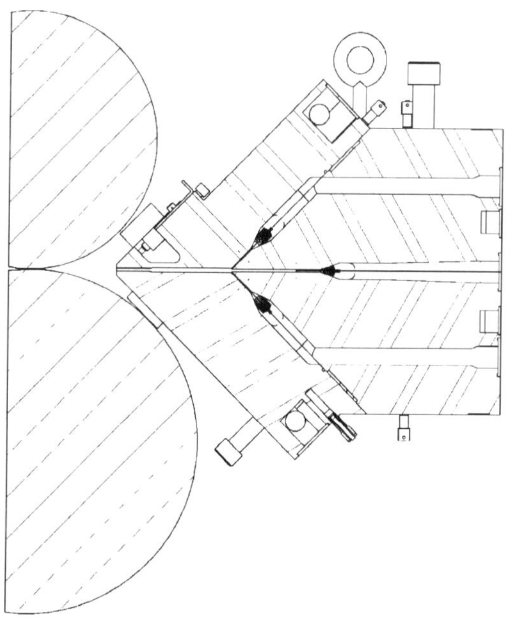

FIGURE 3.54 Combining layers in a multi-manifold die. (Courtesy of EDI, Chippewa Falls, WI.)

rates that are extremely low in comparison. When combined with high differential melt temperatures, these varying shear rates yield a combined product structure that will distort uniformity to an unacceptable degree. An example of this is to consider a basic three-layer structure that consists of a relatively high core layer viscosity combined with a lower cap layer viscosity. The core layer will tend to be thick in the center of the sheet, and the cap layers flow heavy to the die ends. In the most extreme cases this causes a total end encapsulation.

Multimanifold dies allow specific manifold designs to be applied to a range of targeted materials. Subsequently, this enables a higher differential melt temperature variation to be tolerated while still achieving good results. A multimanifold die typically permits melt temperature differentials between adjacent materials to be as high as 75° F (41.6°C). At higher temperature differentials, isolating individual manifold temperatures from neighboring manifolds becomes an insurmountable challenge for the die designer. In any event, if large temperature differences are to be encountered, the multimanifold die approach would be favored over a simple feed block/single-manifold die approach, but limitations do exist and sometimes cannot be overcome easily.

Examples of materials that cannot be combined in normal circumstances can be cited. A classic example is nylon 6 or nylon 6,6. It possesses a relatively high melt temperature of over 500° F (260°C). This would not be a suitable candidate to co-extrude with polyvinyl chloride (PVC), which possesses a correspondingly low melt temperature, which ranges between 360 to 400° F (182 to 204°C). If this combination were attempted, the PVC would degrade due to the heat that is transferred as a result of the exposure of the higher temperatures that are characteristic of processing nylon, thus making the product unsuitable for use.

Resin blending and employing techniques to vary the resulting molecular weights are used to overcome apparent adverse melt temperature differentials. Many material suppliers have addressed this problem using relatively sophisticated polymers that permit processing structures that by previous standards were considered incompatible.

3.4.1.3 Adhesive Effects

It is absolutely imperative in any co-extruded structure that the adhesive forces between adjacent layers have adequate bonding to maintain structural integrity throughout the life of the product. Layer-to-layer adhesion in the co-extrusion process is not due to primary forces such as chemical bonding. This can be defined further as not relating to a typical chemical reaction present in the extrusion process, even though crystalline materials go through a phase change from a liquid to a solid as they are heated and cooled and amorphous materials go through a glass transition temperature. There are unique forces that cause adjacent layers to bond together. They are referred to as *secondary forces or van der Waals forces*. [10].

Van der Waals forces are commonly found in polymeric chains and usually exhibit themselves in the co-extrusion process as polar or dipolar effects. These forces are quite small compared to chemical bonds, but in a polymeric network where longer molecular chains are present, these forces are quite significant. These intermolecular forces are very prevalent in polymeric families that have similar molecular backbones. Following this supposition in terms of adhesive effects in the co-extrusion process, it is reasonable to expect that materials will bond to similar materials. Consider olefins, for example. They adhere readily to each other without the need of an adhesive or bonding agent. Included in the olefin family are low-density polyethylene (LDPE), polypropylene (PP), linear low-density polyethylene (LLDPE), thermoplastic olefins (TPOs), and high-density polyethylene (HDPE). It is found that they all adhere to each other readily without the need of an intermediary agent. There are possible exceptions. Additives can challenge these natural tendencies and make it difficult to achieve good layer-to-layer bonding due to their specific characteristics. Slip agents in high concentration are one of several examples of additives that can have a negative impact on the naturally occurring bonding that takes place between many materials.

Similarly to the olefins, styrenic compounds exhibit mutual attraction to each other without the aid of an intermediary ingredient. The example cited earlier of HIPS–GPS is an excellent example of members of the styrenic family showing mutual affinity toward each other in a co-extrusion structure. Acrylonitrile–butadiene–styrene (ABS) and styrene–butadiene–styrene (SBS) are other styrenic compounds that generally adhere quite nicely to each other without the need of an intermediary ingredient. The one notable exception involves ABS. When the acrylonitrile content exceeds 20%, mutual adhesion with a styrenic compound is retarded. This can be explained by the fact that the acrylonitrile molecule acts as an inhibitor to adhesion in concentrations over 20%.

Vinyl compound adhesive behavior is, as one would expect, that rigid polyvinyl chloride (RPVC) would adhere readily to flexible polyvinyl chloride (FPVC) materials. As expected, there are many examples of RPVC–FPVC in the sheet industry, and examples of these combinations are also prolific in many industrial profiles. Vinyl compounds present some unique problems with respect to co-extrusion because of their heat sensitivity and because of this fact that commercial applications are limited to compounds within the vinyl family.

When the naturally occurring adhesive forces between resins are not sufficient to promote adequate bonding for structural integrity or if additives and/or fillers are present that alter the normal polymeric bonding affinity, an adhesive material may be used to create the necessary bond that otherwise would not exist. There are numerous materials that act as adhesives to enable this bond to occur. These materials generally fall in three polymeric groups: the ionomers, ethylene methacrylic acids (EMMAs), ethyl vinyl acetates (EVAs) and the SBS family of materials.

Table 3.9 gives a comprehensive breakdown of adhesive characteristics used in many commercially available structures. Materials are grouped into their basic chemical families: styrenics, olefins, polyesters, vinyls, and bonding or adhesive resins.

3.4.2 Commercially Available Structures

Co-extrusion use is extremely prevalent in the food-packaging industrial market segment. The most common structures run the gamut from two to seven layers. There

TABLE 3.9 Guide to Adhesion in Co-extruded Structure

Material[b]	LDPE	LLDPE	HDPE	PP	TPO	HIPS	GPS	ABS	PET	PETG	RPVC	FPVC	PC	PMMA	EVOH	PVDC	PA	EVA	E/MA	SBS
Olefins																				
LDPE	G	G	G	G	U	U	U	U	U	U	U	U	U	U	U	U	U	G	G	G
LLDPE	G	G	G	G	U	U	U	U	U	U	U	U	U	U	U	U	U	G	G	G
HDPE	G	G	G	G	U	U	U	U	U	U	U	U	U	U	U	U	U	G	G	G
PP	G	G	G	G	U	U	U	U	U	U	U	U	U	U	U	U	U	G	G	G
TPO	G	G	G	G	U	U	U	U	U	U	U	U	U	U	U	U	U	G	G	G
Styrenics																				
HIPS	U	U	U	U	U	G	G	G[c]	U	U	U	U	U	U	U	U	U	G	U	G
GPS	U	U	U	U	U	G	G	G[c]	U	U	U	U	U	U	U	U	U	G	U	G
ABS	U	U	U	U	U	G[c]	G[c]	G	U	U	U	U	G	G	U	U	U	G	G	G
Polyesters																				
PET	U	U	U	U	U	U	U	U	G	G	U	U	U	U	U	U	U	G	G	U
PETG	U	U	U	U	U	U	U	U	G	G	U	U	U	U	U	U	U	G	G	U
Vinyl																				
RPVC	U	U	U	U	U	U	U	U	U	U	G	G	U	U	U	U	U	?	?	?
FPVC	U	U	U	U	U	U	U	U	U	U	G	G	U	U	U	U	U	?	?	?
Miscellaneous																				
PC	U	U	U	U	U	U	U	G	U	U	U	U	G	G	U	U	U	U	U	U
PMMA	U	U	U	U	U	U	U	G	U	U	U	U	G	G	U	U	U	U	U	U
Barriers																				
EVOH	U	U	U	U	U	U	U	U	U	U	U	U	U	U	G	U	U	G	G	G
PVDC	U	U	U	U	U	U	U	U	U	U	U	U	U	U	U	G	U	G	G	G
PA	U	U	U	U	U	U	U	U	U	U	U	U	U	U	U	U	G	G	G	?
Adhesives																				
EVA	G	G	G	G	G	G	G	G	G	G	?	?	U	U	G	G	G	G	?	?
E/MMA	G	G	G	G	G	U	U	U	G	G	?	?	U	U	G	G	G	?	G	?
SBS	G	G	G	G	G	G	G	G	U	U	?	?	U	U	G	G	?	?	?	G

Source: Ref. 11. Courtesy of PTi, Aurora, IL.

[a]G, good adhesion; U, unsatisfactory; ?, questionable or unknown.
[b]TPO, thermoplastic olefin. Other abbreviations are defined in Table 3.7.
[c]Good with acrylonitrile content below 20%.

are materials that use layer multiplier technology for iridescent flms, which can possess over 100 layers. But this is indeed a specialty product limited to iridescent films for decorative packaging. The vast majority of applications for sheet f all into the two-to seven-layer category.

3.4.2.1 Two-and Three-Layer Structures

Two-layer structures are perhaps the simplest form of co-e xtrusion available and are used widely for placing a cap layer onto a base material. A prime example of a tw o-layer structure is that of the HIPS–GPS in beverage cups, the HIPS being used for structural integrity and formability and the GPS being used to produce a glossy layer for appearance. Outdoor signage applications use either polymethyl methacrylate (PMMA) on ABS or polycarbonate (PC) on ABS. The ABS is used because of its strength; the PC and PMMA with ultra violet stabilizers are used for their weatherability . Other widely used applications include plastics bo wls and dishware. Figure 3.55 shows typical structures of two-layer products, with cap layers usually representing only 5 to 10% of the entire structure.

Two and three structures can be manuf actured by using either tw o or three extruders. Setups with two extruders can produce monolayer, A–B, B–A and A–B–A structures (where the A and B designation refers to one of each of the tw o extruders). This structure can be created using either a multimanifold die or a co-extrusion feed block/single-manifold die approach. This can be seen quite readily in the typical arrangement shown in Figure 3.56. By adding a third e xtruder and a properly configured feed block or multimanifold die, an A–B–C structure can be made. The advantage of this system (Figure 3.57) is that the skin (or cap) layers can be of tw o distinct materials versus that of an A–B–A structure.

3.4.2.2 Three-Layer Classic Structure

Three layer structures represent a classic form of co-e xtrusion and are e xtremely practical with today's emphasis on using regrind. A simple, yet extremely important structure is the utilization of tw o skin layers o ver a regrind layer. The primary benefit of this structure is the utilization of materials in the core that cost less mone y or require reuse due to their abundance as a by-product in a process, as is the case with many thermoforming applications. The use of a three-layer structure utilizing regrind as a center component with tw o distinctly different skins is used in man y food-packaging applications, with polypropylene, polystyrene, or polyester being the principal ingredient. One skin can be used as a food-contact layer with properties designed to retain fla vor, and another outside skin layer can be designed to

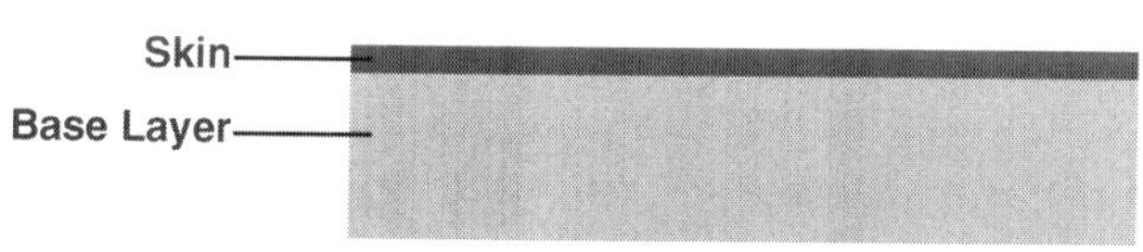

FIGURE 3.55 Two-layer structure, (Courtesy of PTi, Aurora, IL.)

FIGURE 3.56 Extruder arrangement for an A–B–A structure. (Courtesy of PTi, Aurora, IL.)

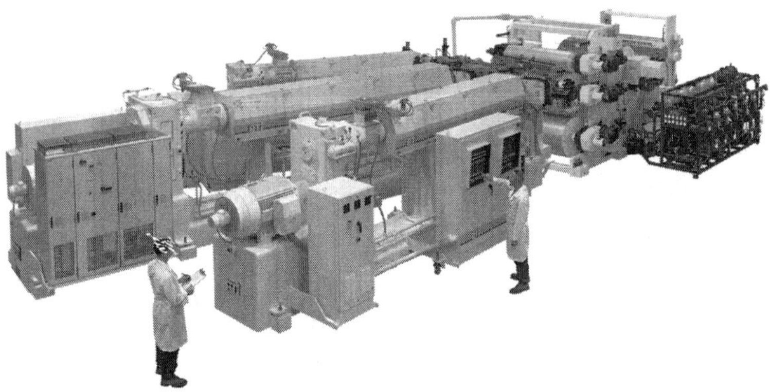

FIGURE 3.57 Extruder arrangement for an A–B–C structure. (Courtesy of PTi, Aurora, IL.)

enhance sealability and the freshness of the product. The center layer is usually a regrind or blend of regrind and virgin material.

Other three-layer structures in volve the tw o materials mentioned earlier with PMMA/ABS/PMMA and PC/ABS/PC, with skin layers containing ultraviolet stabilizers for outdoor furniture such as stadium seats and similar applications. The food-packaging industry uses similar structures with polyester , polypropylene, and/or high-impact polystyrene as the main ingredient for packaging aesthetics.

3.4.2.3 Four Layer Structures

Four-layer structures are not as common as three-layer structures, by virtue of their demand and market potential. A classic example of a four-layer structure is that used to create a disposable beverage cup. The four-component structure is a HIPS structure consisting of a white HIPS inner skin layer that will come into contact with the beverage liquid, a core layer consisting of approximately 50% regrind, another decorative colored HIPS layer, and a glossy outer skin made from GPS for gloss effect. This four-layer product is favored by many manufacturers, particularly those with inline thermoforming operations. Beverage cup forming creates a great deal of skeletal scrap, due to the unique mold geometries and the respective mold cavity center-to-center distances that are required. The scrap rates can be as high as 50% with all of the material going back into the co-extrusion process as the core layer, making the economics of this structure very attractive. Most applications for this product use a four-layer feed block. However, there are a number of processors that produce this product using more sophisticated manifold dies. The layout for this arrangement can utilize either a three-or-four extruder complex (Figure 3.58).

3.4.2.4 Five-, Six-, and Seven-Layer Structures

Structures of this nature typically involve a unique performance objective for the end-product materials. This can be a moisture or oxygen barrier, or both. Seven-layer co-extruded barrier sheet is made using a specialty barrier resin that consists of ethylene vinyl alcohol (EVOH). Due to the lack of naturally occurring bonding characteristics, an adhesive must be applied to both sides of the EVOH. Usually, this is an ethylene vinyl acetate (EVA) family of products. This is then followed by two layers of regrind (one on each side of the adhesive–EVOH–adhesive sandwich) and an inside and an outside skin layer (Figure 3.59). The inside layer is the food-contact layer, which is usually white for aesthetic value. The two skins can be the same resin or different, depending on the food being packaged.

Six-layer structures use the same basic ingredients as the seven-layer structure. However, the regrind is typically placed all on the same side of the structure

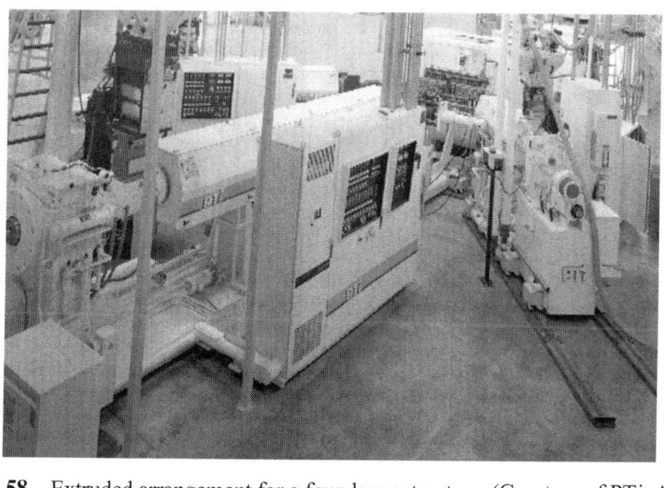

FIGURE 3.58 Extruded arrangement for a four-layer structure. (Courtesy of PTi, Aurora, IL.)

adjacent to the outside skin layer, forming an asymmetric structure. The reason for this is that retort packaging requires the use of steam for cooking the product contents in the formed container. This asymmetric placement shifts the barrier layers such that the moisture from the steam will have minimal effects on the EVOH barrier performance. Placing the regrind as shown in Figure 3.60 accomplishes the results desired.

One unique product which deserves special mention that is not used for food packaging is the automotive gas tank. This product must have exceptional barrier properties, due to the volatility of the gasoline while remaining very functional and strong. The co-extruded sheet uses an ultrahigh-molecular-weight HDPE as a carrier material. This six-layer structure, (Figure 3.61) has two skins of the ultrahigh-molecular HDPE with one regrind layer and an EVOH barrier with adhesive on both sides. This is a prime example of how properties of co-extruded products can be utilized in the most demanding of applications.

The extruder arrangement for a six-and seven-layer structure can vary considerably given the many combinations of possibilities. However a classic co-extrusion complex arranges the extruders in a fan pattern as depicted in Figure 3.62. In practice, it is common to have the barrier extruder in the center position, with other extruders fanning out on either side. Products other than EVOH can be used as barrier materials.

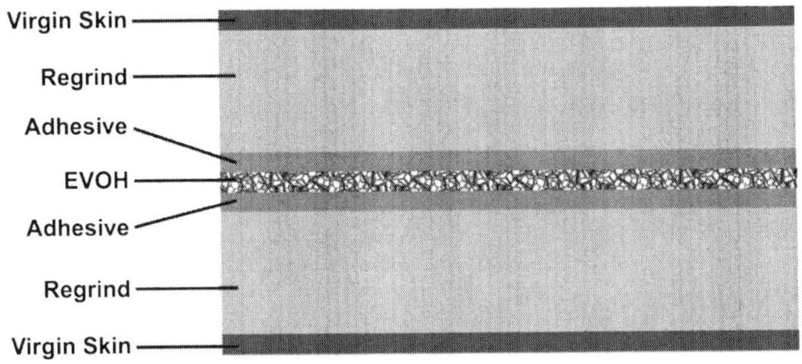

FIGURE 3.59 Symmetric seven-layer structure. (Courtesy of PTi, Aurora, IL.)

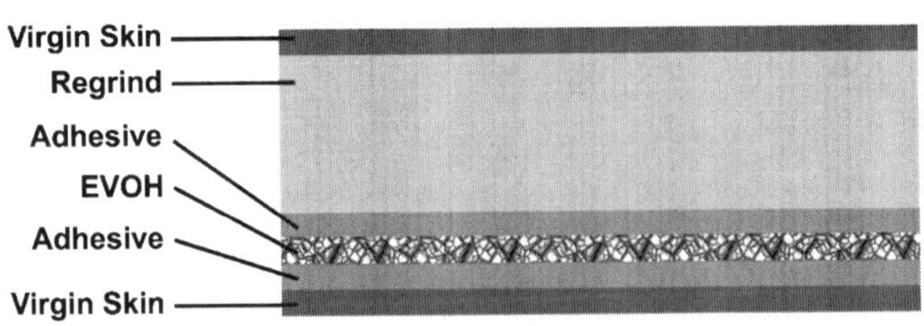

FIGURE 3.60 Asymmetric six-layer structure. (Courtesy of PTi, Aurora, IL.)

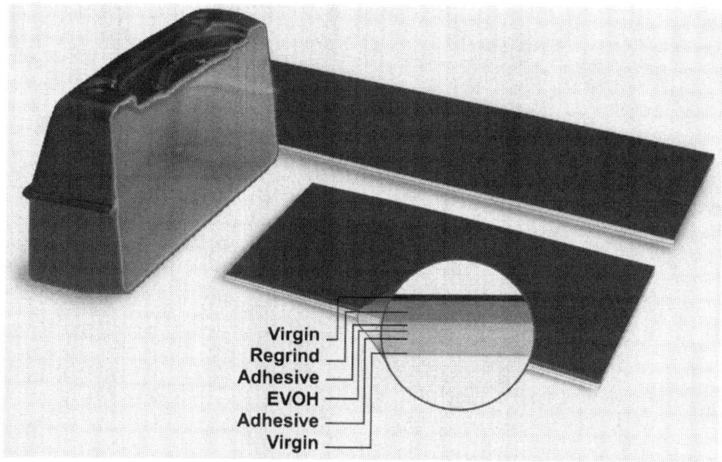

FIGURE 3.61 Six-layer HDPE and EVOH structure for automotive gas tanks. (Courtesy of PTi, Aurora, IL.)

FIGURE 3.62 Classic co-extrusion complex. (Courtesy of PTi, Aurora, IL.)

Five-layer barrier structures in the sheet industry have been produced with polyvinylidiene chloride (PVDC) as a barrier with either PP or HIPS as skins. Naturally, EVA is used as a bonding agent. Although technically a very good barrier, product limitations in recycling PVDC have curtailed its acceptance in the marketplace. PET with EVOH as a barrier is being manufactured as a five-layer structure on a very limited scale. This composite cannot be reclaimed effectively because of temperature disparity between the EVOH and the PET.

Because of cost ineffectiveness due to a lack of reclaim utilization, the PVDC and PET composites just referred to are not economically sound products. It is for this reason that most barrier products favor EVOH as a barrier in either six-or seven-layer configurations with either PP or HIPS skins. PVDC also possesses properties that are exceptional in terms of serving as a barrier against carbide dioxide, nitrogen, oxygen, and water vapor transmission. However, it remains a very difficult material to use because of the problems associated with processing the regrind. However, due to its exceptional barrier performance and ability to utilize the by-product regrind materials back into the co-extruded structures, EVOH has been accepted as the standard in the industry for barrier resin applications.

Nylons (polyamides) are used as a barrier in processed meat and cheese containers. This structure is usually produced in three or five layers. Due to the extreme process temperature differentials required between nylon and the other commonly used resins (i.e., usually, a polyethylene or an ionomer), regrind cannot effectively be reintroduced as a layer.

Other applications for barrier food packing run a wide spectrum of end uses for structures of six or seven layers. Apple sauce, pudding, meats, cereals, microwavable products in self-serve containers, and numerous snack containers are all by-products of this technology. In addition to these examples, co-extruded nonbarrier products are prolific in disposable cups, bowls, dishware, and numerous nonperishable food-packaging applications. The array of products displayed in Figure 3.63 demonstrate the scope and importance of co-extrusion in the food-packaging arena.

FIGURE 3.63 Examples of barrier food packing. (Courtesy of PTi, Aurora, IL.)

3.5 IN-LINE SHEET EXTRUSION: DIRECT COUPLING TO THE THERMOFORMER

The term *conventional thermoforming* usually refers to sheet being fed into a thermoformer from rolls of thermoplastic sheet that are mounted on an unwind stand located to the rear of the thermoformer. The sheet is threaded into transport rails on the forming machine and indexed forward into an oven and subsequently into the forming station to be formed into parts. The oven is provided with heaters that heat the sheet on both top and bottom to bring the sheet to the proper forming temperature. The heaters are zoned to apply the correct amount of heat to all areas of the sheet. The heater panels are typically quartz, ceramic, or infrared panels. The oven is laid out to provide stops as the sheet passes through it. When the machine goes through the forming cycle, the stops bring the sheets to a halt in the mold while being formed and cooled. The length of the stops is sized to match the forming tool length for uniform temperature control. An example of a roll-fed thermoformer is displayed in Figure 3.64.

In a conventional arrangement such as this, the sheet extrusion and thermoforming operations are separate and require the sheet extrusion line to include a winder to produce the rolls. The edges of the sheet need to be trimmed to eliminate the rough edges and beads that are produced by the sheet extrusion process. The trimmed edges produce wound rolls that are very flat and will feed smoothly into the thermoformer. A negative characteristic of this method is that the rolls need to be stored for a period of time to allow the sheet temperature to stabilize throughout the roll. Temperature variations in the roll will raise havoc in the thermoforming reheat process.

To avoid constant roll changeovers, the rolls of sheet are usually wound into very large diameters, referred to as *jumbo rolls*. The labor required for handling of the rolls and the space required to age them can be costly and inefficient. Such systems

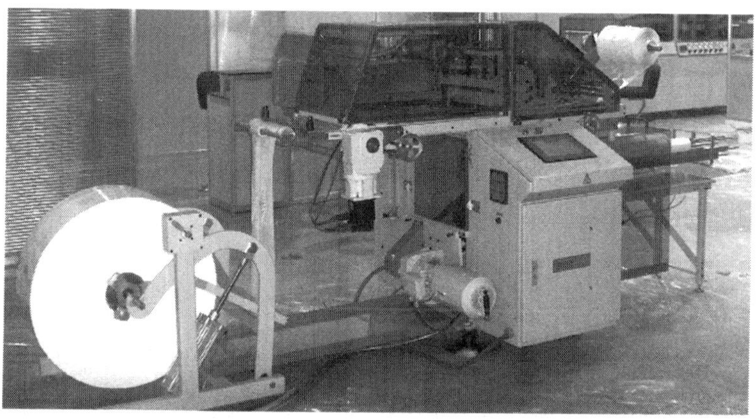

FIGURE 3.64 Roll-fed thermoformer. (Courtesy of G. N. Plastics Company Limited, Chester, Nova Scotia, Canada.)

are practical for short production runs, where either the type of sheet or the thermoforming tools will be changed frequently. One extrusion line can be sized to produce rolls for several thermoformers.

An alternative to the conventional thermoforming system is in-line thermoforming. In-line forming is the marriage of the sheet extrusion process to the thermoformer. The sheet is fed directly into the forming machine from the sheet system, completely eliminating the roll stock and winder. In-line thermoforming is used for long dedicated production runs of one particular part and usually consists of large thermoforming machinery for high throughputs. Products made on such a system are typically packaging items such as cups, lids, plates, bowls, takeaway containers, and containers for prepared foods. The extrusion plant is sized to the capacity of the thermoformer so that the production rate is not limited by the extrusion output.

The advantages of in-line thermoforming are energy savings, reduced labor, storage of rolls, control of scrap, and consistent operating conditions. In-line thermoforming is done by three basic methods. One is a tandem or compact sheet system, a second is a hot sheet system, and the third is the rotary vacuum wheel or drum.

3.5.1 Compact Sheet Systems

The compact sheet system uses a three-roll sheet takeoff. The three rolls are temperature controlled and are used to cool the sheet to below its softening temperature and to impart a surface to the sheet. Some residual heat is kept in the sheet for a savings in energy, but the thermoformer will have a multistop oven similar to the oven used for conventional thermoforming. The compact sheet extrusion system is displayed in Figure 3.65.

The conveyor ends at a set of pull or draw rolls that feed the sheet to the thermoformer. The sheet is drawn along an ambient cooling conveyor, which is of a compact design so as not to overcool the sheet. The draw rolls consist of two rubber-covered rolls that form a nip point to pull the sheet. They are driven and speed synchronized to the three-roll takeoff unit. The draw rolls can increase or decrease tension on the sheet between them and the sheet takeoff and also isolate the sheet from the takeoff, providing a relaxed sheet for feeding into the transport rails on the thermoformer.

FIGURE 3.65 Compact sheet extrusion system. (Courtesy of PTi, Aurora, IL.)

Since the sheet extrusion system is continuous and the thermoformer is cyclic, a loop of sheet is collected between the draw rolls and the thermoformer inlet. This droop provides an accumulation of the sheet between the two machines, and the loop is sensed optically for excessive material. The optical sensor sends a signal to speed up the thermoformer if the loop is excessive and to slow the thermoformer down if the sheet is pulled taut.

After the parts have been formed in the thermoformer the sheet is conveyed to a trim press. The trim press punches the parts out onto a stacking table. The trim from the sheet, often referred to as *skeletal scrap*, which consists of the sheet that represents the space between the formed parts, is stripped away and fed into a granulator. In the granulator the sheet is ground into particles and is then conveyed back to the extruder hopper to be fed back into the extrusion process. Since round or oval parts will generate a 50 to 60% scrap rate,refeeding it back into the process is critical.The particle-conveying system is closed, to protect the material from dirt and other contaminants. Some polypropylene lines have what is called *trim-in-place*, where the parts are trimmed in the forming station as a second operation in this press.The trim-in-place skeletons are handled in the same way as the trim press scrap.

The compact in-line system is the easiest to thread up and operate and can be used for virtually all extrudable thermoplastics. It is the best system to use for sheet material that has low hot melt strength, such as PP or PET.

3.5.2 Hot Sheet Systems

The hot sheet system eliminates the three-roll sheet takeoff. The sheet is fed from the sheet extrusion die directly onto small-diameter conditioning rolls which are mounted at the entrance to the thermoformer transport rails. The conditioning rolls cool the sheet only slightly while giving it dimension and enough mechanical strength to be transported by the thermoformer transport rails without sagging and falling into the bottom oven. The surfaces of the conditioning rolls have much less effect on the sheet surface that in the compact system. Figure 3.66 shows a hot sheet thermoformer system.

Ideally, the sheet in a hot feed system would not have to be reheated in the thermoformer. Due to the cyclic operation of the thermoformer, stoppage for the forming cycle will cause the sheet to lose some of its heat in a nonuniform fashion. A short oven is therefore required, although it will be only a one-or two-stop oven, as opposed to the compact line, which uses three or four stops. The energy savings are significant, due to the fact that so little heat must be put back into the sheet. To compensate for the sheet stoppage during the form cycle, a mechanical accumulator is used to provide sheet storage. This accumulator consists of linked rolls to provide a fully supporting loop. It is typically mounted within the thermoformer, and the accumulation and roll speeds are timed and driven to match the thermoformer cycle rate.

The sizing of the extrusion plant to the downstream thermoformer is critical. The extrusion system output must be sized to the maximum capacity of the thermoforming system. This requires communications between the supplier of the extrusion system and the thermoformimg toolmaker as well as the thermoforming

FIGURE 3.66 Hot sheet thermoformer. (Courtesy of PTi, Aurora, IL.)

system supplier. The most common factors used in sizing the system are the cycle rate of the thermoformer, the width and length of the forming tool, and the thickness of the material. When the cycle rate and tool length are known, the line speed of the sheet extrusion system is calculated, and it is then a simple process to determine the output by using the thickness of the sheet and the specific gravity of the resin being run. It is also necessary to inform the sheet extrusion supplier of the surface finish required on the finished part since this will affect the finish of the middle roll on the sheet takeoff unit, where the sheet will be polished or textured to the effect desired.

3.5.3 Drum Systems

The third type of in-line system is a vacuum rotary wheel or drum. In this scenario a driven wheel is installed immediately below the sheet die. There are a set of small conditioning rolls just prior to the wheel to support the sheet exiting the die. The wheel has individual forming tools installed around the circumference. As the wheel turns and the sheet is draped onto it still hot and somewhat fluid, a vacuum is applied to the tool, drawing the sheet down into it for forming. As the wheel continues to turn, the sheet is stripped away and fed into a trim press, where the parts are punched out and stacked. This continuous method of thermoforming is used for high production rates of thin-walled, shallow-drawn parts which don't require a great deal of cooling. These are used primarily for lids and shallow trays. The production rates of this type of system are limited only by the capacity of the trim press. Figure 3.67 shows an image of a drum-style thermoformer.

Tooling costs are high for large in-line thermoforming systems. However, it remains the preferred method of producing high-quality parts at increasingly higher

FIGURE 3.67 Drum-style thermoformer. (Courtesy of PTi, Aurora, IL.)

throughput rates. There are systems in production today making parts in-line at rates of up to 10,000 lb/hr, and lines that produce over 6000 lb/hr have become commonplace. For long dedicated production runs the savings with an in-line system are significant given the advantage of reduced heat cycles, elimination of costly capitol equipment such as winding units, reduced labor and potential injury from handling large heavy rolls of sheet, and elimination of the need of expensive production floor or warehouse space for roll storage.

3.6 THERMAL ANALYSIS USED IN MACHINERY DESIGN

One of the primary functions of the roll stack is to remove the heat from the sheet once it is extruded into the primary nip point from the sheet die. Predicting the temperatures at which the sheet leaves the roll stack is beneficial information for engineers to have during the design phase of the machinery. It can also be used in forecasting the performance of a new product and/or resin from the end-user perspective.

Not every sheet process has the same overall thermal objective. The desired temperature of the sheet at different points along the line may vary from application to application. There are three basic alternatives for producing sheet products. Sheet can, be wound, shear cut, and stacked or extruded in-line close coupled to a thermoformer. Each requires different end thermal objectives with respect to the target goals for the sheet temperature. There are several design elements that need to be evaluated to evaluate a sheet extrusion process properly. The information required to assess the equipment's design thoroughly includes the following:

- Base polymer to be processed (PP, PET, PS, PMMA, PVC, etc.)
- Extrusion melt temperature
- Extrusion capacity (lb/hr, kg/h)
- Thickness of sheet (in. or μ/m)
- Width of sheet (in., mm)
- Diameter of all cooling rolls (in., mm)
- Configuration of roll stack to determine amount of wrap (vertical up-stack or down-stack, J approach, inclined, and horizontal)
- Roll temperature profile (top, middle, bottom, and auxiliaries)
- Quantity of rolls

Incorporating this information into a thermal analysis for the process will yield the exit temperatures for each roll. The thermal properties of the material will determine the polymer's ability to have heat removed. The extrusion rate, thickness, and width of the sheet are used to determine the line speed of the web as it passes through the roll stack. The time that the sheet remains on the roll is commonly called the *residence time*. The longer the sheet spends on a roll, the greater the ability to remove heat from the roll. Larger-diameter rolls possess longer residence times than those of smaller rolls. The faster the web passes through the roll stack, the shorter the time will be to remove heat from the sheet. The residence time is related directly to the line speed of the web and the size of the rolls. The geometric path the web takes as it passes over the rolls also plays a contributing role in determining residence time on the rolls. Additionally, there are several other variables to consider when reviewing the thermal model of a sheet system. By using a numerical analysis and the aforementioned variables, the web temperatures can easily be modeled such that the required roll diameters may be determined.

For purposes of comparison, consider thermal models for two-roll stand geometries for a given sheet application. The first roll stand features three rolls arranged vertically with each roll being 18 in. in diameter The second roll stand features three rolls also arranged vertically, each roll being 24 in. in diameter. To evaluate the sensitivity of residence time versus cooling performance for a roll stand, each analysis will assume a web width of 60 in., a sheet thickness of 0.025 in. a melt temperature of 435°F, roll temperatures of 120°F, and an overall rate of production of 2500 lb/hr. The thermal properties for the resin will also assume polystyrene for each analysis.

Since each thermal model is for a roll stand that is arranged vertically, the proportional web wrap for each will also be the same when comparing roll-to-roll sheet web surface exposure. The top roll is nominally 5° of web wrap, the center roll is 180°, and the bottom roll is 200°. Using the web wrap figures, the amount of length the web is in contact with the roll can be determined. The length can then be converted into an area of exposure to the roll's surface. It is this exposure to the roll surface that is instrumental in permitting an exchange of heat to occur through conductive heat exchange principles. For other roll stand geometries the web wraps can be manipulated to examine the comparisons for different roll sizes, roll counts, and roll stack arrangements (i.e., vertical down-stack, J-stack, inclined stack, etc.) as well as five-roll systems.

In summary, Figure 3.68 and Table 3.10 display a graphical depiction of the analysis results and a numerical overview. In referencing the graphical depiction, the

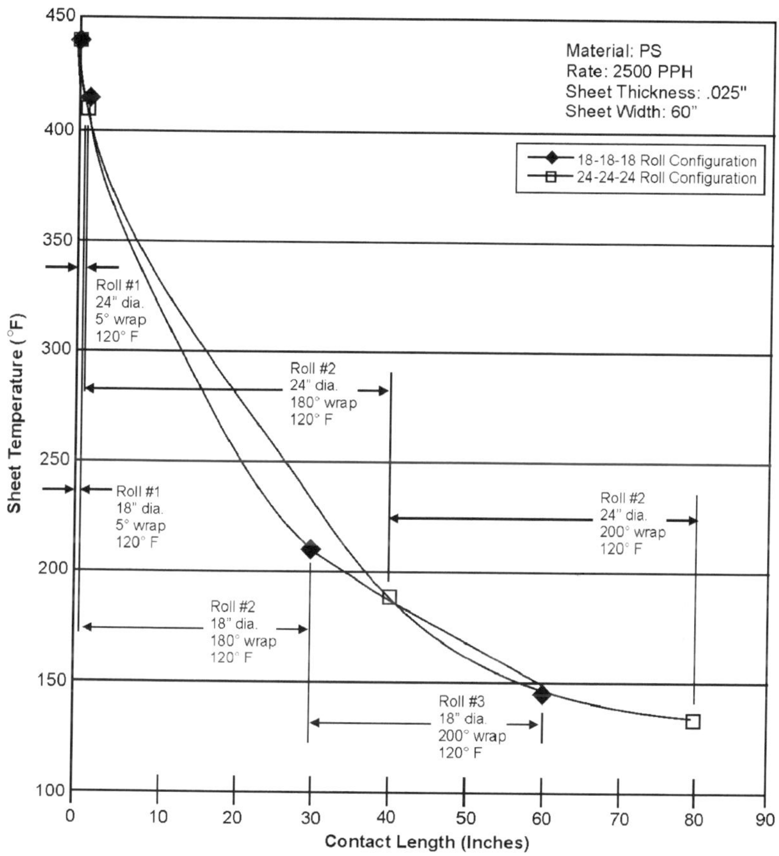

Web Cooling Analysis
2500 PPH, .025" Thick

PPH	2500.00
DENSITY (# / cu ft)	65.500
SPECIFIC HEAT (Btu/lbm F)	0.320
THERMAL CONDUCTIVITY (btu/hr ft f)	0.060
SHEET THICKNESS (in)	0.025
LINE SPEED (fpm)	61.063
SHEET WIDTH (in)	60.000

FIGURE 3.68 Graphical web cooling analysis results. (Courtesy of PTi, Aurora, IL.)

TABLE 3.10 Numeric Web Cooling Analysis Results

	Extrusion Temperature °F (°C)	Roll Temp. °F (°C)	Deg. Wrap	Diameter [in. (mm)]	Btu/hr Removed	Average Sheet Temp. (°F)	Temp. difference Across Roll (°F)	Water Flow Req'd (gal/min)
ROLL (18-18-18)								
1	435 (224)	120 (49)	5	17.75 (450.85)	13,184	419	2	13
2		120 (49)	180	17.75 (450.85)	163,067	214	2	164
3		120 (49)	200	17.75 (450.85)	54,759	146	2	55
ROLL (24-24-24)								
1	435 (224)	120 (49)	5	23.75 (603.25)	16,942	414	2	17
2		120 (49)	180	23.75 (603.25)	183,271	184	2	184
3		120 (49)	200	23.75 (603.25)	42,045	132	2	42

Source: PTi, Aurora, IL.

analysis shows a cooler exiting sheet temperature for the larger roll diameters, along with a higher amount of web wrap exposure on the rolls themselves. At each roll location the web temperatures are lower for the larger-diameter rolls than for the smaller rolls. The exiting temperature leaving the three rolls is also much lower for the larger rolls.

Furthermore, examining the numerical results shows the web temperature leaving the top, middle, and bottom rolls for the 18-in.-diameter stack as 419, 214, and 146°F, respectively. The web temperature leaving the top, middle, and bottom rolls for the 24-in.-diameter stack is 414, 184, and 132°F, respectively. Since the results of these analyses are limited to conductive heat exchange principles versus combined convective and conduction modes of heat exchange, the field results will be lower, due to the missing convective component. As a general guideline, the convective results for the primary roll stand represent approximately 6 to 8% of the conductive component. Omitting the convective constituent is conservative for purposes of performing a thermal analysis, and accordingly, sizing systems, will yield conservative results—but not overly conservative.

Additional information that is gained from these analyses includes the cooling water flow rate requirements for each roll. For the 18-in.-diameter stand, the top, middle, and bottom roll flow rate is 13, 164, and 55 gal/min, respectively. For the 24-in.-diameter model, the top, middle, and bottom roll flow rate is 17, 184, and 42 gal/min, respectively. It is the practice of PTi to apply a 2°F temperature differential across the face of the roll. In the limit as the temperature differential goes to zero across the face of the roll, the flow rate required to provide this differential performance goes to infinity. Conversely, in the limit as the temperature differential goes to infinity across the face of the roll, the flow rate required to provide this differential goes to zero. If the results of the analysis were to yield an impractical flow rate through the roll, then through iterative analytical methods the temperature differential figure can be increased incrementally until an acceptable flow rate figure is achieved. This by no means compromises the integrity of the production system. It simply means that it may require 3°F versus 2°F across the roll face. In other words, there is a balance in the analysis to maintain between practical sizing of system components and targeted roll face temperature differentials. For design purposes a 2°F differential is conservative.

3.7 PLC/COMPUTER CONTROL SYSTEM

The control system is an essential part of the sheet extrusion process. The PLC/computer control system enhances the performance and capabilities of the sheet extrusion system for full-line coordination. A sheet extrusion system is a complex system, and coordinating its many parts for efficient and safe operation is very important. The items that are commonly integrated into the PLC control system include the material feed and blending system, extrusion complex, sheet takeoff system, winder and/or shear and stacker, and many other auxiliary pieces of equipment. It can centralize the control of many different aspects of the sheet extrusion process to one or a number of control interface screens. This can eliminate or minimize the need for discrete controllers such as pushbuttons, selector switches, temperature controllers, pressure controllers, potentiometers, encoders, etc. These screens or interfaces are usually referred to as HMIs (human–machine interfaces) or MMIs (man–machine interfaces). The interfaces are commonly touch screens, and the operator can select from a number of pages for the particular portion of the machine that the operator needs to control.

On a sheet system that utilizes it, the PLC is the hub from which information flows. The PLC consists of a main processor and a number of modules for input, output, and communications. The communication modules are used to transfer information from module to module, rack to rack,or module to controlled component. These items are mounted onto one common rack or on several racks and networked together. Inputs may be from many different sources, such as a motor drive, pressure transducer, limit switch, thermocouple, pressure switch, potentiometer, encoder, pushbutton, and/or selector switch. Outputs may go to many different destinations, such as solenoid valves, motor drives, motor starters, and many others. Several communication protocols and methods are used commercially to enable real-time interaction between the PLC and major components. These include DeviceNET, Profibus, Ethernet, Modbus, and so on. Typically, these communications protocols occur using a specialty cable or cables which would run from component to component to network the system together. For example, this may be between the PLC and drive, between two separate racks mounted in different pieces of equipment, or from the processor to the interface screen. Sometimes only one protocol is needed, but in some cases there may be several.

Examples of the information processed by the PLC are speeds (i.e., rpm ft/min, m/min, etc.), temperatures, pressures, safety interlocks, roll gaps, thicknesses, and diameters, to mention a few. The PLC can be used in conjunction with a motor's drive or starter to turn on motors, set the motor's speed, and coordinate a number of drives to run at the same time and speed (i.e, extruder drives, gear pump drives, roll drives). Pumps are commonly controlled by use of the PLC (i.e., barrel cooling pumps, lube system pumps, hydraulic power unit pumps, gas injection pumps, roll temperature control pumps, silicone, or cooling bath pumps). Coordinating the display, set point, and calibration of the roll stand gap adjustment system is another example of what can be done through a PLC/computer control system. Additional examples include actuating solenoid valves for opening and closing rolls or pull rolls for either hydraulic-or pneumatic-based scenarios, and importing the set points for controlling PID (proportional, integral, and differential) temperature control zones for barrel heaters, downstream zones (i.e., adapters, screen changers, gear pumps, dies, rolls, etc.), and many other items.

The PLC can simplify the task of setting up and coordinating interlocks on an extrusion system, thus eliminating the need for many relays, timers, and so on, as is the case for discrete manually controlled systems. An example would include the interlock of the barrel cooling pump, which should be running before the barrel heaters can be turned to their designated set point. Another example might be that there is flow or that a certain temperature setting is not exceeded in the extruder gearbox lubrication system before allowing continued operation of the extruder. Failure to meet the interlock criteria could shut the machine down or not allow its startup, for example.

Operators are able to gain access into the inner workings of the PLC by using an interface screen device. One common version of this device is Allen Bradley's Panelview. The pages displayed on the screen are for the particular pieces of equipment or section of it. These screens may be of a predetermined format or customized for particular systems. Alarm indication can also be put on these pages to greatly increase the effectiveness of troubleshooting a problem on sheet extrusion equipment. Examples of this include a drive fault, an emergency-stop chord or pushbutton that has been tripped, a pressure switch that is activated, and/or a number of other operational and safety-related items.

Advanced systems for integrated PLC/computer control would be with the use of a central processing computer. This type of system utilizes software such as Wonderware, Siemens, Allen Bradley's RS View, Intellusion, and many others. Each of these software packages allows programming with enhanced graphic capabilities and reporting options. PTi offers such a system, the TITAN control system. The computer communicates with the PLC and receives data to process it for the display on the MMI or HMI. The Titan system greatly enhances fundamental aspects of operating and controlling a complex system such as a co-extrusion sheet system. The TITAN system offers many specific features, including:

- Multiple security levels
- Alarm logging
- Recipe manager
- Diagnostics and remote modem capability (communications on several different levels, drives, temperature controllers, PLCs)
- Real-time and historical trending (pressure, temperatures, amperage, etc.)
- Full machine control and coordination
- Remote video monitoring
- Integration of thickness gauging systems for flat sheet and thickness adjustment
- Overviews and reports for process
- Data logging
- Layer control (for barrier sheet and to controll the percentages of each layer in a structure)
- Ramp features
- Emergency-stop interlock strings, including logic sequencing
- Battery backup in case of loss of power
- Temperature and pressure control and deviation reporting

Extruder(s) and roll stack temperature zones can be controlled either by standalone (discrete) controllers or by PLC PID (proportional–integral–differential) loop instructions. A system of hardwired discrete controllers without PLC support is now obsolete and very rarely built. It has been replaced by one of the following systems:

- Hybrid system comprised of discrete controllers and PLC controls. Temperature loop control is performed by discrete controllers networked with TITAN. Setting the PID loop parameters and displaying process parameters is done conveniently from TITAN HMI but can also be done from individual controllers.
- Total PLC/TITAN control, where all temperature loops are controlled by PLC PID instructions. Controlling the process by PLC provides for simplicity of system, cuts down on hardware and wiring, and simplifies programming and troubleshooting.

Both systems provide sufficient accuracy of 1°F. Error between the set point and process variable measured by a thermocouple connected to the controller/PLC is evaluated continuously by PID loop controller/PLC PID instruction. Through the PID control a calculation is made of the actual error that exists between the set point desired and the actual run condition such that an action is invoked to correct the error until convergence on the set point is reached. The loop output will switch on respective zone heating or cooling for a duration (within a predetermined period) proportional to the controller output.

Head and discharge pressure transducers are connected to pressure controllers and networked with the TITAN system for monitoring and data collecting. Digital outputs of pressure controllers are connected directly to the extruder drive or to a PLC for quick shutdown of the process in case of excessive value. Most extrusion systems require a constant and stable output such that high product quality is achieved and maintained. A constant-head pressure and/or pump discharge pressure must be maintained to accomplish this. The actual (measured) pressure is then fed into PID loop for controlling the extruder speed (rpm).

Typical sheet extrusion lines utilize ac vector drives. These types of drives are capable of delivering high torque in a wide range of speeds. They also provide high accuracy of speed control, matching the performance of the once-dominant dc drives. The motors of cooling rolls and pull rolls are equipped with encoders connected to respective drives to provide precise speed control. The extruder main drive and gear pump drive can be of the type that is considered a sensorless vector type. All drives and local and remote PLC racks are networked together. Utilizing commercial protocols, a communications network can be established to interconnect the drive to the PLC while minimizing wiring, improving the drive diagnostics,and increasing the amount of data from/to the drive itself.

The sheet stack in its simplest form is comprised of a roll stand equipped with three cooling rolls and a traction system. Depending on the process, it can also consist of a variety of auxiliary equipment, such as an additional cooling station and/or cooling belt, pull roll, winder, edge trim knives, gauge scanner, shear, and stacker. The rotation of each roll is coordinated precisely by the control system. The equivalent line speed must be coordinated between adjacent components on the system to prevent excessively high or low web tension. Tension control devices may be incorporated to permit a means of coordinating the downstream components with the main roll stack.

Figure 3.69 is a photograph of a typical TITAN computer control console. This represents the primary interface between the operator and the machine. The controls must be intuitive in nature and easy to maneuver around the primary control functions and informational screens. The console serves as an interface device that must be readily accepted on the plant floor for it to truly be a functional and beneficial production tool.

3.8 SUMMARY

In summary and in review of the many topics presented herein on the subject matter of sheet extrusion, we note especially that our work as educators, processors, and machinery manufacturers is far from over. In fact, it will never be over. There will

FIGURE 3.69 TITAN computer control console. (Courtesy of PTi, Aurora, IL.)

always be new challenges to be met. These challenges will continue to emerge year to year. It is our objective to stay ahead of the curve and meet these challenges head on. The best way to achieve this is to maintain a comprehensive and well-organized educational program within our respective organizations in order to benchmark current technologies and methods as well as being able to identify specific shortcomings that play havoc on the production floor.

The tools that we have referred to are the production tools that processors have available to them. To survive in the global marketplace as manufacturers of plastic sheet goods we must maintain strict control of our cost structures. To control something, it must be well-defined and identified. Raw material and labor costs are both well-defined components of our overall cost formula. However, the day-to-day cost impact of operating old, low-efficiency machines is not so well defined. Our closest competitors are now emerging around the world and are participating in the global markets. The tools that we need to keep our production facilities highly efficient may not presently exist on our plant floors. There is a need to keep our production facilities equipped with the necessary machinery items such that our quest to achieve high levels of production efficiency becomes a reality. The low-producing antiquated sheet extrusion machinery of the 1970s and 1980s must now be put out to pasture and replaced with advanced technologies that characteristically have relatively high production rates and maintain excellent "up-time" performance, thus satisfying our objective to increase overall production efficiency as well as help us maintain a global competitive edge.

ACKNOWLEDGMENTS

This chapter represents a collaborative effort from members of PTi's senior management staff, engineering department, and technical sales group. Additional recognition is given to NDC Infrared Engineering of Irwindale, California and EDI of Chippewa Falls, Wisconsin for their editorial input on gauge scanning systems and sheet dies and feed blocks, respectively. Many special thanks go out to each of these individuals, whose dedication and related expertise have made this work possible.

REFERENCES

1. American National Standards Institute, *American National Standard for Plastics Machinery*, ANSI/SPI B151.7-1996, ANSI, New York, 1996.
2. Wall Colmonoy Corporation, Colmonoy wear-resistant products, http://www.wallcolmonoy.com/products/coating/index.htm, retrieved December 27, 2004.
3. Xaloy, Inc., Proven to perform: genuine Xaloy barrels, http://www.xaloy.com/barrels.html, retrieved December 27, 2004.
4. National Electrical Code 1999, *NEC Handbook*, Section 110-16.
5. Extrusion Dies Industries, e-mail message to author, EDI, Chippewa Falls, WI, July 21, 2004.

6. American National Standards Institute, *American National Standard for Plastic Sheet Production Machinery*, ANSI/SPI B151.20-1996, ANSI, New York, 1999.
7. American National Standards Institute, *American National Standard for Plastic Film Sheet Winding Machinery*, ANSI/SPI B151.5-2000, American National Standards Institute, ANSI, New York, 2000.
8. Buyers' guide to thermoplastics, in *Plastics Technology Processing Handbook and Buyers' Guide, 2004/2005*, October 2004, pp. G-3 to G-279.
9. NDC Infrared Engineering, e-mail message to author, NDC, Irwindale, CA, August 16, 2004.
10. F. W. Billmeyer, Jr., *Textbook of Polymer Science*, Wiley, New York, 1971, p. 16.
11. J. E. Johnson, Co-extrusion, *Plastics Technology*, February 1976.

CHAPTER 4

Thermoforming

SCOTT MACDONALD
Maryland Thermoform, Baltimore, Maryland

4.1 INTRODUCTION

Today's modern manufacturing facility mimics a process that some credit back to the days of the Egyptians. The Egyptians were thought to utilize tortoise shells, the oldest known thermoplastic material. Like most thermoplastic materials, the tortoise shell, becomes moldable and softens when exposed to a heat source. The Egyptians were thought to manufacture food and beverage containers from these shells, utilizing hot water rather than convected or contact heat to make the shell moldable.

In more modern times, thermoforming applications included topographical maps, used to illustrate terrain, and early aircraft windshields, used to protect an airplane's occupants and allow viewing of the surroundings. These are just two of a kaleidoscope of applications in the first half of the twentieth century. After World War II, the United States led the application of the thermoform process to develop products and packaging for commercial, industrial, and consumer use. Today, thermoform products are widely accepted and utilized by billions of people daily. No more cost-effective process can be found for many applications, such as food storage, pharmaceuticals, and medical devices. Future applications and uses are limited only by the materials themselves. Individuals and companies considering an option for packaging, products, and career opportunities should not hesitate to utilize thermoforming.

Simple Guide to This Chapter

Thermoform is defined as follows according to Merriam-Webster's online dictionary:

Main entry: **ther·mo·form**
Pronunciation: 'ther-mo-"form

Handbook of Plastic Processes Edited by Charles A. Harper

Function: *transitive verb*

: to give a final shape to (as a plastic) with the aid of heat and usually pressure

The act of thermoforming, for our purposes, is to give shape to plastic, but today we use several variations to reach the desired shape. They include vacuum, pressure (air), and occasionally, just the force of two individual surfaces crushing the plastic. Products of the thermoforming process, again for our purposes, are divided into two groups, light gauge and heavy gauge, roughly along the lines of the equipment style generally utilized in making the thermoformed product.

Light-gauge items are typically packaging, but some are also products in themselves. Thickness, or gauge, typically is 0.003 to 0.100. Products are usually manufactured on roll-fed in-line thermoformers. Heavy-gauge items usually are products, but some packaging applications appear in this group as well. Thickness, or gauge, typically is 0.100 to 0.500 and thicker. Products are typically manufactured on cut-sheet or sheet-fed equipment such as a single-station thermoformer, shuttle thermoformer, and rotary thermoformer.

Are You in the Right Process?

Typically, the uninformed designer or purchaser of any given plastic product is drawn to thermoforming, due to low up-front costs in product development. But are you in the right process? This question should not only be asked at the inception of a product, but also throughout the product's life times, as design changes, product requirements, and even volume requirements can offer improvements or even require a process change to injection or several other types of processes to produce the desired result through the most cost-effective and most efficient processing method.

4.2 THERMOFORMING PROCESS

As described previously, the process in its simplest expression is forming with heat. As practiced, this heat source causes the plastic to deflect or distort from a flat sheet into a draped sheet or sagged sheet. When it reaches this stage it is then drawn over or into a mold cavity. The most common technique used is a plastic sheet draped over a simple male mold cavity. It is from this simple technique that most variations occur and give the user a superior part or product for a specific application. After each different technique is completed the product is cooled, then removed or ejected from the tooling in preparation for the next sheet of plastic or cycle of the machine.

Let's review the simplest commonly used techniques in thermoforming.

- *Male mold drape* (Figure 4.1*a*): plastic sheet heated to heat deflection temperature and then formed over a male mold. Vacuum, pressure, or force must be applied to form the sheet to the mold cavity.

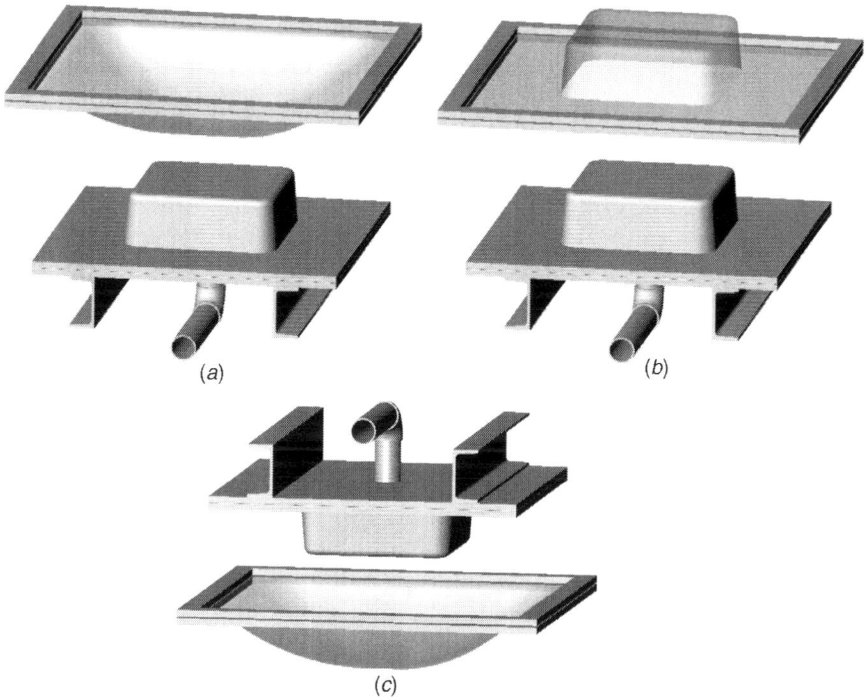

FIGURE 4.1 Male mold: (*a*) drape; (*b*) drape formed; (*c*) top platen.

- *Male mold drape formed* (Figure 4.1*b*): plastic sheet formed and cooled enough to be removed from the mold. Any vacuum pressure or force utilized to form the product must be removed in this step to allow the product to be removed or ejected.
- *Male mold top platen* (Figure 4.1*c*): variation on the male mold drape by placing the mold on the top platen instead of the bottom platen. Vacuum, pressure, or force must be applied to form the sheet to the mold cavity.
- *Female mold drape* (Figure 4.2*a*): plastic sheet heated to the heat deflection temperature and formed over a female mold. Vacuum, pressure, or force must be applied to form the sheet to the mold cavity.
- *Female mold drape formed* (Figure 4.2*b*): plastic sheet formed and cooled enough to be removed from the mold. Any vacuum pressure or force utilized to form the product must be removed in this step to allow the product to be removed or ejected.

More advanced but commonly used techniques include two closely associated techniques called billow forming and snapback forming. We describe first the snapback method with a male mold. Figure 4.3*a* shows the male mold, plastic sheet heating, sources of vacuum and pressure, and the immediate surrounding tooling to

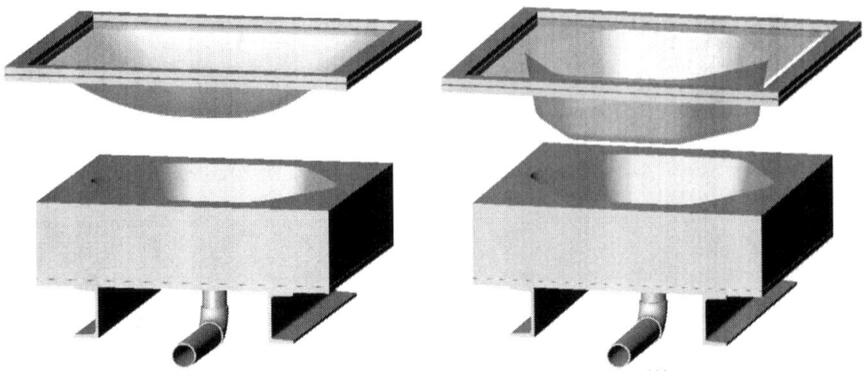

FIGURE 4.2 Female mold: (*a*) drape; (*b*) drape formed.

support technique. The plastic is at heat deflection temperature and about to be thermoformed. The vacuum source pulls the now flexible sheet upward against gravity, pre-stretching the sheet (Figure 4.3 *b*). The prestretched material is then pushed and/or pulled into a part of tooling (difficult to illustrate) that forces the plastic into a form similar to the mold in size and configuration called an *assist* (Figure 4.3*c*). Finally, the formed part is removed from the mold (Figure 4.3*d*).

There is also the ability to process using the snapback method with a female mold. The first step again is to set up the female mold, plastic sheet heating, sources of vacuum and pressure, and the immediate surrounding tooling to support the technique. The plastic is depicted as at the heat deflection temperature (Figure 4.4 *a*). Next the vacuum source pulls the now flexible sheet upward against gravity, prestretching the sheet (Figure 4.4*b*). Third, we show the prestretched material pushed and/or pulled into a part of the tooling (difficult to illustrate) that forces the plastic into a form similar to the mold in size and configuration called an assist. Notice that the center of the sheet is the last area of the sheet to be pulled up from the prestretched material by gravity position (Figure 4.4 *c*). Finally, we illustrate the part, now formed female, cooled and removed from mold (Figure 4.4*d*).

4.3 LIMITATIONS OF THE PROCESS

As with all plastic processing, each process has strengths and weaknesses, and thermoforming is no different. What initially draws a designer or user to thermoforming is usually tempered by a lack of uniform wall thickness in the process, limitations of design surrounding the materials, problems in depth-of-draw applications, and the necessary secondary operations to utilize a thermoform part as a product itself. Whenever a thermoformed item is used a product (e.g., enclosures for electronic components), the product must be CNC (computer numerically controlled) trimmed to size or height, holes must be drilled, and standoff or support blocks must be located. Fastening methods include either bonding or mechanically fastening to the

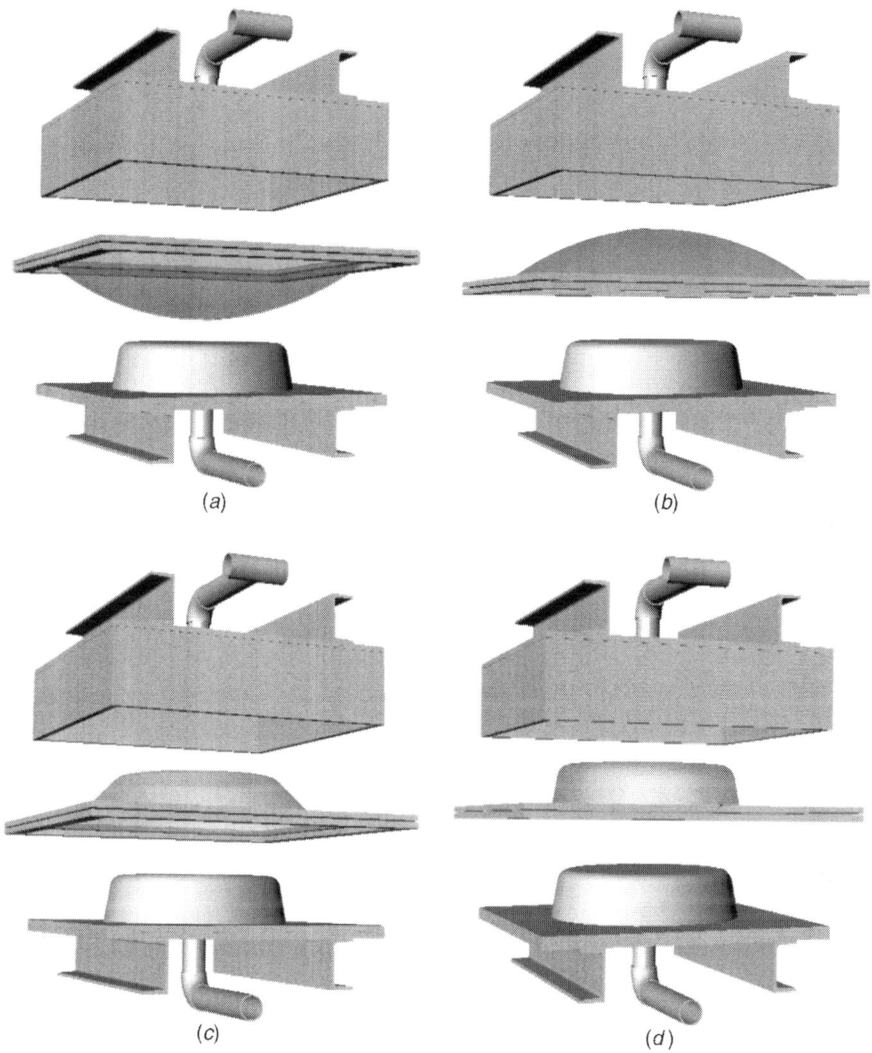

FIGURE 4.3 Snapback male mold.

product. Not only are these extra steps time consuming, they are generally costly and result in thermoforming usually being utilized in low-volume product applications.

4.4 MATERIALS FOR THERMOFORMING

Thermoforming utilizes a huge range of plastics, and most of these fall into the category of thermoplastics. Thermoplastics are a family of materials characterized by several simple properties, the largest being their ability to be reprocessed into new

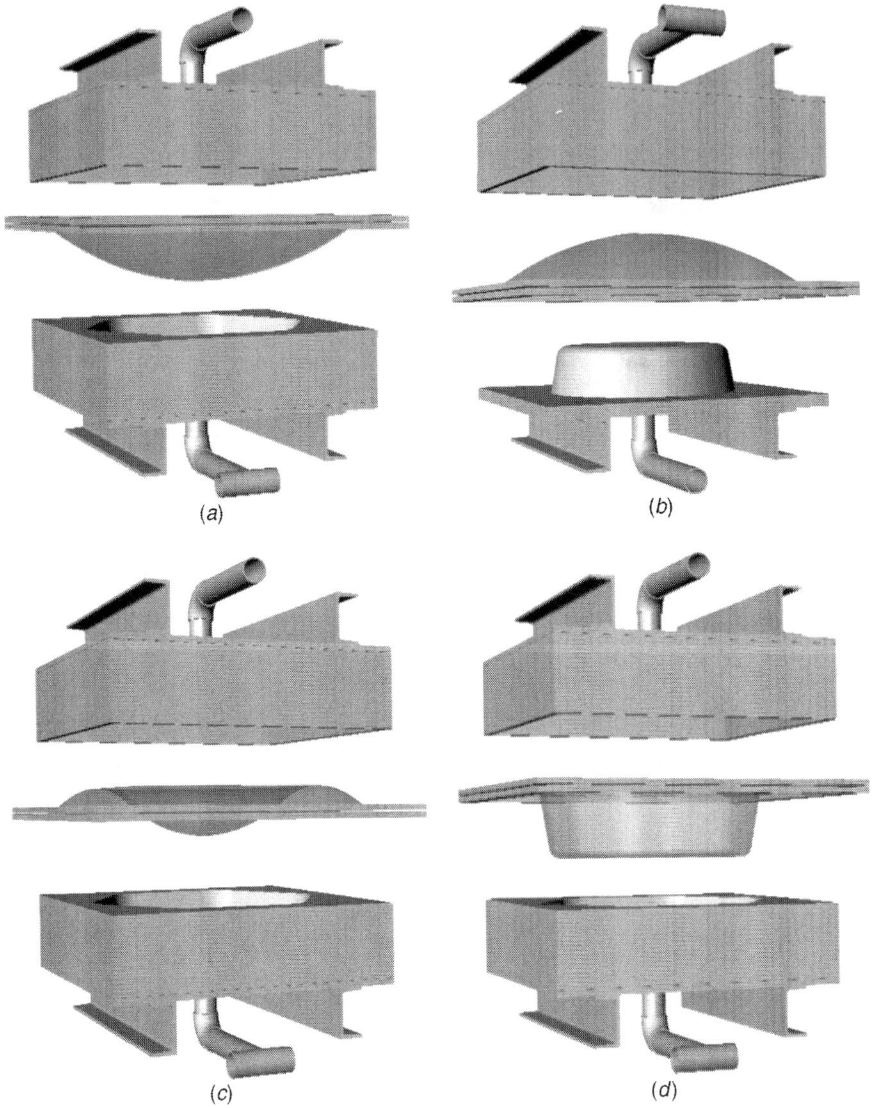

FIGURE 4.4 Snapback female mold.

products exhibiting the same characteristics as the previously manufactured product. In contrast, thermoset plastics can only be processed into a product once. After that it must be used as filler or in some other utilitarian application. They cannot simply be reground and reprocessed as thermoplastics. Under the f amily of thermoplastics fall many well-known plastic materials, such as polystyrene, polyvinyl chloride (PVC), polyester, polyethylene, and polypropylene. These are only a few examples, as many other materials exist in various grades and alloys.

4.4.1 Choosing the Correct Material for the Project

When choosing a material, cost, although a consideration, should not be the driving factor in material selection. One must consider the following variables and environmental issues:

- *Chemical resistance:* Will the product be subjected to chemicals in liquid or gaseous form?
- *Heat:* What will be the highest temperature to which the product will be subjected? This is known as the *heat deflection temperature*. It must be avoided in a finished product so that the product will retain its size, shape, and other physical properties after manufacture.
- *Cold:* What will be the lowest temperature to which the product will be subjected? This is known as the *cold crack temperature*. This temperature should be avoided to prevent product failure while in use.
- *Outdoor or indoor use:* Will the product be facing ultraviolet radiation from the sun? What will be the product life expectancy in an outdoor or indoor environment?
- Will the product be bonded mechanically or using a solvent, fastened to mechanically like a bolt, screwed or inserted, welded, heat sealed, radio-frequency sealed, or sonic welded or sealed?

Each of these factors must be addressed prior to final material selection. Only then can the additional level of material grade, such as utility or regrind grade, virgin grade, co-extruded or capped sheet, and color issues be addressed.

Most thermoplastic materials can be modified by the addition of other materials to create plastic alloys. These alloys are very common and readily available as products in themselves from the extruder of plastic sheet and film. They have been developed to answer specific product and package needs as the market requires. An example of this is adding polyvinyl chloride to acrylonitrile–butadiene–styrene (ABS), improving the flammability characteristics of ABS. ABS is of the styrene family and is a self-perpetuating material for fire. By adding PVC, the flammability resistance of the ABS increases, altering the final product from a self-perpetuating plastic to a self-extinguishing plastic. Most materials are also modified with additives which can improve specific material characteristics. An example of this is adding a rubber modifier to polystyrene to improve the impact resistance of the polystyrene. This is very common, as the impact resistance of polystyrene, unmodified, is low in relation to that of other plastic materials.

4.5 COMMON PACKAGING AND PRODUCT PLASTICS

The most common materials utilized in packaging contain several key elements: low cost, readily available, and contain the correct properties for the product's specific use. The following are some common examples of material uses.

Packaging

- *Polyethylene:* milk jugs and detergent containers
- *Polypropylene:* single-issue and reusable food containers
- *Polystyrene:* salad and bakery food containers
- *Polyvinyl chloride:* blisters and clamshells
- *Polyester:* soda and water bottles

Products

- *ABS:* enclosures for electronics and automotive and transportation parts and components
- *Nylon:* automotive and medical components
- *Polycarbonate:* aircraft and transportation vehicle windshields; medical components
- *Polypropylene:* medical components and medical disposables
- *Polystyrene:* novelty items and children's toys
- *TPO (thermoplastic olefin):* transportation vehicle external components

4.6 THERMOFORM TOOLING REQUIREMENTS

Thermoforming requires some of the least inexpensive tooling of all of the common plastic processing methods. Other processes, including injection, extrusion, and blow molding, all require steel or aluminum tooling to produce even the smallest number of parts. Basic thermoform tooling consists of a mold from which to manufacture the product and a method to trim the product. In-line thermoforming of light-gauge products also requires a method of quick removal of the parts from the processing equipment. This, in most applications, is a stacker or stacking device, but other solutions also exist. In-line thermoform tooling components typically include a mold, the die station or cutting station tooling, and the stacker. In some cases the molds may need to be internally water cooled as well.

Single-station and rotary cut-sheet tooling components typically include a mold and a fixture or series of fixtures to allow quick CNC trimming and finishing. Thermoforming low-volume product or prototypes can be manufactured from wooden and other inexpensive tooling components to offer quick solutions. This is one of the many attractive features of thermoforming. These solutions work for both light-gauge packaging and heavy-gauge product applications. The single largest factor when choosing a tooling solution is how many parts are required, and what tolerance requirements must they meet? Thermoform tooling solutions can be broken down into several categories based on volume of products required: (1) prototype and limited production, (2) limited and low production, and (3) medium- and high-volume production.

4.6.1 Prototype and Limited Production

In prototyping, the typical designer is wishing to manuf acture a sample product as quickly and cheaply as possible. Tolerances typically are not the determining factor at this volume level. When developing the prototype,the final material selection and finished product quality reflect the quick and dirty attitude as well as the limited money usually earmarked for prototyping.

Until 10 to 20 years ago,wood was used extensively. Closed-cell solid woods such as maple were materials of choice in developing wooden patterns. Open-cell woods were utilized, but typically last only se veral cycles. By coating wooded patterns of all kinds with high-heat synthetic materials, patterns can even be utilized in lo w-volume production. Caution here is necessary , as glue joints and w ood contain a great deal of moisture, which allows expanding and distortion of the pattern under heat, which makes wood a poor choice for limited production.

Synthetic materials, made from small glass beads suspended in a casting resin, replaced wood as the material of choice for prototyping and limited production. These materials have high heat resistance, limited moisture absorption,and excellent dimensional stability. Many close-tolerance prototypes or products can be produced from this material. Due to its good insulation properties, there is a lack of ability to transfer heat quickly and efficiently from plastic to mold. This causes extremely long cycle times, which are the single largest drawback for this material.

An important concept to keep in mind is that the quality of the prototype tool will be clearly reflected in the prototypes themselv es. Do not become so o verly concerned with development costs that your prototypes become unacceptable for the use originally intended.

4.6.2 Limited and Low Production

Several solutions exist at this production le vel. As noted earlier, synthetics can be utilized for limited runs, but their inability to cool quickly mak es them of limited value. High-temperature epoxy filled with aluminum makes an excellent alternative to synthetic tooling, but the tooling requires extra steps to manufacture. These extra steps and low transfer/conduction of heat, even though higher than that of synthetics, make this a choice that most thermoformers now avoid. The low production rate, coupled with electric and labor costs, have reduced the use of this once-popular tooling method,due to other ever-increasing costs of manufacturing.

The method most utilized today is a single or small number of production cavities, which allow for excellent part quality and reasonable tooling costs. The construction of this tooling must be of aluminum or an alternati ve material that has the ability to transfer heat quickly without damaging the mold. Although most in ventors and finance people wish to avoid these up-front costs, a project's successful completion is strongly dependent on high-quality tooling.

It is important to remember to be realistic in your actual product v olume needs and the quality required at this production le vel. A poorly constructed lo w-volume tool will create a small number of poorly manufactured products, and commonly results in missed deadlines and unhappy customers.

4.6.3 Medium- and High-Volume Production

At these production levels, the key questions are: How many parts are required, and specifically, in what time frame must they be produced? Failure to deliver in a regular, timely fashion can be the death of even the best designed, manufactured,and marketed product. High-volume light-gauge packaging products will require multiple cavities with integrated water cooling, and in many instances into the cavities themselves, to meet production and cycle-time requirements. High-volume heavy-gauge products will require the same design as the light-gauge products, but the added dimension of secondary operations and overall physical space required could overwhelm the production facility if not tooled properly. Several criteria must also be considered, including physical mold size, material construction, cooling integration, and how and where the molded product's secondary finishing operations will be performed.

In comparison to many other processing methods, thermoform tooling does not require, or utilize, steel-fabricated mold tooling. Most thermoform molds are constructed of aluminum for several reasons. First, and foremost, is ease of machining, which more than offsets the higher material cost of aluminum compared to steel. Second, the lack of actual wear in the thermoforming process versus other plastic processing methods makes steel unnecessary as a tooling material. Most thermoforming tooling will never be repaired for wear, and only in the highest-volume applications does noticeable wear occur. Specialty engineering and filled plastic materials may cause unusual wear and should be considered separately from most other common thermoplastic materials.

The high transfer rate of heat from plastic to aluminum molds is also extremely important when considering volume production runs. Other materials are utilized occasionally, but their exponentially higher material costs make them an uncommon choice. Integrating water and being creative in how water flows under and through cavities will make dramatically higher product output rates.

4.7 ASSISTING THE PROCESS MECHANICALLY

Although thermoforming is commonly called *vacuum forming* or *pressure forming*, it also relies on mechanically assisting the plastic to get a desired shape, finished wall thickness, or even removal of plastic features that not would result without their use. The term *plug, plug assist*, or simply *assist* refers to a mechanical application of force to control the plastic sheet during the thermoform process. Materials utilized include syntactic foam, aluminum, delrin, or other high-heat materials that will not require replacement due to heat damage from contact with hot plastic. They are commonly mounted on the platen opposite the thermoform tooling. Controlling the opposite platen during the process can result in many changes to the finished plastic product.

4.8 TOOLING FOR SECONDARY OPERATIONS

All thermoformed products, regardless of size gauge and complexity, require secondary trimming operations. This, of course, requires more tooling as well as being specific to the part being manufactured. Tooling for trimming and finishing in thermoforming again breaks into the broad categories of light and heavy gauge and their specific needs and requirements. Light-gauge products and packaging can be steel-rule-die cut in prototype to high-volume applications. This method of cutting is also commonly found in the printing, folding carton, and corrugated box industries as well. Steel rule dies are inexpensive and in most cases disposable tooling, requiring replacement depending on use and treatment when stored. When utilized with in-line thermoforming equipment, steel rule dies are capable of tolerances of ± 0.031 or $\frac{1}{32}$ in. When utilized off-line and cut manually, closer tolerances can be held but should only be used in low-volume applications or prototyping.

Punch and die, or match metal cutting, is also commonly utilized, but up-front costs are exponentially higher for manufacture of the cutting tools. This in itself makes them cost prohibitive. These tools require fabrication of high-quality steel and a secondary press when the product is manufactured. Some presses are integrated into the thermoforming machines, while others are physically separated from the thermoforming machine. This is a case-by-case situation. Examples of products utilizing punch and die tooling include medical and cosmetics or high-volume items such as food applications. These products require punch and die tooling when particulate contamination issues downstream in the finished product become undesirable or unacceptable, or when extremely close tolerance products are required.

Trimming a few heavy-gauge parts can be accomplished by simple hand routers used in woodworking applications. Accuracy here is truly in the hands of the operator of the power tool. Heavy-gauge products are normally trimmed by CNC three-or-five-axis routers in production. These routers are highly accurate cutting machines but do require precision jigs and fixturing to complete the secondary trimming process. Poor-quality jigs and fixturing will result in a poorly trimmed part. Programming of these CNC routers is very similar to any machine shop utilizing CNC machining centers. Many of the programs utilized by the CNC machines are the same as the machine shop or very closely related.

4.9 APPLICATIONS

Applications of thermoforming abound, and each of us touches thermoformed products daily. The largest consumers of plastic to be thermoformed are the food-processing and packaging industries. Simply walking through a grocery store is an eye-opening experience to the food-packaging industry. Fresh fruits such as apples are not only shipped in thermoformed trays but are also then placed in thermoform display racks for easy access by the consumer. Clear and colored packages for mushrooms, herbs, and tomatoes can easily be found. Farther into the store we can find prepackaged items such as yogurt, sour cream, and cottage cheese in single-use or reusable thermoformed cups and containers. Prepackaged and precooked products include soups, frozen dinners, and gourmet food items. Thermoformed plastic cups and disposable plates constitute a product in themselves.

Entering a department or big-box retailer, one notices tabletop thermoformed point-of-purchase displays, large thermoformed end-of-aisle displays and printed signs, and consumer products in blisters and clamshells. Many of these styles of packages are also heat sealed and radio-frequency sealed, not only for ease of assembly by the manufacturer but also because intentionally oversized thermoformed packages reduce theft of expensive but small items such as cellular telephone accessories and printer cartridges. Just look at your automobile. As you look around at all of the plastic items in your car, you notice the cup holder for your coffee and the housing for your gearshift, and perhaps the plastic covering your odometer and speedometer. These can all be thermoformed parts.

Maybe you noticed a few items during your last visit to the doctor. The plastic display holding the pamphlets about the latest and greatest medical breakthrough drug, as well as the packaging in which the drug is stored, are both examples. As the visit progresses you notice the sterile, individually packaged selections of other diagnostic devices and medications. These are all thermoformed items.

One only has to take a second look around in our world to find examples of thermoformed products and packaging. The list is as diverse and varied as the plastic materials themselves. Every day, everywhere, thermoforming applications are an indispensable and necessary part of our lives.

4.10 CAREERS IN THE THERMOFORM INDUSTRY

With over 200 thermoforming companies publicly listed, as well as a large number of companies utilizing thermoforming machines within their internal or secondary processes, career opportunities are still in abundance nationally. Although the national issue of outsourcing to other nations has been felt in the thermoform industry, skilled workers are still difficult to find and retain. This correlates to well-paid career positions and demand for any skilled person willing to update skills continually and be eager to work. Many jobs have been lost nationally and much consolidation of the industry is yet to occur, but some specific skill sets seem to be more resistant to downsizing in total job numbers. They include positions in thermoforming within the following industries: food packaging, medical packaging, pharmaceutical packaging, and promotional and specialty products. Innovations in materials, processes, and newly created industries seem to assure thermoforming a place in American manufacturing and industry.

The following is a short list of some of the job titles to be found in the thermoforming industry:

- Thermoform tool designer
- Thermoform tool programmer

- Thermoform part designer
- Thermoform packaging designer
- Thermoform packaging engineer
- Thermoform manufacturing engineer
- Thermoform mechanic
- Thermoform technician
- Thermoform pattern maker
- Thermoform steel rule die technician

There are also a huge number of support staff positions in the fields of, but not limited to, quality, production control, scheduling, marketing, and product development.

Following are a few companies that provide a sense of the range of businesses and industries manufacturing thermoformed products:

- Pactiv Corporation: www.pactiv.com
- Solo Cup Drinking Cups: www.solocup.com
- Alcoa Packaging and Consumer Products: www.alcoa.com

Plastic product manufacturers tend to be much smaller than the food, institutional, and consumer packaging companies listed above, and most gross less than $50 million in sales annually. Typical of these is Maryland Thermoform Corporation, www.mdthermo.com.

4.10.1 Finding a Position in the Thermoform Industry

To find a job in thermoforming, as in any industry, a friend in the business or word of mouth about one's work reputation is generally the best way. However, each of us at one point or another has had to consider the possibility of finding a job on our own. Several well-know periodicals in the thermoform industry, such as *Plastic News, Plastic Distributor*, and *Fabricator Magazine*, have help wanted sections. Also search for companies on the Internet, using keywords such as *plastic thermoform, vacuum forming, pressure forming*, and *plastic packaging*. Apply online to a specific company through its Web site. Some of the larger, more sophisticated companies list a human resources department at their site. Many also have specific departments or positions that list contacts and e-mail addresses. Always chose a specific department rather than the human resources or personnel department—the contact of last choice. Others simply direct you to send your résumé to the "Contact Us" or "Info @" e-mail address. This should always be addressed to a specific department or person, if possible.

When finding an advertiser in a periodical or a company on the Web in which you are interested, submit your résumé even if the job or position listed is not in your background or even if openings are not listed at all. Highly skilled employees in the field are hard to find, and even when a job is not advertised, one may be available to the right applicant.

ACKNOWLEDGMENTS

A special thank you to Peter Turk, SPE Baltimore–Washington Chapter, for his assistance in developing this chapter. A warm, personal thanks to Jeff Fish for providing the graphics and to Deanna Lewandowski for proofreading and editing.

BIBLIOGRAPHY

1. Baird, Ronald J., and David T. Bair. *Industrial Plastics: Basic Chemistry, Major Resins, Modern Industrial Processes.* South Holland, IL: Goodheart-Willcox Company Inc., 1986.
2. Cherry, Raymond. *General Plastics: Projects and Procedures:* McKnight & McKnight Publishing Company, 1967.
3. Florian, John. *Practical Thermoforming: Principles and Applications.* New York: Marcel Dekker, 1987.
4. Mooney, Peter. Plastic custom reports. in *Yearly Report.* NC: Advance, 2002.
5. *Plastic Distributor and Fabricator.* Riverside, IL: KLW Enterprises, Inc., 2000–2005.
6. *Plastic News Magazine Weekly.* Detroit, MI: Crain Communications, 1990–2005.
7. Plastic processing: guide to methods, machinery, and auxiliary equipment, in *Modern Plastics Encyclopedia.* New York: McGraw-Hill, 1969.
8. Rosen, Stanley R. *Thermoforming: Improving Process Performance.* Dearborn, MI: Society of Manufacturing Engineers, 2002.

CHAPTER 5

Blow Molding

NORMAN C. LEE

Consultant, Greensboro, North Carolina

5.1 INTRODUCTION

Blow molding is a process for producing hollow objects, primarily from thermoplastic materials. Bottles and packaging are the primary applications of blow-molded parts. As an industry segment, approximately 80% of polyethylene (PE) and a major share of polyethylene terephthalate (PET) materials are used for bottles and packaging by the blow molding industry. Additionally, blow molding is used to produce many other objects, such as components used under the hood and on the exterior of cars. Gas tanks, structural panels, toys, and double-walled cases are also blow molded. A wide range of materials is used in these applications, including polypropylene, polyvinyl chloride, polycarbonate, and fluoropolymer.

5.1.1 Plastic Process

To visualize the blow molding process, think of an inflatable plastic tube that is closed on both ends except at one point. If air enters at this point, the tube will expand, causing the walls of the tube to become thinner. It is much like blowing up a balloon (Figure 5.1). The blow molding process begins with a hot, soft plastic tube called a *parison* or *preform*. The parison is placed between two halves of a mold that has one or more hollow cavities (Figure 5.2). The mold halves are then clamped together. Air is blown into the parison, expanding it against the inside walls of the mold to form a hollow plastic part in the shape of the cavity. The mold is cooled, usually with water, and the hot plastic cools by contact with the mold. Once the part has cooled, the mold opens and the part is ejected.

Handbook of Plastic Processes Edited by Charles A. Harper

FIGURE 5.1 Inflating an expandable tube.

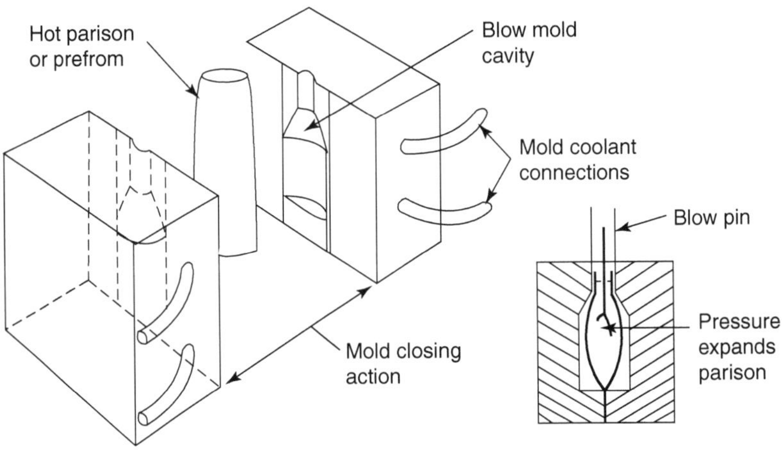

FIGURE 5.2 Basic blow molding process.

5.1.2 Brief History of Blow Molding

The plastic blow molding process is derived from the ancient art of glassblowing and became a production reality with the availability of low-cost plastic materials or *resins* in the early 1940s. Shown in Table 5.1 are some key events in the development of the process. Today, blow molding has developed in two different but related directions: extrusion blow molding and injection blow molding. These are described in detail in Chapter 2.

5.1.3 Typical Blow Molding Markets and Product Applications

5.1.3.1 Markets

Bottles are the primary application of blow-molded plastic parts; however, the blow molding industry also makes products used in the aircraft, automotive, building and construction, electronics, furniture, lawn and garden, medical, and recreation fields.

TABLE 5.1 Key Blow Molding Industry Events

- The first patent, issued in the 1850s for blow molding with a material other than glass, was issued to Samuel Armstrong. Items were made from natural rubber latex, and were mainly novelty items that were unique because of their soft feel.
- The next major advancement came in 1869 with the commercialization of celluloid, which is considered to be the first true thermoplastic material.
- In the 1880s, cellulose nitrate was introduced and was used to produce novelties and toys. This material was softened by steam; one disadvantage was its high flammability, which kept it from being widely used.
- In 1919 a material that was much more stable, cellulose acetate, became available and gained much greater acceptance as a commercial material. By 1930 it was being used for making squeeze bottles.
- In the early 1930s, Plax Corporation developed the first blow molding machine. This early machine was crude and produced only small quantities. However, from this early beginning, a machine to make 25,000 bottles a day was developed.
- In 1939, low-density polyethylene (LDPE) was introduced by I.C.I. (Imperial Chemical Industries, England). This material was much friendlier to the blow molding process and opened up the way to further materials, process, and market development.
- In the 1940s, Plax introduced the first LDPE bottle.
- The first injection blow molding machine was introduced in 1942. This hybrid machine allowed the production of plastic bottles with precise neck dimensions and tolerances comparable to glass bottles.
- Continental Can was issued a patent for a continuous extrusion blow molder in 1950.
- The industrial blow molding industry can be said to have started in 1953 with the development of high-density polyethylene (HDPE). HDPE was developed simultaneously by Phillips Petroleum Co. in the United States and in Germany by Professor K. Ziegler.
- By 1956 the commercialization of the blow molding process was complete, and explosive growth had begun.
- During the 1950s, industrial blow molding machines were developed in Europe, since Plax and Continental patents where tightly held in the United States.
- European machines became available in the United States by 1958. Empire plastics bought the first machine to make toy bowling pins.
- Several individuals and companies built machines. Norman Lee and others converted the Reed injection molding machine to make a blow-molded baseball bat.
- In the early 1960s, Ideal Toy ordered six machines built by Walden-Hartig with a die head design from Empire Toys. They were used to make dolls' bodies.
- By 1960, with U.S. patents freed, Zarn, Inc. began making milk bottles on Uniloy machines for Borden Dairy in High Point, North Carolina.

5.1.3.2 Products Made by Blow Molding

Blow-molded products are many and varied, because the process and the materials it uses are continuously being developed. Applications include balls, toys, bellows, and car bumpers (fenders). Many of the bottles and jars are used in the food industry to package both solid and liquid products, such as plastic containers for salt and sauces, are blow molded. Often, such containers are designed to use a reusable seal (a lid or cap) that is commonly produced by injection or stretch blow molding. Most food-packaging containers are small, that is, less than 1.75 pints (1 liter) in capacity.

However, it must not be thought that blow molding is restricted to such small containers. Surfboards 13 ft (4 m) in length have been made commercially, and tanks of over 2200 gal (10,000 liters) in capacity have been produced. A common, large container is made from high-density polyethylene (HDPE) and has a capacity of 45 gal (180 liters). See Figure 5.3 for various blow-molded products.

Containers The primary functions of a blow-molded container are to protect the product from spoilage and to keep the contents from escaping.

Barrier Properties The permeation (gradual movement) of gases and liquids through the walls of a plastic container is one of the major factors that determine the shelf life of food products. The transmission rates of oxygen, carbon dioxide, and flavor constituents are usually of the greatest interest. Permeation is affected by the materials, design, and processing method used to make the container.

Leakage Container leakage is controlled by the design and fit of the closure, the bottleneck finish, and the dimensional tolerances. The thread type, the number of threads per unit of length, and the length of the thread also influence leakage. Buttress threads (sawtooth shape) are often used, as they pull the closure down uniformly. If the closure contains a soft, resilient material, this will also help to seal and prevent leakage.

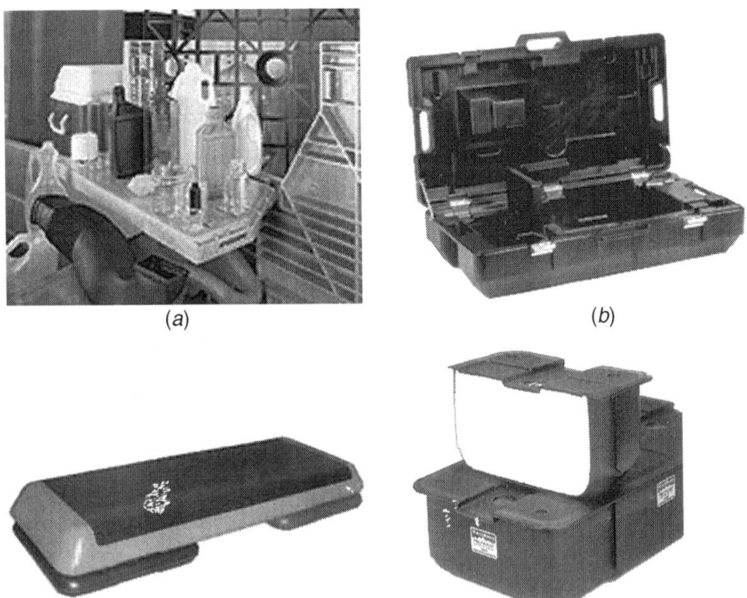

(a) (b) (c) (d)

FIGURE 5.3 (*a*) Assorted industrial and consumer blow-molded parts; (*b*) intricate blow-molded case; (*c*) aerobic step; (*d*) dock floates, foam filled for buoyancy. [(*a*) From Uniloy Milacron; (*b*) and (*d*) From WhiteRidge Plastics, L.L.C.; Reidsville, NC; (*c*) from WEK South Corp., Reidsville, NC.]

5.1.4 Blow Molding Materials

5.1.4.1 Materials Used

The blow molding industry originally developed around low-density polyethylene (LDPE). The industry then modified its machinery and processes to suit other materials. Now, another polyolefin, high-density polyethylene (HDPE), dominates the blow-molding industry. Polypropylene (PP) is also becoming more widely used for blow-molded applications. Other materials that are important to the blow molding industry are unplasticized (rigid) polyvinyl chloride (UPVC) and polyethylene terephthalate (PET). Because of interest by the automotive industry, the use of engineering plastics such as nylon (PA) and polycarbonate (PC) in blow-molded applications is also on the rise.

Thermoplastic materials are often suitable for many applications without extensive material or component modification, but in other cases they are not suitable. One reason is that many plastics are permeable; that is, a liquid product or part of the product can migrate through the walls of the container, or something from the environment, such as oxygen, can enter the container through the plastic walls. For a given plastic material, there are several ways of overcoming this problem. They include increasing the product thickness; coating the product with an impermeable material such as glass, cross-linked PE, or more commonly, polyvinylidiene chloride (PVDC); incorporation of a platelike filler (e.g., glass flake); and using a combination of plastic materials. Such combinations, in which one parison is surrounded by another, are usually produced by co-extrusion. Orientation (stretching the parison) also yields improvements.

5.1.4.2 Chemical End Use and Physical and Processing Requirements

HDPE is impervious (unaffected by, or resistant) to water, water vapor, inorganic chemicals, and inorganic materials in aqueous (water-based) media, especially at the ambient temperatures normally encountered during storage and transport. Typical materials in these categories are:

- Most inorganic acids
- Most bases (alkaline)
- Minerals and inorganic chemicals as dry solids and powder
- Minerals and inorganic chemicals in solution or aqueous suspension

Its high moisture resistance makes HDPE an ideal container material, keeping moisture out as well as in. However, with organic chemicals and fluids, HDPE must be used more selectively. Some liquids pose no major problem under ambient conditions. These include:

- Most alcohols
- Most glycols
- Glycerin
- Most organic acids and their derivatives

For hydrocarbon fluids, oils, and solvents such as those listed below, absorption and permeation through HDPE become very important factors.

- Straight-chain hydrocarbons (aliphatics: butane, pentane, hexane, heptane, octane, etc.)
- Other hydrocarbons (aromatics: benzene, toluene, xylene, etc.)
- Chlorinated and oxygenated products of the above (chlorobenzene, chlorinated solvents, ethers, ketones, esters)

Most of these compounds are flammable and would not normally be contained in HDPE containers without special consideration. Their use as an ingredient in any product being considered for packaging or storage in HDPE containers requires proper testing to determine whether or not excessive permeation and absorption will occur.

Another aspect of chemical end-use requirements that must be considered involves the property of certain chemicals, in either aqueous or organic media, to cause "environmental stress cracking," which weakens the material and may cause early failure. In general, certain detergents and aliphatic chemicals stress-crack HDPE. To resist this effect, high-performance HDPE resins, primarily copolymers, have been developed.

5.1.4.3 Physical and Processing Requirements

The HDPE resins used in extrusion blow molding have a high molecular weight (MW) to provide the high melt viscosity required for proper extrusion and melt strength in the parison. They also have a broad molecular weight distribution (MWD) to enable easier flow at processing temperatures and flow rates.

5.1.5 Resin Considerations for Blow Molding

5.1.5.1 Melt Viscosity Ratings

The HDPE resins commonly used in extrusion blow molding usually have a fractional (less than 1) melt index (MI) of 0.9 per 10 min or lower. However, low MI values, indicative of high MW, are misleading when comparing the melt processability of one resin with another. MI measurements are made on a simple laboratory capillary instrument called an *extrusion plastometer* (Figure 5.4). They represent the weight of extrudate (grams collected in 10 minutes) that is forced through a standard orifice under a static pressure of 43.3 lbf/in^2 (0.290 MPa) at a temperature of 374°F (190°C). Under these conditions, the shear stress in the melt passing through the orifice is only 0.019 MPa. Although 190°C may represent a realistic processing temperature for HDPE, the applied pressure and the resulting melt shear stress are considerably lower than those encountered in actual processing.

The test conditions above are specified by ASTM D1238, Method E. To provide a better view of how HDPE blow molding resins will behave under actual processing conditions, modifications to this procedure are provided by other procedures specified in D1238 using the same instrument at the same melt temperature but with higher applied static pressures. Table 5.2 lists these conditions. Flow determinations via

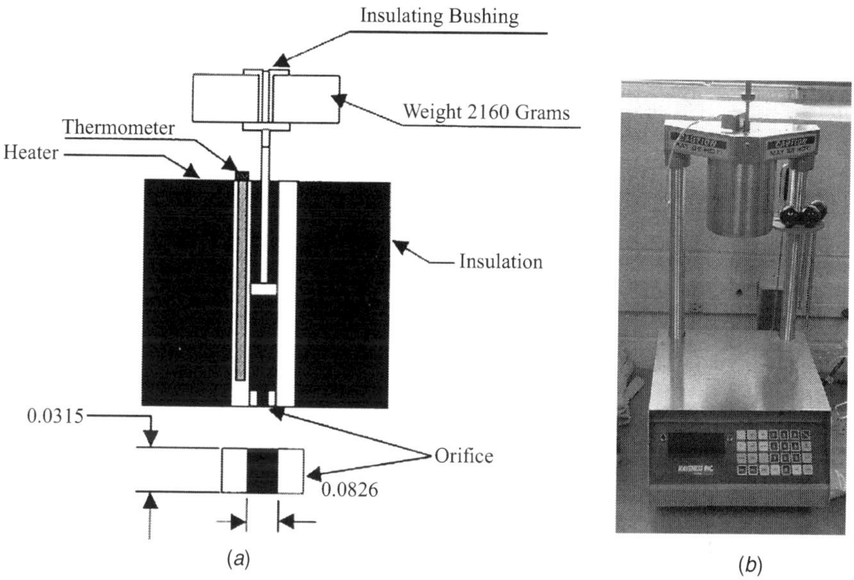

FIGURE 5.4 (*a*) Melt indexer; (*b*) Kayness melt index tester. [(*b*) (Courtesy of Davidson County Community College, Lexington, NC.]

TABLE 5.2 ASTM D1238 Test Method Conditions[a]

Method	Weight (g)	Static Pressure (MPa)	Shear Stress (Pa)
E	2,160	0.30	19,000
N	10,000	1.38	89,000
F	21,600	2.98	190,000

[a]All at 190°C.

method N, called *melt flow* (MF), and method F, called *high load melt index* (HLMI), are often provided on specification sheets, particularly for higher-molecular-weight HDPE resins with MI measurements of less than or equal to 0.1 g per 10 min.

5.1.5.2 Melt Rheometer

Melt fluidity measurement can also be made using a gas-pressurized rheometer under conditions established by Canadian Industries Ltd. [190°C, but with an effective pressure of 10.3 MPa (1500 lbf/in^2)]. The orifice dimensions of this instrument are smaller than those on the melt index device (diameter is 0.489 mm and land length is 4.470 mm). The pressure above induces a shear stress of 0.283 ×106 dyn/cm^2. This rating is called the *CIL flow index*, and like the preceding methods, measures the grams of extrudate collected in 10 minutes.

5.1.5.3 Basic Melt Rheology

Three commercial HDPE resins with equivalent MIs of 0.2 g per 10 min were rated by the HLMI method and found to vary over a wide range, as indicated in Table 5.3. Such differences in melt properties are of great significance in blow molding performance and equipment design. The higher HLMI/MI ratio indicates greater melt flow under equivalent pressure. Although the MIs indicate equivalent average molecular weights (MWs), resin X is more shear sensitive due to a broader molecular-weight distribution (MWD), especially at the low end of the MWD spectrum. This increased proportion of shorter, linear molecules acts as a lubricant. The ratio of HLMI/MI may thus be used as a guide in rating MWD; the higher ratio always indicates a broader distribution.

The melting and pumping actions of the extrusion process, maintained by heat and pressure, acts on the resin via two rheological mechanisms: viscous and elastic. Viscous energy is dissipated in working against the resistance of the system, whereas elastic energy is stored. This stored component causes an overall "swell" in the extruded parison. The capacity of molten HDPE to undergo elastic deformation, hence to store energy, increases as MWD broadens, especially with an increased proportion of larger molecules at the high end of the MWD spectrum.

5.1.5.4 Melt Swell

As already noted, the swell effect is the result of the elastic recovery of stored energy in the melt. The long molecular chains are stretched and straightened during extrusion and attempt to return to their favored, coiled and tangled, condition as soon as the forces acting on them are removed. Melt swell, rather that exact die and mandrel dimensions, establishes the final dimensions (both diameter and wall thickness) of the parison. Control of swell, by resin selection or rheology, and its prediction are essential in precision blow molding operations. Data on melt swell can be developed using laboratory instruments such as the Melt Indexer (ASTM D1238) and employing different weights in the range 2160 to 21600 g. The gas rheometer used in the CIF flow index method and the Instron melt rheometer can provide guidance in predicting melt swell characteristics at temperatures and shear conditions matching those experienced on commercial blow molding equipment. Basic differences in the *L/D* ratios of the rheometer orifice and equivalent parison tooling must be taken into account in developing such comparative data.

The *swell ratio* (extrudate diameter divided by orifice diameter) increases with shear rate. Some resins tend to level off in swell in certain ranges of shear rate, while

TABLE 5.3 Melt Properties for Three Resins

Resin	MI	HLMI	HLMI/MI
X	0.2	32	160
Y	0.2	20	100
Z	0.2	16	80

for others swell continues to increase as shear rate increases. The former types provide better parison dimension control and are preferred.

5.2 STAGES AND TYPES OF BLOW MOLDING

5.2.1 Stages of the Blow Molding Process

There are many ways to make blow-molded plastic parts, but they all have the following stages in common:

1. Plasticizing (melting) the resin
2. Production of the parison (extrusion) or preform (injection)
3. Inflation of the parison or preform followed by cooling in the mold
4. Ejection of the part from the mold
5. Trimming or finishing the part (the trim step is frequently performed while the other four steps are cycling)

It should be noted that when several pieces are made at one time in multiple molds on one machine, the first four steps may overlap. Also note that step 3—inflation of the parison or preform and cooling of the part—takes the most time and controls the machine cycle.

5.2.2 Types of Blow Molding

The types of blow molding include extrusion, injection, and stretch. Figure 5.5 describes these types and subtypes.

5.2.3 Extrusion Blow Molding

There are two basic types of *extrusion blow molding*: the continuous method and the intermittent method.

5.2.3.1 Continuous Blow Molding

In the continuous method, the parison is extruded continuously from a head or die unit. The extruder produces an endless parison, which subsequently is pinched and/or cut by the closing mold halves. In one variation of the continuous method, a shuttle press carries one or more molds, which may be of the multicavity type and are supplied with one or more parisons. As soon as the mold closes, it moves away and an open mold moves into place as the parison or parisons continue to be extruded. When the mold closes, air is blown into the parison to inflate it into the shape of the cavity. In some molds, the bottom side holds a blow pin. In the case of bottles, air is blown through a hole from the top. In other cases, a blow needle is mounted at the side or in the mold and is actuated to penetrate the parison. Once the parison has expanded, air pressure holds it against the mold as the part(s) cool.

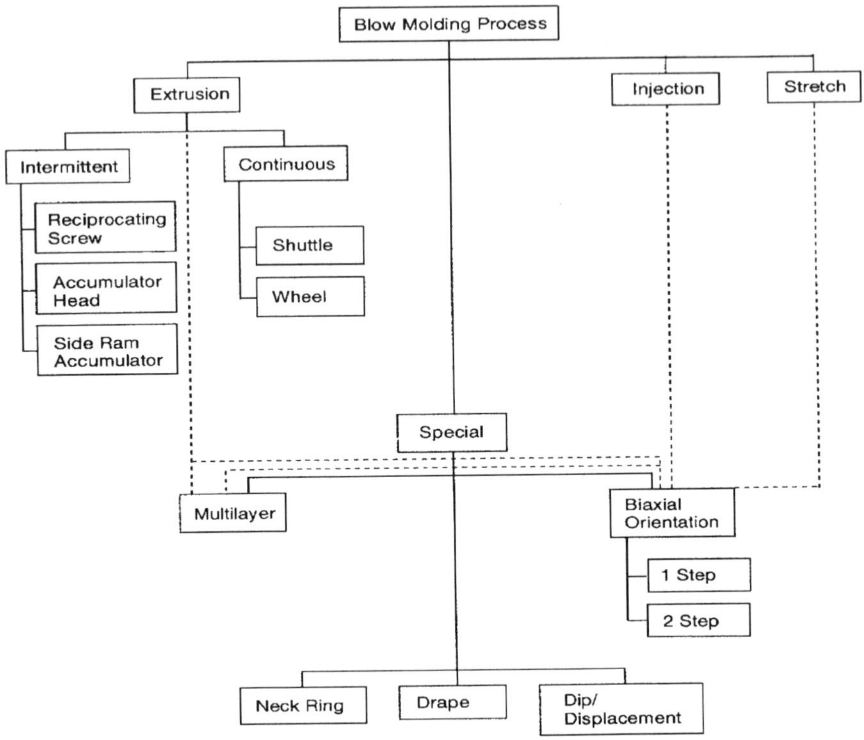

FIGURE 5.5 Types of blow molding processes.

5.2.3.2 Intermittent (Accumulator) Method

In this method, the extruder runs continuously. A chamber, called an *accumulator*, gathers and retains (accumulates) a substantial volume of well-plasticized melt that is delivered by the extruder as the previous part is being blown and cooled. Once the required amount of hot plastic has accumulated and the mold is ready, a plunger or ram forces the melt out through the die head to form the parison (Figure 5.6).

The intermittent or accumulator method is widely used to make large industrial parts, ranging from municipal waste containers to leaf blower housings (Figure 5.7). A benefit of the intermittent or accumulator method is that it allows the delivery rate of hot plastic from the die head to be independent of the delivery rate of the extruder. The extrusion of resin into the accumulator chamber is relatively slow compared to the extrusion rate of the parison by the ram. Accumulator capacity determines the maximum size of large blown parts. A large accumulator may hold enough plastic melt to make a parison for a 150-lb (68-kg) part.

The advantages of an accumulator are:

- It holds a large volume of melt for large items requiring very long (up to several minutes) molding and cooling cycles.

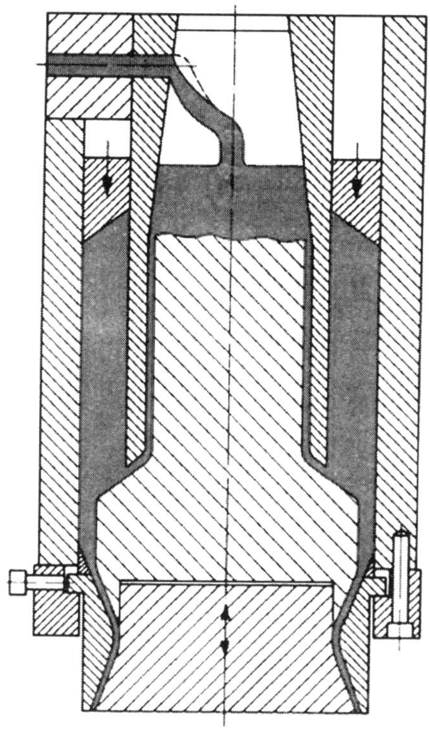

FIGURE 5.6 One type of accumulator.

FIGURE 5.7 John Deere lawn blower.

- It permits high production rates.
- It permits the fast extrusion of large parisons (short parison formation time) and consequently a short suspension time for the parison, allowing comparatively little sag and better control of wall thickness.
- It provides more uniform shot size (weight of plastic), which minimizes waste.
- It decreases idle mold time to a minimum.

5.2.3.3 Co-extrusion Blow Molding

Co-extrusion refers to the technology used to make products that contain multiple layers in their wall structures. Such products are said to be co-extruded. The layers may be made of the same or different materials, colored or uncolored material, or recycled and virgin materials. Packaging of various types is the primary application of co-extruded products, as better barrier properties are the main reason for the multilayer structure. The multilayered structure of co-extruded products is created by combining two or more melt layers in a die head before their extrusion as a parison. The main difference between multiple-layer and single-material extrusion blow molding is in the extrusion system. In co-extrusion, each material is extruded from its own extruder. Examples of products made from this process are ketchup bottles and automotive gas tanks.

Arrangement of Extruders for Co-extrusion An arrangement of extruders to produce co-extruded, multilayer structures is illustrated in Figure 5.8.

Multilayered Structures A co-extruded, multilayered structure (Figure 5.9*a*) may be created to provide one or more characteristics that cannot be provided by a single-layer product. These may be based on a physical requirement: for example, a better heat barrier or increased resistance to permeation. Also, cost considerations may require that a virgin wall material be replaced by reclaimed material, or that a costly color be used in only one layer of the structure instead of throughout the entire wall thickness.

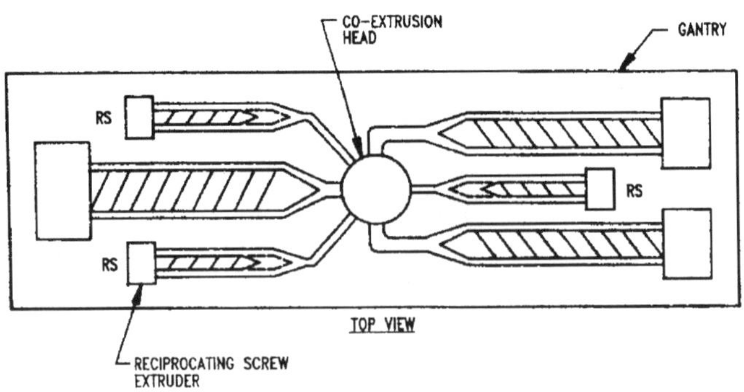

FIGURE 5.8 Co-extrusion blow molding.

Co-extrusion Systems The extrusion system in co-extrusion blow molding must supply several streams of melted material to the die simultaneously . Some streams are smaller, by design, in volume than others, in order to produce thinner layers. The processing conditions may also differ from one material to another. See Figure 5.9*b* which shows a typical co-extrusion die head, and Figure 5.9*c*, which illustrates the multiple layers in a packaging container.

5.2.4 Three-Dimensional Blow Molding

Three-dimensional blow molding w as developed several years ago in Japan and refined further in Germany in recent years. Parts with irregular shapes conventionally blow molded create lar ge areas of flash, whereas a seamless molded part has flash only at the top and bottom of the part (see the example of an air duct in Figure 5.10). The advantages of the reduced flash are that it allo ws the use of a smaller e xtruder, less power, and granulating equipment, thus less regrind handling. Other advantages include even wall thickness; improved mechanical strength; no finishing at the outer diameter, resulting in high product quality; and the use of lo wer clamp forces.

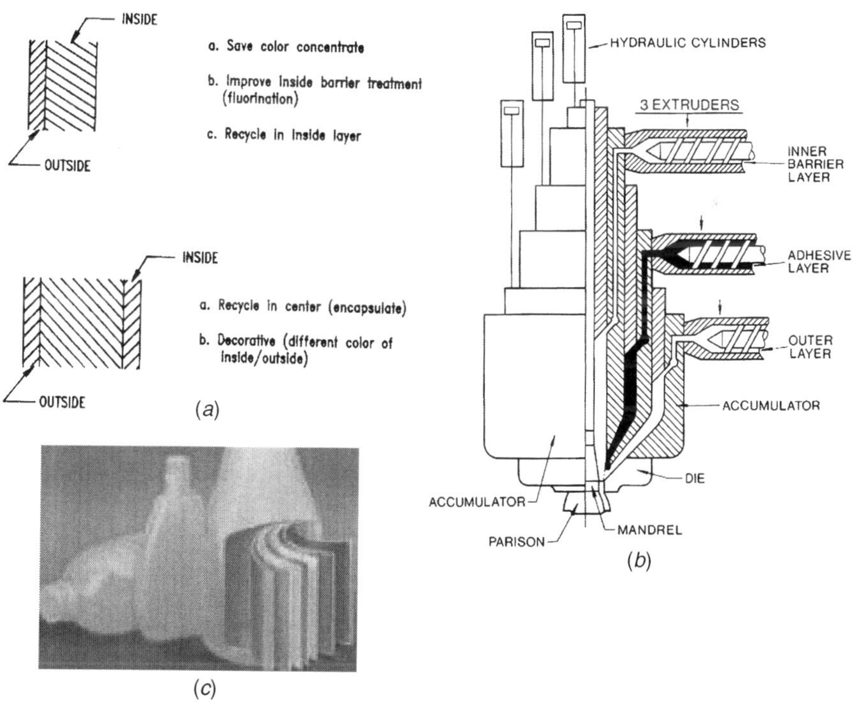

FIGURE 5.9 (*a*) Multilayer structures; (*b*) multilayer extrusion die and manifold deli very system; (*c*) multiple layers used in packaging applications. [(*c*) Courtesy of Bekum America Corporation, Williamston, MI.]

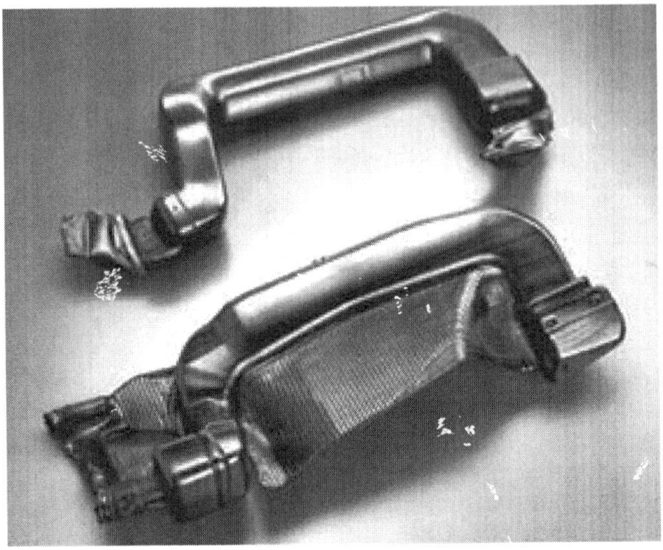

FIGURE 5.10 Air duct parts. Top, three-dimensional blow-molded part; bottom, conventional blow-molded part. (Courtesy of SIG Kautex, Inc., North Branch, NJ.)

5.2.4.1 Three-Dimensional Extrusion Processes

Three types of three-dimensional technology methods are a vailable: (1) the suction blow module, (2) vertical clamp parison manipulation, and (3) horizontal segmented mold parison manipulation.

Suction Blow Molding Figure 5.11 shows the four phases of the suction blo w molding process.

Parison Manipulation In the vertical clamp method the mold opens vertically, the lower half slides out, the parison is placed in the cavity, and the mold slides back and blows (Figure 5.12). This method is ideal for multilayer applications: for example, fuel filler pipes. Conventual mold halves provide a lower mold cost. The mold is also very accessible, and double-clamp stations can be used for higher v olumes (Figure 5.13).

In the se gmented mold/horizontal clamp process (Figure 5.14) the parison is extruded, and as the mold be gins to close, the parison is inflated. Then the parison is manipulated to the ca vity configuration and the mold completes closure, cools, and opens with part ejection. The parison manipulation is accomplished with a six-axis robot. Complex shapes can be produced by using se gmented mold technology, with the incorporation of v alue-added design features through the use of a clamp function and pinch-off for twin tubes, brackets, and so on. The part quality is high due to the minimum contact between parts and mold.

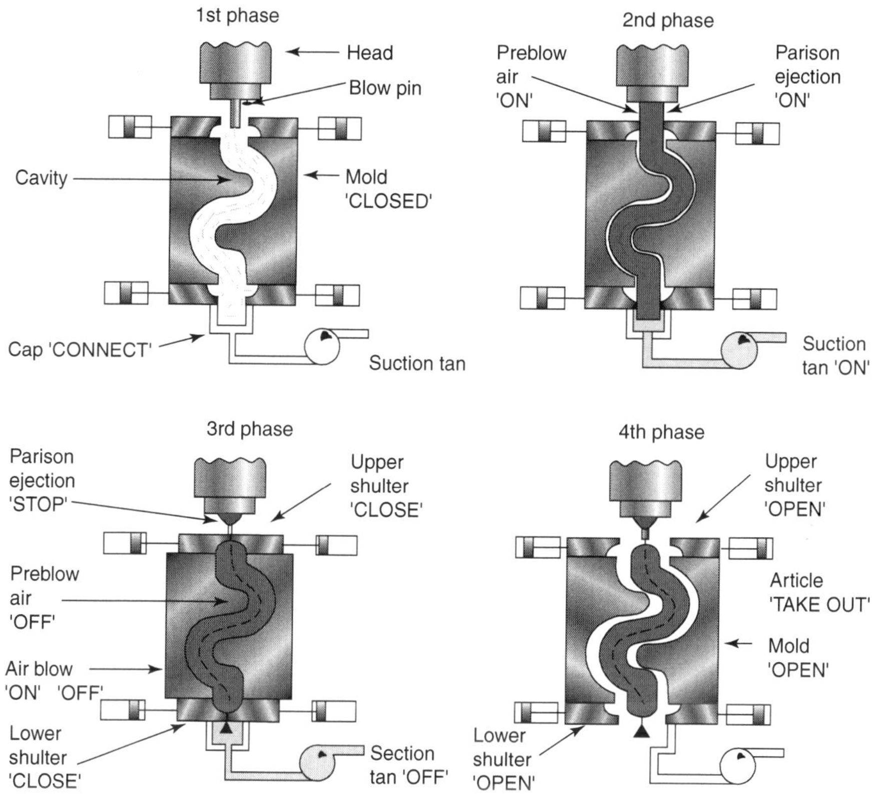

FIGURE 5.11 Suction blow molding.

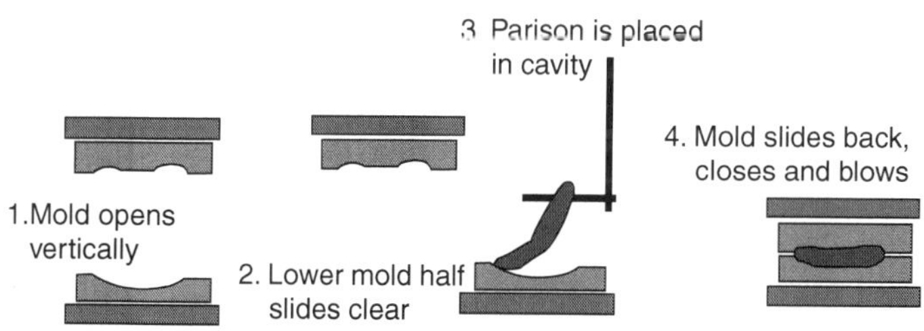

FIGURE 5.12 Vertical clamp method. (Courtesy of SIG Kautex, Inc., North Branch, NJ.)

5.2.4.2 Double-Walled Parts and Containers

Double-walled parts, mostly cases with shallo w draft angles, were developed in the early 1970s with pivots, latches, and a modular mold system as well as a special blo w machine with patents issued to Peter Sherman, who uncovered several molders in different regions of the United States and Europe. A typical part is shown in Figure 5.15*a*.

FIGURE 5.13 Integrated mounting. (Courtesy of SIG Kautex, Inc., North Branch, NJ.)

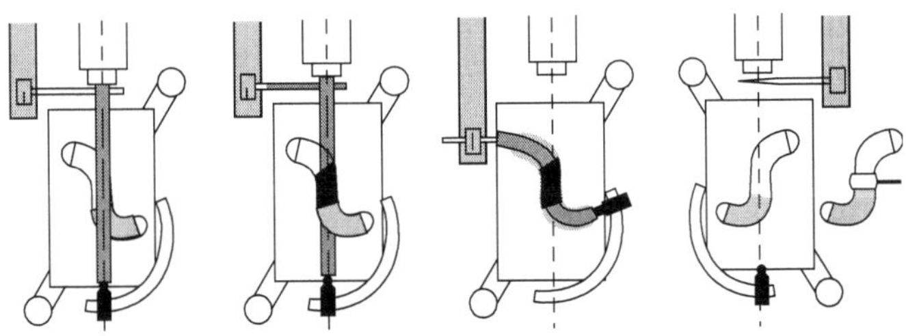

FIGURE 5.14 Horizontal clamp.

The parison is usually produced on a continuous extrusion machine with a transfer parison into a mold or a shuttle press (Figure 5.15 *b*). In Figure 5.15*c* the parison has been prepinched on the bottom. It has been preblown with low pressure to form a pillow. As the mold closes, the parison bulges and forms over the male form. The side and top of the parison begin to be trapped around the edges, essential for the successful double-walled part (Figure 5.15*d*). The mold closes further, with air preventing the walls from collapsing (Figure 5.15*e*). Cooling and high-pressure air are used to conform the mold core and cavity (Figure 5.15*f*). The high-pressure air is usually blown through a hollow needle.

5.2.5 Injection Blow Molding

Injection blow molding is used to produce a molded parison called a *preform.* Injection blow molding is usually preferred over extrusion blow molding for making small parts that require high production volumes and closer control of dimensions. The injection blow molding process occurs in two steps: (1) injection molding a preform onto a support pin or core, which provides neck threads that are already formed

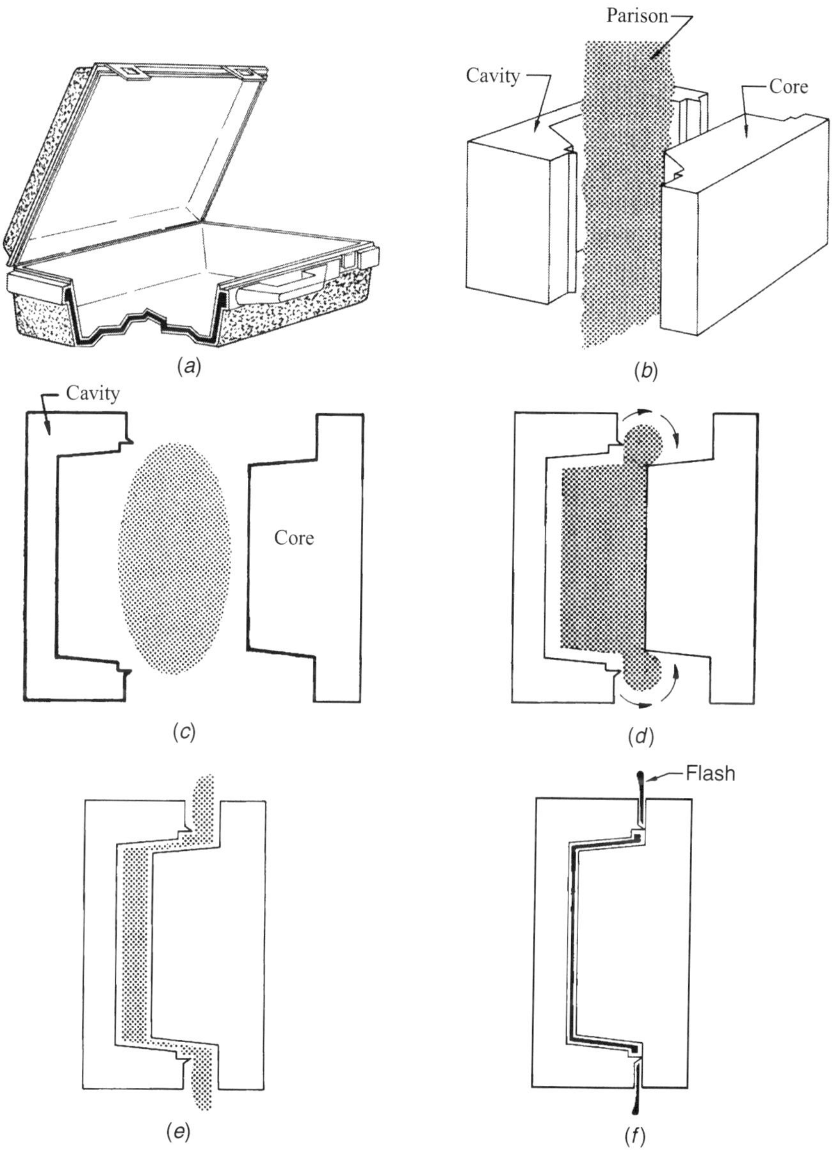

FIGURE 5.15 Process sequence double-walled case.

to their required dimensions; and (2) blowing the preform, still on the support (core) pin, to its final shape in a separate mold (Figure 5.16).

5.2.5.1 Injection Blow Molding Machine

Injection molding machines with screw plastication units that incorporate space for melt accumulation at the front of the barrel were developed in the mid-1950s. The

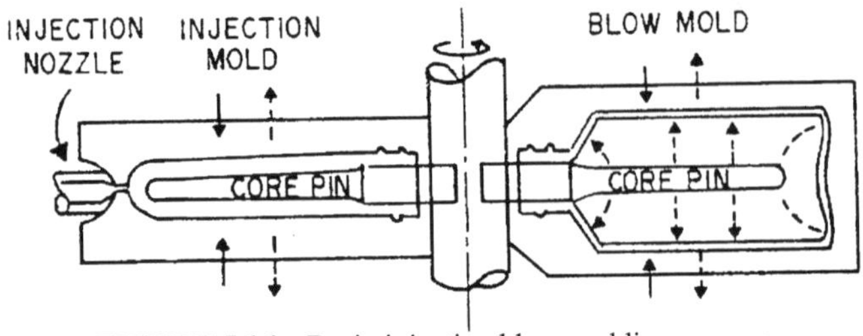

FIGURE 5.16 Basic injection blow molding process.

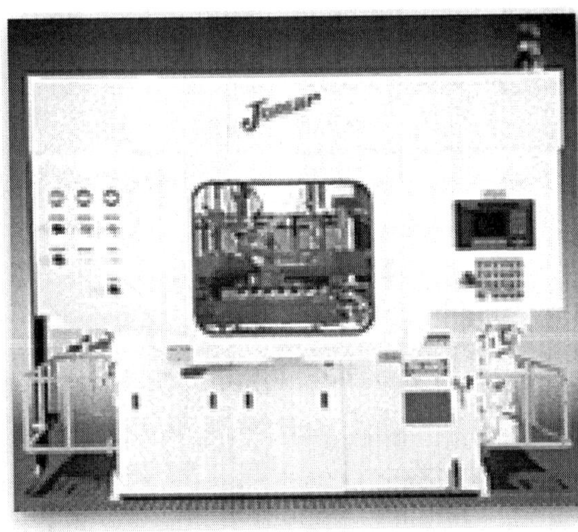

FIGURE 5.17 Injection blow molding machine. (Courtesy of Jomar Corporation.)

adaptation of these units to the blow molding process led to the reciprocating screw blow molding machine (Figure 5.17). The plasticating screw is designed for reciprocating (back and forth) movement in the barrel. At the start of plastication, the screw tip is close to the end of the barrel. As plastic is melted, a cushion of melt builds up which, as in reciprocating screw injection molding machines, gradually forces the screw backward. When the correct amount of plastic melt has accumulated in the barrel ahead of the screw, the screw stops turning and the melted resin is forced under high pressure through a sprue and gate opening into the mold cavity.

In phase 1 of this process, the preforms are molded by injecting the plastic material into a matched metal mold, consisting of top and bottom split cavities, over a core that forms the inside of the tubular preform. The preforms are allowed to cool only long enough to hold their shape. In phase 2 the mold opens and the hot, semiviscous preforms are indexed to the next station (Figure 5.18), where split cavities in the shape of the part are closed over the preforms. Here, the hot preforms are blown to the shape of the cavities, then cooled. In phase 3 the blow molds open and the parts are indexed to the next station for ejection. In this three-phase process, all three phases take place at the same time (Figure 5.19).

FIGURE 5.18 Uniloy Milacron transfer turret, indexing injection-molded performs on cores to blowing station.

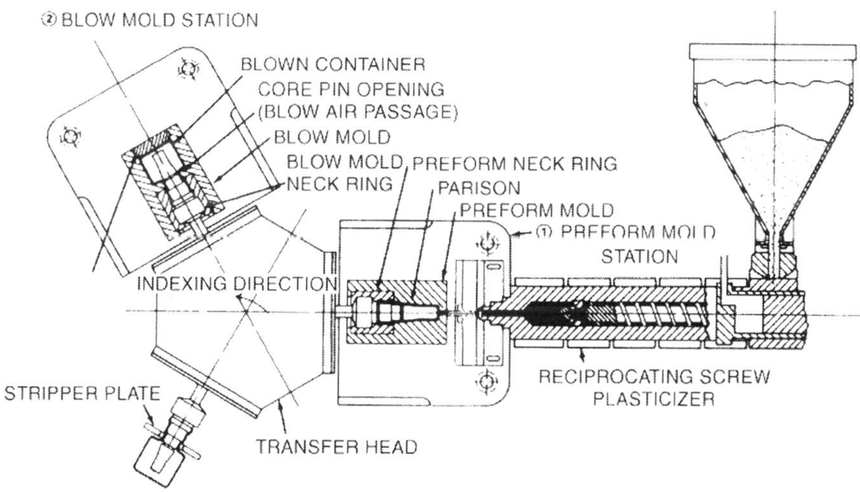

FIGURE 5.19 Three-station injection blow molding machine.

In this type of machine, the core rods are mounted on a turret mechanism, turning in sequence from injection station to blow station to ejection station. Cost dictates the use of matching multicavity injection and blow molds. The majority of machines are of the three-station type described, with 120° indexing and matching injection and blow molds. Special four-station machines are indexed on 90° turns, with the fourth station being used for special conditioning of the parison core rod after stripping (pick-off) and before blowing (Figure 5.20). Four-station machines are often used for two-color or multicolor bottles.

5.2.5.2 *Stretch Blow Molding*

Stretch blow molding is a modification of either injection blow molding [referred to earlier as injection stretch blow molding (ISBM)] or extrusion blow molding. It involves conditioning (heating) a molded and cooled preform to a specific temperature, closing it in the blowing mold, then very quickly stretching it in two directions, length and diameter. This biaxially orients the polymer molecules. Often, a rod is used to stretch the hot preform in the axial direction, with air pressure then used to stretch it in the radial direction. Doing this improves impact strength, transparency, surface gloss, gas barrier, and stiffness properties. The four plastics commonly used are polypropylene (PP), polyvinyl chloride (PVC), polyethylene terephthalate (PET), and polyacryonitrile (PAN), with the most common application for stretch blow molding being the soda bottles made of clear or tinted PET (Figure 5.21). An injection stretch blow molding machine is shown in Figure 5.22.

5.2.6 Control System

Modern control systems are based on microprocessors, and blow molding machines are no exception. Such systems are capable of performing not only the control function but also troubleshooting and statistical functions. Microprocessor-based controls do not necessarily produce better products, but they do produce them more quickly and may produce them more consistently (Figure 5.23).

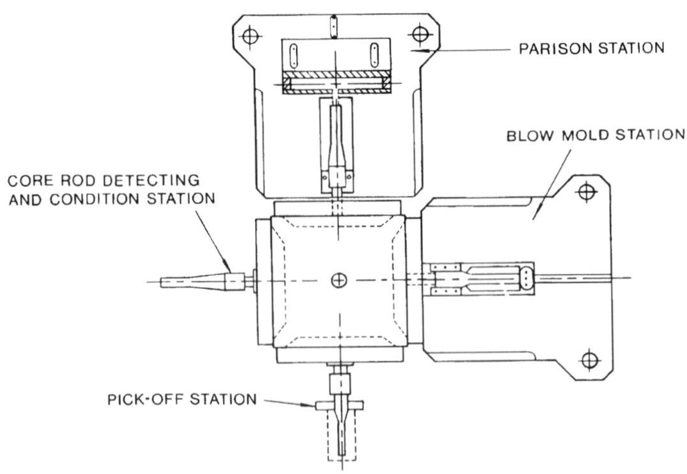

FIGURE 5.20 Four-station injection blow molding machine turret.

FIGURE 5.21 PET carbonated beverage bottles made by Amcor Twinpak, Canada.

5.2.6.1 Reprogramming

With microprocessor-based controls, reprogramming machine sequences to suit special conditions is easier than it would be with solid-state or relay controls. Machine sequences can be altered by a field programming device to incorporate special machine movements or sequences. Users can modify the programs to adopt modifications and even new processes. Machine set points, process set points, and process sequences can be loaded into the controller by magnetic card, for example, making setting up the machine easier and more consistent. Even though the machine will not produce parts that are identical to those produced in a previous run, because machines and materials differ, the new run will be as close as possible to the original production. Only fine tuning will be required.

5.2.6.2 Universality

The machine manufacturer can fit the same controller to all machines in their range, variations are obtained by varying the program. Manufacturing processes are simplified and costs are reduced by doing this. A more versatile machine is obtained even if the costs are similar to those of alternative control systems.

FIGURE 5.22 Injection stretch blo w molding machine manuf actured by Jomar Corporation, Pleasantville, NJ.

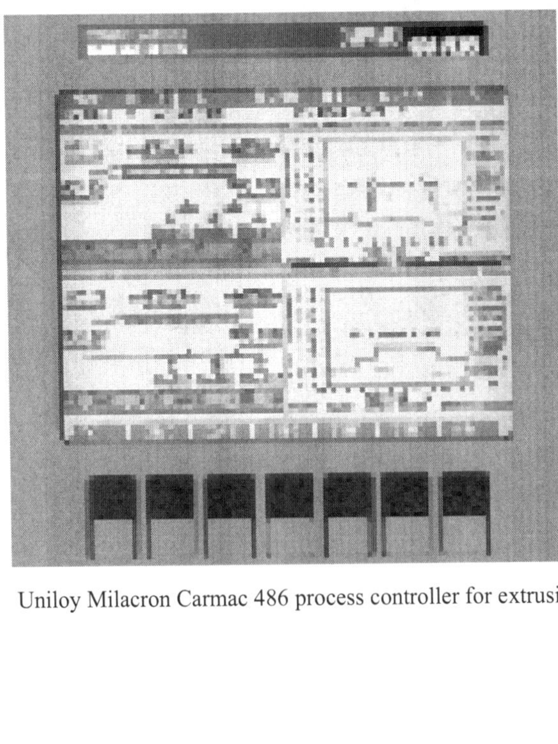

FIGURE 5.23 Uniloy Milacron Carmac 486 process controller for extrusion blow molding machines.

5.2.6.3 Safety

This is a very important benefit of microprocessor control, as more safety features can be incorporated. The status of guards, interlocks, and other safety devices can be determined, and the machine will be allowed to start only if it is safe to do so. What is more, the system can be updated to comply with new safety requirements as they become effective.

5.2.6.4 Troubleshooting

A microprocessor-based controller can troubleshoot both itself and other components in the system. The results of this survey can be displayed in plain language, or on a machine diagram, together with suggested solutions.

5.2.6.5 Ancillary Control

The memory of many microprocessor systems is large enough to accept the sequencing of ancillary equipment. Robots, trimmers, and material handlers, for example, can be brought into the machine cycle, thus making downstream equipment synchronous with the molding machine and less costly to control. Even the simple temperature controller is now able to communicate with the control system of the molding machine and with other computers. For example, the master computer can set the temperature, alter the control characteristics or terms of the temperature controller, and maintain a record of the actual temperature.

5.2.6.6 Open- and Closed-Loop Control

A process is said to be in *closed-loop control* if when a command is given, the control system monitors what happens and readjusts the strength of its response, if necessary, to achieve what is desired. *Open-loop control* is the term given to a control system that does not monitor what happens when a control command is given. In blow molding, the term *closed loop* is often applied to some function of the hydraulic system: for example, to parison programming speed.

5.2.6.7 Statistics

A modern control system is capable of accepting large amounts of data, rearranging and presenting the data in a way that informs the machine user of what is happening or what is going to happen. This can only be done by a sensible application of statistics. Such statistical procedures could be done by hand,but by the time the calculations were finished, the molding cycle would be over. For online testing to be successful, a decision must be made as to what will be measured, what calculations will be performed, and what will be done with the answer. By accepting the results of various measurements (e.g., screw speed, parison speed), the microprocessor can perform the statistical calculations extremely quickly and display the answer just as quickly.

5.2.6.8 Automatic Quality Control

Because of the power of the microprocessor, it is now relatively easy to incorporate features in a molding machine that make it possible to verify product quality during the molding cycle. Verification of product quality at the point of manufacture can be documented in statistical process control (SPC) records. The average or set value of a monitored value may be displayed automatically on a VDU (video display unit), together with appropriate control limits: for example, an upper control limit and a lower control limit. During the actual molding cycle, the data are gathered and displayed on the VDU. By doing this, production drifts or trends can easily be spotted and out-of-specification parts rejected completely or diverted for inspection.

5.2.7 Advantages of Extrusion and Injection Blow Molding

5.2.7.1 Extrusion Blow Molding

The blow molding process is a natural production process for containers and hollow parts. It is the preferred process for high-volume containers for packaging applications such as food, personal care items, and household products, as well as for industrial high-strength applications such as automotive and agricultural tanks, pressure vessels, and air ducts Figures 5.24 to 5.27 show several examples of blow-molded parts. Capital investment in machines and tooling is largely dependent on quantities of production.

5.2.7.2 Injection Blow Molding

The main advantages of injection blow molding are:

- There is no scrap or flash to trim and reclaim.
- The neck finish and details are very accurate and of high quality.
- There is no process weight variation.
- Injection blow molding typically offers the lowest part cost for high-volume bottles weighing 12 oz (37 g) or less.

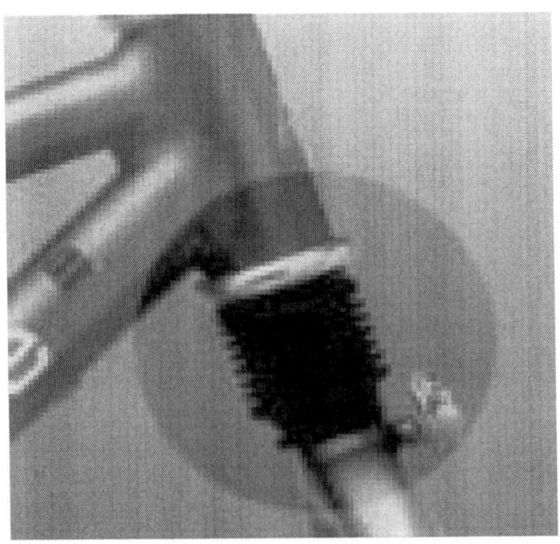

FIGURE 5.24 Cannondale bike bellows. (Courtesy of Blow Molded Specialties, Providence, RI.)

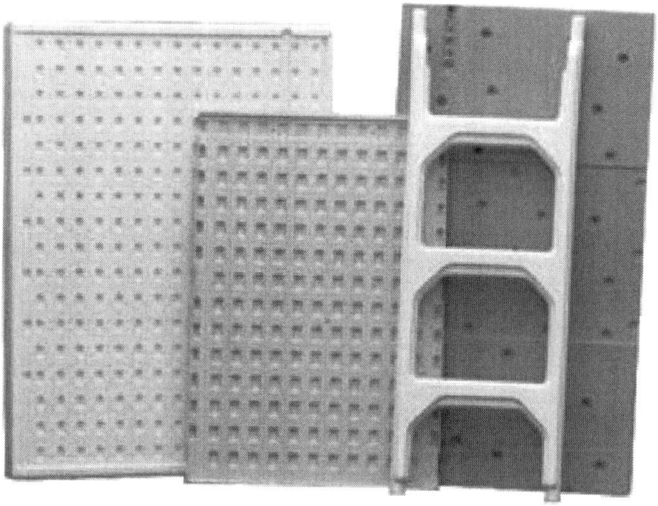

FIGURE 5.25 Panels. (Courtesy of Blow Molded Specialties, Providence, RI.)

FIGURE 5.26 Assorted bottles and containers. (Courtesy of Blo w Molded Specialties, Providence, RI.)

It is important to understand that man y different resins are processed by this method. Among these are LDPE and HDPE (lo w- and high-density polyethylene), polypropylene, polyethylene terephthalate, polyacrylonitrile, styrene–acrylonitrile, and others.

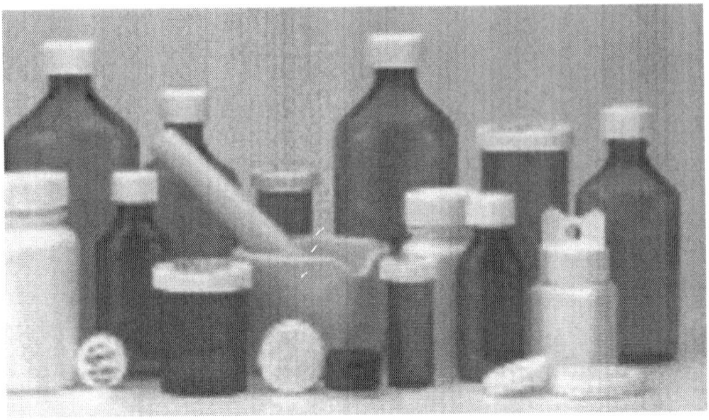

FIGURE 5.27 Medical prescription containers. (Courtesy of the Kerr Group.)

5.2.8 Disadvantages of Extrusion and Injection Blow Molding

5.2.8.1 Extrusion Blow Molding

A disadvantage of blow molding is uneven wall thickness. Although this may be minimized with programming, the wall is usually thicker at pinch-off areas and thinner in corners. The tendency of thicker walls to shrink more in the center of the part has to be compensated for in the product design. For example, rectangular containers usually have outwardlycurved surfaces. Another disadvantage is that close dimensional tolerances are difficult to achieve, the exception being bottle threads produced by injection blow molding. The accuracy of surface finishing details on extrusion blow molded products is also relatively low.

5.2.8.2 Injection Blow Molding

The disadvantages of the injection molding process are:

- The cost of tooling is higher than for extrusion blow molding.
- Bottle sizes and shapes are limited to an ovality ratio of 2 : 1 and a blow-up ratio no greater than 3 : 1.
- Slightly offset necks are possible with this process, but handles are not.

5.3 BLOW MOLDING EQUIPMENT

The basic components of the extrusion blow molding process are shown in Figure 5.28.

5.3.1 The Extruder

Extrusion is the process of applying heat and pressure to the resin to melt it and force it through an accurately machined die orifice or opening to form a desired shape. For extrusion blow molding, the shape of the die orifice that forms the parison is a circular

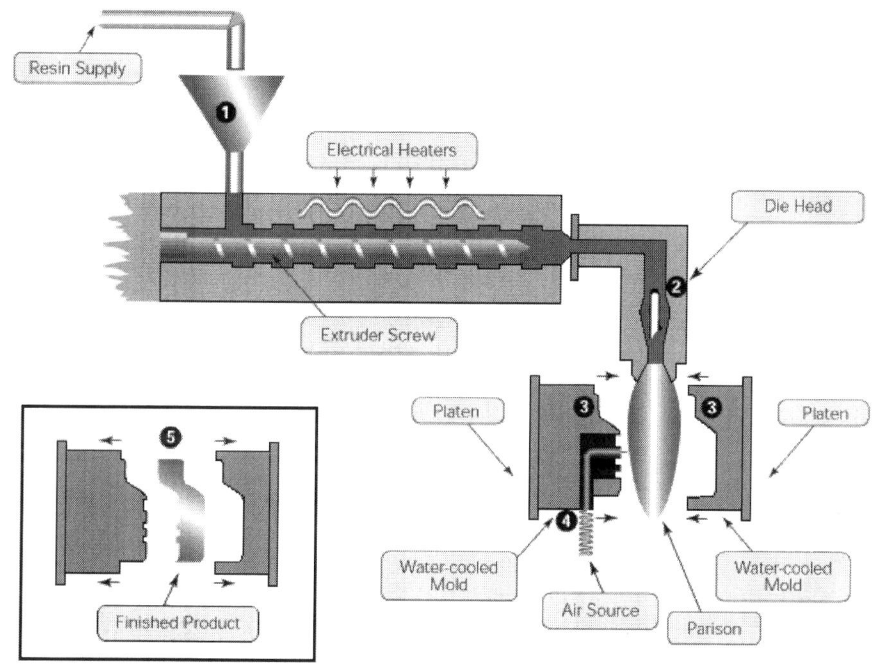

FIGURE 5.28 Components of the extrusion blow molding processes. 1, Feed hopper; 2, die head; 3, mold; 4, air source; 5, finished part. (Courtesy of Delta OEM Division, Raleigh, NC.)

slot. There are several parts to an extrusion blow molding machine and die head assembly, including (1) the screw (with feed, transition, and metering sections) and barrel, (2) the screw drive, (3) the material feed hopper, (4) the die head, and (5) the die core or mandrel and die tip. Typical components of a single-screw extruder are shown in Figure 5.29. The plastic material is melted (plasticized) in the extruder barrel and delivered to the die at a uniform rate by rotating the screw at the proper speed (rpm).

The size of an extruder is given by its barrel bore size and by the ratio of the length (L) of the screw to the screw diameter (D). This ratio of length to diameter is called *L over D* and is written as either L/D or $L:D$. In blow molding, the screw L/D ratio is usually 20 : 1, 25 : 1, or 30 : 1, depending on the application. To feed the extruder, resin is fed into the funnel-shaped hopper. Hopper loading is usually accomplished with an automatic vacuum feed device. The hopper is kept filled with resin, and an observation window on the side of the hopper allows the operator to monitor the level. A cover on the top of the hopper keeps out such contamination as dust or dirt, as well as other foreign matter, such as tools, screws, or bolts that could damage the screw, barrel, or die. In addition, at the base of the hopper there is a valve or gate that stops the resin flow when the feed to the barrel must be stopped or the hopper emptied. A set of bar magnets with spaces between them is often fitted inside the hopper near the feed throat, to attract and catch steel objects that somehow got into the resin supply or fell into the hopper.

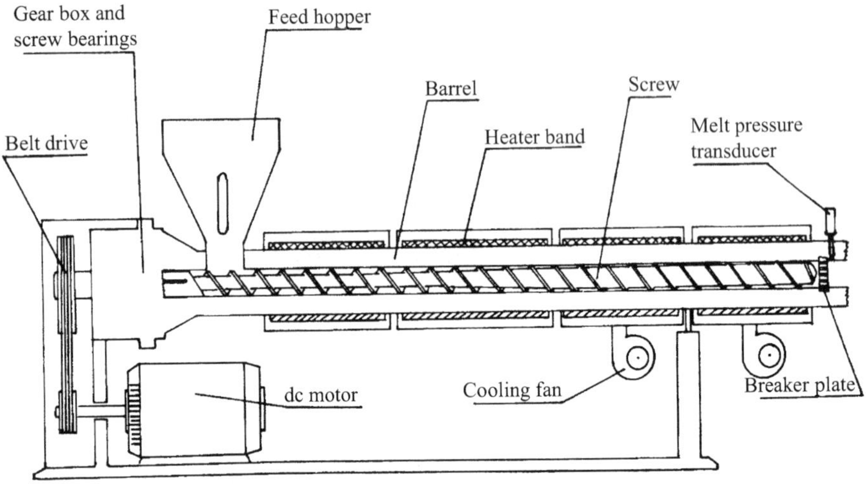

FIGURE 5.29 Single-screw extruder.

The feed throat section of the barrel, immediately below the hopper, is water or air cooled to prevent resin from melting too soon, possibly sticking together and "bridging" in the feed throat before reaching the screw. Bridging can cause a production shutdown, which can result in oxidation or thermal degradation of the resin in the barrel. If degradation occurs, black specks and off-color material will appear in the parts. Degradation usually requires a thorough cleaning of the extruder screw and die head to completely remove the burned plastic.

Note: Degrading plastic that is trapped in the barrel can present a serious hazard to operating personnel. It can give off a large volume of gas in the confined space of the barrel, resulting in a rapid and dangerous buildup of pressure. If allowed to continue, this pressure buildup can develop enough force to shear the die or the hopper mounting bolts and literally blow the die or hopper off the machine and into the air. At the same time, hot plastic and gas will spray out. These events can cause serious damage or injury to anything and anybody that happens to be in the way. Preventing an accident of this type requires careful operator attention to be sure that the feed throat cooling water is on and cooling, that die head pressures are normal, and that the plastic melt is being conveyed forward through the barrel and out the die.

In operation, the resin enters the feed throat and falls onto the screw. The screw is turned by a variable-speed drive and rotates within a hardened liner inside the barrel. As the screw rotates, the resin is conveyed forward and is compressed, melted, and mixed. A well-designed high-*L*/*D* screw will improve melting and mixing, along with providing better parison appearance and less variation in wall thickness. With higher-*L*/*D* screws, the melt temperature is also easier to control and the production rate is improved. The disadvantages of high-*L*/*D* screws are higher cost, greater power requirement, and a longer material residence time in the barrel, which must be avoided with heat-sensitive materials such as PVC which are easily degraded.

The rotating screw creates a large amount of frictional heat. If too much heat is developed, the melt temperature may become too high for good processing, or the plastic resin may thermally degrade. To prevent this from happening, the excess heat is removed by cooling the barrel. On small extruders, approximately $2\frac{1}{2}$ in. and less, cooling is usually done with fans that operate with a temperature controller. On larger extruders, excess heat is removed more effectively by circulating temperature-controlled water or oil through barrel channels and also sometimes through the center of the screw.

External heating by the barrel heater bands is always required during startup to melt the material, but once the screw begins to turn, frictional (shear) heat becomes the primary source of heat for maintaining the melting process. During normal operation, about 80% of the heat to melt the plastic comes from the frictional or shear heating action of the rotating screw, with about 20% coming from the heater bands. After the plastic melt reaches the end of the barrel, it passes through the screen packs, breaker plate, and die. The screen packs serve to filter out any foreign matter that may have entered the extruder with the plastic feed. To prevent "heat streaks" in the finished part, which are caused by the thermal degradation of trapped material, the interior surface of the head end is designed to be smooth, with no sharp corners or edges. To maintain the proper melt temperature, the head and die assembly is heated from the outside by heater bands.

5.3.2 Extrusion Blow Molding Head and Die Unit

The function of the extrusion blow molding head and die unit is to maintain the melt at a constant temperature and viscosity and to form the parison consistently at the desired rate and wall thickness. (*Note*: The rate and wall thickness may be varied intentionally as the parison is being formed.) In most blow molding, the die head unit diverts the melt flow through a 90° angle from horizontal to vertical, extruding the parison downward. In the case of some "wheel machines," the parison may be extruded upward. There are two types of parison dies: *center-feed* and *side-feed*.

5.3.2.1 Center-Feed Die

In the center-feed die, the flow is vertically downward around the core (Figure 5.30). The main advantage of a center-fed die is that the melt flows uniformly downward all around the core. There is no reason for one part of the melt to flow faster than another. A disadvantage of the center-fed die is that the core must be supported inside the die head unit by either a perforated support or a "spider" configuration. The perforated support or spider holds the core centered inside the die, giving the parison a uniform wall thickness as it exits the die. The perforations in the support or the openings between the spider arms permit the melt to pass through. The perforated support is usually the method of choice because it keeps the flow of melt more uniform.

Both types of support cause flow lines (weld lines) in the parison, because the melt stream must split to pass around the solid portions of the support and then be

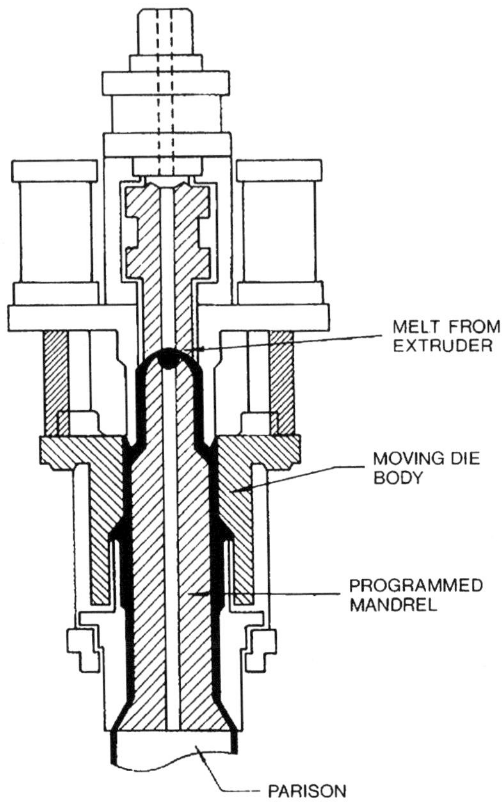

FIGURE 5.30 Center-feed die.

rejoined. Flow lines can result in a blow-molded part with poor appearance and reduced strength in the area of the flow lines. Certain center- and side-fed die designs, called *spiral flow dies*, can greatly reduce or eliminate flow or weld lines in the parison and are widely used for this reason.

5.3.2.2 Side-Feed Dies

With the side-feed die head, the melt flow enters at one side of the core (Figure 5.31) and is guided around the core through channels to form a uniform tube to be extruded. There are no multiple weld lines, but it is more difficult to achieve a uniform rate of flow all around the die opening because a portion of the melt must pass around the core while another portion flows directly to the die opening. Each die maker uses proprietary designs to achieve uniform flow.

5.3.2.3 Wall Thickness

Wall thickness around the circumference of the parison is adjusted by a set of screws at the die orifice. Control of the wall thickness along the length of the parison is possible with a suitably shaped die and a mandrel that can be moved axially within the die body.

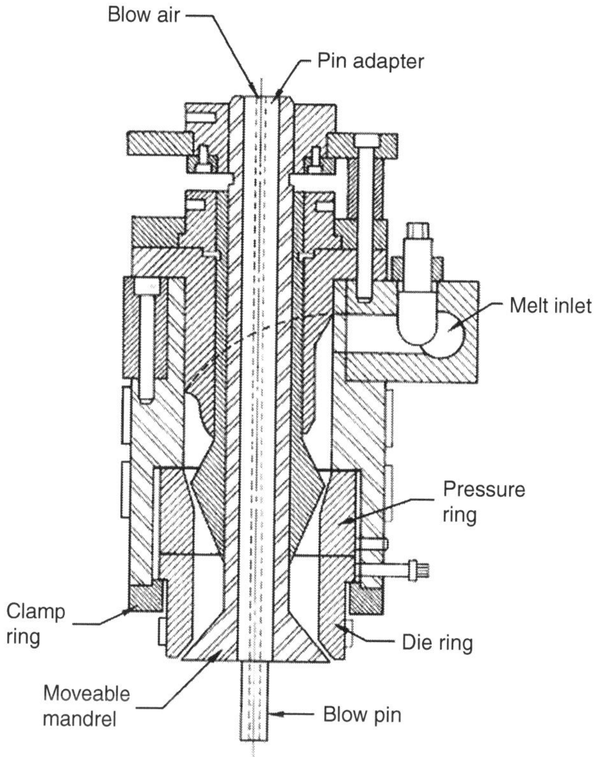

FIGURE 5.31 Side-feed die.

5.3.2.4 Accumulator Head

Accumulator head machines are used for blow molding large containers or parts. Resin is melted in the extruder and pumped into the accumulator, where it is held in readiness for the next cycle, as discussed in Section 5.2.3.2. The reciprocating screw machine, in which the melted resin is accumulated and held in the barrel, is shown in Figure 5.32.

5.3.3 Die and Mandrel

The die and mandrel (sometimes called a *pin*) are sized according to the desired parison diameter and wall thickness. At the die face, the melt should flow at a consistent rate all around the die to provide uniform wall thickness. The die land is the ring-shaped (annular) section at the end of the die. It is also the working area where the volume is kept constant. The die land length should be between 10 and 40 times the die opening (or die slot) dimensions. A relatively long die land will reduce parison swell and surface defects caused by melt fracture, but it will increase back pressure on the melt in the extruder. This can improve mixing, but it can also increase the melt temperature and may reduce the extruder's maximum output, possibly limiting the

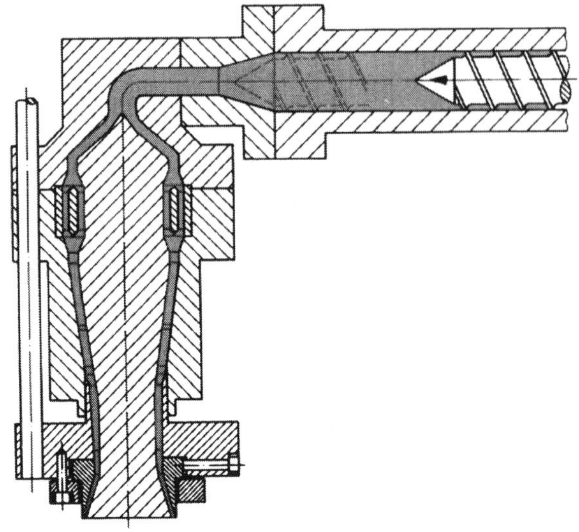

FIGURE 5.32 Reciprocating screw transfer.

production rate or increasing the drop time (parison formation time). A shorter die land, in conjunction with a high extrusion rate, can result in outer surface roughness in both the parison and the blown part. Determining the right die land length requires a combination of experience and trial and error. Figure 5.33 shows the two types of mandrel and die: converging, which tapers inward, and diverging, which tapers outward. Whether a converging or diverging design is used depends on the design of the die head and on the type of machine.

5.3.3.1 Die Swell

As the hot resin leaves the die opening, it swells, growing thicker but shorter. The term for this phenomenon is *die swell*, which is a bit of a misnomer since it is the hot plastic and not the metal die that does the swelling. The amount of swell depends on resin type, machine type, melt temperature, die head temperature, die design, and the rate of parison formation. The size and shape of the die opening or slot must be designed to compensate or correct for die swell.

5.3.3.2 Parison Adjustment

The purpose of the adjustment ring in a die head is to adjust for an uneven plastic flow from place to place around the parison. One cause is an off-center mandrel; another is variations in melt or die surface temperatures or variations in the material. The adjustment ring is a movable ring at the base of the die head. It can be moved with the adjustment screws. Moving this adjustment ring changes the die head gap through which the plastic flows. If the melt flow through the die gap is uneven, the parison wall will be thicker on one side than on the other (as will be the blown part).

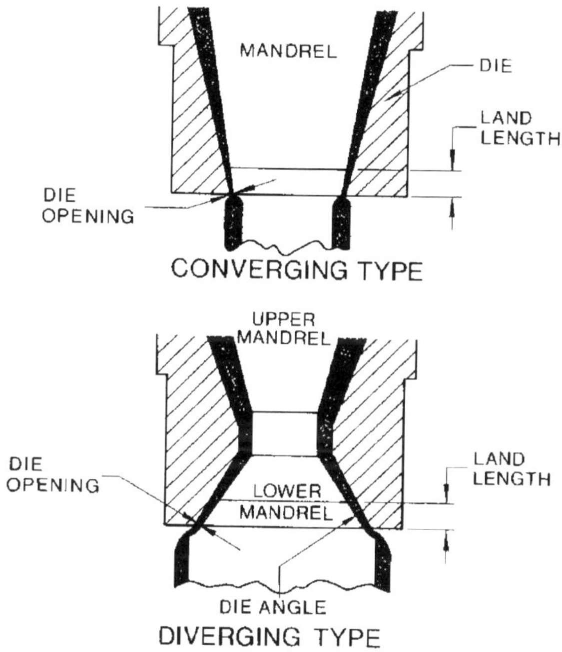

FIGURE 5.33 Converging and diverging dies.

This may cause the parison to curve or swing toward the hotter side, which will be the thicker side. To correct for (i.e., to achieve) even flow and uniform wall thickness, the adjustment ring screws should always be tightened on the side opposite the direction of swing (Figure 5.34).

5.3.3.3 Die Shaping

Die shaping is a technique frequently used to improve the circumferential wall thickness distribution of a blow-molded part. Die shaping is also referred to as *ovalization* or *profiling* and may have other descriptive names. Die shaping consists of opening the head tooling die gap at the orifice to increase the thickness of some longitudinal portion of the parison. The thicker section of the parison blows in the problem area of the part, which is usually where it has been difficult to maintain adequate wall thickness and increases the thickness. Opening the die orifice is done by machining an area at the die face and "running out" the pin or mandrel part of the way through the land of the die, as shown in Figure 5.35.

5.3.4 Parison Programming

Unless the temperature is too high, the parison will stay in one piece as it is formed, but as it extrudes it is subject to gravity and therefore thins out at the top. This

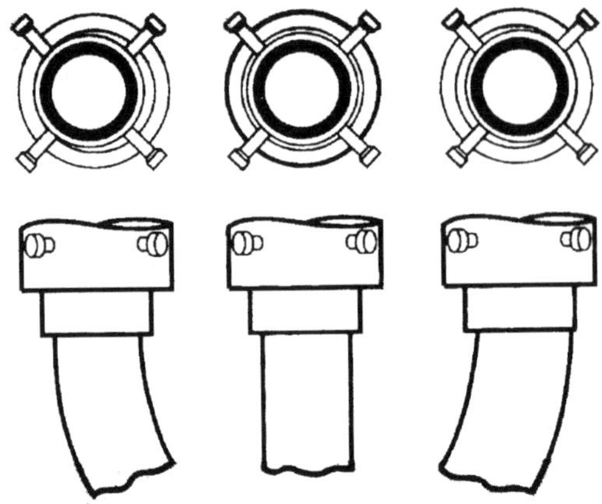

FIGURE 5.34 Head adjustment to achieve uniform parison wall gauge.

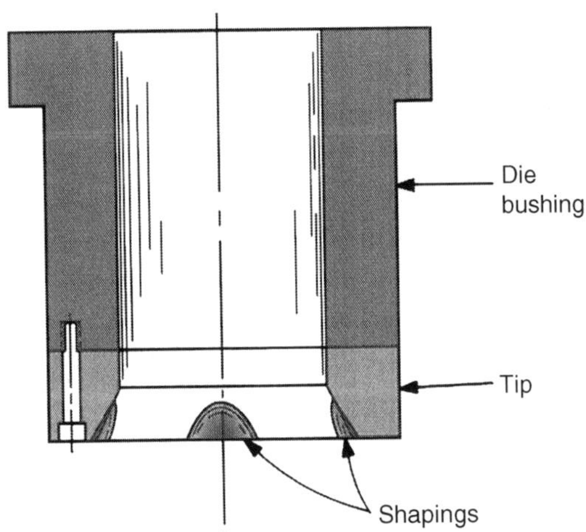

FIGURE 5.35 Die shaping.

thinning is referred to as *sag, draw down, neck down,* or *stretch out.* Such a parison (and final part) will end up being thicker at the bottom and thinner at the top, resulting in a part with nonuniform wall thickness. Parison sag is affected by the same factors as those that affect swell, that is, by resin type, machine type, melt temperature, die head temperature, die design, and the rate of parison formation. While swell and sag tend to work against each other (i.e., the extrudate expands and shortens with swell while it thins and stretches with sag), the net result is usually a parison (and a final part) that is thicker at the bottom and thinner at the top. Sag becomes even more of a problem when parison formation takes a long time, since gravity has more timc to act on the parison, resulting in more stretching and thinning.

When the parison reaches some critical length it appears to be extruding faster, indicating that the upper part is stretching and thinning out. This condition can be overcome by gradually increasing the wall thickness during extrusion by moving the mandrel inside the die, a process called *parison programming.* It is done with an automatic timing device that raises the mandrel some predetermined distance at the proper time, to extrude more melt while the parison is being formed. When the mold closes, the mandrel returns to its original position (Figure 5.36).

Parison programming can also be used to alter the wall thickness of the part deliberately in selected locations. By moving the mandrel up or down during extrusion, parison wall thickness can be increased or decreased to compensate for irregular part configurations in addition to sag. The die head gap distance is the most important factor affecting the parison wall thickness (Figure 5.37).

Modern parison programming devices use a microprocessor to control the up-and-down movement of the mandrel (controlling up to 100 points is common). A further method of controlling wall thickness is to vary the extrusion pressure through a fixed die opening by changing the screw rpm. Higher screw rpm increases pressure and melt output; lower rpm reduces pressure and output. An example of a parison programming controller where pressures can be monitored is shown in Figure 5.38.

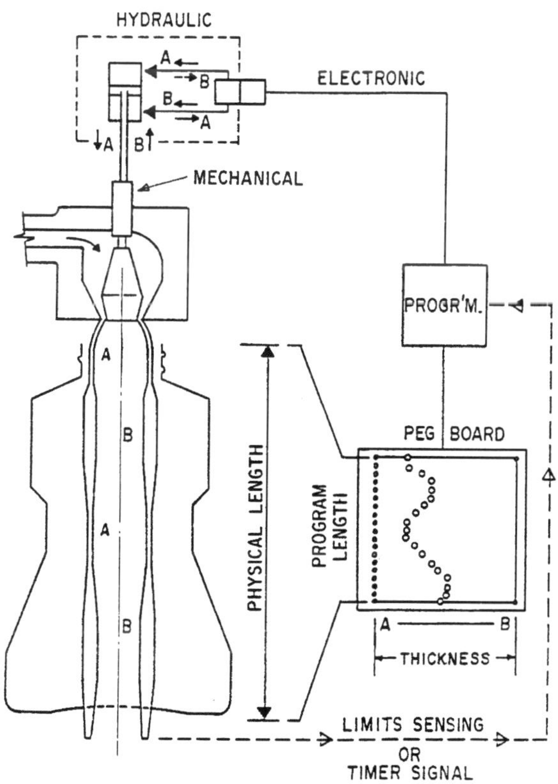

FIGURE 5.36 Parison programming.

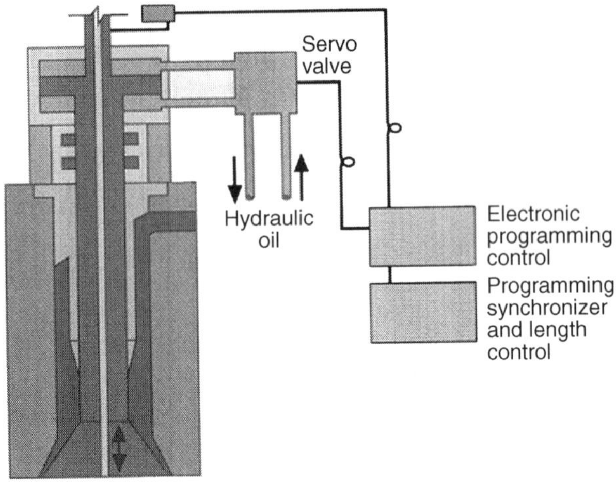

FIGURE 5.37 Programmed mandrel mo vement to achie ve desire parison and part w all thickness.

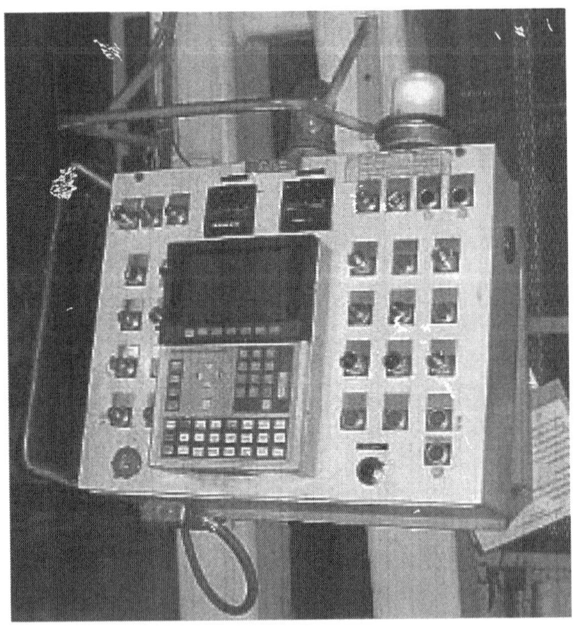

FIGURE 5.38 Parison programmable controller . (Courtesy of WEK South Corp., Reidsville, NC.)

5.3.4.1 Blowup Ratio

The *blowup ratio* is defined as the ratio of the average diameter of the finished product to the average diameter of the parison (Figure 5.39). The maximum blow-up ratio for applications with a thick-walled parison is 5 : 1. For most applications, 3 : 1 is preferable.

5.3.5 Mold Clamping Systems

The blow molds are mounted on platens, and the platens are moved by the clamping system. Aluminum platens are often used. They are light and stiff and can be moved quickly by the clamping system. Clamping systems may be of the pneumatic, toggle, or hydraulic type. No matter what system is used, the clamping pressure must be applied consistently and evenly to obtain consistent, reliable, and uniform opening, closing, and clamping. If the closing pressure is not consistent, there may be problems (e.g., in the pinch-off trim in the handle area of a container). To obtain the control now required, many users prefer machines that have clamping systems based on proportional hydraulics.

Clamping systems vary based on part configuration. Basically, there are three types. The L-shape type has the parting line at an angle of 90° to the centerline of the extruder. The T-shape type has the parting line in-line with the extruder centerline. Mold opening is perpendicular to the machine centerline. The third type, is the gantry type, in which the extruder/die head unit is arranged independently of the clamping unit. This arrangement permits the clamp to be positioned in either the L or T shape without being tied directly into the extruder assembly. Regardless of the arrangement, sufficient daylight in the mold platen area is required to accommodate parison systems [i.e., single or multiple parisons (multiple-head systems), unscrewing equipment, etc.].

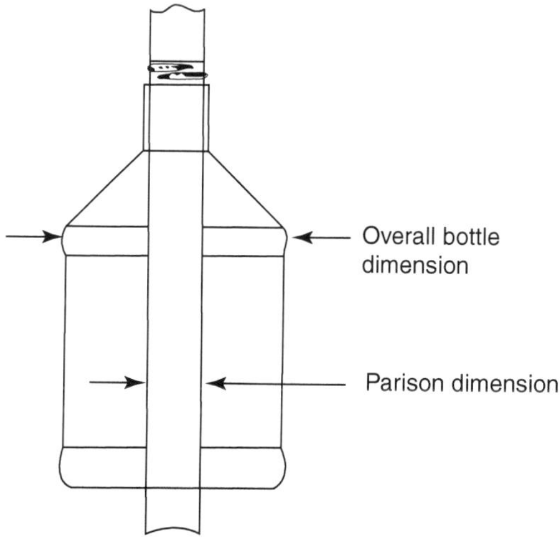

FIGURE 5.39 Dimensions used to determine the average blow-up ratio.

5.3.5.1 *Clamping System Requirements*

The clamping system is required to do several things. For example:

- To close and clamp the mold. Even on large molds, the mold closing speed should be 10 in./sec (250 mm/s). The amount of clamp force required for holding the mold closed during the blowing process will depend on the projected surface area of the part and the blowing pressure needed to expand the hot parison completely inside the mold cavity. In general, for every square inch of part surface area, approximately 145 lb/in^2 (1 MN/m^2) of clamping force is normally needed for clamping.
- To pinch the material at the base to form a weld and (almost) sever the flash. From 2 to 20 tons per foot (0.6 to 6 MN per linear meter) of pinch section may be required.
- To move the mold at different speeds to minimize shocks, reduce product tearing, and give stronger welds. Final high-speed closing can improve the weld strength.
- To mold and shape the top of the product.
- To open the mold so that the product may be ejected.

5.3.5.2 *Clamp Operation*

It is extremely important that the daylight, maximum, and minimum opening be adjusted readily and simply to compensate for various mold sizes and shut heights. Platen closing must be synchronous and parallel to produce acceptable parts. If this is not the case, there will be excessive wear on the bushings, and unacceptable parts will result. The overall closing of the platens should be smooth, with no hesitation or bounce-back, especially during final closing, as a bad pinch-off will result.

5.3.6 Presses

The accumulator method generally uses a stationary press, sometimes on a rail track that is moved only for mold setup. In the continuous method, there are several alternative press systems.

5.3.6.1 *Shuttle or Reciprocating Press*

The shuttle or reciprocating press, frequently used with multicavity molds, has one or more parisons being extruded from the die head while the mold halves are rising upward or moving sideways. A blow pin or hollow needle enters the parison while the mold halves are rising or sliding sideways.

5.3.6.2 *Rising or Vertically Moving Press*

A hydraulically operated cylinder on the platens clamps a mold around the parison and separates it from the extruding parison. The mold drops back down for the blowing and cooling cycle. After ejection of the part, the clamp rises again to receive another parison that had been extruded in the meantime (Figure 5.40).

5.3.6.3 Sliding or Horizontally Moving Press

In this type of machine the clamp assembly moves a mold to one side for blowing and cooling while a second clamp assembly closes another mold over the continuously extruding parison. After cooling, the first part is ejected, and the clamp assembly moves back to pick up another parison as the second clamp assembly moves aside (Figure 5.41).

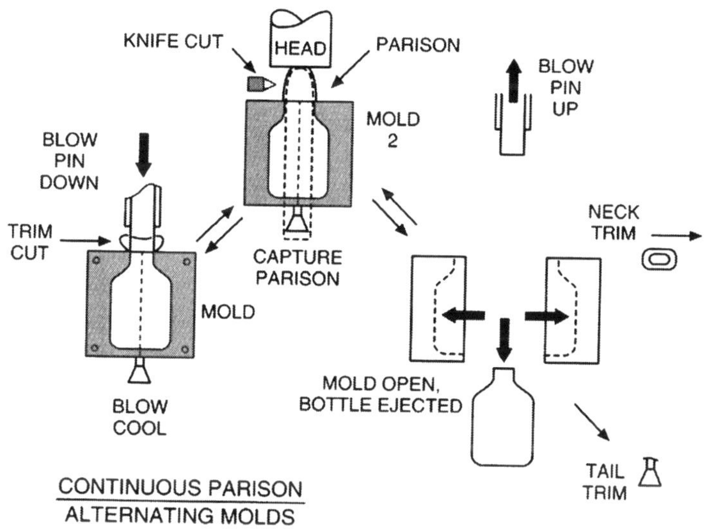

FIGURE 5.40 Rising (vertical) press.

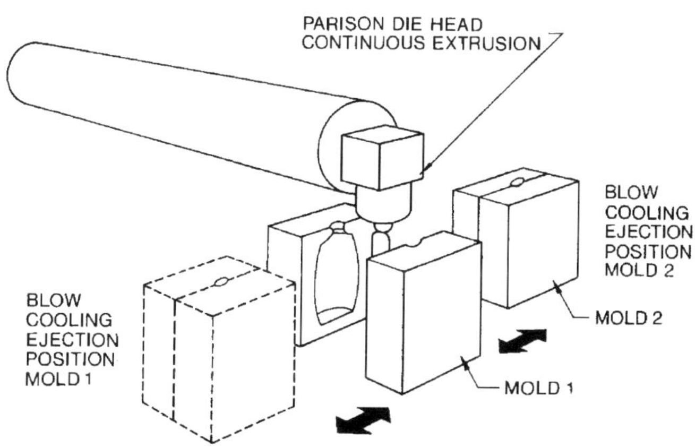

FIGURE 5.41 Sliding horizontal moving press.

Both the rising and sliding methods reduce the cycle time since no time is lost while blowing and cooling are completed. The parison extrusion rate is adjusted so that it matches the lag time required for the blow-cooling stage of the cycle.

5.3.6.4 Directional Valve

In this type of machine, a directional valve mounted at the end of extruder diverts the melt stream from a die head assembly on one side, to a die head assembly on the other side. One side is therefore blowing–cooling as a parison is being extruded on the other. As with other continuous methods, the extrusion time has to be matched to the blowing–cooling time. These die heads may also have several parison streams so that multiple cavities can be used to produce several parts at the same time. A blow pin needle is positioned in the mold, preferably near the pinch-off, to penetrate the parison effectively for blowing. In the case of bottle molds, a blow pin is mounted in the clamp assembly.

5.3.6.5 Continuous Rotary Wheel

For very high production and long run parts such as personal care products, a rotary wheel is the preferred production method. A vertical wheel is mounted next to the extrusion die. Multiple split blow molds are mounted on the wheel. Each of the molds is operated independently, either hydraulically or pneumatically. While the wheel turns, one mold at a time passes under the die to receive and close around the parison. The motion is continuous. The blowing air is introduced through a hollow needle and penetrates the parison when the mold closes or shortly afterward (Figures 5.42 and 5.43). The air is fed through the wheel axle, as is the water used to cool the molds. Rotary cams, located in the wheel axle, control the mold opening, mold closing, blowing air start, and blowing needle insertion.

Machines of this type may extrude the parison either downward or upward, where it is pulled along as needed. These wheels may carry up to 20 or more molds. Rotation of the mold carrier is arranged to provide ample time for each mold to blow, cool, and eject the product. The extrusion speed of the parison is coordinated with the rotation speed of the wheel. The ejected parts are connected by sections of the parison, which requires trimming and finishing steps at both ends of the parts. The weight, and therefore the wall thickness of the blown parts, can be reduced or increased (within limits) by varying the extruder speed or the rotation speed of the wheel.

5.3.6.6 Shuttle Rotary Wheel

This type of machine combines the advantages of the wheel and shuttle technologies and may have four or more stations. The parison is extruded continuously. The mold closes and the parison is cut; then the wheel indexes to bring the next mold into position to clamp over the parison. This process is repeated continuously, with blowing and cooling taking place in nonactive stations. In the last station, the part is ejected. This method is used for high production rates with larger containers and is more flexible for mold changes (Figure 5.44).

FIGURE 5.42 Uniloy Milacron vertical wheel blow molding system.

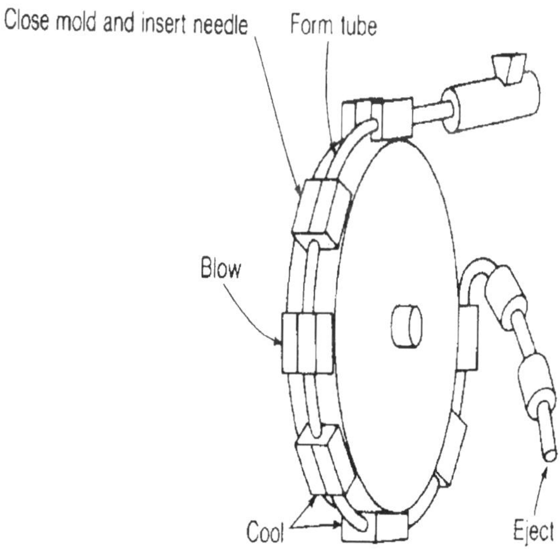

FIGURE 5.43 Continuous rotary wheel.

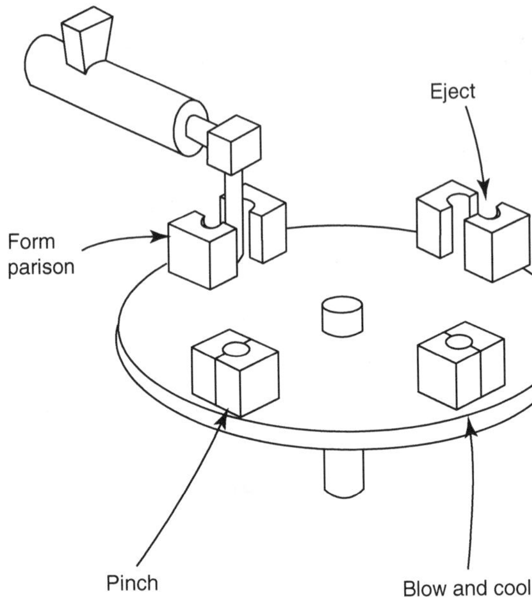

FIGURE 5.44 Shuttle rotary wheel.

5.4 BLOW MOLDS

A blow mold consists of several parts, counting its various inserts, but it is usually made in two halves. When the mold is closed, these halves form one or more cavities, which enclose one or more parisons for blowing. The two halves are usually alike for containers, but for industrial parts they are often different. For example, a mold for a double-walled case would have a female cavity on one half and a male core on the other.

5.4.1 Mold Materials

Because of the comparatively low clamping and blowing pressures, a blow mold need not be made of a high-tensile-strength material, with the possible exception of molds for very long production runs, say, hundreds of thousands or even millions. The predominant raw materials for blowing molds are beryllium copper, cast aluminum alloys, zinc alloys such as Kirksite, and occasionally, bronze. Beryllium is a light metal that offers excellent thermal conductivity. Great care must be taken in machining beryllium copper alloys because of the toxic nature of beryllium metal.

Of the alloys above, aluminum is the softest. Aluminum molds wear easily and are the most easily damaged in the shop. On the other hand, they are the easiest to machine. Aluminum and beryllium–copper molds may be slightly porous. Occasionally, blow molders have experienced some permeability of such molds to the viscous resin. This may affect the appearance of the blown part. The remedy is to coat the inside of the mold halves with a sealer (such as radiator sealer) that will not affect the heat transfer between the resin blown against the mold and the mold walls.

While steel molds last longer and are less prone to wear or damage than light-alloy molds, they are heavier, more expensive, and more difficult to machine than those made of nonferrous alloy. Higher weight means more setup time in the molding shop. Moreover, the heat conductivity of steel is inferior to that of the three main nonferrous mold materials. This results in a slower cooling rate and a correspondingly longer cooling cycle and consequently, a lower production rate for steel molds. The heat transfer rates and several other properties of various mold materials are given in Table 5.4.

5.4.2 Extrusion Blow Molds

A blow mold consists of several parts, counting its various inserts, but usually consists of two halves (Figure 5.45). Pinch-off edges are generally provided at both ends of the mold halves. A blowing pin may have the additional function of shaping and finishing the inside of the neck. Both mold halves must have built-in channels for cooling water. Sets of guide pins and bushings or side plates in both mold halves ensure perfect cavity alignment and mold closing. Accurate guiding devices in both mold halves reduce setup time.

TABLE 5.4 Properties of Common Mold Materials

	Thermal Conductivity (Btu-in./ft^2 hr-°F)	Density (lb/in^3)	Hardness[a]	Tensile Strength (lb/in^2)	Composition
Beryllium–copper alloy					
CA 172	770	0.298	37 RC	150,000	1.8% Be
CA 824	750	0.304	34 RC	135,000	1.65% Be
Aluminum alloy 7075 T6	900	0.101	150 BR	72,000	1.6% Cu, 0.23% Cr, 5.6% Zn, 2.5% Mg
AISI P-20 steel	200	0.282	30–60 RC, 290–330 BR	120,000	0.35% C, 1.0% Ni, 1.0% Cr
AISI 420 stainless steel	166	0.28	50 RC	213,000	0.38% C, 0.8% Si, 13.6% Cr, 0.5% Mn, 0.3% V
Kirksite A	640	0.25	100 BR	35,000	Contains Cu, Al, Mg

[a]RC, Rockwell C; BR, Brinell.

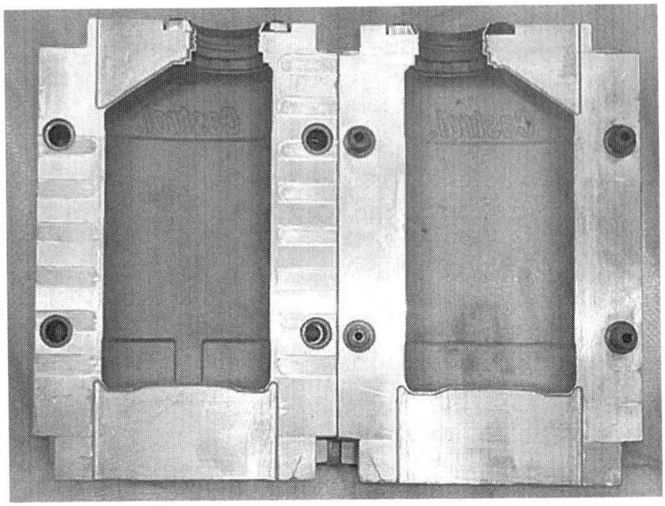

FIGURE 5.45 Quart-size bottle mold cavities for Castrol motor oil.

In some blowing processes, mold closing is carried out in two steps, first at high speed with lower pressure to perhaps $\frac{1}{4}$ to $\frac{1}{2}$ in. ($6\frac{1}{2}$ to 13 mm) of separation ("daylight"), and then slower, with higher pressure. Low-pressure, low-speed close helps to protect the mold from being damaged by tools or anything else that might have fallen between the halves (which, however, should never happen in a well-kept shop!). Although slow-speed, low-pressure close can prevent damage to the mold, it *does not* protect an operator from being severely or fatally injured if he or she caught between the closing mold halves. Getting caught between closing mold halves is usually a result of someone deactivating a safety gate or other safety device during mold setup, machine repair or troubleshooting, or press operation. One should *never* deactivate a safety device or enter or operate a machine that has one or more non-working or deactivated safety mechanisms. Besides the personal injury hazard, deactivation of machine safety mechanisms is illegal under the Occupational Safety and Health Act (OSHA) and may result in very large fines to the company.

Molds are not necessarily positioned to be vertical, that is, in line with the parison. Occasionally, they may be tilted intentionally. This will result in a nonuniform distribution of resin, which may be helpful, for instance, when such irregular pieces as a pitcher with a handle are being blown. It may also result in some saving in parison length.

5.4.2.1 Pinch-off

Because of the comparatively high pressure and mechanical stress exerted on a mold bottom during closing as it pinches one end of the parison together, the pinch-off in a nonferrous-metal mold is frequently an insert made of hard, tough steel. The effect on the blown part always shows in a *weld line*. The pinch-off section does not cut off the excess parison "tail." Its protruding edges cut nearly but not completely through to create an airtight closure by pinching the hot parison along a straight line. This makes it much easier later to break off or otherwise remove the excess tail piece (Figure 5.46). A high-quality pinch-off in a thick-walled parison is more difficult to obtain than in a thin-walled parison. However, much depends on the construction of the pinch-off insert.

The pinch-off should not be a knife-edge but, according to some molders, should be formed by lands about 0.05 to 0.015 in. (0.1 to 0.5 mm) long. The total angle outward from the pinch-off should be acute, up to 15°. These two features combine to create a weld line that is rather smooth on the outside and forms a flat elevated line or a low bead inside rather than a groove (Figure 5.47). A groove, which weakens the bottom along the seam, may be formed when these two features of the pinch-off are missing. One method of obtaining more uniform weld lines is to build "dams" into the mold halves at the parison pinch-off areas. The dams force some of the molten resin back into the mold cavities to produce strong, even weld lines (Figure 5.48).

5.4.2.2 *Importance of Fast Mold Cooling*

Fast heat transfer through the material of which the mold is made is of utmost importance because, as mentioned, the cooling step controls the length of the blow molding cycle. Cooling takes up roughly two-thirds or 66% of the entire blowing cycle. Good heat transfer means faster cooling, and faster cooling means more items blown per hour, that is, less expensive production. This is the main reason that for blowing molds, the above-mentioned alloys are generally preferred to steel even though steel is usually more durable.

Considering only their heat transfer rate, the principal blow mold materials follow each other in order from highest to lowest thermal conductivity: (1) beryllium–copper, (2) Aluminum, and (3) Steel. Because the different mold materials have different heat transfer rates, a blowing mold, with the exception of the steel pinch-off inserts, should ordinarily be made of only one material. Different materials with consequently

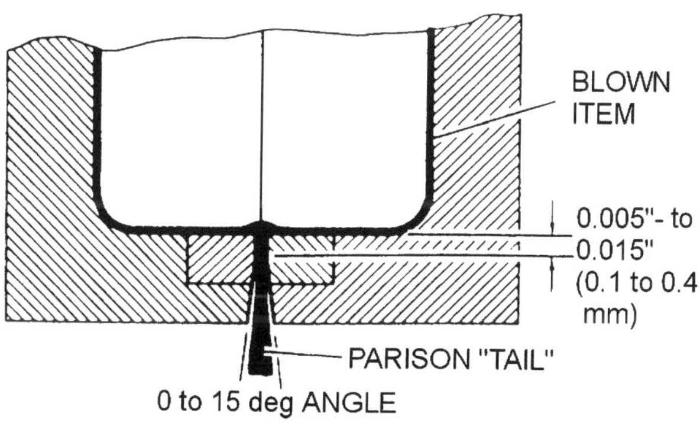

FIGURE 5.46 Pinch-off.

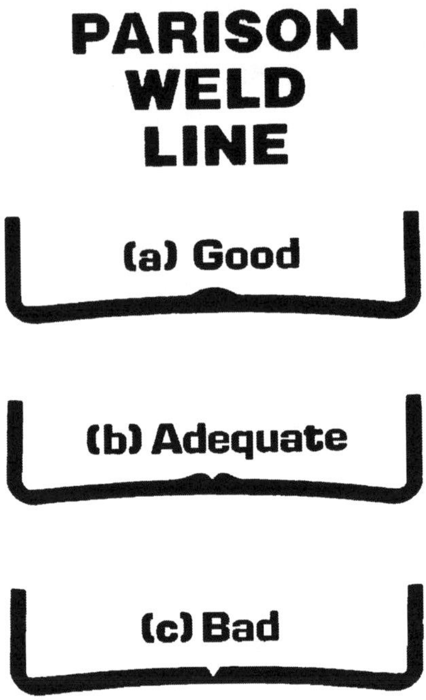

FIGURE 5.47 Weld lines.

different heat conductivity at various points of the mold will result in nonuniform cooling. This, in turn, might set up areas of stress in the finished piece, yielding parts that are more susceptible to warping, splitting, or failure in use. However, some parts will benefit from different cooling rates in certain areas of the part, so several different alloys may be used in the same mold.

The blowing mold halves must always be cooled adequately to solidify the part quickly and immediately after the parison has been blown out against the mold walls. The cooling water may be tap water. If the water has a high mineral content that may allow sediments to settle in the narrow cooling channels, a closed system for circulating purified water should be used. Unless the cooling water is cold enough, as in winter, it should be chilled to 40 to 70°F (4 to 20°C) by a heat exchanger. Such a low temperature may, however, cause water condensation on the outside mold walls. Some molders use noncooled tap water. Usually, the cooling water is recirculated, that is, reused time and again for a long period. Sometimes, it is partly recirculated and mixed with fresh tap water, to maintain the desired temperature and to economize.

The water usually circulates through the hollow mold halves. Sometimes, a copper tubing system is cast into the mold. However, to create the most useful flow, water channels are machined into the mold halves. Well-placed channels will

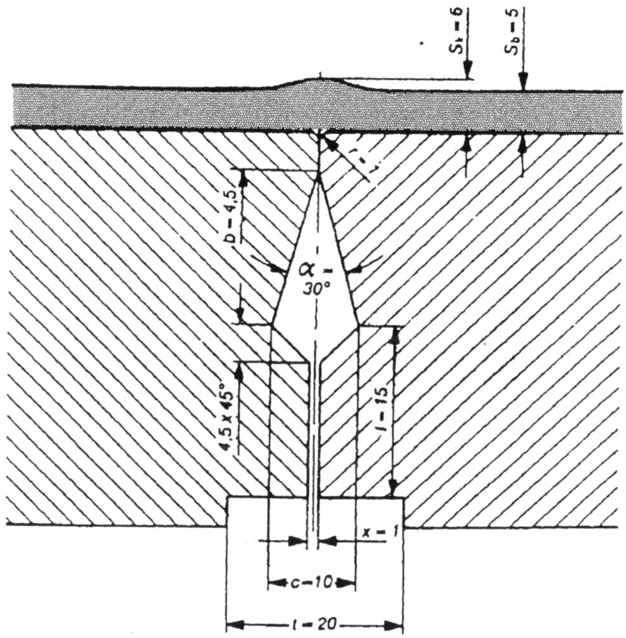

FIGURE 5.48 Pinch-off dam.

ensure that the cooling water comes as close to the mold cavity as feasible. Cooling channels should also be as close (lengthwise or other) as possible to the parting lines caused by the separation lines of the two halves or inserts. Parting lines will nearly always show along mold separation lines. Cooling these areas will result in better surface finish of the piece along the parting lines.

Generally, cooling channels are located in the top or bottom areas, that is, around a bottle neck or the bottom pinch-off, or both, and in mold areas where greater masses of resin are required than in other areas. Such areas, as well as thicker wall sections, often require additional cooling. Otherwise, the plastic in these sections may still be viscous when the piece is ejected, while the thinner wall sections have solidified. This may result in a deformed piece or one that the customer will reject because of nonuniform shrinkage and resulting warpage. To obtain uniform cooling and shrinkage, a simple cooling system is used for each half. Typical mold cooling line patterns are shown in Figure 5.49.

Occasionally, the blowing pin is also cooled. Sometimes, air cooling from the outside is provided for the pinched tail of the parison sticking out of the mold bottom. The tail is much thicker than the wall and cools correspondingly slower. Air may also be circulated inside the blown part to speed up its cooling. Cooling time is strongly affected by the parison melt temperature. The higher the melt temperature and the thicker the walls, the longer the cooling time required to make a part.

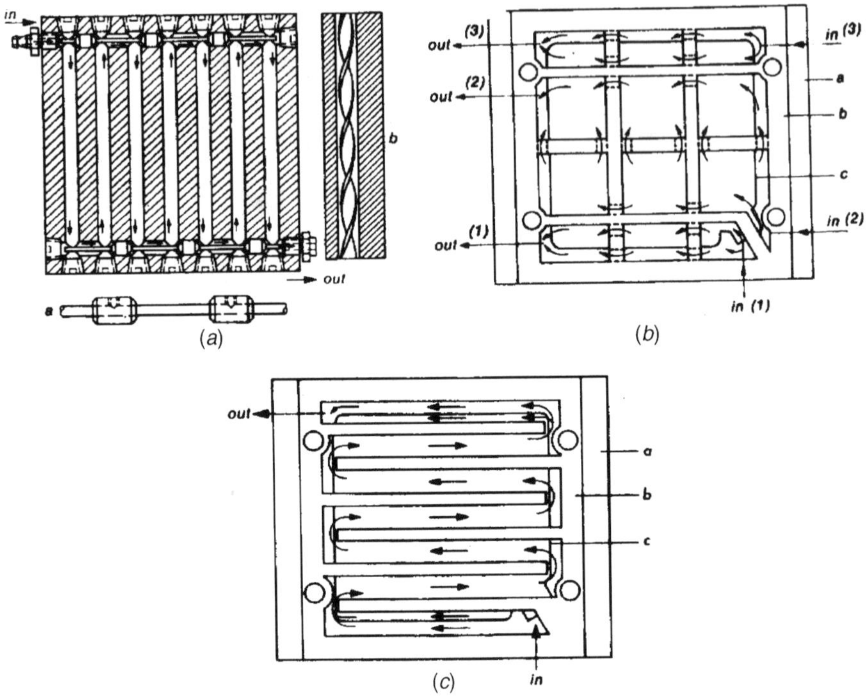

FIGURE 5.49 Mold cooling line patterns.

5.4.2.3 High-Quality Undamaged Mold Cavity Finish

A high-quality mold cavity finish and undamaged inside surfaces are essential in blow molding to avoid surface imperfections in the end product. If the highest possible gloss for the end product is desired, the mold cavity should be sandblasted with 100-grit flint sand and have vacuum assists for the removal of trapped air. If other end-product finishes are desired, the mold cavity should be finished accordingly. Even a first-class machining job inside the mold cavity cannot prevent the occurrence of parting lines, especially if the blown item has a very thin wall.

5.4.2.4 Effects of Air and Moisture Trapped in the Mold

In highly polished molds, air may become trapped between the mold walls and the hot, still-soft piece, marring the surface of the piece. This can happen especially when large thick-walled pieces are blown. In such cases the mold must be vented either by sandblasting (with a coarse rather than fine sand)—resulting in a matte outer surface—or by grooves in the separation lines, or, in extreme cases, by valves in the mold.

Generally, about one-half of the parting line periphery is vented to a depth of 0.002 to 0.004 in. In venting difficult areas, such as handles or thread inserts, holes are usually drilled into these areas so that they vent to the atmosphere. The vent holes are normally about 0.008 to 0.010 in. in diameter. Particular care must be taken when drilling these holes to avoid drilling into the mold cooling channels (Figure 5.50). Moisture in the blowing air may result in marks on the inside of the blown part. These may result in high reject rates, especially if the part is transparent or translucent. Moisture in the blowing air can be removed by means of a heat exchanger that cools the compressed air, or by using moisture traps and separators in the pipelines.

5.4.2.5 Injection of Blowing Air

The injection of blowing air can be done by various means, such as downward through the core, through a blowing needle inserted sideways through the mold wall, or from below through a blowing pin that moves up into the end of the parison that will become the (frequently threaded) neck of a bottle (Figure 5.51). Sometimes, different blowing devices are used in combination. Like every step in the blow molding cycle, blowing time and duration must be well coordinated with all other parts of the cycle. Compared with cooling time, the blowing time is very short.

To obtain rapid inflation of the hollow piece, the volume of injected blowing air should be as large as possible. The opening through which the air enters the mold must, of course, be adequate. The thinner the wall and the lower the melt and mold temperatures, the faster the blowing rate should be and the higher the blowing pressure, up to about 150 lb/in^2 (10.5 kg/cm^2) for very cold molds and thin-walled parts. High blowing pressure requires correspondingly high clamp force to keep the mold tightly closed during the blowing step.

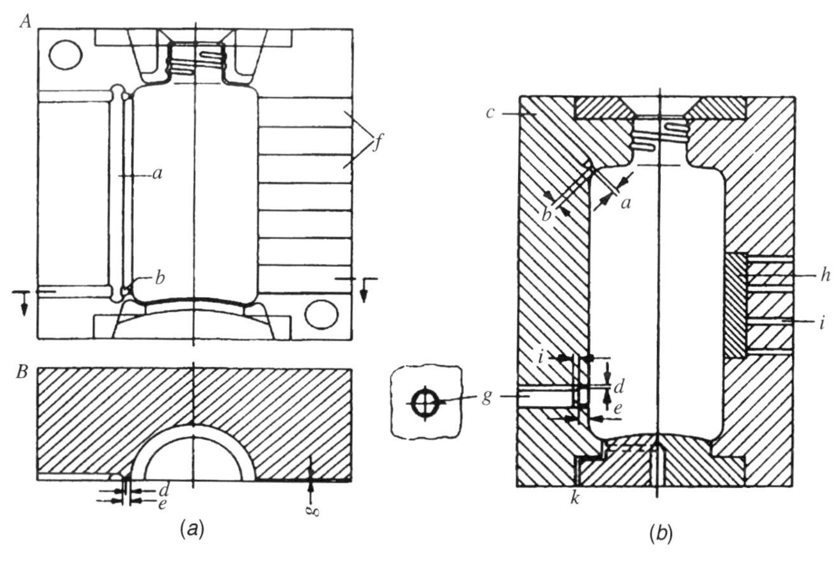

FIGURE 5.50 Mold venting methods.

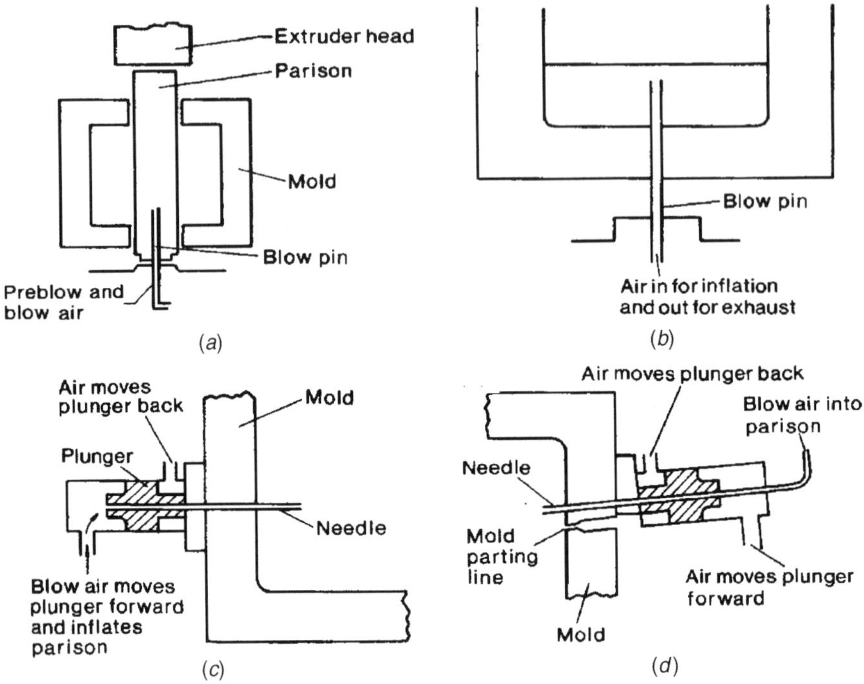

FIGURE 5.51 Blowing methods.

5.4.2.6 Ejection of the Piece from the Mold

Ejection of the blown piece can be effected forward between the mold halves or downward, provided that the press is built in such that a free fall from between the open mold halves is possible. Many machines have an automatic ejector, or stripper assembly. Ejectors, or knockout pins and plates, push the piece out with a rapid motion, preferably hitting it on a trim area so that the piece itself will not be distorted. Ejection of the blown piece is part of the automatic blowing cycle.

If the piece is not too large, it can be blown forward out of the mold by an air jet from behind. Very large pieces are manually removed from the mold by the operator. To reach the part, the operator must first push aside the protective gate or shield, which stops every mold or platen movement automatically so that the operator is in no danger of being hurt by a moving mold.

Automatic stripping, being a very fast operation, requires a pneumatic (air) pressure system, as do the blowing process, the cutoff at the die, the operation of automatic valves in the die head, and other functions. Because so many moving parts are actuated by air, it is sometimes recommended that the air in the pneumatic system be lightly oiled. Hydraulic (always oil) systems are generally used for clamping and for moving the molds, the platens on which they are mounted, and other heavy parts of the molding equipment (such as rams) that are used in some molding processes.

5.4.3 Understanding Injection Blow Molds

5.4.3.1 Tooling

Tool components for injection blow molding must be machined to precise tolerances because of their critical combined function in the overall molding cycle. The preform (parison) molds must be more accurate and robust than the blow molds since they are subjected to high hydraulic clamp pressures and high plastic injection pressures. The core rods are subjected to the same conditions as the preform molds and must be concentric to provide uniform wall gauge and good blowability in the preforms. The blow molds are subjected to only a third of the clamping force of the preform cavity. Because several component parts make up a complete injection blow molding tool, they must all be precisely machined. Excessive dimensional variations between tool components cannot be tolerated.

5.4.3.2 Parison or Preform Neck Ring

The preform neck ring forms the finished shape of the threaded or neck section of the container. It also centers and retains the core rod securely inside the parison to minimize core rod deflection during the injection part of the cycle. The neck ring is a separate component or mold segment that is precisely fitted into the injection mold parison cavity. A close, nick-free fit prevents melted plastic from entering the spaces between mold segments, which, if permitted, would result in visible surface ridges or even flashing. Poor-fitting or damaged mold segments are likely to result in objectionable parts or scrap. The parison neck ring has its own independent thermal control channels for achieving the molded neck finish and is usually made from nondeforming air-hardened tool steel.

5.4.3.3 Parison or Preform Cavity

The injection preform or parison mold consists of stationary and movable halves, with cavity dimensions determined by the desired weight and wall thickness of the preform and the type of resin to be molded. The number of mold cavities that will fit into a specific machine is limited by platen size, machine clamp tonnage, barrel capacity, and the size of the individual cavities (Figures 5.52 and 5.53).

Temperature control (conditioning) is an essential part of the cavity design. Temperature zones are created by channels drilled into the mold body for the circulation of a liquid conditioning medium. These channels are connected to form individually controlled temperature zones. The number and location of the zones within the parison cavity are critical design considerations as they directly affect the blowing efficiency of the parison shape. Each temperature control channel is positioned along the parison cavity profile in a way that permits the temperature to be varied along the parison profile, thus ensuring uniform blowing of the parison wall. Depending on the type of plastic material being injection molded, preform cavities are made from air hardened tool steel or prehardened steel.

5.4.3.4 Core Rod Assembly

The core rod forms the inside diameter of the neck finish and the inside shape of the parison. It also incorporates the air channel and valves used to blow the plastic melt

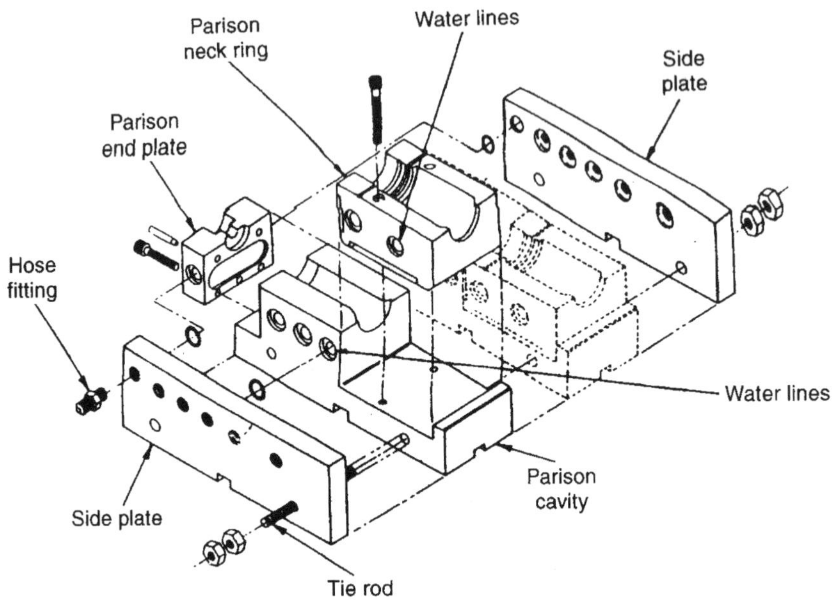

FIGURE 5.52 Injection perform mold assembly.

FIGURE 5.53 High-cavitation high-productivity 96-cavity Husky PET preform mold-matched with 96 cores on each of four indexing mold faces. (Courtesy of Husky Corporation.)

into the final shape of the bottle. Core rods are usually made of oil-hardened tool steel. A crown nut, spring, and star nut assembly at the rear of the core rod valve permits blowing air to flow but pulls the valve shut prior to injection (Figure 5.54). Non-temperature-controlled core rods are heated by the melt to provide blowability; however, they do not maintain a consistent temperature profile along the parison surface as do core rods with temperature control. With the advent of heat-sensitive and high-melt-temperature engineering thermoplastic resins, temperature-controlled core rods provide a more accurate and consistent temperature profile along the core rod–molding surface.

5.4.3.5 *Injection Manifold Systems*

The manifold assembly is fastened to the stationary preform die set. It is made up of the manifold base, nozzle clamps, and nozzles. Nozzle clamps with retaining screws hold the injection nozzles and are accessible from the rear of the manifold for ease of assembly and disassembly. The hot melt is injected from the machine nozzle into a manifold that performs the same function as a hot runner manifold in regular injection

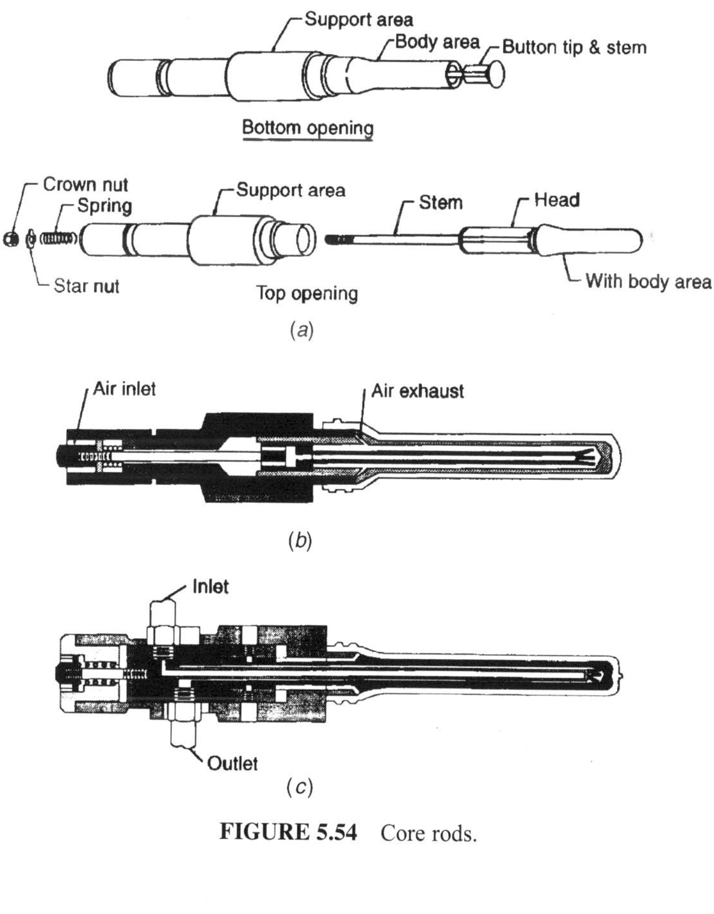

FIGURE 5.54 Core rods.

molding. Manifold designs range from simple, for molding easily processed plastic materials such as polyethylene, to complex, as in the case of higher cavitation (multiple cavities) molds and when molding heat-sensitive materials or high-temperature engineering thermoplastics such as polycarbonate.

The injection nozzles allow passage of the melt from the manifold to the preform cavities. Nozzle orifices are generally varied in diameter from cavity to cavity in a given mold set to provide a balanced fill of all cavities, that is, to ensure that each cavity fills at the same time and at the same pressure. Balanced fll is required to produce injection-molded performs that have similar characteristics: for example, similar weights and shrinkages rates. Variations in these characteristics from one preform to another can result in variations in the dimensions, appearance, and strength of the final blown parts.

5.4.3.6 Parison or Preform Design

A crucial element of injection blow molding is the configuration of the core rod and the injection-molded preform or parison. Each container shape has its own unique core rod and parison design. If the initial design of the preform and core rod is not suited for the particular bottle to be molded, a production processing problem will continue until the design is corrected.

The designer should always keep in mind maximum processibility or blowability when selecting the preform and core rod shapes. Under no circumstances should processibility be sacrificed for maximum cavitation. Consider injection clamp tonnage relative to the projected area of the parison shape, the physical limitation of the machine platen area, the machine trigger bar length, and the total shot capacity of the machine. Always insist on building a single- or unit-cavity tool before going into production tools. Much can be learned about the molding characteristics of a particular core rod and parison design through single-cavity sampling. If the design is difficult to process during the single-cavity stage, you can be sure that the process window will be narrow in production, which is not a desirable situation. Design changes are normally made during the single-cavity stage to produce parts in specification with as broad a process window as possible. Normally, the single-cavity tool can become part of the production tool.

5.4.3.7 Injection Blow Mold Cavities

Injection blow molds are used to shape the final form of the container. Because blowing is a low-pressure process, they are not exposed to the high clamp and injection pressures seen by the injection preform mold. The blow mold is subjected only to the press clamping pressure needed to withstand the 150-lb/in^2 (1034-kPa) air pressure necessary to blow the parison shape (Figure 5.55). A very important consideration in blow mold design is to ensure adequate cooling. Since cooling time is the largest part of the process cycle, time spent on designing the mold to provide efficient cooling is time well spent in terms of cycle-time reduction and productivity gains. Here again, as in the parison mold, peripheral channels should be provided for

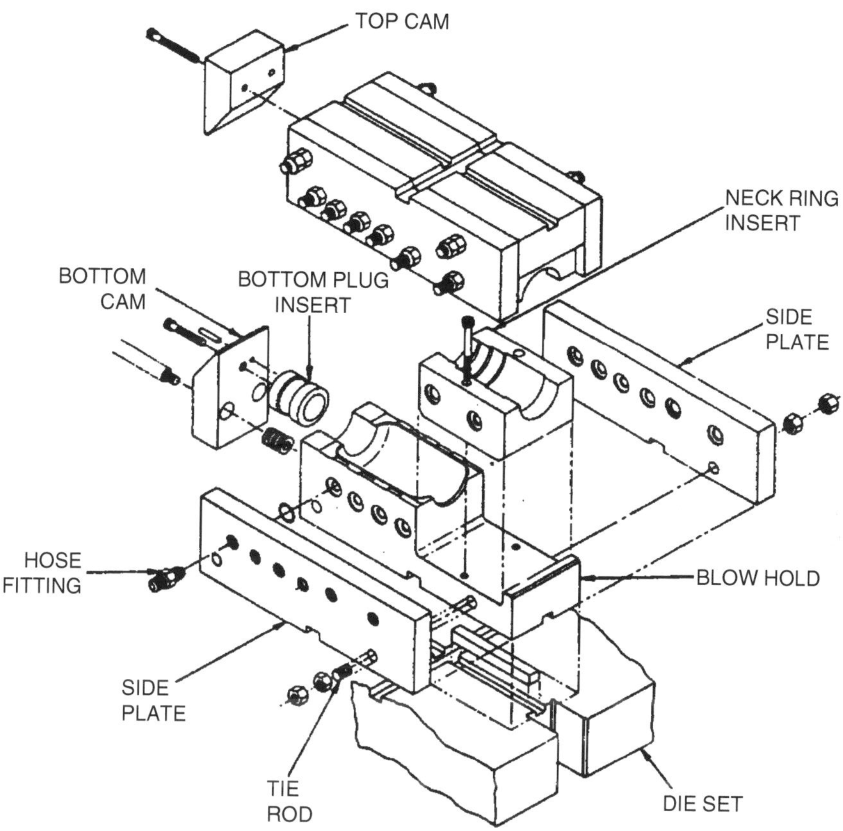

FIGURE 5.55 Injection blow mold assembly.

maximum cooling to help reduce overall cycle time. Blow mold cavities for making 1-liter water bottles are shown in Figure 5.56.

5.4.3.8 Bottom Plug

The bottom plug is an independent part of the blo w mold cavity and forms the bottom configuration of the container (Figure 5.57). It is usually made in halv es, may be fixed in position or retractable, and may be water cooled. Most polyolefin (e.g., polyethylene or polypropylene) containers are flexible and can be stripped o ver the fixed bottom plug when the mold is open. With polystyrene, polycarbonates, and other rigid plastic materials, a retracting bottom mechanism is necessary . When a retracting bottom mechanism is used, the upward movement of the top half of the mold permits a blown bottle, for example, to be removed from the blow mold cavity without stripping or distortion. The retracting mechanism can be operated by a cam, or in difficult cases, by air or hydraulic cylinders.

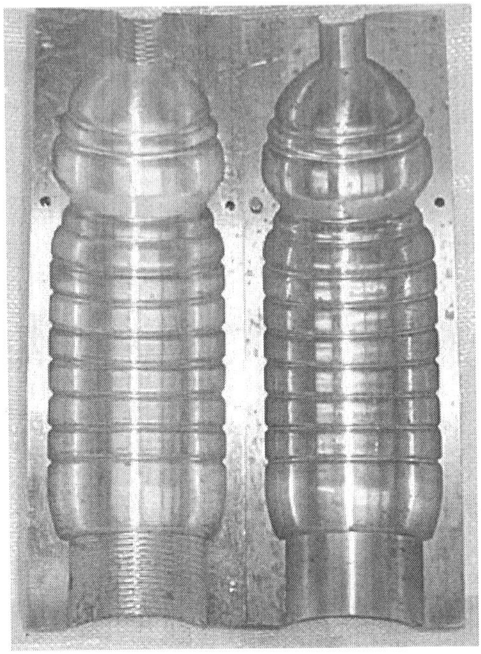

FIGURE 5.56 Mold cavities for a 1-liter water bottle. (Courtesy of Roa Designs International, Schiller Park, IL.)

FIGURE 5.57 Bottle cavity inserts with independent bottom plugs. (Courtesy of Ryka Blow Molds Ltd., Mississauga, Ontario, Canada.)

5.5 BLOW MOLDING OPERATION

5.5.1 Heating and Heating Input

Plastic materials are poor conductors of heat, and they have high specific heats. This means that it is difficult to remove heat from plastics even by placing them against a cool surface. This fact has a strong effect on how plastics must be processed if a fast cycle time is important, which is almost always.

5.5.1.1 Melt Temperature

For a given machine and material, the actual temperature of the plastic melt depends on the barrel zone temperature settings and also to a great extent on the screw rotational speed and melt pressure. The melt temperature should be checked periodically, as it has a significant effect on melt viscosity and die head pressure. Plastic melt temperature is measured in the die or nozzle or is taken from an extrusion or injection of material into the air. When measuring melt temperature by pushing material into the air to obtain a melt sample, take particular car to avoid being burned. Plastic melt is hot enough to cause severe burns, and it also adheres to the skin. Wear the required safety gear, which usually includes gloves, long sleeves, and a face shield, when measuring melt temperature or when there is a danger of being splashed with plastic melt during startup or purging.

5.5.1.2 Product Temperature

Both melt and product temperatures should be measured and recorded. Product temperature will vary from place to place, so the hottest point, usually the tail in extrusion blow molding, should be selected as the reference point. During production, that point should be checked with an infrared pyrometer or a probe. As with all temperature measurement, take particular care to get accurate and consistent readings, because processing decisions may be based on them.

5.5.1.3 Zone Heating

The barrel of an extruder or injection molding machine is divided into zones or regions. Each of these is fitted with its own heating or heating and cooling system. The zones are controlled with temperature sensors and associated equipment (e.g., a microprocessor-based controller). Accurate temperature control is important to the process, so calibration must be performed. Ammeters should be fitted to the heater circuits so that heater failure may be spotted; that is, when a heater fails or burns out, the ammeter will show that the proper amount of current is not flowing. Heater failure can cause several problems, and it may go unrecognized if ammeters are not placed in the heater circuits.

5.5.1.4 Temperature Settings

With a new part or material, it is advisable to start off by setting the barrel temperatures at the lowest settings recommended for a particular grade of material until process experience is obtained. Although each material and each process has its own requirements, it is common to set the first zone temperature at the lowest value, while the temperatures of the other zones are set progressively higher. As the product is molded, the temperatures may be adjusted to achieve the desired cycle time and part quality. Once good parts are made and the process has settled down, the temperatures are recorded for later use.

5.5.1.5 Temperature Checks

Always be sure that the machines are set and running at the temperatures specified on the record sheet. This is very important because barrel and melt temperatures affect dimensions, surface finish, and output. In extrusion blow molding, the internal surface temperature of the die is also important. It should be measured using a temperature probe. Die and melt temperatures should be checked during the actual extrusion–molding operation and the set temperatures of the heaters adjusted to maintain the desired temperature.

5.5.1.6 Heat Input from Screw Rotation

During startup, or anytime the screw of an extruder or injection machine is not turning, all of the heat needed to melt the plastic and maintain melt temperature comes from the heater bands. When the screw is turning at production speeds, about 80% of the heat comes from frictional heat generated by the rotating screw, with the rest coming from the heater bands. On many continuously operating extruders, the heaters never come on once the extruder is up and running at operational speed. In fact, as mentioned earlier, cooling fans are often required to keep the barrel and melt temperatures from rising above the set points. This means that the heater bands never or rarely need to come on. This is not the case when injection molding preforms. Injection molding machines do not have cooling fans. During operation, the heaters will come on whenever the barrel temperatures fall below the set point of the temperature controllers.

5.5.1.7 Barrel Residence Time

The thermal (heat) decomposition of resin depends on both temperature and time. A resin may be degraded by short exposure to a high temperature or by longer exposure to a lower temperature. The length of the time that resin is exposed to high temperature in the extruder or injection barrel is therefore important. If the residence time is too short, the material may not be fully melted and at the desired temperature. If the residence time is too long, it may degrade. The aim, therefore, must be to keep the residence time uniform by maintaining a consistent screw rpm and a consistent cycle, and by preventing unplanned production stops.

5.5.2 Safe and Efficient Setup, Startup, Operation, and Shutdown

Note: These procedures represent good practice. In an actual production environment, always follow the plant's procedures.

5.5.2.1 Starting Up

During the startup of a blow molding machine, precautions should be taken such as that no one stand in front of the die or nozzle and the hopper be firmly in place so that the screw cannot be accessed. Startup is the most hazardous time in the process because material left in the barrel from the last run can overheat and degrade, spewing hot gases and degraded plastic from the nozzle or die at high pressure. Another potential hazard is that many steps must be taken, and often quickly.

Startup Preparations An understanding of the settings is needed to start up a blow molding machine. Obtain advice from someone who is very familiar with the equipment, or better yet, consult the written procedures. A general procedure follows:

1. Turn on the main power switches and then select or set the temperatures.
2. Ensure that the cooling water is on and check to see that it is flowing through the feed throat.
3. Preheat the hydraulic oil to its correct operating temperature. This may be done either by pumping the oil back into the tank or by using a preheater fitted for this purpose.

Once the machine has reached the required temperature, it should be allowed to settle down before any material is introduced into the barrel. The settling-in (equilibration) time, sometimes called the *soak time*, is the time needed for the barrel, screw, breaker plate, and die or mold temperatures to stabilize close to the temperature set points. The equilibration or soak time will depend on the size and type of machine. It may take 20 minutes for a small machine and it may take several hours for a larger machine. This time should be used to prepare for the production run.

Other startup steps include:

- Check the nozzle/die and molds to see if they are clean and operational.
- Review the production order for color, quantity, and other requirements.
- Check for necessary tools and equipment and be sure they are in place and working properly.
- Make sure that all auxiliary equipment is clean and operational, to include hopper loaders, conveyors, grinders, vacuum pumps, and leak testers.

Melt Temperature Two methods are commonly used to measure melt temperature in a blow molding machine: (1) extrusion or injection of the material onto a suitable surface, then measuring the temperature of the plastic mass with a thermocouple probe; and (2) direct reading by a thermocouple that is placed in the barrel and is in direct contact with the melt. When the temperature of the melt is measured with a probe, care should be taken during the measurement to ensure that the purging of hot plastic does not cause an accident. As has been mentioned, molten material will cause serious burns because it is very hot, it adheres to the skin, and it is very difficult to remove. Burns are a common injury in molding operations, so long sleeves, gloves, and face shields should be worn when handling hot material or where there is danger of being splashed with hot plastic melt, particularly during startup or purging. As with other situations requiring personal protective equipment, the plant's requirements must be followed.

Warming Up an Empty Machine The machine's warm-up cycle should be programmed so that thermal overshoot does not occur and heating times are kept reasonably short. Once the machine is at the set temperature, it should be allowed to equilibrate (*heat soak*) before material is introduced into the barrel. It is advisable to keep this time as short as possible so that any resin left in the barrel after purging does not degrade. Check the machine for correct temperatures by briefly rotating (jogging or "inching") the screw. If the screw requires excessively high motor currents or will not rotate, allow the machine to equilibrate further. The setup sheets should show the normal heat soak time.

Before starting the machine, be sure that the set conditions are satisfactory by purging a few pounds of resin out of the die (or nozzle) at slow screw speeds. Check the melt temperature with a melt probe and check the general appearance of the melt. It should be smooth and free of dark specks, streaks, bubbles, or other signs of degradation. If no material is delivered when the screw is turned, check to see that plastic feed is available to the screw, check for faulty heater bands, and determine if the feed has bridged in the hopper.

Note: Checking for bridging or material blockage in the machine feed throat requires caution and the immediate attention and expertise of trained personnel wearing the appropriate safety equipment. This is because of the potential for serious personal injury from the hot gases and degraded and overheated plastic that may spray violently back through the hopper.

Warming Up a Full Machine When material is in the barrel or die head, the machine is said to be full. A full cold machine might result when there is a power failure or when the machine is deliberately shut down because of deterioration of the material by oxidation or depolymerization. The machine must be heated in a safe way because decomposition produces gases under pressure and can cause serious accidents. To warm a cold machine, set all temperatures just below the melting temperatures of the material, for example, at 275°F (135°C) for LDPE. Allow the machine to reach and equilibrate at these temperatures, then raise the temperature of the die (nozzle) to the process set point. To reduce the potential for dangerous pressure buildup in the barrel, wait for a period of time to allow the plastic in the die or nozzle to melt, then raise the other barrel temperatures to the set point. Allow the machine to equilibrate to these temperatures before beginning to purge.

Initial Operation and Purging When the machine is fully up to temperature, put a small amount of material in the hopper, make sure that the hopper lid is in position and the hopper gate is open, and start the screw at 10 to 15 rpm. Do not allow the screw to turn in an empty machine, because this can damage the barrel and screw. Check to see that the set process conditions are correct by running for a minute or so on a extruder or by running a few purge cycles on an injection machine. A machine with a grooved barrel may require careful hand feeding of material by trained personnel, because it is easy to overfeed the machine.

After a short period of operation, check the melt temperature with a melt probe and also the general appearance of the melt. Dispose of the hot, sticky plastic melt in a safe way after being satisfied that the material is feeding well and the melt looks satisfactory. On an extruder, also check the drive motor current. It should be with the normal startup range. If it is too high, there is probably unmelted material in the barrel. If it is too low, there may be feeding problems. On an injection molding machine, material should flow freely from the nozzle. If it doesn't, the nozzle may be blocked by unmelted plastic. Do not attempt to clear a blockage by turning the screw or injecting under high pressure. If all is well, fill the hopper to the normal level for running. Check to see that the monitoring equipment is working, and in extrusion blow molding, when material starts to extrude from the die, turn on the screw cooling if it is required.

Commencing Molding: Manual Operation When purging is complete and satisfactory melt is being produced, molding may begin. The molding process usually begins with manual operation, in which the operator initiates each part of the molding cycle by pushing buttons according to the operation sequence. When the mold is closed, the clamping pressure should be checked. To begin extrusion blow molding, produce a parison and check its temperature and appearance. The parison must be long enough to reach the bottom of the mold for pinch-off. Start molding, and adjust conditions and settings as needed to obtain a satisfactory part. Increase screw rpm gradually until it reaches the normal operating speed, while constantly checking the parts. A periodic check should be made to ensure that the hopper has enough material and that the melt is not leaking or weeping from around the nozzle, die, or adapter areas. Product quality should be watched closely to see that parts are free of froth or unmelted resin particles, and when the cycle is stable and the parts are good, production can begin.

Note: If for some reason a different material was used for purging, it must be removed, from the barrel before molding begins, according to procedures required by the material manufacturer.

Commencing Molding: Automatic Operation Automatic operation is started only after a satisfactory melt is made, after the purging procedure and after the machine settings have been established based on experience or process records, or as determined from the manual operating conditions if this is a new product. Begin molding on an automatic or a semiautomatic cycle (in which the operator opens and closes the safety gate to start the cycle) using predetermined cycle times. These may be calculated, based on experience, or determined from the manual operating conditions. Adjust conditions gradually until the product of the required quality is obtained at an optimum rate. After each adjustment, allow the machine to settle down for a reasonable time (approximately six cycles) before making further adjustments. With intermittent extrusion machines, adjust the cooling time until the molding can be ejected without distortion. Screw start delays and screw speed (rpm) can be adjusted to fill this time.

Changing Conditions and Dimension Verification Any changes must be well thought out in advance and should be made gradually. As an example, any increase in screw rpm may cause not only an increase in output but also an increase in temperature. Changes must be made one at a time. The machine must be allowed to settle down and the effect of the change noted; otherwise, no one would know what is going on.

Frequent or incorrect changes in the process can cause time to be lost and lage amounts of scrap to be made.

Recording Production Conditions The object of molding is to make moldings (parts) to the required specification (quality) and at the quoted cost. To do this, it is essential to keep accurate records. On many machines, data are recorded by computer. The data should be preserved. Critical parts such as those for medical applications are bar coded and the data stored. When this is not possible, an appropriate record sheet should be completed initially and then updated periodically throughout the run. It is also good practice to keep sample moldings. Logs of key events, reasons, and observations are also useful.

5.5.2.2 Safety in Normal Machine Operation

Operation Once the machine has settled in, controls and heaters should operate between an upper and a lower limit. This allows parts to be made to specifications. Most machines have process controls that warn when a condition is moving outside a limit. The operator should advise the process technician so that the cause can be found quickly and the problem corrected, both to minimize the production of bad parts and to reduce the likelihood of hazardous overheating or excessive melt pressure.

Safety Considerations Machines that are set to run automatically usually eject the parts onto a conveyor for finishing operations so that the operator does not have to reach into the press. For some large industrial parts, an automated picker or robot picks the part from the press and delivers it to the finishing operations. Often, particularly on small runs and when using a semiautomatic press cycle, the operator removes the parts by reaching into the press and removing the part from the mold. Redundant safety switches and devices must be in place and working to prevent the press from closing on the operator inadvertently.

In all cases, the operator must wear gloves ("cooled" plastic parts are extremely hot to the touch when ejected), safety glasses, and usually, earplugs. In plants with several different types of equipment running, the noise generated could damage hearing if protection is not worn. Entrances into plant areas that require eye, hearing, and sometimes helmet protection are usually marked with signs at the entrance indicating the type of safety protection required to alert personnel before they enter the area.

In September 2000, the American National Standards Institute (ANSI) issued a new safety standard, ANSI/SPI B151.15-2000 (effective September 2003) on extrusion blow molding machines operating in the United States (see Table 5.5). The numbered clauses in the table are from this specification.

5.5.2.3 Shutting Down

A great deal of money can be saved by using the proper shutdown procedures. For example, if the material could be prevented from degrading or burning, a large amount of purging could be eliminated. Additional money would be saved if a complete machine shutdown and cleanout were unnecessary, and startup would certainly be easier.

Temporary Stops It is a good idea during a temporary stop to purge the cylinder or barrel periodically by passing material through the machine and/or making air shots. If the plastic material starts to look a bit discolored, increase the frequency of purging. When a minor machine repair is required, set the heaters on the plasticizing cylinder to low values [about 302°F (150°C)] to minimize thermal degradation.

TABLE 5.5 Safety Standards for Extrusion Blow Molding Machine

Clause	Safety Caution
6.2.1 Operator's gate	Operator's gate, window, and mounting hardware to keep the operator away from hazards associated with moving parts and hot parison(s), including electric, hydraulic, and pneumatic interlocks.
6.2.1.1 Power operated gates	Leading edges mounted with pressure-sensitive switches to stop or open the gate. Closure of the gate should not initiate cycle start.
6.2.2 Operator's gate electrical interlock with monitoring	To prevent all clamp, carriage, calibration, or takeout motions when the gate is open.
6.2.3 Operator's gate hydraulic and pneumatic interlock with monitoring	To prevent hydraulic- or pneumatic-powered motions when the gate is open, including monitoring and alarm.
6.2.4 Emergency stop button	At least one emergency button is provided near the point of operation.
6.2.5 Reset	Resetting a safety interlock should not directly initiate a cycle.
6.2.6 Rear guard	A fixed guard for the molding area opposite the point of operation.
6.2.7 Top guard	A fixed guard to prevent reaching over another gate or guard.
6.2.8 Additional safety requirement for large machines only	Presence-sensing device; mechanical latch; double acknowledge system.
6.2.8.1 Emergency stop	At least one emergency stop button in a walk-in mold area.
6.2.8.2 Blow air release	Monitoring of blow air to prevent mold opening under full blow pressure.
6.2.9 Part discharge opening	Guarding required near conveyor openings.
6.2.10 Windows to molding area	All windows to conform to ANSI Z97.1.
6.3.1 Guards	Fixed guards (or movable guards with interlocks) at all other hazardous points.
6.3.2 Guarded feed throat opening	Guarding where access to the rotating feed screw is a hazard.
6.3.4 Extruder barrel covers	Cover or barrier to prevent inadvertent contact with high voltage or high temperature.
6.3.5 Window	All windows to conform to ANSI Z97.1.
6.4 Safety signs	Safety sign kit to current standard.

Overnight Stops For overnight stops with thermally stable plastics such as polyethylene at blow molding temperatures, close the gate at the base of the feed hopper and turn off the barrel heaters. With the nozzle–die heat on, purge the barrel clean by pumping the screw dry. As soon as nothing more comes from the die or nozzle, stop the screw drive and set the barrel cooling (if equipped) to maximum. When the machine is cool, shut everything off.

High-Temperature Work When extrusion blow molding at very high temperatures (with a material that does not melt until 265°C), oxidation and material removal can be a problem. Depending on the material type, it may be purged with a wet HDPE (approximately 2% of water is added to the HDPE before use). The water reduces viscosity. In condensation polymers such as nylon and polycarbonate, the water reduces the viscosity further because it causes depolymerization. Water also acts as a lubricant with nylon.

To purge, shut the gate at the base of the feed hopper and run the machine until it is free of high-temperature material. With an accumulator extrusion blow molding machine, open the die gap and keep the machine and cylinder temperatures at a high value [518°F (270°C)]; run wet HDPE through the system and fill the accumulator. The melt will foam and there will be crackling and spitting noises. Keep purging, reduce temperatures to 419°F (215°C), and open the die gap fully. Introduce dry HDPE and then purge the barrel clean by pumping the screw dry. Turn off the heaters when no more material comes through the die, set barrel cooling at the maximum, and then when machine is cool, turn everything off. [*Note*: A 7-kg (15-lb) accumulator machine may require 90 kg (200 lb) of wet HDPE and 45 kg (100 lb) of dry HDPE for proper purge of a high-temperature resin.]

Heat-Sensitive Materials A major problem with heat-sensitive materials is decomposition ("burning") of the plastic in the machine. The results may be discoloration and rejection of the molded part. When decomposition occurs, a complete shutdown is usually necessary, although it may be possible to purge the heat-sensitive material with another, more heat-stable material to clean the machine of contaminated resin.

Purge Materials Purge compounds are materials used to clean the cylinder (barrel) and may be purchased for this purpose. Instead of a commercial purge compound, a resin such as LDPE (low-density polyethylene) may also be introducedinto the barrel to push out a thermally unstable material such as PVC. Many purge materials do not melt or flow as ordinary resins do, so to prevent blockage it is advisable to remove the die assembly before purging. The die should be cleaned thoroughly. Once the purge compound has come through, the shutdown procedure should be followed.

When PE is used as a purge, it may be stored in a small hopper alongside the main hopper, from which it can be introduced rapidly into the machine by a power-operated valve. When PVC degrades, the rapid introduction of purge material is often necessary. With a stoppage of more than $\frac{1}{2}$ hour when running PVC, the barrel should be purged with PE. Stripping, cleaning, and purging should be done when restarting a PVC run after a power failure or some other unplanned shutdown.

Shutting Down an Injection Blow Molding Machine For injection blow molding machines, retract the barrel away from the parison mold's sprue bushing. Run the screw dry by allowing the screw to turn just until no more melt exits the nozzle. Do not allow the screw to continue turning because it may cause unnecessary and possibly damaging wear to the screw and barrel. Follow with a small amount of purge material, the type of which will depend on the type of material that had just been run. In the case of PE, purging is not usually necessary. It is generally safe simply to leave the screw in the forward position and turn off the barrel heats.

Other materials may require different purging materials and procedures. For PETG, a material used to make soda bottles and many other types of beverage bottles, polystyrene, low-melt-index HDPE, or cast acrylic are the typical purging materials. Polycarbonates are generally purged with a low-melt-index HDPE or a cast acrylic resin. Polyetherimide resins, which are molded at very high temperatures [in the range 700 to 750°F (370 to 398°C)], are purged in either a one- or a two-step process. In the one-step process, extrusion-grade HDPE (with a low melt index, in the range of 0.3 to 0.35 g per 10 min) is run through the machine after as much of the resin as possible has been pumped out. The barrel temperatures are reset to the normal melt processing range for HDPE once the HDPE begins to exit the barrel. HPDE is run through the machine until the purge exiting the nozzle is clear and clean. The screw is left in its forward position inside the barrel, the heaters are turned off, and the machine is shut down.

In the two-step process, a material that is intermediate in melt temperature between the high-temperature material and the lower-melt-temperature purge material is used. For example, a polycarbonate, which normally processes in the range 560 to 590°F (293 to 310°C) may serve as an intermediate temperature material that is used for an initial purge. Once the PC begins to exit the nozzle, the barrel temperatures can be reduced to that for molding PC. The next step is to purge the PC, again either with a low-melt-index HDPE or a cast acrylic resin. Acrylics and polystyrenes should not be used as purge materials for resins that are processed at high temperatures, that is, at temperatures above 590°F (310°C).

Chemical purging compounds that are designed to work with certain families of materials are often used, usually in conjunction with a plastic purge material. Always check with your supervisor or team leader for the material manufacturer's recommendations regarding purging materials and the safe procedures for their use.

Note: When purging materials from an injection barrel, always wear a full-face safety purge shield, always move the barrel (injection carriage) to a rearward position, never inject or purge through an open mold, and always make sure that the purge safety guard is functioning and closed to avoid a serious burn accident.

Check the Recommendations It is important to fully understand the correct procedures before a machine is started or shut down or before a resin change. Material suppliers issue process brochures and material data sheets with specifications and a great deal of other information. These should be studied and used to develop plant procedures for each material, particularly for high-temperature materials and those that are extremely heat sensitive or thermally unstable.

5.6 FINISHING

The finishing of a blow-molded part should be considered in the product design, mold engineering, and process planning phases. A good way to do this when in the initial stages of part design is to imagine that the mold has opened but that the part is still hanging in the mold and the flash is still attached. Depending on the part and process, the following secondary operations may need to be considered.

- Removing a dome or other sections from the part body.
- Removing the flash and performing drilling operations (Figure 5.58).
- Decorations: (hot stamping, heat transfer, and serial numbering by the hot stamp method).
- Automatic weighing and recording equipment.
- Safety (ergonomics and noise control).

In the following sections we outline items to be considered when planning the finishing processes.

5.6.1 Product Design

1. Radii, no square corners either inside or outside.
2. Wherever possible, include orientation and register features for positioning and holding in downstream tasks.
3. Lay out flash pockets in the product drawing.
    a. Design pockets wide enough to accept normal flash variations.
    b. Circumferential flash is to be avoided if at all possible.

FIGURE 5.58 Finishing an extrusion blow-molded traffic safety barrel. (Courtesy of Crocker Ltd., Three Rivers, MI.)

c. When circumferential flash is una voidable, compensate for it in the mold design.

5.6.2 Mold Engineering

1. Can the part be molded together or "siamesed" (Figure 5.59)?
2. Is the application suited for "family" molds?
3. Can container handles or similar components be molded-in as inserts instead of being added later (Figure 5.60)?

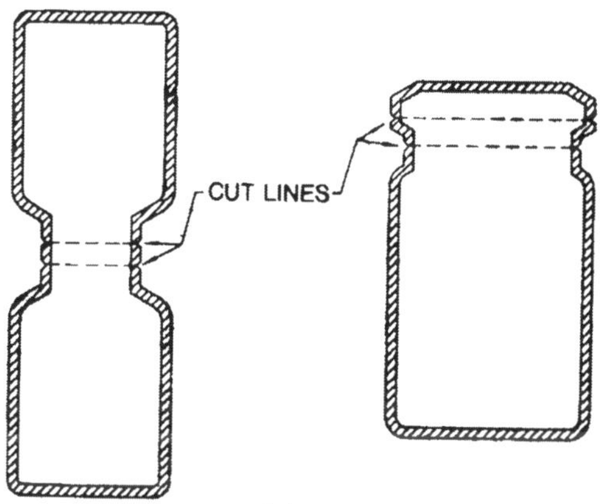

(*a*)

(*b*)

FIGURE 5.59 (*a*) Two-up molding; (*b*) container molded "Siamese," or two-up, on a machine specially designed for automatic separation and trimming. [(*b*) Courtesy of WEK South Corp., Reidsville, NC.]

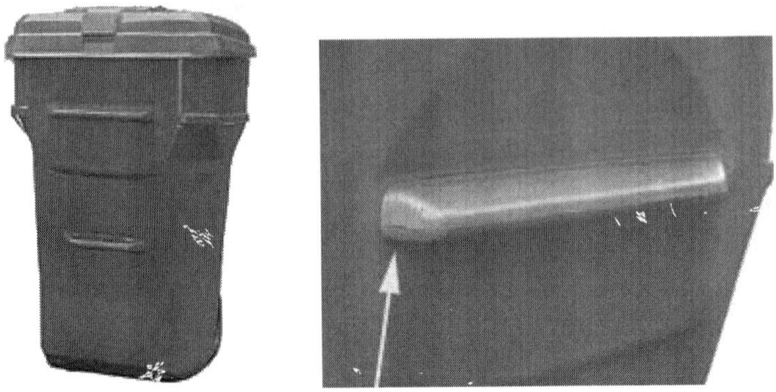

FIGURE 5.60 Container with premolded handle inserted into the mold as an insert to produce a container with a strong integral handle. (Courtesy of WEK South Corp., Reidsville, NC.)

5.6.2.1 Flash Pockets

1. Calculate the depth and plan to cut "metal-safe." Don't count on the old rule of thumb: 0.070 in. deep per side.
2. Keep the flash pocket depth tight enough to get cooling contact so that the flash is rigid. (Molten wavy flash, in addition to encouraging burn accidents, enlarges the potential for rejects due to parisons and/or parts welding together, and minimizes the potential for using deflashing dies.)
3. Consider "corrugating" the flash, especially head and tail tabs. (This is more expensive but provides superior results and can decrease the cycle time.)
4. Evaluate "blowing the flash," which provides a dimensionally stable, cool flash. (Again, a fringe benefit can be decreased cycles.)
5. Accommodate circumferential flash. Consider employing flash breaker pinch-offs so that flash can be removed in sections. This is especially important when deflashing is performed by machine. A full ring flash is usually removed by hand and is very difficult to automate.

5.6.2.2 Pinch-offs (Extremely Important)

1. *Width:* keep to a minimum; 0.015 in. is recommended.
2. *Hardness:* beryllium–copper preferably, or steel, is preferred over aluminum, for long life and maintenance of sharpness.
3. Incorporate in-machine replaceable inserts whenever and wherever possible.

5.6.2.3 Part Ejection

1. Configure the mold to eject the part so that it clears the mold *without human intervention.*
2. Utilize repinch mechanisms as strippers.
3. Preplan to include mechanisms to retain the part in the preferred mold half, and then include ejection features.

5.6.3 Process Planning

First, lay out the entire process, step by step. What has to be done, when, and where? Whenever possible, finish the part to the maximum degree possible, possibly even to include packaging, in the molding room and adjacent to the blow molding machine. Doing this promises the best economics, as it eliminates the labor involved in warehousing, transport, and storage. An additional benefit is that quality problems can quickly be communicated back to the molding department.

A common problem in blow molding plants is moving parts to separate finishing departments which work on completely different schedules or only on day shift, whereas molding is often an around-the-clock 24/7 operation. Careful planning and an efficient plant layout are required when designing for the following elements:

- Labeling
- Decorating (e.g., labeling and hot stamping)
- Drilled holes
- In compression-molded areas, consider punching in mold.
  - In blown areas, consider a "bubble," which can be "snare cut" rather than drilled (i.e., is chipless).
- Threads
  - Internal threads can be formed by using unscrewing cores.
  - External threads can be formed by threaded cavity sections that are split along parting lines.
- Deflashing
  - In entirety
  - Partial, in stages

5.6.3.1 *Specific Considerations*

Investigate making the flash productive during its limited life. Can it be configured to serve as a temporary handling, positioning, or registration device?

Part Handling: Removal Automatic removal is preferred since it enables consistent gate-closed operation. This can be achieved through stripper "pop and drop," side takeout, or bottom takeout. Parts may also be removed manually, but unless the operator is exceptional, the cycle time in some operations can vary from shot to shot, creating inconsistencies in thermal history, which leads to shot-to-shot variation in product weight, location of programming points, and dimensional variations. This is demonstrated by the readily visible symptom of variation in the tail length and the overall parison length. The operator will say that "those tails keep changing on me." Actually, the operator is changing the tail length by running an inconsistent cycle.

Part Handling: Secondary Cooling

- Parts up to 0.090 in. thick, if molded in a well-cooled mold with reasonably designed flash pockets, should not require secondary cooling.

- Medium-weight wall parts are sometimes conveyed through fan cooling.
- Heavy wall parts (0.375 in. and heavier) may be passed through chilled water spray or conveyed through a chilled water bath.

Having to hold the parts until they cool, prior to deflashing, is more often than not due to the retained heat in the flash. Deflashing early rather than later minimizes the hazard of the retained heat migrating back into the part, which, in turn increases the possibility of warpage.

Part Handling: Warp Control or Shrink Fixturing Not all parts will require shrink control. For those that do and depending on part shape, place them in shrink fixtures for whatever amount of time is necessary. Thicker parts tend to require more time in the fixture than do thinner parts. There are no real rules of thumb—experimentation with the actual part is the norm. Shrink fixture dwell time is usually stated on the shop floor by the number of shots: for example, "retain in fixture for five shots." Control of shrink fixture dwell time can be handled in the following ways:

- Shrink table, with manually operated clamps. Operator keeps track of where the coolest part is, replacing it with a fresh one from the mold part. Least desirable.
- Carousel or turntable shrink fixture. Operator indexes the table each shot, removing and replacing parts in rotation.
- Indexing conveyor, as part of an automatic finishing line.

Part Handling: Conveying The first consideration is whether or not the part to be conveyed has been deflashed. Deflashed parts generally can be moved with simple single-speed constant-running belt conveyors without problems. Conveying "fresh from the mold" parisons with hot flash tends to cause rejects because the hot flash sticks to other parisons, or worse yet, to other parts. Depending on flash configuration, when conveying parisons fresh from the mold with hot flash must be done,conveyors with support rails for the flash are recommended. When parts are flashed circumferentially, conveyors with separators are required to guarantee that the hot flash from one part cannot touch an adjacent part or its flash.

Trimming and Deflashing These terms are reasonably synonymous in industrial and large-part blow molding and refer to flash removal. *Trimming* is more descriptive in container blow molding where blow domes are cut away. In some plants and for some products, computer numerically controlled (CNC) routers are used to trim flash and remove unwanted material automatically rather than versus manually when part volumes are high enough to justify the cost of automation (Figure 5.61).A router will cause serious injuries if it makes contact with any part of the body while it is rotating. CNC machines move rapidly and must never be approached while in operation.

An alternative to the router is the laser . Laser cutting uses a powerful enough beam of light to cut plastic by heating and vaporizing a very small area. A laser is used to make highly accurate intricate holes and complex patterns. The laser power can be adjusted just to etch the surface of the plastic or to penetrate deeply, even drill

FIGURE 5.61 Using a CNC router to trim blow-molded panels automatically. (Courtesy of Agri-Industrial Plastics, Fairfield, IA.)

through. Laser cutting provides a clean, finished appearance and is more precise than conventional cutting methods. Never look into a laser when it is on or when it could be turned on at any time. Laser protection goggles must be worn at all times to avoid serious eye damage when working with high-powered laser cutting equipment. As with CNC robotic routers, stay away from the machine while the head is active, even if it is not moving at the moment.

A third alternative for making complex, precise trimming cuts is the water jet. Although it may seem as impossible to cut with water as to cut with light, water jet cutters use a thin jet of water at pressures up to 55,000 lb/in^2 to quickly cut the toughest plastics. Water jet cutting is very accurate, creates no smoke or dust, and produces a relatively smooth cut edge. As with a laser, most water jet cutters are mounted on a robot or a rotating head. Personal protective equipment *will not protect* against close contact with a high-pressure jet. Do not go near the water jet cutter nozzle unless the machine has been turned off and locked out. Even with the introduction of robotic devices for flash removal and trimming, the most common method of flash removal with large industrial parts is manual, using a knife to trim and a hammer to remove the parison tail.

Increasingly in the high-speed production of containers, machine deflashing is used. Although each application has its own requirements, most employ an assortment of pneumatic and hydraulic deflashing presses. A common deflash press uses female bottom nests into which the parts are placed. For the most part, these presses are semiautomatic, operator loaded, and unloaded. The majority are simple designs with nests mounted on tables. A few have channels for the scrap to drop to a conveyor below. Parts with circumferential flash cannot ordinarily be trimmed in this way.

Where interior flash webs must be removed, the bottom nest includes a die or cutout. The top tooling includes a punch. Deflashing presses may be either bottom or top acting. The part, cradled in the nest, contacts the deflashing die, which overstrokes slightly. A top-acting stabilizer/ejector member is generally required. For head and tail flash, as well as circumferential flash, the tooling will vary depending on whether or not the parting line is flat or irregular. Flat parting lines are well handled by a die plate. The flash to be removed rests on a shard, smooth surface, and a sharpened blade in the shape of the area to be trimmed is pressed with considerable force against the plastic, cutting through it and coming into contact with the die plate below. Irregular parting lines are better handled by "opposed chisel" tooling, which works by shearing the plastic between two sharp edges.

A feature common in deflashing machinery is the guillotine, which is used to chop off blow channels and necklike areas. Guillotine cutters work like the ancient capital punishment devices but usually have an air or hydraulically assisted fall (moving blade) to ensure proper cutting every time. Guillotine cutters are available in manual or automatic models. Because of their size and the difficulty of guarding the blade adequately, they are designed carefully and include several safety devices. Some include photo eyes that detect the thickness of what is under the guillotine. If the thickness is too great, the machine will not function. All systems guard the blade area, and to ensure that the operator is safely away from the cutter, some machines have pressure mats that must detect a human's weight before the blade will move. Some manual systems require two distant buttons to be pushed to ensure that the operator's hands are safely away from danger.

There is a category of secondary equipment referred to as *fixtures*. Generally, fixtures are job specific for a specific part and designed for secondary machining. Some fixtures include deflashing, both partial (e.g., an interior section needs to be punched out) and complete exterior deflashing.

5.6.3.2 Removing Domes and Other Sections

Rotating saw blades are often used to remove domes or other sections from the part body. Disadvantages are:

- The process is very noisy.
- Clips and fluff made by the saw blades when cutting a part create a mess in the work area and may require the wearing of respirators.
- There is danger from exposed saw blades and chips being flung that may get into the eyes.
- Finished cuts are not very smooth and require extra trimming.

The better way is to use a knife cut when possible. Figure 5.62 shows a groove for knife cut. Groove size will depend on part size and wall thickness. Generally, the height and depth of the groove should be at least twice the wall thickness, with shaped corners to prevent the knife from wandering. As with all cutting devices, the knife must be guarded during operation. Always turn off the machine and lock out

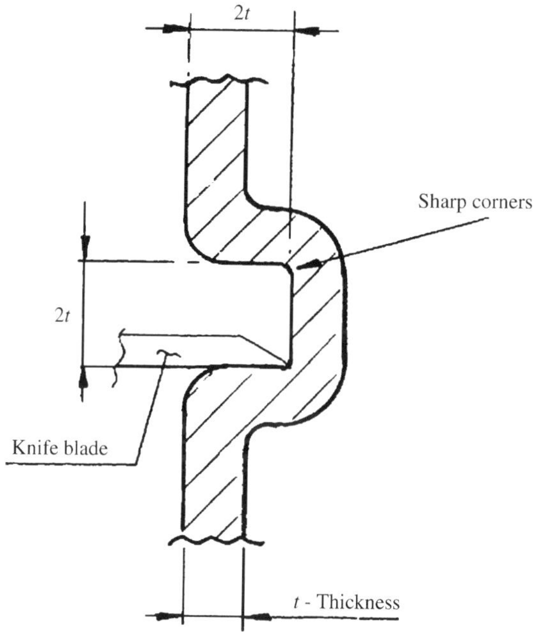

FIGURE 5.62 Grove for knife cut.

the knife movement mechanism before working on it, and wear cut-resistant gloves when working on or around trimming knives.

5.6.3.3 Flash Removal

Round parts can typically be cut in a machine, which rotates the part. The blade may be brought in manually or automatically. Squared parts that are rotated need some type of control to maintain the depth of the blade into the part, something to bring the knife in and out automatically, and something to start and stop the machine in the same orientation each time. Other operations can be added to the cutting machine if needed.

5.6.4 In-line Inspection and Testing Equipment

Because blow molding can be a high-volume process, in-line inspection and gauging equipment play an important part in monitoring product quality. There are several basic types:

- Laser beam for outside dimensions
- Ultrasonic for inside dimensions and wall thickness
- Vision equipment for visible defects
- Sliding, rolling, and other mechanical contact

5.6.4.1 Laser Devices

Very accurate, fast, and repeatable measurement of outside dimensions is possible with laser measurement devices. Laser light shines across a narrow gap to a set of sensors. The product runs through the gap, and the shadow that it casts is read by the machine as a measurement. To measure in more than one direction at once, more than one laser device is used. Laser systems can take many measurements in a second and can update a display once a second, an important ability with any high-speed production.

5.6.4.2 Ultrasonic Testers

Wall thickness in hollow products is something that cannot be tested by laser, visual, or mechanical gauges unless the testing is done at a cut end. If an unintended change occurs in a process between measurements and goes undetected, it can mean many dollars in scrap and lost production. Ultrasonic testers can measure the wall thickness of hollow products as they run by sending high-frequency sound waves at the moving product. The waves will bounce back differently as they hit different materials and different thicknesses. The controller of the device measures the time it takes for the signals to return and how they return, and from that it can determine the wall thickness. These gauges work very well for single-layer products and for products with distinctly different layers. When the materials being used are not significantly different from one another, the gauge finds it difficult to differentiate between the layers.

These units are small and can be made to fit most places. Some gauges will give printed readouts whereas others give a digital display that can either show the value of the wall thickness or inner diameter or can display it graphically on a screen in the form of raw data, a control chart, or a bar graph. As with other on-line measuring devices, ultrasonic gauges can be connected to alarms so that if the product is varying or approaching unacceptable dimensions, the operator can be alerted before the products become unusable.

5.6.4.3 Vision Systems

Vision systems are a major component of many automated systems. Typical components of a vision system are a controller and software for machine and user interface, video cameras, and one or more lighting sources. Camera-based systems (Figure 5.63) are used to detect such defects as black specs, flow lines, and gels. Vision systems are used increasingly to detect missing, wrinkled, or skewed labels, bar codes, and other product identification media as well as to detect the presence or absence of a component or feature such as a lid, untrimmed flash, formed threads, or proper fill height of the liquid product in a beverage bottle. A computer is used to capture, analyze, and archive the data while a video monitor displays the screen images. Typical defects detected by vision systems include:

- Holes
- Thickness variations
- Gels
- Contaminants
- Shape and dimension

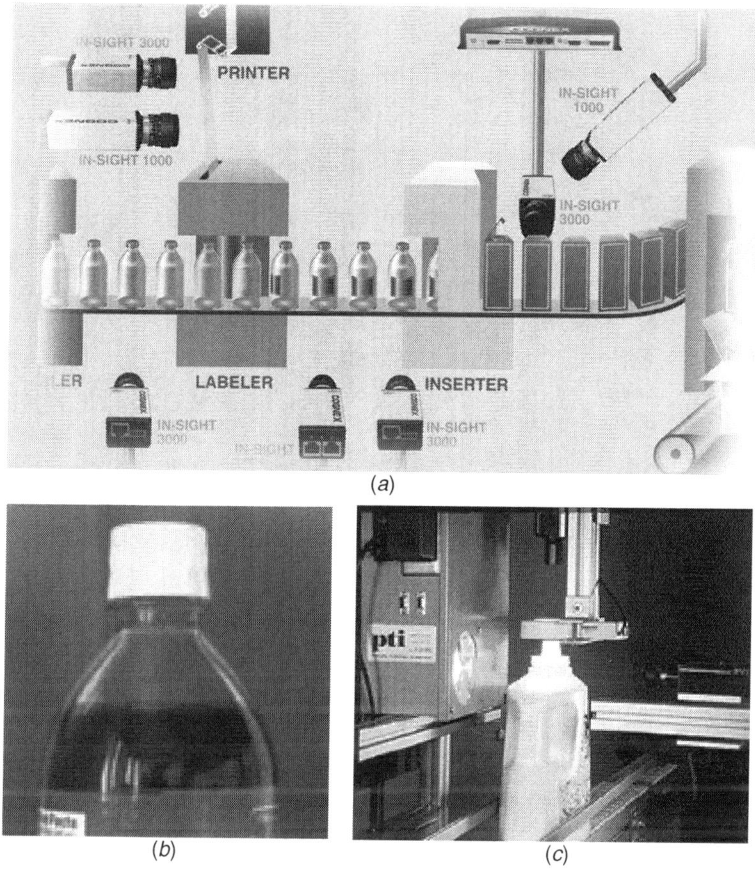

FIGURE 5.63 (*a*) Camera-based vision inspection system for plastic containers; (*b*) scanning bottle labels for unacceptabe skew; (*c*) scanning bottlenecks for defects related to thread profile, neck slant, flash, and neck diameter, and height. [(*a*) Courtesy of Cognacs Corp., Natick, MA; (*b*) and (*c*) courtesy of Packaging Technologies and Inspection, Tuckahoe, NY.]

- Wrinkles or creases
- Streaks
- Tears
- Printing defects
- Missing or skewed tabs, labels, bar codes

5.6.4.4 Mechanical Testing

Sliding, rolling, and other mechanical contact can be used,for example, to check dimensions or to test for bottle or container leakage using pressurized air to detect holes or weak weld lines. Figure 5.64 shows blow-molded containers being tested automatically for leaks. Defective containers are automatically di verted from the inspection line to a holding area or container for defecti ve or otherwise nonconforming product.

FIGURE 5.64 Container inspection systems using pressurized air to detect holes or weak spots in blow-molded containers. (Courtesy of Packaging Technologies and Inspection, Tuckahoe, NY.)

5.6.4.5 Conveyors

Conveyors are commonly used in blow-molding. The operation of conveyors is usually automatic. Check to be sure that the conveyor is running properly. If a blow-molded part or piece of trim or flash should jam, it can stop the conveyor and cause premature failure of the conveyor belt or motor, cause product to back up and jam the prior operation or station, or even stop the line. Regardless of the cause of a jam, if something is stuck in the conveyor, *always stop the conveyor* before removing the stuck object unless it can be done without risk. Conveyors may not look like a piece of equipment with a lot of power and speed but do not be deceived, because they can cause serious injury if not properly operated. Stay away from drive chains, belts, or gears, and never step on a conveyor. A high-speed bottle conveyor is shown in Figure 5.65.

5.6.5 Granulators

Because plastic materials are valuable, most blow molding facilities must be able to reclaim startup waste, defective products, edge trim, flash, and (sometimes) purgings. The equipment of choice is the granulator (Figure 5.66). The main function of the granulator is to reduce these items into pieces that are small enough to go back into the machine to be molded once again.

5.6.5.1 Grinders

Most of the time, this recovered material is called *regrind* and the machine is referred to as the *grinder*. This implies that the plastic is ground up, but that is a bit misleading. The granulator actually cuts or chops the pieces with a series of very sharp blades. One set of blades is carried on a rotating drum or axle, driven by an electric

FIGURE 5.65 High-speed overhead bottle converyor. (Courtesy of Graham Packaging, York, PA.)

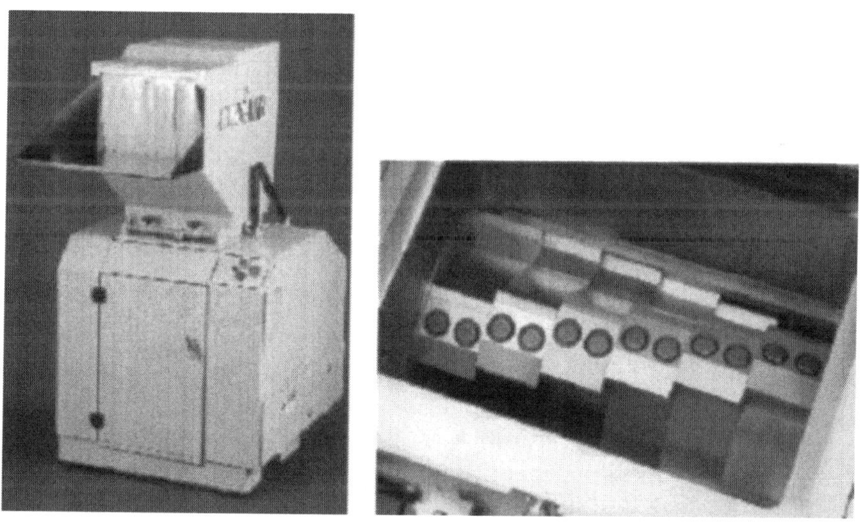

FIGURE 5.66 Conair granulator and view inside the cutting chamber.

motor and belt-drive system, while the other knife or set of knives is fixed. The moving blades pass by one or more stationary bed knives at a clearance of only a few thousandths of an inch. The plastic is cut and recut into smaller pieces. A heavy-duty metal screen with holes of a specified size keeps the plastic in the cutting chamber until the pieces are small enough to fall through the screen. The smaller the desired size of the material, the smaller the holes. Purge blobs are handled by a machine that actually shaves them into pieces. It takes a great deal of force to cut plastic, and granulators are strongly built and very powerful.

5.6.5.2 Safety Considerations

There are some important things to remember about the operation of granulators:

- Always wear the proper safety equipment. Granulators can be loud, so hearing protection must be used. Pieces of plastic can be ejected from the granulator, so always wear safety glasses. Some plants may require that the operator wear protective clothing, depending on the type of material that is being reduced.
- Never overload the granulator by stuffing it full. Feed it at a steady rate. Overloading can clog the feeding section of the granulator, stopping the reduction process until it is stopped and the excess material removed. This causes downtime that could have been avoided.
- Check the material collection drawer frequently to be sure that it is not overfilled. An overfilled drawer or box will cause the granulator to clog and eventually heat the material to the point at which it will melt, creating a very costly mess. Some molding facilities use automatic unloaders on the granulators to vacuum the material continuously from the collection drawer.
- **Caution**: The blades are sharp and the granulator is a powerful piece of equipment. It is meant to cut pieces of plastic that are very tough, much tougher than the finger or arm of an operator. *Do not* reach into the granulator for any reason when it is running. Stay clear of the blades.
- When cleaning the granulator, unplug it or lock it out before any work is done. Even though modern granulators have safety switches and interlocks, the safest way to avoid injury by a granulator is to ensure that it has no power.
- Never work on a granulator without cut-resistant gloves, and don't let anything—tool or body part—get between the moving knives and the bed knife when the drum or axle is turned by hand.
- Before starting the granulator, check to be sure that it is clean. Just like the rest of the material handling system, one piece of foreign material can contaminate many pounds of reground material. Checking to ensure that the granulator is clean includes making sure that the granulator catch bin is clean.
- Be sure that there is no metal or other contamination in the material to be granulated. Containers of startup waste or bad product are attractive as garbage containers. If anything other than the correct plastic material gets into the granulator, *stop*. Discard what's in the granulator and catch drawer, lock out the granulator, and clean it carefully. If an automatic unloader is in use, it will be necessary to inspect or discard a significant amount of material.
- Remember that despite guarding and feed chutes, the granulator may occasionally eject a piece of plastic, or even a large chunk, with lightning speed. Eye or

facial injuries are possible. Stand to one side if possible, and do not look directly into the granulator.

5.7 TROUBLESHOOTING

The blow molding process looks simple, but there are many variables, machine controls, materials, and process conditions, and many of these interact. Therefore, a thoughtful approach to any problem is best. Random changes or guesswork rarely solve a molding problem. Good problem solving (troubleshooting) requires four steps:

1. Collect the data. Collect as much useful information as possible, but don't jump to conclusions.
2. Interpret the data. What does it mean?
3. Take action. Repair the machine, change a setting, try a different material.
4. Check the results. Did that action solve the problem?

It is important to let the process settle down after making a change. It usually takes time for change to take effect. Record the action and effects of changes so that you know where you have been and where you are now, to help in solving the next problem. A basic list of container defects and processing irregularities is presented in Table 5.6, which also outlines corrective actions.

TABLE 5.6 Container Irregularities and Corrective Actions

Observation	Action	
	Machine Running	Machine Stopped
	Finished Container	
Rough surface	Check for moisture condensation in the mold; blast with an air hose and raise the coolant temperature.	Add mold venting.
Excessive shrinkage	Increase the blow cycle;. increase the blow pressure; lower the melt temperature.	Check the die-mandrel concentricity.
Warpage	Check the mold cooling; increase the blow cycle; lower the melt temperature.	Same as for shrinkage.
Weld line breaks	Same as for warpage.	Same as for shrinkage; inspect the pinch-off areas.
Thin wall at parting line	Increase the mold clamp pressure.	Inspect the mold alignment; inspect the mold venting.

TABLE 5.6 (*Continued*)

Observation	Action: Machine Running	Action: Machine Stopped
	Forming	
Parison blow out	Lower the melt temperature; reduce the blow pressure.	Check the mold for hot spots; check the parison alignment; check for contamination inside the tooling.
Container sticking	Lower the melt temperature; lower the mold coolant temperature.	Check the mold design.
	Parison	
Excessive stretch	Lower the melt temperature; lower the die temperature; increase the extrusion rate.	—
Rough surface	Reduce the extrusion rate; raise the melt temperature.	Clean the die tip; clean the tooling; change the tooling.
Uneven parison	Reduce the extrusion rate.	Align the die and mandrel; inspect for contamination; inspect for heater band failure.
Fisheyes (bubbles)	Lower the melt temperature; lower the feed section temperature.	Check the resin for moisture and for contamination.
Streaks	Raise the extrusion back pressure.	Inspect the tooling for contamination or damage; check the tooling design.
Curl	Increase the extrusion rate.	Check the tooling temperature profile; check the tooling alignment.
Wrinkles	Lower the melt temperature.	Check the tooling temperature profile; check the tooling alignment.

ACKNOWLEDGMENTS

During the last 10 years or so I have conducted blow molding seminars for the Society of Plastic Engineers. I have had a number of experts in the various subspecialties in the field of blow molding participate in these seminars. I have found that no one person can be conversant with the daily details of each of these specialties. The work in this chapter has been greatly influenced by them and I wish to acknowledge them since this chapter would not have been possible if I had not had their support.

John Bergerhouse	Resin materials	Equistar Chemicals
Gary Carr	Blow molding machines	Bekum America Corp.
Lewis Ferguson	Engineered materials	Parisons
Robert Jackson	Automation	Jackson Machinery
Jerry Jinks	Injection blow molding	R. J. Abramo Associates, Inc.
Joe Krupp	Three-dimensional blow molding	SIG Kautex
John McBride	Controls	Hunkar Laboratories, Inc.
John McNamara	Finishing	Blow Molding Automation Specialist, Inc.
Richard Morgan	Extrusion machines	Cincinnati Milacron
Robert Slawska	Extrusion machine heads	Proven Technology, Inc.
David Tekamp	Stress analysis	Stress Engineering Services, Inc.
Bruce Vogel	Extrusion blow molds	Portage Casting and Mold, Inc.
Ron Walling	Materials testing	Advanced Material Center, Inc.

I wish to thank John Bartolomucci, who worked with me on a project for Davison Community College in North Carolina and helped locate on the Internet some of the diagrams and pictures used. This work would not have been possible without my right-hand assistant, Loretta Lee, quality engineer for White Ridge Plastics, who helped with the word processing, editing, and insertion of pictures, tables, and graphs.

RECOMMENDED READING

1. Norman C. Lee, *Blow Molding Design Guide*, Hanser/Gardner Publications, Cincinnati, OH, 1998.
2. *Blow Molding of Thermoplastics*, company literature, Hoechst, Frankfurt, Germany.
3. Norman C. Lee (editor), *Plastic Blow Molding Handbook*, Chapman & Hall, New York.
4. *Polyolefin Blow Molding*, operations manual, company literature, Equistar Chemicals.
5. *The Bekum Blow Molding Handbook*, Tony Whelan, company literature, Bekum, Bodenewich, Germany.
6. Norman C. Lee, *Understanding Blow Molding*, Hanser/Gardner Publications, Cincinnati, OH, 2000.

CHAPTER 6

Rotational Molding

PAUL NUGENT
Consultant, Reading, Pennsylvania

6.1 INTRODUCTION

Rotational molding, or *rotomolding*, is a relatively young industry, with its roots found in polyvinyl chloride (PVC) plastisol molding of the late 1950s. With fewer than 2000 companies worldwide even today, it has remained a small but exciting niche within the plastics world that reaches almost every conceivable market yet typically enjoys growth rates above the industry average. It offers much in the way of design flexibility and scale of product, with products ranging from simple bulk storage containers to sophisticated automotive, medical, and aerospace applications. Regularly used to produce parts of great complexity with a constantly growing range of applications, rotational molding lends itself to hollow complex forms as does no other process. Small parts such as dolls' heads can be made in essentially the same manner as that used for huge 22,000-gallon (80,000-liter) storage tanks. Intricate parts such as fuel tanks and components for aircraft ducting are now more common as rotational molding is recognized by a broader group of designers and engineers.

Rotational molding, however, remains in something of an infancy in terms of development. The apparent simplicity of the process and the low-tech markets commonly pursued have caused it to be overlooked somewhat. Current market trends to consolidate components and shorter product life cycles with lower total volumes have brought an increased level of interest. Importantly, this interest by the market has attracted the interest and development efforts of major material suppliers, a key factor in the future development of the industry.

In this chapter we take a look at the basic process of rotational molding, the markets that it serves, the materials, molds, and machinery that are used, and some of the design basics that can help to make molding easier. For those who seek a deeper

Handbook of Plastic Processes Edited by Charles A. Harper

and broader treatment of the subject, *Rotational Molding: A Practical Guide*, also by the author, is the most comprehensive book available on the process, details of which can be found at www.paulnugent.com.

6.1.1 Rotational Molding Markets and Applications

With its ability to produce large hollow parts, the most commonly molded parts worldwide are water tanks. This may account for as much as 70% of the global activity in rotomolding, with much of it concentrated in developing regions or areas where water is in short supply. The largest general markets in monetary terms are to be found in North America and Europe, and in these regions, without the dominance of water tanks, rotational molding is applied to a much wider range of applications with an ever-increasing degree of sophistication. Estimates of molders and the consumption of material around the world are shown in Figure 6.1.

Ideal nontank applications for rotational molding are typically complex hollow forms in relatively low quantities (up to approximately 10,000 parts). There are obvious

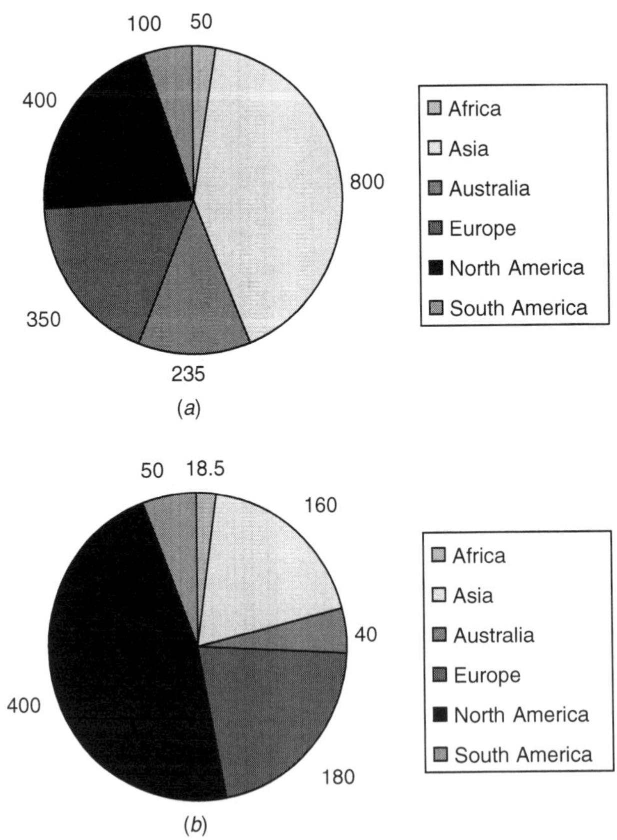

FIGURE 6.1 Estimated number of (*a*) molders and (*b*) material consumption (kilotonnes) by continent.

exceptions to this, children's toys for example, but for the most part, rotomolding offers benefits to end users with its low upfront costs in tooling and equipment and the ability to render almost any design that the molder can conceive.

The process is inherently low pressure and is closely tied in many respects to the processing attributes of polyethylene. Material choices outside the polyethylene palette are limited by comparison to other processes, and thus rotomolding does not always suit applications that have more demanding performance criteria. Tolerance control, surface finish, and part stiffness are among the most common challenges.

Typical Applications The versatility of rotational molding is constantly demonstrated in a wide range of products in an equally wide range of markets. Some typical market sectors and the applications within them include:

- *Agriculture:* storage tanks, spraying equipment tanks
- *Automotive:* interior panels, fuel tanks, ductwork, air-intake systems
- *Building–construction:* water tanks, septic tanks, highway barriers
- *Electrical–electronic:* aboveground pedestals, belowground chambers
- *Floor care:* vacuum cleaner parts, floor cleaner tanks
- *Industrial:* chemical tanks, cases, shrouds and housings, corrosion and pollution control equipment parts
- *Lawn and garden:* mower shrouds, fuel tanks, ductwork
- *Marine products:* dock floats, hulls, fuel tanks, seating, fenders, live wells
- *Material handling:* tanks, drums, barrels, hoppers, IBCs, pallets
- *Medical equipment:* spine boards, anatomical figures, inflatable masks, implants
- *Playground equipment:* slides, climbing frames
- *Signs and displays:* point-of-purchase displays
- *Sports and recreation:* toys, play balls, kayaks, canoes, helmet linings, pet products
- *Transportation:* road traffic barriers, cones, signage, aircraft ductwork

6.2 ROTATIONAL MOLDING PROCESS

6.2.1 Four Basic Steps

Rotational molding is a simple process. It utilizes high temperatures, thin-walled metal or composite molds, biaxial rotation in two perpendicular axes, finely divided powder or liquid polymers, and cooling using air and/or water to produce hollow, seamless, low-stress parts. Rotational molding has four basic steps (Figure 6.2):

1. *Loading.* A preweighed amount of powdered or liquid plastic is placed in one half of a thin-walled hollow metal mold that is mounted on the arm of a molding machine. The mold is then closed using clamps or bolts.

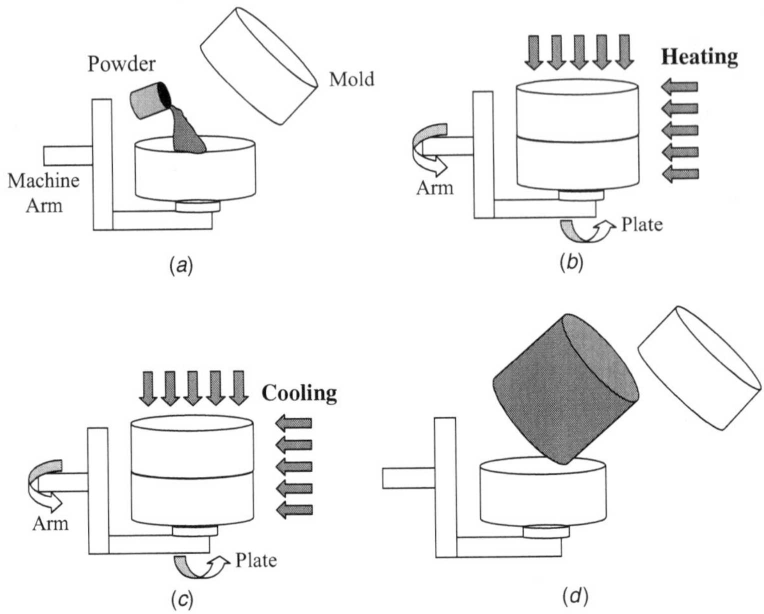

FIGURE 6.2 Four basic steps of rotational molding: (*a*) loading; (*b*) heating; (*c*) cooling; (*d*) unloading.

2. *Heating*. The mold then begins to rotate biaxially about two perpendicular axes while being moved into an oven where heat is applied. The metal or composite mold becomes hot and the powder–liquid tumbling inside rises in temperature. Hot powder material sticks to the mold in successive layers to form the part, while liquid materials typically react as they form the part shape.

3. *Cooling*. When the material has melted and has been consolidated, the mold is moved to a cooling station, where forced air, water mist, or a combination is used to bring the part temperature down to a point below the crystallization or solidification point of the material. Uniaxial or biaxial rotation continues to prevent the molten material from sagging.

4. *Unloading*. Once the part is cool, the mold is moved to the unloading station, where the part is removed. The mold is then ready to begin the process again.

Stages 1 and 4 are often combined into a single operating station (mold servicing) in machine design so that the most basic of machine configurations typically consist of three workstations: heating, cooling, and servicing. This apparent simplicity belies the complex interaction of heat transfer and material distribution that occurs within the mold during the process. Rotational molding is unique among plastics processes in that heating, forming, and cooling of the material all occur in the mold without the use of pressure; until recently, once a mold entered the oven, nothing more was known other than that the powder melted (or liquid reacted) and was then cooled to form the final part. Nowadays, it is possible to measure temperatures inside the mold during the cycle, scan the surface of the mold continuously for temperature readings, or even place a videocamera inside to view the formation of the part.

During the process, a number of key factors must be considered to ensure that the process is effective and economic: Vents are used to ensure that pressure inside the mold is in equilibrium with the external environment; release agents often must be used to ensure that the material does not stick to the mold and surface; rotation settings must be selected carefully, as the relationship between the primary and secondary axes will affect the way in which material is distributed in the final part; only the external surface of the part is in contact with the mold, and particularly in the case of semicrystalline materials such as polyethylene, cooling rates must be controlled to minimize distortion while accelerating the process.

6.2.2 Advantages and Limitations of Rotational Molding

The main advantages of rotational molding can be summarized as follows:

- It is ideally suited to the manufacture of hollow, complex shapes ranging in size from table tennis balls to industrial tanks over 20,000 gallons (75,700 liters) in capacity.
- Both molds and machines are simple and relatively low in cost. It is a low-pressure process that allows thin-walled low-strength molds to be used. Small production runs can be cost-effective.
- Low pressure and low shear rates lead to parts that have low levels of molded-in stress.
- Parts have good wall thickness distribution compared to processes such as blow molding and thermoforming. External corners tend to thicken, which can be an advantage in applications where wear is critical.
- Parts can have thin walls relative to their size and volume (i.e., large tanks).
- Different sizes of parts can be produced simultaneously on the same machine at the same time.
- Parts made of different materials can be molded simultaneously on the same machine and even on the same arm at the same time.
- Large metal inserts and graphics can be molded directly into parts. A wide range of surface textures and details can be reproduced.
- Color changes can be made quickly and easily. There is no purging process such as that found in injection molding or extrusion processes—parts can be molded in a new color without loss of material or parts.
- Multiple-layer parts can be formed using the same low-cost molds. Multicolored parts and parts with foamed layers can be produced using simple techniques.
- All material placed in the mold is used to form the part. Scrap is limited to those areas that are removed from the part during finishing. Insulated sections of the mold can be used to minimize this runout.

The main limitations of rotational molding can be summarized as follows:

- The process is typically not well suited to very large production runs of smaller parts. For smaller parts, blow molding or even injection molding may be more competitive.
- The number of materials that are available for rotational molding is limited in comparison to other processes.
- Material costs are higher, due to the need for grinding pellets of raw material to a fine powder for molding. Developments in micropellets may help to reduce this penalty.
- Cycle times are long in comparison to other processes, as both the mold and material must be heated and cooled. The materials used require more thermal stabilization and therefore cost more than for other processes.
- Loading of molds and unloading of parts is labor intensive by comparison to other processes, especially for complex parts.
- Release agents are required to ensure that the material does not stick to the mold during demolding. This demands constant attention by the machine operator.
- Bosses and ribs for stiffening cannot easily be molded into parts; designers must depend more on part geometry and design to produce stiff parts.
- Large flat surfaces are difficult to produce, due to warpage. Designers will typically use ribbing and surface details to avoid them in parts.
- The inner surface of a part is formed freely during molding. This means that dimensions cannot be controlled with the same degree of accuracy as, for example, injection molding.
- A part is free to shrink during cooling; dimensional accuracy can therefore be more difficult to predict.

6.2.3 Analyzing the Process

Figure 6.3 shows typical data measured during a molding cycle for polyethylene. In this case, two thermocouples were used, one measuring the external environment of the mold and the second measuring the internal temperature of the mold (the enclosed inner air volume). Examining the external temperature profile first, at the beginning of the process the internal and external temperatures are similar. As the mold progresses into the oven, the temperature of the environment rises rapidly. The profile and rate of recovery of the oven temperature depend on the size and performance of the oven burner and the design of the oven itself. The environment temperature settles around the set point of the oven. Variation in the temperature during this stage of the cycle is due to the rapid movement of air in the oven and the rotation of the mold past the hot-air inlet. Once the oven cycle is over, the environment temperature drops rapidly to the temperature of the cooling bay. Temperatures in the cooler are dictated by ambient conditions and any special cooling systems employed. For enclosed coolers, heat radiating from the mold can raise the surrounding temperature initially, but this falls over a period of time as the mold cools.

Looking at the internal air temperature profile in more detail, as shown in Figure 6.4, a number of consistent transition points can be observed for polyethylene moldings. Heat transfer through the mold and energy absorption by the material affect the internal air temperature profile at key points in the cycle:

A Powder begins to adhere to the mold surface.
B Powder has adhered completely to the mold surface.
C Peak internal temperature that can be related to part properties.
D Crystallization of polymer.
E Demolding point for solidified part.

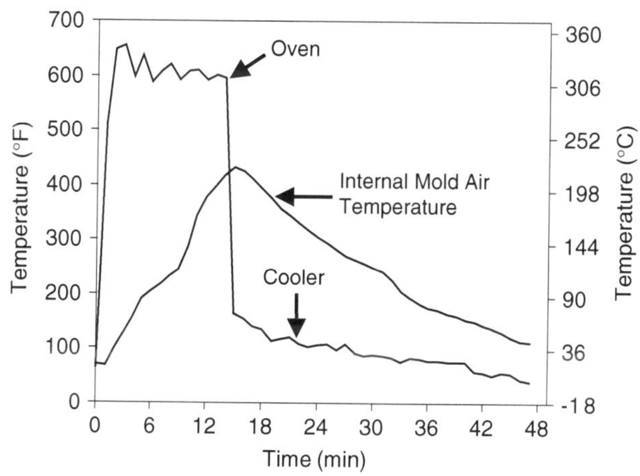

FIGURE 6.3 External and internal temperature profiles during a typical molding cycle for a polyethylene part.

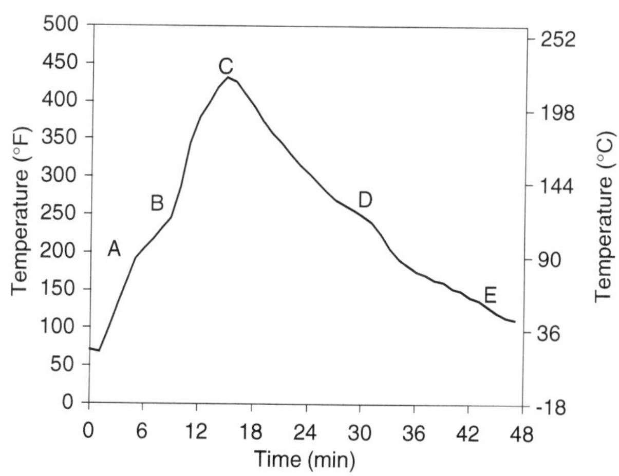

FIGURE 6.4 Internal air temperature profile during a typical molding cycle for a polyethylene part.

At the start of the process, the material and air inside the mold are at ambient conditions dictated by the residual heat left in the mold from the previous cycle and the temperature of the material as it is placed in the mold. As rotation and heating begin, energy is transferred through the mold and into the air and material masses within the mold. The material does not stick immediately but flows as a powder until the material mass and the mold surface are sufficiently hot to allow smaller particles in the material to begin sticking to the surface. This occurs at point A.

As heating continues between points A and B, material fuses to the mold in successive waves. The absorption of energy by the melting process (as material is converted from a powder to a solid molten mass) causes the rate of increase in temperature of the internal air to slow down, producing a plateau. At the end of this plateau (point B), when all the material has adhered to the mold surface, energy is once again transferred directly to the internal air and the temperature rises again, more rapidly.

Heating continues until point C. This peak temperature can be related to the final properties of the final molded part; the ultimate degree of "cure" or "cook" can be related to the peak internal temperature observed within the mold during the cycle. This peak will occur some time after the part leaves the oven, depending on the thermal inertia in the mold and part wall thickness. Increases of 30 to 40°F (16 to 22°C) into the early stages of cooling are not uncommon for thick parts.

As cooling of the mold begins, the temperature of the part inside falls. The rate of temperature drop of the air inside the mold lags the external surface temperature of the mold. This rate is dominated by the cooling method employed (typically, fan cooling only during the beginning of the cycle) and the ambient conditions (cooling in summer can be dramatically slower than in winter). Cooling continues until the material reaches crystallization. At this time, energy released during the formation of a crystalline structure maintains the temperature within the part, producing a second plateau similar to the melting plateau found during the heating stage of the cycle. This occurs at point D.

Beyond the crystallization point, cooling continues using water and/or air until the part is safe to handle. Recent work has shown how temperature changes in this region can be related to separation of the part from the mold surface due to shrinkage.

The characteristic shape of this curve is consistent for almost all polyethylene parts, only the relative times and temperatures at which the transitions occur will change, depending on mold material, part material density, part thickness, and size. For other materials, the characteristic shape of the internal temperature profile is similar but will vary according to the melting temperature of the material. Amorphous materials such as polycarbonate do not exhibit a crystallization plateau during cooling.

Thus, the rotational molding process can be divided into six distinct phases as shown in Figure 6.5:

1. *Induction.* Powder inside the mold picks up heat through the mold wall but remains free until point A.

2. *Sintering.* The powder and mold reach a sufficiently high temperature to begin a process of successive layers of powder attaching to the mold inner surface. Absorption of energy during melting of the polymer slows down the increase in

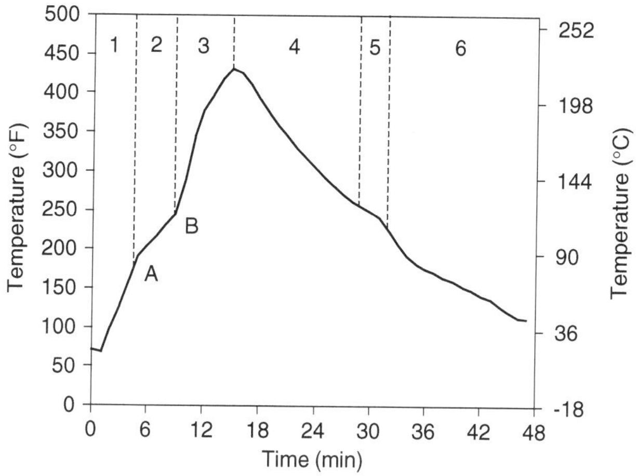

FIGURE 6.5 Six phases of rotational molding.

temperature within the mold. There is a natural separation of particle size during sintering—smaller particles adhere first, followed by progressively larger particles.

3. *Consolidation.* Once the powder has adhered to the mold completely, heating continues to fuse and consolidate the material into a solid homogeneous mass.

4. *Melt-phase cooling.* The polymer is molten and must be kept moving to prevent sagging.

5. *Crystallization.* Once the polymer is cooled to its crystallization temperature, energy is released, which maintains the internal temperature on a second plateau. This will not occur for amorphous materials.

6. *Solid-phase cooling.* Beyond the point at which the material has crystallized, the part is solidified and may pull away from the mold surface.

Rotolog is a radio-frequency-based temperature measuring system that is an ideal tool for diagnosing cycles and problems during molding. The data that it provides give a true insight into the molding process. However, it is limited by the use of an insulated container to protect it from the heat of the oven, which means that it must be cooled after several cycles (depending on duration). Several other approaches to gathering temperature data from the mold have been developed, including noncontact infrared systems and directly hardwired systems employing high-temperature slip rings.

6.3 ROTATIONAL MOLDING MATERIALS

6.3.1 Desirable Characteristics

There are a number of desirable properties that a material should have in order to be suitable for rotational molding. They include the following:

1. *Thermal stability* is required to allow for a wide processing window during molding and to prevent loss of properties during grinding into powder. Most rotational

molding materials are stabilized with antioxidants to reduce the effects of the long heating cycle typically used during rotational molding and to provide good service life, especially in outdoor applications.

2. *Low shear melt viscosity* must be low enough to allow the material both to wet the internal surface of a mold and to fuse to itself in a homogeneous mass. The common measure used as a guide to the melt flow properties of materials is the Melt Index (MI), measured in grams per 10 minutes. Most commercially available rotational molding polyethylene materials have an MI value between 2 and 8 g per 10 min (2.16 kg at 190°C). Higher flow rates may be required for complex molding details.

3. Good *impact strength* is most often critical in low-temperature applications. Materials will usually be expected to exhibit good performance down to −40°F (−40°C).

4. *Environmental stress crack resistance* (ESCR) is the ability of a material to resist long-term exposure to stress cracking agents. Materials with ESCR values in excess of 1000 hours are typically considered acceptable based on one of the common test procedures.

5. *Flexural and tensile strength* must be sufficient for the final application for which the material is intended; polyethylenes have relatively low strength values, engineering materials such as nylon and polycarbonate offer better levels of performance.

6. *Elongation to failure* should be high for most applications.

7. *Free-flowing physical properties with good heat transfer properties* are necessary to allow the material to distribute evenly and quickly during the heating cycle. Poor flow properties will significantly affect the final appearance of a part.

Recent research has investigated the effect of rheology and thermal properties on sintering and bubble removal in polymers. The techniques developed are a major step forward in the process of evaluating materials prior to molding. Conclusions from this work show that melt rheology and surface tension play a primary role in the earliest stages of material densification, while diffusion of gas from within entrapped bubbles is very important during the latest stage of densification. Materials such as linear low-density polyethylene melt slowly with a gradually decreasing viscosity. This creates a three-dimensional network of particles connected at their boundaries, trapping large amounts of air. Materials with lower viscosities tend to fuse more quickly, trapping less air and therefore consolidating faster.

6.3.2 Range of Materials Available

Of the many polymers available to all plastics processes, there is only a fairly narrow group of materials that can be processed easily by rotational molding. This group is gradually growing as more material suppliers and end markets recognize the potential of the process. From a market perspective, the principal materials in use today are ranked as follows:

Polyethylenes (LDPE, LLDPE, HDPE, XLPE, mLLDPE, EVA)	95%
Polyvinyl chloride	3%

Nylons (nylon 6, 6,6, 11, 12)	<0.5%
Polypropylene	<0.5%
Polycarbonate	<0.2%
Fluoropolymers (ETFE, ECTFE, PFA, MFA, PVDF)	<0.1%

There are, however, a range of polymers that have either been successfully molded over the past 30 years or are being used in small quantities for specialist applications. These include:

- Acetal copolymer
- Acrylic
- Acrylonitrile–butadiene–styrene (ABS)
- Cellulosics
- Epoxy
- Ionomers
- Phenolic
- Polybutylene
- Polyester
- Polyester elastomer
- Polystyrene
- Polystyrene, impact modified
- Polyurethane
- Silicone

6.3.2.1 Polyethylene

Polyethylenes offer an excellent combination of ease of processing, thermal stability, and cost with which they continue to dominate the market. The development of a broad range of alternative materials for rotational molding has been slow for two main reasons:

1. The ability of a material to re-form into a homogeneous solid after being pulverized into a free-flowing powder without the need for pressure eliminates many engineering materials from the rotational molding palette.
2. Specialist materials tend initially to be more expensive than the benchmark polyethylene and are often hampered by low market demand to encourage material development.

While polyethylene satisfies the need for many varied applications, structural requirements, scratch resistance, chemical resistance, and an expanded temperature performance range are just some of the attributes that molders and end-users alike need in order to expand into new markets. The demand for these properties will grow as the overall scale of the rotational molding market grows and will hopefully drive further research and development in new materials.

6.3.2.2 Polypropylene

Polypropylene has been used in rotational molding for many years. However, it has always been hampered by poor processability, poor low-temperature impact strength, and sensitivity to overcure. A number of commercially available materials are currently on offer, but none has yet been able to fully utilize the potential benefit of polypropylene. Recent grades have improved the processability, and additive packages have improved the mechanical properties of finished molded articles. Applications for polypropylene include hot (not boiling) water storage, autoclavable containers, and chemical containers where the chemical resistance of polyethylene is not sufficient.

6.3.2.3 Nylon

A broad range of nylon combinations can be used for rotomolding. The most common commercial grades are nylon 6, nylon 11, and nylon 12, with typical melting points of 419°F (215°C), 367°F (186°C), and 352°F (178°C), respectively. Nylon 6,6 can be molded but has lower impact strength when processed at the high temperatures required for rotomolding. There are many other grades, such as nylon 4,6, nylon 6,9, and nylon 6,10, that have been tested but are not produced commercially. Special grades have been developed with enhanced flame retardancy properties for use in aerospace applications such as those being used to mold ductwork on Boeing aircraft.

Nylons require some care in processing, but when properly molded are strong, abrasion resistant, self-lubricating, and have good impact, chemical and fatigue resistance, and excellent continuous-use service temperatures. They retain their strength properties at high temperatures; values fall off as the material rises in temperature but do not fall away dramatically until the glass transition temperature (T_g) for the material is passed. Nylons 11 and 12 have better resistance to oxidation that results in longer retention of properties at elevated temperatures than that of nylon 6.

Nylons need additional antioxidants or processing with an inert atmosphere (typically, nitrogen or carbon dioxide) to prevent degradation. Heat stabilizers such as copper salts can be added to improve the heat stability of the final molded product for applications where the service temperature exceeds 165°F (75°C). Nylon 6 can be molded without nitrogen if lower impact and inferior surface appearance is acceptable. Nylons 11 and 12 do not require nitrogen atmospheres, although using a nitrogen bleed during molding can improve color retention and impact strength.

Liquid nylons in a caprolactam-based reaction and reaction injection molding materials such as Nyrim can also be molded rotationally. Raw materials, usually in two streams, are heated to a liquid state and then mixed just before being added to the mold. Rotation deposits the material on the inside surface of the mold. Control of the reaction rate is critical in ensuring that the material is evenly distributed during the curing stages of molding. The reaction generates heat that accelerates the process—an oven may not be required for some applications.

Nylon is commonly used in applications requiring high–temperature resistance while retaining good strength properties (e.g., car and truck ductwork). It is also used where good abrasion resistance is required and where chemical resistance beyond that of polyethylene is required. High flexural strength and toughness makes it a good choice for pressure vessels, and excellent permeation resistance makes it ideal for solvent and fuel containment. Both powder and liquid nylons are used for fuel tanks for motorcycles and military vehicles.

6.3.2.4 PVC (Liquid)

Polyvinyl chloride (PVC) is available in two distinct forms for rotational molding; either liquid or dry. Liquid materials in the form of plastisol and organosol have been the mainstay of PVC molding since their introduction in 1947. Dry powders or vinyl micropellets were introduced in the early 1980s. Vinyl plastisol is the most common form of PVC used for rotational molding.

Plastisols are typically low cost, have a wide range of hardness, are easy to automate for dispensing, can have high clarity and transparency, allow high mold surface reproduction, and are easy to color. They can be produced in Food and Drug Administration (FDA) grades, have good chemical resistance, and are very forgiving in process. Altering the ingredients of the plastisol controls the durometer, or surface hardness, of a molded plastisol product. Rotomolding grade plastisols range from a Shore durometer of about 30 A to about 70 D. The inherent hardness is controlled by the properties of the materials used and the ratio of solids (materials) to plasticizer.

6.3.2.5 PVC (Powder)

Both vinyl powders and micropellets are fairly recent additions to the PVC palette for rotational molding. They offer a wide range of hardness, do not require a pump for dispensing, are clean, can mold with uniform wall thickness, have a lower bulk density and somewhat higher cost, but are in only limited use in the industry. The powdered form of PVC has been in use for some time in slush molding; a process most closely akin to the Engel molding process of the mid-1950s. In slush molding, a heated mold is filled with powder, allowed to rest for a fixed amount of time, and then dumped. The mold temperature and the residence time of the upright mold determine the amount of powder that sticks to the mold walls. After the excess material is dumped out of the mold, the coated mold is heated until the PVC powder sinters and fuses together. This process works very well for thin-walled parts. In continuous production, the powder that has been dumped out of the mold increases in temperature, thereby speeding up the overall molding process. Heavy-walled parts are impractical, since gravity can overcome the capillary attraction of the molten PVC to the static mold wall.

6.3.2.6 Polycarbonate

Polycarbonate was introduced in the 1950s and was first molded rotationally in the late 1960s. It is hygroscopic and requires drying before molding and should have less than 0.02% moisture before being charged directly into the mold. Improper drying will produce parts containing gray matter, bubbles, a yellow color, a reduction in transparency, and a loss of impact strength. Moisture decreases the average molecular weight of the polycarbonate during molding and therefore degrades physical properties.

As higher temperatures are required to process polycarbonate, heat transfer through the mold to the material can be a problem. The best results are achieved using oil-jacketed molds and finer powder than the 35 mesh (500 μm) commonly used for polyethylene. However, bubbles can still occur in parts, and it is common to see textured mold surfaces used to disguise their presence. The use of a nitrogen atmosphere can reduce the yellow color caused by degradation.

The advantages of polycarbonate are its higher stiffness, toughness, clarity, and high-temperature resistance [275°F (135°C) versus 122 to 158°F (50 to 70°C)] over polyethylene. Weatherability and dimensional stability are good and the material can be ductile even at low temperatures. FDA-approved and flame-resistant grades are available. It is most commonly used for applications such as street lighting and display covers (Figure 6.6). The disadvantages are that it is high in cost, is more difficult to process, must be dried and has low chemical resistance.

6.3.2.7 ABS

Acrylonitrile–butadiene–styrene (ABS) is an amorphous thermoplastic terpolymer. It has represented something of a "holy grail" for rotational molding materials in recent years, being seen as a potential scratch-resistant material available at a reasonable cost. It has been molded in the past commercially and has been the subject of several recent research projects at the university level. Studies have shown that ABS (along with polystyrene and acrylic) is sensitive to degradation and that could sinter to produce a porous and friable structure with low tensile strength and high surface porosity. Increased rubber content in ABS leads to increased difficulty in densification.

ABS offers a range of potential advantages, including good impact strength, rigidity, chemical resistance, good adhesion properties, paintability, and resistance to solvent cracking. Molded parts can be painted using a primer and topcoat combination for applications such as mannequins, toys, sanitary ware, auto components, sports equipment, and furniture.

FIGURE 6.6 Polycarbonate display globes. (Courtesy of Konan Tokushu Sangyo Co., Ltd.)

6.3.2.8 Fluoropolymers

Fluoropolymers offer superb chemical and temperature resistance for demanding applications. They exhibit low surface tension (nonwettable and antifouling surface), excellent chemical resistance to many aggressive chemicals, and good temperature resistance. They also offer good aging resistance, as they are not affected by ultraviolet rays, do not undergo oxidation, and do not absorb water.

There are two distinct families of fluoropolymers: perfluorinated (e.g., PTFE) and partially fluorinated (e.g., PVDF) polymers. The major types that are rotationally molded are polyvinylidene fluoride (PVDF), ethylene–tetrafluoroethylene (ETFE), ethylene–chlorotrifluoroethylene (ECTFE), and perfluoroalkoxy (PFA).

Fluoropolymers are used for demanding applications where purity, chemical resistance, and high-temperature performance are required. Typical products include high-purity chemical storage vessels and processing vessels. They are also commonly used to line tanks and in piping components for transportation and processing of aggressive chemicals: pump housings, vessels, columns, elbows, tees, and pipe sections with unusual shapes.

6.3.2.9 Foamed Materials

Foaming of molded parts is used for increased stiffness, insulation, floatation, impact dampening, and even sound absorption. In rotational molding, two primary techniques are used: (1) polyurethane foam filling, and (2) foaming the part material during molding using a chemical blowing agent.

Polyurethane foam is commonly used to fill the void inside a hollow part for strength or insulation. The foam density can be in the range 2 to 4 lb/ft^3 (0.03 to 0.06 g/cm^3) and is often used to provide buoyancy in flotation devices. The urethane foam is created using a two-component liquid system that is metered into the part using a special mixing head. The amount of foam added is typically measured in seconds. The liquid reacts inside the part and rises to fill the void. As the foam reaches the filling point, this is blocked to allow the pressure inside the part to increase and fill the part. The pressures reached can easily distort a part, so a fixture is often used to support it.

The downside of polyurethane is that there is almost no adhesion between regular polyethylene parts and the foam (some adhesion with cross-link polyethylene occurs). A number of techniques are used to develop a mechanical bond, but over time the interface between the foam and the part will break down. The initial reinforcement and insulation properties can be reduced significantly. Recycling of parts is not easy, and the process of filling is messy and potentially an environmental problem.

Using blowing agents to foam the material of the part directly offers improved strength and recyclability. It can also be performed during the molding process, thus removing the need for secondary operations and fixtures. The most common material to be foamed in rotational molding is polyethylene, although work has been done using polypropylene. The main disadvantage of foamed polyethylene is that the insulation properties are much lower than for polyurethane.

6.3.2.10 Reinforced Materials

Reinforcement or fillers can be added to rotationally molded parts either during molding as a dry blend or by compounding them into the raw material before molding. The process of mixing fillers with powder is quite simple, but the results are rarely good. Fibers (glass and carbon), glass spheres (solid, hollow, and gas filled), wood chips, metal powders, and other fillers have all been investigated. If the filler is similar in size and mass to the polymer particles, the filler can be dispersed through the wall relatively uniformly. However, if the intended filler is larger or heavier, a natural separation of the powder particles and the filler will occur and the filler will tend to congregate at the inner surface.

6.3.3 Material Preparation

The rotation process exerts negligible shear on the materials used for rotational molding. As a result, molding material must be free flowing enough to reach every surface detail and have low enough melt viscosity properties to form a smooth finish. To achieve this, the majority of materials are molded as finely divided powders or pellets, although an increasing number of liquid materials are also used.

Materials are most commonly ground to a 35-mesh (500-μm) standard that is defined as the size through which 95% of the material will pass. The average size of the powder particles is typically around 50 mesh (297 μm), although a full range of particle sizes from very fine dust to 600 μm may be present. Powders ground to finer and coarser mesh sizes are used for specialized applications and materials. Nylon materials have been molded directly as pellets, due to their ability to flow easily beyond their melting point. Surface definition can still be a problem, and it is common for blends of powder and pellets to be used to produce a smooth part.

More recently, micropellets in the range 0.020 to 0.060 in. (500 to 1500 μm) have been used successfully for a range of applications. They offer good moldability and eliminate the need for grinding. This has the potential for reducing cost and eliminating the shearing action of grinding from the heat history of the material. However, extruder throughput levels are lower than for regular-sized pellets and this tends to offset some of the potential cost savings. As they are used directly as a pellet, they eliminate the potential for cross-contamination during grinding and reduce the amount of dust produced during molding. Micropellets are very free flowing and can in some cases produce uneven wall thickness distribution, due to short residence time during rotation on large flat surfaces or internal corners. Again, some molders will combine 10 to 20% powder with the pellets to aid in producing a smooth surface. In situations where filling a mold with powder is difficult, the higher bulk density of micropellets can allow more material to be placed in tight spaces.

6.3.3.1 Grinding Process

The bulk of materials ground for rotational molding are polyethylenes (95%+ of the market). Polyethylene is a relatively tough material that can be difficult to divide. The most common method of pulverizing is performed using high-speed attrition mills that grind pellets approximately $\frac{3}{16}$ to $\frac{1}{4}$ in. (5 to 6 mm) in diameter to the particle size distribution required.

An attrition mill uses a stationary and a rotating disk with a series of radially serrated teeth machined into them within a mill housing. The disks are positioned opposite each other with a narrow gap that tapers down from the center of the plates to the outer edge. Figure 6.7 shows the inside of a large mill, and Figure 6.8 shows the layout of a two-stage grinding mill. Modern mills have trended toward horizontal operation for more uniform powder production and reduced wear on the cutting teeth. Production units use single-, double-, and triple-mill configurations according to the throughput required.

High speeds are required to cut the material efficiently at the interfaces between the two plates. A flywheel is attached to the rotating disk spindle and is balanced to eliminate vibration at high speeds and also to reduce deflection of the rotating disk during cutting. The stationary disk is attached to a water-cooling jacket used to control temperature at the plate during operation. The stationary disk can be moved in and out from the outside of the machine to allow the gap between the cutting faces of the disks to be adjusted.

Gap Size The separation between the rotating and stationary disks is very important to optimize particle size and distribution. These in turn affect the bulk density and dry flow rate of the powder. Wider gap settings will produce larger particle sizes; narrower settings can generate too much heat and affect overall system throughput. In two-mill systems, typical gap settings for the first mill can be between 0.010 and 0.020 in. (0.25 and 0.5 mm); the setting for the second mill is typically around 0.010 in. (0.25 mm) less than the first, but the disks should be no closer than 0.005 in.

FIGURE 6.7 Static and rotating disks inside a large attrition mill. (Courtesy of Pallman GmbH.)

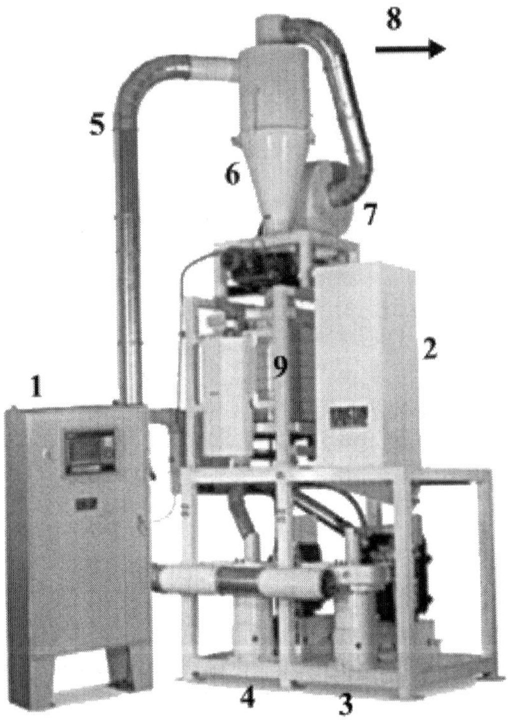

FIGURE 6.8 Two-stage attrition grinding mill. (Courtesy of Reduction Engineering, Inc.)

(0.125 mm). The settings are affected by initial pellet size, material toughness, overall throughput rate, and so on. Optimizing the settings for a single mill will difer from a dual-mill system, as the single mill acts as a secondary mill also,resizing larger particles from the first cut.

Cutting Teeth The number and condition of the cutting teeth on the wheels will affect grind quality. A 360-tooth disk is coarser and will produce larger particle sizes—a 480-tooth wheel is finer and a smaller particle is produced. The design of the cutting edges is also important, as the valley between the teeth affects the amount of material sheared during cutting—the larger the angle, the larger the particles produced. A sharp edge on the cutting blades is important for a clean shearing action. Dull cutters generate more heat as pellets roll through the blades producing more fine particles.

Grinding Temperature Monitoring temperatures during grinding is important for smooth operation and final powder quality. Overheating at the mill face can cause a meltdown situation that can cause the mill to stop rotating. Mills will typically have a thermocouple sensor in this area to prevent this from happening. Also, as particles are drawn away from the mills for sifting, the air temperature plays an important role in "polishing" the particles. A sufficiently high temperature is required to soften the particles and allow tails and hairs formed during cutting to retract into the main body of the particle. Operating temperatures are often in the range 194 to 212°F (90 to 100°C). Too low a temperature will affect the quality of powder, while maintaining a relatively high temperature will improve bulk density and dry flow. Dual- and triple-mill grinding requires a balance between the mills to ensure that too much work is not being done in one over another. Monitoring feed rates, oversize production, and mill amperage can help to balance overall throughput.

Sifter (Sieve) The sifting unit on the mill is typically a stacked sieve (one long continuous surface folded into a number of levels for maximum use of space) selected for the final size of powder required. The efficiency of the sieves is affected by the presence of poor-quality particles with tails, too many fines that can clog the screens and reduce efficiency, and the presence of static.

Cryogenic Grinding Cryogenic grinding uses liquid nitrogen to freeze material prior to feeding it into the mill and to maintain a low temperature throughout the system. It is used for soft or very tough materials that cannot be ground at normal temperatures. By freezing the pellets, they are shattered as they pass through the mill. Cryogenically ground materials tend to feel somewhat coarse by comparison with polished polyethylene powders.

6.3.3.2 Powder Quality and Assessment

Powder particle size, shape, and size distribution are important factors in determining the moldability of a material. Heat is transferred to the powder by conduction with other particles and the mold and by convection with the surrounding air.

Particle Shape Figure 6.9*a* shows a magnified view of an improperly ground polyethylene powder. Many of the particles have tails and "hairs" attached. These can lead to a number of problems, including reduced bulk density, poor flow characteristics, and unevenness during molding. Figure 6.9*b*, on the other hand, shows a sample of powder that has been polished to remove tails and hairs that can interlock to cause problems. Note that although the particles are not uniform in shape, they are rounded and therefore able to flow much more easily. Also, a range of particle sizes is present; this is essential in promoting even flow and good surface reproduction during molding.

Particle Size Distribution (ASTM D1921) Particle size distribution (PSD) is measured using a set of sieves stacked vertically, with mesh sizes typically ranging from 100 mesh (150 μm) to 30 mesh (600 μm). A sample of material (typically 100 g) is shaken, vibrated, or tapped through the sieves for a fixed time period (10 minutes) and the quantity retained on each sieve is measured. A typical size distribution is shown in Figure 6.10. A broad range of distributions can produce quality moldings; the main aspects requiring control are the level of fine particles (<100 mesh) and the level of coarse particles (>35 mesh). Typical quality guidelines look for a PSD with 95% <35 mesh (500 μm) and a maximum of 15% <100 mesh (150 μm).

Sieving is a dynamic process that may create a static charge, and an antistat (e.g., carbon graphite) may be necessary to prevent a coarse bias in the results (i.e., too

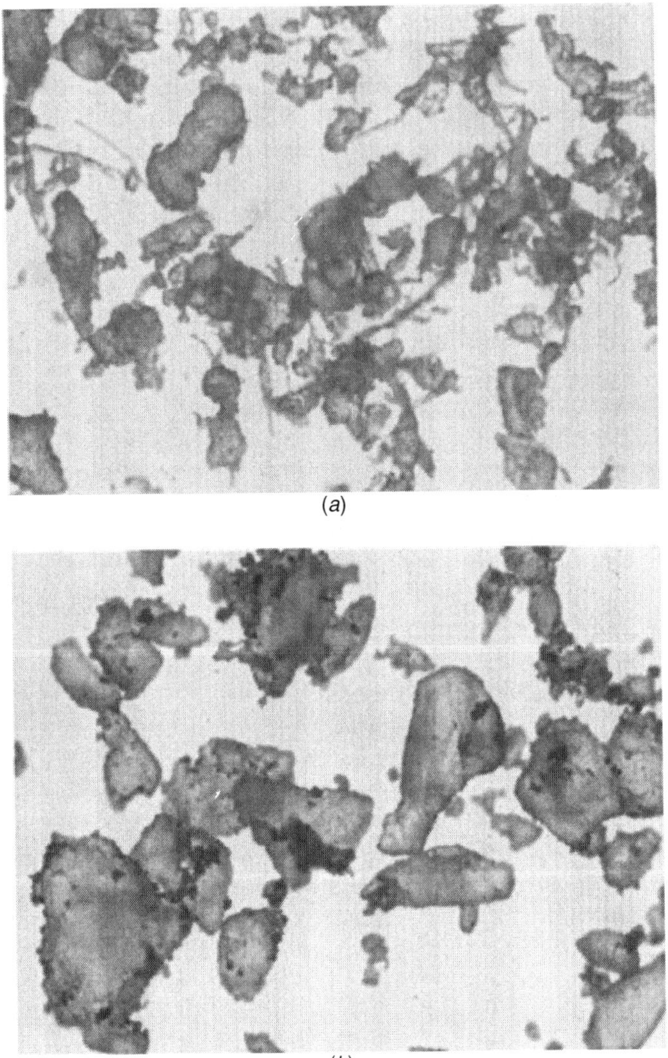

(*a*)

(*b*)

FIGURE 6.9 (*a*) Poor and (*b*) good powder particle shapes. (Courtesy of Wedco, Inc.)

many particles are unable to pass freely through the upper sieves). Polyethylene powders of 35 mesh (500 μm) do not typically need an antistat. Also, interference can be caused by too much material on a sieve screen that causes mass binding, resulting in a coarse bias, as material cannot pass through.

Dry Flow (ASTM D1895) The shape of the particles will affect the way in which the material will flow during molding. This flow, called the *dry flow rate*, is measured using a funnel of specified shape and dimensions. Powder that has been ground properly will flow through the funnel smoothly and steadily as shown in Figure 6.11. If the powder has not been ground properly and the particles are hairy or have tails attached, it will not flow well or even not at all in some cases. Flow rate is important since easy-flow powders will produce parts of more uniform wall thickness than that of powders that are tacky, sticky, or that tend to bridge.

Since powder can settle during shipment, to reduce errors the lot should be mixed prior to testing and samples for taken from several locations. A recommended

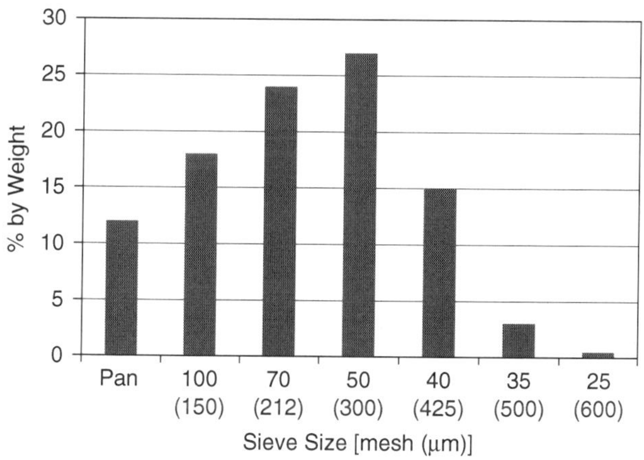

FIGURE 6.10 Typical powder particle size distribution.

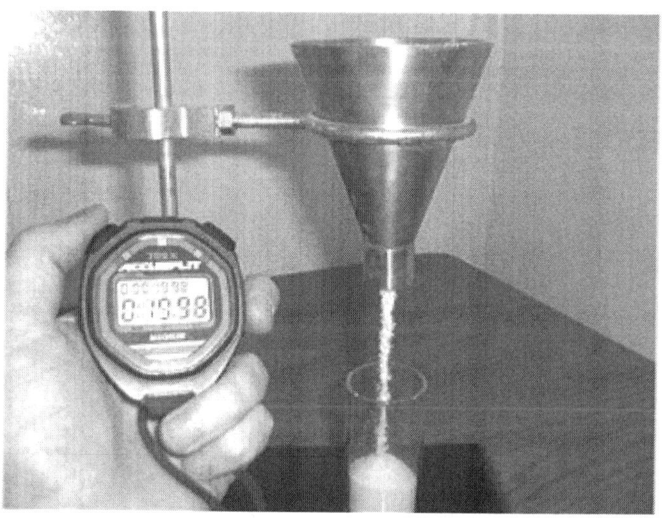

FIGURE 6.11 Free-flowing powder during a flow rate test.

powder flow rate should lie between 25 and 32 seconds for a 100-g sample. Higher flow rates can be tolerated for large simple part shapes that do not have a lot of fine surface detail.

Bulk Density (ASTM D1895) The bulk density of the powder is measured using a cylinder of known volume placed under the dry flow test funnel as shown in Figure 6.12. The powder is leveled off after filling the cylinder (taking care not to tamp it and cause settling) and weighed. The weight of the powder is then divided by the volume of the cylinder to calculate bulk density. Typical powders will have bulk densities in the range 0.320 to 0.400 g/cm^3. Poorly ground materials with many tails will have lower bulk densities, as the powder does not pack as well.

6.4 MOLDS

6.4.1 Mold Design

Molds for rotational molding are hollow, thin-walled, and lightweight with good heat transfer characteristics and must be sufficiently strong to withstand repeated handling. They are relatively low in cost compared with injection or blow molding tools. The choice of mold material and method of manufacture used will depend on the size, complexity, surface finish, and the number of molds required for the production run anticipated for the product.

For all their apparent simplicity, molds for rotational molding are often subjected to more rigorous demands than those used in other processes. Injection, blow, and thermoforming molds are used to shape molten or softened plastic. Rotational molding

FIGURE 6.12 Bulk-density cylinder of known volume (40 cm^3) filled with powder during a flow rate test.

tools are used to heat the raw material from ambient conditions, shape it, and then cool it back to close to ambient conditions. This thermal cycle presents a dichotomy for the mold maker in that the mold must be thin enough to allow heat to pass quickly into and out of the part and at the same time be sturdy enough to withstand the rigors of repeated opening and closing. A very thin mold will heat quickly but will distort during demolding; a very thick mold will maintain its shape well but may heat at an uneconomical rate.

The range of materials that have been used for rotational molding molds includes:

- Sheet carbon steel [sheet is defined as up to $\frac{3}{16}$ in. (4.76 mm) thick]
- Sheet stainless steel
- Plate aluminum [plate is defined as above $\frac{1}{4}$ in. (6.35 mm) thick]
- Cast and machined aluminum
- Cast and machined nickel
- Electroformed nickel and nickel–copper
- Cast and machined Be/Cu
- Sprayed metal
- Glass- or carbon fiber–reinforced epoxy
- Cast rubber

The main commercial processes used to produce molds are:

- Cast aluminum
- Fabricated carbon steel
- Fabricated stainless steel
- Fabricated aluminum
- Machined aluminum
- Electroforming

Other methods used for specialist parts and prototyping include:

- Oil-jacketed molds
- Glass- or carbon fiber–reinforced molds
- Sprayed metal

The most common types in use are fabricated and cast molds, as shown in Figures 6.13 and 6.14.

6.4.2 Choosing a Mold Type

Four major factors must be considered when choosing the type of mold to use for a particular part: size, complexity of form, quantity of molds required, and the appearance of the molded part.

FIGURE 6.13 Fabricated sheet metal mold. (Courtesy of Rivers Metal Molds, Inc.)

6.4.2.1 Size

Physical size may eliminate some methods of tool manufacture. Most large molds are fabricated from sheet metal, although large castings such as 18-ft (5.5-m) kayaks can be produced in single pours. Aluminum casting is often limited to the size of individual cast pieces by the capacity of the furnace used, but larger molds can be made by combining several castings. The cost of manufacturing a very large pattern for casting must also be taken into account.

Strength and durability of the mold will also affect the choice. Stainless steel molds do not rust and are tough. Cast and machined aluminum molds are inherently softer and require more care during production. Molds are often grouped together in what are known as *spiders*. Total mold weight can be a factor in multiple-mold spiders. Cast molds will often be considerably heavier than comparable fabricated molds. The weight capacity of the machine may be a limiting factor.

6.4.2.2 Complexity of the Form

The geometry of the part will also affect the decision. Features such as corner radii, inserts, holes, threads, and so on, may affect the choice of mold materials and how a part must be mounted or how a parting line is constructed.

Where is the parting line likely to be? How complex is it? Are there aesthetic requirements that will direct the path of the parting line? Can it be made in one plane or are there areas above or below the general plane that need to be at the parting line to allow the part to be withdrawn? Are there any special features that the parting line must pass through? Complex parting lines in small to moderate-sized parts will favor casting and possibly machining.

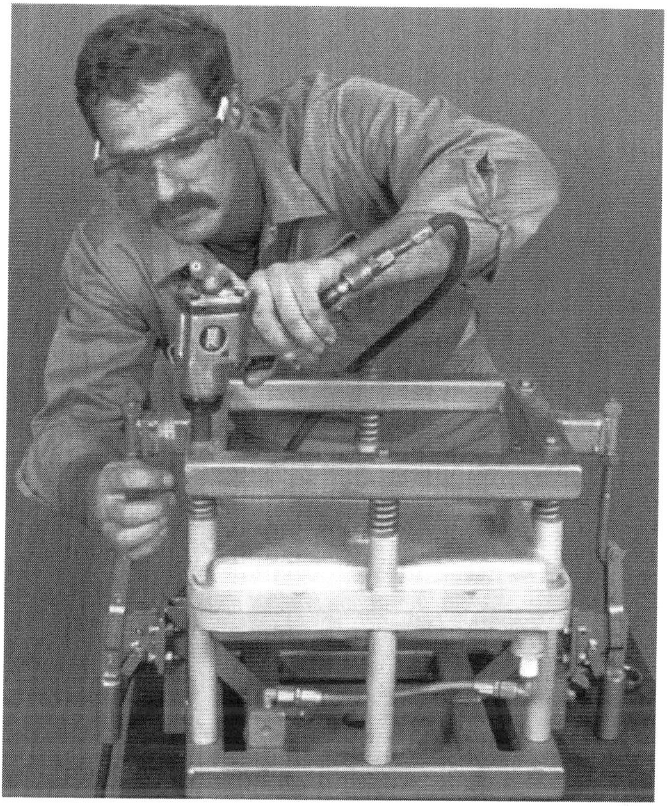

FIGURE 6.14 Fitting and adjustment of a frame for spring tension on a cast aluminium mold. (Courtesy of Wheeler Boyce, Inc.)

How many pieces are required to make the mold? The more complex the parting line and number of pieces, the more expensive and difficult to operate the mold will be. Maintenance during operation should be a major consideration in the design process.

Are there tight areas that might affect the flow of material? Do they extend more than 6 in. (150 mm) into the part? If so, they may be difficult to finish.

Is the mold a form that can be readily translated to a lay-flat image or developed by a machine tool? How many changes of cross section are involved? Spinnings and pressings can reduce the amount of handwork required for fabricated molds; the mold maker will probably use a standard gauge of material that will dictate the limits of what is typically subcontracted.

Mold inserts (cores, etc.) that are removed each time the mold is opened will affect mold construction by introducing wear points. Fixed inserts that must be interchanged less frequently are preferred. Molded-in inserts typically require bushings to prevent mold wear and must be readily accessible for servicing during operation.

Does the molded part have areas that must be thicker or thinner than others? Will cores or plates of more or less conductive material have to be incorporated?

Shielding of the mold to reduce heat or vectors (air amplifiers) to increase heat can be used. Detailed molds that require a high degree of precision will favor machining, electroforming, or casting over fabricated methods. Chemical etching can be used on steel molds to create fine surface details and patterns.

6.4.2.3 Quantity of Molds

When multiple molds are required, the level of repeatability that can be achieved must be taken into account when choosing a mold-making method. Sheet metal molds have improved considerably, but casting, electroforming, or machining offer better and more repeatable accuracy. Repeat cast molds are lower in cost, due to the fact that the same master pattern can be used.

High-volume production runs that can use either steel or aluminum molds will often be made in cast aluminum due to repeatability. Lower-volume production runs will often take advantage of the ability to use lower-cost sheet metal tools or even composite tools to produce small batches. As volume increases, the project can be converted to cast molds.

6.4.2.4 Appearance of the Product

The required surface finish for a part may also dictate which process is used to produce a mold. Some surface finish and details can be added during casting or fabrication; others will be added after the mold is finished (peening, etching). Aluminum or electroformed molds can produce a wider range of surface finish than can sheet metal molds; intricate castings and engraved surfaces can easily be incorporated onto the pattern. Sheet metal can be lightly textured after forming and can even use patterned plates and chemical etching to produce more complex finishes.

Parts requiring a highly polished finish will favor machined aluminum or nickel-plated molds. Cast aluminum and steel can be burnished to a polished finish but not a mirror finish. Designers should note that polished finishes require many hours of hand work in a mold and can add considerably to the cost of a mold. Also, finishes on cores should be considered carefully—highly polished surfaces can create a vacuum, whereas rough textures may provide enough grip to lock a part in the mold.

As a rule, very complex parts with fine surface detail (e.g., statuary) are typically not suitable for sheet metal molds; cast or electroformed molds are better choices.

6.4.3 Mold Elements and Features

The basic components of molds are similar regardless of the manufacturing process chosen. Figure 6.15 shows the construction of a typical cast aluminum mold. Each mold requires a mounting plate (1) to enable it to be attached to the machine, a frame to support the mold cavity (2), posts to connect to the mold (3) (often, spring loaded), a parting line flange (4), the cavity itself (5), clamping or bolting mechanisms (6), pry points (7), a vent (8), and lifting points (9). Inserts, removable cores, and other ancillary devices may be added as necessary.

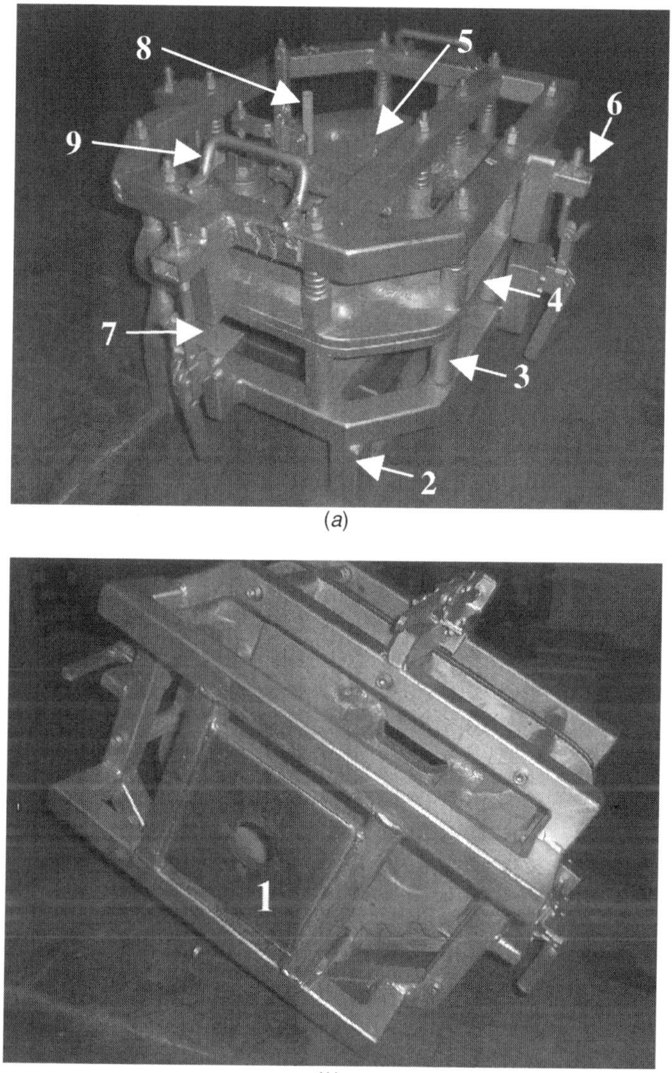

(a)

(b)

FIGURE 6.15 Typical mold elements.

6.4.3.1 Mounting Plate

Typically, a heavy steel plate is attached to the bottom of the mold frame to allo w the mold assembly to be fitted to the machine arm. The pattern of holes drilled in the plate will be dictated by the machine(s) on which the mold is intended to be used. In the situation where several different sizes or types of molds are options, adaptor plates may be required to provide alignment with the machine arm bolt pattern. Machinery manufacturers are discussing standardization of machine-head mounting patterns.

6.4.3.2 Framing

Framing should be designed for each mold application, as it has a number of important functions to fulfill. Typically made from steel box tubing, the frame will closely match the parting line contour to provide uniform support for clamping. The frame must be strong enough to support the weight of the mold, resist the torque applied for closure, allow for ease of handling, and withstand the rigors of repeated heating and cooling. The frame will also be used as the attachment point for mounting plates, clamping, pry points, and other components. As the frame will typically extend beyond the mold cavity, it must be considered when determining how molds will fit on the machine and in the oven.

Multiple molds can be mounted in a single frame in a spider arrangement. The molds are connected to the frame individually and are opened simultaneously when the frame halves are separated. Ideally, mold frames should be stress relieved by heating to 1825°F (1000°C). Heating the frame to 500 to 600°F (260 to 315°C) in a standard rotomolding oven will not allow the stresses in the frame to be fully relaxed. Figure 6.16 shows a tank mold supported at the flanges by a steel frame. If the mold is fabricated aluminum and the framework is carbon steel, bolting is generally used. Allowances are made for the difference in thermal expansion rates by using enlarged bolt holes and/or springs that allow the mold to float.

6.4.3.3 Support Posts

The cavity is suspended within the frame via the parting line, flanges, and support posts. These posts, or stand-offs, should be as short as possible to prevent distortion

FIGURE 6.16 Stainless steel mold supported in a steel frame at parting lines. (Courtesy of Rivers Metal Molds, Inc.)

and cracking due to thermal expansion and contraction—6 in. (150 mm) is typical. For aluminum cast molds, the support posts are commonly used as spring loading points to allow clamping pressure and forces created due to thermal expansion to be more uniformly distributed across the mold. One half of the mold is supported on springs and floats while the other is fixed. Some molders prefer no springs, ensuring that the parting lines are well matched and that the mold is sufficiently thick to allow clamping forces to be applied directly to the cavity. Fabricated steel and machined aluminum molds may be clamped directly at the parting line. There may be additional support posts, depending on the size of the mold and whether or not internal pressure will be used in the mold during production.

6.4.3.4 Parting Lines

The parting line is a critical part of the design of a mold. This is normally chosen about the largest dimension of the part to assist with removing the part from the mold and ideally should be flat in a single plane for ease of construction. Parting lines that are complex introduce additional cost to the mold and may also cause problems in ensuring consistent clamping force where a change of slope or angle occurs.

Several designs for the flanges along the parting line have been developed, such as tongue and groove, offset (stepped), and flat, to ensure good sealing and ease of location. Alignment pins along the parting line or a tongue-and-groove parting line can be useful to maintain parting line alignment, as shown in Figure 6.17. Spider

FIGURE 6.17 Flat bar parting line with alignment pins. (Courtesy of Rivers Metal Molds, Inc.)

arrangements will normally include some form of protection, such as one or more guide pins, to prevent inadvertent clashing of parting lines. These can help correct assembly of molds but can also create damage to the cavity surface if they swing out of alignment. In the case of multipiece molds, hinging of smaller sections of the mold can be helpful in reducing wear on parting lines.

The location of the parting line should be chosen to allow one piece of the mold cavities to hold the total charge of material if a fill port is not acceptable. This larger mold portion is best for mounting to the machine arm. Also, it is not a good practice to locate a parting flange along a corner, as porosity and excess flashing may occur.

6.4.3.5 Mold Cavity

The simplest cavity consists of two mold pieces with a simple flat parting line. However, multipiece cavities may be required for complex parts to allow a part to be removed. Six- or eight-piece molds are not uncommon but usually require more care and maintenance during use.

6.4.3.6 Clamping Mechanisms

There are a number of different types of clamping methods that can be used, including over-center clamps, quick-acting toggle bolts, spring-loaded bolts (at the parting line or through the spider), captive bolts and nuts, C-clamps, and even Vice Grips. The objective of the clamping system is to apply uniform pressure along the parting line—the spacing of clamps or bolts is critical in ensuring minimal flash.The clamping point where force is applied should be as far inbound on the flange as possible to ensure that force is applied as close as possible to the parting line.

6.4.3.7 Pry Points

Pry points are incorporated into a mold frame to provide leverage points where mold halves can be separated without damaging parting lines. Part design and the condition of the release agent in a mold will have a large influence on how readily mold halves will come apart during demolding. If material has built up on the parting line or a part is sticking slightly (even in just one area), molds can require an initial levering in one corner to separate them. Pry points allow substantial forces to be applied and distributed through the frame without damaging the mold.

6.4.3.8 Vent

As mentioned above, the thicknesses used in practice for molds are affected by their ability to withstand the repeated thermal cycling found in rotational molding. They are not normally designed to withstand pressure, so a vent pipe is usually incorporated to allow equilibrium between the atmosphere and the inside of the mold and to allow unwanted emissions to escape and cooling air to enter the mold. The most common vent tube materials are Teflon tubing, steel or stainless tubing, and silicone tubing. Ideally, a vent should be in the line of draw, reach to the center of the part volume, and be located in an area that does not affect the final function of the part.

6.4.3.9 Lifting Points

Ergonomic and safe handling of molds must be considered when attaching lifting points. When using a hoist, consider using four lifting points for balance.

6.4.3.10 Inserts and Cores: Mold Sections

Mold inserts and cores are removable or interchangeable areas of a part. These can be used to form threads, undercuts, handles, holes, and so on. For cores (e.g., core pins) that are removed every pass, the removable piece should be protected if possible by hinging or by mounting with guides on the frame so that it cannot be dropped or damaged during handling. For steel bushings the parting lines of cores are recommended to resist wear.

6.4.3.11 Fill Ports

Some molds have a shallow cavity half that prevents all the material from being placed in the mold at the beginning of the cycle. A fill port is essentially a removable area of the mold that is opened after the cavity is closed to allow material to be poured into the mold. This can also be used for double-shot processes when material is added during the molding cycle. Once again, steel bushings at the parting line are recommended as well as spring loading to ensure good seating.

6.4.3.12 Hinge Mechanisms

Opening and closing molds can create wear on mold components. Removing the mold halves and setting them down may create further damage at parting lines. One way to reduce this problem is to hinge mold pieces so that they remain on the machine at all times. The hinge mechanism must open only enough to allow the part to be removed (Figure 6.18). This can be a major timesaving feature for mold servicing.

FIGURE 6.18 Mold hinging mechanism. (Courtesy of Wheeler Boyce, Inc.)

6.4.3.13 *Roller Guides*

Another technique used to keep mold pieces together on the machine is the use of roller guides. Mold pieces are mounted on guides that allow the pieces to be separated for part removal. The guides and roller system need to be robust to withstand repeated heating and cooling cycles. Figure 6.19 shows roller supports used as guides on a machined aluminum mold.

6.4.3.14 *Drop Box*

A drop box (or dump box) is used to add material during a cycle. This may be to produce a multiple-layer part or to allow extra material to be added that could not fit into the mold at the start of the cycle. The drop box is insulated to prevent the second charge of material from melting before it is added to the mold. It is connected to the mold through a Teflon plugged opening [typically, 2 to 4 in. (50 to 100 mm) in diameter]. A pneumatic actuator connected through the arm of the machine is used to open the box.

6.4.3.15 *Airflow Amplifiers (Vortex Generators)*

Airflow amplifiers (vortex generators) work by using high-pressure compressed air through a specially designed nozzle to entrain large volumes of air from the surrounding environment. When in the oven, this can increase the flow of heat to areas of the mold, thereby increasing the wall thickness of the molded part. In the cooler they can help to increase cooling in hard-to-reach or thick areas of the part.

FIGURE 6.19 Roller guides used on a machined aluminium mold. (Courtesy of Wheeler Boyce, Inc.)

6.4.3.16 Thermal Pins

Thermal pins can be used to direct heat into difficult areas on a mold. These can be passive designs that utilize pipes or tubes (all-thread) with fins attached to them to increase heat flow. The fins attract heat and concentrate it down the pin to the attachment point. They can also be active designs that utilize a condensing vapor contained within a wicked copper tube to transfer heat into difficult areas on molds. The core pin has threaded sections that allow aluminum fins to be screwed onto the outer surface to adjust the amount of heat flow. The heat transfer rate can be up to 5000 times more than that of steel pipe.

6.4.3.17 Insulated or Shielded Panels

Heat can be excluded from areas of the mold by several methods:

- Metal grids to break up airflow to the mold surface
- Insulated blankets with glass or ceramic wool encased in a sheet metal box next to the mold surface to reduce heat flow
- Teflon sheet (usually, on the inside surface of the mold)

Care should be taken, as insulated mold surfaces will not expand as much during heating as surrounding surfaces—this can lead to distortion and cracking. Also, wool can trap water and cause variable insulating values.

6.4.4 Mold Release Systems

During molding, the plastic part is formed at the inner surface of the mold. The interface between the plastic and the mold is critical, as this will have an impact on part size, finish, and ease of removal from the mold. A mold release system is used to prevent plastic from sticking to the mold surface. There are a number of types available, falling into three main categories:

1. *Nonpermanent:* soaps, stearates, or silicones that are either dusted or wiped onto the mold almost every time a part is made. They can also be added directly to the plastic powder.
2. *Semipermanent:* principally siloxane and silazane polymer compounds that bond to the surface of the mold, providing release for a number of moldings before needing to be reapplied. These compounds are supplied in solvent- and water-based formulations—they are the most common form of release used.
3. *Permanent:* fluorocarbons that are bonded to the mold surface before molding takes place. The surface must be shot blasted, Teflon material sprayed on, and the mold baked to cure the Teflon in place. Teflon provides a uniform coating that can have a high-gloss finish. The term *permanent* is slightly misleading, as these coatings do wear out after a number of moldings.

The choice of which type of mold release to use is most often based on cost, ease of use, and the ability to control part size. Part design may also affect this decision, as simple rounded shapes are ideally suited to Teflon coatings. Teflon can last for many molding cycles before needing to be replaced (up to 30,000, although up to 5000 may be more common) but cannot be repaired in production. Semipermanent releases require a reasonable degree of skill to be maintained properly, but in some cases can offer more control for parts that require close tolerances to be maintained. Recently, high temperature paint coating systems have been developed that release in a fashion similar to Teflon. The paint can be repaired in production if scratched. It is available in satin and gloss finishes and has the advantage of being able to be buffed like car paint.

6.5 MACHINERY

Until the early 1980s, the main commercially available machines were carousel-style machines in the turret configuration, a few shuttles, and single-station styles. The choices were limited, thereby making machine selection relatively easy. Today, there are numerous configurations of carousels, shuttles, and single-station machines from which to choose. This makes machine selection more exacting and more complicated than before, which makes it increasingly important that the way in which the machine will be used is defined; each style has distinct advantages and disadvantages based on its use.

6.5.1 Basic Requirements of a Rotomolding Machine

The rotomolding process consists of three stages: (1) heating, (2) cooling, and (3) servicing. The basic rotational molding machine must be able to satisfy the three stages of the rotomolding process. A fundamental requirement of the machine is that it must rotate the mold continuously during the heating and cooling stages.

6.5.1.1 Heating Stage

The primary goals of the heating stage are:

- To raise the interior temperature of the molds to the sticking point (tacky temperature) of the molding material as quickly as possible.
- To provide uniform heating of the mold during the fusion period (sintering phase).
- To control the rate of heating and the point at which the heating cycle ends to prevent degradation of the molding material.

A number of methods of heating have been used, including:

1. *Forced-hot-air convection.* Mainstream rotomolding machines have an enclosed oven with temperature control provided by a closed-loop system that operates within 1% of the set temperature. The heat source is usually natural gas, although liquified petroleum gas, fuel oil, diesel, and electric are also in use. The burners are

specified oversize for the volume to be heated to ensure fast recovery so that cycle times are kept to a minimum. Ovens have circulation fans that produce high-volume high-velocity air that is recycled within the oven. These fans generate a turbulent airflow within the mold-residence area of the oven to ensure efficient heat transfer to the mold by creating a scrubbing action on the outside surface of the mold. The extremes of operating temperatures are typically in the range 350 to 900°F (176 to 482°C), although more practically, 500 to 650°F (260 to 343°C) is used (higher temperatures can affect final part properties and increase maintenance costs).

2. *Direct (open)-flame heating.* Direct (open)-flame heating uses naked gas flames directly on the outside surface of molds for heating. It is a relatively simple technique, although careful positioning and adjustment of flame quality is necessary to ensure even and controlled heating rates. It can be effective for simple shapes such as large storage tanks and is still relatively common.

3. *Molten salt spray heating.* Molten salt heating is a little used method that involves spraying a noncorrosive eutectic mixture of hot inorganic salts onto the mold to transfer heat. By using a liquid medium,the heat transfer coefficient is much improved over convective heat transfer, which allows the use of lower temperatures, typically from 446 to 554°F (230 to 290°C). The system has limitations due to difficulty in recovering, recirculating, and reheating the salt mixture and the need for more accurate mold parting lines to prevent salt from entering the mold.

4. *Hot-oil heating: direct conduction.* Hot-oil heating methods use jacketed (twin-walled) molds or molds with piping attached to or embedded in the mold surface; hot or cold oil is pumped through the cavity or piping to provide heating or cooling as required. Good control of the process can be achieved, as the temperature ramp is rapid. These are most popular when a long production run warrants the higher initial cost of the mold and are used principally for engineering-grade materials.

5. *Infrared Heating.* Infrared heating uses electrical or gas-fired radiation heaters to provide quick and efficient heating. This method is clean and can apply energy into deep pockets and areas that are traditionally difficult to mold, and it can be more effective than other methods at *not* heating specific areas of the mold. However, complex molds can cause shielding of areas of the mold by casting shadows, resulting in nonuniform wall thickness distribution.

6. *Electric heating.* New developments are taking place in the use of electrical heating elements embedded in carbon or glass fiber molds and also in the use of shaped metal heating elements (akin to those used in injection molding). Cooling channels can also be embedded into the molds. This is an efficient and simple process in that the machine concept is not complicated by having to move or be enclosed.

7. *Microwave heating.* Development work in Europe has produced a machine that uses microwaves to heat liquid materials inside a microwave-transparent mold. The system allows fast heating of the material and shorter cycles. Development work is continuing to expand the range of materials that can be used.

When using hot-air heat transfer, the air delivery system in the heating area (oven) should be engineered to promote uniform temperatures within the mold-residence area to ensure that the mold heats up evenly, to assure even distribution of the material inside the mold. The most efficient design will deliver the air at 90° to the mold-carrying arm assembly to promote uniform heating between spiders of tightly packed molds. Air is typically changed in the oven 25 to 30 times per minute to provide effective uniform heating. Figure 6.20 shows how airflow in one style of oven might circulate.

Oven exhaust fans are employed to evacuate noncombustibles, and other products of combustion, from the factory. Efficient oven designs have a modulated (adjusting) exhaust fan that accounts automatically for the burner setting and extracts only as much air as necessary to maintain a safe atmosphere within the factory. Nonmodulated exhausts must be sized for the maximum energy input of the burner; this means that they will extract more heat than necessary when the burner modulates (slows down) at the point when the oven has reached its set point.

6.5.1.2 Cooling Stage

The primary goals of the cooling stage are:

- To cool the material within the mold uniformly.
- To control the rate of cooling to produce good part properties and dimensional stability.
- To reduce the mold and part temperature to a safe handling temperature.

Cooling is normally carried out in a separate chamber, although single-station machines combine both heating and cooling in a single chamber; hot oil and electrical

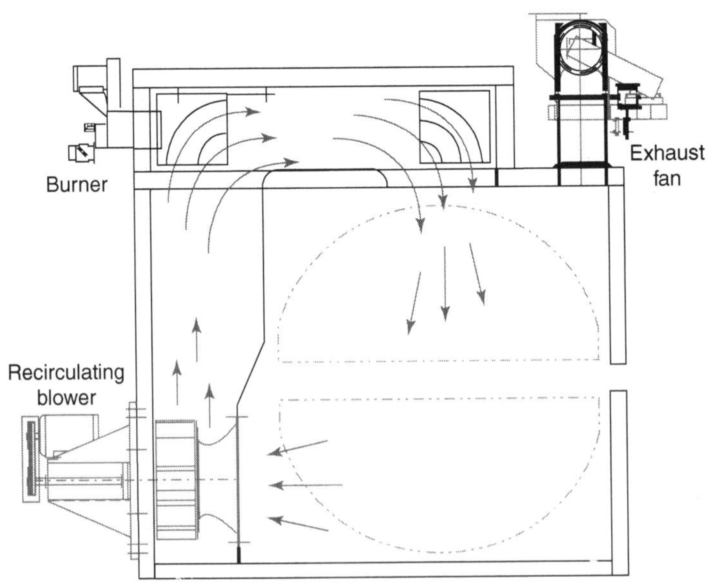

FIGURE 6.20 Typical air circulation pattern in an oven during molding. (Courtesy of Ferry Industries, Inc.)

heating systems typically use a single station also. Coolers are commonly equipped with water sprays, cooling fans, and exhaust fans to remove the heat from the area. Water spray is generally a very fine mist to envelop the molds to maintain uniform cooling throughout the mold-residence area in the cooler. Large water droplets tend to agglomerate at different areas of the molds, promoting uneven cooling and creating stresses in the molded parts that can cause warping or uneven shrinkage. Large water droplets can also enter the mold and cause water spots or streaks in the molded part.

The same principle as that used in the oven-circulating fan is employed in the cooler fans. High-velocity high-volume fans are utilized to create airflow and turbulence throughout the mold-residence area in the cooler to promote efficient removal of heat. Cooling fans should generate high-velocity airflow that will penetrate uniformly through tightly packed molds to assure even heat dissipation from the entire surface of individual molds.

The major problem with cooling rotomolded parts is the effect of changes in the ambient temperature of the air drawn into the mold-residence area of the cooler by the fans. Seasonal changes, as well as morning-to-afternoon ambient temperature differences, can negatively affect cooling rates. The ideal situation would be to manage the ambient temperature. However, standard refrigerant cooling is generally cost prohibitive and is not used in rotomolding. Evaporative coolers can efficiently condition the cooling-air temperature to some extent by providing cooler-air during the warmest periods to bring more consistency to the process, but cannot maintain total control of the temperature.

Internal cooling can be used to provide more uniform cooling between the inner and outer surfaces of the part. This can help to speed up cooling by removing heat more quickly and also help to control warpage by balancing the shrinkage rate between the two surfaces. This may be particularly useful in obtaining more uniform material properties in thick-walled articles. The typical method of accomplishing this is to supply a stream of cold air through the arm of the machine. This airstream is forced or drawn into the mold to remove heat from the internal surface of the part.

6.5.1.3 Servicing Stage

The primary goals during this stage of the process are to remove the finished parts and to efficiently charge the molds with raw material. The servicing stage of the molding cycle is often the most time-consuming stage of the process and can theresfore affect overall production output. Ideally, the mold-carrying arms are prepositioned before exiting the cooler so that the machine operator need not waste time getting the mold into position to begin demolding. Multiple positions can be preprogrammed that will automatically reposition the molds during demolding for the operator's convenience. Figure 6.21 shows an arm that left the cooling station in the horizontal position ready for the operator to begin demolding.

Platforms can be provided to raise the operator to an acceptable working height. These platforms traverse in and out to locate the operator conveniently at a satisfactory height for efficient servicing of the molds. Figure 6.22 shows an elevated platform used to allow an operator to work at the height of the molds. Working at the proper height can be an important factor in avoiding back injuries related to lifting or reaching.

FIGURE 6.21 Servicing (demolding) station at ground level. (Courtesy of Ferry Industries, Inc.)

FIGURE 6.22 Platform used to raise an operator to the level of a mold. (Courtesy of Ferry Industries, Inc.)

Semiautomatic material handling equipment can relieve the operator's responsibilities for weighing and delivering raw material to the molds. Material dispensers can automatically identify the mold and accurately deliver the correct color and weight of the charge of raw material to prevent scrap or waste due to underweight or overweight parts. The system can improve housekeeping and improve cleanliness around the machine.

6.5.1.4 Rotation

During the heating and cooling stages, the molds must be rotated to ensure that even material distribution takes place inside. Rotating the molds is required to disperse the raw material to all areas on the inside surface of the mold. The raw material begins as a pool of liquid or powder that remains at the bottom of the mold under the force of gravity [rotation speeds are rarely high enough to generate any centrifugal (centripetal) forces that push the materials outward]. Constant rotation on two axes causes the lowest point of the mold to be in continuous transition, thereby allowing all surfaces of the inside of the mold to pass through this pool. As the mold heats up, the raw material melts and adheres to the inner mold surface. The ever-decreasing pool of molding material is ultimately fully absorbed into a homogeneous layer on the inside surface of the mold.

The speed of rotation must be fast enough to allow all inner surfaces of the mold to pass through the pool during the fusion phase of the process, yet slow enough to allow a free flow of material with no centrifugal forces. The required ratio of rotation between the horizontal axis and the vertical axis is a function of the part shape and size. The ratio must be such that all inner surfaces of the mold pass through the pool of raw material to create the desired wall thickness characteristics in combination with heat transferred through the mold wall. Rotation can be accomplished by two principal means: Biaxial rotation and rock-and-roll rotation.

Biaxial Rotation Biaxial rotation is created by turning the mold 360° simultaneously in the horizontal (major) and vertical (minor) axes. The mold is mounted on an arm assembly that is geared to allow rotation on both axes. Some rotomolding machines are designed to operate with a fixed 4 : 1 ratio and only the major axis shaft is driven. To accomplish this, the gear head must have a 4 : 1 ratio. Turning the major axis shaft against a stationary minor axis shaft creates a fixed 4 : 1 ratio. Generally, the major axis shaft speed is variable, but a ratio of other than 4 : 1 is not possible on a fixed-ratio machine, limiting its molding capability.

Two styles of arms are commonly used for creating biaxial motion:

1. *Straight arm.* This configuration allows two molds or spiders (a framed group) of molds to be mounted directly opposite each other on either side of the central concentric shafts. This can allow a large number of small parts to be mounted simultaneously within the two rotating volumes shown in Figure 6.23. The volume available for mounting a mold or spider is defined by the plate diameter and the spherical swing diameter. Note that the maximum diameter of the mold to be mounted decreases as

the height of the mold above the mounting plate increases. The maximum size or swing designated for a machine is rarely usable, as the spherical dome reduces the volume available.

2. *Offset (dropped or cranked) arm.* The arm is offset (dropped) to allow a larger mold to be mounted on the machine. The main arm (horizontal axis) is rotated and power to the minor rotation (vertical axis) is transferred via a set of gears and chain drive down through the arm. Figure 6.24 shows the available volume for mounting a mold(s) on an offset arm; Figure 6.25 shows a typical offset arm arrangement. An offset arm can be programmed to operate in a manner that simulates a rock-and-roll machine. The dropped or offset arrangement allows a much larger mold or spider to be mounted on the arm. The volume available is defined by the plate diameter, the height to the center of rotation from the plate, and the spherical swing radius.

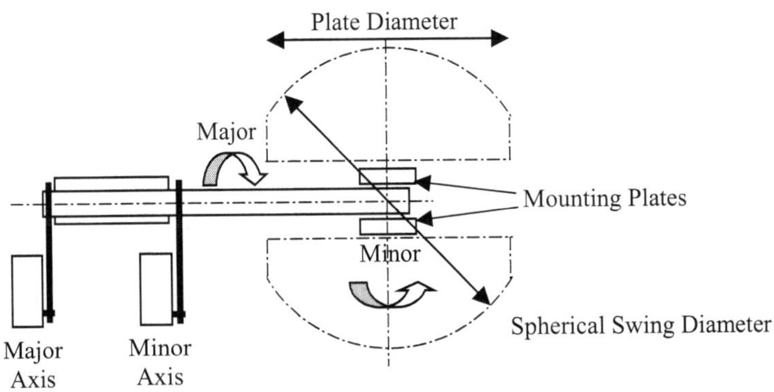

FIGURE 6.23 Straight arm configuration showing spatial volume available for two mold spiders.

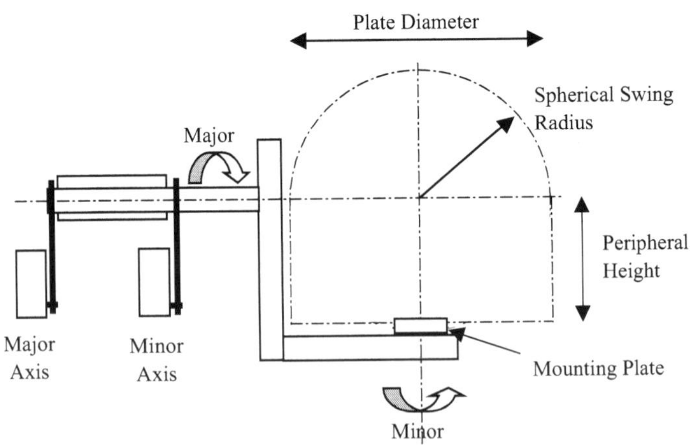

FIGURE 6.24 Offset (dropped) arm arrangement showing the spatial volume available for large molds.

FIGURE 6.25 Typical offset arm arrangement. (Courtesy of Ferry Industries, Inc.)

Rock-and-Roll Rotation Rock-and-roll rotation is accomplished by continuously rotating the minor axis 360° while tilting the major axis back and forth at least 45° in two directions. The major or horizontal axis is the "rock"; the minor or vertical axis is the "roll." The speed of the roll is variable to help the flow of the material, and the speed and frequency of the rock is variable to assure the desired wall thickness distribution.

6.5.2 Machine Styles

A number of different styles of machine are commonly used in rotational molding:

- Open-flame rock and roll
- Rocking oven
- Shuttle
- Clamshell
- Vertical oven
- Carousel
- Oil-jacketed molds

Each style may be suited to different sizes of parts and different volumes of production. Simple open-flame machines offer the lowest-cost entry to the process; single-station shuttle machines and clamshells offer efficient use of limited space. Rock-and-roll machines dominate water tank production, and large shuttle machines are common for the more extreme sizes of tanks produced. As production demands increase, multiple-arm shuttles become more useful and offer flexibility in terms of multiple molds in process.

However, with three functional stages to the rotomolding process, the most efficient machines for the majority of small to moderate-sized applications will consist of three stations to allow all three phases of the process to be performed simultaneously: (1) the oven station for the heating stage, (2) the cooler station for the cooler stage, and (3) the servicing station for the servicing stage. Generally, the three stations are arranged in a circular pattern with three mold-carrying arms indexing from station to station around a common center point. This style of machine is called a carousel style because it operates similarly to carousels at amusement parks. Figure 6.26 shows the center arrangement of three arms for a small machine, and Figure 6.27 shows how this fits within the final machine assembly.

The three-arm carousel-style machine has one mold-carrying arm for each of the oven, cooling, and servicing stations. Contemporary machines provide for independent programming of the process variables for each of the arms on the machine. Molds that are mounted together on an individual arm must have compatible cycle times, temperatures, cooling rates, rotation speeds, and so on. However, this compatibility need not extend from arm to arm. Products with different oven processes, cooling processes, and servicing requirements may be run on a common machine by varying

FIGURE 6.26 Center assembly for a three-arm turret machine. (Courtesy of Ferry Industries, Inc.)

FIGURE 6.27 Three-arm turret machine showing the oven (1), cooler (2), and servicing (3) stations. (Courtesy of Ferry Industries, Inc.)

oven temperature, cooling rates, rotation speeds, and so on, for the individual arms; however, differences in cycle times will affect production throughput.

Most rotomolding materials have a processing window that will allow acceptable properties of a finished part without optimizing the process exactly. Therefore, products with differing wall thickness, shapes, and mold types can still be run on the same arm if the compromise-for-compatibility does not cause the final product to go beyond the processing window and greatly reduce the desirable properties.

If the process dictates that each stage of the process takes an equal time, the mold-carrying arms can all index to the next station at the same time; for example, all arms require 10 minutes of oven residence, 10 minutes of cooling, and 10 minutes to service. This is an ideal situation for a turret configuration of a carousel-style machine. A turret configuration requires that all mold-carrying arms be attached to a common base. When the base indexes, all arms are advanced to the next station simultaneously. In this example, every 10 minutes all arms can advance to the next station to begin the next stage of the process. The net result is that every 10 minutes an arm with molded parts is presented to the machine operator.

If one stage of the process takes longer, however, the frequency of indexing is decreased. For example, if the oven phase can be completed in 10 minutes and the cooling phase can be completed in 10 minutes but the servicing phase takes 20 minutes, the index frequency must decrease to once every 20 minutes. One way of dealing with this is accomplished by reducing the oven temperature to allow a longer oven residence time without overcuring the parts and decreasing the cooling rate to allow a longer cooling time. The net result is a reduction in the index frequency by half, reducing production by 50%, which is typically unacceptable.

Another way to address this problem is by using a turret configuration with four mold-carrying arms and four stations (Figure 6.28). The additional station would be

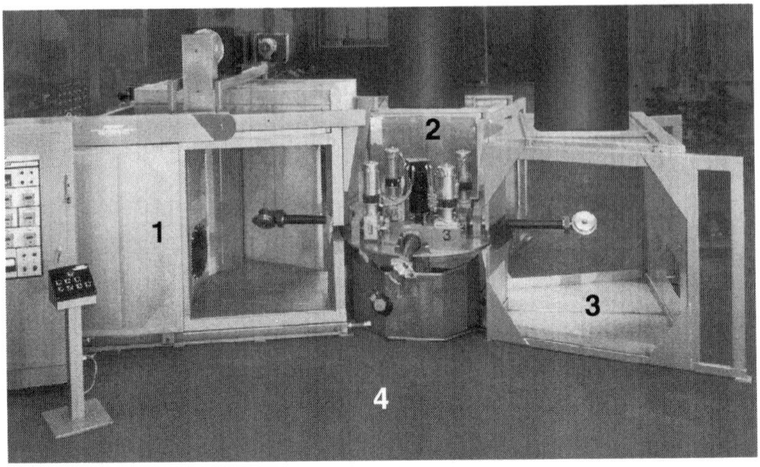

FIGURE 6.28 Four-arm turret machine showing the oven (1), cooler A (2), cooler B (3), and servicing (4) stations. (Courtesy of Ferry Industries, Inc.)

dedicated to servicing the molds. Now, with one oven, one cooler, and two servicing stations, the frequency of indexing in the example above can return to 10 minutes, resulting in 10 minutes in the oven, 10 minutes in the cooler, and 10 minutes in each of two servicing stations, resulting in 20 minutes for servicing the molds. The frequency of indexing and production remains at the higher levels. Additionally, a fourth arm is available on the machine to mount even more molds. The addition of the fourth arm and station can apply to any of the three stages of the process. If the heating stage takes longer, a second oven can be added; if the cooling stage takes longer, a second cooler can be added.

Another configuration of the carousel-style machine is the independent-arm configuration. The independent-arm configuration allows more flexibility in the molding process by allowing each arm to index to the next station while not disturbing the position or process stage of the other arms. This configuration typically has five stations, consisting of one oven, one intermediate station (or wait station), one cooler, and two servicing areas, and can be equipped with three or four arms. Figure 6.29*a* shows a three-arm five-station layout; Figure 6.29*b* shows the same configuration with four arms.

Arms on an independent machine are not, of course, truly independent; for an arm to advance, the station ahead of the current station must be free. On a five-station machine with three arms, there will always be two levels of redundancy that can allow two arms to advance at any time. In situations where the cycle settings for each arm are similar, this is not a problem; however, as the differences between cycle settings on arms increase, the availability of redundant stations becomes more important. A five-station four-arm machine has only one level of redundancy and can therefore be very sensitive to delays or variations in cycles; if any arm is delayed for longer than the cycle step of the next arm, arms can be held in their station. Ideally, for a four-arm configuration there should be six stations.

(a)
(b)

FIGURE 6.29 Independent (*a*) three-arm and (*b*) four-arm machine layouts. (Courtesy of Ferry Industries, Inc.)

6.6 MOLD DESIGN

In the past, many rotationally molded products were developed with very little or no documentation. Product ideas were sketched on paper and then, in the case of cast molds, a wooden pattern was made. Much of the design review process was done with a wooden pattern. Once the pattern was finalized, it was used to make the first mold. After sample parts were made from the first mold, another design review was performed. This review included any processing problems. It was common practice to consider the first mold as a prototype mold, since changes were expected. These changes usually resulted in revisions to the wood pattern, and another mold was made to verify that all problems were eliminated. This process was adequate because the tooling was inexpensive and few projects required rapid execution.

As projects have become more complicated and costly, the development process requires more up-front thinking and more documentation. Most companies have realized that the up-front cost of a complete design with full documentation can be much less than the cost of changes. Also, time-to-market has become more important, and time delays for mold changes are not acceptable. The use of computer-aided design, solid modeling, and structural analysis tools has vastly improved the process of design development.

6.6.1 Basic Design Guidelines

In the following sections we deal mainly with the design and moldability of polyethylene parts. Some modification may be required for other materials. As with any plastic process, one must understand the basic part requirements to have a successful design. There have been relatively few publications on the design of rotationally molded parts despite the fact that rotational molding offers considerable design freedom with generally fewer restrictions than other processes. It can produce both open and closed shapes ranging in size from 1 to 2 in. (25 to 50 mm) in diameter to tanks 20 ft (6 m) long by 7 ft (2.2 m) in diameter. Parts can range from simple shapes such as balls and containers to complex fuel tanks and machine housings. These can vary in thickness from $\frac{1}{16}$ in. (1.6 mm) to 2 in. (50 mm), and within the limits of powder volume, the part wall thickness can be varied using the same mold. Single- or double-walled parts for insulating properties can be produced. Parts can be colored or transparent, flexible or stiff.

The decision to use rotational molding over other processes is usually dictated by the following elements:

1. *Material.* Rotational molding involves a limited number of common materials that are relatively straightforward to mold; polyethylene, nylon, polycarbonate, and PVC are the most common. Specifying a material will usually depend on strength requirements, how rigid or flexible it needs to be (surface properties), operating temperature, color, and ultraviolet requirements. Other materials may require special handling or processing controls.

2. *Number of parts.* Rotational molding is ideally suited for small runs but can be justified for runs of up to 50,000 to 100,000 parts per year. Beyond this (depending on part size and form) a more automated process may be a better alternative. However, if rotational molding is the only process suitable, high volumes can be achieved by using multiple molds and machines (e.g., the toy industry). Throughput rates for rotational molding must be considered, as they are lower than for blow molding or injection molding. This means that multiple tools might be required, narrowing the gap for justification.

3. *Complexity of shape.* The complexity of shapes that can be produced by rotational molding is almost unlimited. It offers the potential to convert complex metal parts consisting of multiple sections into single parts. Fuel tanks for vehicles are an

excellent example of how traditional parts that require many steps to form can be converted to a single molding operation, resulting not only in cost savings but also in a better design with fewer operations to monitor and fewer connections that can potentially fail.

4. *Size.* Large parts are typically the domain of rotational molding, although the capabilities of injection molding, blow molding, and thermoforming are constantly increasing in size and thickness. Large tanks between 500 and 20,000 gallons (1890 to 75,000 liters) are eminently suited to rotational molding, but other processes are increasing their capability in terms of volume. Parts the size of trashcans are now commonly produced by injection and blow molding where volumes justify the investment in tooling and equipment. Parts required in low quantities are often more easily justified for rotational molding due to the lower cost of tooling versus other processes.

6.6.2 Wall Thickness

Rotational molding has the advantage that the wall thickness can be varied after the mold is made, but it is best to have a nominal wall thickness in mind when the part is designed. Parts are formed against the outer surface of the part (inner surface of the mold) without cores that allows wall thickness to be varied easily. Similar changes in other processes could mean substantial mold modifications. However, it should be remembered that thicker parts tend to shrink more than thinner parts and may therefore cause fit problems. Also, thicker parts will require longer oven cycles and at higher thicknesses may be limited, due to thermal degradation at the outer wall.

The distribution of wall thickness is affected by heat transfer rates, the ratio and rates of rotation, mold wall thickness and material, powder characteristics, and release agent. The major factors are heat transfer and ratio of rotation. Maintaining a uniform wall thickness is very important, as it aids in uniform cooling rates and reduces the possibility of warpage. Despite the fact that the internal surface of a part is free-formed in rotational molding, the distribution of material within a molded part will generally be quite uniform. Wall thickness variation is normally quoted at ± 20% for most commercial applications. However, careful adjustment of the mold and molding parameters can achieve ±10%, with a corresponding increase in costs. When specifying wall thickness on a rotationally molded part, it is best to set a nominal wall thickness along with a minimum thickness. Also, the part weight should be specified with a tolerance since it controls the wall thickness. Figure 6.30 shows how external corners (E) attract more material and are thicker than internal corners (I). Away from corners, wall thicknesses will generally be more uniform, except at the center of large flat surfaces where parts tend to be thinner.

Variation of wall thickness has a direct impact on the way in which rotationally molded parts are dimensioned. Since the wall section may vary and in many cases cannot be predicted accurately before a part is made, the part must be dimensioned to an external feature. If an internal dimension is required, allowances for the upper tolerance of wall thickness variation must be made. Designers used to specifying injection-molded parts with controlled surfaces both inside and outside the part can find this difficult to deal with. However, extrusion blow molding and twin sheet

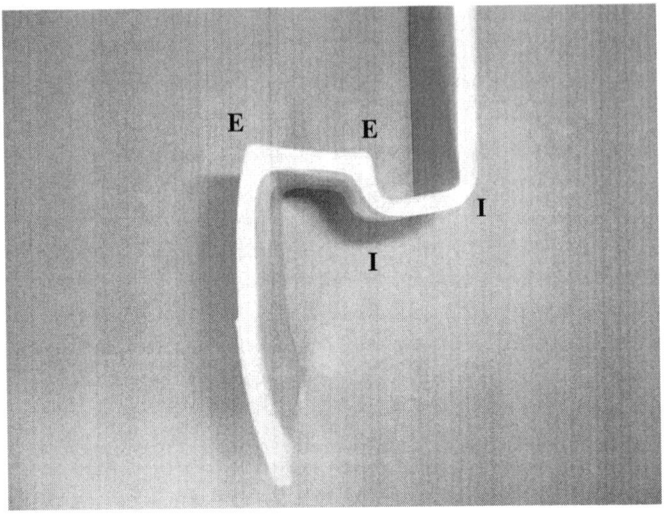

FIGURE 6.30 Cross section of a molded part comparing b uildup of material at internal (I) and extranal (E) corners.

thermoforming have the same limitations. Rotational molding actually maintains more uniform wall thicknesses than either one of the other competiti ve processes.

Typical molded parts ha ve wall thicknesses in the range 0.0625 in. (1.6 mm) to 0.5 in. (12.5 mm). Parts as thin as 0.020 in. (0.5 mm) have been produced for aircraft applications but require special controls; thicknesses up to 2 in. (50 mm) can be found in large tank applications. Liquid systems such as PVC can be produced in v ery thin layers by controlling the rotation of the mold; parts as thin as 0.010 in. (0.25 mm) can be produced. The typical range of thicknesses is 0.0625in. (1.6 mm) to 3/8 in. (10 mm).

Controlling the flow of heat to a particular area can deliberately induce variations in the wall thickness of a part. This can be achieved by utilizing shielding material (wire mesh) or insulation (ceramic or glass w ool) to slow down heat transfer, or by increasing heat transfer by preheating of the mold or by the use of airflo w amplifiers. Painting the outer surface of the mold either black or silver will also help to attract or deflect heat.

6.6.3 Draft Angles

It is possible to produce parts with zero draft in rotational molding, as the material shrinks away from the mold wall during cooling. However, in most cases it is best to have draft on a part to help in remo val from the mold. F or a large hollow part, draft may not be needed on the outside surf ace, but if 1° of draft will not af fect the function of the part, draft should be used. The actual angle will depend on the material being used. Materials with high shrink le vels (crystalline) such as polyethylene and nylon will mo ve away from the ca vity much more than lo w-shrink-level materials (amorphous) such as polycarbonate. Also, materials such as nylon and polycarbonate are much more rigid than polyethylene and therefore require more draft to allow easy removal. Table 6.1 lists recommended draft angles for a range of materials. For recesses or holes, 3° of draft is a minimum and 5° is better.

Figure 6.31 shows a detail that is shrinking onto a mold core instead of away from it; greater draft angles are needed to allow the part to be removed from the mold. The forces generated by plastic as it shrinks onto cores can be quite significant. In general, the use of larger draft angles is recommended. Adding surface texture and other features to a mold may influence the demolding process and require a larger draft angle. A rule of thumb for texture is that one additional degree of draft is required for each 0.001 in. (0.025 mm) of textured depth.

6.6.4 Corner Radii and Angles

6.6.4.1 Corner Radii

Corner radii are an extremely important consideration for rotomolded parts. The perfect rotationally moldable part is a ball. Smooth corners allow powder to flow and form uniform wall thickness distributions; sharp corners can promote stress concentrations and may be difficult to fill during molding. Sharp outside corners will be thicker and potentially incompletely filled; sharp inside corners will be thinner, as shown. Corner radii should be as generous as design limits will allow (Figure 6.32).

TABLE 6.1 Recommended Draft Angles (deg) per Side for Various Materials

	Outside Surfaces		Inside Surfaces	
Material	Minimum	Better	Minimum	Better
Polyethylene	0	1	1	2
Polyvinyl chloride	0	1.5	1	3
Nylon	1	1.5	1.5	3
Polycarbonate	1.5	2	2	4

Source: Ref. 1.

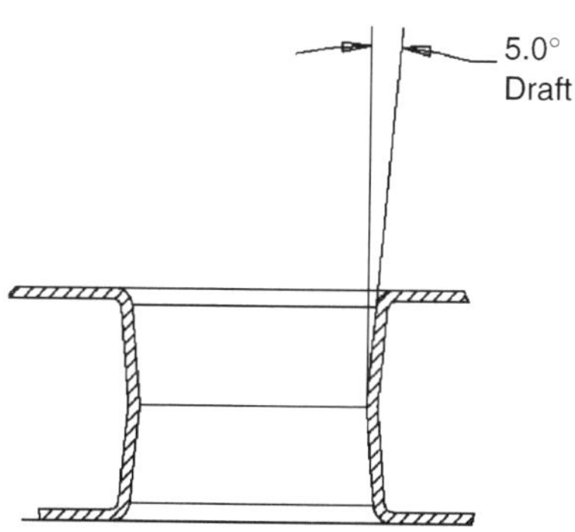

FIGURE 6.31 Core cross section showing increased draft.

Generous corner radii help to improve the flow of material inside the mold and produce more uniform wall thicknesses. They also help to distribute stresses over a broader area, thereby increasing part strength. Rotationally molded parts benefit from thicker external corners, points at which stresses and wear are usually highest. Internal corners tend to be thinner however, as powder does not remain in contact with these surfaces as much as with external corners; internal radii should therefore be more generous than external radii.

Target radii for polyethylene are 0.25 in. (6.5 mm) for external corners and 0.5 in. (12.5 mm) for internal radii. Stiffer materials such as nylon and polycarbonate require larger radii to allow corners to form properly; radii up to 0.5 in. (12.5 mm) and 0.75 in. (19.0 mm) are recommended. Table 6.2 lists recommended radii for various molding materials. Two situations require larger radii. For wall thicknesses greater than 0.1875 in. (4.75 mm), the minimum outside radius should be at least equal to the wall

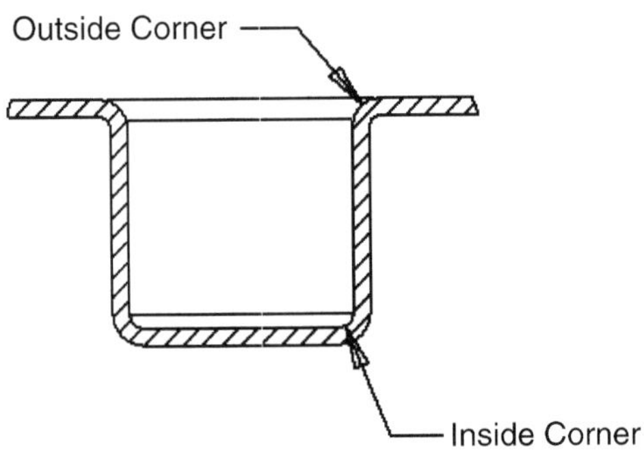

FIGURE 6.32 Outside and inside corner radii.

TABLE 6.2 Recommended Radii for Various Materials

Material	Standard[a]	Inside Radius [in. (mm)]	Outside Radius [in. (mm)]
Polyethylene	Ideal	0.500 (12.7)	0.250 (6.35)
	Commercial	0.250 (6.35)	0.125 (3.18)
	Minimum	0.125 (3.18)	0.060 (1.52)
Polyvinyl chloride	Ideal	0.375 (9.52)	0.250 (6.35)
	Commercial	0.250 (6.35)	0.125 (3.18)
	Minimum	0.125 (3.18)	0.080 (2.03)
Nylon	Ideal	0.750 (19.05)	0.500 (12.7)
	Commercial	0.375 (9.52)	0.375 (9.52)
	Minimum	0.187 (4.75)	0.187 (4.75)
Polycarbonate	Ideal	0.500 (12.7)	0.750 (19.05)
	Commercial	0.375 (9.52)	0.375 (9.52)
	Minimum	0.125 (3.18)	0.250 (6.35)

Source: Ref. 1.

[a]Ideal tolerance, minimum care required; commercial tolerance, possible with reasonable care.

thickness and the minimum inside radius should be increased equally. For corner angles of less than 45°, the radius should be increased to as much as 0.5 in. (12.5 mm).

6.6.4.2 Corner Angles

A target angle is 90° or greater for most plastics. For polyethylene, corners as sharp as 30° are possible (depending on wall thickness), but a 45° angle is better. Materials such as nylon and polycarbonate require more open corners to prevent bridging; a minimum of 45° is recommended. A limiting factor on corner angles is that there must be an adequate radius at the point where the two walls meet. The process of molding a corner is related to the same rules as for closely spaced parallel walls—bridging occurs when the material fills the profile before material at the point of the corner has had a chance to flow out.

6.6.5 Stiffening of Parts

6.6.5.1 Ribs

Ribs are routinely added to rotationally molded parts to stiffen them. This is because many parts are large, yet thin-walled. They may be incorporated into the design of a part to enhance the stiffness of flat sections or areas that will be expected to support extra loads (e.g., vertical walls in a part that stacks). Most ribs must be designed as a double-walled feature, as shown in Figure 6.33. This is because the material must flow into the mold cavity that forms the rib. Ribs should be designed so that the width of the rib is equal to or greater than the height, and this should be a minimum of four times the wall thickness, *t*. This guideline is derived from the bulk density of the powder versus solid plastic (approximately 3 : 1) and the ability of the powder to flow into and out of the rib. Keep in mind that the rib strength depends mainly on its height. Figure 6.34 shows the use of ribs around a tank to add strength during operation.

Single-walled ribs can be used, but the rib height must be limited to two or three times the wall thickness, *t*. Single-walled ribs can present a molding problem in that the material may bridge over the mold cavity and not form a solid rib. Also, increased material thickness in this area can produce higher shrinkage, increased warpage, and longer cooling cycles. Single-walled ribs have several uses unrelated

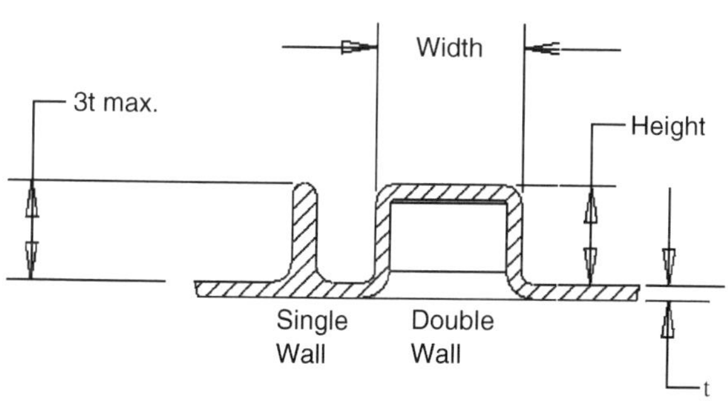

FIGURE 6.33 Stiffening rib designs.

FIGURE 6.34 Rib pattern on a tank. (Courtesy of Gregstrom Corporation.)

to stiffening. They can add a wear surface to the bottom of containers such as garbage cans. This can extend the life of the product. Another use is to tighten the fit between two parts. It can be difficult to control the tolerance of rotomolded parts so that a snug fit is achieved between mating parts. However, the parts can be designed with clearance, and after sample parts have been evaluated, "crush" ribs can be added at key locations to tighten the fit.

Standard engineering design formulas for the cross-sectional stiffness of ribs can be used to compare the relative merits of using one cross section over another. In general, a number of shorter smaller ribs can often be better than a single large rib. An experimental study used a series of corrugated plates to mold polyethylene sections. These were tested for bending resistance and tensile properties in both perpendicular and parallel directions. The tests concluded that a corrugation with equally spaced peaks and troughs, with the trough depth being 30% of the pitch of the corrugations, gave the best overall performance in terms of property per unit weight.

6.6.5.2 Recesses

Like ribs, recesses are often added for stiffness as shown in Figure 6.35. Recesses are also used as cosmetic features and to break up laze flat surfaces. One of the most

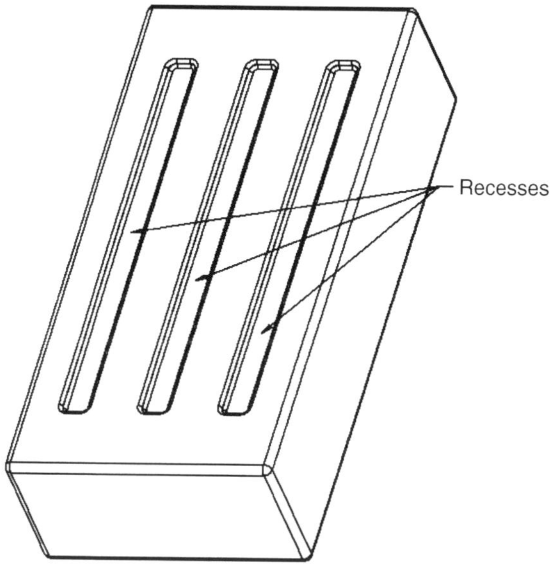

FIGURE 6.35 Stiffening recesses.

useful part features for rotomolded parts is the recess/kiss-off combination. This combination involves using long recesses to strengthen a flat w all and adding a series of kiss-offs along the length of the recess. This part feature is common on panel type parts. Sometimes recesses and kiss-offs are added to both sides of the part.

6.6.6 Holes

Holes can be molded completely through a rotomolded part by a variety of methods, including:

- Cores
- Through-pins
- Insulated panels
- Knife edges
- Dual cavities

The two main concerns with molded-through holes are draft angle and heat transfer. Since the part is shrinking onto the mold cores that form the hole, 3° should be the minimum draft that is used, but 5° is better, as shown in Figure 6.36.

To improve heat transfer, the hole should be formed from both sides of the mold, and if possible, the mold should also ha ve a hole through it to impro ve airflow. Molded through holes are used for axle holes, rotating assembles, drain holes, the ultimate stiffening rib, and many other functions. Figure 6.37 sho ws a series of through-holes molded in an insulated backboard.

Countersinking the opening to a hole will help to minimize surfce distortion on the flat surface immediately adjacent to the hole. Insulated panels on the mold can reduce the buildup of material in an area to create a hole or opening. This area can use Teflon blocks or mineral wool insulation and is usually designed to cover an area smaller than

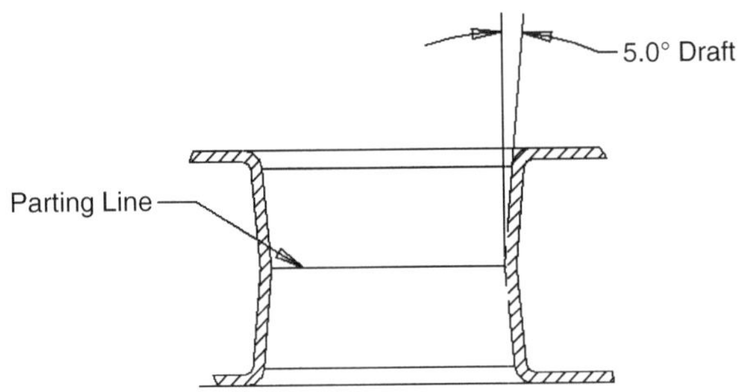

FIGURE 6.36 Molded through-hole cross section.

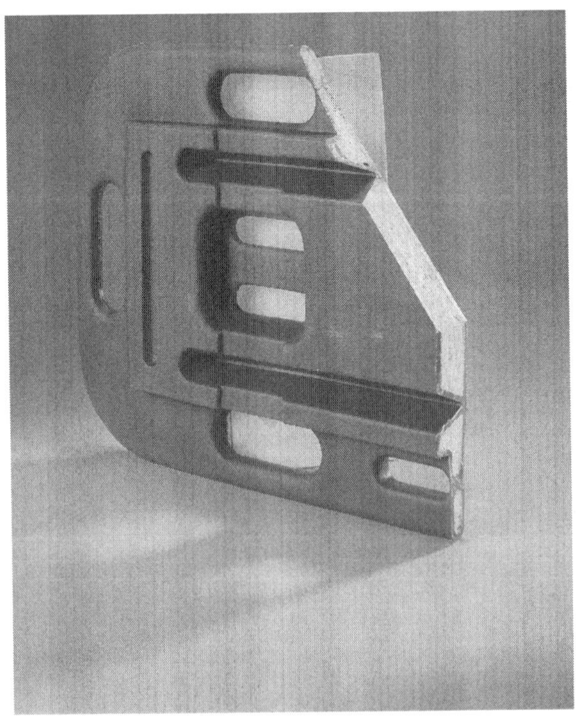

FIGURE 6.37 Through holes molded into a part. (Courtesy of Gre gstrom Corporation.)

the desired opening, as thinning will occur at the edges of the opening. A knife-edge feature in a mold can also be used to form holes. Figure 6.38 shows a mold and part cross section and how plastic forms around the feature. The central panel can be knocked out or cut out to form the required opening.

Dual cavities (they can be the same part or two different parts using the same material and wall thickness) can be molded together simultaneously. They can be connected and require separation by a secondary operation, or they can be divided by a Teflon ring in the mold. This produces two separate parts, but depending on the final requirements of the edge of the opening of the parts, often requires a trimming or finishing operation.

6.6.7 Bosses

Internal and external bosses are used for a variety of purposes but should not be confused with screw bosses in other plastic parts. Screws are not recommended for attaching directly to rotomolding parts because of their poor holding power. Extended bosses are used as pivot pins for many parts that rotate in assembly. External bosses can also be used as spacers between parts. The ends of external bosses can be cut off to allow hoses to be attached to containers for liquid filling or removal. It is important that the length of an external boss does not exceed its diameter, as shown in Figure 6.39. Internal bosses are used to increase the support at holes used for axles and pivot pins. A hole drilled through a single wall is not a good wear surface. An internal boss can give two to three times the wear surface. Both internal

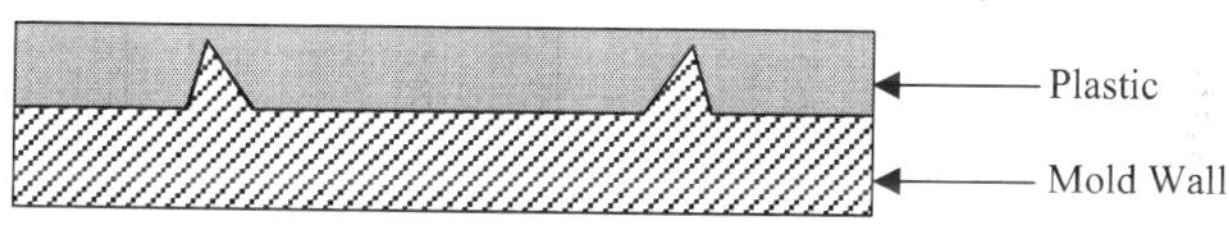

FIGURE 6.38 Cross section of knife-edge detail.

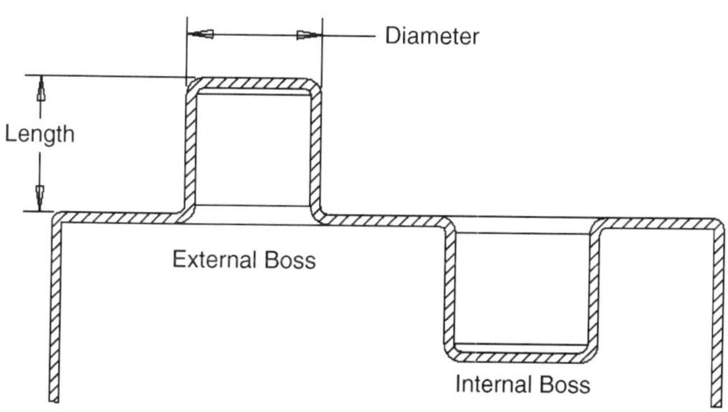

FIGURE 6.39 Cross section of rotomolded bosses.

and external bosses can be used to attach threaded fittings or caps. This gives flexibility when designing tanks and containers.

6.6.8 Tolerances and Shrinkage

Tolerances for rotationally molded parts are generally given as a percentage of the dimensions or an inch per inch (centimeter per centimeter) value. The designer should endeavor to use the broadest tolerance possible that can be tolerated by the final application. Overspecifying tolerances inevitably leads to higher costs and rework of molds or parts.

Two design variables have a major effect on final part dimensions. The first is the quality and accuracy of the pattern; this is being minimized by the use of computer numerically controlled (CNC) machined patterns. The second variable is the shrinkage value that is used for making the pattern. Determining this value requires an understanding of the material being used, the size and thickness of the part, and the complexity of the part design. It can be difficult to anticipate the actual part shrinkage, as variables in the molding process, such as color, wall thickness distribution, release agent, and cooling cycles can vary. Large parts typically have greater amounts of shrinkage and will therefore be more difficult than small parts to hold to close tolerances. Thicker parts will also vary more, due to higher levels of shrink than in thinner parts. The best approach is to use shrinkage data gathered for similar -shaped parts using similar parameters. Note that the inside surface of rotationally molded parts are free-formed and should not have dimensions with applied tolerances.

Crystalline materials such as polyethylene and nylon will shrink most and will also be susceptible to variations in the molding cycle. Amorphous materials such as polycarbonate and PVC are able to withstand cycle variations with lower variation in part dimensions. The design of the part can be critical in determining how a part will shrink during cooling; cores, changes in geometry, and any feature that restricts the free shrinkage of the part will distort it to some extent. Recommended free-shrink and restricted shrink tolerances are given in Table 6.3 for a range of common materials. Adding reinforcement to materials will affect the shrinkage of parts. Fibers or glass spheres can significantly reduce shrinkage (down to around 0.5% for polyethylene). Higher draft angles (up to 7 to 8°) are also recommended.

6.6.9 Warpage and Flatness

Large flat surfaces are a major problem in rotomolded parts, as they tend to distort during cooling due to uneven wall thickness and uncontrolled release. The problem is most apparent in highly crystalline materials such as polyethylene. If possible, flat surfaces should be broken up with additional design details such as that shown in Figure 6.40. Creating steps or recesses within a large flat surface reduces the part distortion by both reinforcing the surface and distracting the eye. When large flat surfaces must be used, a slight crown of around 0.015 in./in. (0.015 cm/cm) should be added to the surface. A crown helps to keep the surface from oil canning and minimizes distortion. Smooth or polished surfaces will exacerbate distortion, while a heavy texture or the use of "stone-effect" materials will improve the cosmetic appearance of a flat surface even if

TABLE 6.3 Recommended Shrinkage Tolerances

Material	Standard[a]	Free Shrink (in./in. or cm/cm)	Restricted Shrink (in./in. or cm/cm)
Polyethylene	Ideal	0.020	0.015
	Commercial	0.010	0.008
	Precision	0.005	0.004
Polyvinyl chloride	Ideal	0.025	0.015
	Commercial	0.020	0.010
	Precision	0.010	0.005
Nylon	Ideal	0.010	0.008
	Commercial	0.006	0.005
	Precision	0.004	0.003
Polycarbonate	Ideal	0.008	0.005
	Commercial	0.005	0.003
	Precision	0.003	0.002

Source: Ref. 1.

[a]Ideal tolerance, minimum care required; commercial tolerance, possible with reasonable care; precision tolerance, possible with difficulty and added cost.

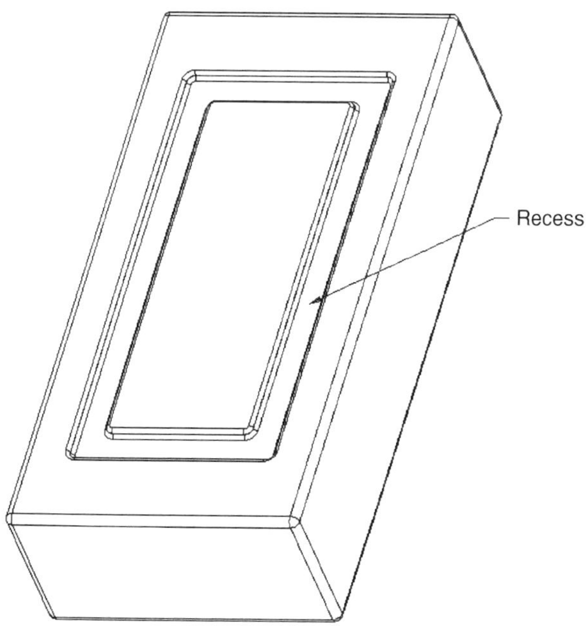

FIGURE 6.40 Recessed panel used to offset warpage.

it is slightly distorted. The use of internal pressurization can also help to control part distortion by forcing the part against the mold during cooling.

As with general size tolerances, the tolerance that should be applied to the flatness of a part will depend on the material used, the size and thickness of the part, and the complexity of the part design. Control of the cooling cycle is also critical. Guidelines for the flatness tolerances are given in Table 6.4 for a range of common materials. Note that stiffer materials such as nylon and polycarbonate are less prone to distortion than materials such as polyethylene.

6.6.10 Double-Walled Parts

A unique design requirement for rotationally molded parts relates to parallel walls as shown in Figure 6.41. They can be closely spaced, which enables double-walled parts suitable for insulated containers or boats to be made. When determining the separation of the walls, a guideline can be derived from the fact that most rotational molding powders have a bulk density approximately three times less than that of the final solid material. The minimum volume necessary, therefore, to allow the powder to fit between the mold halves physically is three times the wall thickness. For practical purposes, the walls can be separated by a factor as little as three times wall thickness in some areas but generally five times or greater is required to allow for easy filling of the mold. If this separation is not possible or the shape of the mold prevents all the powder from being added when the mold is open, powder can be added via a fill port. This can be a slow process, as the mold may have to be rocked back and forth or even vibrated.

TABLE 6.4 Recommended Flatness Tolerances

Material	Standard[a]	Flatness Tolerance
Polyethylene	Ideal	0.050
	Commercial	0.020
	Precision	0.010
Polyvinyl chloride	Ideal	0.050
	Commercial	0.020
	Precision	0.010
Nylon	Ideal	0.010
	Commercial	0.005
	Precision	0.003
Polycarbonate	Ideal	0.010
	Commercial	0.005
	Precision	0.003

Source: Ref. 1.

[a]Ideal tolerance, minimum care required; commercial tolerance, possible with reasonable care; precision tolerance, possible with difficulty and added cost.

Solid walls cannot be formed by moving the walls closer together, as material will tend to bridge and not flow completely into the cavity. Powder flow properties can be critical in forming these cross sections: If the powder has reduced flow characteristics, it can bridge across the narrow gaps and form unwanted voids. High-flow materials such as cross-linked polyethylene can allow narrower cross sections in certain areas to be produced. Figure 6.42 shows a cutaway section on a double-walled part. Double-walled containers, particularly deep ones, will typically have thinner internal walls, due to a combination of reduced heat transfer rates at the bottom of the core and lower residence times for powder on the inner surface. Extra heat can be applied by the use of airflow amplifiers or baffles.

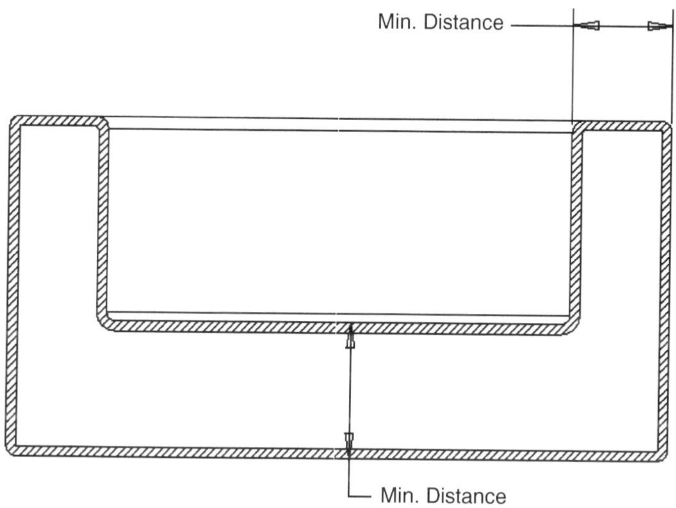

FIGURE 6.41 Double-walled part.

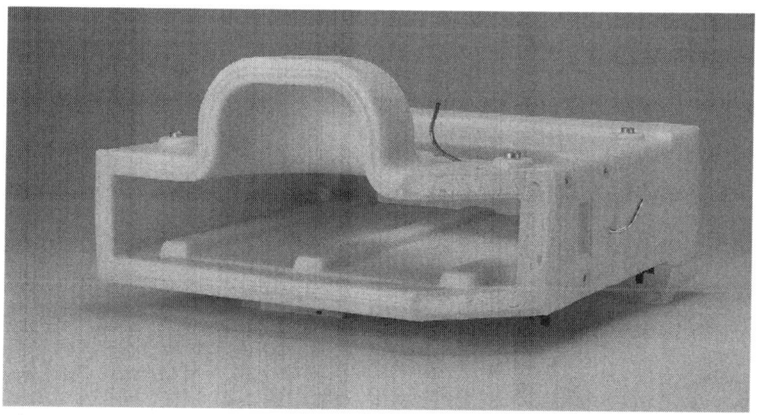

FIGURE 6.42 Double-wall construction. (Courtesy of Gregstrom Corporation.)

6.6.11 Kiss-offs

Kiss-offs are a part feature that causes two walls of a part to form together in small areas, as shown in Figure 6.43. Kiss-offs can increase part stiffness even more than ribs by forming a beamlike structure by combining opposing surfaces. They can be used individually or in a series or pattern. Two types of kiss-offs are used; a cone-shaped kiss-off forms a solid single wall in a circular shape at a single point; an elongated kiss-off forms a solid wall in an oval shape or continuous line.

The wall thickness, or separation, of the kiss-off should be 1.5 to 1.75 times the part wall thickness. This creates a small gap in the mold. It is better to start with a small gap (leaving the mold "metal-safe") and increase it as necessary by machining or grinding. This guideline is reduced for kiss-offs at the bottom of deep cores; actual wall thickness on the core is lower, so that the gap should be reduced to between 1.0 and 1.25 times the target wall thickness. Figure 6.44 shows a kiss-off that has been formed properly—the separation between mold surfaces is 1.75 times wall thickness. Kiss-offs need to be spaced so that there are 2 in. (50 mm) or more between the kiss-off areas. This allows the material to flow through and around the kiss-offs to maintain an even wall thickness (rules similar to forming ribs apply).

One disadvantage of kiss-offs is that they can cause a blemish on the opposite wall. This can be objectionable for cosmetic parts. This leads to the development of an alternative kiss-off called the "almost" kiss-off. An almost kiss-off leaves a slight gap in the molded part. A good starting point is to use a kiss-off spacing of two times the wall thickness plus 0.125 in. (3.2 mm). Since the two walls are not formed together, there is no blemish. This design is almost as strong as a regular kiss-off since the opposite wall can flex only 0.125 in. (3.2 mm) before it is supported by the kiss-off. This technique is used in playground slides because regular kiss-offs can also cause a small hump in the part at that point.

6.6.12 Undercuts

An *undercut* is defined as a protrusion into or out of a part that prevents it from being removed from a mold, as shown in Figure 6.45. Rotational molding is more tolerant of

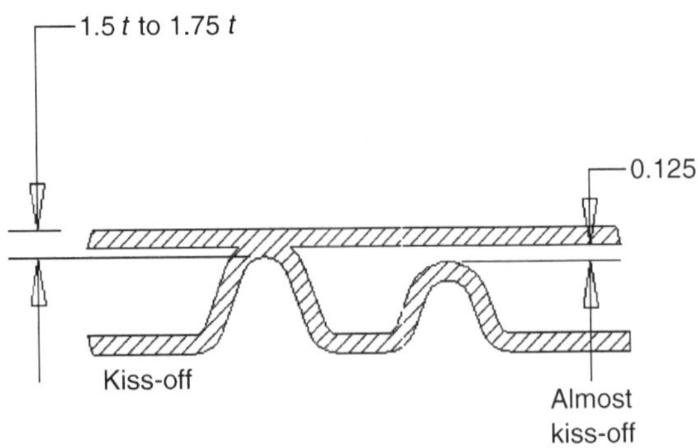

FIGURE 6.43 Kiss-off designs.

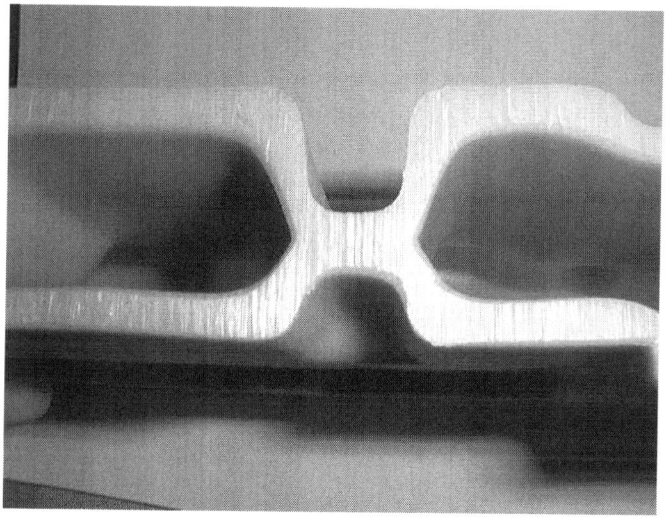

FIGURE 6.44 Properly filled kiss-off.

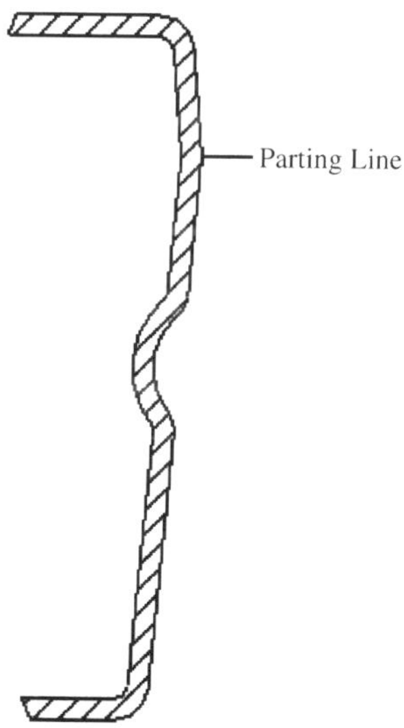

FIGURE 6.45 Cross section of an undercut in a part.

undercuts than are most plastic processes. Since the molded material is free to shrink away from the mold wall, the part may shrink an amount that is greater than the undercut. In general, for an unrestricted shrink part in polyethylene, undercuts equal to 50% of the linear shrinkage of a part in that area are acceptable. Fle xible materials such as PVC can tolerate large undercuts, producing parts such as dolls' heads that must actually collapse to be removed through the neck area. More rigid materials such as nylon and polycarbonate have lower tolerances, as they shrink much less. Undercuts on mold cores should be avoided since the part shrinks onto the core and the plastic must be stretched to demold the part. Sometimes, large undercuts can be accomplished if the part can be rotated out of the mold half with the undercut.

6.6.13 Threads

Both internal and external threads can be molded in rotomolded parts. Figure 6.46 shows common buttress thread profiles. External threads not on the parting line and all internal threads require additional mold sections that must be unscrewed before removing the part from the mold. Only coarse threads should be molded into the part. The most commonly used thread is a modified buttress. This is used for most containers, and a variety of standard caps are available to fit this thread. Small fine threads should be handled with metal inserts molded into the part or by spin-welding plastic fittings onto the part in a secondary operation. Machining fine threads into rotomolded polyethylene parts will typically produce a ragged edge and a poor finish.

6.6.14 Inserts

One of the advantages of rotational molding is the ease with which metal or plastic inserts can be molded into a part. It is common for threaded metal inserts, metal bosses, and bushings to be molded into parts, allowing other components to be attached directly to the molded part in a finished assembly. The most common types are:

- *Aluminum:* low cost; good conduction; poor thread strength
- *Brass:* chemical resistance; slightly better thread strength; more expensive
- *Steel or zinc-plated steel:* excellent thread integrity; least expensive; will rust
- *Stainless steel:* corrosion resistant; can bind and break drills and taps

During the molding process, plastic forms around the insert, embedding it in the wall of the part. For best adhesion, the insert must be clean and free of oil and dirt. Figure 6.47 shows a typical spring-loaded insert assembly. The inserts are located at the appropriate point in a mold and can be held in place by bolts, spring-loaded

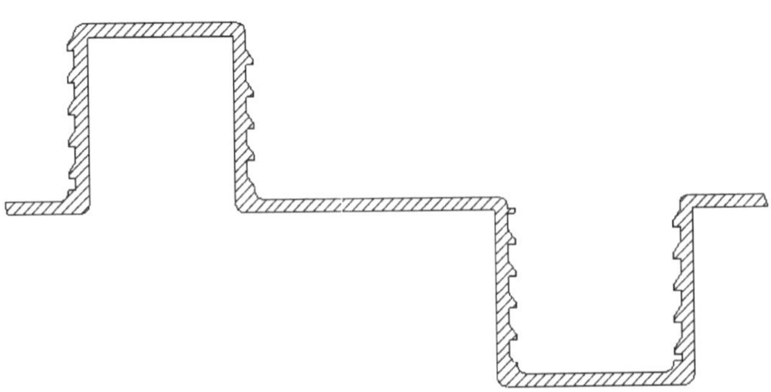

FIGURE 6.46 Internal and external thread forms.

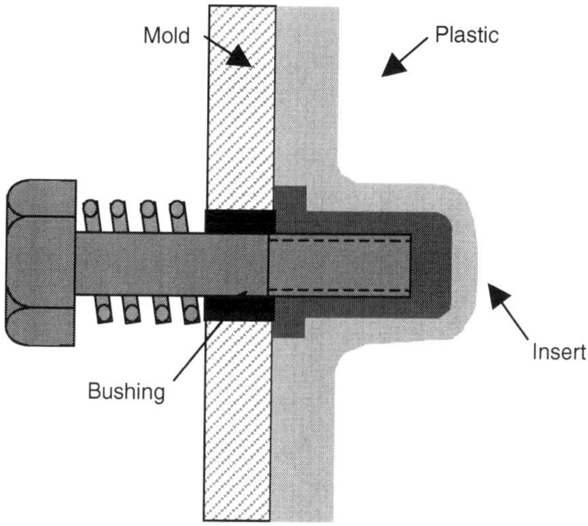

FIGURE 6.47 Spring-loaded insert assembly.

mechanisms, or magnets. Spring-loaded mechanisms allow the insert to move with the part as it shrinks. When the spring acts in the direction of shrinkage, this can reduce or eliminate distortion associated with locking an area of the part while the surrounding material free-shrinks. A bushing in the wall of the mold helps prevent distortion due to the forces created by the plastic as it shrinks. These forces can also distort a part that has multiple inserts and can make it more difficult to be removed from the mold.

Since inserts are commonly used for attachments, torque resistance is an important factor. Hex-shaped or "T-nut" inserts have the best jack-out strength resistance; knurled finishes are better than a smooth surface but will not provide as good a resistance. A double-hex profile with a turned-down center is the most popular configuration; the center turn-down allows material to flow around the hex and bind the insert mechanically in place. A square profile improves torque resistance; a rectangular one will improve it further. Check for torque using a torque wrench. Buildup of material around the insert is critical, so heat transfer and powder flow must be considered. Aluminum and brass inserts have better heat transfer so plastic will mold around them better than around steel inserts. Large inserts may require a heat pin external to the mold to attract more heat into this area.

Inserts should not be located close to a corner because they may restrict the flow of material. Guidelines for the distances to be used are similar to those for forming ribs. Inserts are not limited to simple fittings and bosses. They can be as complex as complete metal grids within racking pallets [up to 48 in. (1.2 m) long] or large metal plates in parts on which to locate machined metal components. Two of the main concerns when dealing with large inserts are heat transfer and differential shrink between the plastic and the insert. Inserts are most typically held in place with a mechanical grip by the plastic, but they must reach a certain threshold temperature for material to build up on them.

6.6.15 Parting Lines

Parting lines can become an aesthetic issue on many cosmetic parts. This can be caused when flash must be removed from the part. One way to minimize the parting line blemish is to simplify and soften the parting line by removing any sharp corners to trim and place a raised bead on the part at the parting line as shown in Figure 6.48. This makes it easier to trim the flash and minimize any surface blemishes. Another option is to place the parting line at the tangent point on a corner radius. For some parts, this moves the parting line to a less visible area of the part and also makes trimming easier.

Determining the location of the parting line is also critical in ensuring that wherever possible, it is always perpendicular to the line of pull. This allows maximum clamping force to be applied—vertical sections of parting lines tend to leak and cause flash in a part. If a step in a parting line is necessary a sloping transition is better than a vertical one. The parting line also determines the number of mold pieces. Two pieces are ideal; additional pieces complicate the mold-building process and add to maintenance during production. Check the part design to see if it can be modified to allow for a two-piece mold.

Thought should also be given to positioning the parting line to aid the operator in pulling parts from the mold. If possible, there should be an area that allows the part to be gripped or rotated from the mold half. Another factor to consider for parting line location is making sure that there is enough volume in the cavity for the powder, ideally, all the powder will fit in the fixed half of the mold attached to the machine. Since the plastic material is added to the mold in a powder form, its volume is approximately three times that of the molded plastic. If there is not enough volume in one half of the mold, the material may be added through a fill port in the mold after it is closed. Sometimes, details can be added to the part to give more volume for the material. Unless a parting line is machined, there will be some mismatch and gaps, particularly on large molds. A typical guideline is a maximum of 0.050 in. (1.27 mm) mismatch and 0.020 in. (0.508 mm) gap or flashing.

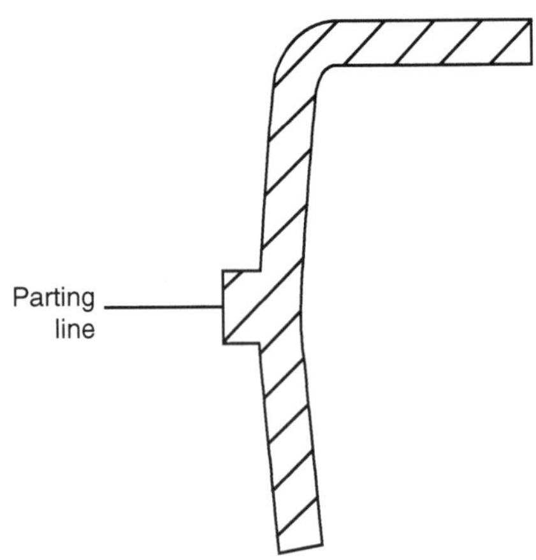

FIGURE 6.48 Raised bead at a parting line.

6.6.16 Vents

Vent tubes are required to relieve pressure in the mold. The most common size is 0.75 in. (19 mm) in diameter and leaves a hole in the part. The vent needs to be located in an area of the part with sufficient distance to the opposite wall. If the vent is close to the opposite wall, it will cause a blemish. From a cosmetic viewpoint the design engineer may want to specify the location of this hole. Ideally the vent should be located so that the internal tip of the vent extends to the center of the largest open volume of the part. Often, this is left to the molder to specify and even install, but the designer can help by building a supporting boss or feature specifically for the vent. Placing the vent hole in a shallow recess [0.125 in. (3.2 mm)] such as that shown in Figure 6.49 may be beneficial. The surface around the vent hole is usually distorted by shrinkage of material during cooling. This distortion is usually not a problem if it is maintained within the recess. If the vent hole must be plugged, the recess allows room for the plug.

6.6.17 Graphics and Texture

6.6.17.1 Graphics

Graphics can be added to rotomolded parts in several ways. The easiest and most inexpensive way is to mold them directly into the part as a raised or recessed detail that is etched or formed in the mold. This can be in the form of lettering, logos, or artistic shapes (common permanent molded in graphics include a part number, a recycling number, and a date wheel). Usually, the height or depth of this detail is 0.060 in. (1.5 mm) or less. This is one place where very small radii or sharp corners can be used; however, deep lettering with sharp radii may not fill completely. Chemical etching of the mold can produce quite complex designs, as the masking process can print almost any image on the mold surface.

Adding colored graphics adds cost to a part but will often transform an ordinary part. Colored graphics are available that can be molded directly into the plastic during the molding process or added to the part after it is molded. Surfaces of the part that are covered with a graphic (transfer) are usually smooth or slightly matte to

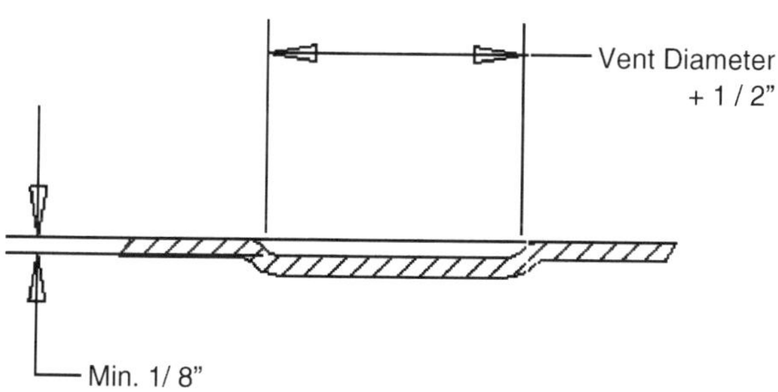

FIGURE 6.49 Recess feature for a vent location.

allow the graphic to adhere properly. Areas on the mold are masked during finishing to prevent texture from being applied.

The typical decorating methods for plastic parts such as painting, hot stamping, silk screening, and labeling require treatment of the part surface to improve adhesion on polyethylene parts. Polyethylene has a relatively inert surface that is resistant to solvents and does not interact with adhesives. Preparation for painting is typically done by flame treatment, corona discharge, or oxidizing chemicals (the latter must be aggressive to affect polyethylene and are not recommended). Polypropylene is more reactive and will respond to adhesives. Materials such as nylon, PVC, and polycarbonate can be painted more readily.

6.6.17.2 Texture

The appearance of cosmetic parts can be improved with the use of textured surfaces. The texture hides minor imperfections in the surface of the molded parts. Texture also aids the mold maker in that the texture hides minor imperfections in the mold and reduces finishing time. Unless parts require a glossy finish, all roto-molded parts should have a matte finish that is accomplished by lightly sandblasting the mold. A heavier texture is then added to selected areas of the part. This also adds a cosmetic feature to the part. Most mold makers have standard texture finishes that will cover most texture needs. Aluminum is much softer than steels and can therefore accept a greater range of textures ranging from mirror polish to "leatherette" or rippled surfaces. Steel will not accept a heavy texture applied by shot blasting.

Texturing the surface of a part will affect the way in which it demolds. A very rough surface will produce more wear of release agent and may cause demolding problems as a part grips the microscopic roughness of the mold. This may require an increase in the draft angles specified for a part [an extra 1° for each 0.001 in. (0.025 mm) of texture depth]. Stiff materials such as nylon and polycarbonate require a smooth mold surface. A very smooth surface for polyethylene parts may cause warpage problems, as the parts will release from the surface more easily.

Typically, mold makers will supply molds in an *industrial finish*, which for most steel molds is a fine shot-blasted surface and for aluminum, a hand-sanded finish. Producing a high polish or very fine finish on a mold is a time-consuming and expensive operation and should only be specified when necessary for aesthetic parts.

6.6.18 Part Assembly

Parts assembled from multiple plastic parts will be affected by molding tolerances. Following are some guidelines for designing the assembly of parts:

- Overlap the joint between mating parts. This conceals misalignment and compensates for variation between parts. The tank and base shown in Figure 6.50 show how an overlapping joint hides any distortion.

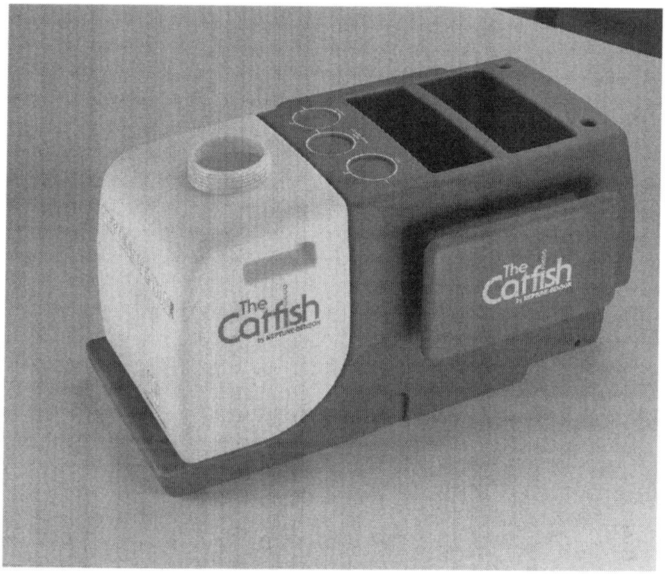

FIGURE 6.50 Tank inset into a base with overlapping joint. (Courtesy of Gregstrom Corporation.)

- Mold in ridges or beads along the edges of mating parts. This generates design features and creates a perception of alignment.
- Simplify the profile of mating parts. This helps to maintain uniform spacing between parts.
- Tighten mating parts by molding in small ribs without draft.
- Make molds that are metal-safe so that modification is easier. This allows for the addition of crush ribs to tighten the fit between parts (small ribs along the gap between two parts determined after the first parts are fitted together).
- Key assembled parts to fit in only one way. Add protrusions or pockets to allow parts to go together in only one way.

REFERENCE

1. Glenn Beall, *Rotational Molding: Design, Materials, and Processing,* Hanser/Gardner Publications, Cincinnati, OH, 1998.

CHAPTER 7

Compression and Transfer Molding

JOHN L. HULL
Consultant, Ivyland, Pennsylvania

7.1 INTRODUCTION

Compression molding and transfer molding are the present-day and historic processes for molding thermosetting plastics [phenolics, melamines, ureas, diallyl phthalates (DAPS), unsaturated polyesters, silicones, epoxies, alkyds, etc.]. Unlike thermoplastics, thermosetting plastics (generally referred to as *thermosets*) experience an irreversible and exothermic chemical reaction during molding, called *polymerization*, brought about by heat (approximately 150°C) and pressure (between 2000 and 5000 lb/in^2). Following polymerization, the rigid thermoset molded parts retain their physical, chemical, and electrical properties through a temperature range from −50°C or more to 150°C or more.

7.2 COMPRESSION MOLDING

Figure 7.1 shows the process of compression molding. The mold bottom half, containing one or more bottom cavities, is bolted to the bottom platen of the molding press. In this diagram an upward-closing press is shown. Compression molding can also be done in downward-closing presses. The mold halves are kept heated to about 150°C, more or less, depending on the plastic being molded. A metered charge of molding compound, granular or preformed, is placed in the open bottom cavities. The press is then actuated to close, generally fast upward movement (200 to 800 in./min) until the molding material contacts the upper mold half. Then the closing speed is reduced (0 to 80 in./min) as the material in the cavities is heated by the mold and becomes fluid. As the mold continues to close, the material is forced to flow so as to

Handbook of Plastic Processes Edited by Charles A. Harper

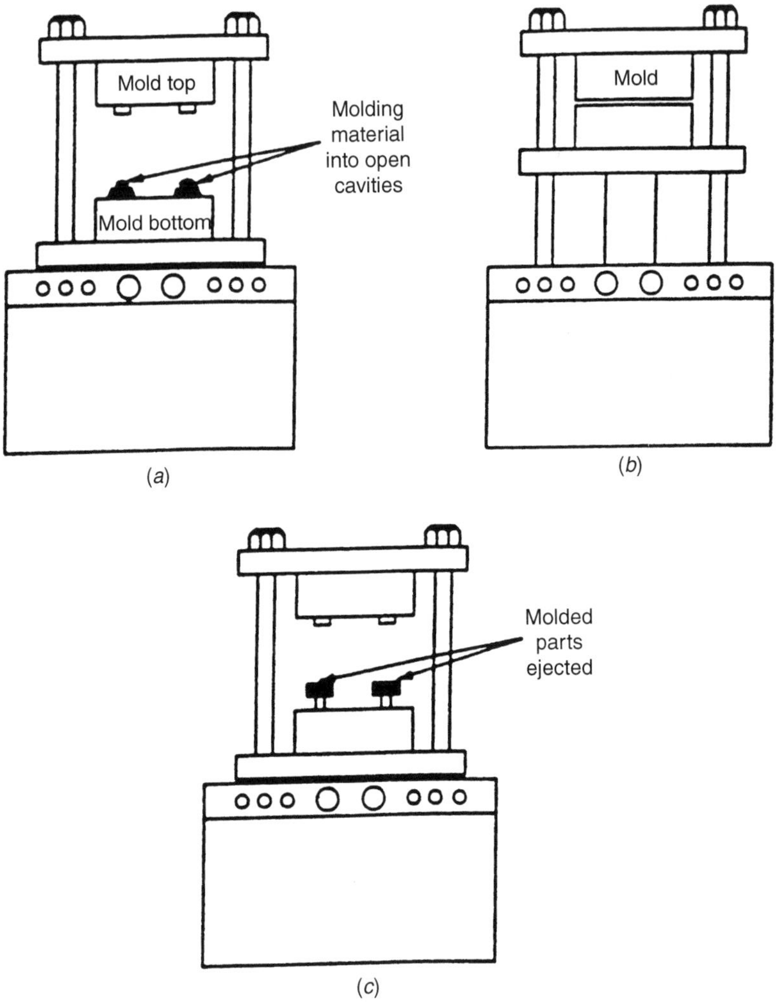

FIGURE 7.1 Compression molding sequence: *(a)* molding material is placed into open cavities; *(b)* the press closes the mold, compressing material in the hot mold for cure; *(c)* the press opens and molded parts are ejected from the cavities.

fill the cavities. The metered charge contains about 3 to 5% more material than is required for the molded parts, including runner and cull. As the mold halves are moving together to fully close the cavities, the slight excess of material is squeezed out along the land surfaces, the flat areas sealing off the cavities and causing the plastic to be compressed for the polymerization or cure. The slight excess of material on the land area cures into a very thin flash, which is readily separated from the molded part following cure and removal of part from the mold (Figures 7.2 and 7.3).

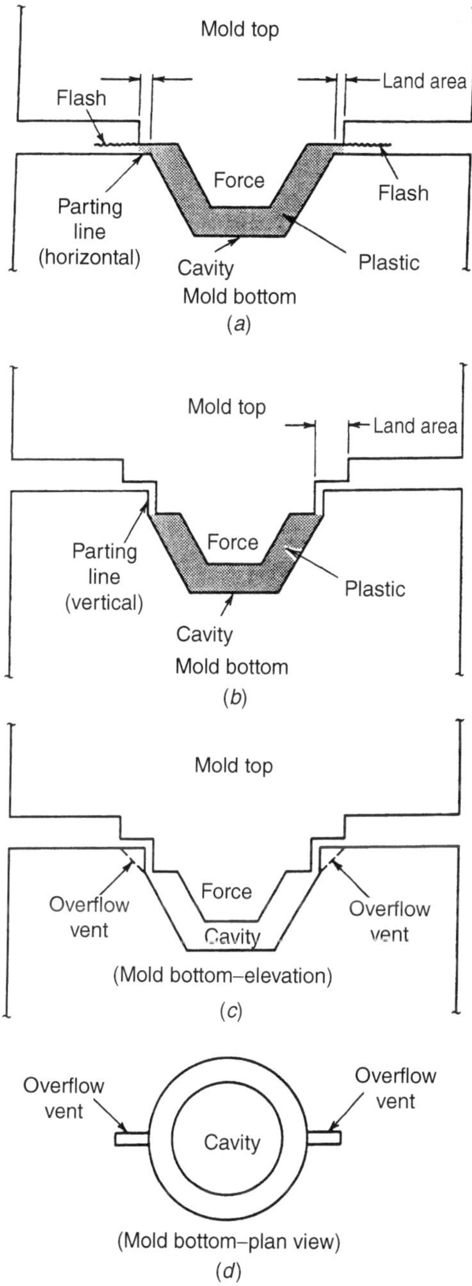

FIGURE 7.2 Compression mold cavity venting options: *(a)* flash-type compression mold; *(b)* fully positive compression mold; *(c)* semipositive compression mold; *(d)* semipositive compression mold.

FIGURE 7.3 Downward-closing compression mold.

7.2.1 Breathe Cycle

Some thermosets require a minor but important variation in the compression molding cycle, specifically the breathe cycle (Figure 7.4). The variation is necessary to allow easy escape of vapor, air, and other gases from the molding compound as the plastic experiences heating from the mold and from the exothermic reaction of the resin and catalyst (in the molding compound) during cure. The breathe cycle is almost always used with phenolics and other formaldehyde molding compounds, and mostly in compression molding.

The breathe cycle or degassing cycle is effected by incorporating, in the press cycle control system, a time point in the close or cure period (often,a second or more before the mold reaches full close) but at a time when the material has not yet reached full hardness. The cycle controller opens the mold to perhaps 1 in. or more

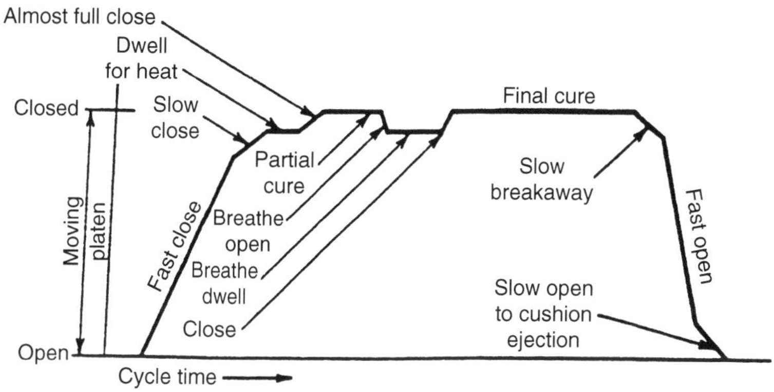

FIGURE 7.4 Typical compression cycle.

("breathe open"), holds the mold in that position for 2 to 4 seconds ("breathe dwell"), and then closes the press under full tonnage for the balance of cure. With some molding compounds and relatively large molded parts, the cycle may call for two or even three breathe cycles before final close and cure. The mold is then held closed under pressure for cure, lasting less than a minute for small parts, or up to several minutes or longer, depending on the size and thickness of the molded part.

Following cure, the press opens and the moving platen moves downward. About 1 in. before the platen bottoms out, a mold fixture, called the *pin plate*, to which are attached one or more metal ejector pins extending up to the bottom cavity surfaces, stops its downward travel, while the platen and mold continue their downward travel for that final inch. During that final travel, the ejector pins are pushed upward against the molded parts, effectively lifting them out of the cavities (Figure 7.4).

At this point, the molded parts are not completely cured but should be sufficiently rigid to be removed from the cavities and from the ejector pins. During the next several minutes, the plastic continues to cure as its temperature gradually returns to ambient temperature. The press operator removes the parts from the mold area manually and proceeds to load the mold cavities with plastic for the next "shot."

7.2.2 Additives

Thermosets are generally formulated with a variety of additives to improve the physical, electrical, and chemical properties of the molded part(s). For physical strength, glass fibers are frequently mixed with the base resin and catalyst. For arcing resistance in molded electrical parts, mica, alumina, or silica may be added. Halogen salts may be added as a fire retardant. Other additives may be used to accelerate or retard the cross-linking reaction. A small amount (3% by weight) of mold release additive such as carnauba wax or zinc stearate is often used to lessen the adhesive characteristics of the molding compound to enable the molded part to be ejected from the mold cavity with minimum resistance.

7.2.3 Cycle Variables

Variables that can be adjusted to optimize the molding c ycle include the mold temperature, higher for shorter cures (more rapid cross-linking). Speed of platen tra vel during press close may need to be reduced to ensure that the molding compound is partially softened by heat from the lo wer mold half before the material contacts the upper half cavities. Speed of travel during the mold final close is sharply reduced to ensure a gentle flow of material into all parts of the ca vity as the mold reaches full close.

7.2.4 Preheating

Molding with preheated po wder or preforms often enables shorter c ycles for larger parts and parts with thick cross sections. A *preform* is made by putting the metered granular charge into a room-temperature preformer to compact the material into a rectangular or hockey puck shape. The molding compound becomes a well-pack ed solid block, containing just the right amount for the ca vity or ca vities. The press operator places the preform into a beside-the-press high-frequenc y electronic preheater before placing the preform into the mold cavity. Once inside the preheater, the preheat cycle is actuated, the plastic material is exposed to alternating frequency of 70 or 100 MHz (frequencies authorized by the Federal Aviation Administration for the plastics industry), which rapidly heats (similar to microwave heating) the preform from room temperature to about 100°C in 10 seconds or more (depending on the weight of the perform) (Figure 7.5). The preform can then be transferred manually (with gloves) from the preheater to the mold cavity by the operator just before he or she actuates the press controls for a molding c ycle. Because the preform is already at 100°C rather than at a nominal 20°C, the time in the mold to reach 150°C is much shorter (and the preheating is more uniform throughout the plastic, rather than heating the plastic only where it is in contact with the mold) (Figure 7.6).

7.2.5 Moisture Absorbtion

Many thermoset molding formulations are hygroscopic in their unmolded granular or preformed state, meaning that they tend to absorb moisture at room temperature when the ambient humidity is high. Temporary adjustments may be needed in the length of the breathe cycle, or in the number of breathe openings each cycle to allow the moisture to escape as v apor, before the mold is fully closed. Phenol formaldehyde compounds are especially hygroscopic. Such materials should be stored in air conditioned (reduced humidity) space, in containers that ha ve plastic film or other barriers to minimize moisture penetration, especially if storage of unmolded material may be prolonged.

7.2.6 Hot Rigidity

Another property of thermosets that can af fect cycle times is *hot rigidity*, referring to the fact that at the time a thermoset molded part is ejected from the mold ca vity,

FIGURE 7.5 A high-frequency preheater, positioned next to the press, enables faster cures for both transfer and compression molding.

it is not fully cured. Even at molding temperature of, say, 150°C, 100% cure may not be reached for hours or longer following removal from the cavity (Figure 7.7). But if a molded part shows signs of warpage after removal from the mold, or shows signs of deformation in some section (often near ejector pin sites), it may be necessary either to lengthen the cure time another 5 to 10 seconds or to raise the mold temperature another 5° or more, or both, to allow the plastic to achieve a higher percentage of cross-linking before opening the mold and ejecting the part. It may even be worthwhile to increase the high-frequency preheating time by 5 or 10 seconds.

On occasion, ejector pins may need to be increased in diameter to lessen the localized pressure on the plastic when the ejector pins are pushing the molded part out of the cavity. And, on occasion, more ejector pins may be required to better

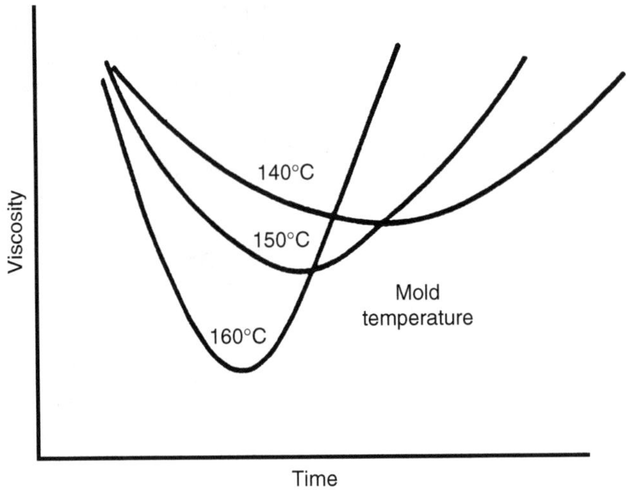

FIGURE 7.6 Higher mold temperatures shorten the flo w time in the mold and accelerate cure, with compression, transfer, and injection molding of thermosets.

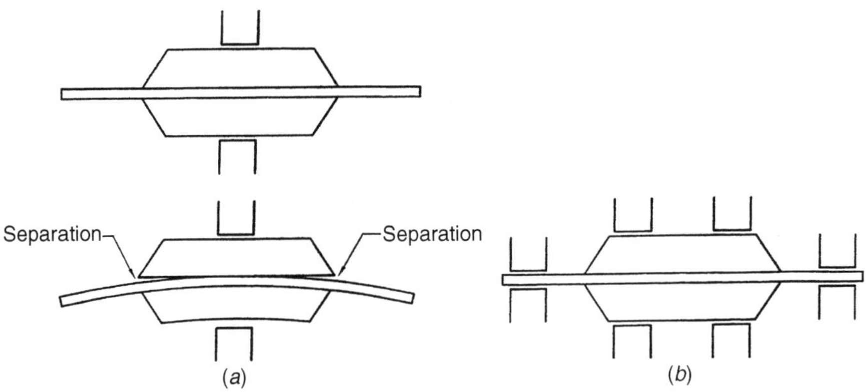

FIGURE 7.7 *(a)* Poor hot rigidity. On ejection of hot molded parts from a cavity, distortion can occur, causing separation of epoxy from leads, breaking epoxy bond. Subsequently , moisture will enter along interf ace between epoxy and leads. *(b)* Adequate ejector pins, top and bottom, acting on lead frames as well as on body of encapsulated de vice, will minimize possibility of separating plastic from lead frame during ejection.

distribute the forces needed to ef fect the ejection without distortion. Furthermore, part and cavity design should be reviewed to ensure that sufficient draft (usually, 5°) has been incorporated on parts and mold sections perpendicular to mold parting surfaces.

7.2.7 Automatic Compression Molding

In fully automatic compression molding, the molding compound is metered and fed from a hopper or preform dispenser into one or more cups on a movable frame. The frame is then positioned automatically under a heating source, often an infrared heating lamp, to preheat the molding compound before dropping the charges into each cavity. Following the preheating, the frame transports the preheated charges directly over the bottom mold half, opens the bottom of the frame under the cup, and allows the preheated plastic to drop from each cup into its respective cavity. The frame then returns to its position below the hopper in preparation for the next cycle. The molding cycle proceeds through press close, breathe if necessary, material cure, and ejection, following which a "comb" or robot will enter the space between the open mold halves, remove the molded part or parts, and transport the parts to a container or conveyor belt. The next cycle is then initiated automatically.

7.3 TRANSFER MOLDING

In 1928, Louis Shaw, founder of Shaw Insulator Company of New Jersey, secured a patent on a method for molding thermosets, whereby following press close (with cavities empty), a measured quantity of molding compound is placed in a chamber in the upper mold half (called the *transfer pot*), then forced by a transfer plunger to flow from the pot, into a runner system, through which the material is caused to flow, under continued pressure from the transfer plunger, into the cavity or cavities until the cavities are full, at which time the plunger stops further advance but holds a pressure (often 2000 to 5000 lb/in^2, depending on the molding compound) on the material as it cures. On completion of cure, the controls signal the press to open, the transfer plunger to travel into the pot another $\frac{1}{2}$ in. to ensure that the cull (cured residual plastic) is pushed downward out of the pot, following which the transfer plunger is retracted from the pot to await the next cycle. The press continues to open the mold and effect ejection of the molded parts as in compression molding (Figures 7.8 to 7.10). Molded parts are removed manually or automatically, and the press and mold are ready for the next cycle.

Advantages of transfer molding over compression molding are several. One advantage is that part dimensions, in direction of mold closing and opening, can be held to closer tolerances because the mold is firmly closed, metal to metal, before the plastic flows into the cavity. Flash at the mold–parting surface is minimal, if any. Another advantage of transfer molding over compression molding is that the mold cavities are exposed to plastic that is already somewhat fluid, enabling longer cavity life. In compression molding, dry granular plastic contacts the mold cavity surfaces, and in time, the hard chrome plating on cavity surfaces will abrade, requiring more frequent mold maintenance.

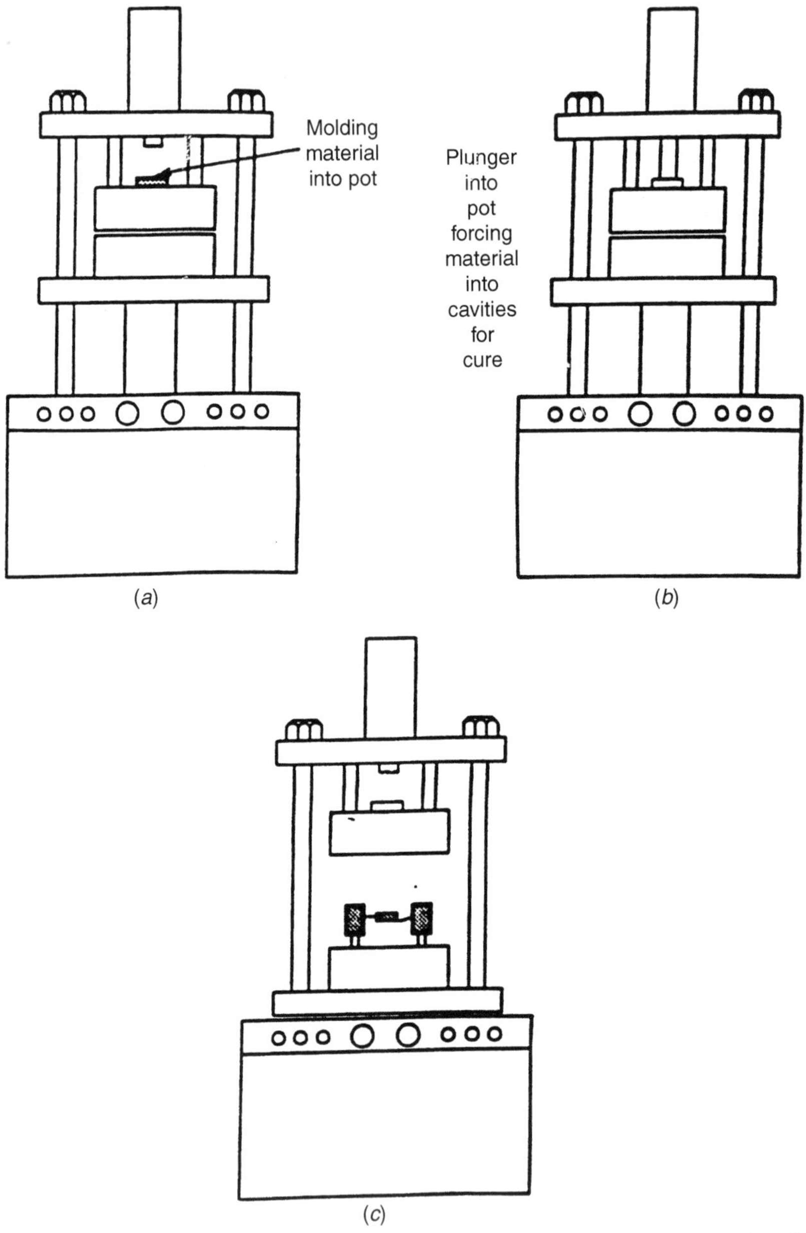

FIGURE 7.8 Tranfer molding sequences: (*a*) the mold is closed and material is placed in the pot; (*b*) the plunger descends into the pot, causing material to melt and flow through runners into cavities; (*c*) after cure, the press opens, the plunger retracts, and parts are ejected with cull and runners.

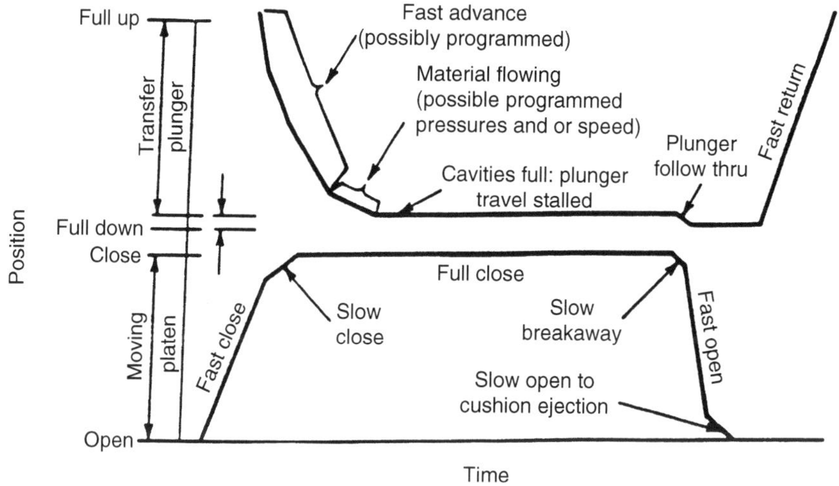

FIGURE 7.9 Typical transfer cycle.

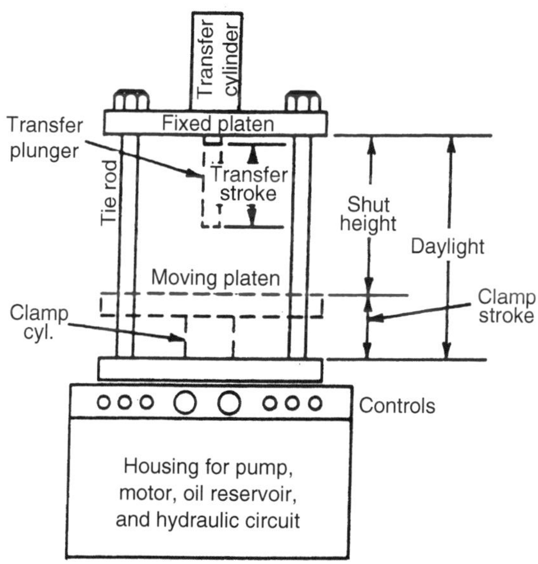

FIGURE 7.10 Typical specifications and terminology of compression and transfer processes.

7.3.1 Insert Molding Using the Transfer Process

Perhaps the major adv antage of transfer molding o ver compression molding is the process of *insert molding,* in which a metal or glass component (the insert) can be placed into the open cavity before the mold is closed, and held firmly in place between the closed mold halves when the transfer of plastic from pot to cavities takes place. The fluid plastic flows around the inserts and adheres to the insert during and following cure. Following cure, the molded part, including the molded-in insert, is ejected from the mold. The process enables lower costs than attaching other items after molding the plastic (Figure 7.11).

7.3.2 Direct Encapsulation of Semiconductor Devices by Transfer Molding

The application of insert molding using the transfer molding process has become the dominant technique to pro vide electrical, physical, and moisture protection for a wide range of both passive electrical and electronic devices (coils, chokes, resistors, capacitors, potentionmeters) and active electronic devices (diodes, rectifiers, transistors, varistors, integrated circuits, light-emitting diodes, etc.). For most encapsulated devices, epoxy molding compounds are used mainly because the viscosity of epoxies in the cavity-filling stage of the molding process is very low, minimizing damage to the fragile inserts and attendant connecting wires. Epoxies adhere well to the wire leads, stamped lead frames, and semiconductor chips, preventing moisture penetration between the interface of the plastic to the metal conductors and other materials of the device. The epoxy molding compounds generally have a high filler content, as much as 50% or more by weight, which assures minimum shrinkage of the plastic during and following cure and enables a very low coefficient of thermal expansion,

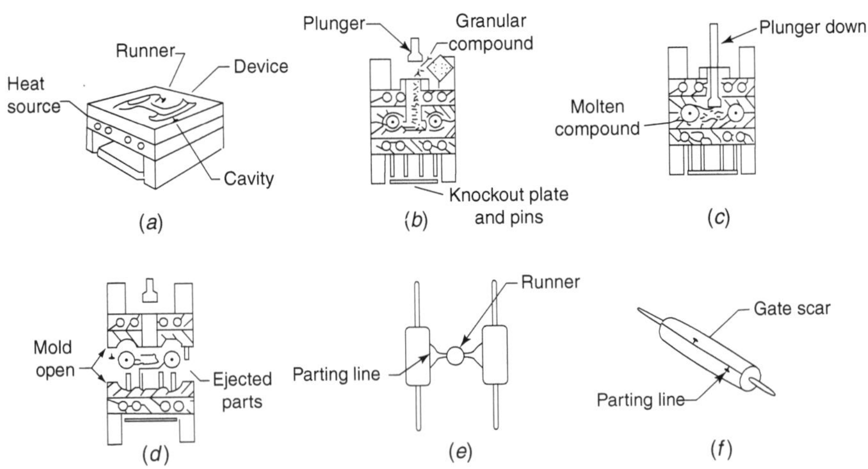

FIGURE 7.11 Tranfer molding with inserts: *(a)* perspective view of transfer mold (bottom half); *(b)* mold closed with devices positioned for encapsulation, plunger retracted, and granular compound fed into heating chamber or pot; *(c)* plunger moves downward, forcing molten compound around de vices in ca vities; *(d)* mold open and knock out pins eject encapsulated devices; *(e)* encapsulated devices as molded; *(f)* encapsulated device showing parting line and gate scar.

a property vital to minimize stresses on circuit elements when exposed to wide swings in ambient temperature of the device in space vehicles, self-directing military missiles, high-altitude aircraft, automobiles, and so on (Figures 7.12 to 7.15).

7.4 RELATED PROCESSES FOR MOLDING OF THERMOSETS

7.4.1 Screw Injection Molding

Screw injection molding, today's preferred process for molding thermoplastic materials, is also adaptable for molding with some thermoset plastics. Because of the proven economics of the thermoplastic injection molding process and the ease of fully automatic production, the thermosetting plastics industry has adapted a variety of thermosets capable of high-volume injection molding, benefiting from the technology developed for thermoplastics molding (Figure 7.16).

The injection molding of thermosets includes preheating, often in reciprocating screw machines, and is similar, in principle, to transfer molding. It is most commonly practiced in horizontal than in vertical presses, when no molded-in inserts are required. When inserts are to be molded in, as in encapsulation of electric/electronic devices, the mold opens and closes vertically, and the transfer plunger or screw feeds the heated plastic horizontally into a runner system through a nozzle entry on the mold parting surfaces, or vertically through the top or bottom half of the closed

FIGURE 7.12 Transfer press for laboratory development and modest production. Note the hand mold for inexpensive pilot production of small parts.

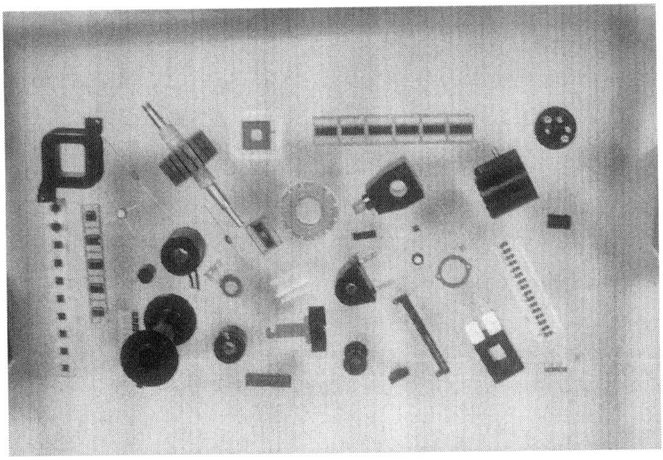

FIGURE 7.13 Typical electrical and electronic components encapsulated by transfer molding (insert molding).

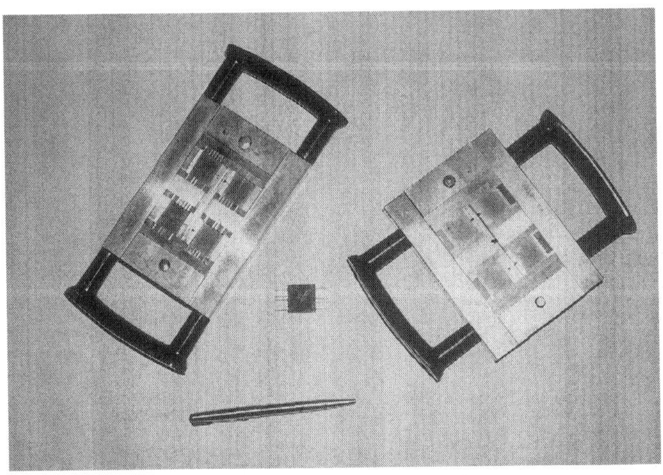

FIGURE 7.14 Hand transfer mold for low-cost product development to encapsulate a ceramic substract with a semiconductor circuit.

mold. The transfer pot and plunger is replaced with a reciprocating screw and barrel. A material feed hopper is positioned above the barrel. In cycling, the press is first actuated to close the mold. The screw is rotated in the barrel, and a measured granular charge of plastic is dropped from the hopper into the barrel. The rotating screw augurs the material, imparting frictional heat from the shearing action of the screw and deriving heat from the heated barrel. As the plasticized material is pushed toward the nozzle end of the barrel by the screw rotation, the screw reciprocates

FIGURE 7.15 Production transfer mold, with work loading fixtures, for encapsulating 160 dual in-line integrated circuits per cycle, producing about 40,000 devices per shift with one press and operator.

FIGURE 7.16 Injection molding press for molding thermoset plastics.

backward in the barrel to a fixed distance where the measured amount of plasticized material is packed into the nozzle end of the barrel.

With the mold closed, the screw then reciprocates forward, driving the molten plastic from the barrel through the nozzle into the runner system and then into the cavity or cavities of the closed mold. The mold temperature is kept at about 150°C to allow the material in the mold to cure sufficiently for mold opening and ejection of the molded part. At this point the screw rotates and reciprocates to prepare the next charge of material, while the mold opens for ejection of part or parts and then closes for the next injection. Operation of injection presses is fully automatic, and production rates are high.

7.4.2 Plunger Injection Molding

A variation of screw injection molding presses for thermosets is plunger injection molding, in which the screw is replaced with a plunger. The cycle is essentially like transfer molding but is often in horizontal configuration. The principal advantage of thermoset plunger injection molding over thermoset screw injection molding is when highly filled molding compounds and fiber-reinforced molding compounds are used. Such thermosets impose substantial abrasive wear on rotating screws, and the fibers are often damaged or broken, causing a loss of strength of the final molded products.

7.4.3 Vacuum-Assisted Venting

The process of closed-mold molding, including transfer and injection molding, presents an actual or potential problem of trapping air in the molded part. When the mold is closed and the molten plastic is being forced into each cavity, the inflowing plastic must displace the volume of air in the cavity. The mold generally has one or more vents machined into the land area, a few thousandths of an inch deep, and $\frac{1}{8}$ to $\frac{1}{4}$ in. wide, which allow air to flow out of the cavity as the plastic fills the cavity. The vents are shallow to minimize the flow of plastic through the vent and to enable the mold temperature (hot for thermosets, and cool for thermoplastics) to allow any plastic flowing through the vent to harden so as to prevent excessive bleeding of pressurized plastic from the cavity.

Positioning one or more vents in a cavity to ensure complete evacuation of air is often difficult if not impossible, especially with parts of complex shapes. If the flowing plastic fills the vent or vents before all air is forced out, a void is created in the molded part, causing a reject. To prevent such rejects, a process called *vacuum-assisted venting* has been perfected which involves machining a vacuum manifold in the mold, and use of a vacuum pump and vacuum reservoir placed adjacent to the press (Figure 7.17). The vacuum manifold requires an O-ring seal completely around the land area between the mold halves, plus a channel machined into the land area of one of the mold halves, often a half-round groove about $\frac{1}{4}$ in. in radius. The vent or vents from each cavity lead into the groove. The groove is connected by a -in.-diameter flexible hose (larger or smaller, depending on the size of the mold) leading from an opening into the manifold groove to a vacuum reservoir or chamber that is kept evacuated to about 2.5 lb/in^2 (absolute) by means of a mechanical vacuum pump connected to the reservoir.

Once installed, the vacuum venting system maintains the 2.5-lb/in^2 pressure in the vacuum reservoir. The vacuum valve in the hose connecting the manifold to the reservoir is kept closed until the injection screw or plunger is activated to move molten plastic from the injection barrel or the transfer pot to the mold. The valve

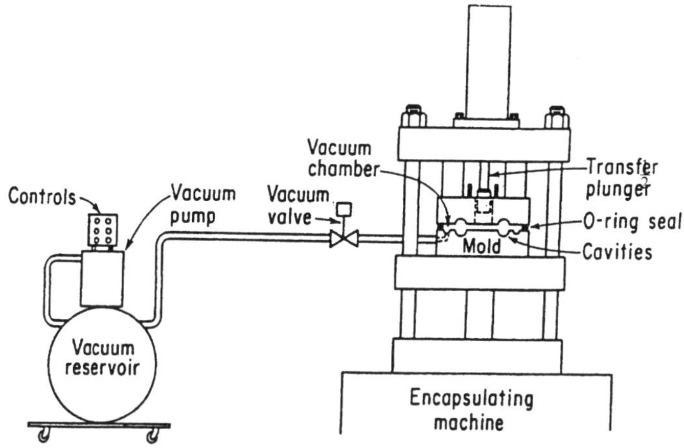

FIGURE 7.17 Vacuum-assisted venting. The vacuum line is connected to the manifold reaching all cavity vents. After the mold is closed, the vacuum valve opens, drawing cavity into the vacuum reservoir. The valve closes when the cavity is full.

opens and air rushes from the mold runner system and cavities through the vents into the manifold and into the reserv oir (and out through the v acuum pump exhaust). Once the transfer plunger or injection screw finishes its travel to fill the cavities, the vacuum valve closes, allowing the mechanical v acuum pump to bring the v acuum reservoir back to 2.5 lb/in^2 to be ready for the next press cycle. What little amount of air that may be trapped in the cavity is generally absorbed into the liquid plastic with the help of the fill pressure and holding pressure during hardening.

In most instances the overall molding cycle time is reduced when using v acuum venting (as compared to the c ycle time *without* vacuum venting), giving the added benefit of more economical processing. The shorter ca vity fill time also reduces molding stresses in the molded part and lessens tendencies to w arp or deform following molding.

7.4.4 Optional Press-Closing Mechanisms

Although most presses used for plastics molding are opened and closed with direct hydraulically driven rams, many use smaller hydraulic rams in conjunction with a mechanical toggle mechanism that multiplies the clamping force of a smaller pump and ram (Figure 7.18).

7.4.5 Retractable Positioning Pins

In transfer molding with inserts, the inflowing plasticized material may exert sufficient forces on the insert to cause it to be displaced from its hoped-for centered position. To prevent or minimize such movement, ejector pins are sometimes used

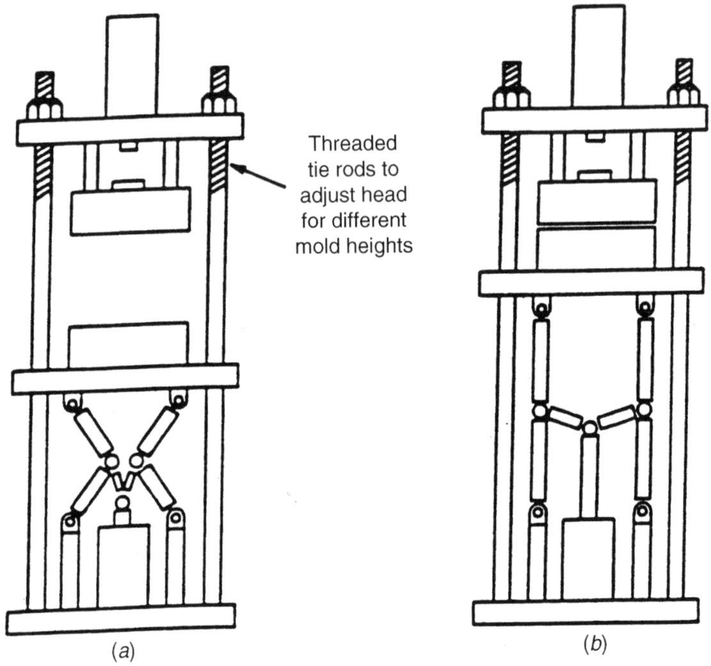

FIGURE 7.18 Toggle press: *(a)* open; *(b)* closed. Threaded tie rods are used to adjust the head for different mold heights.

as retractable positioning pins (Figure 7.19). In such cases the pins will be extended and retracted by one or more dedicated hydraulic pistons and cylinders. The pins will be extended a precise distance so as not to quite touch the insert during cavity fill, and then retracted so the ends of the pins are in the same plane as the cavity surfaces. Retraction takes place before cure of the plastic when the cavities are essentially filled with plastic. At this stage of the cycle, the plastic surrounding the insert will generally give adequate resistance to any significant movement of the insert as the pins are retracted and the still-fluid pressurized plastic flows into the space where the extended pins were. Following cure, the retractable pins are extended by their dedicated hydraulic system at the same time as any additional ejection pins are moving, to ensure uniform distributed force on the cured plastic during ejection.

7.4.6 Electrically Driven Molding Presses

In recent years several molding press manufacturers have replaced hydraulic pumps and cylinders with linear electric motors to drive the moving platen and to drive the reciprocating screw or plunger for its linear travel to move plasticized material from the injection barrel to the cavities. Advantages include absence of hydraulic leaks

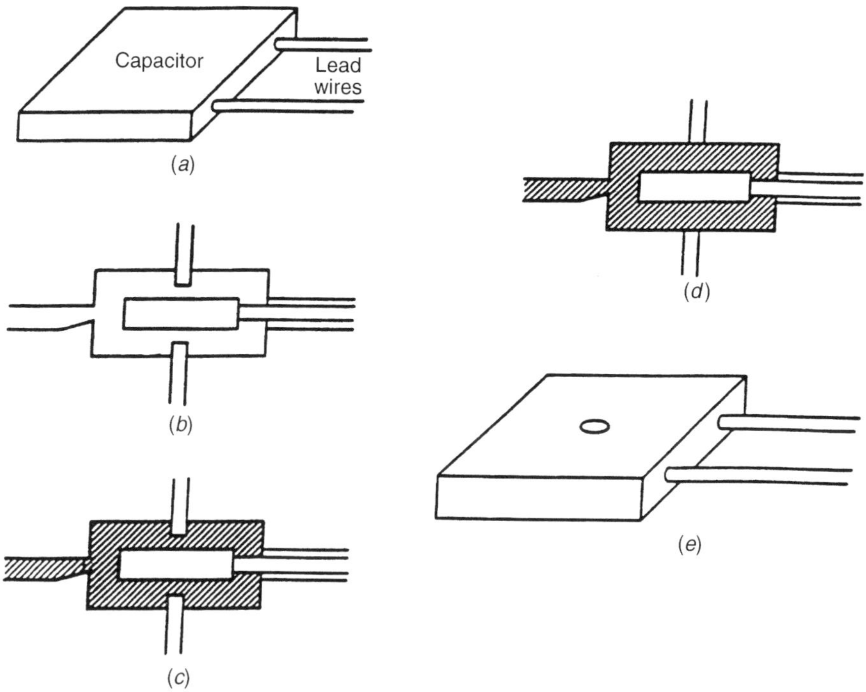

FIGURE 7.19 Retractable positioning pins: *(a)* insert and lead wires, for example, a ceramic capacitor before plastic is molded around it; *(b)* ejector pins are extended into cavity before cavity is filled with molted plastic; *(c)* pins minimize movement of capacitor during cavity fill; *(d)* pins are retracted promptly after cavity fill, but before plastic has hardened, and while transfer or injection pressure is maintained on material to enable it to fill voids as pins retract; (e) encapsulated capacitor after cure, showing the flush ejector (retracted) pin scar.

and oil vapors, reduction in operating noise level, and more accurate control of velocities and moving forces during the various phases in mold closing and opening and ejecting.

CHAPTER 8

Composite Processes

DALE A. GROVE

Owens Corning Corporation, Granville, Ohio

8.1 INTRODUCTION

The purpose of this section is to remind readers that the choice of ra w material and equipment are based on part performance, sales volume, and desired manufacturing cost. Following this important point, the remainder of the section concentrates on raw material selection, quality control, and to some degree the interdependencies of some processes to raw materials. This is then followed by a series of relatively short descriptions and an introduction to a series of composite processing techniques. The purpose of this composite processing chapter to acquaint ne w composite professionals with the myriad of processing technologies that are a vailable today.

8.1.1 Economics of Composite Processes

All part design projects be gin by examining raw material selection and manuf acturing methodology. Recently, composite-based materials are often being selected o ver unreinforced plastics, which suffer from lower mechanical properties, or over metals due to similarity in mechanical performance and fabrication ease. A good way of seeing this change is to lift the hood of an automobile and look at the changes in ra w materials from the 1970s to today. There are many more plastic and composite parts under the hood that are subject to harsher en vironmental conditions and higher temperatures than e ver before. Although some of the dri ving influence for mo ving to plastics in the automoti ve industry are dri ven through better gas mileage or CAFE regulations, most of the material shift is based simply on performance. When added performance value, such as flame retardanc y, environmental resistance, antistatic characteristics, wear performance, or possibly the processability ease of one material

Handbook of Plastic Processes Edited by Charles A. Harper

over another, are not relevant, it is possible to rank raw materials based on several standard formulas that are given below.

Only volume considerations: $$\rho C \tag{8.1}$$

Tensile strength performance: $$\frac{\rho C}{\sigma_T} \tag{8.2}$$

Flexural strength performance: $$\frac{\rho C}{\sigma_F^{1/2}} \tag{8.3}$$

Tensile deflection performance: $$\frac{\rho C}{E_T} \tag{8.4}$$

Flexural deflection performance: $$\frac{\rho C}{E_F^{1/3}} \tag{8.5}$$

where ρ (rho) is the density, C is the cost in monetary units over weight or mass, σ_T (sigma) is the tensile strength, σ_F is the flexural strength, E_T is the tensile modulus, and E_F is the flexural modulus. Formulas (8.1) to (8.5) are derived by calculating the volume of the part that is required to pass a particular test, finding the appropriate thickness, and multiplying the volume term by the density and unit cost [1]. These equations are fundamental in material selection when all other testing considerations and processing costs are equal. The main reason for using composites is captured in equations (8.4) and (8.5). Composites offer an inexpensive way of achieving modulus at relatively low densities compared to metals.

After selecting a few possible material candidates, the next consideration is processing costs combined with the number of parts required. There are many choices and considerations here, depending on the shape of the part, performance considerations, and material selection to name a few. Typical questions include, but are not limited to, the following:

- Is the sales volume sufficient to invest in a full-scale compression molding machine?
- Is a third-party toller warranted?
- At what volumes should the conversion occur between hand layup, resin transfer molding, and sheet molding compound?
- Does the sales volume suggest that direct long fiber thermoplastics technology is more appropriate than pellet-based long-fiber thermoplastics?
- Is used equipment available that could be fixed with only minor investment?

Clearly, many good decisions need to be made at an early starting point based solely on economics or presently available equipment, if any. With this in mind, production benefits and investment costs are listed in Table 8.1. Note that higher production rates normally come with the price of a higher initial investment costs and

TABLE 8.1 Composite Manufacturing Methods and Costs

	Matrix[a]	Reinforcement[b]	Part Shape[c]	Labor[d]	Investment[e]	Production Rate
Thermoset Processes						
Hand lay-up	TS	CM CSM V PRE	Open 3D surface	H	1–20K	1 part/5–24 h
Spray-up	TS	R	Rough 3D	M	1–20K	1 part/5–24 h
Vacuum bag	TS	CM CSM V PRE	Smooth 3D	M–H	20–50K	1 part/5–24 h
Preform/plenum method	TS	R	3D	M–H	30–50K	1 part/0.5–3 h
Resin transfer molding (RTM)	TS	CM CSM V PRE	3D	M–H	50–100K	1 part/2–10 min
Bulk molding compound (BMC)	TS	CS R	$2D^+$	M–L	500–1000K	1 part/30–120 s
Sheet molding compound (SMC)	TS	CS R V	$2D^+$	M–L	800–1200K	1 part/30–120 s
Mold for RTM, SMC, BMC	—	—	$2D^+ - 3D$	—	5–60K[f]	—
Pultrusion	TS/TP	R, CSM, CM, V	Linear – $1D^+$	M	200–1200K[g]	5–40 ft/min
Filament winding	TS	R, CSM, CM, V	Hollow parts	M–L	150K–3000K[h]	1 part/0.5–10 h
Thermoplastic Processes						
Long fiber thermoplastic (LFT)	TP	CP REC	3D	L	500–1000K	1 part/15–60 s
Direct LFT injection molding (DLFT)	TP	R REC	3D	L	~1500K	1 part/15–60 s
Glass mat thermoplastic (GMT)	TP	CSM CM V R	$2D^+$	M–L	~2000K	1 part/30–120 s
Extrusion/compression molding	TP	R REC	$2D^+$	M–L	~1500K	1 part/30–120 s
Mold for LFT, DLFT, GMT	—	—	$2D^+ - 3D$	—	5–60K[3]	—

[a]TS, thermoset; TP, thermoplastic.

[b]CM, continuous fiber mat (the mat may be needled or held together by binder); CSM, chopped stand mat (the mat may be needled or h eld together by binder); V, veil; PRE, preform; R, roving; CS, chopped strand; CP, chopped pellet; REC, recycled composite materials.

[c]$1D^+$, one-dimensional part with possibility of changing normal area with length; $2D^+$, two-dimensional part with some three-dimensional possibilities; 3D, Nearly any three-dimensional shape is possible.

[d]Labor level: H, high; M, medium; L, low.

[e]Investment cost is based on obtaining all necessary equipment unless otherwise noted. Amount based on 2004 U.S. dollar.

[f]Mold costs only; the rest of the process is already present. Note that lo wer investment costs usually create fewer parts per time.

[g]Pultrusion costs are technology and part sensiti ve. A typical LFT pultruder line will be some what expensive, thermoset single cavity less so.

[h]This process is very part-size dependent. Large diameters/vessels will be expensive: typical average cost around \$750,000 to \$1,000,000.

that processing methods are sensitive to selected raw materials. At times it can be advantageous to begin with a less expensive initial process and run to a full-scale investment as sale volumes increase.

8.1.2 Thermosets Versus Thermoplastics

Composite processing techniques depend on the choice of matrix resin, length of fiber in the finished product, and desired design and cost targets. Thermoset processes take full advantage of their initial low viscosities to fully wet out their reinforcing agents, whereas thermoplastic processes normally involve high temperatures and pressures to reduce the viscosity to a point where the material can be pumped/pushed. The major advantages for thermoset resins, such as polyester, vinyl ester, epoxies, and phenolics, are (1) ease of achieving good appearance, (2) good wet out as a result of low viscosity, and (3) low creep rates; whereas thermoplastics, such as polypropylene, polyethylene, nylon, polyethylene terephthalate, and polybutylene terephthalate, offer (1) the ability to recycle material, (2) the ability to reheat and flow the material over several cycles, and (3) added toughness. It is possible to find exceptions to the high thermoplastic viscosity rule, especially if lower-molecular-weight thermoplastic polymers, oligomers, or even monomers are polymerized during the selected process; and it is also possible to toughen thermoset materials by adding some impact modifiers. Most of the generic pros and cons between thermosets and thermoplastics are captured in Table 8.2, whereas Table 8.3 describes the major resins and their thermal property transitions and physical properties in terms of modulus and density. Modulus and density both represent integral properties with regard to micromechanic composite models.

It is not the purpose nor the scope of this chapter to provide a full description of addition polymerization, condensation polymerization, effective catalysts, initiators, co-initiators, inhibitors, or binder additives for various effects, such as flame retardancy, ultraviolet resistance, nor other chemistries or inexpensive additives or fillers. Readers interested in these topics should consult Refs. 1 to 7 as a starting point.

8.1.3 Reinforcement Theory

A unique combination of high strength, modulus, and impact resistance is possible by employing high-modulus fibers and plate-like materials of sufficient length and

TABLE 8.2 General Advantages and Disadvantages of Thermosets and Thermoplastics

Material	Advantages	Disadvantages
Thermosets	Low viscosity, low creep rates, good appearance	Viscosity varies, pot life, cure effects T_g
Thermoplastics	Reprocessable, impact strength, low cycle time	High viscosity[a], creep

[a]Exceptions exist e.g., cyclic resins, nylon 6 in monomer form.

aspect ratio within a matrix resin. Mechanical properties are highly dependent on fiber volume concentration, fiber length/aspect ratio to an asymptotic limiting point, interface chemistry (how well the reinforcement sticks to the resin), and particularly to the orientation of the fibers versus testing direction. Maximum moduli benefits occur with longer lengths and aspect ratios (length/diameter) reinforcements, and when the testing direction matches the direction of the fiber. These benefits are reduced as the difference in testing direction and fiber direction increases until a low point occurs at 90°. Similarly, testing angle dependencies hold true for impact or crack propagation–related properties, except that here the best properties are obtained at fiber orientations normal to the crack direction. Composite materials, which exhibit angular property dependencies, are classified as anisotropic materials.

Equations predicting property behavior are complex in nature; people unfamiliar with matrix algebra or tensor transformations are likely to find the mathematics overwhelming. Since it is not the purpose nor the intent of this chapter to cover composite micromechanics, readers are referred to Refs. 13 and 14. Readers interested in digging deep into all theoretical details, without the mathematical background, may first wish to consult Refs. 15 to 17, which have concise explanations of the calculus, matrices, and tensors.

Some readers may be asking at this point: Why is this important? The answer is that theories pinpoint what went wrong with a preexisting part or what can go wrong with future part designs. Typical part failures include resin-rich areas, fibers pointing the wrong direction, weld lines, where fibers cannot point across the join line, insufficient matrix cure (when applicable), lack of fiber flow (when applicable), wrongly sized fiber input, or possibly fibers getting ground to dust (when applicable). The point here is that theories or predictive programs exist that can predict the problem with a particular part and/or design. Understanding these concepts can be particularly important during early design and fabrication stages.

8.1.4 Reinforcement Forms

The best way of quickly becoming familiar with the differences in various reinforcement forms is to see them. Figures 8.1 to 8.4 display the major reinforcement input forms. Table 8.4 lists the major reinforcing fibers employed in the composite industry and their associated properties.

8.1.4.1 Chopped Strands

There are a variety of chopped inputs that can be employed in composite processes. The main considerations for selecting a chopped strand is to achieve a good interface and to obtain a critical aspect ratio, fiber diameter, and "size" without running into processing issues. The finish or sizing influence, particularly in materials formed in injection molding of greatly reduced fiber length, should not be overlooked. Size or finish formulations can significantly aid desired property performance, and they are normally constructed to obtain the following effects: (1) bond the

TABLE 8.3 Major Composite Resin Physical Properties

	T_g[a] (°C)	T_m (°C)	Processing Temperature (°C)	HDT (°C)	Density[b] (g/cm^3)	Tensile Strength (MPa)	Tensile Modulus[b] (GPa)
Thermosets							
Flexible epoxy	Varies[c]	NA		70	1.2	40	1
Epoxy	Varies[c]	NA		175	1.2	70	2.6
Melamine	Varies[c]	NA	160[liq]		1.48	40	Filled
Melamine, phenolic	Varies[c]	NA		169–202	1.6–1.7	53–77	Filled
Phenolic	Varies[c]	NA	150[liq]	160–226	1.3–2.1	54–60	Filled
Flexible polyester	Varies[c]	NA			1.1	10	0.3
Rigid Polyester	Varies[c]	NA	150[liq]	130–204	1.2–2.0	34–70	3.1
Polyimide	Varies[c]	NA			1.6	18	4
Polyurethane	Varies[c]	NA			0.4–0.6	11–21	0.5–0.8
Silicone	−110	NA	25–230[liq]		1.0–1.7	1.6–30	Filled
Vinyl ester	145–160	NA	105–260[liq]	235–260	1.6–2.0	70	2.1–4.1[d]
Thermoplastics							
ABS	90–111	NA	100–245	70–88	1.0–1.2	30–65	1.8–3.2
Acetal	−82[e]	160–171	170–196	66–123	1.3–1.4	37–66	1.4–3.2
Acrylic	100–105	200[e]	243–250	41–100	1.2	47–79	2.3–3.8
Nylon 4,6	43[e]	295[f]	295–310[f]	90–160	1.1–1.4	38–99	1–3.3
Nylon 6	52[e]; 60	220[g]	230–335	70–85[f]	1–1.2	48–100	0.3–4

Nylon 6,6	50[e]	255[g]; 265[e]	279–300	50–164	1–1.2	40–86	0.7–3.3
Polyamide–imide	280	NA	360	275–280	1.4–1.5	69–160	
Polybutylene-terephthalate	17, 80[g]	220–267[f]	239–260	47–175	1.2–1.5	35–75	1.5–5.2
Polyether ether ketone	140	343[h]	360–370	154–295	1.3–1.4	90–150	3.1–8.3
Polyethylene (HDPE)	−125[e]	110–135	190–274	41–93	0.9–1.4	10–50	0.2–1.6
Polyethylene terephthalate	73–78	250–265[g]	280–300	63–100	1.3	50–57	2.5–3
Polyphenylene sulfide	88	285–290[f]	309–350	104–260	1.3–1.8	70–124	2.2–5.5
Polyphthalamide	90	310[f]	320–350[f]	138	1.2	85	3.7
Polypropylene	−17[g]	130–168; 162–185[g]	202–252	37–148	0.9–1.2	20–80	0.5–1.6

[a]T_g is the glass transition temperature. The temperature where a second-order transition occurs with re gard to heat capacity, density, and modulus.
[b]Standard properties are from Ref. 8 unless otherwise noted.
[c]Varies with copolymer types, hardener types, as well as processing conditions (such as a time–temperature–transformation diagram) .
[d]From Ref. 11.
[e]From Ref. 7.
[f]From Ref. 10.
[g]From Ref. 9.
[h]From Ref. 12.

FIGURE 8.1 Milled, dry chopped strand, and wet chopped strand forms. (Courtesy of Owens Corning. Used with permission.)

FIGURE 8.2 (*a*) Single- and (*b*) multiended rovings. (Courtesy of Owens Corning. Used with permission.)

glass surface directly to the material through coupling agents; (2) allow fibrous bundles to disperse (come apart) through dispersant additives; (3) provide slippage or lubrication during feeding operations; (4) provide protection from fiber–fiber abrasion, which can lead to notches and corresponding property reduction; and (5) enable easier cutting during manufacture through use of a film former. Although there are occasional applications where it is advantageous to select fibers that are not compatible with the matrix resin, such as added impact performance, in which case formulation testing should be performed, it is generally wise to follow the manufacturer's recommendations. If a wet chopped strand is employed over a dried form, some provision must exist to remove or vent off the water. Wet chop is less expensive than dry chop, but there is a processing penalty with regard to water removal that can often become a significant barrier.

FIGURE 8.3 Typical roving unwind stand. (Courtesy of Owens Corning. Used with permission.)

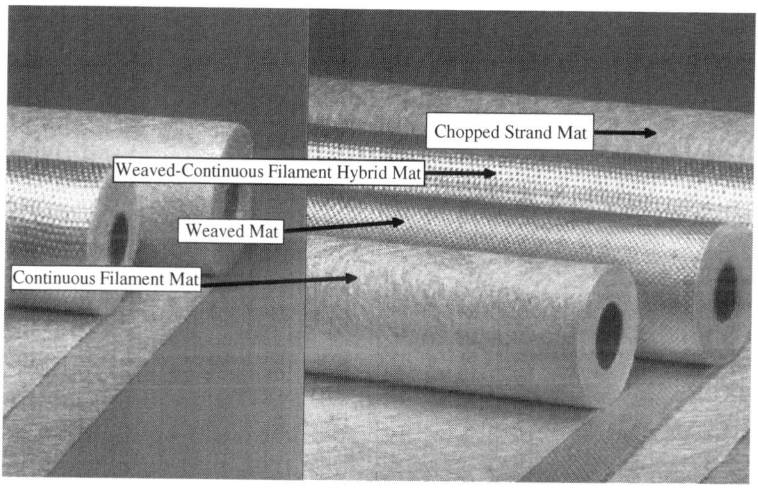

FIGURE 8.4 Various weaved and nonwoven fabrics. (Courtesy of Owens Corning. Used with permission.)

8.1.4.2 Rovings

Continuous filament forms can be purchased in either single-end, multiple-end, or even commingled glass–plastic filament forms such as Twintex[1], depending on what is desired. Roving packages can come in inside or outside payout forms. Outside payout offers the advantage of zero twist, whereas inside unwound packages, which is the normal mode for fiberglass products, allow for easier splicing. Splicing methods include

[1]Twintex is a registered trademark of Vetrotex France Corporation.

TABLE 8.4 Major Composite Fibers and Associated Properties

	Specific Gravity (g/cm^3)	Tensile Strength (GPa)	Tensile Modulus (GPa)	Failure Elongation (%)
Boron[a]	2.4–2.6	2.3–2.8	365–440	1.0
Cellulose[b]	1.5	1.1	2.4	
Cotton[b]	1.5	0.4	1.1	
Flax[c]	1.5	0.9	50	1.8
E-glass[a]	2.5–2.6	1.7–3.5	70	3
S-R glass[a]	2.5	4.8	85	5.3
PAN-based carbon fiber[a]	1.8	5.6	295	1.8
Pitch-based carbon fiber[a]	1.96	1.9	520	0.4
Kevlar 29[a]	1.44	2.6	59	4
Kelvar 49[a]	1.45	2.6	128	2.4
Kraft paper[b]	1.0	0.9	72	
Nomex 430[d]	1.4	0.6	12	28
Polyester (PET)[a]	1.4	0.8		15
SiC[a]	2.7–3.3	0.3–4.0	45–480	
Silk[b]	1.3	0.4	6	
UHMWPE (spectra)[a]	1.0	3	117	3.5

[a]Value from Ref. 14.
[b]Value from Ref. 13.
[c]Value from Ref. 18.
[d]Value from Ref. 19.

glue (inside unwind only), air splices, which intermingle the fibers together, and a sewn splice, which adds a foreign filament to hold the rovings together.

As with chopped strand input, rovings come with sizing to enhance the interfacial strength of these materials to desired resins. Because size is usually added through a dilute aqueous slurry, during drying, size migration can occur, which causes outer package edges to contain more size than the inner sections. Depending on the process, size and cure variation can potentially lead to differences in wet-out and durability within a single roving package. One method of mitigating size migration is to use dielectric drying during the drying operation or to purposely wind the package to leave diamondlike vent holes to hasten the drying process. Sizing ingredients also dictate how firmly each fiber is held to its nearest neighboring fibers. Tightly bound filaments can be advantageous until a point is reached where unwinding fibers stick to the fiber layers below (blocking). Well-bound filaments tend to process better from a processing/mechanical perspective, especially if a thermoset ingredient is present, whereas less bonded filaments will be easier to wet out in some processes. The best roving sizing for a given process is a question of chemical compatibility to the supplied matrix and runnability.

Although partially controlled by sizing ingredients, fuzz generation can be reduced significantly by using ceramic guide eyes. The rate of fuzz generation (or wear rate) is directly related to how hard the roving is pushed against another surface, the speed with which the roving is dragged over the other surface, and material tribological factors, such as roughness, friction coefficients, and wear rates. Selecting the wrong guide eyes or tensioning bars (if used) can and will ruin roving processability.

The final important processing consideration in some roving-related processes is tension uniformity. This is particularly true where wet-out is a consideration or when fiber breakage is an issue. High tensions in some processes will lead to better wet-out at the expense of fiber breakage, whereas low tensions could cause wet-out issues or breaks due to snagging. To fully understand roving operations where tension affects quality, regular tension quality checks should be performed. Tension meters can be purchased through specific company Web sites (e.g., Refs. 20 and 21). All processing lines should come equipped with some measure of tension control such as those suggested by Roisum [22].

8.1.4.3 Mats and Veils

There are a wide array of weaved and nonwoven fabrics of various fiber materials. The material is either in roll form, with the necessary special splicing precautions, such as use of an accumulator, floating looper, or a flying splice, or in box form, where the "tail" is available for splicing at any time and the remainder of the material is gently guided back and forth within the box. Splicing may be accomplished through glue, splicing paper, or sewing methods.

Weaved single-ended yarns or roving materials can achieve tremendous properties in the weaved (machine direction), weft (cross machine direction), or other specially designed angular directions. These materials are used when very high moduli are required in a few critical directions. The weaved fabric will be imparted with the roving sizing chemistry unless other finishes are applied. One important characteristic of weaved materials is their conformability or drapability. This property is linked directly to the weave pattern, with looser weaves being more drapable. In other words, the fewer roving crossovers between each weave, the more it can be shaped.

Mats or veils formed from continuous fibers may be constructed by blowing or placing continuous filaments in regular straight lines or swirl patterns or by forming the filaments and blowing them pneumatically onto a moving conveyor wire. Preformed continuous fibers are then either bonded together, through the application of an applied binder, or adhered mechanically through sewn fibers or needles.

Chopped strand mats may be formed either by chopping roving input or by applying dry chopped products directly. Longer chopped fibers can be processed into veils through dry-laid techniques such as the Rando–Webber process or possibly by carding, whereas shorter fiber length veils would be likely to use the wet-laid process, which uses wet chopped fiber as an input. Dry-laid webs are capable of flowing more than continuous filament mats in glass mat thermoplastic (GMT)-type processes while still providing a fairly long fiber length. Smooth veil products are often used to provide the best fiber flowing capability in GMT, or they are used to enhance aesthetical attributes while still providing some physical property enhancement. The main purpose of veils in the composite area is to hide unsightly weave lines or patterns in the material residing below the outer layer(s).

8.1.4.4 *Pre-pregs*

Weaved or nonwoven fabrics are sometimes further processed through various thermoset baths, thermoplastic solvent baths, powdered thermoplastic, or even melt-impregnated techniques to form pre-pregs. Pre-pregs have the advantage of nearly complete resin saturation and are usually less irritable to skin and easier to place on a part; the main disadvantages associated with pre-pregs are the added raw material costs in purchasing or manufacturing them, combined with the loss of conformability around curves and corners.

8.1.5 Fillers and Additives

It is not possible to leave the realm of raw materials without a very brief mention of fillers and additives. Readers interested in a broader perspective of these benefits are referred to textbooks on the subject [1,23]

8.1.5.1 *Fillers*

Fillers are used primarily to reduce costs, and secondarily to achieve modest modulus increments, flame retardancy, wear resistance, hardness, antiblocking, flatness, and other enhancements. Typical examples of cost-reducing fillers include clay, calcium carbonate, talc, wood flour, ground shells, wollastonite, and mica. Reinforcment is possible in those materials that have a sufficient aspect ratio, such as needle-formed wollastonite and platelike mica. Flame retardancy benefits are achievable with such fillers as calcium carbonate, aluminum trihydrate, magnesium hydroxide, and bauxite. Talc and mica have been used in the past for their non-blocking properties (i.e., preventing materials from sticking to itself), whereas alumina is often used in the thermoset lamination field to provide hardness to laminated flooring or decorative laminates. One of the predominant uses of certain grades of mica is to act as a flattening or leveling agent to achieve automotive class A surfaces in sheet molding compound (SMC) formulations. Clearly, fillers have their role in composite formulation and performance value. Readers interested in further details are referred to Ref. 23 or to vendors that work in that area.

8.1.5.2 *Nanofillers*

The most interesting new filler on the market are nanofillers or nanoreinforcements. Nanofillers are comprised of extremely small particles, typically measured in nanometers, which have high aspect ratios. The most common nanofiller is montmorillonite, a silicon-based clay comprised of ultrathin platelets, although nanographite and nanocarbon tubes are also available [24,25]. With the right combination of mixing energy, residence time, and charged dispersants such as dimethyl dihyrogenated tallow quaternary ammonium chloride, octadecylamine, or ionomers, it is possible to take the initial montmorillonite platelets, widen the spacing between the platelets with the introduction of bulk polymer layers, which is termed an *intercalated structure*, and ultimately exfoliate the ultrathin platelets to form a well-dispersed composite structure [24,26–28]. The small size, yet high aspect ratio of these particles permit very good reinforcing properties at levels of 3 to 5% that can match many normal extruded composite or mineral-filled materials used in the 20 to 40% weight range [27]. The advantages of these materials are flame retardancy, higher crystallinity levels (nucleating agent), vapor barrier properties in films, reinforcing properties [24,25], and in the case of carbon-based forms, electrical conductive properties [24,25]. Methods of following the intercalation and ultimate exfoliation process normally involve wide-angle x-ray diffraction, which follows the spacing between the clay layers, electron microscopy, and in the case of conductive fillers, resistance measurements [24,25]. Although nanocomposite materials represent very exciting promising new materials, they are processed through traditional extrusion processes and are not discussed further in this chapter. The reader is referred to Chapter 3 of this book and Refs. 24 to 29 on nanocomposites and extrusion parameter variables.

8.1.5.3 Additives

There are a host of additives to improve properties such as impact performance (while sacrificing tensile strength): initiators, activators, retarders, flame retardants, ultravoilet stabilizers, thermal stabilizers, and many more. When a desired or special effect is required, it is usually best to consult textbooks and vendors that work in those areas [1,23].

8.1.6 Quality Control Methods

Molded parts should be visibly checked for any discrepancy that the final customer may find objectionable. Typically, this involves quick part examination, an occasional physical test, a cure check when thermoset resins are employed, and other specific customer-related tests. One thermoset quality control measure includes the use of differential scanning calorimetry (DSC), differential thermal calorimetry (DTA), thermal mechanical analyzers (TMAs), and dynamic mechanical analyzers (DMAs) to measure changes in the glass transition temperature of a part [2,5,14]. The *glass transition temperature* is defined as the temperature where a material shows an abrupt modulus (DMA), density (TMA), or heat capacity (DTA, DSC) slope change with temperature, and it is also an indication of the extent of cure, with higher-modulus thermoset materials achieving greater cure levels. Measurements are performed by heating the material and examining heat capacity, density, or modulus changes. The differential scanning calorimetry instrument measures the heat difference between a blank cell and a sample while maintaining a user-specified temperature profile. The principle of the differential thermal analyzer is exactly the opposite of DSC, with the supplied energy being controlled and temperature differences recorded. Thermal mechanical analyzers can be used to observe density changes as the volume of the material is observed with changes in temperature. Classical dynamic mechanical analyzer methods involve ramping the temperature up from temperatures as low as liquid nitrogen and measuring the modulus differences as experienced through a dynamic torsional pendulum [14]. Unfortunately, the glass transition temperature is a blurry temperature that is sensitive to measurement method and temperature rate [2], so any quality control method that employs these techniques must be performed consistently.

In some thermoset materials the dielectric constant varies sufficiently with cure to permit an indirect cure measurement. By aligning the cavity or surface with dielectric sensors, it is very feasible to measure cure levels across the part during processing. Clearly, this is a very powerful technique that can be used to determine directly what is happening online [14]. Whether the composite matrix is a thermoplastic or thermoset resin, temperature and pressure transducer measurements during molding are other methods of judging differences in viscosity, cure characteristics (when applicable), and possibly even mold fill characteristics if the transducer is located at the end of a flow front [30].

Fiber and filler content can be determined in several different ways. If the fibers or fillers are not combustible, the part or a section of the part may be placed into a high-temperature oven or furnace to burn off the resin. Provided that there is good resin decomposition, all that will remain are the fibers and fillers of interest. The fillers can then be removed by shaking the materials on a sieve that is sufficient to catch most fiber sizes. Note that this technique would not work well with very short fiber materials which contain fillers of the same length as the fibers. If the burn-off and sieving steps work, the part content can be broken down quickly, to the percentage of glass, filler, and resin. A slight modification of these technique with the replacement of nitrogen for air, enables similar measurements with carbon powder or fibers. The burn-off method does not always work. Sometimes the bulk resin chars; other times the fibers themselves burn. When this situation occurs, the next best option is to find solvents to separate the matrix resin from the fiber and fillers. Given enough time, heat, and solvent selection, it is often possible to physically separate the various ingredients. Analytical methods such as near-infrared and x-ray spectroscopy work reasonably well with some charable thermoplastic resins. On the other hand, many thermoset-based resins will not dissolve, due to their very high molecular weight. Sometimes it is possible to use very acidic or caustic systems to degrade away thermoset materials, but at this point there is a concern about the fibers and fillers also degrading as well. When all else fails, it may be necessary to create materials of known composition to use as a fingerprint with infrared and other fingerprinting techniques.

When it is possible to quickly separate a part into base components, it is possible to calculate a theoretical density, the real density, and the void content:

$$\rho_t = \frac{1}{\Sigma(w_i/\rho_i)} \tag{8.6}$$

$$\varepsilon = \frac{1-\rho_a}{\rho_t} \tag{8.7}$$

where ρ_t is the theoretical density when no voids are present, Σ represents a summation operation, w_i is the weight fraction of a given component, ρ_i is the density of a given component, ε is the porosity fraction, and ρ_a is the actual measured density. Density measurements may be performed through ASTM D792.

The presence of voids may also be detected through acoustical, ultrasonic, and if the fibers employed are conductive, electrical or magnetic methods. Sound travels faster through higher-modulus materials, which is why approaching trains can be detected much faster by listening to metallic railroad tracks rather than by sight or sound. By the same principle, voids slow sound through a medium, and this slower rate of sound velocity can be detected through acoustics and ultrasonic methods. Electrical and magnetic methods can be helpful for those circumstances where a conductive fiber is present. As with the case of sound, voids are high-resistance areas, and this will be detected through the passage of electricity [14].

Other quality checks often are based on aesthetics, such as unwet fibers, and part dimensional stability checks, such as warping. Destructive testing should be done in a manner consistent with random sampling and with consistent environmental conditions, particularly for those materials whose physical properties depend on moisture intake. Six-sigma methodology can be particularly helpful in employing adequate sampling and in pinpointing accurate test methods versus poor testing procedures [31,32].

8.2 COMPOSITE MOLD PROCESSES

The equipment that most composite processes have in common is the mold. As with unreinforced plastics, many composite processes, such as resin transfer molding (RTM), sheet molding compounds (SMCs), bulk molding compounds (BMCs), long fiber thermoplastic injection molding (LFT), and glass mat thermoplastic compression molding (GMT), use closed molds, whereas others, such as hand-layup and spray-up techniques, employ an open mold. In the sections below we cover mold-making commonalities between these processes prior to delving into specific composite processes. Figure 8.5 shows some general mold differences and generic terminology.

8.2.1 General Molding Practices

8.2.1.1 Materials of Construction

Molds are formed from numerous materials, depending on raw material cost, part volume, machining costs, part material selection, ease of maintenance and repair, and to

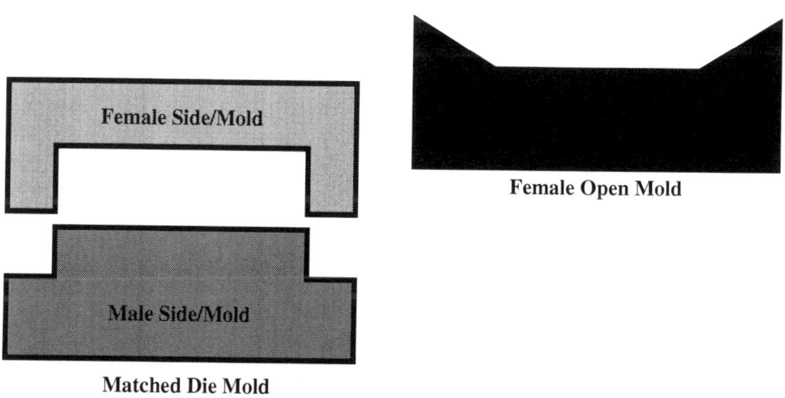

FIGURE 8.5 General mold terminology.

some degree, thermal property performance. The advantages and disadvantages of a number of common mold materials as well as their cycle life is captured in Table 8.5. Note that many of the mold materials listed can support coatings,such as chrome, that significantly increase the number of mold c ycles to failure. Another important consideration is the thermal properties of the mold material and its af fect on cycle time; materials such as beryllium–copper of fer fast cycle times at the e xpense of shorter tool life through its associated thermal characteristics.

Part sales conditions strongly influence material selection.As an example, suppose that a desired part has limited sales volume, employs low-pressure methods, and uses such thermosets as polyesters, epoxies, polyurethanes, or low-melting-temperature thermoplastic matrices. Under these conditions it may be very wise to select silicones, metal-filled epoxy, epoxy composite, or room-temperature vulcanates as the mold material. If, on the other hand, a reinforced, higher-temperature thermoplastic is the material of choice and part sales volume is high, the proper selection may be P-20 or hardened H-13 steel, due to their expected tool life of over 1 million cycles.

8.2.1.2 General Mold Design Guidelines

Avoid sharp bends and angles in parts by using smooth radii with a minimum radius to thickness of 0.25/1 around part bends and 0.5/1 around ribs. These criteria come from stress concentration factors versus radii curves observed in mechanical engineering textbooks [39] or in injection molding textbooks [40]. Composites add a further complications in that reinforcements may not extend to the very corner of a tight turn which can readily lead to a weak point. High resin contents are particularly a problem in processes like hand layup or resin transfer molding where it is very easy not to place the reinforcing media near a turn in the part. This problem also arises in materials with reinforcing media that cannot easily flow with the resin, such as compression molding of longer length glass mat thermoplastic materials or in applications involving weaved fabrics. The opposite problem of not having sufficient resin can arise when a resin must wet out a f abric such as in a R TM process. Very high fiber loadings can lead to poor performance as well since insufficient resin will lead to a material that is not fully consolidated.

A significant draft angle must be applied to enable better part release. Typically composites require larger draft angles of two to three degrees for shorter sections up to as much as five degrees for buried long length sections, are recommended due to the higher friction/wear characteristics of many composite based materials [40, 41].

Undercut sections, when applicable, require particular attention in enabling part release from the mold. This can be as easy as a mold modification change, if the customer can accept it, or as complicated as purposely releasing one section of the part prior to another, or possibly e ven worse, if it requires sections of the mold to rotate/swivel during the release portion of the molding c ycle [40,42].

The mold must be designed so that the part will shrink to the desired f inal shape without warping. This can be accomplished by either purposely matching the mold to part coefficient of thermal expansion, or if this is not possible, resorting to published nomigraphs from ra w material v endors. Thermoset based materials and semicrystalline thermoplastic materials are known to shrink substantially more than amorphous

TABLE 8.5 Materials of Mold Construction

	Cost ($/lb$_m$)	Density (lb$_m$/ft^3)	Heat Capacity [Btu/(lb$_m$-F)]	Thermal Cond. [Btu/(ft^2-hr-°F/in.)]	CTE [10^{-6}(1/°F)]	Fab. Ease	No. of Mold Cycles[b]	Advantages
Aluminum 7075	Medium	170[c]	0.23[d,e]	840[d,e]	12.9[d]–13.6[c]	Medium	250–1000K	Thermal properties; cheaper than steel
Berylium–copper	Medium–high	520–550[d]	<0.1[d]	1450–1800[d]	9.8[d]	Medium	5K	Excellent thermal properties
Brass (Cu/Zn)	Low–medium	550[d]	0.09[d,e]	1100[d,e]	9.8[d]	Easy	5K	Fabrication ease
Glass/epoxy	Low	100–125[c]	0.22[f]	2–7[f]	13.1[c]	Easy	0.1–2K	Low cost, castable
Kirksite[g]	Low–medium	415[h]	0.10[h]	65[h]	15.2[h]	Easy	5–10K	Castable, fab. ease
RTVs[a]	Very low	—	—	—	—	Very easy	0.01–0.2K	Inexpensive, ease of fab.
Steel (H-13)	High	485[i]	0.11[i]	120–170[i]	5.8–7.3[i]	Difficult	>1000K	Long life, strong, resists wear
Steel (P-20)	High	490[j]	0.11[j]	202–215[j]	7.0–7.5[j]	Difficult	>1000K	Long life, strong, resists wear
Wood (mahogany)	Low	44[c]	—	220[c]	12[c]	Easy	~0.1 K	Light, inexpensive, fab. ease

[a]Room–temperature vulcanates, castable rubbery mold substance.

[b]Mold cycles to failure are very dependent on compressional strength, mold alignment, hardness treatment (if any), coatings (if any), corrosion resistance (if applicable), thermal properties/effects, and molded material. Note that $K = 1000$ in the table.

[c]From Ref. 33.

[d]From Ref. 17.

[e]From Ref. 34.

[f]From Ref. 8.

[g]Kirksite is a zinc-based alloy.

[h]From Ref. 36.

[i]From Ref. 35.

[j]From Ref. 37.

thermoplastics, which in turn may shrink more than a well designed low profile added thermoset part. Composite formulations can make shrinkage matters worse. The orientation of reinforcements and added fillers significantly reduce shrinkage in the direction parallel to the particle's main axis. The anisotropy of these materials can make shrinkage very difficult to predict which can easily lead to some prototyping work or mold rework. General rules of thumb for shrinkage are shown in Table 8.6. Ideally shrinkage should be solved by correctly sizing the mold first, using all processing variables other than cycle time secondly, and lastly adjusting cycle time as a final resort [42,43]. The mold forces the part to withstand some internal stresses that could relieve themselves under higher temperature conditions. Warpage increases with larger tool sizes, greater crystallinity levels in thermoplastic semicrystalline materials, and as the temperature difference between ambient and the cure temperature increases [33,40].

Considering the influence of temperature uniformity on shrinkage, the greater the thickness uniformity of the part and mold temperature uniformity, the less prone the part will be to warp. Cooling or heating lines are preferred over serpentine flow paths since the temperature of the cooling lines must change through the mold.These lines are ideally fed from a manifold system that partitions the flow evenly. Conformal cooling channels, although more expensive to machine, can be worthwhile in the long run, due to better dimensional control and possibly, shorter cycle times. Typical cooling–heating line dimensions are 16 to 19 mm in diameter and are usually located one to two cooling–heating line diameters away from the part. When two molds are employed, two cooling systems must be employed, and these systems should use either water or oil. If a given press is going to switch from SMC to GMT, as an example, the cooling–heating lines should be able to work for both mediums, and these lines are preferably short between the mold and thermal unit to avoid heating or cooling losses outside the mold [41,44].

Sink marks and void regions occur in thicker part sections than compared to thin part sections. Since it is often necessary to design molds of varying thicknesses, the problem of sink marks, indentations in the surface near the thick section, and voids (holes) in the center of these thick sections can be problematic. One way to resolve this issue is to purposely employ an insert in the thick section [42]; a second possible way out of the dilemma is to add blowing agents to the formulation.

Weld and meld lines should either be avoided entirely or be placed at an unobjectionable point from either an aesthetic or physical property perspective. In the case of

TABLE 8.6 General Shrinkage Rules of Thumb

Variable	Effect
Pressure	Higher pressure results in less shrinkage.
CTE difference	Larger difference will result in more shrinkage/warping.
Cycle time	Shorter cycles lead to hotter part release and more shrinkage.
Low-profile additive (thermoset)	LPAs reduce shrinkage in thermosets.
Crystalline level (thermoplastic)	Higher crystallinity means more shrinkage.
Mold temperature	Lower temperatures lead to less shrinkage.

Source: Refs. 33 and 40.

composites, weld lines are like kryptonite, since only the strength of the resin will be achieved in this area. Sometimes it is possible to remove weld lines entirely where a small opening occurs. It is better in those cases to postcut the holes than to try to mold a weld into a part that will serve as a structural member. In other cases the weld line can be placed at a noncritical stress region through blank placement (SMC/BMC/RTM/GMT) or gate placement (LFT) [41,45,46], or the weld line can be eliminated through proprietary equipment that can cycle the flow fronts into each other [47,48].

8.2.1.3 Matched Die Mold Equipment Measures

When employing matched mold techniques for processes such as SMC, BMC, GMT, or blob LFT methods, it is important that the matched die molds have stop blocks, shear edges, guidance blocks, ejector pins (if needed), and venting measures. Stop blocks are short metallic blocks located around the periphery of the mold, and these should only come in contact when there is no material in the mold or when the initial blank size is too small. Molding on stops should be avoided with most composite materials since this places the molding pressure on the blocks instead of on the part where compaction forces are necessary to achieve full consolidation with minimum void content [41,45]. The exception to this rule occurs with RTM, where a fluid is introduced into the mold [49] (note that RTM normally does not employ a full compression molding machine) or when a purposely foamed article, which normally has blowing agents in the formulation, comes with a desired thickness outcome [50]. Shear edges prevent excessive material bleed-out in the molds. They work by bringing together two vertical edges, which engage roughly 12 to 19 mm before mold filling, that remain in close proximity (a 0.05- to 0.08-mm gap) to each other. Guidance edges engage slightly earlier (50 mm) than the shear edges and prevent mold damage due to mold shifting. If part removal is an issue, ejector pins can be used to push parts out of their cavities. These are normally located in thinner sections of the part, since these sections will solidify first. The size of the pin is such that it will not cause significant part damage while pushing the part. Venting should exist, especially in ribs or boss sections, which can cause dieseling, and burn marks on GMT parts and throughout the rest of the part. The actual tonnage of a typical compression molding machine depends on the part thickness, initial mold coverage, designed press closing rate, and material flowability, which includes viscocity, filler type, filler level, and mold temperature. It is not a simple function of part coverage unless all of the other variables are fixed. Figure 8.6 shows a number of these important features [41,45].

8.2.1.4 Mold Release Measure

A material should *never* be molded without either prior knowledge that the material contains a mold release agent or application of a mold release, which can be in the form of polyvinyl alcohol (PVA) films or releases, PVA sprays, silicone sprays, fluorocarbon sprays, wax sprays, and the like. There is nothing worse in the processing world (other than breaking a mold for a foolish reason) than spending hours carefully removing glued raw materials from a mold [14,40,42,51]. A second consideration is whether the molded part will undergo a subsequent painting operation. If it will, it may be wise to use a mold release that does the job while not severely

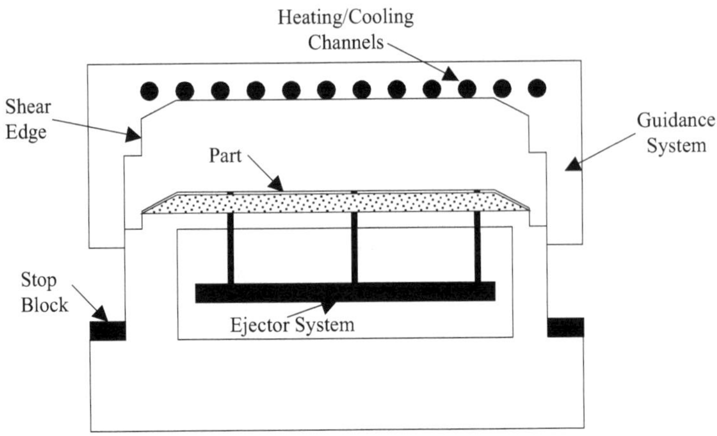

FIGURE 8.6 Important matched die mold features.

reducing the surface energy of the part, as happens with the application of silicones and fluoropolymers [40,51].

8.2.1.5 Rapid Prototyping

Modern technology has caught up to the world of rapid prototyping. Today it is possible to make a computer-assisted drawing and use laser cutting technology, stereo lithography, rapid computer numerical control, or other techniques to create a mold out of materials such as room-temperature vulcanates (urethane/silicone), epoxy-based molds, and others. Although these molds will not last for many cycles, they will enable rapid assessment of customer needs, especially where lower-temperature thermosets are the material of choice [52–58].

8.2.1.6 General Safety Measures

From a safety perspective, all injection and compression molding machines should come equipped with a light curtain, a beam of light, or lasers that prevent last-second adjustments to the press as it is lowering [59–62]. There have been several fatalities in the compression molding industry related directly to operators turning off the safety equipment and making last-minute adjustments. It is absolutely imperative to prevent this from happening, through proper maintenance and discipline. When high pressures are applied, lock and tag-out procedures should be employed before anyone climbs under or between two mold surfaces [63,64]. Thermoset materials often require safety and environmental restrictions. Although environmental emissions are beyond the scope of this chapter, peroxide initiators and other highly combustible materials are dangerous if not handled appropriately [65]. Emissions from some materials require the use of masks or breathing apparatus.These should always be placed in an area of the plant away from general operations and in a chilled environment [65].

8.2.1.7 Buying Molds

Buyer beware is the best motto in buying molds. Everything should be negotiated up front, including who owns the mold, the terms of the lease program (if applicable), who will store the mold, what tolerances are expected, what warpage is acceptable, and so on. Be extremely cautious and put things in writing.

8.2.2 Open-Mold Methods

8.2.2.1 Hand Layup

The hand-layup or open-mold approach is one of the oldest composite processing methods in existence. It is used to create large parts in the aircraft and marine industries and in common everyday parts such as spas, bathtubs, and sinks. Weaved or knitted fabrics, continuous strand mats, and chopped strand mats are set above optional fine weaves or veils for smooth surface attributes, which normally sit above a gel coat layer that makes direct contact with the open-mold surface. Fabrics normally consist of higher-modulus material such as glass, carbon fibers, Kevlar, and the like, or pre-pregs, fiber assemblies preimpregnated with a resin. Assemblies are glued together with the application of a thermoset binder that can be applied manually or through semiautomated equipment. After resin is applied, which can be done by applying sufficient resin under pressure through grooved rollers or a following roller, the assembly is either left alone to cure or placed in a high-temperature environment. High-temperature ovens are an absolute requirement if fully polymerized thermoplastic pre-pregs are going to be employed. The final part may be cut either manually or robotically with routers, high-velocity water jets, or possibly grinders. Figure 8.7 illustrates the essentials of this technique. Figure 8.8 shows the process in actual use. The open-mold section of the part will be the side of higher surface quality. It is this side where a gel coat can be prepared first and allowed to harden somewhat prior to adding an optional veil or fine-weaved fabric. If smoothening the opposite side of the mold is desirable, cull plates, films, or veils or fine fabrics may be employed [14,66].

This particular process may seem very straightforward, but as unsophisticated as it seems, there are plenty of ways to go astray.

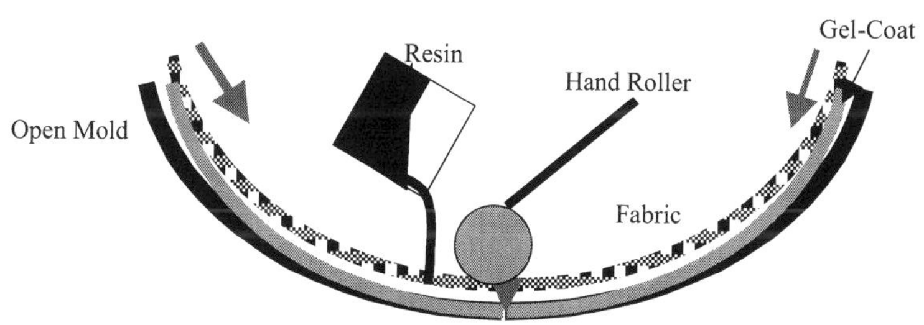

FIGURE 8.7 Hand-layup method.

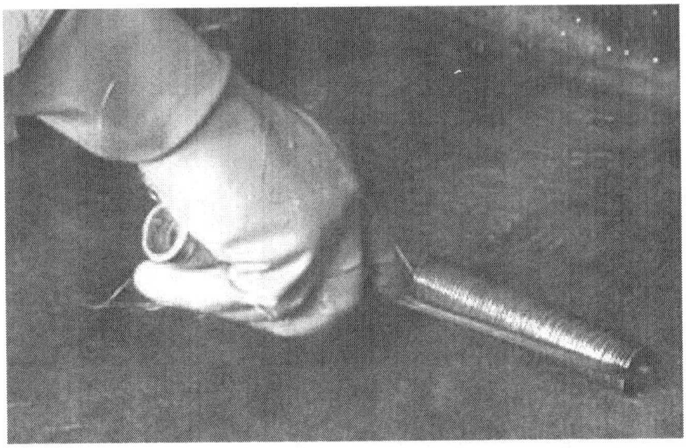

FIGURE 8.8 Hand layup in practice. (Courtesy of Owens Corning. Used with permission.)

1. The mechanical properties of the final part need to be addressed. Longer fibers will provide for better properties than shorter fibers. Reinforcement orientation also plays a role. Fibers oriented in the longitudinal direction will provide longitudinal strength; fabrics with orientation in the transverse direction will provide properties in the normal direction; fibers oriented in the ±45° direction provide the best shear strength. If high bending forces are likely, it would be wise to place some of the longer directional fiber fabrics in the outer regions since these regions support the structure under bending moments.

2. Symmetrical fiber assembly placement is important in avoiding warp. Unsymmetrical parts will almost certainly warp.

3. The part should avoid any tight bends or corners since these points will be high stress concentration regions and because the fabrics probably won't bend or drape into the corners or angles.

4. Avoid large weld or meld lines. Fabrics stacking sequences should be staggered a minimum of 2 in. to avoid large-scale weld or knit lines.

5. Thickening resin viscosity and settling are both issues. Depending on the initiators and catalyst package, the resin may solidify as quickly as a few hours. This will require the laborers to finish thick parts quickly, before the resin becomes so thick that fiber wet-out becomes impossible. On the other hand, low-viscosity resins will have the tendency to puddle in nonhorizontal regions and settle in the lower mold regions. The effect of wet-out versus wet-through is strongly influenced by the fabrics involved. Open fabrics will not wet-through as easily and may lead to more uniform fiber wet-out, whereas less open fabrics will be less likely to wet-through. The arrows in Figure 8.7 roughly illustrate the magnitude of resin settling in a boat hull part. One method of avoiding poor wet-out and resin puddling issues is to use pre-pregs, which will also improve the mechanical performance of the final part.

6. Some resins will require long cure cycles with specific temperature profiles, which consist of a slow heat-up, a hold cycle for flow, a secondary heat-up, a secondary hold at higher temperatures to ensure cure, followed by a slow cool-down [14,66].

The major advantages of this method are the low investment cost (only an open mold and a roller are needed) and the ability to produce very large parts. Limitations to this technique are gentle curvatures, the need to fit inside an open mold, timing limitations brought on by resin pot life, high labor costs, slow production rates, and the fact that the performance of the part will depend greatly on the laborer's skill level in achieving a well-wet-out part under low-resin-content conditions [14,66].

8.2.2.2 Spray-up

The spray-up or gun roving method offers a faster way of placing fibers onto an open mold, framework, or in some cases dirt (outdoor swimming pools). As with hand layup, this method has virtually no part-size limitations. Essential features of the chopper gun are displayed in Figure 8.9. Here a roving is pulled into a chopper gun, chopped to a desired fiber length, and sprayed with thermoset resin prior to being blown on to an open mold or fine wire frame. The advantages of this technique versus the hand-layup method is the manufacturing speed, more uniform resin placement, a material that is more drapable, and less skilled labor involved. The main disadvantage is the loss in mechanical properties associated with a finer fiber cut and the fibrous appearance of the part if veils, fine-weaved fabrics, or gel coats are not employed [14,66–70].

FIGURE 8.9 Spray-up technique. (Courtesy of Owens Corning. Used with permission.)

8.2.2.3 *Vacuum Bag Molding/Pressure Forming/Autoclave*

The last set of open-mold techniques consolidate parts by applying either atmospheric pressure (vacuum bag molding), higher pressures (pressure forming), or a combination of high temperatures and high pressures (autoclave). All of these techniques have one thing in common: an outer bag that squeezes the various hand-laid fabrics or pre-pregs into a final part that can achieve void contents as low as 0.5%. In many ways this process is similar to techniques employed in thermoforming, but here there must be a way to remove air and volatiles from the fiber–resin assembly. Like the preceding open-mold processes, vacuum bag molding, pressure forming, and autoclave methods are amenable to lar ge parts such as aircraft wings, aircraft tail sections, satellites, and boat hulls in the marine industry . Large sizes apply equally as well to the autoclave method, which is a sophisticated name for a vessel whose temperature (up to 500°F) and pressure (up to 500 lb/in^2) are carefully controlled. Autoclaves units as large as 10 m in diameter by 30 m in length exist [14,66, 68–70]. Figure 8.10 shows the essential features of this process.

Each of these processes be gins by applying a mold release to the mold. Mold releases can be w axes, fluoropolymers, silicones, and the like, but care should be taken to ensure compatibility of the gel coat (or resin) with the rest of the resins and fibrous assembly. Depending on the part design, there are times where a nonporous mold release layer or edge dam is placed at the edges of the assembly to a void edge leakage. Raw materials for this process are then put into place over this layer. In vacuum bagging or pressure forming the materials can be essentially the same as with hand layup or spray-up, whereas in the higher -technology autoclave process, raw materials typically consist of about 60% f iber volume content pre-pregs, tapes, or pre-resin-impregnated fabrics to achie ve the lowest void contents possible. When using B-stage thermoset pre-pregs, resin life issues exist. Most thermoset pre-pregs have a lifetime of 8 days without refrigeration. If the pre-preg was placed into a cool environment, it should come to room temperature prior to opening it to the outside temperature to avoid moisture blisters. Pre-pre gs should exhibit some tack, which can be useful for holding inserts securely in place. There are no life issues with thermoplastic pre-pregs, but the use of high temperatures no w becomes mandatory unless the thermoplastic material is being polymerized during the process step. General design rules discussed in the previous open-mold sections apply here as well with regard to fiber orientation and general assembly . A separator film is placed

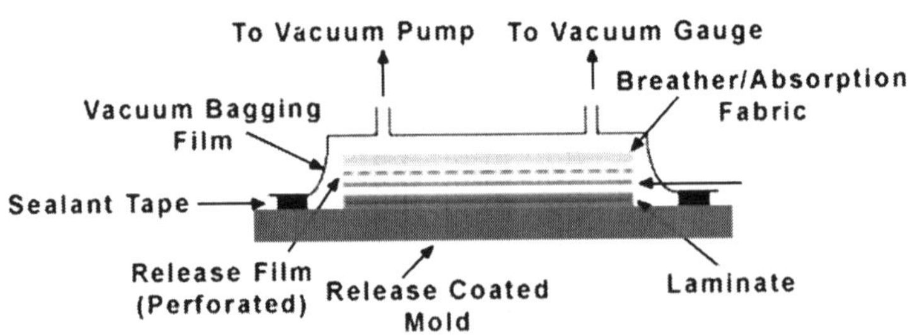

FIGURE 8.10 Vacuum bagging method. (Courtesy of Owens Corning. Used with permssion.)

around this assembly next. This layer should be slightly wider than the composite structure, and it normally consists of a high-elongation conformable perforated film.

The separator must be capable of releasing from the composite structure below while allowing vapor and typically some resin to pass through it. Ideally, the resin passing through this layer is captured in a bleeder fabric, which is normally a nonwoven polyester, treated glass, or cotton fabric. The bleeder ideally absorbs the excess resin while being sufficiently conformable to avoid transmitting wrinkles from the bag to the part. The next possible/optional layer is the barrier film layer. This is another film layer, but here the film has many fewer perforations. This layer of perforated Mylar or Teflon film should, under ideal conditions, be the final barrier to resin flow. This layer is then followed by an optional breather layer to distribute the air evenly and to avoid transferring bag wrinkles below this level. Essentially the same type of materials as the bleeder material can be used here, only these materials tend to be more porous and not collapsible under pressure. Finally, sealant tape is applied around the outer periphery of the part, thermocouples are carefully placed and sealed, and the final bag assembly is placed around the outside of this assembly. Vacuum bags can be polyvinyl alcohol, fluoroelastomers, nylon, coextruded nylon, or possibly reusable silicone-based materials, depending on temperatures and pressures applied. If holes occur, especially around the vacuum port, patching is possible through the employment of patching RTVs. There are many possible options as to the assembly of these layers. Sometimes there will be no barrier or breather layers and a resin trap and/or resin dam is employed to prevent the resin from getting too far in potentially clogging the vacuum lines [14,66].

When an autoclave is employed, a typical heat-up cycle consists of a slow warm-up, a hold temperature, and pressure that initiates resin flow, a secondary higher temperature heat-up, which is the high-flow region, a longer hold at a higher temperature, where flow ends and the thermoset cross-links, followed by a slow cool-down. Note that these cycles can easily last for hours [66].

General design rules for all bag methods are similar. The outcome of these processes is to achieve uniform resin–fiber ratios and low voids while maintaining the desired part geometry. The following rules apply:

1. Sharp corners and angles should be avoided. These will be stress concentration points for both materials as well as the vacuum bag itself. Gentle curves in the vicinity of 0.187 to 0.5 in. are recommended [66].

2. Rapid thickness changes should also be avoided by changing thickness gently as with a card deck that is spread across the table such that each card is in contact with the prior and subsequent card.

3. Rubber inserts can be employed when holes reside within a material. It is easier to use inserts when employing pre-pregs since these materials will be less drapable and stiffer than fabrics.

4. Bridging can be avoided by using "ears" or longer flaps to enable the bag to move into these regions without puncturing, or by using either foam or rubber inserts as stress intensifiers in a specific point. The intensifiers should be placed somewhere between the barrier film and the bag assembly.

5. As mentioned with other open-mold methods, cull plates can also be applied both as stress intensifiers and as a means of achieving a smoother inner surface away from the mold [66–70].

It is imperative with these processes to achieve good sealing. Leaks occur predominantly at the interface between the bag and the vacuum line. Proper care should be taken to cause the resin flow to either stop, as in the case of dams, or to be contained, as with resin traps. The vacuum lines should be checked for resin and pinch points whenever insufficient vacuum is observed, and the initial design of these lines must avoid large pressure drops associated with loss of pressure around elbows, fittings, or a long tube length to the vacuum pump [14,66]. Finished parts often need trimming. This is accomplished either manually or robotically through routers, water jets, and on high-end lasers. Holes can readily be drilled using normal methods.

The primary advantage of this technique is the automatic removal of VOCs through the vacuum lines, the lower void contents that are possible with this technique compared to other open-molding methods, and the large part size capability. The disadvantages reside in poorer surface quality of parts compared to closed-mold processes, very high costs of the autoclave process (when used), long cycle times, and the distinct possibility that thick parts will require multiple cycles to obtain desired porosity/strength levels [14].

8.2.3 Closed Mold Methods

8.2.3.1 Resin Transfer Molding

Resin transfer molding (RTM) is a low-pressure process that involves injecting a low-viscosity thermoset resin into a closed cavity containing fibrous preforms. Medium to large parts, including boat hulls, automotive parts, truck parts, seats, saunas, shower enclosures, and bathtubs, are typical applications. The biggest hurdle to overcome is to partition the injected resin evenly to avoid unwet sections, high resin contents, and trapped air. When done properly, fiber contents as high as 50 to 60% have been achieved through this technique. Ideally, the resin can be quickly partitioned throughout the part with a manifold system, and then the resin fills the mold such that the resin-free volume or area of the part is constantly being diminished. This approach purposely forces the air from the mold in a symmetric manner, such that the last filled section of the part through a vent or vacuum. Vacuum-assisted RTM (VARTM) can be very beneficial in pulling the resin through the part and in removing the air pocket. The choice of fabric or wet chop sizing chemistry can be advantageous in achieving full wet-out through a wicking mechanism. Figure 8.11 captures the essential details [14,49,68,71,72].

The preform comes from either manually combining continuous or chopped strand mats, spraying preform roving to form a preform, or better still, spraying fibers and resin onto a screen that pulls vacuum. To avoid potential knit/weld lines, the preform must fill the entire mold, and because of its bulk, it should remain stationary as the resin fills the part. There are really three methods for creating a preform. The most primitive

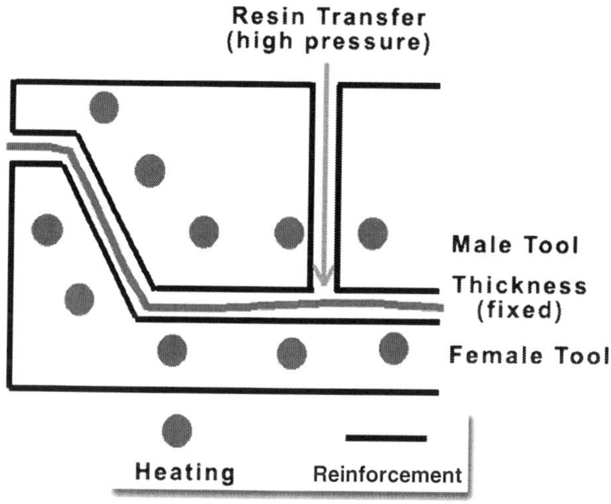

FIGURE 8.11 RTM process in operation. (Courtesy of Owens Corning. Used with permission.)

method is to assemble the entire assembly by hand and optionally, to sew it together. This method has the drawback of higher cycle time and less uniformity than is possible through other methods. One advancement of this method is to use the direct fiber method, where chopped rovings and binder are blown onto a preform mesh. The third method involves using a plenum where fibers are chopped and resin or powder is sprayed onto a mesh that holds the assembly in place through vacuum pressure. Both the direct fiber and plenum methods can be done robotically to save on labor and to achieve better preform uniformity [14,49]. Figure 8.12 shows a typical plenum process, the P4 process, in operation. The same rules of why a particular fabric is employed and why the fabrics orientation should be symmetrical in hand layup apply to RTM.

Resin viscosity should be at 1 Pa·s or less during the mold process, whereas the lifetime of the resin should be fairly fast (2 hours or less) to reduce the molding cycle. Because of the low viscosities employed, fully polymerized thermoplastic materials are unlikely ever to be employed in RTM. Only prepolymerized thermoplastics such as cyclic poly butylene terephthalate or in situ polymerized thermoplastics are possible. Polyester, polyurethane, and epoxy resins are commonly employed. Mold releases should be applied initially for good part release. If a good exterior appearance is desirable, a gel coat should be added prior to addition of the remaining materials [14,49,68,71,72].

From a tooling perspective, the RTM or VARTM process are low-pressure processes that require a match mold die optionally equipped with a vacuum system. Since the compression mold equipment is not utilized, it is a much less expensive than the SMC/BMC process. Molds made out of composite tooling are not uncommon. Tooling and dimensional criteria discussed in the molding section are applicable here. If optional vacuum methods are to be employed, the outer mold rim should employ an O-ring to seal the mold firmly against air links. The resin pumping

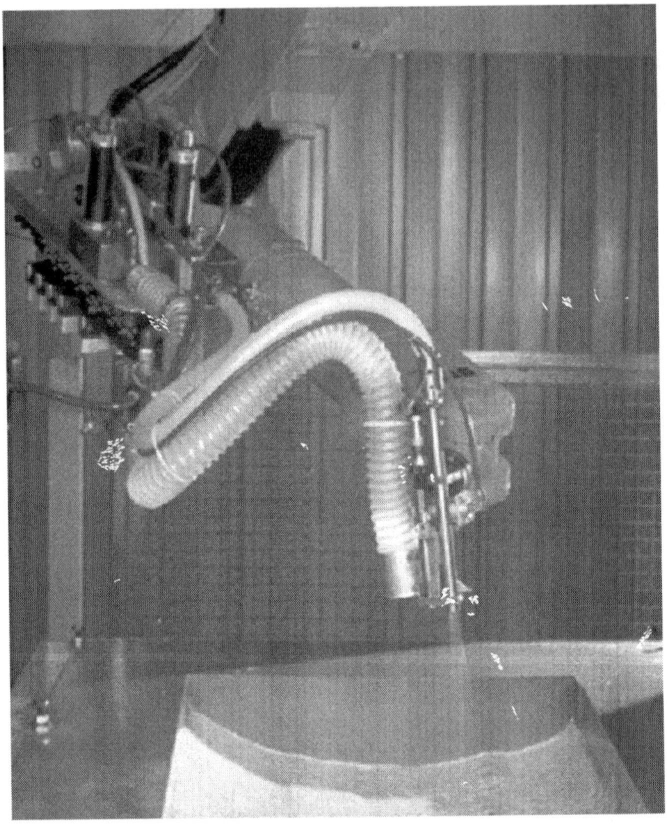

FIGURE 8.12 Plenum preform method (Owens Corning's P4 process). (Courtesy of Owens Corning. Used with permission.)

method should employ the single valve as opposed to the multivalve for easier maintenance and fewer operator mistakes [14,49].

The major advantages of this method are the ease of day-to-day operation, faster cycle times than by hand layup if nonmanual preform methods are emplo yed, good surface quality compared to open-mold methods, low exposure to chemicals, and long fiber lengths compared to the SMC/BMC operations. The disadvantages of this method are the need to ha ve skilled mold craftsman and resin-rich areas in ribs and exterior portions of the mold [14,49].

8.2.3.2 Sheet Molding Compound

The sheet molding compound process delivers the best-appearing parts that the composite industry can of fer. Parts produced through this technique f ind heavy use in class A exterior automotive parts, such as hoods, doors, trunks, and in home articles such as bathtubs, sinks, showers, appliances, electronic components, and business machines. The long fibers present in these material provide good stiffness and impact resistance [73,74].

This process begins by forming a doughy sheet. Roving input is chopped to a length of 0.75 to 2 in., although there is really no reason why mat input, other than known flow issues, could not be used, as well as either chopped or unchopped input, along with fillers, low-profile additives, catalyst, catalyst aids, and a thermoset resin, normally a polyester-based thermoset, or possibly epoxy and vinyl ester resins have been used. This assembly is then compressed between a series of compaction rolls to work the resin into the reinforcing medium. The initial paste has a viscosity near the range 40,000 to 100,000 cP. Figure 8.13 displays the essentials of the SMC sheet- making process, and Table 8.7 provides rudimentary formulation details. The roving should ideally be uniformly distributed between two plastic polyethylene carrier sheets which are laden with the full thermoset chemistry package. A slight alteration to this process can be used to create a thick molding compound (TMC). Here the roving and binder are mixed together intermittently through an early nip, and a very thick sheet can be created. Whether it is SMC or TMC, material from this process is then wound into a package and placed into storage for 1 to 2 days until the resin cures to the B stage. The thickening during this time period is usually assisted with such additives as magnesium

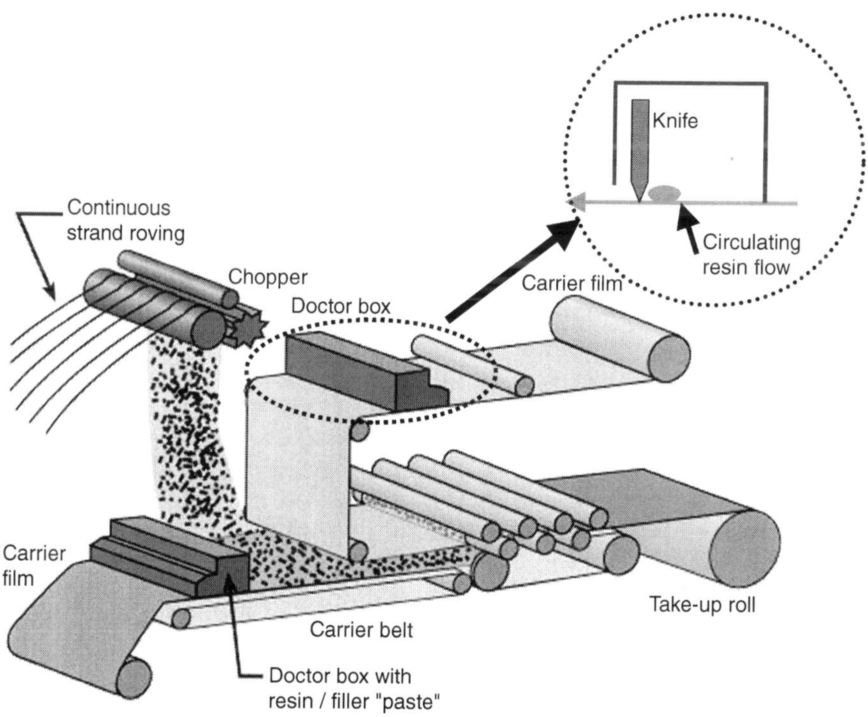

FIGURE 8.13 Sheet molding compound (SMC) process. (Courtesy of Owens Corning. Used with permission.)

oxide, magnesium hydroxide, or calcium hydroxide to achieve B-stage viscosities in the range 60,000 to 750,000 poise. Properly B-staged material are then staged (time-wise), preweighed, and placed into a heated compression mold (250 to 350°F), which under pressure (500 to 2500 lb/in^2) forms a composite part that can subsequently be painted to create world-class surface quality parts. There is also some recent technology that uses lower temperatures and pressures to achieve faster part cycling. Typical compression molding cycle times of 30 seconds to several minutes are possible, with thicker parts requiring more dwell or curing time [14,68,74,75].A typical compression molding apparatus is shown in Figure 8.14.

Low-profile additives (LPAs) are an integral part of SMC technology. If the formulation does not contain this ingredient, it is virtually impossible to achieve automotive class A parts via profilometer (roughness meter) readings. Without the use of a low-profile additive or low-shrink additive, thermoset polyester resins shrink on the order of 7%. Typical LPAs are made from polymethyl methacrylate, polyurethanes, polycaprolactones, or, in particular, polyvinyl acetate with 1% acid functionality. It is important that the LPA not be fully compatible with the polyester resin such that phase separation occurs, but compatible enough to avoid huge domain differences [74].

TABLE 8.7 Rudimentary SMC Formulation

Material	Percentage (%)
Resin (typically, polyester)	25–30
Fibers (typically, glass)	25–30[a]
Fillers ($CaCO_3$, ATH)	40–50
Other (LPAs, initiators, thickeners)	3–5

[a]Structural parts will use more fiberglass, which results in less surface quality but more structural performance.

FIGURE 8.14 Compression molding operation. (Courtesy of Owens Corning. Used with permission.)

One important consideration is the ability to fill the mold with a few well-placed blanks. Most of the time, blanks are placed in the center of a heated mold and flow occurs outwardly to fill the part. If continuous roving and/or matting is used, the continuous fibers will not flow as far as the chopped rovings and the initial blank sizes will have to have more mold coverage or be placed purposely in the part where added structural enhancement is necessary. It is easily possible to run into viscosity issues with SMC. Low viscosity results in less glass flow and parts with a high resin fraction in the outermost region. On the opposite extreme, high viscosity can easily lead to unfilled parts. Clearly, fiber length, proper resin mixing/formulation, temperature control during storage, and inventory/staging control are important process aspects. When in doubt, several programs exist, such as Cadpress,[2] that can provide theoretical predictions for mold flow, cycle time, and fiber orientation [76].

There are two very common defects associated with SMC: blistering and paint pops. To avoid blistering, SMC materials must remain in a fairly dry humidity-controlled environment prior to molding, and the forming part must be cured sufficiently throughout. Poor temperature uniformity can promote blister formation. The dreaded paint pop defect arises during secondary painting processes. These are typically due to an entrapped gas, whether air, water, or solvent, heating up, and eventually, due to the higher pressure, breaking the surrounding material to form a pop. One way to minimize this defect is to try to provide pathways for whatever gases are being emanated. A second alternative is to apply an in-mold coating or in-mold coating film over the original SMC part during the SMC compression step. The problem with the film method is that the part cannot have too much three-dimensional structure (e.g., deep draws, deep ribs) or the coating film will not drape well over the preexisting part. Paint pops can be removed by postpatching, but this is a secondary process that should be minimized [14,77].

In summary, SMC has some advantages whenever large parts of high surface quality levels are required. The main disadvantages are the high investment cost and part size limitation compared to vacuum molding or hand layup.

8.2.3.3 Bulk Molding Compound or Dough Molding Compound

Bulk molding compound (BMC) or dough molding compound (DMC) employs a somewhat similar process to SMC, except that the input material comes from an extruded log that contains 5–10% less fiber with shorter fiber lengths ranging from 0.125 to 0.5 in. The log can either be made on site or purchased from a BMC compounder. Figures 8.15 and 8.16 display typical BMC mixers and molding equipment. As with SMC, the log is allowed to proceed to the B stage and is then placed into a hot compression mold, or it can be injection molded. Typical mold temperatures run from 250 to 350°F with application pressures of 350 to 2500 lb/in^2. Cycle times of 30 seconds to several minutes are possible, depending on final part thickness, with larger thicknesses requiring more cure or dwell time. The same issues of viscosity control apply from SMC along with the need to avoid water absorption, which will

[2]Cadpress is a registered trademark of the Madison Group.

FIGURE 8.15 Typical BMC mixer. (Courtesy of Owens Corning. Used with permission.)

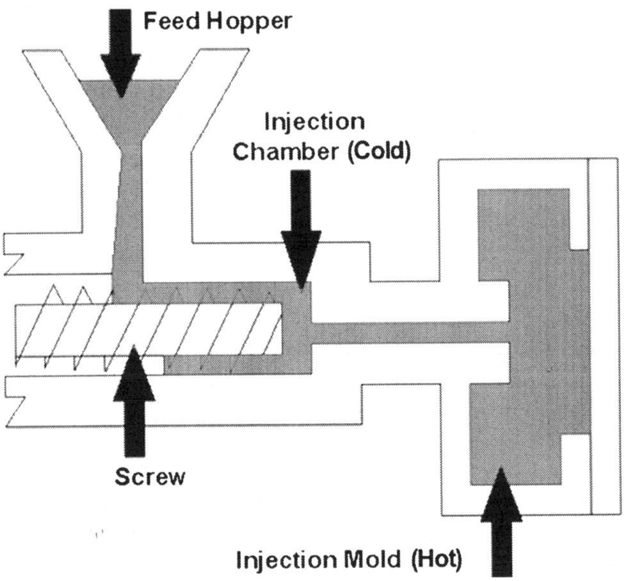

FIGURE 8.16 Typical thermoset or BMC injection molding. (Courtesy of Owens Corning. Used with permission.)

result in waviness or blistering in an extreme case. Typical parts consist of electrical housing, appliance housing, and air-conditioning housings [5,14,68].

The major advantages of the BMC process match those of the SMC process: high production rates, low labor skill level, and excellent part appearance. The main disadvantages of this technique are high investment costs combined with part size limitations. BMC also has the added disadvantage of less fiber content at a lower fiber glass length such that the mechanical performance of a BMC part will be inferior to that of an SMC part [14].

8.2.3.4 Glass Mat Thermoplastics

Glass mat thermoplastics (GMT) is the thermoplastics counterpart of the sheet molding compound process. Blanks, which are formed through a prior thermoplastic lamination process (see Section 8.3.1), are preheated through several zoned ovens, placed into a cold mold, and then compression molded to form a two-dimensional compression molded part. The main disadvantage of GMT compared to LFT is fiber uniformity, while the main advantages reside in the longer glass length, which provides impact strength and added stiffness. The vast majority of GMT materials are formed from polypropylene–glass laminates which employ either random continuous needled webs, unidirectional glass fiber webs, short chopped fiber wet-laid web, weaved materials, or commingled fibers. The stiffness and added impact energy of a long fiber polypropylene product is difficult to match with any other materials. The most famous GMT application is automotive bumpers. Other known applications include basketball backboards, automotive consoles, automotive door modules, lines, hockey sticks, snowboards, doors, and other relatively flat articles [41,43,45,78,79].

GMT begins by preheating blanks through a three- or four-zone infrared oven or air-recirculation oven. The blanks loft and become "flowable" in the subsequent compression molding operation. During the preheating step it is important that uniform temperatures be achieved throughout the blank. One trick that is normally done toward this goal is to use a slightly cooler final oven zone such that the hot exterior of the lofted blank will cool due to the surrounding cooler air and heat the cooler center of the blank. Typical blank temperatures should reach the range 200 to 230°C. Hot blanks must be transferred quickly to the compression molding machine. Slow transfers can easily lead to unfilled parts since the viscosity of this material will rise rapidly as the thermoplastic begins to crystallize or solidify [45,80].

Blank placement within a compression mold is critical. Unless it is desirable to fill a rib, blanks should never be placed side by side with each other. This will result in a weak meld or weld line that could easily be avoided by using a larger blank size. Initial blank sizes typically cover 80% of the mold. The type of blank should be based on the mechanical forces or other outside influences on the final part. If strong properties are needed in a given direction, unidirectional fibers may be required, but if a part needs to get fibers into a thin rib, a shorter chopped version is a better choice since the chopped fiber versions will flow better than the continuous versions [80].

Molding machines are normally set to 25 to 70°C with mold pressures of 10 to 14 MPa [45]. Closure rates range from about 900 mm/s for fast closure down to 30 to 70 mm/s for the actual part compression speed [81]. Lower mold temperatures are normally associated with faster cycle times. Cycle times of 20 seconds to several minutes are not uncommon. Many of the essential mold features were described in the earlier matched die mold section. Weld or knit lines should be avoided by cutting holes or adding inserts during secondary operations unless the part can survive the critical stresses with a well-placed weld line.

There are several common defects to be wary of in the GMT process. Bubbles or blisters can occur, but unlike in the SMC process, these are absolutely not due to moisture. They are due instead either to mold hot spots (improper cooling design) or are due to attempting to remove the part from the mold prior to full thermoplastic crystallization or solidification, especially in the thickest mold sections. The solution to this problem is either to live with a longer dwell time or to redesign the mold cooling section. A second common problem with GMT is higher resin content in ribs or the exterior part periphery. Longer, continuous fibers will inherently flow less than their chopped fiber counterparts. There are times when the chopped alternative must be employed or when a hybrid loading is used, with the more continuous materials in the center of a stacked blank input and the chopped fibers toward the outside. A third common issue is postpainting. If this is desired, the material should be run through a flame treatment, corona treater, or other methods of raising the material surface energy prior to painting or applying decoration. In-mold coating technology can be used for parts that do not have heavy draws or ribbing. The fourth issue with regard to GMT is to be wary of part shrinkage and warp. Polypropylene is a highly crystalline thermoplastic that can undergo large shrinkages. One way of mitigating this affect is to have part fixtures near the mold that force the cooling part to maintain the final desired shape at the expense of having higher internal stresses [45,78–80].

8.2.3.5 Long Fiber Thermoplastic or Direct Long Fiber Thermoplastic Processes

Long fiber thermoplastic (LFT) composites represent a rapidly growing thermoplastic composite business segment where either (1) long fiber pellets or (2) rovings combined with plastic pellets are used to create injection- or compression-molded parts of longer fiber lengths. These methods differ from GMT since the starting material comes from either pultruded long glass pellets or directly from a combination of thermoplastic pellets and fibrous reinforcements. Thermoplastic resins have included polypropylene, various nylon resins, polyphthalamide, acetal, polyurethane, and thermoplastic polyesters (PET/PBT), to name a few. Fiber inputs are mostly glass rovings followed by carbon fiber, natural fibers, and stainless steel (stainless steel is used mostly for electrical properties). The value of fiber length resides in the unique combinations of greater stiffness and impact resistance that are unattainable in normal thermoplastic extrusion-grade materials of the same matrix material.

LFT mechanical properties are very dependent on process methods, part thickness, gating design, injection molding variables, and even injection molding equipment itself. In all cases, preserving fiber length is absolutely critical in achieving optimum performance. The fastest way to check fiber length retention is to burn off a molded part and observe whether a skeletal structure remains. If the original part turns to dust, which is a common mistake when applying this technology, or if it cannot be picked up without disintegrating, the following quick checklist should be consulted [82]:

1. Use a compression ratio screw from 2 : 1 to 2.5 : 1 combined with a free-flow screw tip.
2. Employ a short nozzle, minimum nozzle orifice of 8 mm, and a sprue orifice of 9 mm.

3. Avoid fiber breakage by using large gates at least two-thirds the part thickness, avoiding 90° bends, feeding at the thickest part location, and using amply-sized runners.
4. Avoid fiber breakage by setting screw speeds under 60 rpm with back pressures under 3.5 bars.
5. Set the molding temperatures higher than for compounded grades to obtain further resin lubrication.
6. Employ sufficient hold pressure and time to reduce voids and sink marks.

Although injection molding is the primary part-forming method, there are a variety of other approaches, including injection compression molding, where a fibrous ingot is initially injected into a partially open cavity and molded [82,83]; structural foam molding, where blowing agents are added to create a foamy composite structure [82]; gas-assisted injection molding, which is used to create hollow composite parts [82,84]; and extrusion compression molding, which is a compression molding variant of the injection compression technique [82]. Many of these techniques, particularly compression molding of a fibrous ingot, may be alternatively done through direct long fiber thermoplastic (DLFT) techniques [85].

Long fiber mechanical properties are naturally difficult to measure since mechanical characterization in each section of the mold will be a consequence of fiber length (be wary of any study that does not measure fiber length in the final part) and fiber orientations in a skin–core structure where fibers in the skin are aligned in the flow direction and core fibers are aligned transversely. Mechanical properties are known to depend on part thickness, mold filling time, wall temperature, and diverging and converging flow fields, to name a few parameters [82,86–88]. Since mechanical properties vary through the part, prototype parts should always be tested sufficiently prior to market introduction.

A major weakness of LFT materials occurs where weld or meld lines exist. Either the part should be designed to place these weak points in noncritical areas, or specialized equipment can be used to vibrate flow fronts back and forth to jiggle fibers across the weld or meld line. In short, weld lines are an integral design consideration that cannot be ignored [80].

The testing and design period can be reduced through fluid dynamic models that are capable of predicting how the mold fills, where the weld lines occur, and rough fiber orientation distribution of the skin and core layers in molded parts [89–92]. The skin always has fibers located in the mold filling direction, whereas the core has fibers oriented in the transverse direction. This comes about due to the fountain flow filling phenomena and cooling and viscosity features [46]. Clever part designs take advantage of some of the fiber orientation predictors to design the mold and flow fronts for optimum part performance. When these techniques are taken into consideration with the criteria described earlier, the time to market can be reduced significantly.

From a historical and marketing perspective, long fiber thermoplastic and more recently, direct long fiber thermoplastic technologies have come a long way since the 1960s, when Wilson Fiberfil introduced wire-coated concentrates to the injection molding community. The early poorly impregnated wire-coated products did not achieve the mechanical performance that was possible by impregnating molten polymers into the rovings. PCI (now Ticona/Celstran[3]) and ICI (now LNP/Verton[4]) discovered that by passing the rovings around "heated spreaders" or "heated bumps" it was possible to impregnate the rovings with a thermplastic melt [93]. Some of the original compositional patents prevented many competitors from entering the field, but with the expiration of these patents, LFT has become increasingly popular and segmented. A number of companies have joined Ticona and LNP, including Sabic, Montech Technologies, RTP, Neste, Vetrotex, and CPI, to name a few [94]. Several companies, such as Baycomp, have elected to go with a powder impregnation technique to achieve long fiber pellets; however, the cost of obtaining a fine powder has restricted this route with some resin systems.

With the growth of this industry, a number of equipment vendors have developed proprietary equipment to handle rovings and resin pellets directly in creating DLFT composites. These methods effectively eliminate the middle compounding step in exchange for equipment investment and operational complexity. Advantages of DLFT include a single heat history and possibly long-term cost savings [95]. Depending on the part volume, it may be worthwhile to consider the equipment investment to save on raw material costs. Companies such as Lawton, Woodshed DLFT, Krauss Maffei, Husky, Coperion, and Berstorff have all developed DLFT proprietary equipment. The concept here is either to use the extruder itself and/or converging rollers to pull the rovings into the extruder, cut the fibers, gently spread and mix the cut fibers with hot thermoplastic resin, release any vapor such as the entrained air or moisture coming in with the roving, and either force the material into an injection molding machine or optionally, create a hot charge for compression molding applications.

Roughly 90% of these materials find their way into the automotive industry; the remainder occur in lawn and gardening, marine, sporting, gears, and construction applications. Typical automotive applications include instrument panels, technical front ends, bumper beams, underbody/heat shields, battery trays, door systems, load floors, and seat structures, to name a few [96]. With all of the newly developing equipment technology in this area, it is somewhat surprising that 90% of the business remains in the form of direct compounded pellets, with the remaining 10% turning to DLFT for large-volume automotive applications. The most economical route for large-volume parts remains open to debate, with the introduction of superconcentrated pellets at 70 to 75% glass concentrate [94,97,98]. Time will tell which route captures the most market share.

8.3 OTHER COMPOSITE PROCESSES

8.3.1 Lamination

Lamination processes intermittently combine/adhere several input materials into a single output material. These processes can employ relatively low, nearly impulse pressure points as found in a single hot nip, consecutive hot nips, or a Meyer laminator, which consists of a double continuous belt wrapped around several heated

[3]Celstran is a registered trademark of Ticona GmbH, Germany.

[4]Verton is a registered trademark of LNP Engineering Plastics/General Electric.

nip sections, or they can employ high pressure as found in continuous double-belt pressures that apply pressure either at numerous small roller points (Sandvik) or in a continous fashion (Hymenn).

8.3.1.1 Low-Pressure or Hot Nip Lamination

The simplest lamination combines various materials through a single or successive hot and cold nip(s) where one or more of the materials acts as an adhesive to bond the remaining materials together. Nips serve as hot pressure pulse points to compress and consolidate materials in the thickness direction. This method can be used to fasten nonwoven facers reinforcement to other materials, to add desired films to other substrates, or to consolidate composites through a successive series of nips with tighter gap settings. Figure 8.17 displays the essence of a typical low-pressure lamination operation.

Roisum [22,99,100] does an excellent job of describing the mechanics of a nip operation. Ideally, both nips should be driven with an independent drive to prevent rollers from fighting each other due to differences in the diameter of the rolls usually caused by uneven wear. Although it is possible to change the nip pressure online, through techniques such as changing the tilt on one roller, a smaller roller on a larger roller, intermediate loaded rollers, or controlled crowned rollers, applying pressure at specific roller points, deflections should be minimized in most operations. Nip materials can vary from metals to various rubbers, depending on how hard the desired nip should be combined with how long it should last. Both rollers should not show uneven diameter differences across their width, unless crowning is desirable (TIR=total indicated readout), should be aligned under regular intervals in the machine direction/cross direction, and should be controlled under either gap or load control methods [22,99,100].

Gaps should be checked regularly with nip impression paper. An excellent source of nip impression paper can be found from Sensor Products. When nip uniformity is in doubt, nip impression paper can quickly resolve the question of whether the nip is a simple one or as complicated as a double-belt laminator [101].

Curl or warp is another very common lamination issue. Either one of this issues is related directly to differences in strain across the final laminate. If curl is related to moduli differences (not temperature), it is sometimes possible to alter one of the

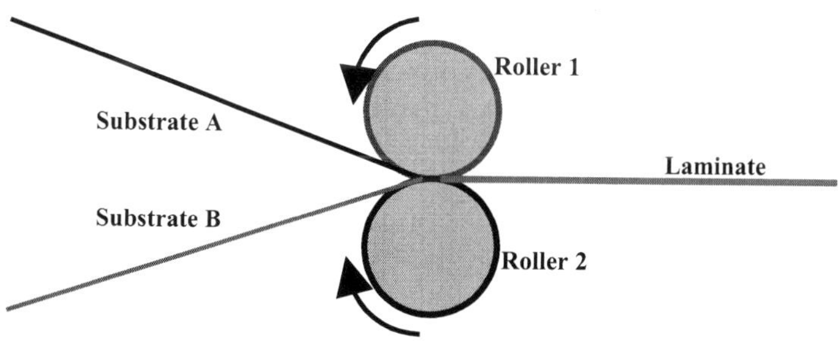

FIGURE 8.17 Low-pressure lamination process.

incoming tension to match strains. If tensions do not work, it becomes necessary to examine other variables, including laminate construction. Equation (8.8) expresses the strain matching relationship to prevent cross-direction curl, and equation (8.9) provides the strain matching relationship in the MD direction:

$$\frac{T_A}{T_B} = \frac{t_A}{t_B}\frac{E_A}{E_B} \tag{8.8}$$

$$\frac{T_A}{T_B} = \frac{t_A}{t_B}\frac{\nu_B}{\nu_A}\frac{E_A}{E_B} \tag{8.9}$$

where T_A and T_B are the tensions of materials A and B, respectively; t_A and t_B are the thickness of materials A and B, respectively; E_A and E_B are the moduli of materials A and B, respectively; and ν_A and ν_B are the Poisson ratios of materials A and B, respectively. Clearly, the Poisson ratio effect can absolutely make unwarped lamination a difficult, if not impossible task [99]. Perhaps the best way of avoiding material-induced warpage issues is to construct laminate structures in an absolutely symmetrical and orientationally balanced fashion [80].

From a safety perspective, higher-pressure nips should have a maximum load response such that the control method switches to load control [12]. Although this mechanism does protect equipment, the inertia of the equipment is unlikely to open or stop the nip point prior to seriously injuring an operator by pulling an arm off. Operators should stay clear of the nip point either by putting equipment into caged areas [102,103] where possible, adding light curtains where necessary [59–62], or as an absolute minimum precaution operating training programs coupled with lock out and tag-out maintenance procedures [63,64].

8.3.1.2 Thermoset Lamination

Translucent or opaque paneling is often made through a thermoset lamination process. This process begins in a very similar manner as the SMC manufacturing method, but now the material is cured directly through some ovens, as opposed to a storage period/compression mold, to form flat or corrugated paneling. Roving can be added to either side of this construction to prevent a limp edge and ease the final trimming process. Corrugation is accomplished through shaping rollers or by gradually conforming the material to the desired corrugation shape within the curing oven [67,68].

To prevent residual stresses or undesired warp, it is important that the material be cured evenly from side to side and that the material be uniform in resin content. One method of checking for this is to run thermocouples along the top, middle, and underside of the material and proving that both sides of the laminate experience the same temperature history [68,80]. Low-profile additives can greatly assist in minimizing shrinkage and warpage issues [67,68].

Thermoset lamination also applies to the decorative high-pressure laminate (HPL) area. Here layers of melamine- and phenolic-treated papers are combined in isobaric high-pressure double-belt laminators to create decorative articles that are used to decorate countertops, laminated floors, office furniture, cabinets, and the like. The outer layer, usually consisting of a high-wear overlay, employs a combination of melamine and aluminum oxide particles to create a highly resistant layer. This is followed by a printed paper layer and a phenolic-treated kraft paper underlayer to balance the laminate [104–108].

8.3.1.3 Thermoplastic Lamination

GMT blanks or thermoplastic paneling are formed by combining thermoplastic resin either in film form or as a hot melt along with reinforced mats under the influence of high temperatures and pressures. This is accomplished by squeezing the materials together between two hot double belts. Pressure may be applied either in a continuous manner through air pressure (isobaric) [109,110], through a series of small rollers that make contact between presses and the double belts (isochoric) [109,111], or through discontinuous nip points that are applied to the double belts [112]. The application of continuous isobaric or isochoric systems is absolutely essential in achieving wet-out of thick thermoplastic materials. For this reason, either the multiroller method (isochoric) or the continuous high-pressure method (isobaric) is recommended. The advantage of the isochoric roller design is the heat transfer efficiency since metal rollers are always making contact between the hot presses and double belts, whereas the isobaric laminator has the advantage of an entirely uniform pressure [80,109–111]. Lamination of GMT or thermoplastic paneling should always apply pressure directly to the consolidating panel to obtain complete wet-out.

Thermoplastic films or extruded melted resins are placed above, below, and in-between the two-dimensional fibrous mats/veils. If resins are pumped onto the belts or the fibrous mats, it is important to use an extrusion die with adjustable lip settings and deckles, adjustable devices that restrict flow in the cross direction [113,114]. Fibrous mats can include a variety of inputs, including weaved mats, randomly swirled rovings which are held together through glass needles, continuous rovings combined with a veil or random mat swirls, short fiber veils formed through a wet-laid process, or dry-laid veils with medium-length fibers formed through a dry-laid process such as the Rando–Webber process [80,110]. Reinforcing input should either come in boxes with the ending tail being available for splicing to the next box, or equipment such as flying splicers, accumulators, or floating loopers to obtain a continuous operation. Light defect detection equipment may be employed to find defects in any unwinding roll stock [116].

The mass of resin and fibrous inputs are then combined and passed through a hot pressurized zone. Sufficient pressure must be applied to obtain low void contents. Although edge dams can prevent the leakage of high flow resins, they may be avoided, and it is important to avoid using such high pressures that the impregnating resin leaks out from the laminator belts. Edge dams have the advantage that resin leakage would never occur at the expense of possibly acting as stops. Proper maintenance should be performed to maintain a continuous flat profle with a flat belt, flat rollers, or any other equipment that affects the conformation of the material between the two belts [80,104]. Carbon paper or special films may be employed to determine if the pressure profile is truly flat [100,101].

After the material passes through the hot zone, it is imperative that it be evenly cooled on both sides during the pressurized cooling section; otherwise, warping will result as a result of one side crystallizing sooner than the other. The best way of resolving this issue is to place thermocouples between the various layers and to ensure that both sides of the composite cool uniformly [80]. It is also possible to obtain warpage if there is extra resin at points across the width; if this is occurring, die adjustments will be necessary to even the resin percentage across the width.

Finally, a gauging system should be used to ascertain basis weight and thickness quickly [117]. This saves an inordinate amount of time in setting the die lips properly. Produced laminates are subsequently wound for paneling applications or slit and cut to desired blank sizes for subsequent GMT compression molding [45].

8.3.1.4 Foam Lamination

Very lightweight structures can be formed through the foam lamination technique. Unlike thermoplastic lamination, where pressure should be applied to material directly, in foam lamination a height is fixed and the foam expands under heat to a set height. Typical materials such as polyurethane or phenolic resins are foamed between two nonwoven substrate such as fiberglass veils or CSM to form insulation boards that are used in the construction industry [50].

8.3.2 Pultrusion

Pultrusion is the composite counterpart to thermoplastic extrusion. The pultrusion process represents an efficient, low-labor, and reasonably good line-speed process for creating fishing poles, golf club shafts, I-beams, rebar for construction, ladder rails, hammers, walkway gratings, antennas, door lineals, window lineals, and oil well suction rods, to name a few applications. Profiled shapes are formed by pulling roving, mats, and/or veils through a thermoset bath or hot thermoplastic melt, impregnating the strands by wrapping the reinforcing media around pins to wet the roving, whether the pins are in a prior thermoset resin bath or a heated chamber for thermoplastic-based materials, and then pulling the wet-out assembly through a die. The die is followed by an optional oven section for a thermoset material, a cooling section for either thermoplastic or thermoset pultruded articles, a large puller, and then either a pelletizer, if long fiber thermoplastic pellets are desired, or a moving carbide- or diamond-tipped saw if large linear parts are desired [14,68,118–121].

8.3.2.1 Basic Pultrusion Technology

Pultrusion technology begins at the unwind stand, where input rovings, mats, or veils reside. Rovings exist in both an outside unwind manner, where there is zero twist but splicing timing issues exist, or inside unwind manner, where normally there is a twist combined with splicing advantages. It is not uncommon to have as many as 1200 to 1500 rolls of materials on creel tables/stands at a time, which forces painstakingly careful setups. When the input comes in roll form, carousel delivery systems are advantageous in providing sufficient space such that all of the rolls can be accessed without disturbing the others. Stands normally are equipped with some form of tension control and guidance system to begin the process of aligning materials into the pultrusion region.

From here the reinforcement media contacts and entrains resin—the exact manner depends on resin type—into a shaping region and then a pultrusion die. Careful zoned temperature control must exist within this region to cure thermoset resins in a manner that avoids overheating or to melt thermoplastic resins sufficiently to obtain good wet-out. The material ultimately enters a pultrusion die where it is shaped, cured (thermosets), and consolidated. The rise in pressure that occurs by dragging fluid into a tighter region helps to obtain reinforcement wetting for both thermoset and thermoplastic resins. Dies should maintain uniform temperature control, should be very smooth to prevent any material hang-ups, and should be hardened to a Rockwell hardness greater than 70. Typical chrome plated dies last for 60,000 to 150,000 m. Unlike thermoplastic extrusion, pultrusion dies are exactly shaped to meet the final part shape. There is little concern about die swell with high filler loadings.

After leaving the die, the material is either cooled (thermoplastic) or possibly further heated (thermoset) prior to cooling before reaching the puller. The ideal puller pulls at constant line speed and normally has a belt or series of clamps that conforms to the shape of the pultruded structure. From a safety perspective, pullers must come equipped with emergency switches at their entrance. It is entirely possible to lose an arm if these devices are absent. Furthermore, pullers should be anchored firmly into the floor. Tension forces can be high in these processes, and it is very possible for the puller to walk backward toward the front end of the line without proper anchoring.

The material is then cut to a desired final shape through a moving saw. Since composite materials often employ abrasive fillers and fibers, the saw tips should either be carbide- or diamond-tipped. Typical line speeds of a thermoset pultruder, which is displayed in Figure 8.18, are 2.5 to 3000 cm/min, depending on the pultrusion thickness and raw material type [14,68,80,119,120].

8.3.2.2 Thermoset Pultrusion Technology

The overwhelming volume of thermoset pultruded parts employ unsaturated polyesters resins followed by phenolic, vinyl ester, and epoxy resins. Thermoset resins are typically applied in a resin bath. Prewetting of the low-viscosity resins (0.5 to 2 Pa·s) is relatively straightforward by using minor roller path changes, proper sizing chemistry on

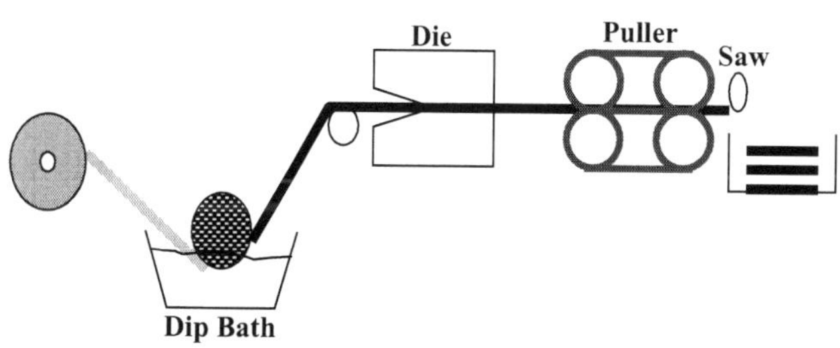

FIGURE 8.18 Typical thermoset pultrusion operation.

the reinforcements, and possibly by wetting agents in the resins. It is advantageous to keep the resin bath area fairly cool while preheating the resin-enriched reinforcing media. There are also some advantages to performing vertical pultrusion since the excess resin will not naturally drop to the lower side of the reinforcing medium, due to the effects of gravity. Line-speed enhancements are possible by using preheating or heating methods such as radio frequency (RF) on insulative reinforcements such as glass, or induction heating on conductive reinforcements such as carbon fibers.

The goal of thermoset pultrusion is to achieve a well-wetted nearly fully cured structure (90% cured) devoid of burn blemishes or rough skin (sloughing). Wetting can be an issue if the reaction is too rapid to allow sufficient flow brought on by the pressure rise that occurs by dragging fluid into the tight final die region. There are some trade-offs between the rise in pressure, brought on by faster line speeds, and the ever-decreasing residence time for flow to occur. At some point insufficient flow occurs, and it is not possible to run faster unless preheating efforts are employed combined with methods that even the temperature profile, such as RF or induction heating.

Voids can still arise due to trapped styrene or moisture. There was a time where phenolic pultrusion was believed to be impossible to run, due to the evolution of water or gases, but as fate would have it, these gases naturally vented from the die on their own, causing no problems other than normal VOC (volatile organic component) concerns.

Curing must be well controlled. Rapid, extremely fast curing causes a large exotherm leading to the outer skin to cure prematurely, whereas the interior material has yet to cure. Such an approach, particularly on thick parts, will result in cracked and burned surfaces. Clearly, increasing temperature is often not the answer to faster pultrusion rates. Instead, it is better to control temperature settings and temperature uniformity such that a nearly uniform gel point occurs near the die exit. In thermoset pultrusion a delicate balance exists among pressure, cure rate, and shrinkage.

The ideal thermoset resin for pultrusion expands on the application of heat, continues to expand during curing, and then shrinks after curing is complete. By expanding throughout the cure step, the pressure will continue to rise in the pultrusion die leading to a uniform surface appearance. Unsaturated polyesters come very close to behaving this way, which is why this material is a preferred resin. Epoxy resins, on the other hand, tend to shrink earlier than polyester resins and can run into processing issues if insufficient filler is present and if insufficient mold release is present. By controlling the expansion or contraction of thermoset resins within the die, it is possible to gain considerable control over the process. Researches have proven that the worst or highest pull forces occur during the curing time period since higher cure rates lead to lower pull forces.

Because of the natural tendencies of the resins to expand, there is a continual pressure buildup until the solid resin shrinks from the die surface. Pulling forces increase constantly during pultrusion. At first the force is due to viscous drag; then as the fluid transforms into a solid, the pressure rises due to expansional or frictional forces, and finally, the pull force increases due to frictional effects alone [14,68,118–120].

8.3.2.3 Thermoplastic Pultrusion Technology

The most important thermoplastic resin for composite pultrusion is polyprop ylene followed by a variety of thermoplastic resins, including polyurethane, nylon, polyphenylene sulfide, PMMA, and ABS [119,120]. These resins offer unique impregnation challenges to the pultrusion process. High viscosities encountered with thermoplastic resins require specialized impregnation equipment such as cylinderical spreaders, rotating cylinder spreaders, or spreading bumps, combined with optional roving/fabric preheating, and high-melt-temperature settings. The cylindrical spreaders (pins) or spreader bump regions serve as resin nip regions, where the entrained resin with the reinforcing media gets pulled into a tight nip region, which is where the fibrous tows wrap around the spreaders, experiences an increase in pressure, and then penetrates the reinforcement media. Figure 8.19 captures the essential of the technique. If the wrap angle or tension is great enough, it is possible for the rovings or mats to touch the curved surface, resulting in potential fiber abrasion damage—the glass sizing chemistry plays a very significant role with regard to the damage extent. Impregnation enhancers do not come without a penalty of higher tension or pull forces on the rovings or fabrics. Control of wrap angles, pretension, number of impregnation zones, amount of resin dragged into the nip regions, and other geometrical aspects of the curved surfaces all play a role [80,114,116,117]. Two recent studies have demonstrated the influence of both the pins and die on thermoplastic impregnation and physical properties [121,122].

Thermoplastic pultrusion employs several alternative methods of introducing the resin to the fibers. The most popular method involves creating a crosshead die, where resin is literally pumped into an impregnation chamber prior to the exit die. This method has the advantage that the starting raw materials begin in their simplest, least expensive pellet form. (Depending on the resin type, the powdered form can often be more expensive than the pellet form.) Other methods involve powdered input, which is typically sprayed on in an electrostatic bed, commingling of thermoplastic or fibrous rovings, or solution methods where the thermoplastic is dissolved in a solvent and the fabric or roving is passed through this lower-viscosity medium [80,116]. The use of pre-pregs is not uncommon in this technology [120].

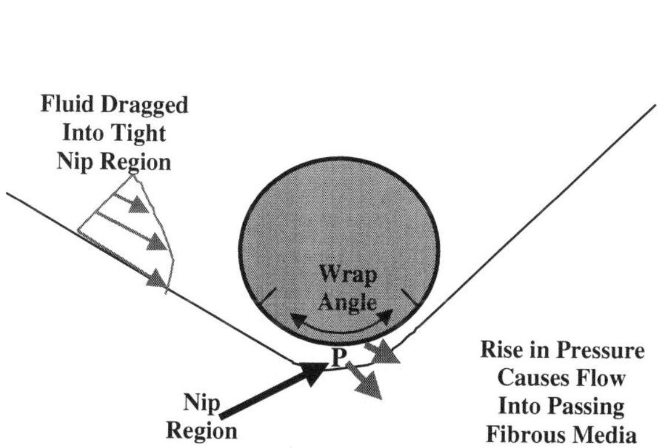

FIGURE 8.19 Thermoplastic resin impregnation around a curved region.

Preheating can be very advantageous since it is very difficult to locally wet-out fibers. Cooling measures should be taken to avoid warping or distortion that is caused in a similar manner, as discussed in the lamination section [80,115].

8.3.2.4 Pultrusion Quality Control

The pultrusion process is particularly sensitive to the linear density of roving or the basis weight of mats or veils that enter it, especially if the die is fixed. Vendors supplying these inputs are normally expected to provide tight linear density or basis weight limits. Depending on the line speed, the output can be so slow that the area occupied by the fibrous matter is related directly to the volume fraction of this material, and hence the final fibrous weight. Faster line speeds will lead to drag effects which make the fiber content relationship depend on geometry and resin rheological properties. Since tows or veils undergoing these melt process are likely to observe differences in shear rate across their heights or widths, particularly in thermoplastic pultrusion, it is important that the sizing and binder chemistries of these materials avoid catenaries (i.e., that the tow or veil remain somewhat together as opposed to coming apart under the application of various stresses across their thickness). The process should be monitored with regard to temperature settings, pull forces, cure (thermoset), and impregnation levels in a daily manner to obtain optimum performance [80,114].

8.3.2.5 Variant Processes

The thermoset pultrusion process is amenable to several notable process variants. Curved parts are manufactured by setting up the dies in a curved manner to induce part curvature. A large curved piece forces the material across the walls of the curved surface, resulting in a curved part in a batch like manner. It is also possible to alter the area or volume either by having a floating die, which enables simple area changes, or by a the *pulforming* process, where it is possible optionally to add BMC material and to compress the material along essentially a multiple-moving-chamber compression molding device (continuous) or a single-chamber compression molding device (batch). A further adaptation is essentially to do a combination of filament winding with pultrusion, termed *pull winding*, to produce such parts as hollow tubes with improved hoop resistance. In this process, added filaments are wound around a mandrel or prior pultrusion assembly prior to getting pulled through a hot die [120].

8.3.3 Filament Winding

Hollow cylindrical articles such as piping, vessels, golf clubs, and baseball bats, or spherical articles such as satellites, domes, and others articles, are all formed through filament winding. In this method, binder-entrained or binder-impregnated rovings, tapes, or webs are circumferentially wrapped around a moving mandrel to create open or partially closed objects as illustrated in Figure 8.20 [14,68,123–125].

The most often employed thermoset technique is wet filament winding. Reinforcing media of zero twist are dipped into a thermoset bath, where binder

FIGURE 8.20 Filament winding apparatus. (Courtesy of Owens Corning. Used with permission.)

impregnation occurs, wound on a mandrel, B-staged, and cured. During B staging the wound material is consolidated further by padding it down as the material is continuously turned. At this point curing can take place in a normal oven, under vacuum conditions, or under the pressure and heat of an autoclave. Thermoset filament winding can equally be accomplished with wet wound, B-staged filaments, or pre-preg media. When B-staged precursors are employed, the wound article is immediately ready for curing [14,68,125].

Thermoplastic filament winding, which still resides in early development stages, is much more difficult than thermoset, due to high resin viscosities. Normally, the reinforcing media must be preimpregnated through powder preimpregnation, commingled filaments, melt impregnation methods, or solvent impregnation methods. Most of these methods employ impregnation bars or bumps in some form to form a nip region that massages the thick thermoplastic melt into the passing reinforcement. If desired, partially impregnated powder or other thermoplastic forms can be further wrapped in a thermoplastic sheathing. The steps in thermoplastic filament winding consist of thoroughly preheating the materials, wrapping the reinforcement under fairly high tension to impregnate the resin further, and employing a consolidation roller directly on the wound fabric to further consolidate the wound media. Low tensions and nip pressures should be avoided because in the case of thermoplastics achieving thorough wet-out is absolutely essential [126–129].

There are essentially three different winding techniques circumferential or hoop winding, helical winding, and polar winding. Circumferential or hoop winding involves wrapping the materials in either a single band or multiple repeating bands in an almost 90° angle to the mandrel. This technique, which is used to create tubes and piping articles, requires the mandrel to turn much more rapidly than the traversing action of the applicator eyelet. Helical winding follows geodesic patterns whereby the winding pattern would actually follow a straight line if the mandrel surface were unwrapped and flattened out like a sheet of paper. (Circumferential winding is actually a subset of the more general helical winding approach.) This method is also used in tubular wound articles. The last technique, *polar winding*, is the only way to form a closed article at one end of the mandrel. This pattern literally goes from end to end such that there is a fairly fast traversing eyelet with a nearly stationary mandrel. As the wrapping medium reaches the end of the article, the mandrel quickly turns 180° prior to the wrapping medium returning to the opposite end of the article. If a part requires a winding motion outside these general methods, the wrapping will undergo slippage in an attempt to meet with the nearest geodesic line. Slippage may be reduced by using pins to force certain curvatures or by applying high-friction material such as silica to increase the friction coefficient of the wound structure [14,125].

Fiber placement is aided further by the high degrees of freedom that can exist in filament winding. These include (1) mandrel revolution, (2) transversing action of the filament placing media in the x direction, (3) traversing action of the filament placing media in the y direction, (4) traversing action of the filament placing media in the z direction, (5) revolution of the filament placing media in the xy plane, (6) revolution of the filament placing media in the xz plane, and (7) mandrel tilting. There are literally seven independent ways of moving and swiveling to obtain many different laydown patterns [125].

Because of the vast array of angles, placement techniques, and degrees of freedom, not to mention the desired mechanical properties, it nearly takes an expert with an advanced degree to design a part with the sophisticated differential equations that arise. Fortunately, these issues have been well studied and tackled through presently available computer software. Modern programs can provide recommended laydown patterns, time to create a part, curing conditions to avoid residual stresses, length of tow required, weight of the final part, thickness of the final part, and so on. Several packages, which include FiberGrafiX[5], Cadwind[6], and PatternMaster[7], exist to perform these tasks [130–132].

From a tooling perspective there are a variety of mandrels that can be employed. These include reusable metallic cylinders; collapsible mandrels that come apart with some disassembly; water-soluble mandrels, which normally contain mixtures of polyvinyl alcohol and sand (these can be dissolved and removed from one end of the finished article); collapsible mandrels, which use air pressure to push firmly against the wound medium; destructible mandrels made of plaster; and inner liner mandrels, which literally become part of the finished article. The inner liner method is often employed in fluid-containing tanks [14,125].

[5]FiberGrafiX is a registered trademark of Entec Composite Machines/Zoltek.
[6]Cadwind is a registered trademark of Material Belgium.
[7]PatternMaster is a registered trademark of McLean Anderson.

Typical issues that arise with filament winding involve good consolidation, low residual stresses, and removal from the mandrel. Consolidation improves with lowering viscosity, turning the mandrel during B staging, applying sufficient tension to consolidate the material during filament winding, and if necessary, curing the material under vacuum or pressurized conditions. Residual stresses naturally arise in this process since the mandrel constantly presses against the innermost wrapped material for dimensional control and since shrinkage forces occur from either thermoset shrinkage reactions or thermoplastic crystallization. Residual stresses may be diminished by using a collapsible, expanding mandrel, applying even-temperature profiles, the layer cured approach (where some of the material is cured prior to adding the outer layers), and sometimes using inhomogeneous layers or unusual temperature profiles. The last place to run into trouble, not to mention the most painful place to lose a part, is during mandrel removal. The destructible plaster method often leads to occasional part loss due to operator error in chiseling out the partA second source of trouble occurs if a taper is absent from some geometries. It is quite possible for wound articles to literally become stuck to a simple metal mandrel [14,125]. From a performance perspective, any manufacturing site should measure and maintain records on roving width, roving tension, resin viscosity, resin mix ratio (via a refractometer), fiber laydown, temperature, and reinforcing fiber content [14]. Changes in these parameters will alter part performance.

8.4 CONCLUSIONS

It is my sincere hope that this chapter acquaints and equips new composite professionals to the wide variety of composite processes available today. Depending on the dimensions and physical performance criteria of the final article desired, there are multiple methods to create a high-modulus composite part that can be used in automotive, aerospace, recreation, and industrial areas. May your composite parts be stiff and well consolidated.

ACKNOWLEDGMENTS

The author is indebted to Owens Corning for providing nearly all of the figures; for providing helpful reviews through the efforts of Jim Dottavio, Y. M. Lee, Len Adzima, and Bob Schweizer; for legal assistance from Jim Dottavio and Todd Hall; for research assistance from Karen Hettler; and for some cost background information from Kevin Spoo and Mark Greenwood. A significant amount of help was also received from several board members of the Composite Division of the Society of Plastic Engineers (contact SPE at www.4spe.org for more information, and become a member yourself!): Klaus Gleich, Enamul Haque, and Tim Simko. Finally, the author wishes to sincerely thank his wife, Gloria, for putting up with a husband who was in the office on far too many weekends.

REFERENCES

1. R. Gachter and H. Muller , *Plastics Additives Handbook,* 3rd ed., Hanser Publishers, New York, 1990, p. 607.
2. Z. W. Wicks, F. N. Jones, and S. P. Pappas, *Organic Coatings: Science and Technology,* Volume 1, *Film Formation, Components, and Appearance,* Wiley, New York, 1992.
3. D. Stoye and W. Freitag, *Resins for Coatings: Chemistry, Properties, and Applications,* Hanser Publishers, New York, 1996.
4. F. A. Cassis and R. C. Talbot, Polyester and vinyl ester resins, Chapter 2 in *Handbook of Composites,* S. T. Peters (editor), Kluwer Academic Publishers, Norwell, MA, 1998.
5. R. E. Wright, Thermosets, reinforced plastics, and composites, Chapter 2 in *Modern Plastics Handbook,* C. A. Harper (editor), McGraw-Hill, New York, 2000.
6. O. C. Zashe and S. H. Goodman, Unsaturated polyester and vinyl ester resins, Chapter 4 in *Handbook of Thermoset Plastics,* S. H. Goodman (editor), Noyes Publications, pp. 97–168.
7. G. Odian, *Principles of Polymerization,* 2nd ed., Wiley, New York, 1981.
8. www.matweb.com/search/searchSubcat.asp.
9. J. Brandrup and E. H. Immergut, *Polymer Handbook,* 3rd ed., Wiley, New York, 1989.
10. *Modern Plastics Encyclopedia,* 71(12), 1995, Mid-November 1994 issue, McGraw-Hill, New York.
11. Plastics, 9th ed., Thermoplastics and thermosets desk-top data bank, *International Plastics Selector,* 1987.
12. www.zeusinc.com/peek_thermal_prop.asp.
13. M. R. Piggott, *Load Bearing Fibre Composites,* Pergamon Press, Elmsford, NY, 1980.
14. S. M. Lee (editor), *International Encyclopedia of Composites,* VCH Publishers, New York, 1991.
15. E. Kreyszig, *Advanced Engineering Mathematics,* 4th ed., Wiley, New York, 1979.
16. J. G. Simmonds, *A Brief on Tensor Analysis,* Springer-Verlag, New York, 1982.
17. R. H. Perry and C. H. Chilton, *Chemical Engineers' Handbook,* 5th ed., McGraw-Hill, New York, 1973.
18. J. Rich and H. G. Fritz, Compounding and characterization of long fiber reinforced composites based on flax and polyprop ylene, *ANTEC 2002 Plastics: Annual Technical Conference,* Volume 2, *Materials,* SPE, Brookfield Center, CT, 2002.
19. L. L. Clements, Organic fibers, Chapter 10 in *Handbook of Composites*, S. T. Peters (editor), Kluwer Academic Publishers, Norwell, MA, 1998.
20. www.tensionmeters.com.
21. www.tensitron.com/quicklinks.html.
22. D. R. Roisum, *The Mechanics of Rollers,* Tappi Press, Atlanta, GA, 1996.
23. G. Wypych, *Handbook of Fillers: The Definitive User's Guide and Databook of Properties, Effects, and Uses,* 2nd ed., Plastics Design Library, 1998. www.4spe.org.
24. R. Stewart, Nanocomposites, *Plastics Engineering,* May 2004, p. 22.
25. B. K. Kandola, Nanocomposites, Chapter 6 in *Fire Retardant Materials,* A. R. Horrocks and D. Price (editors), Woodhead Publishing, 2001.
26. M. K. Doglovskij et al., Effect of mixer type on the exfoliation of PP nanocomposites, *ANTEC 2003 Plastics: Annual Technical Conference,* Volume 2, *Materials,* SPE, Brookfield Center, CT, pp. 2255–2259, 2003.

27. D. Merinska et al., Properties of polypropylene nanocomposites prepared by dif ferent ways of compounding, *ANTEC 2003 Plastics, Annual Technical Conference,* Volume 3, *Special Areas,* SPE, Brookfield Center, CT, pp. 2744–2750, 2003.

28. G. D. Barber et al., Dispersion of nanoscopic clay particles in thermoplastic composites, *ANTEC 2000 Plastics: The Magical Solution,* Volume 3, *Special Areas,* SPE, Brookfield Center, CT, 2000.

29. S. J. Dahman, Melt processed polymer clay nanocomposites properties and applications, *ANTEC 2000 Plastics: The Magical Solution,* Volume 2, *Materials,* SPE, Brookfield Center, CT, 2000.

30. L. E. Nunnery, Tooling innovations for thermoset molding, *ANTEC 1996 Plastics: Racing into the Future,* Volume 1, *Processing,* SPE, Brookfield Center, CT, 1996.

31. G. Eckes, *Six Sigma for Everyone,* Wiley, New York, 2003.

32. T. Pyzdek, *The Six Sigma Handbook Revised and Expanded: A Complete Guide for Greenbelts, Blackbelts, and Managers at All Levels,* McGraw-Hill, Norwell, MA, 2003.

33. J. L. Cadden and P . F. Sadesky, Tooling for composites, Chapter 25 in *Handbook of Composites,* S. T. Peters (editor), Kluwer Academic Publishers, Norwell, MA, 1998.

34. J. R. Welty et al., *Fundamentals of Momentum, Heat, and Mass Transfer,* 2nd ed., Wiley, New York, 1976.

35. www.imsteel.com.

36. www.eazall.com.

37. www.ess.elwd.com.

38. www.tooling.vantico.com, for Renshape SL5185.

39. W. C. Young (editor), *Roark's Formulas for Stress and Strain,* 6th ed., McGraw-Hill, New York, 1989.

40. D. V. Rosato (editor), Molds to Products, Chapter 4 in *Injection Molding Handbook,* Kluwer Academic Publishers, Norwell, MA, 2000.

41. H. Giles, Compression molding tooling for thermoplastic composites, *ANTEC 2001 Plastics: The Lone Star*, Volume 2, *Materials*, SPE, Brookfield Center, CT, 2001, 1-58716-098-6.2

42. Design of plastic products, Chapter 8 in *Plastics Processing Data Handbook,* (2nd ed.), Kluwer Academic Publishers, Norwell, MA, 1997.

43. H. Giles, Effect of flow in glass reinforced thermoplastic composites on part w arpage and shrinkage, *ANTEC 1996 Plastics: Racing into the Future,* Volume 2, *Materials*, SPE, Brookfield Center, CT, 1996.

44. P. F. Jacobs, New frontiers in mold construction: high conductivity materials and conformal cooling channels, *Rapid Technologies for Plastic Tooling Applications,* December 10–11, 1999.

45. E. Haque and P. Bristow, Processing of glass fiber mat reinforced thermoplastic composites, *ANTEC 2001 Plastics: The Lone Star,* Volume 2, *Materials,* SPE, Brookfield Center, CT, 2001.

46. J. E. Travis et al., *Proceedings of the SPE Reinforced Plastics/Composites Institute 42nd Annual Conference,* 1987.

47. B. B. Jensen, J. Graae, and J. Christiansen, The influence of push–pull process on weld lines, fibre length, and fibre orientation, Aalberg University, http://iprod.auc.dk/forsk/poly/student/proj/10-97-2.htm.

48. www.plasticstechnology.com/articles/200208cul.html.

49. Resin transfer moulding, Chapter 6 in *Component Design for RTM,* Kluwer Academic Publishers, Norwell, MA, 1997.
50. A. H. Landrock (editor), *Handbook of Plastic Foams,* William Andrew Publishing/ Noyes, Mendham, NJ, 1995.
51. S. T. Peters (editor), *Handbook of Composites,* Kluwer Academic Publishers, Norwell, MA, 1998.
52. S. Schermer, The development of medium to long run tooling with a view to rapid manufacturing, *ANTEC 1999 Plastics: Bridging the Millenia,* Volume 3, *Special Areas,* SPE, Brookfield Center, CT, 1999.
53. M. A. Kotnis, Composite material for rapid tooling produced by high speed CNC machinery, *ANTEC 1999 Plastics: Bridging the Millenia,* Volume 3, *Special Areas,* SPE, Brookfield Center, CT, 1999.
54. K. Bussard, Building soft tooling molds using rapid prototyping and dif ferent methods of construction in a uni versity environment, *ANTEC 2000 Plastics: The Magical Solution,* Volume 3, *Special Areas,* SPE, Brookfield Center, CT, 2000.
55. B. J. Arnold-Feret, Working draft—Rapid tooling—What w orks and what doesn' t, update 2002—Will be updated and re vised soon, *ANTEC 2002 Plastics: Annual Technical Conference,* Volume 1, *Processing,* SPE, Brookfield Center, CT, 2002.
56. F. Lindner and W. Michaeli, New rapid tooling concept, *ANTEC 2000 Plastics: The Magical Solution,* Volume 1, *Processing,* SPE, Brookfield Center, CT, 2000.
57. M. J. Janaes and J. E. F olkestad, Rapid tooling: the Rapid-to-market advantage #663, *ANTEC 2000 Plastics: The Magical Solution,* Volume 1, *Processing,* SPE, Brookfield Center, CT, 2000.
58. B. A. Ferret, Rapid tooling—it's not rock et science, *ANTEC 2003 Plastics: Annual Technical Conference,* Volume 1, *Processing*, SPE, Brookfield Center, CT, 2003.
59. www.sti.com/curtains/cudatash.htm.
60. http://world.keyence.com/products/safety/slc.html.
61. http://oeiweb.omron.com.
62. http://content.honeywell.com/sensing/prodinfo/safetysensors/.
63. www.safetytrainingnetwork.com/products/lockout.html.
64. www.lockout-tagout.com.
65. *Safety and Handling of Organic Peroxides: A Guide,* prepared by the Organic Peroxide Producers Safety Di vision of the Society of Plastics Industry , Publication AS-109, August 1999.
66. D. R. Sidwell, Hand lay-up and bag moulding, in *Handbook of Composites,* 2nd ed., S. T. Peters (editor), Kluwer Academic Publishers, Norwell, MA, 1998.
67. Society of the Plastics Industry Inc., *SPI Plastics Engineering Handbook,* 5th ed., Kluwer Academic Publishers, Norwell, MA, 1991.
68. *Kirk–Othmer Encyclopedia of Chemical Technology,* 4th ed., Volume 21, Wiley, New York, 1997.
69. S. T. Peters, Composite materials and processes, Chapter 5 in *Handbook of Materials for Product Design,* C. A. Harper (editor), McGraw-Hill, New York, 2001.
70. Reinforced plastics, Chapter 12 in *Plastics Processing Data Handbook,* 2nd ed., Kluwer Academic Publishers, Norwell, MA, 1997.

71. P. Hubert et al., Dimensional stability of complex shapes manufactured by the VARTM process, *ANTEC 2002 Plastics: Annual Technical Conference,* Volume 2, *Materials*, SPE, Brookfield Center, CT, 2002.

72. D. J. Michaud, Curing behavior of thick sectioned R TM composites, *Journal of Composite Materials,* 32(14), 1273–1296, 1998.

73. M. C. Rios and J. M. Castro, Multiple criteria optimization studies in in-mold coated SMC, *ANTEC 2003 Plastics: Annual Technical Conference,* Volume 2, *Materials,* SPE, Brookfield Center, CT, 2003.

74. K. E. Atkins, Low profile additives in thermosets, in *Plastics Additives: An A–Z Reference,* G. Pritchard (editor), Chapman & Hall (London), Kluwer Academic Publishers, Norwell, MA, 1998.

75. L. M. Abrams and J. M. Castro, Practical methods for predicting steady-state cure time during SMC compression molding, *ANTEC 2000 Plastics: The Magical Solution,* Volume 1, *Processing*, Brookfield Center, CT, SPE, 2000.

76. M. Wacker and G. W. Ehrenstein, Rheological and thermoanalytical in vestigations on a "Class A" LP-SMC paste, *ANTEC 2002 Plastics: Annual Technical Conference,* Volume 1, *Processing*, SPE, Brookfield Center, CT, 2002.

77. M. C. Rios and J. M. Catro, SMCstuff: multiple criteria optimization studies in in-mold coated SMC, *ANTEC 2003 Plastics: Annual Technical Conference,* Volume 2, *Materials*, SPE, Brookfield Center, CT, 2003.

78. www.azdel.com.

79. www.quandrant.com.

80. Personal experiences of D. Grove (author).

81. Based on an e-mail from K. Gleich.

82. S. Gottgetreu, Injection molding long glass f iber reinforced thermoplastic composites, ANTEC 2001 *Plastics: Annual Technical Conference,* Dallas, TX, Paper 505, SPE, Brookfield Center, CT, 2001.

83. M. Miklos, Updated process builds reinforced plastic parts, *Machine Design,* 71(5), 204, 1999.

84. O. Becker, D. Karsono, K. Koelling, and T. Altan, Gas-assisted injection molding of long glass f iber reinforced thermoplastics, *ANTEC 1997, Plastics: Annual Technical Conference*, SPE, Brookfield Center, CT, 619.

85. D. Schwendemann, New developments in co-rotating twin-screw extrusion for production of long glass f iber composites, *ANTEC 2002, Plastics: Annual Technical Conference,* San Francisco, CA, Paper 544, SPE, Brookfield Center, CT.

86. F. Gonzalez and C. Chassapis, Fracture toughness variations of injection molded long fiber reinforced thermoplastics, *Journal of Reinforced Plastics and Composites,* 20(10), 810–820, 2001.

87. M. Pechulis and D. Vautour, The effect of thickness on the tensile and impact properties of reinforced thermoplastics, *ANTEC 1997, Plastics: Annual Technical Conference,* SPE, Brookfield Center, CT, pp. 1860–1864.

88. F. Gonzalez and C. Chassapis, Experimental investigation of fracture toughness in injection molded long f iber reinforced thermoplastics, *ANTEC 1999, Plastics: Annual Technical Conference,* New York, SPE, Brookfield Center, CT, pp. 2171–2175.

89. J. Linn, On the characterization of the phase space of the Folgar–Tucker equation, presented at the SIAM-EMS Conference: AMCW 2001, Berlin, September 2–6, 2001.
90. A. Rios, B. Davis, and P. Gramann, Computer aided engineering in compression molding, presented at the Composites 2001 Convention and Trade Show, Tampa, FL, October 3–6, 2001.
91. P. Carreau, Rheological properties of concentrated fiber suspensions in polymers, presented at the 74th Annual Meeting of the Society of Rheology, Paper HS23, October 2002.
92. S. G. Advani, University of Delaware, www.me.udel.edu/~advani/index.html.
93. P. Mapleston, Process technology upgrades long-fiber compounds, *Modern Plastics International,* 27(1), 78–83, 1997.
94. K. Gleich, Extrusion/compression of long fiber thermoplastic composites, *ANTEC 2001, Plastics: Annual Technical Conference,* Dallas, TX, SPE, Brookfield Center, CT, Paper 506.
95. M. Voelker and C. Weber, CPI in-line compounding systems, *ANTEC 2001, Plastics: Annual Technical Conference,* Dallas TX, SPE, Brookfield Center, CT, Paper 507.
96. K. Gleich, New developments in long fiber reinforced thermoplastic composites, *ANTEC 2002, Plastics: Annual Technical Conference,* San Francisco, CA, SPE, Brookfield Center, CT, Paper 185.
97. A. Bricout, Processing and benefits of commingled glass fiber reinforced thermoplastic composites, *ANTEC 2001, Plastics: Annual Technical Conference,* Dallas, TX, SPE, Brookfield Center, CT, Paper 508.
98. F. Esposito, LNP, Vetrotex partner to market LFRT's, *Plastics News* (*USA*), 12(25), August 2000.
99. D. R. Roisum, Web Handling and Converting Course.
100. D. R. Roisum, *The Mechanics of Web Handling,* Tappi Press, Atlanta, GA, 1998.
101. www.sensorproducts.com.
102. www.cesphepherd.com/utility_mesh.html.
103. www.gabions.net.
104. www.wilsonart.com.
105. www.formica.com.
106. F. S. Merritt and J. T. Richetts (editors), *Building Design and Construction Handbook,* 6th ed., McGraw-Hill, New York, 2002.
107. R. J. Porter, Paper for resin bonded paper laminates, in *Plastics Additives: An A–Z Reference,* G. Pritchard (editor), Chapman & Hall, London, 1998.
108. Roger M. Rowell, Specialty treatments, Chapter 19 in *Wood Handbook: Wood as an Engineering Material,* U.S. Department of Agriculture, Washington, DC, 2001.
109. www.hymmen.com.
110. www.berndorf.co.kr.
111. www.smt.sandvik.com.
112. www.meyer-machines.com.
113. Extrusion, Chapter 16 in *Polypropylene: The Definitive User's Guide and Databook,* William Andrew Publishing, Mendham, NJ, 1998.
114. E. B. Gutoff, E. D. Cohen, and G. I. Kheboian, *Coating and Drying Defects,* Wiley, New York, 1995.

115. A. F. Turbak, *Nonwovens: Theory, Process, Performance, and Testing,* Tappi Press, Atlanta, GA, 1993.

116. www.cognex.com/pdf/downloads/Smartview_Nonwoven.pdf.

117. www.acs.honeywell.com/ichome/rooms/DisplayPages/LayoutInitial? pageName=chem_filmsconv&CategoryName=chem_filmsconv&ParentCalatogName=ICHome&ParentName=ROOT

118. A. F. Freed and R. K. Gupta, Minimizing voids in pultruded polymer composites, *ANTEC 2003 Plastics: Annual Technical Conference,* Volume 2, *Materials,* SPE, Brookfield Center, CT, 2003.

119. S. M. Moschiar, M. M. Reboredo, and A. Vazquez, Pultrusion processing, Chapter 6 in *Advanced Polymer Processing Operations,* William Andrew Publishing, Mendham, NJ, 1998.

120. B. A. Wilson, Pultrusion, Chapter 23 in *Handbook of Composites,* S. T. Peters (editor), Kluwer Academic Publishers, Norwell, MA, 1998.

121. I. Ekhator and P. Bates, Pultrusion compounding of commingled glass and polyprop y-lene fibers, *ANTEC 2002 Plastics: Annual Technical Conference,* Volume 2, *Materials,* SPE, Brookfield Center, CT, 2002.

122. P. J. Bates et al., Pressure build-up in staggered pin compounding system, *ANTEC 2000 Plastics: The Magical Solution,* Volume 2, *Materials,* SPE, Brookfield Center, CT, 2000.

123. www.owenscorning.com.

124. Reinforced plastics, in *Plastics Processing Data Handbook,* 2nd ed., Kluwer Academic Publishers, Norwell, MA, 1997.

125. Y. M. Tanopul'skii, S. T. Peters, and A. I. Beil, Filament winding, Chapter 21 in *Handbook of Composites,* 2nd ed., S. T. Peters (editor), Kluwer Academic Publishers, Norwell, MA, 1998.

126. A. Lutz and T. Harmia, 2nd Impregnation techniques for f iber bundles and to ws, in *Polypropylene: An A–Z Reference,* J. Karger-Kocsis (editor), Kluwer Academic Publishers, Norwell, MA, 1998.

127. Processing of advanced thermoplastic composites, Chapter 5 in *Handbook of Materials for Product Design,* C. A. Harper (editor), McGraw-Hill, New York, 2001.

128. J. P. Nunes et al., Advances on filament winding technology to produce composites from thermoplastic tow pregs and coated tapes, *ANTEC 2002 Plastics: Annual Technical Conference,* Volume 2, *Materials,* SPE, Brookfield Center, MA, 2002.

129. J. P. Nunes et al., Production of structures from thermoplastic composite to wpregs, *ANTEC 2001 Plastics: The Lone Star,* Volume 2, *Materials,* SPE, Brookfield Center, MA, 2001.

130. www.entec.com/modules.shtml.

131. www.material.be/products/cadwind.html.

132. www.mcleananderson.com/controls/PMRWIN.htm.

CHAPTER 9

Liquid Resin Processes

JOHN L. HULL[1]
Consultant, Ivyland, Pennsylvania

STEVEN J. ADAMSON[2]
Asymtek, Carlsbad, California

9.1 CASTING, POTTING, AND IMPREGNATING

9.1.1 Resin System Description

Liquid reactive resin systems are two-component thermosetting polymers which are generally applied, as liquid mixes, by pouring or spraying at room temperature and low viscosity and allowed to cure (cross-link) in place, sometimes with added heat, or sometimes relying on the heat of the exothermic reaction to effect the cure. In some instances, liquid resins are used for low-cost development of new plastic products with the full expectation of converting the process to plastics molding when the new product is deemed ready for higher-volume production and marketing. Most applications, however, are for insulating electrical and electronic components by coating current-carrying devices such as resistors, capacitors, semiconductors, and compact circuits.

9.1.2 Proportioning

The two components are generally referred to as component A, which is the basic resin, and component B, which is the catalyst or reactant. The required proportions for ideal stoichiometry are critical to achieving the required chemical and physical properties of the polymerized resin system (Figure 9.1). The required proportions are generally given as percentages by weight (gravimetric), as for example, component

[1]John L. Hull, Section 9.1.
[2]Steven J. Adamson, Sections 9.2 and 9.3.

Handbook of Plastic Processes Edited by Charles A. Harper

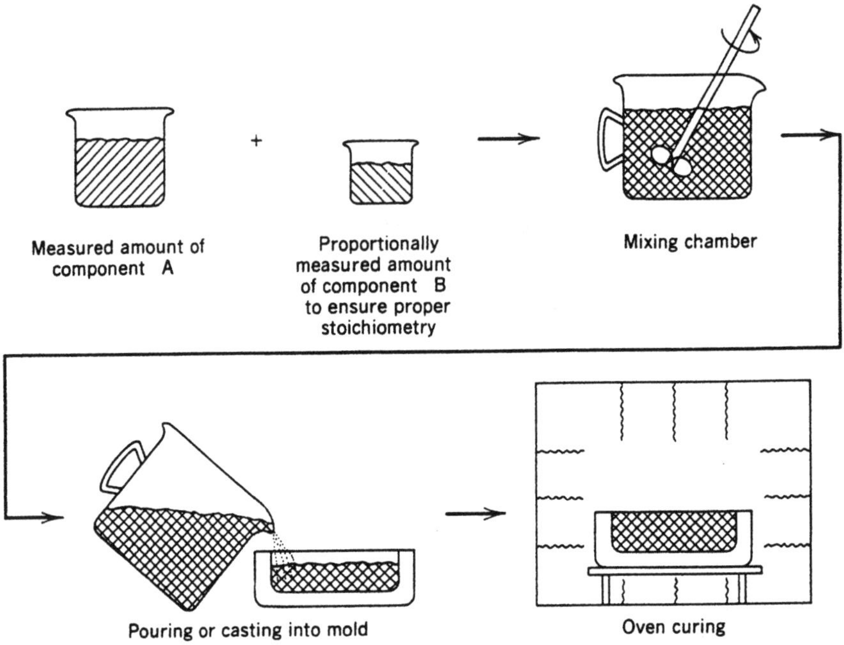

FIGURE 9.1 Hand proportioning, mixing, casting, and curing with a two-component liquid reactive resin system.

A, 60% (by weight) resin, and component B, catalyst, 40% (by weight). A filler material may be included in component A, for improving physical and/or electrical properties of the mix. Components A and B, before mixing, are traditionally quite stable at room temperature. Their shelf life at room temperature may be several months or longer without degradation. For storage of more than a few days, each component should be tightly sealed to prevent moisture from being absorbed by the components. Reduced temperature is recommended for storage of the separate components, often about 40°F.

9.1.3 Mixing the Two Components

In preparation for use, the components should be at room temperature, and they should be carefully measured (in accordance with the ratio stipulated) and then mixed thoroughly by stirring, before casting, to ensure homogeneity of mix. Another technique for mixing the two components of the resin system is to use a mixing head (often part of a liquid resin dispensing machine) (Figure 9.2). A *dynamic* mixing head uses a moving (usually rotating) mixing blade in the mixing nozzle or dispenser, which effectively achieves a uniform mix during a short 2- or 3-in. travel of components through the mixing head. Component A is piped, at modest pressure, into the mixing head on one side, and component B enters the mixing head on the other side of the head, at the same hydraulic pressure. The mixed resin exits the mixing head through a nozzle,

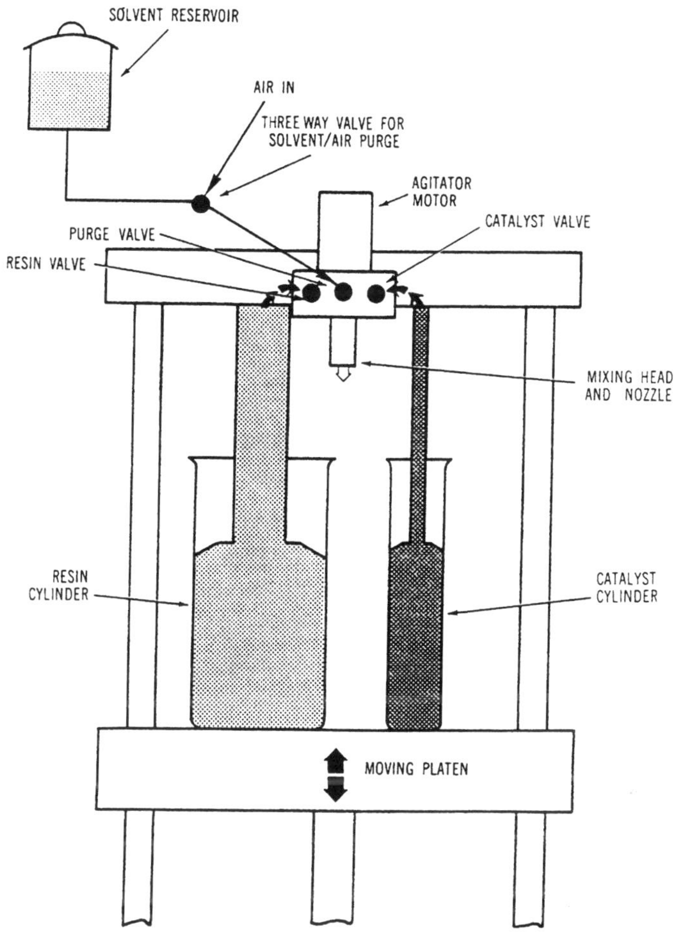

FIGURE 9.2 Automatic proportioning (volumetric), mixing (dynamic), and dispensing (foot pedal- or timer-controlled continuous or intermittent shot size) machine with a manual solvent-type purge system.

which feeds directly into the top of the mold or casting chamber. The rotational speed of the mixing blades should be regulated to ensure adequate mixing without adding significant heat to the mix, due to the mechanical shearing action of the blades.

A variation of the dynamic mixing head is a static mixing head (Figure 9.3). In a static head, a *flow splitter*, consisting of several segments of a twisted metal blade, added together, is inserted into a metal tube and prevented from rotating in the tube. It allows a wide opening for the resin components, under moderate pressure, to progress through the static mixer, with each element cutting the flow stream in half and rotating that segment of the stream 180°. As the resin components complete the flow through the six or more flow splitters, the resin and catalyst are thoroughly blended as they exit the mixing head and are directed into the mold (Figure 9.4).

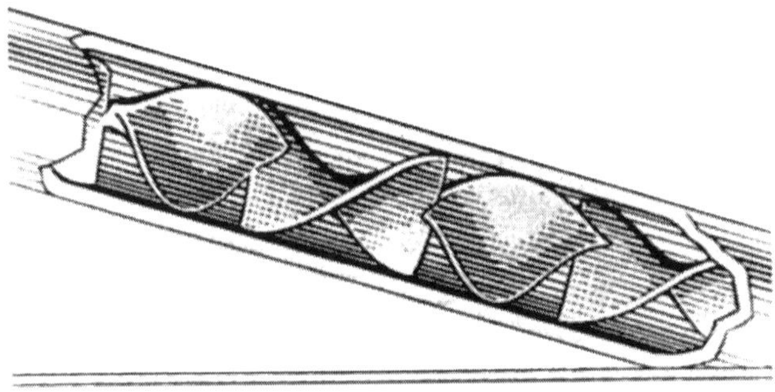

FIGURE 9.3 Typical flow-splitter elements in a static mixing head.

FIGURE 9.4 Dispensing machine with a static mixing head and nozzle fed by flexible resin and catalyst lines for dispensing remotely from resin and catalyst reserv oirs. (Courtesy of Packaging Systems, Inc.)

Whichever means are used to mix the tw o components, it is important to minimize the possibility of trapping air into the resin system as it goes through the mixing head and fills the mold or container holding the electrical or electronic de vice, which is to be surrounded by the resin. Air in the mix, as the resin cures or hardens, will create a bubble or void, in the insulating material.

9.1.4 Curing the Resin System

From the moment that components A and B are in contact, the irreversible exothermic chemical reaction begins. Initially, the heat of reaction raises the temperature of the resin and catalyst mix, resulting in a reduction of viscosity in the mix. This reduction in viscosity is beneficial in two ways. First, it enables the liquid resin system to flow more easily during cavity fill. Second, the lower viscosity enables better and faster wetting of the mold and part surfaces and ensures a better bond between the resin and the embedded component.

The heat of the reaction of resin and catalyst encourages the cross-linking, or polymerization, of the resin system and accelerates the cure. As the cure progresses, the liquid resin–catalyst mix thickens to a nonflowing gel, and stiffens and hardens rapidly to a solid. Certain additives may be added to the resin system formulation, before mixing occurs, to accelerate, or to retard, the cure. The aim is to achieve a complete fill in the cavity before the resin system loses its liquidity. Heating the mold before the resin flows in is another way to accelerate mold fill but may cause premature hardening and lack of complete fill. Conversely, cooling the mold below room temperature will prolong the flow time and may prove necessary to achieve complete fill. Should problems arise in achieving complete cavity fill in an acceptably short cure cycle, contact the resin systems supplier, who will probably be able to help to optimize the process.

9.1.5 Mold Shrinkage

It is important to understand that the mixed and cured polymer will experience some mold shrinkage before the cure is complete. The shrinkage is expressed in thousandths of an inch per inch, perhaps between 0.001 and 0.005 in./in. For example, assume a rectangular mold cavity, say, 1 in. wide, 2 in. long, and 1 in. deep. If the resin system has a mold shrinkage of 0.003 in./in., the finished part, when the plastic has cured, would measure 0.997 in. wide, 1.994 in. long, and 0.997 in. deep.

It is also important to realize that any components in the cavity that are to be embedded in the finished part may reduce the amount of shrinkage in the finished part. Glass, ceramic, and metal inserts molded in the part will prevent the shrinkage from occurring throughout the molded part. Depending on the size of the rigid insert in the finished part and the configuration of the rigid insert, it is possible that the resin shrinkage will fracture the insert. A ceramic substrate, with a thickness of 0.010 in., holding a variety of electronic components, may fracture on cure. Adding a filler of, say, fine grains of silica, to a percentage (gravimetric) of 30% or more, may well reduce the shrinkage sufficiently to prevent breakage of the substrate.

Other damage that can be caused by mold shrinkage is the possible separation of small lead wires between components mounted on a substrate or connected to one or more silicon chips in the mold. It is also important to recognize that plastic (like most metals) will expand with heat and shrink with cold. Selection of components to be surrounded by plastics, positioning of inserts to be encapsulated, and design of the molded plastic are all matters for consideration by the designers.

9.1.6 Voids in the Cast Liquid Resin Part

Most electric and electronic devices to be embedded in plastic must also be void-free: no air bubbles, no fissures, no possibility of air–moisture ingress between plastic and leads extending beyond plastic, and so on. With liquid resin systems, air may be inadvertently mixed into the liquid as it is poured from one container to another or as the two components are mix ed together. Also, moisture may be found on the mold or insert prior to molding because of a fingerprint or an unblocked sneeze or drop of perspiration. To ensure no rejects, such moisture or air must be pre vented. In molding, casting, and embedding with liquid resin systems, especially in expensive electric and electronic applications, "cleanliness is next to godliness."

9.1.7 Use of a Vacuum in Processing with Liquid Resin Systems

In casting and potting liquid resins for applications demanding the ultimate perfection, significant successes have resulted from the use of vacuum in every step of the process. The series of procedures include the following:

1. Vacuum "bakeout" of molds or cups, shells, and cases prior to filling
2. Vacuum degassing of the measured quantities of components A and B, and mixing of the components under v acuum with controlled temperature of the liquid blending, with full visibility of the mix to confirm complete degassing
3. Casting the fully degassed and mixed resin system into each mold, under controlled vacuum, successively, on a simple turntable (or a planetary turntable if the mold size is relatively small, and if the number of molds for each cycle can be justified), and with the mold temperature controlled (Figure 9.5)
4. Allowing return to atmospheric pressure (or e ven to positive pressure up to several hundred pounds per square inch), while the cast polymer is still liquid, to ensure penetration of liquid resin into small interstices of the mold and/or parts, before gel and cure of the cast polymer

9.1.8 Behavior of Liquid Resin Systems Under a Vacuum

Liquid resin systems contain not only the base polymer , which is an engineered blend of liquids designed for extensive cross-linking, but also one or more catalysts (reagents) to ensure a rigid polymer structure for the cured product. Each of the liquids making up the resin system has its individual liquid vapor pressure–temperature (LVPT) curve for the range of temperatures to which the resin components and the eventual resin–catalyst mix will be exposed during the process.

A typical LVPT curve is shown in Figure 9.6. We see that at normal (sea le vel) atmospheric pressure, the liquid remains stable at about 20°C (68°F) as long as the atmospheric pressure remains at a minimum of 3 mmHg. Should the temperature of the liquid rise abo ve 80°C, the curve steepens and some of the liquid(s) v aporize, giving off the "lighter ends" of some of the resin and/or catalysts, thereby losing some of the intended properties of the resin system. In short, the pressure (vacuum) and temperature le vel of the base resin, the catalyst, and the resin–catalyst mix should be controlled to ensure that the system is not de graded by excessive temperature and too low a vacuum (pressure). The formulator of the resin and/or catalyst is generally able to provide a suitable LVPT curve for any specific formulation.

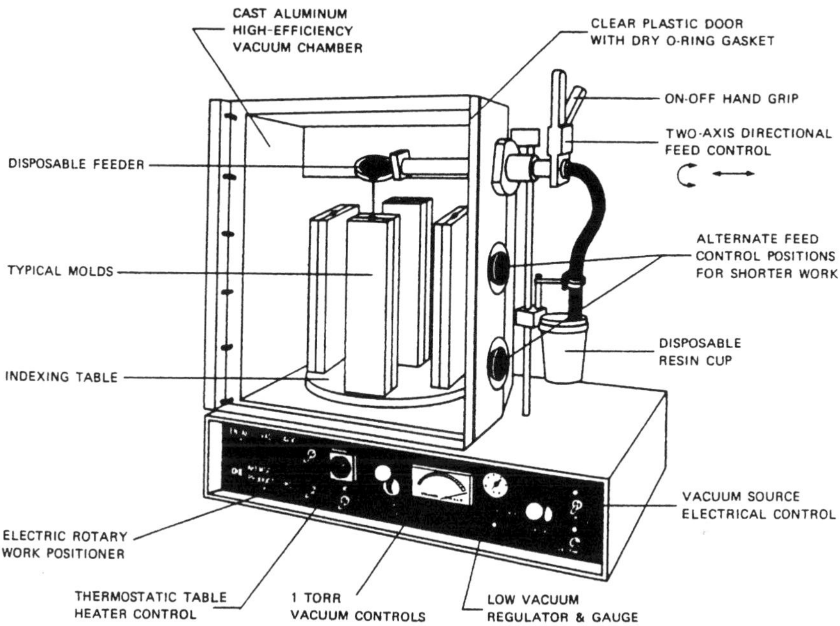

FIGURE 9.5 Simple bench-type vacuum potter for casting catalyzed resin under vacuum into an evacuated mold. Degassing occurs as the mix flows over a heated plate prior to flowing into the open mold. (Courtesy of Aremco Products, Inc.)

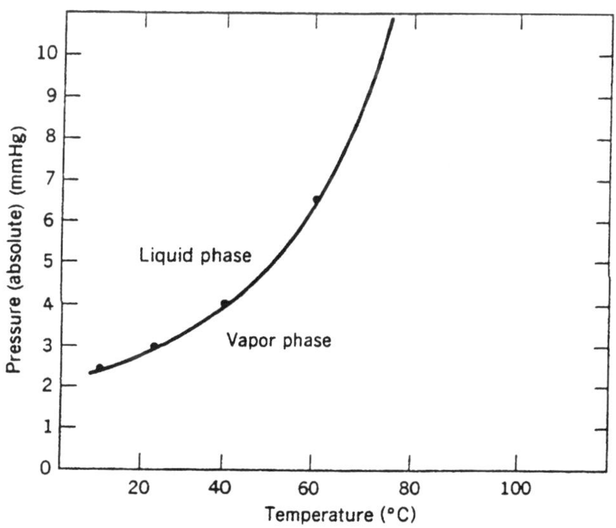

FIGURE 9.6 Vapor–pressure curve for catalyst component of a liquid reactive resin system. Degassing and casting pressures (vacuums) must be controlled so that resin system components stay in the liquid phase.

9.1.9 Complete Vacuum–Pressure Potting Process

The complete vacuum–pressure potting process explains the steps required to achieve a complete process for void-free parts. Figure 9.7 shows a vacuum potting machine with the hinged door on the casting chamber open ready to start the cycle. Figure 9.8 is a four-step schematic of the process. Step 1 is to arrange the several open top molds on the rotary table at the bottom of the casting (lower) chamber, such that as the turntable is rotated by operator, molds will be directly under the director funnel nozzle. Any inserts that are to be an integral part of the finished product are

FIGURE 9.7 Vacuum–pressure potter showing planetary turntables and a director funnel in a casting chamber. Note the handle (in front of the upper timer on the control cabinet) to enable the operator to move the stirrer shaft down to puncture the bottom of the disposable resin/catalyst container in the upper degassing chamber, then to meter flow of mix through the director funnel into molds. Planetary turntables are rotated with handles beneath the casting chamber, positioned conveniently for the operator to manipulate while sitting in front of the sight glass and viewing the funnel and molds to ensure a complete fill. (Courtesy of Hull-Vac, Inc., Warminster, PA).

positioned in the mold cavities. The full-opening isolation valve separating the casting chamber from the degassing (upper) chamber is closed, as is the door to the casting chamber. The several threaded T-bolts are swung into place and tightened against an O-ring to seal the door for vacuum and later pressure. Many users of these systems place aluminum foil on the planetary tables underneath the molds to

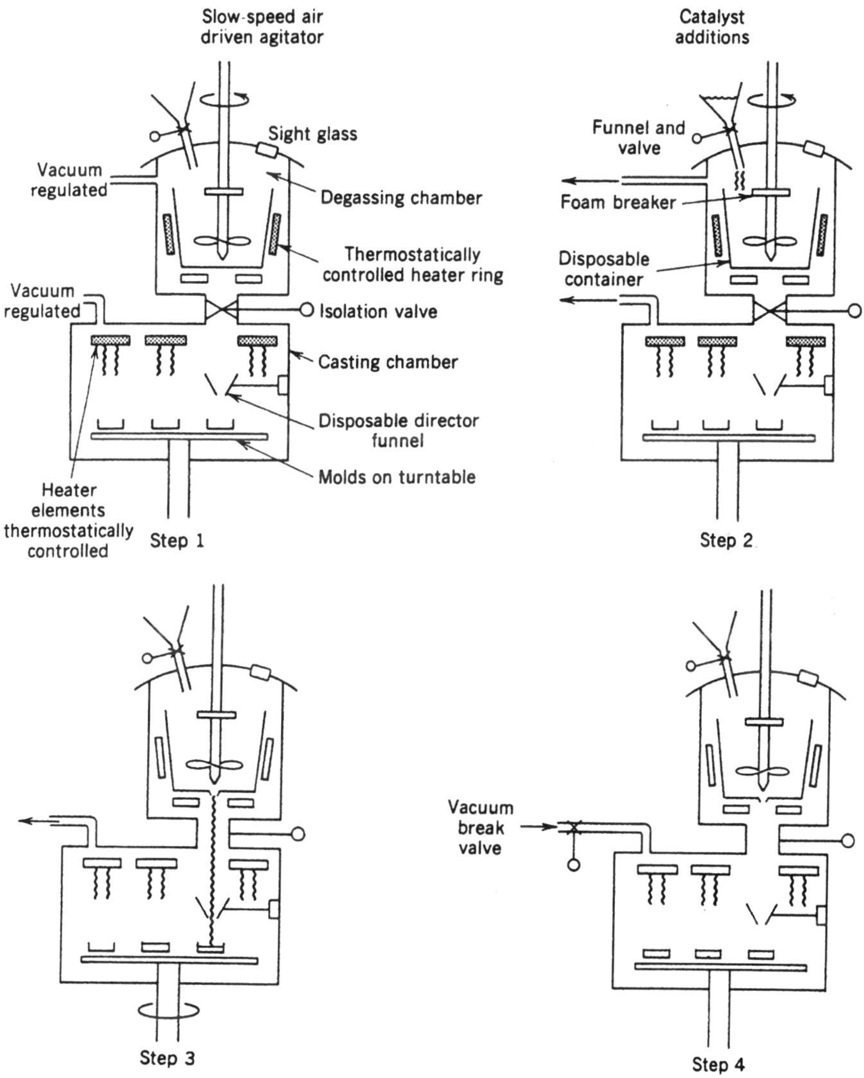

FIGURE 9.8 Vacuum–pressure potter showing the four steps of the degassing and casting process, from blending resin components through cure.

simplify cleanup in event of resin spillage during casing. A vacuum line connects the casting chamber to the mechanical vacuum pump.

Electric heaters with temperature controls are mounted to the casting chamber either internally or externally to maintain the mold chamber temperature modestly elevated to accelerate the vacuum bakeout of the chamber walls and of the molds. The disposable director funnel has been positioned directly beneath the full-opening isolation valve, with the funnel outlet 1 to 2 in. above the openings to the mold cavities. Finally, an empty disposable cup has been placed on the turntable into which any excess casting resin will be dispensed after all molds are full.

Once the vacuum bakeout cycle has begun in the lower chamber, a disposable mixing container with adequate capacity to hold the required amounts of components A and B (plus a slight excess to ensure complete filling of each mold to be processed) is placed into the heater ring, which will support the container during resin and catalyst degassing. The measured amount of component A is then poured into the disposable mixing container. With the disposable container with component A in place, the degassing chamber lid, with an air motor mounted on the lid and the stirrer shaft fastened to the air motor above the chamber lid and its shaft passing through an O-ring rotating seal, is fitted with an impeller blade which will be positioned above the bottom of the mixing container . The rotating shaft has a sharp point below the impeller, which will be $\frac{1}{2}$ in. above the bottom of the container. The chamber lid is then tightened in place with T-bolt clamps. A line from the vacuum pump is then connected to the vacuum port in the lid. Heat is applied to the container heater ring, and temperature of the ring is controlled to the desired level for degassing.

Attached to the degassing chamber lid is a catalyst additions funnel with a manual valve tightly closed. A regulating valve on the vacuum line from the vacuum pump to the degassing chamber is slowly opened and the pressure drops to the set level. The air motor is then started, and the stirrer shaft starts to rotate. Watching the action through the sight glass in the degassing chamber lid, the operator adjusts the vacuum valve to minimize foaming and to ensure that degassing component A is not so violent that the resin is spattered out of the container A foam breaker on the rotating shaft keeps the foam to a minimum as the foam breaker sweeps through the foam, neatly breaking the bubbles as they start to rise. The degassing of component A may require several minutes, but rarely longer than 10 minutes, with the operator constantly monitoring the pressure (vacuum) gauge (usually in the lid). As the bubbles diminish in intensity, indicating that component A is reasonably free of trapped vapor, component B may be added.

Inside the additions funnel, near the exit nozzle, a mark has been scribed which can easily be seen by the operator . The operator gently pours in a small amount of component B until its level is at the scribed mark, filling the space between the scribe mark down to the funnel valve closure. With the catalyst additions funnel valve still closed tightly, the operator starts pouring into the funnel the exactly measured amount of component B to create the correct proportions of component B to component A (already in the degassing chamber). As the operator slowly pours, he or she slowly opens the catalyst additions funnel valve to allow component B to flow into the degassed component A. As component B flows slowly into the degassing chamber, its flow is aided by the atmospheric pressure above the funnel and the lower pressure at the exit end of the funnel nozzle. As the catalyst drops from the funnel into the disposable container, degassing of the thin stream occurs as soon as it flows out of the funnel. The stirrer and the foam breaker continue their slow rotation, and degassing of the component B continues as it is mixed into component A.

At the instant the scribe line appears, the operator closes the additions valve. There will still be a small amount of component B that fills the space between the scribe line and the funnel valve. The amount of component B that is not allowed to flow beyond the scribe line is exactly the amount of component B that was introduced into the funnel while the valve was closed, and which reached up to the scribed line when the valve was opened. The specified component B ratio to the Component A volume has thereby been maintained, and the mixed resin has the correct stoichiometry.

The operator continues to observe (through the sight glass) the mixed resin and catalyst in the disposable container until the bubbles stop rising out of the mix. At that point the degassed resin–catalyst mix is ready for casting and the molds in the casting chamber are thoroughly degassed, ready to receive the polymer. The operator then slowly opens the isolation valve, which until now has remained tightly closed to keep the pressures (vacuum) in the two chambers at the appropriate levels for (1) the vacuum bakeout part of the process, and (2) the resin degassing part of the process. Slowly opening the isolation valve allows the pressures to equalize. The operator closes either one of the two regulator valves and adjusts the other regulator for the desired pressure level for the casting and mold filling part of the process, generally about the level of the resin degassing operations.

The operator checks through the sight glass in the door of the casting chamber to be sure that a mold is positioned directly below the nozzle of the director funnel. He or she then uses the handle controlling the height of the stirrer shaft to force the shaft down until the pointed lower end of the shaft penetrates the bottom of the disposable container holding the degassed resin–catalyst mix. Once the pointed shaft has punctured the bottom of the disposable container, the operator can control the flow of mixed resin and catalyst by slightly raising and lowering the stirrer shaft, causing it to function like a needle valve.

The operator observes the mold filling visually, and stops flow when the resin reaches the top of the fill opening of the first mold. He or she then manipulates the rotating shaft of the turntable, which causes the turntable to rotate in a horizontal plane until the next mold is positioned under the director funnel. The operator then repeats the mold-filling process successively until the final mold has been filled. At that time, he or she rotates the turntable such that the aforementioned overflow cup is directly beneath the director funnel. Opening the needle valve then allows the remaining resin to flow, emptying the disposable container in the degassing chamber and completing the casting operation.

At this point in the process, while the resin is still fluid, the operator may shut down the vacuum pump and open the vacuum break valve, allowing atmospheric air to enter the chambers. The increased pressure will tend to force the liquid resin to flow into any unfilled interstices of the mold. Some applications of casting and potting may require higher pressure during this final part of the cycle. Systems have been furnished for pressures up to 500 lb/in^2. Needless to say, such pressures require ASME code design and construction of the chambers and components to withstand such pressures with safety. Thoroughness in carrying out all the foregoing procedures will assure top quality castings.

9.1.10 Vacuum Impregnation

A process incorporating some of the steps of vacuum casting and potting is used increasingly to improve the performance of electrical devices containing wirewound coils, such as solenoid coils, rotors and stators of electric motors, inductors, and transformers. The wire used for such devices is generally coated with a magnet wire coating, usually transparent and very thin. The coating insulates each wire from its neighbors in the winding. The coils are machine-wound with considerable tension on the wire to ensure getting the maximum number of turns of the wire in as small a space as possible.

Such wires, in application, are energized electrically to perform their functions in motors and coils for radios, computers, automotive applications, and so on. As current is passing through the wires, the wire experiences a tightening, imperceptible to the eye. The frequent tightening and relaxing causes a minute friction between adjacent (and touching) wires. Eventually, such friction causes the magnet wire coating to abrade, eventually causing a short circuit between two adjacent wires. A short circuit generates localized heat, further aggravating the erosion of the wire coating.

Traditionally, such windings were soaked in an insulating varnish, made fluid by a solvent. The varnish would penetrate the windings by capillary action, and the solvent would eventually evaporate (helped by heat). The varnish would reduce the wire movement, prolonging the life of the winding. With the advent of reactive resin systems, formulations were developed that allowed the catalyzed resin to flow into the interstices of the windings, then to cross-link (polymerize), firmly keeping the wires from moving and enabling such windings to survive decades of service. Today, vacuum impregnation (and often, vacuum pressure impregnation) has become the optimum way to prolong the life of products that utilize such windings.

9.1.11 Two-Vessel Impregnation System

The process is generally carried out using a two-vessel impregnation system. The reactive impregnant contains a catalyst that is essentially nonreacting at ambient temperature [68°F (20°C)] and is kept in the storage vessel. The coils to be impregnated are generally placed in a basket that is lowered into the empty impregnating vessel, which is then evacuated to remove essentially all air from the coils in the empty impregnating vessel. Following the vacuum soak, the impregnating vessel remains evacuated, and a valve is opened in a pipeline connecting the two chambers at the bottom.

Atmospheric pressure causes the impregnant to flow out of the storage vessel into the impregnating vessel until the coils in the basket are covered (flooded) with the liquid, which is kept under modest pressure to speed up the penetration by the catalyzed resin now covering the coils. There being essentially no air in the windings of coils, and so on, in the basket, the liquid impregnant flows into the winding interstices, aided by capillary action of the liquid and positive pressure in the chamber and the essential vacuum in the interstices.

Following a soak of about 5 minutes to ensure full penetration of impregnant, positive air pressure is maintained in the impregnating chamber, while the connecting valve is opened, allowing the air pressure to force the remaining liquid impregnant to flow back into the storage chamber through the connecting pipe at the bottom of each chamber. The unused impregnant is then ready for another cycle, following removal of the basket of impregnated windings. The basket is then removed from the impregnating chamber and placed in a heated curing oven, where it is held until the impregnant cures and hardens, completing the process.

In practice, the excess impregnant, which flows from the impregnating vessel back into the storage vessel, has not been heated during the process and therefore has not undergone significant cross-linking. Adding fresh impregnant (resin and catalyst already blended) to the residual impregnant pumped back into the storage vessel from the impregnating vessel generally keeps the impregnant satisfactory for subsequent batches of coils. The operator monitors the impregnant periodically to prevent premature cure of the resin system in the storage vessel.

9.1.12 Trickle Impregnation

Another impregnation process for magnet wire devices, considered somewhat less effective than vacuum pressure impregnation described earlier, is called *trickle impregnation.* In this process the aim is to fill the interstices of windings with reactive impregnants, but without the use of vacuum, relying on capillary action to cause the liquid impregnant to penetrate the winding interstices. The windings are slowly rotated under a thin stream of the impregnant trickled onto the windings for a measured length of time. No vacuum or pressure is utilized in this process, and the equipment to perform the process is considerably less expensive than a typical vacuum pressure impregnation system.

9.2 AUTOMATED LIQUID RESIN PROCESSES

9.2.1 History of Modern Automated Dispensing

In the early 1980s, programmable liquid dispensing equipment did not exist. Meter and mix machines for two-part systems with resin and hardeners were available, but these were not automated robotic systems. At this time the only way to use liquid encapsulants was simple handheld dispensing equipment or to build a mold around the part and pour epoxies, silicones, or urethanes over the part.

Nearly all commercial production of integrated circuit (IC) packages used transfer molding for encapsulation. Transfer molding gave a good-looking hard protective coating around the IC and leads. The finished surface was flat and a logo could be printed onto it. But most important, it was a mass production process. ICs were die bonded onto lead frames, and wired bonds connected the IC bond pads to individual pins or legs on the finished component. Dual in-line (DIL or DIP) and later, quad flat packages (QFPs) were the state of the art.

Strips of lead frames could be placed on racks and loaded into a mold. A single operator could process several hundred parts in one step, and the entire process would take only a few minutes. However, transfer molding was still a manual process. As package designs evolved, new methods of encapsulation were required. Dam and fill encapsulation of wire bonds in ball grid array (BGA) packages utilized liquid dispensing technology. Flip-chip and chip-scale package (CSP) underfill started to become more widely adopted, and this created the need for underfill with liquid encapsulation materials. Figure 9.9 shows a time line for package outlines and when they became widely accepted.

Automated dispensing was made possible by the introduction of low-cost control systems based on personal computers. The early 1980s saw the introduction of personal computers from Apple and IBM [1]. Engineers started to use these computers to control processed PDPs and computer-aided design (CAD) was moved from larger computers such as DEKs to PCs. Pen plotters became standard office equipment to print CAD drawings.

In 1985, PCs started to become commonplace in factories and offices. The PC, for the first time, provided an inexpensive method of controlling x,y,z motion. As can be seen from Figure 9.10, dispensing systems of this time period had a lot of similarities to pen plotting systems. An off-the-shelf dispenser was added to the plotter, such as the EFD unit shown in the figure. However, the simple pen plotters had to be improved to carry higher loads and make them more robust for a factory environment. Concurrent with introduction of the PC and the demand for more flexible encapsulation systems, tooling and automation companies started to offer automated dispensing equipment.

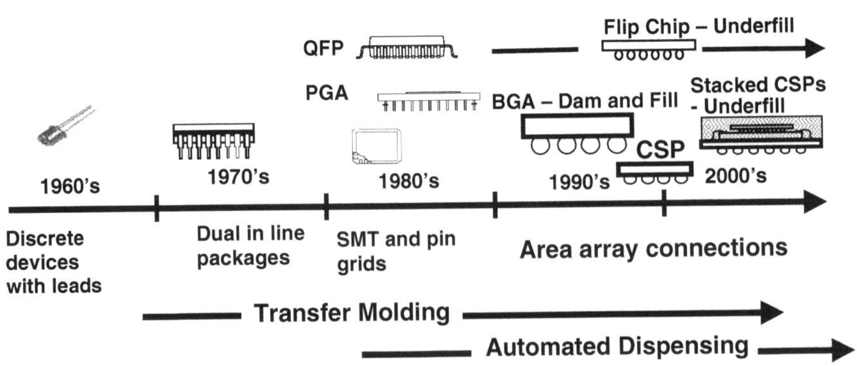

FIGURE 9.9 Time line for package outlines.

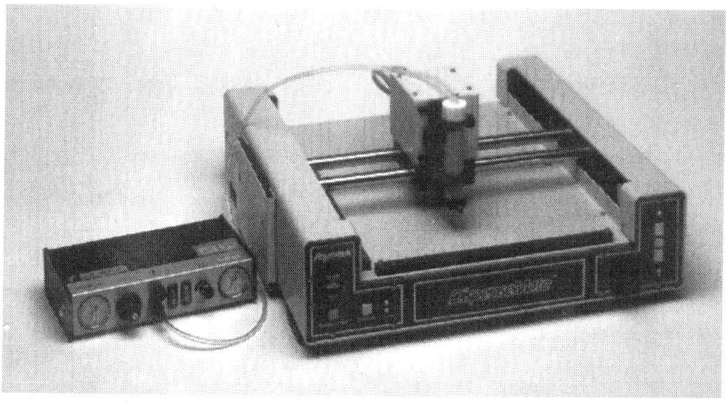

FIGURE 9.10 EFD dispenser and Asymtek benchtop system. (Courtesy of Asymtek.)

Over the last 20 years, dispensing has progressed from simple benchtop systems to fully automated in-line robots with bulk feed apparatus so they can run for a week without attention. Valve technology has also progressed from simple air pressure on a syringe to auger screws, linear positive-displacement pumps, and more recently, jet dispensing.

9.2.2 Why Use Liquid Dispensing?

There are a number of reasons that people chose to dispense liquid resins rather than use molding. The first consideration is usually, the cost. Transfer molding equipment can be expensive, and molds can have lead times of many months. Liquid resin dispensing systems are relatively inexpensive in comparison from an initial investment prospective. Over a period of years, molding equipment will usually prove to be more cost-effective. Considering that products change rapidly, with the drive to continually put more functions in smaller spaces, the life of a package design may not warrant the investment in setting up a molding line.

Dispensing is also seen as a simple step from a prototype, where hand assembly of a product used simple air over syringes to encapsulate the first product samples. The automation of a liquid dispenser is seen as a similar process with a higher throughput. There are also technical reasons for using liquid dispensing rather than to molding. Pressure and temperature are the first two factors in considering if a liquid will be a better choice for encapsulation. Molding equipment uses heat to soften the mold compound and allow it to flow as a liquid through the mold runners. To aid the encapsulation flow, a ram is used to force fluid into the mold. Hence the parts being encapsulated are experiencing high temperatures (120°C+) and high fluid pressures. The reader can understand that if a part is heat or pressure sensitive, these factors must be considered in their decision on what type of equipment to use for encapsulation.

Pressurized fluid in a mold can also cause problems, such as sweeping of wire bonds used to connect an integrated circuit to a lead frame or package. Present-day wire bond wires are now as fine as 18 μm in diameter on 40-μm pitches. Kulicke & Soffa

Industries, a manufacturer of wire bonders, supplies a system with both a wire bonder and a dispenser built onto the same platform. The wire bonds are encapsulated and then cured with an ultra violet light to protect the wires from sweep during a subsequent molding operation.

Other factors can contribute to the choice between using liquid resin encapsulants and mold compounds. For a number of years, liquid epoxy encapsulants have been used for underfilling flip chips in single-chip packages. Several major IC manufacturers have published technical papers on molded underfills [2], proving the feasibility of the underfilling with a molding process. However, the main component cost of underfills materials is the price of fine filler materials that are used to adjust fluid expansion properties of an underfill. These fillers need to be small and round so as not to affect the fluid flow under a flip chip. The size and tight distribution on particles has a large impact on the cost of the underfill. Transfer molding encapsulation not only underfills but also covers the component with typically 0.25 to 0.5 mm of material. Today, it is more cost-effective to underfill the die with liquid encapsulants and over-mold the entire package with lower-cost molding encapsulants rather than trying to use an expensive molding compound for both underfill and encapsulation over the die.

9.2.3 Applications

There are many dispensing applications for epoxy and other materials: attaching components to circuit boards, encapsulating components for environmental protection, and conductive epoxies for electrical contact. What makes all of these applications challenging is that most new dispensing applications are usually attempting to build components in ever-diminishing spaces or are looking to double throughput from last year's target. Dispensing as an industry segment is smaller than stencil/screen printing or molding for encapsulation. However, in many situations the only solution is a discrete dot of something on a nonflat surface or in a cavity.

9.2.4 Glob Top Encapsulation

Glob top is one of the oldest encapsulation applications after potting. In this case, a circuit such as a thick-film hybrid on ceramic or semiconductor chip bonded to a printed circuit board substrate and wire bonds is covered in an epoxy, silicone, or urethane material. Glob top encapsulation provides mechanical and environmental protection to the parts. For many years, watch modules with one die on a printed circuit board used glob top encapsulation.

The materials for glob top encapsulation are usually epoxies and it is recommended that the cured encapsulation has a glass transition (T_g) temperatures above the operating limits of the device. For example, if a device has a maximum operating temperature of 100°C, it is advisable to have a glass transition temperature 110°C, or higher. This means that the component is always under compressive forces up to the T_g and is less likely to be damaged by thermal expansion and contraction under normal operating conditions.

At first, glob top may look like simply pouring epoxy over a die and wire bonds. The goal is to cover the die and bonds with a minimum fluid thickness, at the same time not allowing the encapsulants to run everywhere (Figure 9.11). Fortunately, the epoxies used for glob top applications are thixotropic. Thixotropy aids this process, as the epoxy or fluid is worked mechanically by shear forces in an auger pump, and the fluid viscosity is reduced. This aids fluid flow through a needle and around wire bonds. The viscosity of the fluid recovers fairly quickly, and this can be used to limit flow out onto adjacent components or pads. If components are susceptible to mechanical stress caused by temperature changes, a soft layer of silicone gel can be used as a stress barrier. In glob top dispensing, the operator is often trying to find a balance between a fluid that covers well and how far it flows.

9.2.5 Dam-and-Fill Encapsulation

Dam-and-fill (D&F) encapsulation is used when a production run cannot justify the expense of molding tools. Figure 9.12*a* shows a dam being dispensed; Figure 9.12*b* shows a fill operation. A finished component has the appearance of a molded part with tightly controlled dimensions for the position of the encapsulation materials around and over a die and wire bonds. Apart from a small amount of rounding of the encapsulation at the edges, the finished part is flat, and this allows manufacturers to print logos and identifiers on top of the encapsulation, the flat surface also enables vacuum pickup tools to be used in a pick-and-place assembly system.

The basic construction of a D&F package is to dispense a high-viscosity epoxy material in a square or picture frame around a die and wire bonds (Figure 9.13). The viscosity of the dam material is on the order of 100,000 cP or higher . The epoxy material can be heated in the needle to lower the viscosity to help it pass through a needle. The pump most often used to dispense a dam is an auger pumpAuger pumps

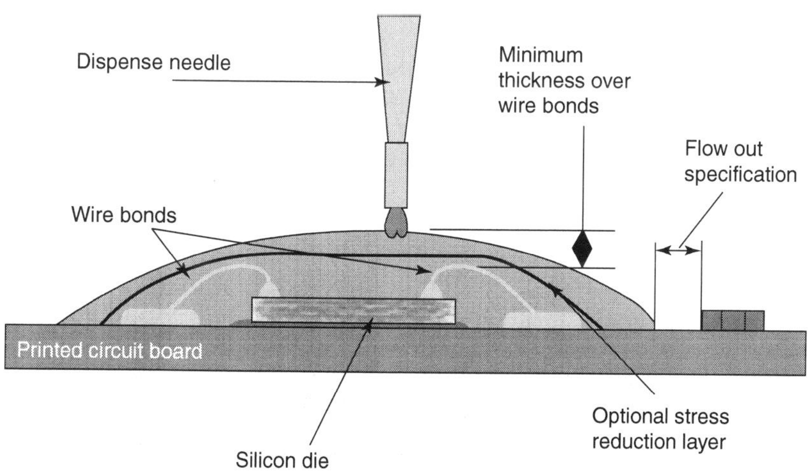

FIGURE 9.11 Glob top encapsulation.

(*a*)

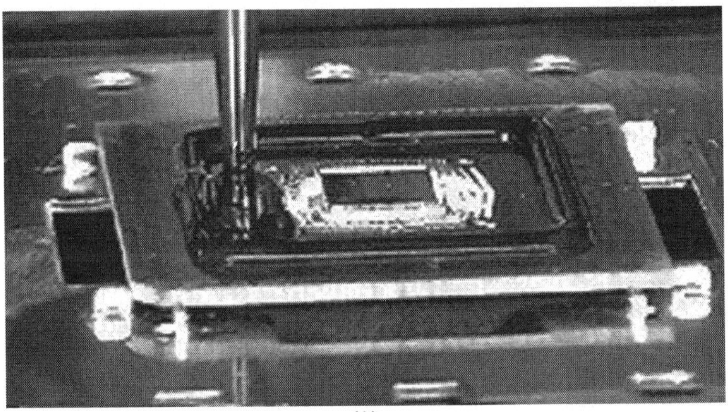

(*b*)

FIGURE 9.12 Dam-and-fill encapsultation. (*a*) Dispense operation, (*b*) fill operation.

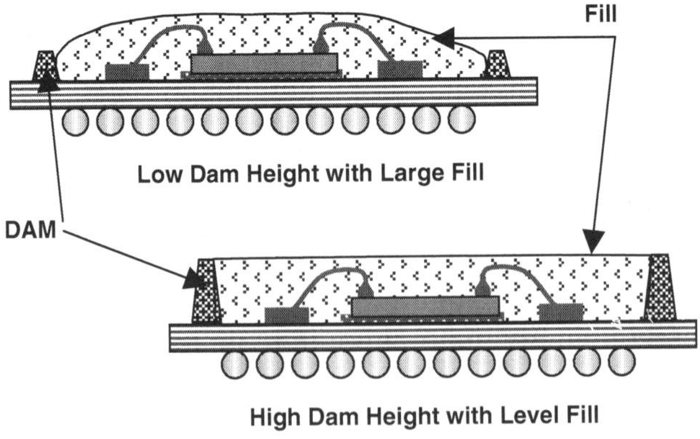

FIGURE 9.13 Dam-and-fill encapsulation.

have a screw feed to pump the fluid through the needle. The auger screw shears the fluid, and this helps to lower the viscosity, as many of the dam fluids are also thixotropic.

Fill fluids are epoxy materials and often have the same basic formulation of resin used in the dam material but with less filler. A dam fluid can have 70% or higher filler content, which increases the viscosity dramatically. There are several fillers that can be used in a dam fluid, and these can also dramatically affect how the fluid flows through a needle and how much the fluid flows after dispensing.

The most difficult part of dispensing a dam is to get the start and end points of a dam line to meet without a gap or bump [3] (Figure 9.14). Highly skilled operators have developed techniques to give a smooth join. Small prestart lines inside the dam are often used. Back tracking of the needle to wipe any fluid tails off the end of the needle has also been employed. This is driven by customers' demand for a flat surface on top of the encapsulation. Specifications of 75 to 100 μm total flatness across the top of a cured liquid encapsulation are not uncommon.

Filling the dam with 5000- to 10,000-cP viscosity material to encapsulate the wire bonds can be done with an auger pump, but a faster pump is a linear positive-displacement pump. Linear pumps can push out fluid at a very high rate, up to 1 cm^3/s. In most cases the requirement is much less than this, but if the geometry of the part permits high flow rates, the pump has the capability. Figure 9.15 show a

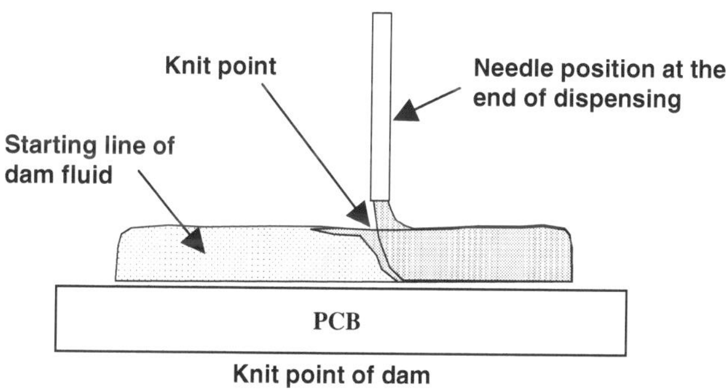

FIGURE 9.14 Dispensing a dam.

FIGURE 9.15 Section through a dam-and-fill encapsulation.

cross section through a D&F package. After curing, it is almost impossible to differentiate the dam area from the filled area, although in this case a small bump in the surface can be seen as the fill material transitions over the edge of the die.

Figure 9.16 shows six profilometer traces over a D&F package. The baseline or zero point is moved slightly in each case to be able to see the traces. Note that the left-hand side of the dam could possibly have used slightly more fill fluid. This can be seen from the slight downturn of the trace at the edge and the fact that there is a slight bump at the die edge on the left side. Both of these artifacts of the dispense process are not apparent on the right side. However, this package still met the overall flatness specification of the customer.

As the equipment evolved for dam and fill, so did the dispensing patterns. Initially, dispensing system were programmed to dispense in a square spiral pattern from either the outside in or the inside out. Through motion studies it was found that a series of lines and dots inside the dam can produce the same results as the spiral pattern's but are much faster to dispense. Fluid flows around the die and wire bonds and forms a bond to the surfaces of the substrate, wires, IC, and the dam walls. It is important that no air bubbles are allowed to form around the wire bonds or in the corners of the dam. To aid adhesion and flow of the fluid, the packages or assemblies are typically heated to 90°C prior to dispensing. The heating also drives off any moisture in the board that could be a cause of voids when the epoxy is cured. When both the dam fluid and the fill fluid are dispensed consecutively on the same machine, a dual-action dispensing tool that carries two pumps is used. Because both of the fluids are wet, they are cured at the same time.

In 2004 there are many packages that still use wire bonds for interconnection. With wire bond pitches at 40 μm and multiple levels in stacked die designs, it is important to know the particle distribution of the filler in the fill fluid. It is not unheard of to find a line engineer struggling with a fill process where the designer used a wire bond process with a fine pitch staggered bond pattern, but did not look at the fluid formulation and check that the maximum particle size in the fluid can pass between the wires. Fine pitch wire bonds can act as a fluid sieve, and it takes

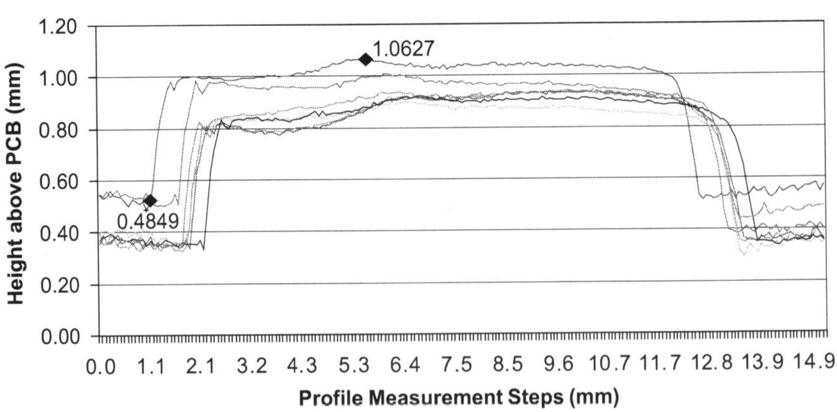

FIGURE 9.16 Profilometer trace of several package encapsulation profiles.

only a few large particles that get stuck going through a field of wire bonds to become a logjam to the encapsulation fill fluid trying to flow around the wires.

The resulting volume of fluid dispensed inside the dam will determine how far the fluid reaches to the crest of the dam lines. Fortunately, linear positive-displacement pumps are highly accurate (1%) in the weight range 30 mg and larger. Linear positive-displacement pumps are not sensitive to viscosity changes in the epoxies.

Many BGA packages (in the millions) have been produced using D&F encapsulation techniques. Most users probably cannot differentiate a liquid-encapsulated part from a molded part. Dam and fill is still widely used for wire-bonded parts, but flip chips and CSP devices are starting to be widely used as the infrastructure to handle these types of devices is put into manufacturing lines.

Dams are also used in some high-reliability aircraft computer assemblies. In this case, large BGA devices are being underfilled on a PC board for improved reliability. Unfortunately, with some of these large BGA devices, the capillary forces to draw underfill under the package are very weak, and this creates a problem to contain the fluid under the BGA. One way around this problem is to use a dam around the BGA. Underfill is dispensed into the dam and under the die. Temporary dam materials can be used to contain the fluid if there is a concern about outgassing of polymers in applications such as space environments.

9.2.6 Underfill

In the late 1960s, IBM [4] realized that peripheral connection to components using wire bonds would limit the number of connections that could be made to a semiconductor die. The concept of an area array connection that could yield many more connections was very attractive. IBM, Motorola, and Delco were some of the first companies to develop what came to be known as *flip chips*. Flip chips make a connection to a package or circuit board using metal bumps on the die to make connections for signals, power, and ground. Some early devices used column structures instead of bumps, as this overcame the thermal expansion and contraction stress problems involved in bump structures. The fabrication of column connections proved impractical for many reasons, including cost. At the same time, many of the problems were found to be mitigated by using underfill. Underfill has a number of benefits, but the two main reasons to use underfill are the reduction of thermal expansion stress and the creation of better adhesion between a package or die and the substrate. Thermal stress problems are primarily a flip-chip phenomenon, as compared to CSPs, where transient mechanical shock is the predominate cause of failure.

9.2.6.1 CSP Underfill

Chip-scale packages are a relatively recent phenomenon (1990s). A CSP is an area array package that looks like a flip chip but is in fact a chip on a substrate (interposer) of PCB material, ceramic, or flex. The CSP substrate usually redistributes the signals to a larger pitch than that found on the original device. The die in a CSP package can be wired bonded or a flip chip. CSP devices are often referred to as BGA devices, and many names are used to describe them (e.g., Micro BGA, Micro SMT). There are many CSP designs, but the devices have common characteristics. The overall size of the package is typically only 20% larger than that of a silicon die. Also, the height of the solder bump for interconnection is on the order of 10 mils high. With larger bumps than flipchips and the introduction of an interposer substrate layer, it was originally believed that CSP devices would be rugged enough to survive use in applications such as cell phones, without the need for underfill. However, in practice it was soon found that mobile devices were failing in the field and underfill was required.

Most mobile devices are battery powered, so a large amount of heat is not generated in a cell phone or PDA. Hence, there are unlikely to be problems caused by differential thermal expansion. By far the largest cause of failure in portable devices is sudden impact shock such as dropping a phone. Most portable devices must pass several drop tests onto a hard surface from 2 m above the ground [5]. Drop testing creates the need for an underfill material that holds the part to the motherboard. With the device bonded to the board, they move together under conditions of vibration and shock forces. Because the board is not moving independent of the board, the solder connections are not subjected to displacement forces and can survive more of the shock forces when compared to a CSP without underfill.

Throughput is a major consideration in the production of inexpensive portable devices, so the underfill operation has to be fast. Ability to stand shock and fast processing affects the design of the underfill materials. Without the need to limit the thermal expansion of the underfill, lower concentrations of filler materials are used. These materials must have high bonding forces and flow out quickly once dispensed.

The underfill process uses capillary forces to pull material under the die. A typical process is to dispense a line of underfill material next to one or two edges of a CSP device or flip-chip die (Figure 9.17). Capillary forces pull the fluid under the

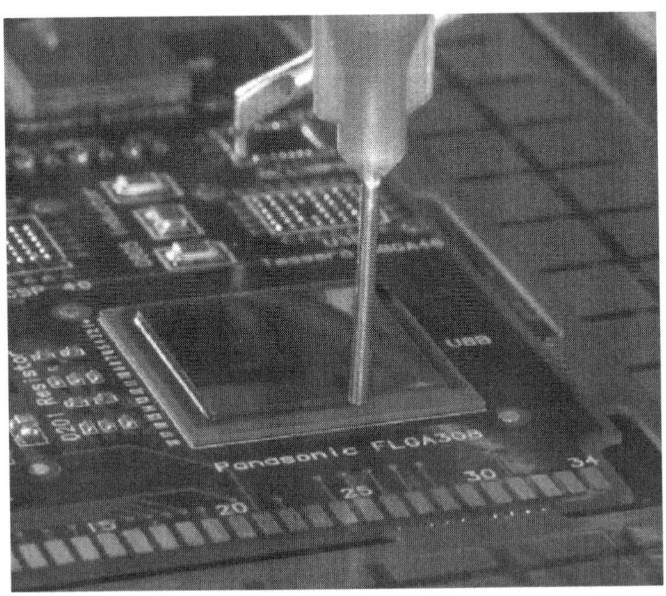

FIGURE 9.17 CSP underfill. (Courtesy of Asymtek.)

package or die. Simple single lines or L-shaped patterns work best (Figure 9.18, 1 and 2). This can be put down in a single pass or in multiple passes, which is often the case in large flip-chip die. Multiple passes are used when other components are placed near the die or package, and the volume of material that would be dispensed for a single pass would start to overflow onto these other components.

Seal passes (Figure 9.18, 3) are employed to get an even fillet of material around the die or package. However, in many cases, self-filleting materials will form fillets on the sides of the device which was not dispensed. Fillets provide more surface area to hold the part down and also remove any sharp corners where stresses can concentrate and cause cracks. The fillet should be halfway up the wall of the chip or device, and some suggest that even 80 to 90% up the side of the package for the fillet is desirable.

Typical CSP underfill operations are three to five times faster than flip-chip underfills for a similar-sized die. Multiple panels of cells phones containing twelve 10-mm^2 die are often processed in 20 seconds. This time includes loading the panel, visually finding all the devices and moving the dispensing needle into position before dispensing. It can be seen that the actual time for dispensing fluid is on the order of 1 second per device. The panels are heated prior to dispensing to 90°C to aid fluid flow. The initial heating process takes place at a preheat station, and the dispensing station is also heated. Sometimes a third heating station is used to allow flow out after dispensing. The amount of time for this process step is limited by the dispensing time, so as improvements are made in dispensing throughput, the amount of time for heating is reduced. There is some concern

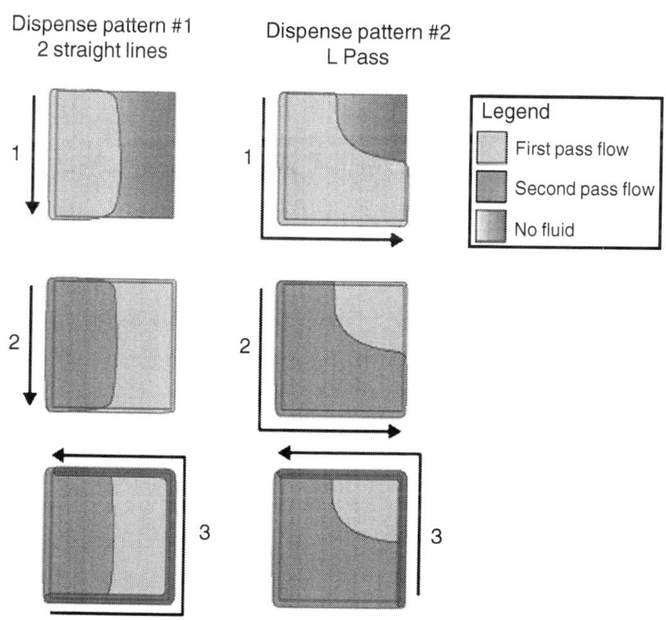

FIGURE 9.18 CSP and flip-chip underfill dispense patterns. (Courtesy of Asymtek.)

that the rate of heating (i.e., 4°C/s) is starting to approach the reflow soldering heat ramp rate and could lead to thermal shock problems. However, no problem has currently been reported.

Many thousands of cell phones use underfill for CSPs. However, due to phones getting smaller and hence having less mass and the motherboards in phones becoming smaller and better supported, the use of underfill in cell phones has been reduced in recent years. Phones with PDAs and camera chips that are often heavier and for which reliability requirements are more strict generally use underfills.

CSP devices have now started to build up in the z direction by adding more die to the package by stacking them on top of each other . This allows the package designer to use mixed semiconductor technologies in one package, such as a microprocessor and memory. Often, a stacked package will use wire-bond and flip-chip technologies in the same package. In a case of mixed interconnection technology, it is often necessary to have a flip chip that is mounted on the top of a die, underfilled. Space limitations for dispensing needles are often a problem, and a new technique where a mechanical inkjet is used to shoot dots of fluid into restricted areas is now being used in the industry. Space limitations are also a problem when a CSP device is mounted to a printed circuit board and other components, such as resistors and capacitors, are mounted in close proximity to the CSP . This can cause a problem where the underfill is pulled away from the CSP toward and around other components. This leaves the CSP with insufficient underfill surrounding all the balls under the package.

Another form of CSP underfill is to put adhesive dots under the corners of a CSP. The theory is that the bumps at the corners are the first to fail, and therefore if these are protected, the device may survive longer than without any underfill. From the published data it would appear that packages assembled with corner dots are more reliable than no underfill. But full capillary underfill can withstand more drops and thermal shock testing than can corner dots. With regard to dispensing time, four dots can be slower to dispense than two lines. However, the four dots are put down before a CSP is placed and reflowed onto a board, and therefore there is one less curing stage in the assembly process.

9.2.6.2 Flip-Chip Underfill

The flip-chip underfill process was developed many years before CSP devices were used (Figure 9.19). The driver for flip-chip underfill is to mitigate stress caused by thermal expansion differences that exist between the silicon die and a package substrate such as a printed circuit board. The expansion coefficient of silicon is 4 ppm/°C, the printed circuit board is on the order of 17 ppm, and the tin–lead solder balls have a thermal expansion coefficient of around 21 ppm. As the die gets larger, the stresses caused by thermal expansion mismatch are exacerbated. It was found by early researchers that the problem could be managed by underfilling the die. This moved the stress force from one where the solder balls were being distorted to one where the forces were being distributed over the entire surface of the die. With underfill materials cured around the solder balls, this also keeps the ball in compression below the glass transition temperature of the underfill. Failure analysis

FIGURE 9.19 Flip-chip underfill on a ceramic PGA package.

following thermal cycling of flip chips usually show that the corner bumps are the first to fail and the failure mode is from crack initiation at the junction of the bump and silicon die. Voids in the underfill often show that the solder metals of the bump flow into the voids: hence, the need for void-free underfills. Underfills can also develop cracks in the fillet that surrounds the die, and this is also an indication that a ball might be about to fail.

Base epoxy resins typically have a thermal coefficient expansion of 70 ppm/°C. This can be reduced by the addition to the resin, of filler materials such as silica. It is not uncommon for underfills to have as much as 70% fillers. The size distribution and shape of the filler particles can have a large impact on how the fluid flows under the flip-chip die. In practice, the desire is to load the underfill with lots of filler, but this slows down the flow of the material under the die. Also, the gap between the die and the substrate is nominally 75 μm but in many cases is much smaller. A second restriction to fluid flow is the bump pattern itself. It is not uncommon to see over 3000 bumps on a 15-mm^2 flip chip, which can act like a sieve to the fluid. Clearly, the filler materials must be small, and to aid flow the shape of the filler particle is important, round being better than flat or acicular. Filler particles are the major cost factor in an underfill system. It has been found that the ideal loading for filler particles should give a resulting underfill when cured to a thermal coefficient of expansion (TCE) of around 21 ppm or matching the expansion rate of solder bumps. Resins have been produced with lower TCEs, and these have been found to be less reliable under temperature-cycling conditions.

9.2.7 Gasketing

Gasketing and sealing of components, enclosures, and so on, uses many different materials, depending on the application (Figure 9.20). Outdoor enclosures need to seal out moisture, and electrical gaskets often have to be explosion-proof. Seals may also need be resistant to chemicals such as solvents, petroleum-based spirits, and acid/alkalis. This means that a number of different gasketing and sealing materials are used for each application. The range can cover cyanoacrylates, epoxies, silicones, and urethanes. These fluids have vastly differing fluid characteristics and as such require different features from a pump or valve.

9.2.8 Solder Masks

Solder masks on printed circuit boards are silkscreen printed. Dispensing of solder masks is used when large components such as capacitors do not permit screening of the mask, or pins are added to a board after screening and these have to be coated for electrical insulation or to keep moisture from the board surface.

9.2.9 Surface-Mount Adhesive and Standoffs

Surface-mount adhesives (SMAs) were used extensively when people moved from through-hole components with legs or pins (e.g., dual-in-line packages) to surface-mountable components with no legs (Figure 9.21). The SMA is used to hold chip components to a printed circuit board through a wave soldering operation. The adhesives are highly filled epoxies that can be screen printed or dispensed as dots on a board between two solder pads (Figure 9.22). Due to the often large number of dots on a board, dispensing the dots has to be done at high speed [6]. Dispensing speeds

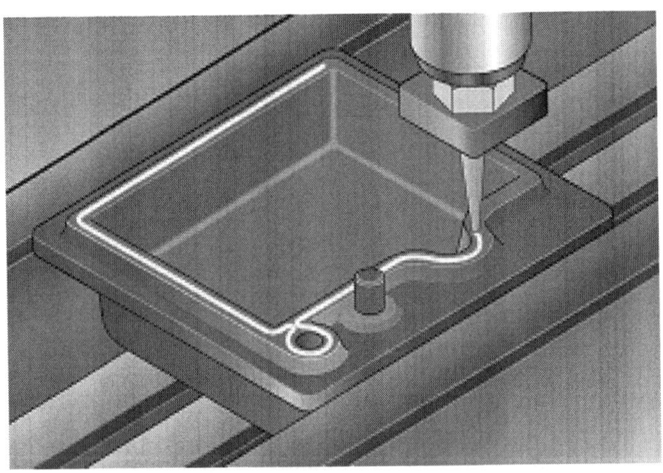

FIGURE 9.20 Gasketing on an electronic housing. (Courtesy of Asymtek.)

FIGURE 9.21 Surface-mount component and adhesive.

FIGURE 9.22 Jet-dispensed SMA dots.

of 30,000 to 50,000 dots per hour on 3-mm centers is a typical rate of dispensing.The dots have to be round, not bridge over contact pads, and not flow after dispensing.

Over the last decade, surface-mount components have reduced in size from 1206 to 0201's and smaller. 1206 components are 0.120 in.×0.060 in. The surface-mount industry has moved from wave soldering to double-sided printing and reflow. Solder pastes have been developed that have tacky fluxes designed to hold components in place up to reflow; hence the use of SMA dispensing has diminished enormously in recent years. One area in which it is still being used is in pro viding standoffs for components.

9.2.10 UV Adhesives and Gaskets

Ultraviolet (UV) activated adhesives are used in many industries; examples are conformal coatings, stacking bonds for recording heads, and dams on lead frames. UV adhesives are normally used where the components cannot be subjected to heat. They are also used where a quick cure is required. One large application for UV dispensing is for sealing flat panel displays.

The viscosity of the fluids ranges from almost waterlike to gel consistency. Some of the UV initiator chemistry can react with metal components of the valves and pumps. Iron is particularly prone to reaction with UV materials, and some stainless steels are also a problem. When specifying UV materials for an application, it is a good idea to discuss this with the fluid vendor. Figure 9.23 shows a jet dispensing a line of UV adhesive for a gasket on a cell phone display.

9.2.11 Conductive Adhesives

Conductive adhesives are made primarily with silver flake materials; some gold and carbon materials are also available. Silver-loaded epoxy materials are used for die bonding and as a replacement for solder. To get good electrical conductivity, flat plate–like silver particles are used. The plates have a large surface area for good conductivity. The problem in dispensing these materials through a needle is that plates do not want to be pushed through a small narrow tube. There is a tendency to clog in the needle or pump and get missing dots. The longer the needle, the more this becomes a problem. One common problem is not complete clogging of the needle but gradual starvation of material, with dots getting smaller, and then a clog can free itself and the operation returns to normal-sized dispensed dots. Of course, the correct amount of material is not being disposed in this case and could cause an electrical problem.

FIGURE 9.23 Jet dispensing of a UV gasket material on a cell phone display.

Auger pumps are the tool of choice for most silver epoxy dispensing applications. These pumps use a screw to move and pressure feed the fluid through the needle. Auger screws can be cut with various thread pitches and depths, and can be made from a number of materials. The tolerance of the screw fit to the auger body can also affect the ability to develop pressure in the pump, and this can affect the cutoff and start of fluid flow.

The fluid and pump combination can have a large impact on the shape and size of a dot or line produced from an auger pump. The smallest dots and lines today are produced from auger pumps. Depending on materials, the smallest dots are approximately 150 μm in diameter and line widths of 175 μm wide. These limits are under attack every day, and many people are trying to get to 75-μm dots to facilitate wafer bumping with conductive adhesives. As said earlier, these sizes are highly dependent on materials. Often, there are other requirements, such as no outgassing for space of disk drive applications. Due to the changes in materials, this will often affect the final dot size to be much larger than these numbers. The dot size is also affected by how much the dot flows following dispensing and/or how the material interacts with the surface on which it is dispensed.

Recently, jet dispensing has demonstrated some very high speed dispensing with silver epoxies (Figure 9.24), particularly for die bond patterns. Jetting can significantly improve throughput but does not give the smallest dots results. Jetting imparts energy to the material that is projected to the substrate. On impact the dot spreads out. Dots can be fired at rates of 200 times per second. One unique advantage of jet dispensing is that it is digital and patterns can be formed with more material at the center than the outside. This makes jet dispensing ideal for silver epoxy die bond patterns and other applications, such as no-flow underfills.

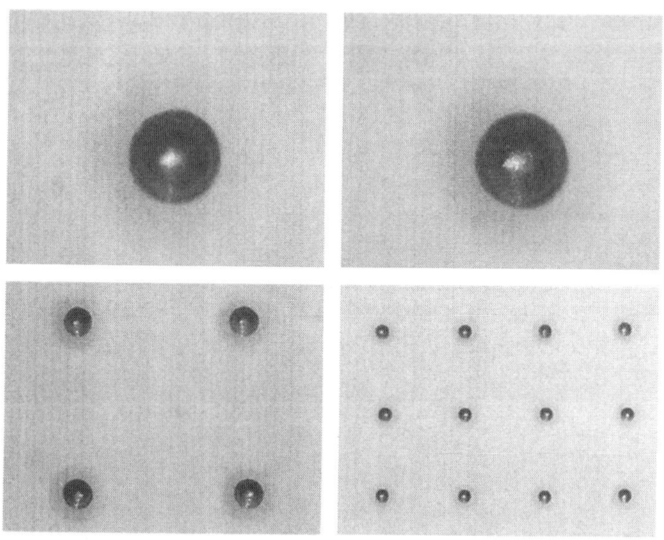

FIGURE 9.24 Jet-dispensed silver epoxy dots (0.33 mm in diameter).

9.3 MATERIALS FOR ENCAPSULATION AND UNDERFILL

Many adhesive, potting, and coating materials are used in the production of electronic equipment: surface-mount adhesives, silver pastes, solder masks, conformal coatings, encapsulations, and underfills. In this section we look at materials for encapsulation and underfill from a packaging engineer's prospective. Packing engineers come from diverse backgrounds: mechanical, electrical, physics, materials, and chemical. What is important is that an engineer faced with the task of selecting materials for a package design understands some of the physical properties of the materials they are about to choose and their impact on the design.

When a new package is being designed, a number of criteria have to be considered. The most common are environmental factors such as mechanical protection, shock and vibration, electrical insulation, damp heat, and thermal environments. Depending on the use environment, some customers may be very worried about mold growth on encapsulations used in tropical regions, although computer manufacturers may be more concerned with having materials that have flame retardants added. If parts are used in an engine compartment, they will have to withstand heat, vibration, and oils. Today, electrical assemblies often have to interface with light pipes or other sensors. Some sensors can be damaged by heat and other physical conditions of the assembly process, which in turn can restrict the material choices and assembly process that can be tolerated.

If a material is going to be used as an encapsulant for integrated circuits (ICs), the extractable ionic contamination levels for sodium, potassium, and chlorine will have to be determined. Similarly, if the encapsulant will be protecting memory devices, the alpha-particle emissions from the encapsulant should be determined,as these can change logic states in MOS-type devices.

Fortunately, the engineer is not alone. Many organizations have developed test specifications for electronics in various use environments. The U.S. military has for many years specified test conditions in its MIL specs, for electronic equipment used in locations from office, to mobile land, to operation on the tip of an airplane wing. Various trade organizations, such the International Printed Circuit Board Association (IPC), Joint Electron Device Engineering Council (JEDEC), Electronic Industries Association (EIA), American National Standards Institute (ANSI), International Electrical Commission (IEC), Semiconductor Industries Association (SIA), Underwriters' Laboratories (UL), and similar organizations in other countries around the world, work closely together to provide input to the national standards bodies. In the United States a number of trade associations develop standards for the American National Standards Institute, which in turn represents the United States at the International Electrical Commission.

9.3.1 Design Plan

Where is the component going to be used? What kind of life is expected under typical life conditions? You have to think about how your product will be used. If it is going to be subject to a lot of thermal cycles, adhesion of the encapsulation to the leads may be a large factor. If a lot of components are to be used in a single compartment in office equipment, what would happen if fire were to break out in an adjacent component, and what would happen if 500 or 1000 components were to be engulfed by fire? Most users of cell phones perform a simple test every day when they drop them onto a hard surface, expecting them to work afterward. So you need to have a design for assembly and a use environment plan. The first consideration is:

- Can I build this component using these materials?

The following are just a few of the considerations that engineers need to review when designing a package or assembly process.

- Is it sensitive to heat, light, or some other physical property?
- Will the materials become part of the circuit (i.e., underfill of radio-frequency devices)?
- Will any of the materials interact with each other or the component?
- How easy are the materials to process, including safety considerations?
- Does this component have to meet any set standards (medical, telecom, space, aeronautic, military)?
- Can these materials be used with existing equipment and operators?

Once you have made your choices of materials to be used and build a few parts, you will need to test the final assembly or components. You may need to rethink some of your material choices and make compromises with some of the material properties to get a final working product that can meet your customer specifications for use environment and reliability.

9.3.2. Material Choices

For encapsulation and underfill, epoxies, silicones, and urethanes are just a few of the choices for the engineer to review. In most cases the choice is paired down to epoxies and silicones. However, urethanes are used in many applications, such as potting transformers and conformal coating applications.

9.3.2.1 Silicones

Silicones are used for encapsulation, die bonding, and wire and tab bond stress relief coatings and are very useful in situations where low or high operating temperatures are going to be experienced, such as in engine modules. At very low temperatures they remain relatively soft and flexible. Glass transition temperatures for silicones can be as low as −120°C and are stable at temperatures up to 250°C with short-term excursions up to 300°C. Silicones have high moisture permeation rates; 2 to 4% is typical but they also have good adhesion to most component parts, so there is less chance of moisture accumulating and forming corrosion sites than there is with epoxies. There are a couple of other drawbacks to the more general use of silicones in electronic encapsulation applications; it is not easy to print package identifications on a finished device, and many engineers do not like to have silicones near a wire bonding operation for fear of contamination of the bond pad surf aces. Silicones are used for stress relief over wire bonds. They are often used to encapsulate modules in boxes and where other materials (e.g., copper) that have a large thermal coefficient of expansion (TCE) are potted in a module.

Silicones have a high thermal coefficient of expansion but a low modulus, which means that they typically expand more than epoxies, but the expansion can cause less damage because the material compresses easily. The thermal expansion coefficients are 125 to 300 ppm/°C compared to 70 to 200 ppm/°C for epoxy resins (Table 9.1). Although the thermal coefficient of expansion for silicones is approximately twice as large as that of epoxies (Figure 9.25), the modulus of silicones is about half that of epoxies [7]. Curing of silicones can be accomplished using heat, moisture reaction, or ultraviolet radiation. They are also used in many medical applications where epoxy materials may not be suitable.

9.3.2.2 Epoxies

Epoxies are used in man y encapsulation applications. Epoxies are typically two-part materials consisting of a resin material and fillers to adjust viscosity and other properties,

TABLE 9.1 TCE Values of Epoxies and Silicones

Material	TCE [10^{-6} (cm/cm)/°C]
Unifilled silicone	320
Filled silicone	125–275
Unfilled epoxy	70–200
Filled epoxy	30–125
PWB(*xy* axis)	12–16
Silicon	2.4

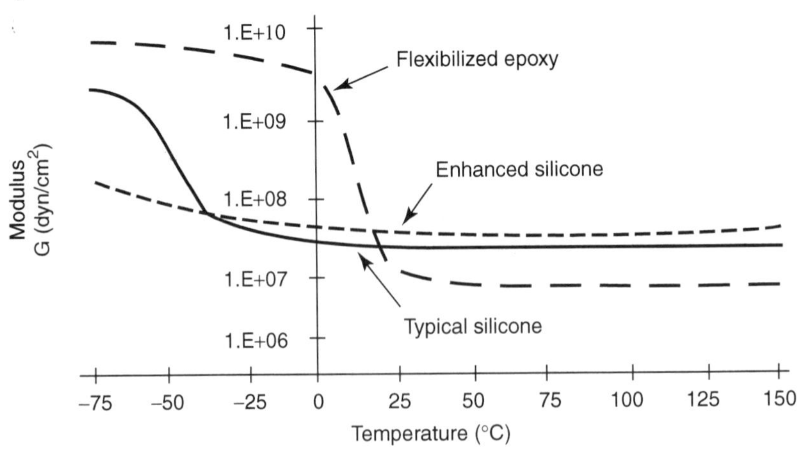

FIGURE 9.25 Silicone and epoxy modulus at temperature. (From Ref. 7.)

with the second part being a hardener system that is used to cure the resin system. For automated dispensing applications the preferred method of purchasing the epoxy materials is premixed, frozen, and degassed.

If a supplier can supply the encapsulants only in two separate containers, resin and hardener, an operator will have to measure out at the correct ratio and mix the two components. Often, the fillers in resin containers fall out of solution, so it may be necessary to keep the containers in motion on a roller or at least turn the can over each day. Mixing of two components can put air into the materials. If the materials are not degassed after mixing, automatic dispensing systems will splutter when an air bubble passes through the dispensing tip. This can result in a missing dot or an air bubble trapped in the encapsulation materials. The air has to be removed, and if the material has a short pot life, a significant portion of the pot life could be used trying to get the air out. It can be seen that mixing of two-part fluids at a manufacturing site can have significant overhead costs compared to having a fluid supplier premix and degas a fluid. Unfortunately, when materials are premixed, they start curing. To inhibit the cure, materials are frozen at −40°C. This will usually give them a storage life of at least six months.

Engineers have to determine if they want to buy the encapsulants for a higher price but with less operator intervention required, or alternatively, use a two-part fluid. Some fluid manufacturers specialize in supplying encapsulants for specialized industries such as the optical industry and do not have the capability to premix and freeze materials. It is possible to find third-party packaging companies that will put materials into a variety of packages: from a vacuum bag with a divider so that the user can remove the divider and mix the materials with minimum handling, to special syringes with membranes that break and static mixing tubes.

Make sure that the pot life is adequate! It is no use having a 1-hour pot life material if the operator is going to spend 30 minutes of the pot life degassing the fluid. If high-accuracy dispensing is desired, an expensive pump will be required. It is not practical to put a 1-hour pot life material into a pump that can cost $20,000. It could end up as an expensive paperweight.

Epoxy systems have been used for many years in the electronics industry. It has been found that overexposure to liquid epoxy systems can cause an allergic reaction in some people. This may not be immediately apparent and may not be obvious for some time. Redness of skin and eyes and dry skin can be signs of exposure to epoxies. Once a person becomes sensitized, there is no other remedy than to remove the person from the environment where the epoxies exist. It is important that all precautions recommended by epoxy manufacturers and by government agencies should be followed. A good report on the epoxy resin sensitivity was a study on epoxies used in the UK construction industry [8].

9.3.2.3 Epoxy Resins and Fillers

The epoxy resin system usually consists of the following constituents:

- Base resin
- Curing agents
- Reactive diluents

- Solvents
- Plasticizers and toughening agents
- Fillers, pigments, and reinforcing fiber

Epoxy Resin The word *epoxy* is derived from two Greek prefixes: *epi*, which means "upon," and *oxy*, which means "sharp/acidic." The invention of epoxy resins is credited to the Russian chemist Prilezhaev (Prileschajew). In 1909 he discovered that olefins react with peroxybenzoic acid to form epoxides. In the mid-1930s, Schlack of Germany applied for a patent for the preparation of polyamines. However, epoxy formulations that an engineer would recognize today are credited to two inventors, P. Castan from Switzerland and S. Greenlee from United States, who applied for patents in the late 1930s. In 1946, Ciba was the first company to sell epoxy resins [9].

Epoxy materials are noted for their versatility, high resistance to chemicals, outstanding adhesion, durability, and toughness. Because of their good electrical properties and the ease with which they can be poured, dispensed, or molded, epoxy resins are especially useful for encapsulating electrical and electronic components.

Hardeners (Curing Agents) Three main curing agents are used with epoxy resins:

1. *Amines:* most common curing agent; short pot life and high exothermic temperatures during cure
2. *Anhydrides:* typically have better electrical properties than those cured with amines; lower dielectric constant
3. *Lewis acids:* long pot life at room temperatures and high T_g values normally used in conjunction with anhydrides; lower-temperature exothermic reaction than amines and less skin irritation

Each of these curing agents will impart different properties on the cured product as detailed above. The resin chemist at the formulating company is often faced with conflicting requirements to produce products with the desired properties. It is through the skill of the formulating company's staff that the right combinations of resins and curing agents are used to achieve the results desired.

Filler Materials Filler materials are added to epoxy encapsulants for a number of reasons, to reduce thermal coefficient of expansion, to increase the strength of the materials, and to increase the thermal conductivity. In some potting applications they may simply be added to reduce the amount of resin used and lower the cost of the material. Other additives are also used to make epoxies fire retardant or to add colors so that they can be identified easily on a production line. Fillers come in many materials, shapes, and sizes. Figure 9.26 shows three commonly used filler material particles. Glass beads make an encapsulation lighter for use in avionic applications, talc imparts a smooth surface on a finished encapsulation, and fumed silica is used in electronic encapsulation and underfill materials to control viscosity. The volume of filler controls the thermal coefficient of expansion. Resins that are designed for flow under flip-chip components, where spacing between the chip and board can be as small as

(a) (b) (c)

FIGURE 9.26 Three common filler materials: (*a*) hollow glass beads; (*b*) talc; (*c*) fumed silica. (From Ref. 10.)

25 μm, require small, fine, round particles to aid flow properties. Unfortunately, the cost of filler materials can dominate the component cost of underfill and some encapsulation materials. It is envisaged that one day the connection or bump height of flip chips and the bump density will be so dense that capillary underfill methods will have to be replaced by preapplied underfill. Also, the smaller a filler particle, the larger the surface area-to-resin volume. The surface area has a large impact on the viscosity, so there is a trade-off between particle size, loading, and how well a fluid flows.

Figure 9.27 shows a cross section through a flip-chip device. Note how the filler materials have settled to the base of the flip-chip joint at the board interface, leaving a pure resin layer next to the silicon die [11]. This would result in higher thermal coefficients of expansion near the bump-to-die interface, which is the point at which most solder balls on flips chips fail and where it would be desirable to have a lower expansion coefficient. In recent years, underfills have improved, and this is not seen as a problem with today's materials.

FIGURE 9.27 Cross section through a flip chip.

It is worth noting that the fillers are not all the same size; there is a distribution of particle sizes. To get a higher packing density of filler you need to have a distribution of particle sizes. In package designs where wire bonds are used to connect a die to a package pin/pad, it is possible on fine wire bond pitch to find that the larger particles will not flow through the gap between the wires. This problem can be overlooked when a wire bond pitch is changed on a design only to find that the encapsulation process no longer works. Filler content can also have an impact on flow rates, as can be seen in Figure 9.28. As filler loading increases, the flow rate is reduced and hence the manufacturing process will take more time.

Ideally, with encapsulants and underfills it is desirable to have a low TCE value. This is achieved by high filler loading of low-expansion particles. Figure 9.29 shows the relationship between filler weight and resin. It is not uncommon to see filler loadings in the 60 to 70% range on encapsulants and underfills. A trade-off has to be made between higher reliability from low TCE, or faster flow throughput rates on the line. In cell phones and other portable products, underfilled flip chips can achieve several thousand temperature cycles from −40 to 120°C before failure. Some manufacturers of cell phones accept 1500 temperature cycles as sufficient to demonstrate the reliability of their product for its planned life in the use environment. This can give the packaging engineer some latitude when it comes to defining and underfill, by allowing a material to be used with a lower filler ratio and achieve faster flow rates and hence improve throughput through the assembly process.

Figure 9.30 shows the impact of filler loading and particle size and how this changes the viscosity of the fluid at different shear rates changes. Note that at 50% loading the differences in viscosity between 8- and 30-μm particles is slight. As the amount of filler is increased to 60%, the viscosity more than doubles; at this loading the influence of the smaller 8-μm particle is significant. As the shear rate is increased, the viscosity of the fluid with 8- μm particles drops. This effect, known as *thixotropy*,

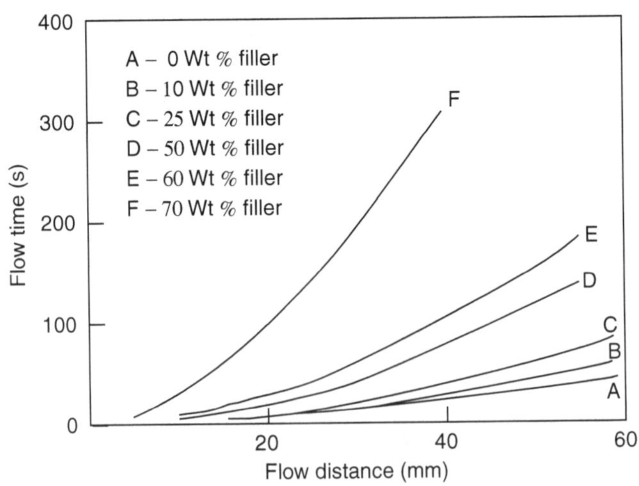

FIGURE 9.28 Impact of filler loading on flow rates for underfills. (From Ref. 12.)

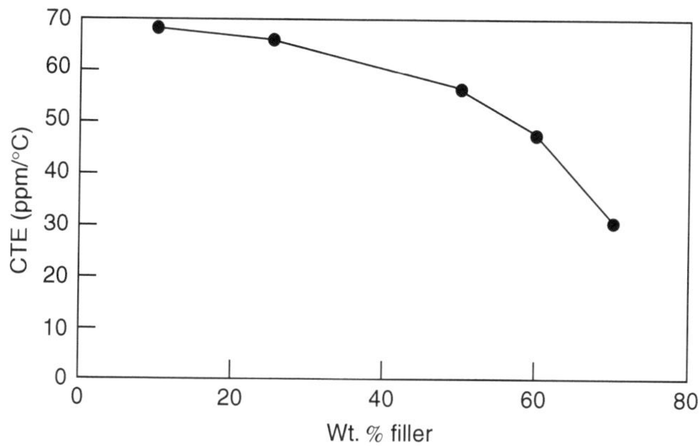

FIGURE 9.29 Percent weight of filler versus CTE. (From Ref. 12.)

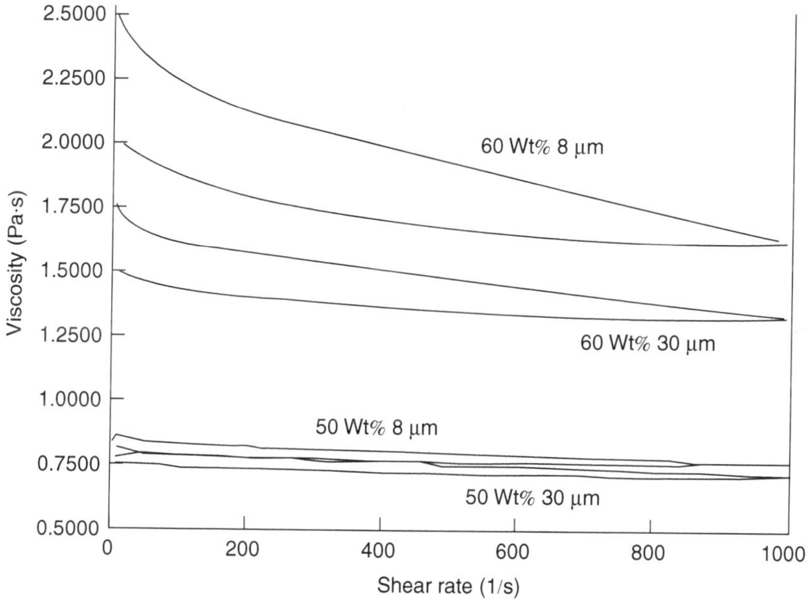

FIGURE 9.30 Impact of filler loading and size on fluid properties. (From Ref. 12.)

can be used in dispensing to lower the fluid viscosity. By using shear forces developed in an auger pump to lower viscosity, after wire bonds have been coated with encapsulants the encapsulation quickly recovers its original viscosity [12].

Filler materials are clearly used to modify the properties of basic resin formulations. There are many different types of fillers, and each is used to impart specific properties. Often, fillers are used in combinations to get a mix of properties in the f inished product. In the case of underfills for flip-chip applications, where the fillers have to be small, preferably spherical to aid flow, and have a tight distribution of particle size, the cost of the filler can be greater than the resin component of the underfill.

9.3.3 Viscosity and Thixotropy

Viscosity is a measure of fluid resistance to flow. This resistance acts against the motion of any solid object through the fluid and also against motion of the fluid through a dispensing needle. Viscosity is affected by the chemistry of the resin, the molecular weight of the resin, or the type and amount of filler added. Lower-viscosity materials are typically thin, unfilled fluids and higher-viscosity materials tend to be highly filled. As shown in Figure 9.30, small particles have a large surface area. Therefore, nanofiller materials will have a larger affect on viscosity than, say, microparticle filler materials, even though they comprise the same percentage of weight in the material.

The temperature at which a fluid is dispensed can have a large impact on the viscosity of epoxy materials. Silicone materials do not change much with increasing temperatures. This effect on epoxy materials can be utilized by adding needle heaters to aid the flow through a needle of a thick epoxy encapsulant.

Some materials experience a viscosity change when being dispensed or worked. These resin systems have shear rate dependence. Materials can be Newtonian (the viscosity remains constant as the shear rate increases) or thixotropic (the viscosity decreases as shear rate increases). Epoxies are usually non-Newtonian and are thixotropic in nature. Thixotropic properties can be utilized in applications such as glob top encapsulation. If the fluid is dispensed with an auger valve, where the screw of the auger shears the fluid, the viscosity of the fluid will drop as it is dispensed through a needle. The fluid will flow over the die and wire bonds. The thixotropic nature of the fluid is relied on to limit the flow of the material or encapsulant.

Viscosity is stated in units of poise or centipoise. Encapsulants range from 1000 to 50,000 cP. Damming fluids used to contain wire bond encapsulation fluids are typically approximately 100,000 cP. In dam-and-fill applications, the dam dispensing pump is usually an auger pump with a heated needle to lower the viscosity of the fluid in the needle, whereas on the same system, a second valve will be used for the wire bond encapsulant and this often requires a linear positive-displacement pump with no needle heater.

9.3.4 Alpha Particles

As silicon devices shrink in size, the logic state of gates in memory devices can change for no apparent reason [13]. No physical damage is observed in a device, and the device appears to work normally after a reset. This is often described as a soft error. Alpha particles cause electron–hole formation, which in turn causes a gate to change its logic state. The source of the alpha particles has been found to be trace amounts of radioactive materials. Lead-based solder materials are generally found to be the main culprit, but some ceramic filler materials can also be slightly radioactive. As a guideline, trace levels as low as 0.01 to 0.5 $\alpha/\text{cm}^2\cdot\text{h}$ are recommend for encapsulants used with sensitive devices. Alpha particle measurement is difficult; very few companies in the world can perform this measurement.

9.3.5 Extractable Ionic Contaminates

The presence of chlorine, bromine, and potassium in electronic packaging materials has been show to have a highly detrimental effect on the life of a component under accelerated life testing. When combined with moisture, they can act corrosively or simply increase the electrochemical reactions, causing electromigration through a process similar to electroplating. The time to failure in plastic-encapsulated materials can be related directly to the chloride and other halides in epoxy encapsulation materials. Very little chloride is required to cause considerable damage to chip metallization. As a general guideline, materials used for encapsulation of integrated circuits should have less than 4 ppm of water-extractable chloride or other corrosive elements. Chloride ions are a by-product of the epoxy resin encapsulation materials. However, the problem is well understood by encapsulation manufacturers, and the engineer should use materials with appropriate cleanliness for the type of component encapsulation.

9.3.6 Thermal Coefficient of Expansion

It is not uncommon to see the acronyms TCE (thermal coefficient of expansion) or CTE (coefficient of thermal expansion) used in papers or on data sheets. Both refer to the expansion of materials when heat is applied to them. The usual form of expressing this quantity is ppm/°C (e.g., 10 ppm/°C or 10×10^{-6}/°C). In either case, using the example given, the material in question would expand 10 μm per meter of material. This does not sound like a lot, but it can be enough to break wire bond welds and bow circuit boards.

The expansion rate can be changed drastically by the use of fillers. However, not all fillers are round particles; some can be acicular particles of strands of glass fibers. In these cases the expansion rate will change, depending on the direction of measurement relative to the filler orientation. A graphic example on this is the TCE of printed circuit boards. The x and y thermal expansion is usually much less than the z expansion rate. Circuit boards are made from woven glass mats in epoxy. The woven glass constricts expansion in x and y but not as much in z.

9.3.7 Glass Transition Temperature

Epoxies and other materials do not always expand at the same rate over a given temperature range. At some temperature point they become flexible, soft, and start to expand at a greater rate. A generally accepted definition for the glass transition temperature (T_g) is where the expansion rate triples. Below the T_g value, materials are in a glassy state and epoxies are hard; above T_g epoxies become rubberlike. This is not a sharp transition point, and obviously, epoxy materials would be starting to soften earlier than the T_g temperature. Some materials, such as silicon materials, can have T_g points at very low temperatures in the subzero range.

There are several methods of measuring T_g: differential scanning calorimetry (DSC), thermomechanical analysis (TMA), and dynamic mechanical analysis (DMA). Unfortunately, these test methods can give slightly different results. T_g values can be expected to vary ±10%, depending on the test method. For electronic packaging applications it is advisable to have the T_g value above the use environment temperature. This will ensure that high expansion rates are not experienced by the component in it normal use. If the temperature does exceed T_g, this can lead to device failure as wire bonds and other joints come under forces that can disrupt welds and solder joints.

Epoxies have to be cured in the 150°C range to get T_g values of 130°C and above. If room cure or low-temperature cures are used, it is not possible to get high T_g values. This becomes a difficult design problem if materials used in the final device are temperature sensitive. Then the designer has to look at low-modulus materials such as silicones. These have T_g values at very low temperatures, so in room-temperature use they will be in a high-expansion region; however, they are soft and can compress under expansion forces.

9.3.8 Cure Shrinkage

Most polymer materials get denser when they cure. This is a result of cross-linking of the polymer chains, and some of the volatiles in the fluid materials boil off during the curing process. Assuming that the polymer or encapsulant cures at a high temperature (e.g., 150°C), when the part returns to room temperature, the epoxy contracts. This can have a bad effect, such as warping PC boards. However, it can also be beneficial: wire bonds coated with encapsulant or a flip chip held down to a board with compressive forces caused by the shrinking of the encapsulant. Also in the flip case, the bump connections under the die are surrounded by compressive isostatic forces that do not allow the solder material to cold flow when subjected to multiple thermal cycles.

9.3.9 Modulus of Elasticity

The ability to compress and return to a predefined shape is elasticity. At some amount of stress the material is unable to return to its original shape and becomes permanently deformed. The results of stress are seen as strain. The ratio of these two is known as the *modulus*. All solids are elastic for small deformation, but if the stress exceeds a certain amount, known as the elastic limit yield point, a permanent deformation is produced. The ratio of tensile stress to elastic strain for a given material is called *Young's modulus*. Figure 9.31 shows the difference between silicone and epoxy modulus characteristics.

Engineers can use their understanding of these modules to overcome problems that they will encounter in component design. For instance, as wire bond pitches get smaller, wire sweep in the molding process can cause the wires to short together. One wire bonding company now produces a combined wire bonder and dispenser. A low-module encapsulant is used over the wire bonds to minimize any transferred forces, but the molding compound offers a rigid exterior for the component. The

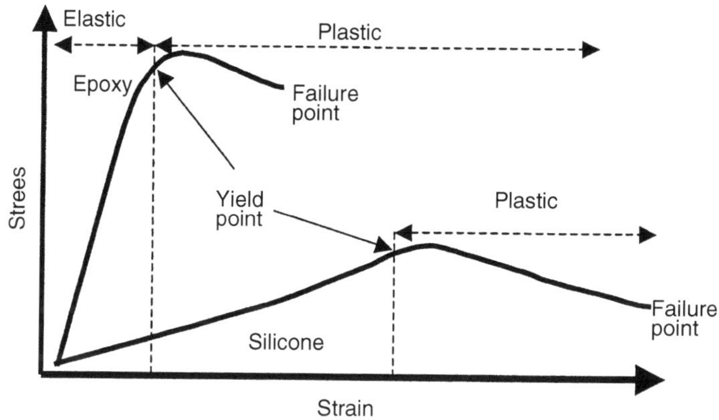

FIGURE 9.31 Modulus example for epoxy and silicone.

hard exterior holds the leads on the component in place, but the low-modulus wire bond encapsulation will allow some movement of the wires but does not allow individual wires to be swept away.

Unfortunately, the most commonly available data on modulus are usually taken at one temperature, and as we have seen with the glass transition temperature, materials change properties with increasing temperature, and the modulus is no different. Lau [14] shows curves for two different epoxy materials' modulus characteristics with regard to temperature (Figure 9.32). As the temperature increases, material F gets softer at a lower temperature than material B. In this case, although a material appears rigid at room temperature, this can soon change. If the expected use environment is at a high temperature, such as engine controls, material B may offer more protection than material F, assuming that the operation stays below 150°C.

9.3.10 Moisture Permeation and Hermeticity

Polymers will absorb moisture over time. Moisture can change the electrical properties of an encapsulant to the point where it will break down electrically and no longer be insulating. The effect can sometimes be subtle and noticed only when measuring properties such as the dielectric constant. Manufacturing engineers on surface-mount assembly lines have experienced a phenomenon know as *popcorning* when plastic packages absorb ambient moisture. In subsequent processing through a solder reflow oven, the rate of temperature changes rapidly and exceeds the boiling point of water, generating steam and subsequently rupturing the plastic encapsulation. The rate of moisture absorption can be from a few minutes to months, depending on the encapsulation type.

Occasionally, someone will state that they have a plastic package that is hermetically sealed. Some plastic package designs do meet a fine and gross leak test criterion that is

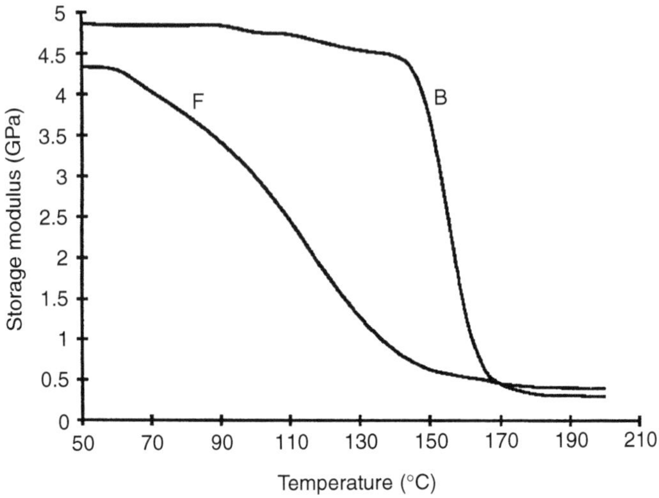

FIGURE 9.32 Change in modules with respect to temperature. (From Ref. 14.)

used to test metal or ceramic packages for hermeticity. However, the basic fact is that polymers are simply membranes, and the forces of nature will try to equalize the moisture content on each side of the membrane over time. This assumes that moisture does not get into a package along leads or around wires that connect the device inside the package to the outside world.

Traeger [15] presents a graph (Figure 9.33) that gives the rate of moisture permeation for silicones, epoxies, glasses, and metals. It can be seen that the exterior moisture level can permeate most polymers in a relatively short time. However, data exist that some combinations of coatings can slow the process to several months.

9.3.11 Test Plan

A component or assembly design and material choices will be based on many factors. Often, a customer will specify the tests they would like to see performed to ensure the reliability of the product in their customer's daily use environment. These specifications can be Military Standards or telecommunications standards such as Telcordia or Underwriters' Laboratories in the United States, which want to ensure that products are safe in the home. Often, customers specify standards from several organizations because their products are not represented fully by one organization. This can get interesting because a flip chip in a package from a component vendor will often be tested using more severe temperature ranges and number cycles than a flip chip mounted to a PC board at a cell phone manufacturing site. There is a hierarchy of reliability testing, and component vendors have to pass more stringent test than do board manufacturers. However, the end product has to meet user expectations, and if it does not, the vendors will hear about it fairly quickly. It is a good

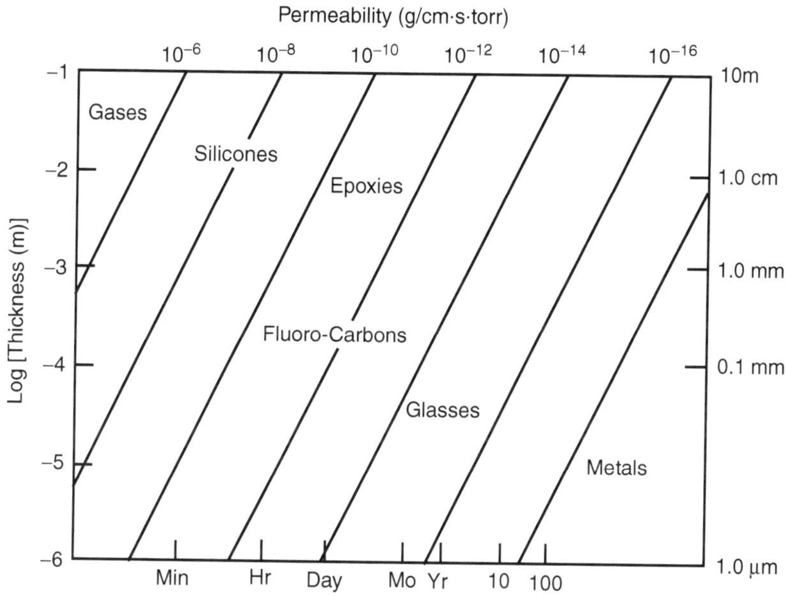

FIGURE 9.33 Time for a package interior to reach 50% of exterior humidity.

idea to have a test plan even if your customer does not. In this way, material choice can be proven before a customer or end user gets to use it. It can be very expensive to have a process established and producing parts only to have a flaw become evident six months after the first device is delivered.

9.3.12 Summary

When designing components or assemblies, start out with a material design plan. Understand how your end product will be a built into the next level of assembly and what it is expected to survive in the use environment. Build a few parts and have a test plan to see that you have been successful in your design choices. This book could never cover every nuance of material properties to make it a recipe for design engineers. Several critical properties have been described and how they affect the design choices available to assembly and package design engineers. Hopefully, these will be found useful.

REFERENCES

1. Mary Bellis, About, History of the IBM PC, http://inventors.about.com/library/weekly/aa031599.htm.
2. S. J. Adamson and D. Gibson, *Precision Dispensing of Liquid Encapsulants for Smart Cards*, IMAPS Europe, Harrogate, Yorkshire, England, June 1999.

3. Tara R. Miles, Louis P. Rector, and Shaoqin Gong, Dexter Electronic Materials; and Tony LoBianco, Amkor Technology, Transfer molding encapsulation of flip chip array packages: technical developments in material design, presented at Semicon West 2000, San Jose, CA, July 12–14, 2000.
4. H. Quinones and A. Babiarz, Flip chip, CSP and WLP technologies: a reliability perspective, presented at Pac Tech, Germany, September 2001.
5. S. J. Adamson and J. J. Klock e, Design considerations for high-speed underf ill of CSP and flip chip packages, presented at the APEX Conference, San Diego, CA, January 2001.
6. A. F. Piracci, Advantages of non-contact dispensing in SMT assembly processes, presented at SMTA/ATE, Chicago, September 2000.
7. Dow Corning, Die encapsulant tutorial, p. 10, retrived September 7, 2004; also Ann M. Norris, Dow Corning, Midland, Michigan and Marlene E. Gladstone, Dow Corning, Fremont, California, Silicone materials for chip-scale packaging, *Chips Scale Review*, 1998.
8. S. M. Tavakoli, TWI Limited, *An Assessment of Skin Sensitisation by the Use of Epoxy Resin in the Construction Industry*, Her Majesty's Stationery Office, Colegate, Norwich, England, 2003.
9. E. N. Prilezhae va, *Reaktsiya Prilezhaeva: Elektrofil'noe Okislenie* (Prilezhaev's Reaction: Electronic Oxidation), Nauka, Moscow, 1974.
10. Darrel Bryant, A matter of choice, *Adhesives and Sealants Industry*, July/August, pp. 29–31, 2004.
11. Krishna Darbha, Juscelino Hozumi Okura, and Abhijit Dasgupta, Impact of underfill filler particles on reliability of flip-chip interconnects, *IEEE Transactions on Components, Packaging, and Manufacturing Technology*, Part A, Vol. 21, No. 2, June 1998.
12. C. P. Wong, Michael B. Vincent, and S. Shi, Fast-flow underfill encapsulant: flow rate and coefficient of thermal e xpansion, *IEEE Transactions on Components, Packaging, and Manufacturing Technology*, Part A, Vol. 21, No. 2, June 1998.
13. Darrel R. Frear, Materials issues in area-array microelectronic packaging, *JOM* (Member Journal of The Minerals, Metals and Materials Society) Vol. 51, No. 3, 1999, pp. 22–27.
14. John H. Lau and Chris Chang, Express Packaging Systems, Inc., How to select underfill materials for solder b umped flip chips on lo w cost substrates, presented at the International Symposium on Microelectronics, p. 693, 1999.
15. R. K. Traeger, Hermeticity of polymeric lid sealants, in *Proceedings of the 25th Electronic Component Conference*, pp. 361–365, 1975.

CHAPTER 10

Assembly

EDWARD M. PETRIE

EMP Solutions, Cary, North Carolina

10.1 INTRODUCTION

As with any other material, plastics components often require assembly to provide practical and commercial products. Frequently, large parts cannot be made efficiently, and smaller parts or subsections need to be joined together. Plastic parts may also require attachment to components made of other materials, including metal, wood, and ceramic. A reasonable demand is that the assembly is as strong as the individual component elements and will last the design life of the product in which it will become a component.

The best system for joining is never better than a well-designed unit that is molded as a single element. However, this often is not possible, and products or subassemblies must be built from smaller components. In this chapter we discuss the major processes that are commonly utilized to join and assemble plastic parts. The fabricator has five broad methods to consider adhesive bonding, heat welding, solvent welding, mechanical fastening, self-assembly, and there are many possible processes defined for each method.

Choosing the optimal method will require a fundamental knowledge of the materials involved, the possible joint designs, the production facilities that are available and practical, and the end-use service requirements to which the assembly will be exposed. In addition to strength and permanence, careful consideration must be given to final appearance, production time, possible disassembly requirements, and all production costs, including all materials, joint formation, labor, and energy.

Since the performance and cost of the final assembly will depend heavily on the plastic used to manufacture the individual parts, this chapter begins with a discussion of how the plastic substrate affects the assembly process and the resulting joint

Handbook of Plastic Processes Edited by Charles A. Harper

performance. In fact, the plastic substrate material, its manufacturing process, and the joint design are often selected to provide the most advantageous overall joining process. This chapter is intended to guide the fabricator to select the correct assembly process and explain how it can be employed most optimally.

10.2 CHARACTERISTICS OF PLASTIC MATERIALS IN ASSEMBLY

Plastics are usually more difficult to join than other materials. The outcome of the joining process may be as dependent on the plastic substrate and how it is manufactured as it is on the process chosen for joining. This dependence is due to the plastic's unique bulk properties and surface characteristics. The bulk properties of the plastic could significantly affect the resulting joint strength using any of the assembly methods indicated above. The plastic's surface characteristics can especially affect joints that are made with the adhesive bonding or welding processes.

The engineering and assembly characteristics of important plastic families of materials are summarized in Tables 10.1 and 10.2, respectively. From this type of information it is easy to see why certain plastics are chosen for specific applications. However, all plastics have certain features that will also affect their assembly properties. These features are manifested in the bulk properties or the surface properties and include the following:

- The nature of the bulk material and the surface material can vary from plastic to plastic and even from lot to lot of the same plastic.
- Many plastics require surface treatment to allow adhesives or coatings to adhere properly.
- The surface of the plastic part must be clean and free of any process lubricants, mold release agents, external contamination, or other possible weak boundary layers.
- Internal additives could migrate to the surface or to the interface; this could happen after molding and before a secondary finishing operation or after the secondary finishing operation once the part is placed into service.
- Aging of the plastic part may make the bond between the plastic and an adhesive or coating weak. Dimensional change resulting from thermal expansion, moisture absorption, and so on, of the plastic substrate relative to the adhesive or to another substrate may cause significant stress on the joint.
- Molded-in stresses within the plastic part may stimulate stress cracking or crazing and reduce the adhesion between the plastic substrate and the adhesive or coating.

It is important to note that plastics can be modified significantly by mixing or combining different types of polymers and by adding fillers and modifiers. These modifications can heavily influence the assembly characteristics. Thus, plastic materials must be specified carefully, and the specification requirements should reflect critical assembly criteria.

TABLE 10.1 Engineering Characteristics of Important Plastic Families

Plastic Material	Type[a]	Important Engineering Characteristics
ABS (acrylonitrile–butadiene–styrene)	TP	Rigid, low-cost thermoplastic, easily machined and thermoformed.
Acetal	TP	Engineering thermoplastic with good strength, wear resistance, and dimensional stability. More dimensionally stable than nylon under wet and humid conditions. Two types are available: homopolymer and copolymer.
Acrylic	TP	Clear, transparent, strong, break-resistant thermoplastic with excellent chemical resistance and weatherability.
CPVC (chlorinated polyvinyl chloride)	TP	Thermoplastic with properties similar to PVC, but operates to a 40 to 60°F higher temperature.
Epoxy	TS	Strong, durable thermosetting material that has broad formulation possibilities. Commonly used in potting, molding, and composite processes. Good high-temperature, chemical resistance, and electrical properties.
Fiberglass	TS	Generic name for a thermosetting composite with high strength-to-weight ratio, excellent dielectric properties, and unaffected by corrosion. The resin matrix can be any thermosetting material.
Nylon	TP	Thermoplastic with excellent impact resistance, ideal for wear applications such as bearing and gears, self-lubrication under some circumstances.
PEEK (polyether ether ketone)	TP	Engineering thermoplastic, excellent temperature resistance, suitable for continuous use above 500°F, excellent flexural and tensile properties.
PET (polyethylene terephthalate)	TP	Dimensionally stable thermoplastic with superior machining characteristics compared to acetal.
Phenolic	TS	High-temperature, rigid thermoset with minimal thermal expansion, high compressive strength, excellent wear and abrasion resistance, and a low coefficient of friction. Used for bearing applications and molded parts.
Polycarbonate	TP	Transparent, tough thermoplastic with high impact strength, excellent chemical resistance and electrical properties, and good dimensional stability.
Polyester	TS	High-strength materials with good chemical and moisture resistance. Often used as a composite material for automobiles, boats, etc. Good electrical properties.
Polyethylene	TP	Good chemical resistance combined with low moisture absorption and excellent electrical properties. Lower-temperature resistance than polypropylene.
Polypropylene	TP	Good chemical resistance combined with low moisture absorption and excellent electrical properties retention up to 250°F.

TABLE 10.1 (*Continued*)

Plastic Material	Type[a]	Important Engineering Characteristics
Polysulfone	TP	Durable thermoplastic, good electrical properties, operates at temperatures in excess of 300°F.
Polyurethane	TP or TS	Can be thermoplastic or thermoset. Excellent impact and abrasion resistance, resists sunlight and weathering.
PTFE (polytetrafluoro-ethylene)	TP	Thermoplastic, low coefficient of friction, withstands up to 500°F, inert to chemicals and solvents, self-lubricating with a low thermal expansion rate.
PVC (polyvinyl chloride)	TP	Resists corrosive solutions and gases both acid and alkaline, good stiffness.
PVDF (polyvinylidene fluoride)	TP	Outstanding chemical resistance, excellent substitute for PVC or polypropylene. Good mechanical strength and dielectric properties.

[a]TP, thermoplastic; TS, thermoset.

TABLE 10.2 Assembly Characteristics of Important Plastic Families

Plastic Material	Type[a]	Important Assembly Characteristics[b]
ABS (acrylonitrile–butadiene–styrene)	TP	Can be heat or solvent welded as well as bonded with adhesives. Many blends are available that can affect assembly properties. Surface preparation for adhesive bonding is generally solvent cleaning and mechanical abrasion.
Acetal	TP	Generally assembled with adhesives (surface treatment will depend on type of acetal) or thermal welding methods. Dimensions of molded parts will vary slightly with RH. Parts should be tried before joining.
Acrylic	TP	Solvent cementing, heat welding, or adhesive bonding can be used. Solvents or certain monomers may attack highly stressed areas, resulting in crazing. To minimize crazing substrate should be annealed at about 10°F below heat distortion temperature.
CPVC (chlorinated polyvinyl chloride)	TP	*See* polyvinyl chloride.
Epoxy	TS	Generally bonded with adhesives. Surface treatment consists of solvent cleaning and mechanical abrasion.
Fiberglass	TS	Generally bonded with adhesives. Surface treatment consists of solvent cleaning and mechanical abrasion. Dimensional change will depend on fiber orientation within the composite.
Nylon	TP	Parts are generally difficult to assemble with adhesives or solvents. Mechanical or self-fastening processes are most often used. For adhesive bonding, primers are often required.

TABLE 10.2 (*Continued*)

Plastic Material	Type[a]	Important Assembly Characteristics[b]
PEEK (polyether ether ketone)	TP	Generally, joined mechanically or with adhesives, but they can also be thermally welded and used in snap fit self-assembly joints. Usually, temperature resistance of the adhesive is not as high as the temperature resistance of the substrate.
PET (polyethylene terephthalate)	TP	Can be joined via mechanical fasteners, self-assembly, adhesive bonding, or heat welding. Surface treatments for adhesive bonding will provide optimum adhesion.
Phenolic	TS	Generally bonded with adhesives. Surface treatment consists of solvent cleaning and mechanical abrasion.
Polycarbonate	TP	Can be assembled using any method. Crazing is possible with certain solvents and monomers.
Polyester	TS	Generally bonded with adhesives. Surface treatment consists of solvent cleaning and mechanical abrasion.
Polyethylene	TP	Difficult material to bond because of low surface energy. Cannot solvent weld because of solvent resistance. Generally assembled using heat welding methods.
Polypropylene	TP	Difficult material to bond because of low surface energy. Cannot solvent weld because of solvent resistance. Generally assembled using heat welding methods.
Polysulfone	TP	Can be joined with all processes.
Polyurethane	TP or TS	Generally bonded with flexible adhesives. Surface treatment consists of cleaning and mechanical abrasion.
PTFE (polytetra-fluoroethylene)	TP	Cannot be joined with adhesives unless substrate is first treated. Sodium-napthalene chemical treatment is often used. Difficult to solvent- or heat-weld because of chemical and heat resistance. Difficult to use self-assembly processes because of low ductility.
PVC (polyvinyl chloride)	TP	Generally, solvent- or heat-welded. When using adhesives, plasticizer migration from the PVC can be a problem. Plasticizer-compatible adhesive should be used.
PVDF (polyvinylidene fluoride)	TP	Difficult to bond with adhesive because of low surface energy, and difficult to solvent-weld because of solvent resistance. Generally, heat-welding processes are used.

[a]TP, thermoplastic; TS, thermoset.

[b]Most plastics can be joined using mechanical fastening processes. Tough flexible plastics can also generally be joined by self-fastening methods.

10.2.1 Bulk Properties of Plastics

The physical and chemical properties of the bulk plastic will affect the quality of the assembled joint. Major elements of concern are the thermal expansion coefficient, dimensional change due to the environment, flexural modulus, and glass transition temperature of the substrate. These elements often cause internal stresses to be generated with the joint. Internal stresses are defined as all stresses on the joint other than those exerted by the normal external loading that occurs in service. Internal stresses generally degrade the theoretical strength of the joint. They are most often encountered with adhesive bonding and heat or solvent welding, but they can also affect the performance properties of mechanically fastened joints.

Significant differences in thermal expansion coefficient between the substrates can cause severe stress at the joint interface. This is common when plastics are joined to metals because of the significantly greater thermal expansion coefficient of most plastics compared to other materials. Table 10.3 shows the thermal expansion coefficients of selected materials. For an equivalent temperature excursion, plastic substrates can grow or contract distances more than 10 times those of most metals. However, many structural adhesives (e.g., heavily filled epoxy) can also have a thermal expansion coefficient that is different from the plastic substrate, and this could also result in internal stress at the substrate–adhesive interface.

Thermal cycling, especially low- or high-temperature excursions during service, also magnify residual stresses that may be locked into the joint. Selection of a resilient joint design (i.e., the use of an elastic adhesive or expandable fastener) or adjustments in the thermal expansion coefficients of the substrate or the adhesive via fillers or additives can help to reduce such stress. Dimension changes in the plastic part, either during the assembly process or after the joint is made and placed in service, will result

TABLE 10.3 Thermal Expansion of Selected Materials

Material	Coefficient of Linear Thermal Expansion (per °C$\times 10^{-6}$)
Metals	
Aluminum	23.5
Brass	18.8
Copper	16.7
Iron, cast	10.5
Steel	10.8
Other materials of construction	
Brick	5.5
Concrete	14.0
Glass	9.3
Granite	8.2
Marble	7.2
Wood, pine	5.5
Plastics	
Epoxy	40–100
Nylon	90–108
Polyethylene	110–250
Polystyrene	60–80
Polyvinylidene chloride	190–200
Polytetrafluoroethylene	50–110
Phenolic	30–45

in internal stress within the joint that could cause joint strength degradation. A significant change in dimension could cause stresses at the joint that result in catastrophic joint failure.

Dimension changes can occur when the plastic part shrinks excessively when coming out of the mold or when it is undercured and continues to cure and shrink in the service environment. Heat aging generally manifests itself as shrinkage caused by the polymer continuing to cure or thermal oxidation. The shrinkage of various thermoset molding compounds as a result of heat aging is shown in Figure 10.1. Dimensional changes may also occur due to nonthermal events such as moisture adsorption in the plastic. Many plastics, such as nylon, will adsorb significant amounts of moisture from the ambient environment. This could cause change in dimensions of the part with differing relative humidities or temperatures.

With adhesive bonding, hydroscopic plastic parts generally should be conditioned at the relative humidity that will be expected in service before bonding. If the bonding operation occurs at high temperatures (e.g., with a hot melt adhesive or an adhesive requiring an elevated temperature cure), the part should be dried before bonding to prevent the moisture from causing the adhesive to foam. Special consideration is also required for additives or modifiers that can migrate to the joint interface during normal aging or on exposure to operating environments. Components within the polymeric bulk material can also migrate to the surface. It is common to find low-molecular-weight polymers or oligomers, plasticizers, pigments, mold release agents, shrink control agents, and other processing aids as well as contaminants adsorbed in the surface region.

The surface regions of plastics are dynamic regions, continuously establishing new equilibrium internally with the bulk material and externally with the surroundings. In

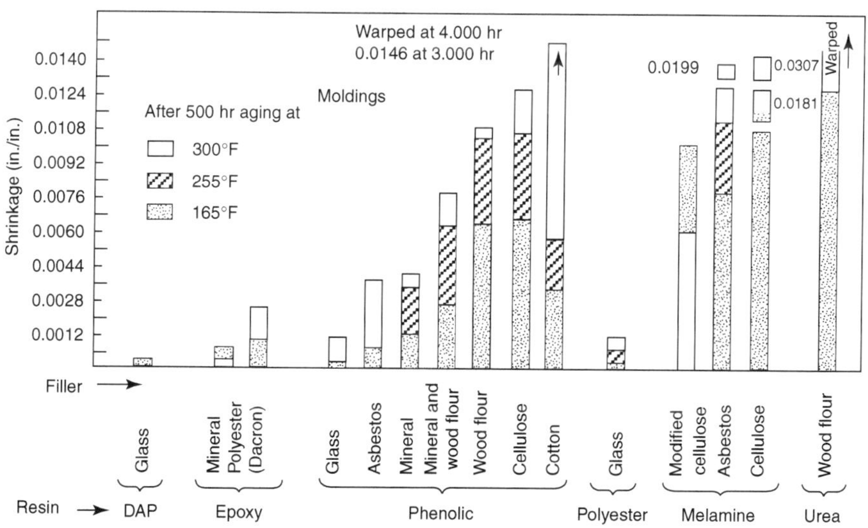

FIGURE 10.1 Shrinkage of various thermosetting molding materials as a result of heat aging. (From Ref. 1.)

flexible amorphous plastics above their glass transition temperature, low-molecular-weight components are able to diffuse out of the bulk and into the surface region, while elements of the surroundings can diffuse into the adherend. The flexural modulus of the plastic substrate can also be a significant factor when it comes to joining processes. This is especially important when adhesive bonding is the process being used. When high peel or cleavage stress occurs, as is generally the case with the more flexible plastic substrates, concentrated stress occurs on the adhesive at the ends of the joint. This requires selection of a tough and flexible adhesive material. More on this subject is provided in Section 10.5.3.

The glass transition temperature of the plastic is an important characteristic in joint designs. At temperatures greater than the glass transition temperature (T_g), the plastic part will show a greater thermal expansion than at temperature below the T_g. This will contribute to residual stresses in the joint as described above. When a material approaches its glass transition temperature, it loses its rigidity and becomes flexible and even capable of flowing. At service temperatures greater than the T_g, the plastic part could show significant creep under prolonged loading. This may dictate changes in the joint design or the base plastic material itself. The glass transition temperatures of selected amorphous and crystalline plastics are shown in Table 10.4.

10.2.2 Consideration of Surfaces

The characteristics of plastic surfaces must be considered carefully, especially when using adhesive bonding or welding processes. These processes rely on the chemical and physical properties of the surface for creating a good bond. Plastic surfaces are

TABLE 10.4 Glass Transition Temperature of Selected Plastics

Plastic	Glass Transition Temperature, T_g (°C)
Amorphous thermoplastic	
ABS	110
Polycarbonate	150
Acrylic	105
Polystyrene	95
Polyvinyl chloride	85
Semicrystalline thermoplastic	
Polyamide	50
Polyethylene	–35
Polytetrafluoroethylene	65
Polypropylene	–10
Thermosetting	
Epoxy, phenolic, polyester	>100 depending on degree of cross-linking

less important with mechanical fastening and self-assembly processes, but they could still affect the performance of the joint as well as the aesthetics of the final product. Plastic surfaces differ from metal, glass, or most other surfaces in often being unstable and thermodynamically incompatible with joining materials such as adhesives. Furthermore, the processing aids used in manufacturing plastic raw materials and in fabricating plastic parts are often not favorable to adhesive bonding.

The term *surface* in adhesive science is usually defined as that portion of the adherend with which the adhesive interacts. The surface is defined by both area and depth of interaction. For a freshly cleaved single crystal, this interaction region might be only one or two atomic layers in depth. For anodized aluminum, a low-viscosity adhesive might reach a depth of several hundred nanometers or more. For a very permeable surface such as plastic, the interaction region may be several millimeters in depth. This larger three-dimensional area that is so critical to adhesive bonding is often referred to as the *interphase* (Figure 10.2).

Thus, with plastics substrates it is not only the surface topology or roughness that must be considered. The physical and chemical nature of the various layers within the interphase can also have a significant affect on the final properties of an assembled joint. To make matters worse, the surface properties could change considerably with the storage, assembly, or service environments. Often, the surface is not the surface that we think it is, and this makes it difficult to provide high performance and consistent bonds with many plastic substrates.

Often, plastic surface materials contain constituents that are very different from the bulk material. For metals and alloys, these surfaces may consist of oxides and adsorbed gases. For plastics, they may be moisture, migrating additives, or adsorbed films, such as shop contaminants. These outer layers can be either loosely bound or tightly adhered to the base material, and they may have high or low cohesive strength.

Certain surfaces also may have weakly attached boundary layers. Examples of these are contaminant films, low-molecular-weight constituents, and oxide layers. A weak substrate boundary layer can provide a "weak link" for reduced bond strength or premature failure, as shown in Figure 10.2. In an ideally bonded assembly, the substrate should be the weakest link. In most assemblies that are properly bonded, however, the adhesive is the weak link because the interfacial adhesive forces are

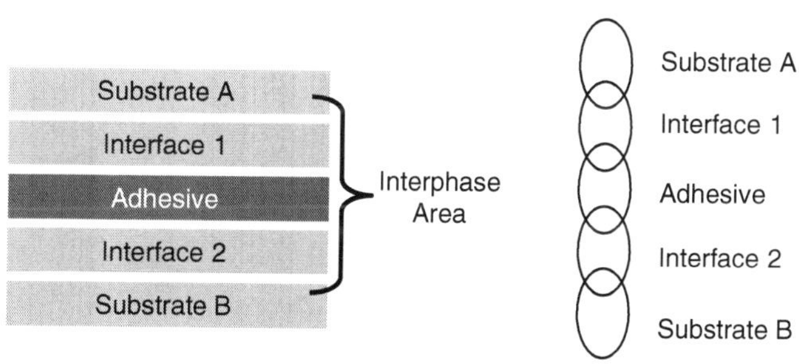

FIGURE 10.2 In an ideal joint the substrate should be the weakest link. The adhesive joint can be divided into at least five regions that are similar to the links in a chain.

greater than the cohesive forces holding the adhesive together. When the surface region becomes the weakest link, it may result in catastrophically low failure strength and an inconsistency in failure values. Thus, it is imperative that these surface characteristics be understood and controlled in some manner.

Plastic substrates can be unstable and thermodynamically incompatible with the joining method. Plastic materials have lower surface energies and lower cohesive strength than most other substrates. Polymeric surfaces are also dynamic—changing constantly with respect to their environment. As shown in Figure 10.3, polymeric substrates have the potential for low-molecular-weight fragments, oxidation products, plasticizers, processing aids, lubricants and slip aids, adsorbed water, and organic contaminants along with various other surprises. For some polymeric materials, there may be greater crystallinity or chemical orientation at the surface, due to the way in which the material was processed. These characteristics could affect the resulting joint strength without having a significant influence on the bulk physical properties or specification values of the substrate material.

A well-known example of this effect is the migration of plasticizer from flexible polyvinyl chloride. The plasticizer can migrate from the bulk adherend to the joint region and then to the interface. If the adhesive is an effective barrier to plasticizer migration, this will create a weak boundary layer at the interface. If the adhesive is not an effective barrier, the plasticizer will migrate through the interface and into the adhesive, thereby possibly changing the physical properties of the adhesive (i.e., acting as a plasticizer for the adhesive).

Many polymeric surfaces slowly undergo chemical and physical change. For example, the plastic surface, at the time of joining, may be well suited to the bonding or welding processes. However, after aging, undesirable surface conditions may present themselves at the interface, displace the adhesive, and result in bond failure. These weak boundary layers could come from the environment or from within the plastic substrate itself. Plasticizer migration and degradation of the interface through ultraviolet radiation are common examples of weak boundary layers that can develop with time at the interface.

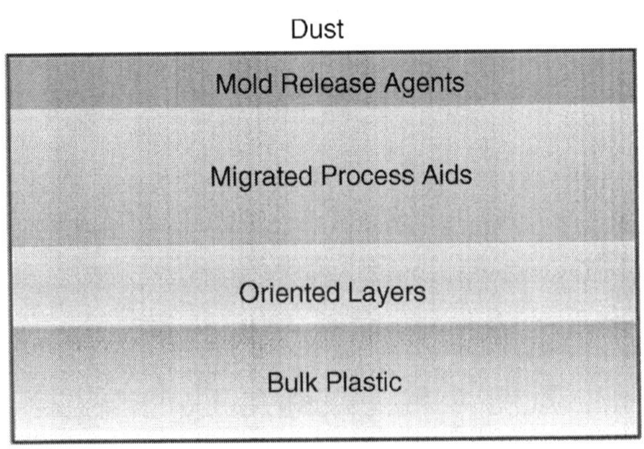

FIGURE 10.3 Surface regions for a plastic substrate.

The nature of the polymeric substrate can change rapidly in response to its surroundings. Even when the bulk of a material is in the glassy state (below its glass transition temperature), the surface region can be quite mobile, owing to the presence of low-molecular-weight polymer constituents and contaminants. Polymers, having both polar and nonpolar regions in their molecular chain, can present different chain segments at the surface, depending on whether or not the surroundings are polar. Wiping a surface with an ionic solution will cause the polar groups to orient toward the surface, whereas the same treatment with a nonpolar solvent such as hexane can bring the nonpolar components to the surface. Exposure to heat after surface treatment could cause fresh, untreated molecular segments to appear on the surface, thereby losing the beneficial characteristics of the surface treatment.

As a result of these dynamic reactions, it is difficult to be confident about the surface of any polymeric material. It is possible that the actual surface to which we are joining is not always the surface that we anticipate. It is also possible that the surface could change once the joint is made and the assembled part is placed into service. Thus, a weak boundary layer that is not there during the bonding process may form during service and contribute to a weakening of the interface. Although these dynamic processes are not always damaging to the integrity of the assembled joint, they need to be considered early in the assembly design process. If such surface interactions are considered possible and especially in the case of adhesive bonding, prototype joints should be made with the anticipated production processes and fully tested with regard to the service environments.

10.3 TYPES OF PLASTICS

The selection of the proper assembly process will depend on the type of plastic being joined. Plastics can be classified in many ways. For the purposes of this chapter, plastics are divided into the following types: (1) thermosetting, (2) thermoplastic, (3) elastomeric, and (4) composite. The considerations that must be given to each type of plastic in specifying an assembly process are summarized in the sections that follow.

10.3.1 Thermosetting Substrates

Thermosetting plastics cannot be dissolved in solution and do not have a melting temperature since these materials are chemically cross-linked. Therefore, they cannot be heat or solvent welded. In some cases, solvent cements or heat welding techniques can be used to join thermoplastics to thermoset materials. However, most thermosetting plastics are not particularly difficult to bond to themselves or to other materials with adhesive systems. Thermosetting plastics can generally be bonded with many different types of adhesives, such as epoxies, thermosetting acrylics, and urethanes. The adhesive should have thermal expansion coefficient and flexural modulus similar to the substrate being bonded. Since thermosetting plastic parts are often highly filled and rigid, this does not generally represent a problem. Most structural adhesives have similar characteristics. With more flexible plastic substrates, the adhesive needs to be tough and flexible. In any case, the adhesive should provide practical and consistent application processes and properties that can resist the ultimate service environment.

Because unfilled thermosetting plastics tend to be harder, more brittle, and not as tough as thermoplastics, it is common practice to add fillers or modifiers to thermosetting resins. These fillers can affect the nature of the adhesive bond (either positively or negatively) and are a possible source of lot-to-lot and supplier-to-supplier variability. Being chemically cross-linked, thermosetting plastics shrink during cure. Sometimes the cure is not entirely complete when the part is bonded. In these cases, cure of the part can continue during the bonding operation or even on aging in service, resulting in shrinkage and residual stresses in the joint (see Figure 10.1). Depending on the nature of the cross-linking reaction, volatile by-products could also generate due to postcuring of the part and provide materials for a weak boundary layer.

The surface of thermoset materials may be of slightly different chemical character than the material beneath the surface because of surface inhibition during cure or reaction of the surface with oxygen and/or humidity in the surroundings. By abrading the surface, a more consistent material is available for the adhesive to bond.

10.3.2 Thermoplastic Substrates

Unlike thermosetting resins, the thermoplastic resins will soften and flow on heating or on contact with certain solvents. They will then reharden on cooling or on evaporation of the solvent from the plastic. This capability is due to the non-cross-linked chemical structure of thermoplastic molecules. As a result, thermoplastic materials can be joined via solvent cementing or thermal welding as well as with adhesive bonding, mechanical fasteners, or self-assembly processes. In the case of heat and solvent welding, the resin from the plastic part being joined acts as the adhesive material.

The following are important characteristics of thermoplastic resins that can affect their joining capability:

- Many thermoplastic compounds are alloyed and really consist of two or more resins.
- Many grades of the same material are available (high flow, high density, etc.).
- Additives, modifiers, and internal mold release agents are commonly employed in the formulation.
- Certain thermoplastics exhibit a relatively high degree of mold shrinkage.
- Dimensional changes due to moisture migration, thermal expansion, and so on, are generally greater than with other materials.
- Properties of the surface, such as surface energy and crystallinity, may be different from the bulk material (this is especially true for thermoplastics that are molded at very high temperatures).

Because there are many forms and chemical variations of the same family of thermoplastic resins, it is important that prototype tests be made on the exact material to be used. Plastics with different grades or from different suppliers, although from the same generic family (e.g., polyethylene), can have significantly different joining characteristics. Thermoplastic materials often have a lower surface energy than thermosetting materials. Thus, physical or chemical modification of the surface is necessary to achieve acceptable bonding. This is especially true of the crystalline thermoplastics such as polyolefins, linear polyesters, and fluoropolymers.

Solvent cementing or thermal welding methods can also join many thermoplastics. Solvent cementing and thermal welding do not require abrasion or chemical treatment of the plastic surfaces. The surfaces must be clean, however, and free of impurities that could cause a weak boundary layer. Bond strengths achieved by solvent or thermal welding are generally as high or higher than adhesive bonding. Bond strengths are often greater than 80% of the strength of the substrate material.

10.3.3 Elastomeric Substrates

The joining of elastomeric substrates is often a difficult proposition. Not only is the viscoelastic nature of the elastomer critically important in achieving a reliable joint, but significant chemical factors must also be considered. These are due primarily to the large number of elastomeric polymers that are available, the degree of freedom that a formulator has in compounding these materials, and the wide variety of curing processes that can be used. Elastomers are unique materials. They differ from other engineering materials mainly by their composition, mechanical properties, and surface characteristics. These unique properties are discussed below in relation to adhesion.

As with other polymers, elastomers can be broadly divided into two types: thermosets and thermoplastics. There are over 30 broad groupings of chemical types of elastic polymers. These are arranged by ASTM D1418 into categories of materials having similar chemical chain structures. The more common elastomer categories and their surface properties are shown in Table 10.5.

There are several challenges in joining elastomeric materials. Many of these challenges are related to elastomer itself. One problem is that significant variation can exist within a given chemical type. This is due to differences in average molecular weight, molecular weight distribution, polymerization processes, variation of structural arrangement, copolymer or terpolymer ratios, and so on. The variations that can exist within a chemical family are generally much greater for elastomers than for thermosets and thermoplastics.

There can be significant differences in the formulation recipe for a given product. This is due to differences and latitude that the compounders have in their choice of additives. Elastomer suppliers have the ability to develop specific elastomer formulations with properties tailored to meet almost any set of requirements. Typical compounding ingredients include cross-linking agents (also called curatives or vulcanizing agents), reinforcements, antidegradants, processing aids, extenders, and specialty additives such as tackifiers, plasticizers, and colorants. Because of the almost limitless

TABLE 10.5 Designations of Common Elastomers[a]

ASTM D1418 Nomenclature	Generic Name	Common Name	Critical Surface Energy (MJ/m^2)	Water Contact Angle (deg)
ACM	Polyacrylate	None	—	—
BR	Butadiene	None	32	—
CR	Chloroprene	Neoprene	38	—
IIR	Isobutylene–isoprene	Butyl	27	115
IR	Isoprene, synthetic	Synthetic natural rubber	31	106
NR	*cis*–Isoprene	Natural rubber	—	120
NBR	Nitrile butadiene	Buna-N or nitrile	37	97
SBR	Styrene–butadiene	SBR	33	95
CSM	Chlorosulfonated polyethylene	Hypalon	37	96
EPDM	Ethylene–propylene dieneterpolymer	EPDM	—	—
EPM	Ethylene–propylene copolymer	EPM	28	105
FKM	Vinylidine fluoride hexafluoropropylene copolymer	Viton, Fluorel, Kel-F, fluorocarbon rubber	19	—
CO	Epichlorohydrin	Hydrin	35	95
AU	Polyester urethane	Ester urethane	—	—
EU	Polyether urethane	Ether urethane	—	—
—	Polysulfide	Polysulfide	—	—
MQ or Si	Dimethyl silicone	Silicone rubber	24	—

[a]Critical surface energy and water contact angle depend on the e xact formulation. Values shown are for selected specific formulations.

number of elastomer formulations, even within a specific family, it can be difficult to find application-specific assembly data.

The large number of compounding ingredients possible for a gi ven generic elastomer such as neoprene or nitrile is a signif icant factor for the joint manuf acturer. Each variation can signif icantly affect the adhesion properties of the substrate. Elastomer material specifications usually do not focus on adhesion characteristics but address mainly chemical and physical properties. Thus, the supplier may ha ve wide latitude within a specif ication to make changes in the compound formulation that could be harmful to the quality of the adhesi ve bond. One solution is to qualify every new lot of material for adhesion as well as the more standard properties. Another possible solution is to ha ve the supplier notify the joint manuf acturer regarding every change in formulation or processing conditions.

Besides curing systems, fillers, and plasticizers, an elastomeric compound may contain protective chemicals such as antioxidants, antiozone agents, waxes, and fungicides. Some of these are purposefully designed to "bloom" or to come to the surface of the elastomer either during processing or on aging. These weak boundary layers often cannot be removed prior to bonding because their supply to the interface is relatively unlimited due to the capacity of the bulk elastomer.

Another problem in joining elastomers with adhesives is that elastomers are relatively deformable viscoelastic materials. It is easy to develop internal stresses at the bond interface. These stresses could adversely affect the bond strength and permanence of the joint. Internal stresses in elastomer bonds generally arise from the pressure on the joint during its formation. Minimal pressure to achieve close substrate contact and wetting is generally all that is necessary when bonding elastomers. Differences in the coefficient of thermal expansion or modulus between the elastomer and its mating substrate can also contribute to significant internal stress.

Because the elastomeric substrate is deformable, the joint will often see localized stress in service. This necessitates a tough, flexible adhesive. Being a viscoelastic material, the properties of many elastomeric materials are highly dependent on the rate of applied stress. Thus, the joint could behave differently when exposed to high rates of stress (i.e., automobile impact or explosive pressure wave) than when it is stressed at lower rates such as is often done in laboratory testing.

Another factor that causes confusion when bonding elastomers is that there are major differences in the ways in which bonding can be achieved. One can bond already vulcanized elastomers using adhesives and processes that are generally available for bonding other substrates. This is often termed *postvulcanized bonding*. One can also create a bond during the vulcanization of the elastomer. This is often termed *vulcanized bonding*. Generally, vulcanized bonding produces somewhat stronger bonds than does postvulcanized bonding, but the processing and materials that are used are more complex and proprietary.

10.3.4 Composite Substrates

Modern structural composites are a blend of two or more components. One component is generally made of reinforcing fibers, either polymeric or ceramic. The other component is generally made up of a resinous binder or matrix that is polymeric in nature, at least for polymeric composites. The fibers are strong and stiff relative to the matrix. When the fiber and matrix are joined to form a composite, they both retain their individual identities and both directly influence the composite's final properties. The resulting composite is composed of layers (laminates) of the fibers and matrix stacked to achieve the desired properties in one or more directions.

The reinforcing fiber can be either continuous or discontinuous in length. The fiber's strength and stiffness are usually much greater than the matrix material. The commonly commercially available fibers are:

- Glass
- Polyester
- Graphite

- Aramid
- Polyethylene
- Boron
- Silicon carbide
- Silicon nitride, silica, alumina, alumina silica

Glass fiber composites are the most common type of composite. However, graphite, aramid, and other reinforcements are finding applications in demanding aerospace functions and in premium sporting equipment such as fishing rods, tennis rackets, and golf clubs. The resin matrix can be either thermosetting or thermoplastic. Thermosetting resins such as epoxy, polyimide, polyester, phenolic, and so on, are used in applications where physical properties are important. Polyester and epoxy composites make up the bulk of the thermoset composite market. Of these two, polyesters dominate by far. Reinforced with glass fiber, these are known as *fiberglass-reinforced plastics* (FRPs). FRPs are molded by layup and spray-up methods or by compression molding from either a preform or from sheet molding compound (SMC). Thermoplastic matrix composites are generally employed where there are high volume and economic considerations, such as in the automotive and decorative paneling industries. Thermoplastic resin–based composites range from high-priced polyimide, polyether sulfone, and polyether ether ketone to the more affordable nylon, acetal, and polycarbonate resins. Practically all thermoplastics are available in glass-reinforced grades.

Resin-based composites are usually defined as either conventional or advanced. *Conventional composites* usually contain glass or mineral fiber reinforcement, and sometimes carbon fiber, either alone or in combination with others. Conventional composites are usually produced in stock shapes such as sheet, rod, and tube. There are many methods of processing composite materials. These include filament winding, layup, cut fiber spraying, resin transfer molding, and pultrusion. *Advanced composites* is a term that has come to describe materials that are used most often for the most demanding applications, such as aircraft, having properties considerably superior to those of conventional composites and much like metals. These materials are engineered from high-performance resins and fibers. The construction and orientation of the fibers are to meet specific design requirements. Advanced composite structures are usually manufactured in specific shapes. An advanced composite can be tailored so that the directional dependence of strength and stiffness matches that of the loading environment.

Designing and manufacturing reliable, economical composite joints are key challenges in making a functional assembly because of high loads and stress concentration and the brittle nature of the composite materials. Most thermoset matrix laminates have limited yielding capability. When the stresses exceed the load limit, the laminates develop microcracks or can fail catastrophically, unlike most metals, which exhibit yield and thereby redistribute the stresses. Thus, there is an especially high concern for the optimal placement and design of joints to ensure reliable operations.

Another characteristic of composite materials that must be taken into consideration is that these materials are generally orthotropic. Dimensional changes will be directional depending on the construction of the composite and direction of the reinforcing fibers. This factor must be taken into consideration when selecting a joining process. Another general limitation that must be considered in composite joint design includes the relatively low strength and stiffness of composites in the out-of-plane direction and poor shear properties. If not accounted for in the design of the joint and assembly, these factors can cause difficulties such as delamination under compressive loading or inadequate out-of-plane load-carrying capabilities.

Thermoset composites are joined by either adhesive bonding or mechanical fasteners. Thermoplastic composites offer the possibility of thermal welding techniques, adhesives, or mechanical fasteners for joining. Composites are also often joined with a combination of mechanical fasteners and adhesives. Many manufacturers distrust adhesive bonds in applications where joints undergo large amounts of stress (e.g., aircraft structures). Mechanical fasteners must be sized to avoid fiber crushing and delamination; adhesives must balance strength and flexibility. The joining of composite materials involves some special problems not faced with other materials.

10.4 TYPES OF PLASTIC ASSEMBLY PROCESSES

In general, joining methods for plastics can be classified as (1) adhesive bonding, (2) welding by direct heating (heated tool, hot gas, resistance wire) or induced heating (spin, induction, ultrasonic, vibration), (3) solvent welding, (4) mechanical fastening, or (5) self-assembly. Other, less common methods of joining (infrared, extrusion, laser, etc.) are also possible, but they generally fall into one of the classifications above. Table 10.6 indicates the most common joining methods that are suitable for various plastic materials. Descriptions of these joining techniques are summarized both in Table 10.7 and in the sections that follow.

The joining of plastics with adhesives is generally made difficult because of the low surface energy, poor wettability, and presence of weak boundary layers associated with these substrates. Adhesive bonding is a relatively slow process that could be a significant drawback in many industries that produce high-volume plastic assemblies. However, with plastic substrates the designer has a greater choice of joining techniques than with many other substrate materials. Thermoset plastics must usually be bonded with adhesives or fastened mechanically, but many thermoplastics can be joined by solvent cementing, heat welding, and self-assembly processes as well as with adhesives or mechanical fasteners.

In the thermal or solvent welding processes, the plastic resin that makes up the substrate itself acts as the adhesive. These processes require that the surface region of the substrate be made fluid so that it can wet the mating substrate. If the mating substrate is also a polymer, both substrate surfaces can be made fluid so that the resin can diffuse molecularly into the opposite interfaces. This fluid interface region is usually achieved by thermally heating the surface areas of one or both substrates, or by dissolving the surfaces in an appropriate solvent. Once the substrate surface is in a fluid

TABLE 10.6 Common Methods of Joining Plastics

Plastic	Adhesive Bonding	Heated Tool	Hot Gas Welding	Resistance Wire	Spin Welding	Induction Heating	Ultrasonic Welding	Vibration Welding	Solvent Welding	Mech. Fastening	Self-Assembly
Acetal copolymer	×									×	×
Acetal homopolymer	×	×	×	×	×	×	×	×		×	×
Acrylonitrile–butadiene–styrene	×	×	×	×	×	×	×	×	×	×	×
Cellulosics	×								×	×	×
Fluorocarbons	×									×	×
Polyamide	×					×	×	×	×	×	×
Polycarbonate	×	×	×	×	×	×	×	×	×	×	×
Polyolefins	×	×	×	×	×	×	×	×		×	×
Polyethylene terephthalate	×									×	×
Polyimide (thermoplastic)	×						×		×	×	×
Polyimide (thermoset)	×									×	
Polymethyl methacrylate (acrylic)	×		×		×	×	×	×	×	×	×
Polyphenylene oxide	×	×		×	×		×	×	×	×	×
Polyphenylene sulfide	×									×	×
Polystyrene	×	×	×	×	×	×	×	×	×	×	×
Polysulfone	×	×	×	×			×		×	×	×
Polyvinyl chloride	×	×	×	×		×	×	×	×	×	×
Thermoplastic polyester	×						×	×	×	×	×
Thermosets	×									×	
Thermoset composite	×									×	
Thermoplastic composite	×	×		×			×		×	×	

TABLE 10.7 Assembly Methods for Plastics

Method	Advantages	Disadvantages	Possible Equipment
Adhesive bonding	Easy to apply, moderate strength, many different substrates	Plastics may need surface treatment, slow	Brush, spray, roller, curing ovens, fixtures, metering and mixing equipment
Heated tool welding	Very fast, high bond strength (approximately 70–85% of substrate)	Stresses may occur, many strength-dependent processing parameters	Simple soldering guns, heated blades and irons, robotic equipment
Hot gas welding	Strong bonds, useful for large parts, effective sealing	Poor appearance, slow, operator dependent	Gas hand gun, special welding tips, gas source, welding rods
Resistance wire welding	Fast, low equipment cost	Stresses occur, poor appearance	Resistance wire, power source
Spin welding	Very fast, strong bonds	Parts must be concentric, high equipment cost, limited substrates	Spinning device, semiautomatic to automatic feeding and pressure equipment
Induction welding	Only bond area is heated, strong bonds	Stresses may develop, induction adhesive must be compatible with substrate, high equipment costs	High-frequency generator, coils, possible automatic feed and fixturing
Ultrasonic welding	Fast, strong bonds, only bond area is heated	Joint design limited, substrate and size limited, high equipment cost, not continuous	Energy converter (20 kHz), stand, horn, automatic feed and timer
Vibration welding	See ultrasonics, can be used on larger parts, continuous sealing possible	Substrate is limited, high equipment costs	Converter (120 Hz) stand, fixturing devices, automatic feed and timer
Solvent cementing	Strong bond strength (up to 80–100% of substrate) economical, minimal equipment requirement	Slow, hazardous materials, may caused stress cracking	Hypodermic needle, brush, tank, drying ovens, holding fixtures may be necessary
Mechanical fastening	Not dependent on surface characteristics of substrate, disassembly possible	Slow, bulky, poor appearance, stress concentrations	Fasteners of general and special types
Self-assembly processes	Not dependent on surface characteristics of substrate, disassembly possible, fast, minimizes number of components required, process insensitive	Design and performance depends on the type of plastic being assembled, does not provide a continuous seal	Specific mold design for plastic components

condition, they are then brought together and held in place with moderate pressure. At this point the molecules of substrates A and B will diffuse into one another and form a very tight bond. The fluid polymer mix then returns back to the solid state, usually by the dissipation of solvent or by cooling from the molten condition.

Thermal welding can also be used to bond one polymeric substrate to another substrate that is not polymeric. In these cases, the molten polymer surface wets the other substrate and acts as a hot melt type of adhesive. Internal stresses that occur on cooling the interface from the molten condition are the greatest detriment to this method of bonding. Thermal or solvent welding is appropriate when the nonpolymeric substrate is porous. For joining polymeric to nonpolymeric substrates, solvent welding can also be used only if the nonpolymeric substrate is porous. If it is not porous, the solvent may become entrapped at the bond line and cause very weak joints. Equipment costs for each method vary considerably, as does the amount of labor involved and the speed of the operation. Most techniques have limitations regarding the design of the joint and the types of plastic materials that can be joined.

10.5 ADHESIVE BONDING

In this section we focus on the adhesive bonding process. It is widely used for joining plastics to themselves, to other plastics, and to nonplastics. The advantages and disadvantages of adhesive bonding as well as the general process requirements are thoroughly defined in most textbooks on adhesives and adhesives. Pocius et al. provide an excellent treatise on the use of adhesive in joining plastics [2], others provide a guide to adhesive bonding of all substrates [3,4], and there are several texts on joining plastics that cover welding and mechanical fastening processes as well as adhesive bonding [5,6].

Adhesives have advanced to a degree where bonding can be considered an attractive and practical alternative to mechanical fastening and other assembly processes discussed in this chapter. Bonding has several distinct advantages and some disadvantages which may make using adhesives impractical in certain applications. The pros and cons of adhesive bonding are summarized in Table 10.8.

There are certain basic process considerations for adhesive bonding that apply to all substrates: (1) wetting of the substrate surface by the adhesive, (2) surface preparation of the substrate, (3) joint design, and (4) selection and solidification of the adhesive. The following discussion concentrates on how these must be considered relative to plastic substrates.

10.5.1 Wetting

There are two important aspects to surface preparation. The first is to ensure that the adhesive spreads over and makes intimate contact on a molecular scale with the plastic substrate. The second is to ensure that there is no weak boundary layer that can interfere with the wetting process and become a weak link in the formation of a bonded joint.

The process of an adhesive spreading over and making intimate contact with a plastic surface is called *wetting*. This molecular-level contact allows the forces of adhesion to develop (Table 10.9), removes entrapped air from the interface, and maximizes the area over which the bond can develop. Without wetting, strong bonds are impossible.

Wetting may be measured by the contact angle that a drop of the adhesive makes with the substrate surface. If the drop beads up with a large contact angle, wetting is

TABLE 10.8 Adhesive Bonding Advantages and Disadvantages

Advantages	Disadvantages
Provides large stress-bearing area	Surfaces must be carefully cleaned and/or treated
Provides excellent fatigue strength	Long cure times may be needed
Dampens vibration and absorbs shock	Limitation on upper continuous operating temperature, due to organic nature of adhesive
Certain adhesives can be cured at room temperature so that heat-sensitive substrates can be joined	Heat and pressure may be required to set the adhesive
Joins all shapes and thicknesses	Jigs and fixtures may be needed
Permits smooth contours	Rigid process control is usually necessary
Seals joints	Inspection of the finished joint is difficult
Joins any combination of similar or dissimilar materials	Useful life depends on the service environment
Often less expensive and faster than mechanical fastening	Difficult or impossible to disassemble
Provides attractive strength-to-weight ratio	Requires careful handling to avoid waste and misapplication

TABLE 10.9 Theories of Adhesion

Theory	Source of Adhesion
Mechanical	Penetration of adhesive into pores and asperities of the substrate
Electrostatic	Electrostatic attraction between interfacial electrical layers—similar to plane condenser
Adsorption	Adhesive has good wettabiliy on substrate and develops interfacial bond based on van der Waals forces
Chemical bonding	Covalent bonds cross-links at interface to prevent propagation of fracture
Diffusion	Interdiffusion of macromolecules
Weak boundary layer	Bulk properties of adhesive dissipate fracture energy away from interface
Rheological	Rupture energy of bond is product of reversible work of adhesion and dissipative factors related to rheological properties of adhesive

poor. If the drop spreads over the surface with a small contact angle, wetting is good. Figure 10.4 illustrates various degrees of wetting on a plastic substrate. To ensure adequate wetting, the adhesive should have a surface tension that is no greater than the critical surface tension of the plastic substrate (sometimes referred to as surface energy, γ).

$$\gamma_{\text{substrate}} > \gamma_{\text{adhesive}}$$

Surface energy of substrates is generally measured by repeated contact angle measurements of liquids with differing surface tension on the substrate surface. The extrapolated surface tension where the liquid completely wets the surface (contact angle = 0°) is referred to as the *critical surface tension* of the substrate. Figure 10.5 lists critical surface tension values of certain common substrates and the surface tension of common liquids, including epoxy resin used in standard adhesive formulations. Clean metal surfaces wet easily with common adhesive because they have high critical surface tension (usually, >500 dyn/cm). Most organic solids and liquids, however, have a surface tension less than 100 dyn/cm. Surfaces with especially low critical surface tension are referred to as *low-surface-energy substrates*.

From the surface tension information provided in Figure 10.5, it is apparent that epoxy adhesive will wet a clean aluminum or copper surface. However, epoxy resin will not wet a substrate such as polyethylene. Epoxies will also not wet a metal surface contaminated with silicone oil or another low-energy contaminant. Therefore, low-surface-energy plastics must be modified prior to bonding to provide surfaces capable of wetting. Along with the surface energy, the topology of a plastic surface will also affect the bond strength of an adhesive or coating on the plastic. When a supposedly smooth solid surface is examined closely under a microscope, it is found to contain irregularities. It is not flat and smooth but contains many surface asperities, such as peaks and valleys, with a certain degree of roughness. The nature of the surface is most important with adhesive bonding and less important for the other assembly processes.

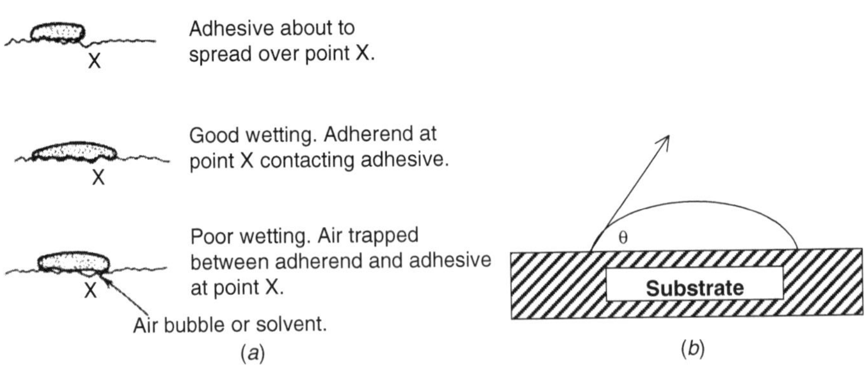

FIGURE 10.4 (*a*) Good and poor wetting of an adhesive on a substrate surface (*b*) Contact angle, θ, made by a drop of adhesive on a substrate (large contact angles indicate poor wetting; small contact angles indicate good wetting).

Substrate	Critical Surface Tension (dyn/cm)
Aluminum	~500
Copper	~1000
Glass	~1000
Epoxy	47
Polycarbonate	46
PPS	38
Polyethylene	31
Silicone	24

Adhesive	Surface Tension (dyn/cm)
Epoxy resin	47
Fluorinated epoxy resin	33
Epoxy resin + DETA	44
Epoxy resin + DEAPA	33

(*a*)

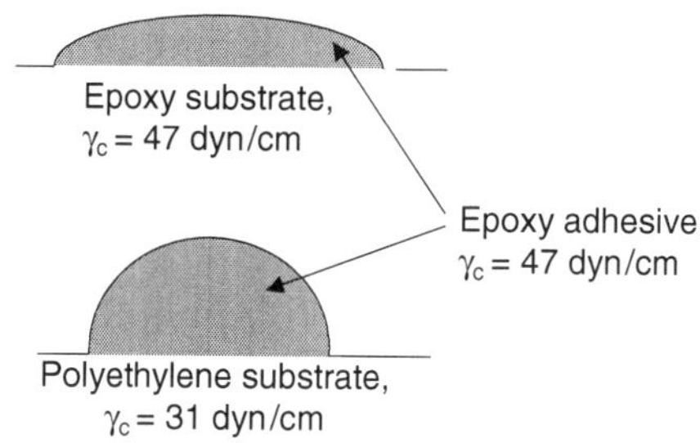

(*b*)

FIGURE 10.5 (*a*) Critical surface tensions for common solids and surface tensions for common liquids; (*b*) contact angle of uncured epoxy adhesive on surfaces of different surface energy.

In most metal surface preparation processes, mechanical abrasion will provide improved adhesion. This operation provides mechanical interlocking and increased area over which the forces of adhesion can develop. For metals, ceramics, and other high-energy surfaces, roughness provides greater bonding area. The greater effective surface area offers a larger area for the forces of adhesion to operate, thereby providing a stronger joint. As a result, abrasion and other surface roughening processes are often recommended as prebond treatments for high-energy surfaces.

However, if the substrate has a lower surface energy than the adhesive, such as the case with many common plastics, surface roughening could actually inhibit the development of high bond strength. The combination of surface roughness and poor adhesive wetting could contribute to air pockets being trapped in the irregularities on the surface. This causes stress concentrations in the adhesive joint, which reduces its ultimate strength. When the voids are in the same plane along the joint interface, as shown in Figure 10.6, they can seriously weaken the bond strength. As a result, prebond surface preparation processes for low-surface-energy materials generally involve changing the chemistry rather than the physical topology of the surface.

10.5.2 Surface Preparation

Chemical or physical surface treatments are generally employed on low-surface-energy plastics to increase the critical surface tension and improve wetting and adhesion. In addition to increasing the critical surface tension, surface treatments are designed to remove contaminants or weak boundary layers, such as a mold release. Modern surface preparation processes provide the ability to bond plastics that were not capable of being bonded previously. This is especially true for the structural bonding and coating of low-surface-energy plastics such as polypropylene, thermoplastic polyolefin, and other olefinic materials. Conventional methods of surface treatments for plastics are summarized below.

- *Abrasion:* surface roughening by high-velocity particle or other abrasive media
- *Primer:* surface coated with a dilute solution chemically similar to the adhesive
- *Chemical etching:* chemical cleaning or removal of surface layers
- *Flame treatment:* surface treated with a controllable gas flame to oxidize surface layers
- *Corona discharge:* surface altered by ionized particles generated by high-voltage electrodes
- *Gas plasma:* surface altered by gas plasma formed by electric current passing through a gas medium at a specified frequency

Table 10.10 summarizes how the surface treating processes affect the plastic surface to improve adhesion. The nature and effectiveness of the process will depend on

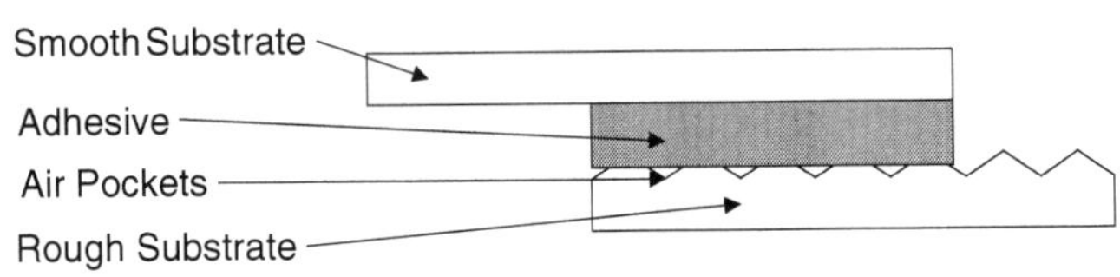

FIGURE 10.6 Effect of surface roughness on interface.

TABLE 10.10 Characterization of Common Surface Treatments for Polymers

Pretreatment Type	Possible Effects of Pretreatment
Solvent	Removal of contaminants and additives that have migratory surface; roughening (e.g., trichloroethylene vapor with polypropylene); weakening of surface regions if excessive attack by the solvent
Mechanical	Surface roughening
Oxidative (flame, chemical, etc.)	Introduction of functional groups. change in topography (e.g., roughening of surface with chromic acid treatment of polyolefins)
Plasma	Removal of contaminants and cross-linking (if inert gas is used); introduction of functional groups if active gases such as oxygen are used; grafting of monomers to polymer surface after activation

Source: Ref. 7.

the substrate being treated and the treatment process parameters, which must be carefully controlled. The degree to which adherends must be prepared is related to the service environment and the ultimate joint strength required. In many low- to medium-strength applications, extensive surface preparation may be unnecessary. However, where maximum bond strength, permanence, and reliability are called for, carefully controlled surface-treating processes are required. Example surface treatments for common plastic substrates are shown in Table 10.11. Details regarding the surface treatment process parameters may be found in ASTM D2093 and various texts on adhesive bonding of plastics. An excellent source of information regarding prebond surface treatments is the supplier of the plastic resin that is being joined.

Abrasion and solvent cleaning are generally recommended as a surface treatment for high-surface-energy thermoplastics and for thermosetting plastics. Frequently, a mold release agent is present and must be removed before adhesive bonding. Mold release agents are usually removed by a detergent wash, solvent wash, or solvent wipe. A clean lint-free cloth or paper tissue is commonly used, and steps must be taken to assure that the cleaning materials do not become contaminated or a source of contamination. Common solvents used to clean plastic surfaces for adhesive bonding are acetone, toluene, trichloroethylene, methyl ethyl ketone (MEK), low-boiling petroleum ether, and isopropanol. A solvent should be selected that does not affect the plastic surface but is sufficiently strong to remove organic contamination. Safety and environmental factors must be considered when choosing a solvent. Solvent cleaning alone can be used in applications that do not require the maximum joint strength.

Solvent cleaning is the process of removing soil and organic contaminants from a substrate surface with an organic solvent. Where loosely held dirt, grease, and oil are the only contaminants, simple solvent wiping alone with a brush or clean cloth will provide surfaces for weak- to medium-strength bonds. There are also more effective means of solvent cleaning. Solvent cleaning is widely used and should precede (and generally follow) any chemical or abrasive surface preparation. However, it is the least effective substrate treatment in that it only cleans the surface of loosely held particulates and organic contamination.

The compatibility of cleaning solvents with plastic substrates is extremely important. Solvents can affect polymeric surfaces and provide unacceptable part appearance or even degradation of properties. Solvents that are recommended for cleaning plastics are shown in Table 10.12. Suppliers of mold release agents are the best source for information on solvents that will remove their materials. Volatile solvents are acceptable; however, the local and most recent safety and environmental restrictions must be consulted before selecting any solvent. These solvents are considered toxic and hazardous and exposure limits as well as safety procedures must be maintained.

Abrasive treatments consist of scouring, machining, hand sanding, and dry and wet abrasive blasting. The abrasive medium can be fine sandpaper, carborundum or alumina abrasives, metal wools, or abrasive shot. Mechanical abrasion is usually preceded and followed by solvent cleaning. The choice is generally determined by available production facilities and cost. Laminates can be prepared by either abrasion or the tear-ply technique. In the tear-ply design, the laminate is manufactured so that one ply of heavy fabric, such as Dacron, glass, or the equivalent, is attached

TABLE 10.11 Typical Surface Treatments to Promote Adhesive Bonding of Plastics

Plastic	Degreasing Solvent	Method of Treatment[a]	Remarks[b]
Acetal (copolymer)	Ketones	Abrasion	For moderate bond strength
		Etch in dichromate–sulfuric acid solution	Note: 1, 2, 3
		Plasma treatment	Note: 2
Acetal (homopolymer)	Ketones	Abrasion	For general-purpose bonding
		Etch in dichromate–sulfuric acid solution	Higher strength, Note: 1, 2
		Satinizing process	Maximum strength
Acrylonitrile–butadiene–styrene	Ketones	Abrasion	Bond strength is usually greater than the part
		Etch in dichromate–sulfuric acid solution	Note: 1
Cellulosics	Alcohols	Abrasion	For general-purpose bonding
		After abrasion dry the plastic at 100°C for 1 hr and apply adhesive before the plastic cools to room temperature	
Fluorocarbons	Chlorinated, ketones, alcohols	Etch in sodium-napthalene solution	ASTM D2093, etching solutions are commercially available
Polyamide (nylon)	Ketones	Abrasion	
		Prime with resorcinol formaldehyde adhesive	Good adhesion to primer coated with epoxy adhesives
Polycarbonate	Alcohols	Abrasion	
Polyolefins	Ketones	Expose surface to flame until the substrate is glossy	Bond immediately after treatment, Note: 2, 3
		Etch in dichromate–sulfuric acid solution	For moderate bond strength, ASTM D2093
		Expose to electric discharge (corona)	Bond within 15 min of treatment
		Plasma treatment	For maximum strength, Note: 1, 2
Polyethylene terephthalate	Ketones	Solvent wipe	For general-purpose bonding
		Etch in sodium hydroxide solution	For maximum strength Note: 2, 3
		Plasma treatment	Note: 2
Polyimide	Ketones	Abrasion	

TABLE 10.11 (*Continued*)

Plastic	Degreasing Solvent	Method of Treatment[a]	Remarks[b]
Polymethyl methacrylate (acrylic)	Ketones or alcohols	Abrasion	For maximum strength, relieve internal stresses by annealing plastic
Polyphenylene oxide	Alcohols	Abrasion	
		Etch in dichromate–sulfuric acid solution	For maximum strength, Note 3
Polyphenylene sulfide	Ketones or chlorinated solvents	Abrasion	
Polystyrene	Alcohols	Abrasion	
Polyvinyl chloride and other vinyl polymers	Ketones or chlorinated solvents	Abrasion	Suitable for rigid plastic. for maximum strength on plasticized material, use nitrile–phenolic
Thermoplastic polyester	Ketones	Abrasion	
Thermoset plastics (epoxy, phenolic, etc.)	Ketones	Abrasion	

[a]Abrasion methods should be preceded and followed by solvent degreasing. Degreasing should also precede other methods of treatment.

[b]*Notes:*

1. A. H. Landrock, *Adhesives Technology Handbook*, Noyes Publications, Park Ridge, NJ, 1985.
2. R. C. Snogren, *Handbook of Surface Preparation*, Palmerton Publishing Co., New York, 1974.
3. E. M. Petrie, Plastics and elastomers as adhesi ves, in *Handbook of Plastics and Elastomers*, C. A. Harper, ed., McGraw-Hill, New York, 1975.

at the bonding surf ace. Just prior to bonding the tear -ply is stripped a way, and a fresh, clean, bondable surface is exposed.

Chemical surface treatments v ary with the type of plastic being bonded. These processes can involve the use of corrosive and hazardous materials. The most common processes are sulfuric acid–sodium dichromate etch (polyolef ins) and sodium-naphthalene etch (fluorocarbons). Both of these processes are described in ASTM D2093. Flame, hot air, electrical discharge and plasma treatments change the surf ace of the polymer both physically and chemically. The plasma treating process has been found to be very successful on most lo w-energy-surface plastics [9,10]. Table 10.13 shows that plasma treatment results in impro ved plastic joint strength with common epoxy adhesive. Plasma treatment requires vacuum and special batch processing equipment.

Better adhesion can be obtained if the parts are formed, treated, and coated with an adhesive in a continuous operation. The sooner an article can be bonded after surf ace treatment, the better will be the adhesion. After the part is treated, handling and exposure to shop en vironments should be k ept to a minimum. Some surf ace treatments, such as plasma, have a long effective shelf life (days to weeks) between treatment and

TABLE 10.12 Common Degreasing Solvents for Polymeric Surfaces

Adherend	Solvent
Acetal (copolymer)	Ketone
Acetal (homopolymer)	Ketone
Acrylonitrile–butadiene–styrene	Ketone
Cellulose, cellulose acetate, cellulose acetate butyrate, cellulose nitrate	Alcohol
Fluorocarbons	Chlorinated alcohol or ketone
Polyamide (nylon)	Ketone
Polycarbonate	Alcohol
Polyolefins	Ketone
Polyethylene terephthalate (Mylar)	Ketone
Polyimide	Ketone
Polymethyl methacrylate, methacrylate butadiene	Ketone or alcohol
Polyphenylene oxide	Alcohol
Polyphenylene sulfide	Ketone, chlorinated solvents
Polystyrene	Alcohol
Polyvinyl chloride, polyvinyl fluoride	Ketone, chlorinated solvents
Thermoplastic polyester	Ketone
Thermoset plastics	Ketone

Source: Ref. 8.

TABLE 10.13 Typical Adhesive Strength Improvement with Plasma Treatment: Aluminum-to-Plastic Shear Specimen Bonded with a Conventional Epoxy Adhesive

	Strength of Bond (lb/in^2)	
Plastic	Control	After Plasma Treatment
Polyamide	846	>3956
Polyethylene	315	>3125
Polyethylene terephthalate	530	1660
Polypropylene	370	3080
Polystyrene	566	>4015
Polytetrafluoroethylene	75	750
Polyvinyl fluoride	278	>1280

Source: Ref. 10.

bonding. However, some treating processes, such as electrical discharge and flame treating, will become less effective the longer the time between surface preparation and bonding. Most optimized surface treatment processes require prolonged production time and provide safety and environmental concerns. Often there is a trade-off in the cost and adhesion quality of surface treatment process, as shown in Figure 10.7. One should be careful not to overspecify the surface treatment required. Only the minimal process necessary to accomplish the functional objectives of the application is required.

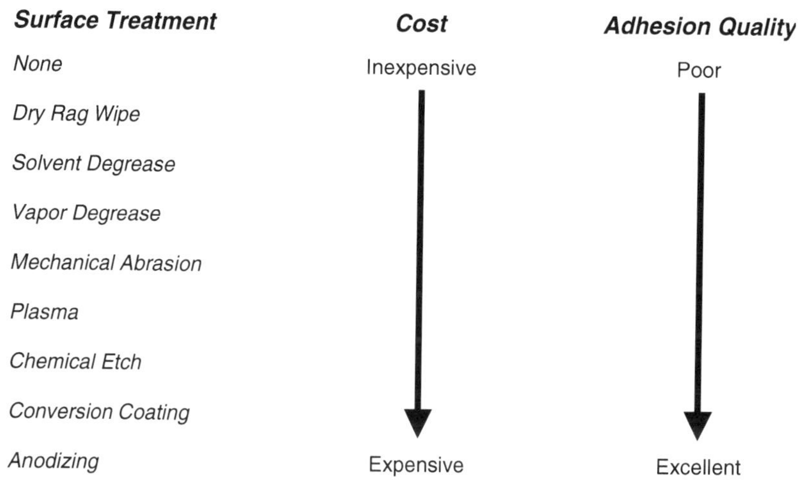

FIGURE 10.7 The objective of surface preparation is to provide a consistent, reproducible adherend surface that when bonded meets the strength and durability required for the application.

Several new surface treatments and modifications of older conventional surface treatments have been introduced over the last few years to provide alternatives to the common processes noted above [11]. The driving factors for these developments have been related primarily to environment and safety. Harsh chemicals and elevated temperature processing associated with conventional chemical and flame treatment methods have inhibited many from using such processes. In addition to providing safer and environmentally friendly processes, these newer surface treatments have also been shown to provide for easier and faster processing. They promise a potentially tremendous positive impact on both manufacturing cost and performance properties. The reduced cost impact can be in the form of equipment costs, implementation costs, operational costs, rework costs, and storage/waste removal costs.

10.5.3 Joint Design

The type of stress that the joint will see in service will play an important role in selecting an adhesive and in designing the joint. Five types of stress, shown in Figure 10.8, are commonly associated with adhesive joints: peel, cleavage, shear, tension, and compression. Most adhesives provide excellent resistance to tensile, shear, and compression because the stress is evenly distributed over the entire joint geometry. Peel and cleavage forces provide the harshest form of stress since the full stress is exerted on a much smaller area of the total joint.

Tensile stress tends to elongate and pull an assembly apart. Compressive stress squeezes an assembly together. Shear stresses pull parallel objects apart lengthwise, causing a sliding motion in opposite directions. Peel stress results when a flexible substrate lifts or peels away from the substrate to which it is bonded. Cleavage stress

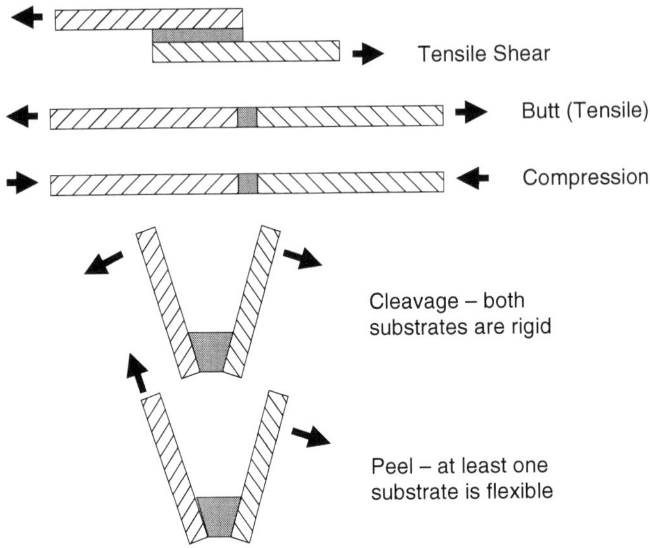

FIGURE 10.8 Joint stresses common in adhesive bonding.

is similar to peel stress, but it arises in rigid substrates when a joint is pried open at one end. In actual application, the stress on an adhesive joint usually consists of a combination of stresses listed above.

Most adhesives provide excellent resistance to tensile, shear, and compressive stress. Unfortunately, plastic substrates are often flexible, and stresses can easily become peel or cleavage. To design the strongest possible joint, one must distribute the loads as evenly as possible over the entire joint area. The goal should be to maximize tensile and compressive stresses. The best joint designs maximize bonding area and provide some mechanical interlocking. Figure 10.9 offers several joint designs that provide improved strength when the substrates are flexible.

The plastic joint may be influenced by the desired appearance of the finished product. It may also be limited by production facilities, allowable costs, or other specific requirements. Plastic adhesive joints should be designed to the following principles:

- Stress the adhesive in the direction of maximum strength (generally shear).
- Provide for the maximum bonded area.
- Make the adhesive layers as uniform as possible.
- Avoid stress concentration.
- Provide an adhesive with approximately the same degree of flexibility and elastic modulus as the substrate.

Many plastics, because of their flexible nature and ability to be formed in thin sections, can easily be subjected to peel or cleavage stress. Rigid adhesives are particularly

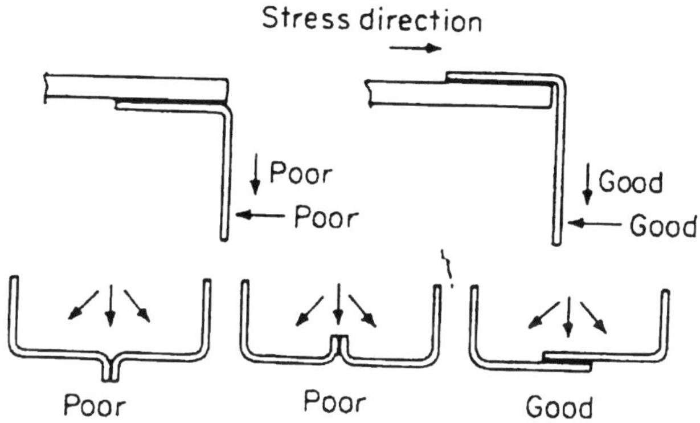

FIGURE 10.9 Joint designs for flexible plastic substrates.

weak in peel because the stress is concentrated at only a thin bond edge. Tough, flexible adhesives distribute the peeling stress o ver a wider bond area and sho w greater resistance to peel. The adhesive should have the same degree of flexibility as the plastic substrate, to avoid uneven stress when the assembly is flexed. For these reasons, flexible polyurethane and epoxy adhesives are often used in bonding plastics.

Typical plastic joint designs for molded plastic parts are illustrated in Fig. 11.10:

- *Butt*: simple design for molded parts. Load can translate into undesirable peel or cleavage stresses. Stronger bonds can be obtained by the designs that follow. Positive positioning is not possible with this type of bond, and joined parts may slip out of position unless locating pins or clamp f ixtures are used.
- *V*: ensures positive positioning, and bonded parts e xhibit good shear and tensile strength. Mating parts must ha ve a close tolerance to ensure surf ace contact.
- *Lap*: provides the greatest joint strength in most instances. It is self-postioning but requires closer tolerance of matching parts than does the b utt joint.
- *Tongue and groove*: postive seating is assured when tolerances are maintained. Joint strength is e xcellent. Required hea vy part thickness may present a problem.
- *Scarfed*: although not frequently used, it does fulfill the requirement of some designs that prohibit lap, tongue and groove, or V joints. It results in a stronger bond than the butt joint.

A scarf, V, or lap joint is acceptable. Butt joints tend to result in clea vage forces when the loads are not aligned perfectly . The tongue-and-groove joint results in good stress distrib ution, but it requires a relati vely thick part. Mold gates should

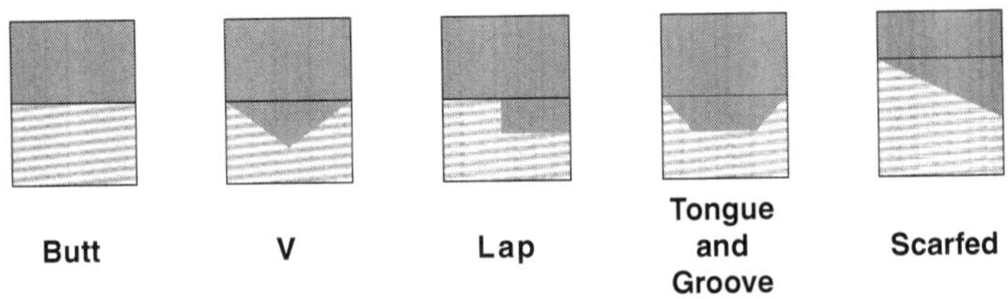

FIGURE 10.10 Common joint designs for molded plastic parts.

not be located in the areas to be bonded. Frozen-in strains are usually found in the gate area. With application of solvent or monomer containing adhesives, stress cracking could result.

Reinforced plastics are often anisotropic materials having directional strength properties. Joints made from anisotropic substrates should be designed to stress both the adhesive and the substrate in the direction of greatest strength. Laminates, for example, should be stressed parallel to the laminations. Stresses normal to the laminate may cause the substrate to delaminate.

10.5.4 Adhesive Materials

There are many types and forms of adhesives that can be used with plastics to provide strong structural bonds. Plastics generally have lower tensile strength than other materials, such as metals, and an adhesive strength can often be achieved that is greater than the strength of the substrate itself. The selection of an adhesive for a plastic bonding application can, at the outset, appear difficult because of the large number of different types of adhesives available. There is no general all-purpose adhesive. Selecting the best adhesive for a particular application depends on (1) the materials to be bonded, (2) the assembly requirements (bond area, production volume and speed, available facilities, etc.), (3) the types of service loads and joint design, (4) the service environment to which the joint will be exposed for its lifetime, and (5) the economics of the bonding process (material, capital, and environmental costs as well as productivity requirements). By using these factors as criteria for selection, the many commercially available adhesives can be narrowed down to a few possible candidates.

The materials to be bonded are prime factors in the adhesive selection process. Generally, the best adhesive is one that will wet the substrate and, when cured, has a modulus and thermal expansion coefficient similar to the substrate or else has necessary toughness and elongation to accommodate stresses caused by thermal movements. Differences in flexibility or thermal expansion between the adherends or between the adhesive and adherend can introduce internal stresses into the bond line. Such stresses can lead to premature failure of a bond. Thus, rigid heavily filled adhesives are often chosen for bonding metals. Flexible adhesives are often chosen for bonding plastics and elastomers.

Figure 10.11 illustrates the strength, modulus, and elongation of several common adhesive systems. Lower-modulus adhesives generally have the flexibility to bond well to plastic substrates. However, these are generally weaker in shear than more rigid adhesives. Fortunately, exceptionally high shear strength is often not required for an adhesive for plastic, since the plastic substrate itself is relatively weak. For many high-surface-energy thermosetting plastics, such as epoxies, polyesters, and phenolics, adhesive bonding is generally easy and can be accomplished with many of the same adhesives that are used on other substrates. For thermoplastics, the surface energy is generally lower, the reactivity is greater, and the thermal expansion is higher than for thermosets. Therefore, when bonding thermoplastics, consideration must be given to the surface energy of the adhesive and the substrate, the compatibility of the adhesive with the substrate, and thermal expansion coefficients.

Adhesives for plastics can be classified broadly into structural and nonstructural types. *Structural adhesives* are those having bond shear strength on the order of 1000 lb/in^2 or greater. This is often sufficient to cause failure of the plastic substrate when the bond is tested. Structural adhesives are generally intended for applications where chemical and temperature resistance are requirements as well as high strength and toughness. *Nonstructural adhesives* are those having bond strength less than 1000 lb/in^2 but sufficient for applications such as pressure-sensitive tapes, labels, laminates, and so on. Nonstructural adhesives are usually employed where production speed, convenience, high peel strength, and resistance to common environments are required.

There are numerous families of adhesives within the structural and nonstructural types. These are generally defined by the chemical composition of the base resin used in the adhesive formulation. These adhesives can also come in many forms,

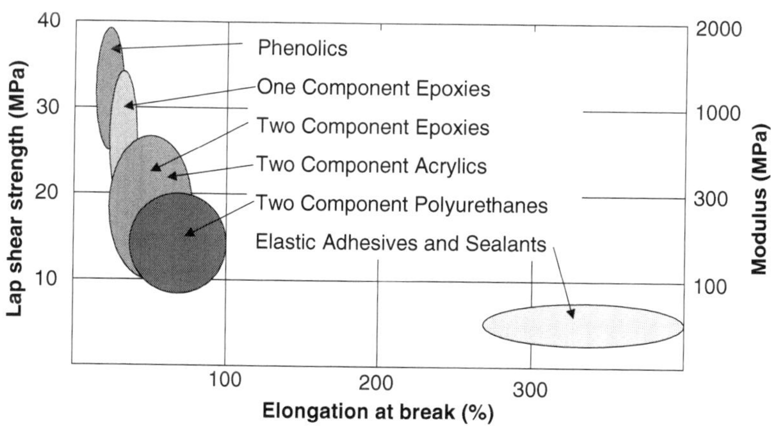

FIGURE 10.11 Common adhesives provide a wide range of physical characteristics. Adhesives with low modulus and high flexibility are generally recommended for bonding plastics.

including 100% solids, solvent-based, and water-based. The most common adhesive families for bonding plastics are identified below.

- Structural adhesives
  - Cyanoacrylate
  - Epoxy
  - Polyurethane
  - Reactive acrylic
  - Light curing adhesive (acrylic and cyanoacrylate)

- Nonstructural adhesives
  - Synthetic and natural elastomers
  - Thermoplastic hot melts
  - Resin latex adhesives
  - Silicone

General properties of the adhesives that are commonly used to bond plastics are presented in Tables 10.14 and 10.15. Each offers a unique combination of performance and processing benefits. Recommended adhesives for specific plastic substrates are listed in Table 10.16. The recommended adhesive is generally determined by its surface tension relative to the plastic being bonded and its flexural modulus and elongation.

TABLE 10.14 General Properties of Adhesives Commonly Used to Bond Plastics

Adhesive	Type Available	Cure Mechanism
Epoxy	100% solids, one- and two-component	Chemical reaction
Reactive acrylic	100% solids, one (with primer)- and two-component	Chemical reaction
Polyurethane	100% solids, one- and two-component	Chemical reaction (parts A and B) or reaction of moisture in the air or substrate
Cyanoacrylate	100% solids, one-component	Reacts with substrate and ambient moisture
Light activated adhesives	100% solids, one-component	Reacts with UV light of various wavelengths
Elastomer	Solvent or water based	Hardens by evaporation of solvent or water
Hot melt	100% solids	Flows by heating and then hardens on cooling
Resin latex	Water based	Hardens by evaporation of water
Silicone	100% solids, one- or two-component	Cures by chemical reaction (parts A and B) or by reaction with ambient moisture

TABLE 10.15 Comparison of Adhesives Commonly Used to Bond Plastics

Adhesive	Advantages	Disadvantages
Epoxy	Usually low priced Good gap-filling capabilities High strength; can be filled with additives Wide range of formulations Versatile Good temperature and solvent resistance	Viscosity reduces on heating Two-component requires metering and mixing One-component requires refrigeration and heat cure Fixturing required during cure Limited pot life can cause waste
Polyurethane	Moderately priced Excellent toughness and flexibility Good flexibility at low temperatures Excellent adhesive for a wide range of substrate materials Varying cure times	Limited temperature resistance Sensitive to moisture in cured and uncured states Two-component requires mixing with metering Toxicity a problem with certain types Short pot life
Reactive acrylic	Moderately priced Good outdoor weathering Good impact resistance and flexibility Good bond strength to many plastics Fast cure at room temperature	Some odor problems Flammable Certain types can be rigid
Cyanoacrylate	Rapid cure at room temperature Single-component adhesive Excellent adhesion to rubber High tensile strength Indefinite pot life Low viscosity	Higher priced Limited gap cure Low solvent and moisture resistance Poor high temperature resistance Poor impact and peel strength Can bond to human skin
Elastomeric	Low price per gallon Many types available for different substrates Easily applied to large surfaces Easily sprayed Long shelf life No special equipment needed	Low strength Poor gap cure Poor creep resistance High shrinkage Slow drying (water based) Solvent can attack certain plastics Flammable (solvent based) Environmental concerns (solvent based)
Hot melts	Moderately priced Good gap-filling characteristics Rigid to flexible bonds available Versatile formulation	Low strength and heat resistance Poor wetting Low solvent resistance Parts must be mated before adhesive cools Short pot life Can burn workers

TABLE 10.15 (*Continued*)

Adhesive	Advantages	Disadvantages
	Fast setting	Requires special dispensing equipment Difficult to automate
Resin latex	Low to moderately priced Good peel strength on various substrates Versatile formulation Easily sprayed to large surfaces No special equipment required	Long cure times Low moisture and solvent resistance Most formulations have limited temperature resistance
Silicone	Moderately to high priced Good gap-filling characteristics Excellent sealant Flexible Good moisture and heat resistance Range of viscosities available Range of formulations available	Low strength Limited solvent resistance Slow curing Most formulations need moisture to cure Some formulations are corrosive Gives off odor while curing Hard to clean

TABLE 10.16 Adhesives for Bonding Common Thermosets and Thermoplastics

Thermoplastics

Acrylic plastics (polymethyl methacrylate)—may also be heat or solv ent welded; with solvent welding, crazing may occur.

- Thermoplastics acrylic (in solvent solution)
- Thermosetting acrylic
- Epoxy
- Neoprene or nitrile synthetic rubber
- Resorcinol–formaldehyde

Cellulose acetate plastic—may also be heat- or solv ent-welded

- Cellulose acetate (in solvent solution)
- Cellulose nitrate
- Epoxy
- Polyvinyl acetate
- Resorcinol–formaldehyde

Cellulose nitrate plastics—may also be heat- or solv ent-welded

- Epoxy
- Resorcinol–formaldehyde

Polyamide plastics (nylon)—may also be heat-welded; not generally solv ent-welded

- Epoxy
- Phenolic blends
- Resorcinol–formaldehyde
- Synthetic rubber-based

TABLE 10.16 (*Continued*)

Polyethylene plastics—may be heat-welded; not solvent-welded
Phenolic blends
Synthetic rubber-based
Epoxy (when treated)
Polystyrene plastics—may be solvent- or heat-welded; low-boiling solvents may cause crazing
Polystyrene in solution
Epoxy
Polytetrafluoroethylene (PTFE, Teflon)—may be heat-welded with difficulty under pressure
Phenolic blends
Silicone
Synthetic rubber-based
Epoxy (when substrate is treated)
Polyvinyl chloride plastics (PVC)—may be heat-welded; not generally solvent-welded
Thermosetting acrylics
Synthetic rubber (neoprene or nitrile)—nitrile for heavily plasticized vinyl
Phenolic–neoprene
Phenolic–nitrile
Phenolic–vinyl
Thermosets
All thermosets, including composites
Epoxy
Furane
Phenolic–neoprene
Phenolic–nitrile
Phenolic–vinyl
Phenol–formaldehyde
Polyester
Resorcinol–formaldehyde
Synthetic rubber-based
Urea–formaldehyde

Source: Ref. 3.

10.5.4.1 Structural Adhesives

Cyanoacrylate adhesives became widely known after their commercial introduction in the 1970s as *superglue* or *instant adhesives*. They are one part adhesives that cure within seconds to minutes at room temperature to form thermoplastic resins when confined between two substrates that contain trace amounts of surface moisture. Cyanoacrylate adhesives have excellent tensile shear strength and good shelf file. However, they are thermoplastic once cured and have limited thermal and solvent resistance. Cyanoacrylates also exhibit several disadvantages, including limited cure through depth (0.010 in.), poor durability on glass, and the potential for stress cracking of plastics.

The discovery of high-viscosity and rubber-toughened formulations and surface primers for cyanoacrylate adhesives is a major development that greatly extends the market potential, especially in the area of plastic bonding for medical and electronic devices. These modern formulations provide the ability to bond to porous substrates, high peel strength and impact resistance, and high bond strength to low-surface-energy plastics.

Epoxy adhesives are widely used for structural bonding. These adhesives are higher in viscosity than cyanoacrylates and acrylics. They can be cured at either room or elevated temperature, and a variety of forms (liquid, paste, film, powder) are commonly available. There are significant variations in epoxy formulations, with a corresponding significant variation in curing conditions and end-use properties. Epoxy adhesives cure to a densely cross-linked structure that has relatively good thermal and chemical resistance. Flexible epoxy adhesives are generally formulated with flexible diluents or elastomeric components to provide higher peel strength and impact resistance. These flexible epoxy versions are commonly used when bonding plastics since they produce less internal stress within the joint.

Polyurethane adhesives are one of the most versatile families of adhesives. They provide good adhesion to a variety of substrates and exhibit a good combination of tensile and flexural properties. They are well suited for bonding plastics because of their flexibility and toughness. They will also provide a chemical bond to many polymeric surfaces, such as the cellulosics and others with hydroxyl groups. They are available primarily in two-part form, for both room and elevated temperature curing. A one-part moisture cure formulation is available that will cure by reacting with ambient moisture either in the air or on the substrate surfaces.

The major disadvantages of polyurethanes are poor hydrolytic and ultraviolet stability for some formulations. They are also very reactive compounds; so that the adhesive is very moisture-sensitive and this can present shelf-life problems. A number of unwanted side reactions can also occur which can reduce the properties of the adhesive.

Reactive acrylic adhesives are two-component, 100% reactive systems based on methacrylate and acrylate monomers which are toughened with elastomers. These adhesives are generally cured at room temperature in minutes to hours; however, elevated temperature cure formulations are also available. When fully cured, reactive acrylic adhesives form thermosetting molecular structures. The adhesives are lower in cost than the cyanoacrylate adhesives. They also provide increased thermal stability, chemical resistance, and peel and impact strength over cyanoacrylates. The presence of an acrylic monomer can provide stress cracking on certain highly stressed plastics. The primary disadvantages of these adhesives are their unpleasant odor and slight yellow concentration. Reactive acrylic adhesives have been recognized to bond very well to many high-energy plastic surfaces as well as moderately contaminated metal surfaces. Recently, reactive acrylics have been introduced that provide very high bond strengths to untreated low-energy substrates.

Light-curing adhesives are generally available from two families: acrylic and cyanoacrylate. Both of these materials bond very well to plastic substrates and have good end-use properties, as indicated above. However, light-cured adhesive cure on exposure to light of the proper intensity and spectral output. These adhesives are 100% solids systems that cure rapidly (seconds) to form thermoset polymer with excellent properties. Secondary cure mechanisms such as heat or chemical activation completely cure the adhesive in shadowy areas.

Light-curing acrylic adhesives cure on command, offering unlimited repositioning time and quick cure once the light energy is activated. They provide high bond strengths on a variety of plastic substrates. Light-curing acrylics come in a wide range of viscosities, from thin liquids to thixotropic gels. This, along with the fast curing speed, make light-curing acrylics especially well suited for integration into automated processes. They provide excellent thermal and chemical stability as well as high peel and impact strength. Light-curing acrylic adhesives are available with flexibility ranging from soft elastomers to glassy plastics.

Light-curing cyanoacrylates is a recently introduced adhesive technology that combines the benefits of cyanoacrylates and light-curing acrylics. These cure via a light-cure process in nonshadowy areas within seconds, and via a typical cyanoacrylate secondary moisture cure mechanism in shadowed areas. This hybrid technology overcomes many of the limitations of cyanoacrylates or light-cure acrylics. The adhesives are surface insensitive and versatile. They offer excellent adhesion to numerous substrates, including rubber and plastic. These adhesives are being used increasingly in high-volume medical and electronic assemblies.

10.5.4.2 Nonstructural Adhesives

Elastomer-based adhesives are based on a variety of synthetic and natural elastomers. They are provided in one-part, solvent, or latex form. They can be activated by evaporation of solvent or water. The elastomers most commonly used are butadiene–styrene, butyl, polyisobutylene, or nitrile rubber. The adhesive formulations are generally highly compounded to provide the degree of tack, peel strength, and curing conditions required for a specific application. Like silicones, these products are generally not used in structural applications but find use as pressure-sensitive or contact adhesives.

Disadvantages of these adhesives are low service temperature and low degree of chemical resistance due to their thermoplastic nature. There is also a negative environmental impact when a solvent system is used. These adhesives are often blended with thermosetting resins such as phenolic or epoxy to provide for higher tensile strength and greater thermal resistance.

Hot melt adhesives for plastics can be produced from four major chemical families: polyethylene, polyamide, polyester, and ethylene–vinyl acetate copolymers. These adhesives are generally available as one-part solids (sticks, powder, blocks) that must be heated to apply and then cooled to form a bond. Polyethylene-based products exhibit good general-purpose performance and wet low-energy plastics very well because of their low surface tension. They offer temperature resistance up to 179°C. Polyamide-based products are used in applications requiring relatively elevated temperature performance (up to 150°C). Polyesters have moderate to high performance properties with temperature resistance to 95°C. EVA-based products have good retention of properties at low temperatures, but their upper temperature limit is only about 50°C.

Hot melt adhesives offer the advantage of setting rapidly, and they provide good adhesion to most plastics. However, caution must be used when bonding to heat-sensitive substrates such as plastic film. They are also relatively tough and flexible. Their major drawback is heat and chemical resistance due to their thermoplastics nature.

Resin latex adhesives are generally formulated from thermoplastic resin emulsions such as acrylic. They are available in a wide range of solids contents and viscosities, and a high concentration of high-molecular -weight material can be used. Water-based adhesives have been widely adopted as an alternative to toxic and flammable solvent-based adhesives. However, they generally have medium to slow cure rates, due to the energy required to evaporate water. Water-based adhesives can also freeze if stored at low temperatures.

Thermoplastic acrylic latex adhesives have excellent adhesive qualities and good weathering characteristics. They are often used in pressure-sensitive adhesive for a wide variety of plastic substrates, including applications such as automobile tags, mounting tape, and protective moldings.

Silicone adhesives have low shear properties but excellent peel strength and heat resistance. Silicone adhesives are often used as either solvent solutions for pressure-sensitive adhesives, or as one- or two-part liquids that can be cured at room temperature. The one-component formulations require ambient moisture to achieve cure. The major advantage of silicone adhesives relative to bonding plastic substrates is their low-surface tension. Silicones will wet many low-energy plastics and thus are often used for bonding and sealing these materials.

Disadvantages of silicone adhesives are their cost relative to other systems and their low tensile strength. They are not generally the adhesive of choice in load-bearing applications, but find significant use in pressure-sensitive tapes, labels, and as sealants. Silicones offer significant temperature resistance (up to 300°C).

10.6 HEAT WELDING

The technique of unifying parts by joining their heated surfaces and allowing them to flow together provides a means of joining for most thermoplastics (i.e., those materials that can be melted or softened by heat without degradation). Certain thermoplastics with very high temperature resistance (e.g., polytetrafluoroethylene, polyimide) cannot be welded with these methods. Materials that have poor heat stability (e.g., cellulose nitrate) can also not be effectively heat welded.

The strongest and clearest heat-welded joints are obtained with thermoplastics having the highest degree of polymerization. These include ABS (acrylonitrile–butadiene–styrene), acetal, acrylic, nylon, polycarbonate, polyolefin, polyphenylene oxide, polystyrene, and polysulfone. Thermoplastic composite parts can also be assembled using heat welding techniques. In these parts, generally the inorganic fillers must be kept to a relatively low concentration to allow for adequate flow of the substrate and subsequent bonding.

In most welding applications the substrates are made of the same plastic materials. However, certain combinations of different plastics can also be heat welded. The bond strengths that can be obtained by welding are generally 50 to 80% that of the tensile strength of the original plastic part. Thus, the choice of heat-welding process generally depends on the production requirements and joint geometry. In the heat-welding processes the molten polymer from the actual part forms a hot melt adhesive, which provides the bond. Plastic heat welding relies on the temperature at the interface being near the glass transition temperature (T_g) for amorphous plastics and above the melting temperature (T_m) for semicrystalline plastics. A welded joint is usually a butt joint. However, thin-film polymers, such as packaging materials, can be welded conveniently using continuous heat-sealing techniques. These are used in producing such objects as envelopes and plastic storage bags.

There are many different heat welding processes. They can generally be classified into two methods, depending on how the heat is applied to fuse the substrates: (1) external heating (hot plate, hot gas, resistance, inductive heating), and (2) mechanical movement or friction (spin, linear vibration, ultrasonic). Heat welding can be accomplished with either no surface treatment or very minor surface preparation (degreasing and removal of mold release), depending on the strength and reliability dictated by the applications. Mechanical abrasion or other prebond surface treatment provides no advantage, since the surface will be melted and a new surface is formed.

Plastics that have a significant degree of internal moisture may have to be dried before heated tool welding, or else the moisture will tend to escape the molten surface in the form of gas bubbles, with a resulting weak and foamy joint. Residual monomer or low-molecular-weight species within the substrate can have the same effect. In all areas of heat welding, equipment systems are continuing to develop toward greater sophistication and automation. Modern production systems have greater electronic control, higher productivity, and new mechanical arrangements to reduce hardware cost and improve safety.

10.6.1 External Heat Welding Processes

Welding using an external source for the heat provides a method of fast, relatively easy, and economical bonds that are high in strength and durability. The principal methods of external heat welding are by heated tool, hot gas, resistance wire, and induction. However, the energy can originate from any source that provides focused and consistent heat to raise the temperature of the interface.

10.6.1.1 Heated Tool Welding

With heated tool welding, the surfaces to be joined are heated by holding them against a hot surface, generally in the temperature range 220 to 350°C. Once the substrate surfaces are sufficiently fluid, they are brought into contact and allowed to cool under slight pressure (5 to 15 lb/in^2) until the interface resolidifies into a structural joint. This process is illustrated in Figure 10.12. Electric strip heaters, soldering irons, hot plates, and resistance blades are common methods of providing heat locally. These can be used for assembling structural parts, and heated rolls or continuous metal bands can be used for sealing films or sheet. When metal plates are used, they are generally either an aluminum–copper or aluminum–bronze alloy. The plates can be coated with PTFE to prevent the plastic from sticking and to reduce contamination of the weld.

Successful heated tool welding depends on the temperature of the heated tool surface, the amount of time the plastic adherends are in contact with the hot tool, the time

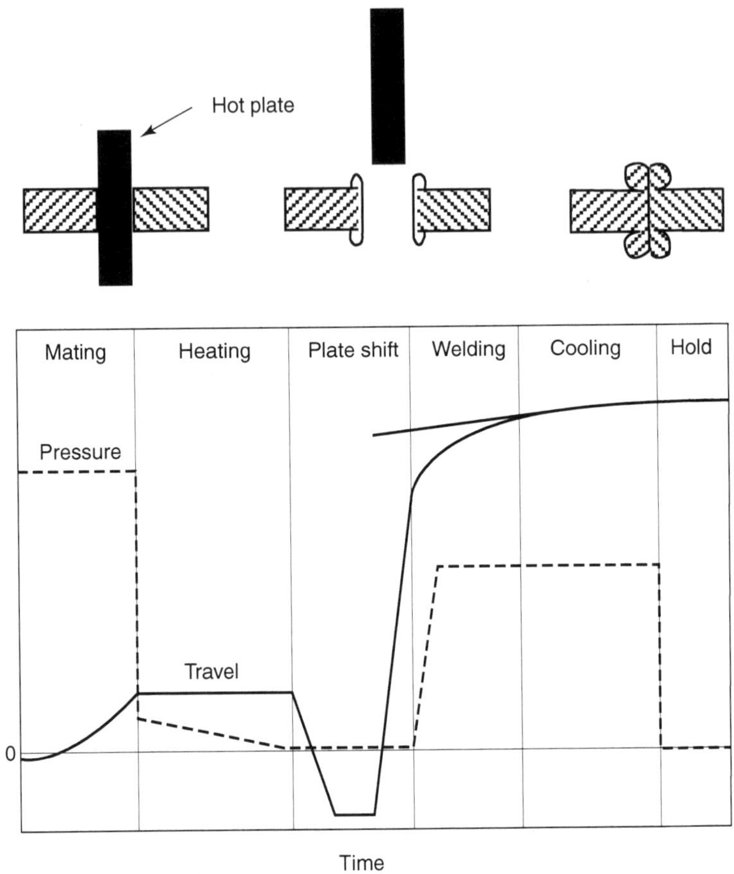

FIGURE 10.12 Hot plate welding process. (From Ref. 12.)

passage before joining the substrates, and the amount and uniformity of pressure during cooling. The conditions vary depending on the size of the joint area and the type of plastic being welded. Temperatures for heat welding a variety of plastics are shown in Table 10.17. Because heat welding requires long heating times and relatively thick cross sections of material, the residual stresses caused by shrinkage of the plastic after welding are relatively low. Weld strengths of up to 100% of the material strength can be met. Heated tool welding is suitable for almost any thermoplastic, but it is most often used for softer, semicrystalline thermoplastics. Common plastic substrates that are suitable for heated tool welding include polyethylene, polypropylene, polystyrene, ABS, polyvinyl chloride (PVC), and the acetals. Usually, heated tool welding is not suitable for nylon or other materials that have long molecular chains.

The method is often used in high-volume operations, where adhesive bonding is objectionably long. It is also often used to join low-surface-energy materials such as polypropylene where the cost and complexity of prebond surface treatment is

TABLE 10.17 Heat Welding Temperatures for Plastics and Polymeric Films

Material	Welding Temperature (°F)
Plastics	
ABS	450
Acetal	500
Polyethylene, low density	360
Polyethylene, high density	390
Polycarbonate	650
Polyphenylene oxide	650
Polypropylene	400
Polystyrene	420
Nylon 6,6	475
Polyvinyl chloride	450
Films	
Cellulose acetate	400–500
Polyethylene	250–375
Polyvinyl alcohol	300–400
Polyvinyl chloride, flexible	200–400
Polyvinyl chloride, rigid	260–400
Polyvinyldene chloride	285
Fluorinated ethylene propylene	600–750

Source: Refs. 13 and 14.

objectionable. Both large and small assemblies can be made easily . Hermetic seals are possible. Heated tool welding does not introduce foreign materials into the part, and as a result, plastic recycling is possible. The heated tool welding process can be completely manual or it can be semi- or fully automated for f ast high-volume production. For automated production, rotary machines are often used where there are independent stations for cleaning, heating, joining, cooling, and unloading.

10.6.1.2 Hot Gas Welding

An electrically or gas (air, argon, or nitrogen)-heated welding gun with an orifice temperature of 200 to 350°C can be used to bond man y thermoplastic materials. The parts to be bonded are be veled and positioned to form a V-shaped joint, as shown in Figure 10.13. A welding rod, made of the same materials as the plastic being joined, is laid in the joint, and the heat from the gun is directed at the rod, which fills the gap. A weld is formed as the rod and parts melt and fuse together.

The hot gas welding parameters include gas temperature, gas flow rate, and speed of transverse. Pressure is applied manually. These parameters are dependent on the plastic being welded. Re gulating the gas flo w or heating source controls the temperature of the hot gas torch. A strong fillet must be formed, the design of which is of considerable importance. Butt joints should be be veled to 60°.

A large difference between the plastic melting temperature and the decomposition temperature of the plastic is necessary for consistent reliable hot gas welding. Hot gas welding can be used to join most thermoplastics, including polypropylene, polyethylene, acrylonitrile–butadiene–styrene, polyvinyl chloride, thermoplastic polyurethane, high-density polyethylene, polyamide, polycarbonate, and acrylic. For polyolefins and other plastics that are easily oxidized, the heated gas must be inert (e.g., nitrogen or argon) since hot air will oxidize and de grade the surface of the plastic. Hot gas welding parameters for certain thermoplastics are shown in Table 10.18.

Hot gas welding is usually a manual operation where the quality of the joint corresponds to the skill and e xperience of the operator. Bond strength of at least 85% of the parent plastic is generally possible. Hot gas welding is a relati vely fast operation that can be used in a wide v ariety of welding, sealing, and repair applications. The most common applications are usually lar ge structural assemblies such as

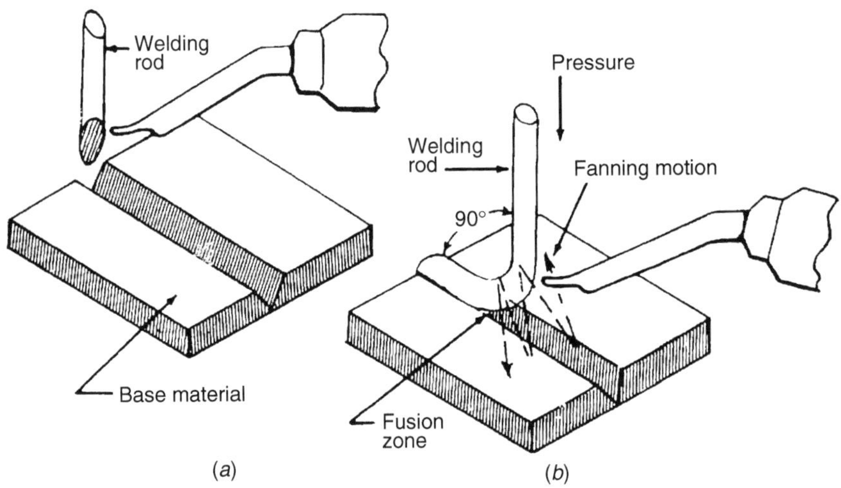

FIGURE 10.13 Hot gas welding process. (Courtesy of Laramy Products Co., Inc.)

TABLE 10.18 Hot Gas Welding Parameters for Selected Thermoplastics

Thermoplastic	Welding Temperature (°F)	Welding Gas
PVC	525	Air
Polyethylene, high density	550	Nitrogen
Polypropylene	575	Nitrogen
ABS	500	Nitrogen
Acrylic	575	Air

extruded sheet. Hot gas welding is generally not recommended for filled materials or for substrates that are less than $\frac{1}{16}$ in. thick.

10.6.1.3 Resistance Wire Welding

Resistance wire welding is a method of assembly that employs an electrical resistance heating wire laid between mating substrates to generate the heat of fusion as illustrated in Figure 10.14. Energizing the wire heats the adjacent polymer and causes a melt area to form. Slight pressure is applied so that the molten polymer fills the joint area completely. The wire is then deenergized, and the molten polymer cools to a structural adhesive. After the bond has been made, the exterior wire is cut off. Resistance wire welding can be used on any plastic that can be joined effectively by heated tool welding. Large resistance-welded parts can require considerable power requirements. Resistance welding has been applied to complex joints in automotive applications, including vehicle bumpers and panels, joints in plastic pipe, and joints in medical devices.

10.6.1.4 Induction Welding

Electromagnetic or induction heating is another method of indirect heat welding. The principle behind this welding method is that magnetic materials become hot when subjected to a high-frequency alternating-current field. In most applications, an electromagnetic induction field is used to heat a metal grid or insert placed between mating thermoplastic substrates. When the joint is positioned between induction coils as shown in Figure 10.15, the hot insert causes the plastic to melt and fuse together. Slight pressure is maintained as the induction field is turned off and the joint hardens.

Electromagnetic adhesives can also be used instead of metal inserts [15]. These are made from metal-filled thermoplastics and are shaped into gaskets or film that will melt in an induction field. The advantage of this method of induction welding is that stresses caused by large metal inserts are avoided. This process usually takes several seconds to minutes, depending on the nature of the substrate and the adhesive, but it

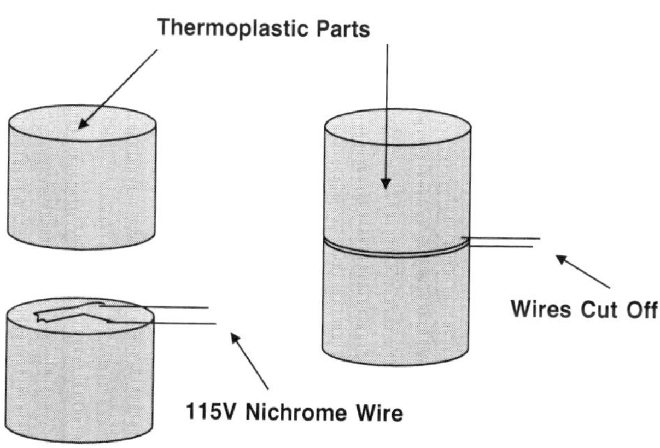

FIGURE 10.14 Resistance wire welding process.

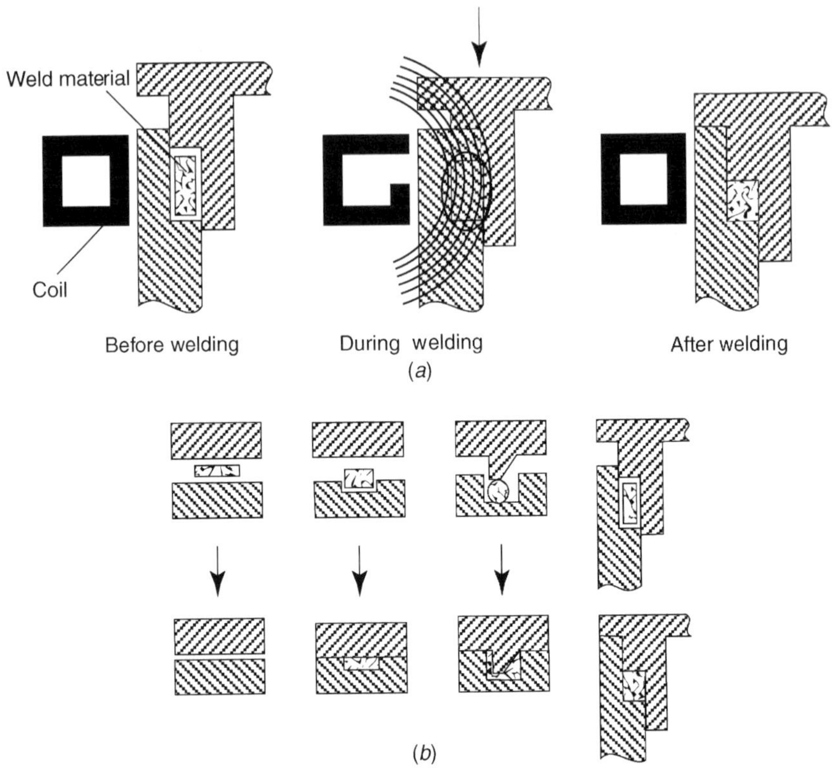

FIGURE 10.15 Induction welding process: (*a*) relationship of part, welding material, and electromagnetic coil; (*b*) common joint designs for induction welding. (From Ref. 12.)

can be done in as little as 1 second. During welding a constant pressure of at least 100 lb/in^2 is applied to the joint to minimize the development of bubbles. This pressure should be maintained until the joint sufficiently sets. When used, metal inserts should be 0.02 to 0.04 in. thick. Joints should be designed to enclose the metal insert completely. Inserts made of carbon steel require less power for heating, although other types of metal can be used. The insert should be located as close as possible to the electromagnetic generator coil and centered within the coil to assure uniform heating. The process is used to join both amorphous and crystalline thermoplastics. Electromagnetic welding is one of the fastest methods of adhering ABS and acetal thermoplastics. This type of welding has also been used on nylon and polycarbonate parts. However, the parts must be dried before electromagnetic welding.

Since the electromagnetic adhesive is simply a hot melt adhesive with magnetic filler, dissimilar plastics and even thermosets can be joined to each other as well as to nonpolymeric materials. Composites with up to 65% by weight of filler and thermoplastic elastomers and foams have also been joined with induction welding processes. Induction welding is often applied when structural assemblies require a hermetic seal.

It is also used when high bond strengths and very fast processing speeds are needed. This welding method can be used to produce a hidden joint and to compensate for irregular surfaces. Induction welding can also be used for joints requiring disassembly since the process is reversible. It is, however, unsuited for parts that have other metal inserts or electronic components that can be heated in an electromagnetic field.

10.6.1.5 Other External Heat Welding Methods

Other heat welding methods that are often used in specialty applications are extrusion welding, laser welding, infrared welding, and high-frequency or dielectric welding. These processes are less frequently used than those described above, but many have found important niche applications. Extrusion welding is similar to hot gas welding except that an extruder is used instead of a hot gas gun. The molten extrusion material is expelled continuously through a heated die and fills a groove in a preheated joint area. The process parameters include extrusion temperature and speed of transverse. A welding shoe follows the application and actually molds a seam into place. The extruded welding process produces a smooth, even weld bead.

Laser welding is a high-speed noncontact process for welding thermoplastics. This method involves directing a laser beam (Nd:YAG, CO_2, and CO) at plastics films or parts, which creates heat and melts the plastic to form a joint. Weld speeds of 500 m/min have been achieved on thin plastic films. Laser welding has been used primarily for small parts and delicate components made of polyethylene or polypropylene. Only recently developed, it is expected to find applications in the packaging and medical products industries.

Infrared welding is the application of infrared energy from a tungsten filament lamp or electrically heated ceramic material to the plastic parts to be joined. Removing the infrared source and clamping the parts together under pressure allows the molten surfaces to fuse and produce a weld. It is considered a noncontact alternative to hot plate welding. Infrared is particularly promising for higher-melting polymers, since the parts do not contact and stick to a heat source. Infrared radiation can penetrate into a polymer and create a melt zone quickly. By contrast, hot plate welding involves heating the polymer surface and relying on thermal conduction to create the required melt zone. Infrared welding is at least 30% fister than heated tool welding. Colorants and pigments can change the polymer infrared absorption properties and will affect the quality of the joint. Generally, the darker the polymer, the less infrared energy is transferred through a melt zone, and surface degradation will possibly occur, due to overheating.

High-frequency welding generates heat in the center of high-polarity polymer films, much as heat is generated in high-electrical loss polymers (i.e., polymers with a high dissipation factor or electrical loss factor) that are placed in a microwave oven. This process is especially effective for sealing highly polar films such as PVC. It cannot be applied to other plastics without the use of polar additives. Microwave welding involves placing a microwave-susceptible material such as a metal or a conducting polymer between two plastic substrates and then exposing the assembly to a microwave energy field of 2.45 GHz. Many plastics are transparent to microwave radiation, so that heat is generated only in the implant material, which in turn melts the parts to be joined.

10.6.2 Frictional Heat Welding Processes

Friction welding uses the heat of friction to cause fusion at the interface. One substrate is rotated or rubbed very rapidly while in touch with the other substrate so that the surface melts without damaging the part. The primary forms of frictional heat welding are ultrasonic, spin, and vibration welding. The major advantage of frictional heat welding methods is that the heating is concentrated at the interface area of the joint. In this way the bulk of the substrates are not heated to high temperatures. Because the heating is localized at the bonding surface, indirect heating processes are very energy efficient.

10.6.2.1 Ultrasonic Welding

During ultrasonic welding a high-frequency electrodynamic field is generated that resonates a metal horn. The horn vibrates one substrate sufficiently fast relative to a second fixed substrate to cause great heating at the interface. The frequency generally used in ultrasonic assembly is 20 to 40 kHz. With pressure and subsequent cooling, a strong bond can be obtained between many thermoplastics. Weld times are generally between 0.1 and 1 second.

The equipment that is used in ultrasonic welding is illustrated in Figure 10.16*a*. The largest possible product that can be welded is physically limited by the principle of ultrasonic welding since the welding horn must expand and shrink during the process. For amorphous materials, the maximum size is ±200 mm and for semicrystalline materials, it is ±70 mm. With ultrasonic welding, the joint design is critical. Typical joint designs are shown in Figure 10.16*b*. The most common design is a butt joint that uses an energy director. The energy director focuses the vibration energy at a localized spot at the interface. It is the material within the energy director that first melts and becomes the "adhesive" for the joint.

Amorphous polymers are easier than crystalline polymers to weld ultrasonically because the molecules are randomly arranged and vibrations pass through the material with little attenuation. Amorphous polymers are also less critical in joint design and in placement of the ultrasonic welding horn. However, excellent results can also be achieved with crystalline polymers if the following guidelines are followed:

- Use of high amplitudes on high-powered equipment is recommended.
- Horn contact should be close to and directly above the joint.
- Shear joints are recommended whenever practical.
- Use an oversized energy director when one is required.
- Rigidly clamp the part in the fixture to prevent loss of welding energy.

Rigid thermoplastics with a high modulus of elasticity are best. Excellent welds are generally obtainable with polystyrene, styrene–acrylonitrile (SAN), ABS, polycarbonate, and acrylics. PVC and cellulose tend to attenuate the energy and degrade the surfaces. Dissimilar plastics can be welded ultrasonically if they are chemically compatible and have similar melt temperatures.

Because heat is generated very locally, the shrinkage stresses are higher with ultrasonic welding than they are with other welding processes. In general, the techniques

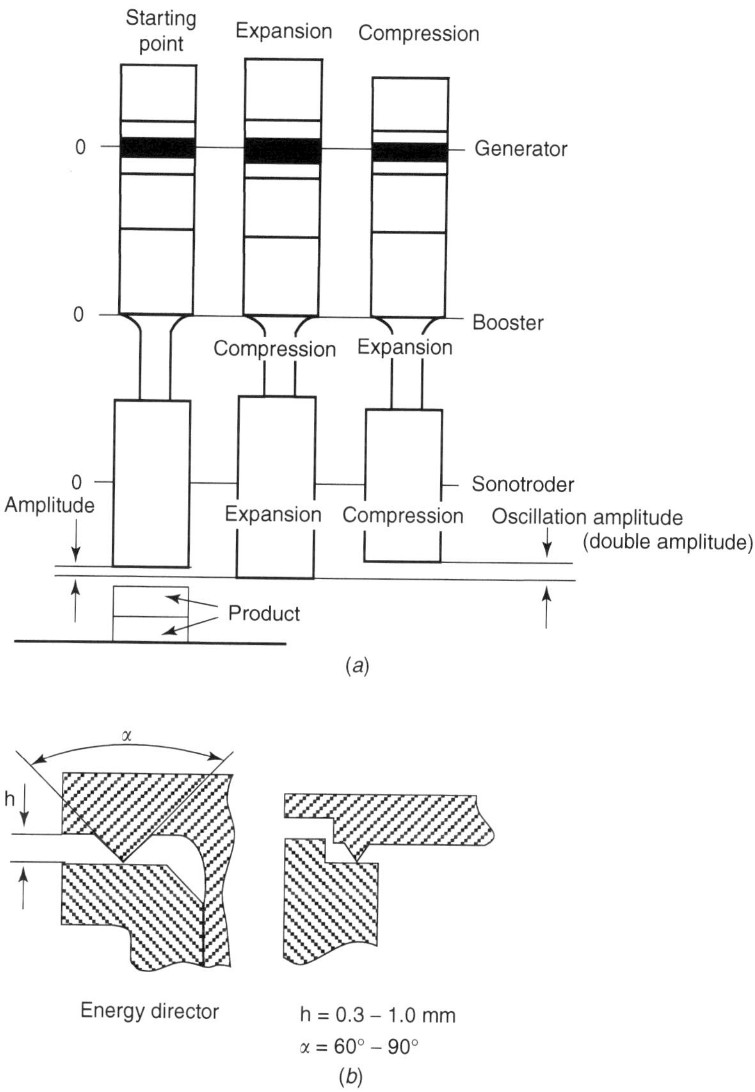

FIGURE 10.16 Ultrasonic welding process: (*a*) ultrasonic welding equipment; (*b*) common ultrasonic joint designs.

employing lower heat generation gi ve high weld strength. Ultrasonics can also be used to stake plastics mechanically to other substrates and for inserting metal parts. Ultrasonic welding is clean and fast. It usually results in a joint that is as strong as the parent plastic material. Because of the power requirements, it is usually applied more to smaller parts. Large parts are generally too massive to be joined with one continuous bond; however, spot welding can be accomplished with the same equipment.

10.6.2.2 Spin Welding

Spin welding is a fast and generally economical method to obtain joints with good strength. With spin welding, one substrate is rotated very rapidly while in touch with the other substrate so that the surfaces melt without damaging the part, as shown in Figure 10.17. Molten plastic forms the weld, and any displaced material forms the flash. Sufficient pressure is applied during the process to force out excess air bubble. The rotation is than stopped and pressure is maintained until the weld cools and sets. Actual welding times for most parts are only several seconds.

The rotation speed and pressure depend on the type of thermoplastic being joined. Practical welding parameters are in the range 1 to 20 m/s measured at the midradius of the part, 0.08 to 0.15 N/mm^2 for frictional pressure, and 0.1 to 0.3 N/mm^2 for holding pressure. The specific parameters will depend on the material being welded.Table 10.19 illustrates the basic parameters that apply for acetal joints that are spin welded.

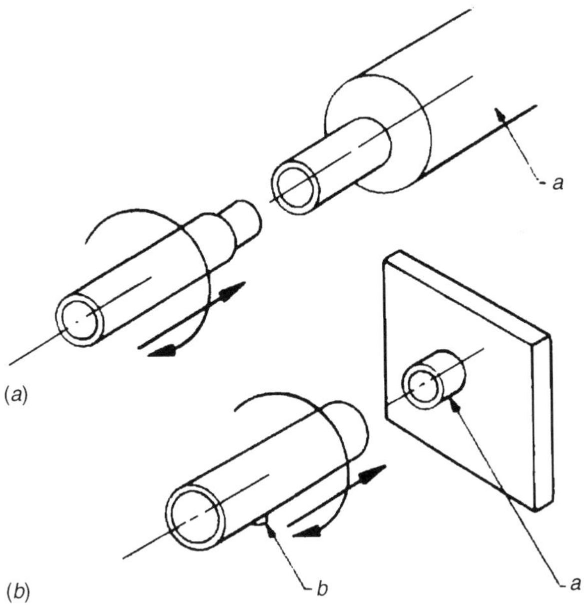

FIGURE 10.17 Spin welding process: (*a*) round to round (*b*) round to square. (From Ref. 5.)

TABLE 10.19 Spin Welding Parameters for Acetal Parts

Parameter	Delrin Acetal	Celcon Acetal
Surface speed range (ft/sec)	10–40	25–30
Spin welding pressure (lb/in^2)	~ 700	40–140
Spin time (sec)	0.25–0.50	0.50

The equipment necessary depends on the material being welded and the production requirements, but spin welding generally can be adapted to standard shop machinery such as drill presses or lathes. A production rate of 300 parts/min is possible on simple circular joints with an automated system containing multiple spin heads. The primary disadvantage of spin welding is that at least one substrate must be circular.

The most important factor in the quality of the weld is the joint design. Spin-welded joints can have straight 90° mating surfaces, or the surface can be angled, molded in a V shape, or flanged. A shallow groove is desirable in one of the parts in order to index the parts and provide a uniform bearing surface. Positioning of the components during spin welding is often considered in the weld design. In addition, a tongue-and-groove type of joint is useful in hiding the flash that is generated during the welding process.

The area of the spinning part should be as large as possible, but the difference in linear velocity between the maximum and minimum radii should be as small as possible. Since the heating depends on the relative surface velocity, the outside edges of circular components will see higher temperature by virtue of their greater diameters and surface velocity. This will cause a thermal differential that could result in internal stress in the joint. To alleviate this effect, joints with a hollow section and thin walls are preferred.

Nearly any melt-processable thermoplastic (even dissimilar polymers with compatible melt temperatures) can be spin-welded. Excellent spin-welded joints can be achieved with ABS, acetal, polycarbonate, and polyphenylene oxide. The plastic should be rigid enough to be able to bear the pressures that are applied during joining. It is also good that the plastic material have a relatively low thermal conductivity to prevent heat dissipation from the bonding surfaces.

Rotation-symmetric components can be welded up to a diameter of 250mm. Spin welding is often used in the assembly of tubes, bottles, and other containers. In addition, small elements can be welded to larger elements. Joints may also be made by rapidly rotating a filler rod around the joint area. A rod consisting of the parent material is rotated at 5000 rpm and moved along the joint as it melts. This type of plastic weld resembles a hot gas weld. It was first developed and patented at TWI in 1991 and is sometimes referred to as *friction stir welding.*

10.6.2.3 Vibration Welding

Vibration welding is similar to spin welding in that it uses heat generated at the surface of two parts rubbing together to produce melting in the interficial area of the joint. However, vibration welding uses linear or axial motion rather than rotational motion. The parts are rubbed together at a frequency of about 120 to 240 Hz. Frictional heat is generated in the joint area, which melts the plastic. Once a molten plastic interface is achieved, the motion is stopped and the joint is then allowed to cool under pressure.

Although similar to ultrasonic welding, vibration welding uses lower frequencies (100 to 240 Hz). Thus, large parts can be bonded because of less reliance on the output of the power supply. Other vibration weld processing conditions are 1 to 5 mm for amplitude, 1 to 5 seconds of weld time, and 0.5 to 2 N/mm^2 welding pressure. The basic joint design used for vibration welding is a butt joint, but unless parts have thick walls, a heavy flange may be required to provide rigidity and adequate welding surfaces. A common solution is a joint design consisting of a flange with a rib on it. Typical joint design for vibration welds are shown in Figure 10.18.

There are two types of vibration welding: linear and axial (Figure 10.18). In both types one of the substrates is clamped so that it cannot move and the other substrate is oscillated using the welding equipment. Linear welding is most commonly used. Axial or orbital vibration welding allows irregularly shaped plastic parts to be vibration welded.

Vibration welding produces a joint strength of about 85% that of the material being joined. Unlike ultrasonic welding, it imposes few restrictions on materials or part size. It is generally preferred when larger parts must be bonded. Vibration welding is often used for joining and sealing of two-part plastic tanks, canisters, and pipe sections. This welding process is also capable of producing hermetic seals.

Vibration welding is ideally suited to parts made from engineering thermoplastics as well as acetal, nylon, polyethylene, and acrylic resin. Almost any thermoplastic can be vibration welded. Unlike ultrasonic welding, with vibration welding it does not seem to matter whether the material is amorphous or crystalline. Vibration welding can also be applied to filled, reinforced, or pigmented thermoplastic materials.

10.7 SOLVENT WELDING

Solvents can also be used to weld many thermoplastics. Solvents function by separating the polymer chains of the substrate and causing their bond surface to soften and flow under pressure. Once the solvent is allowed to evaporate, the plastic hardens and a joint is formed. Solvent welding or *solvent cementing*, as it is sometimes called, is the simplest and most economical method of joining many thermoplastics. Solvent-cemented joints are less sensitive to thermal cycling than joints bonded with adhesive, and they are as resistant to degrading environments as the parent plastic. Bond strength equivalent to 85 to 100% of the parent plastic can generally be obtained by solvent cementing. The major disadvantage of solvent cementing is the possibility of stress

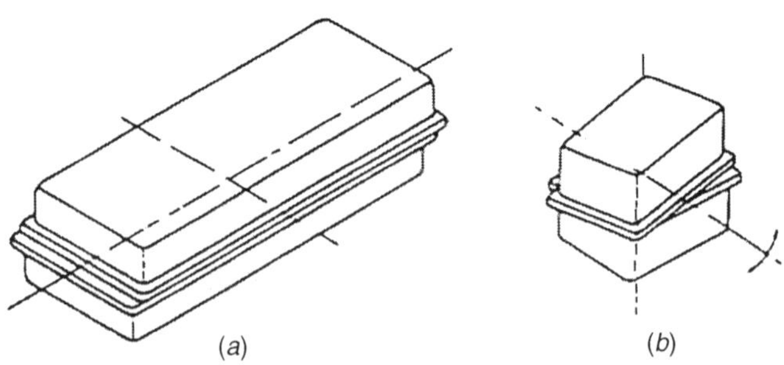

FIGURE 10.18 Vibration welding process. Linear vibration (*a*) is employed where the length-to-width ratio precludes the use of axial welding (*b*) where the axial shift is still within the width of the welded edge.

cracking or crazing. Also, when two dissimilar plastics are to be joined, adhesive bonding is generally desirable because of solvent and polymer compatibility problems.

Solvent cements should be chosen with the same solubility parameter as the plastic being bonded. It is common to use a mixture of fast-drying solvent with a less volatile solvent to prevent crazing and to establish longer contact time of the solvent on the plastic. Table 10.20 provides typical solvents for solvent cementing of common plastics. The solvent cement can be bodied to about 35% by weight with the parent plastic to provide a high-viscosity cement that can fill gaps and reduce shrinkage and internal stress during cure.

When cementing with solvents, the parts to be bonded should be unstressed and, if necessary, annealed. The solvent cement is generally applied to the substrate with a syringe or brush. In some cases the surface can be immersed in the solvent. After the area to be bonded softens, the parts are mated and held under pressure until dry. Pressure should be low and uniform so that the finished joint will not be stressed. After the joint hardens, the pressure is released, and an elevated temperature cure

TABLE 10.20 Typical Solvents for Cementing Thermoplastics

	Solvents																		
Plastic	Acetone	Cyclohexanone	*N,N*-Dimethyl formamide	Dioxane	Ethyl acetate	Ethylene dichloride	Formic acid	Glacial acetic acid	Methyl ethyl ketone	Methyl cellosolve	Methylene chloride	*N*-Methyl pyrrolidone	*O*-Dichlorobenzol	Perchlorethylene	Tetrahydrofuran	Toluene	80 Toluol–20 Ethanol	Trichlorethylene	Xylene
ABS						×			×		×			×					
Acrylic						×		×			×								
Cellulose acetate	×				×				×	×									
Cellulose acetate butyrate	×	×			×	×			×	×							×		
Ethyl cellulose					×	×													
Nylon							×												
Polyaryl ether	×				×														
Polyaryl sulfone			×									×	×						
Polycarbonate					×			×		×									
Polystyrene		×		×	×			×		×	×	×	×	×		×			×
Polysulfone		×				×			×		×		×	×		×			×
Polyvinyl chloride	×	×								×			×						
PPO base (Ndryl)						×					×			×					
Styrene–acrylonitrile	×							×				×	×						
Vinylidene chloride		×	×								×		×						
Vinyl chloride–vinyl acetate	×	×						×		×			×						

Source: Ref. 16

may be necessary, depending on the plastic and desired joint strength. The bonded part should not be packaged or stressed until the solvent has adequate time to escape from the joint. Most solvents are both toxic and flammable. Processing areas must be properly vented and storage areas secured accordingly.

Solvent bonding is suitable for all amorphous plastics. It is used primarily on ABS, acrylics, cellulosics, polycarbonates, polystyrene, polyphenylene oxide, and vinyls. Solvent welding is generally not suitable for semicrystalline thermoplastics, although certain types can be solvent welded at elevated temperatures. It is not affective on polyolefins, fluorocarbons, or other solvent-resistant polymers. Solvent cementing is only moderately effective on nylon and acetal polymers. Thermosets cannot be solvent welded. Plastics that are dissimilar can be solvent welded provided that they are soluble in the same solvent. If there is no common solvent for two resins, solutions can be created composed of two different miscible solvents.

10.8 MECHANICAL FASTENERS

A large variety of mechanical fasteners can be used for joining plastic parts to themselves and to other materials. Generally, fasteners that are used for assembly of metal and other parts can also be used with plastics. However, there are certain fasteners that are designed specifically for plastics. These specialty fasteners include screws, rivets, and spring fasteners and clips.

As with other methods of assembly, mechanical fastening requires that special considerations be given to the nature of the plastic parts that are being joined. Creep of the plastic part due to overloading could also result in loss of preload in poorly designed systems. High-strength metal fasteners can overstress plastic parts, so torque-controlled tightening or special design provisions are often required. Where torque cannot be controlled, various types of washers can be used to spread the compression force over a larger area.

Advantages and disadvantages of mechanical fastening as an assembly process for plastics are summarized in Table 10.21. The main advantage of mechanical fasteners is that they can provide for repeated disassembly and reassembly of the product. In addition to providing high joint strength, mechanically fastened plastic joints should prevent slip, separation, vibration, misalignment, and wear of the parts. The main disadvantage of mechanical fasteners is that they are often used with disregard to the properties of the plastic. This could result in premature failure of the assembly. Besides the fundamental strength of the plastics, important properties that must be considered include creep, relaxation, craze resistance, notch sensitivity, and stiffness. Another disadvantage of mechanical fasteners are that they are relatively heavy and bulky and may not provide a positive aesthetic appearance on many plastic products.

Mechanical fasteners are generally made from metals; however, plastic fasteners are also available. Plastic fasteners are lightweight, corrosion resistant, and impact resistant. They do not freeze on threads of screws and require no lubrication. Metal fasteners provide higher strength and are not affected by exposure to extreme temperatures or chemical environments.

TABLE 10.21 Advantages and Disadvantages of Mechanical Fastening as an Assembly Process for Plastics

Advantages	Disadvantages
Disassembly can be achieved repeatedly	Concentration of stress and crazing
Low technology requirement	Loosening of fasteners due to environmental service
Capable of being used in low- or high-volume applications	Notch sensitivity
Commonly available components	Differentials in coefficients of thermal expansion can cause stress or loosening on thermal cycling
Ability to join thermosets and non-polymeric materials	Cost
Ability to join different materials	Not recyclable generally without disassembly
High, consistent strength	Bulky; not aesthetically pleasing
	Possible need to access both sides of the part

Whatever mechanical fastener is used, particular attention should be paid to the head of the fasteners. Common head styles for screws, bolts, and rivets are shown in Figure 10.19. Conical heads called flat heads produce undesirable tensile stress and should not be used with plastics. Bolt or screw heads with a flat underside, such as pan heads and round heads, are preferred because the stress produced is more compressive. Flat washers are also suggested for the same reason and should be used under both the nut and the fastener head. It is also important that whatever the mechanical fastener, that sufficient diametrical clearance for the body of the fastener be maintained. Without sufficient clearance, expansion of the plastic part relative to the fastener could cause objectionable stress levels around the fastener and part cracking. This clearance can nominally be 0.25 mm (0.010 in.).

10.8.1 Screws

Screws that are inserted directly into a plastic part represent the type of mechanical fastener that is most commonly used. Figure 10.20 shows the pitch diameter and other important dimensions that characterize a specific screw. Screws used in plastic assembly processes can be either self-tapping or machine screws.

Self-tapping or *self-threading screws* consist of thread cutting or thread forming types. They are generally forced into holes that are either molded or drilled into the plastic part. Thread forming and thread cutting screws are generally used for assembly when the screws are not intended to be removed for disassembly of the part. *Machine screws*, on the other hand, can be attached to an insert that is molded or embedded into the part. Standard machine screws commonly used in metal assembly are generally used for plastic parts. However, washers and nonconical heads are commonly used to distribute stress across a larger area. Parts fastened with machine screws can generally be disassembled repeatedly.

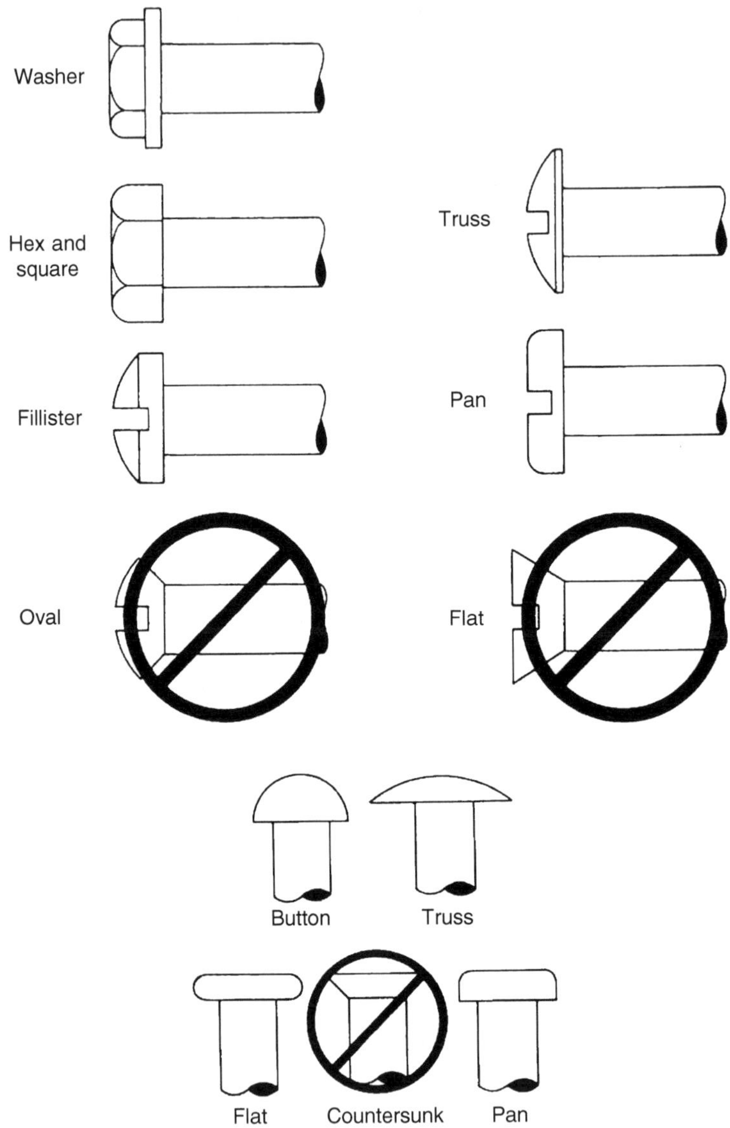

FIGURE 10.19 Common head styles for scre ws, bolts, and rivets used to join plastics. (From Ref. 17.)

10.8.1.1 Thread-Forming Screws

Thread-forming screws displace and compress material as they are forced through a hole that is usually predrilled into the plastic part. The compression and displacement of the plastic combined with its inherent memory will ensure that a tight f it

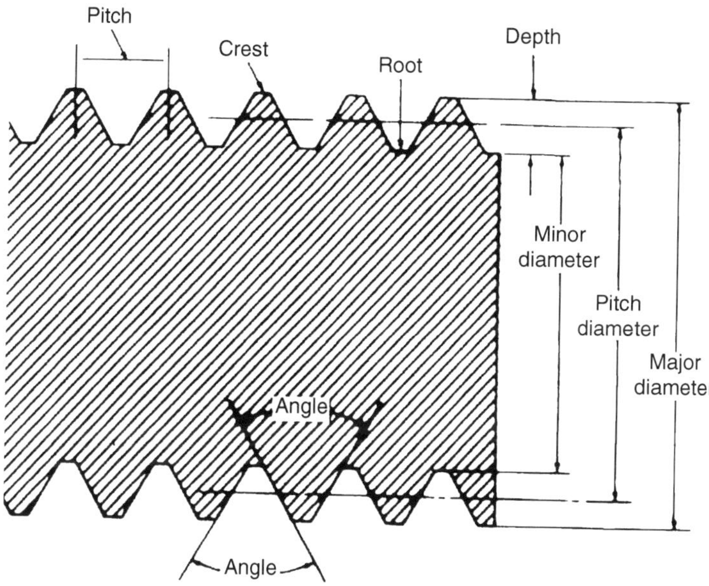

FIGURE 10.20 Terminology for screw threads.

occurs. Thread-forming screws are used in softer, more ductile plastics with moduli below 200,000 lb/in^2. They are usually employed only on thermoplastic parts. With thermosets the material will not flow around the screw threads, and this could result in cracking around the hole.

The assembly strength of thread-forming screws can be increased by reducing hole diameter in the more ductile plastics, by increasing screw thread engagement, or by going to a larger-diameter screw when space permits. Since thread-forming screws induce high stress levels in the part, they are not recommended for parts made of weak resin. The most common problem encountered with thread-forming screws is boss cracking. This can generally be minimized by increasing the size of the boss, increasing the diameter of the hole, decreasing the size of the screw, or changing the thread configuration of the screw.

Several standard thread-forming screws that are commonly used on thermoplastics are shown in Figure 10.21.

- Type B is used for general assembly of thermoplastics; it is a blunt-pointed fastdriving screw with tapered threads.
- Type BP is also used for general assembly of thermoplastics, but it is easier to insert than type B because of the 45° pointed end, which speeds screw alignment during assembly.
- Type U is a drive screw used where disassembly is not intended; it is a blunt-pointed screw intended for permanent assembly.

- Type L is a special combination thread-forming and thread-cutting screw used primarily for nylon.

Types B, BP, and L can be removed and reinserted up to five times. Type U should never be removed.

10.8.1.2 Thread-Cutting Screws

Thread-cutting screws remove material as they are installed, thereby avoiding high stress. Thus, these types of screws are often used on harder, less ductile plastics. Thread-cutting screws are commonly used on thermosetting materials. The six most common types of thread-cutting screws are illustrated in Figure 10.22. They differ mainly in the torque required to drive the screw.

Thread-cutting screws can cause problems if they are installed and removed repeatedly because new threads can be cut each time they are installed. For thread-cutting screws the hole should be tapered to match the screw taper. Holes for thread-cutting screws should always be longer than the screw to allow space for the cutting chips. Screw bosses must be designed carefully. The wall thickness of a mounting boss should be at least equal to the screw diameter. Small boss diameters reduce the tendency for sinks in the molded part: however, the boss may not be sufficient to withstand assembly hoop stress.

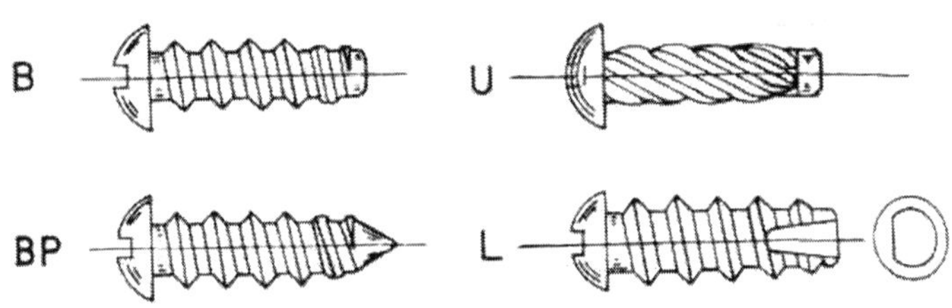

FIGURE 10.21 Thread-forming and thread-cutting screws most often used in thermoplastic materials.

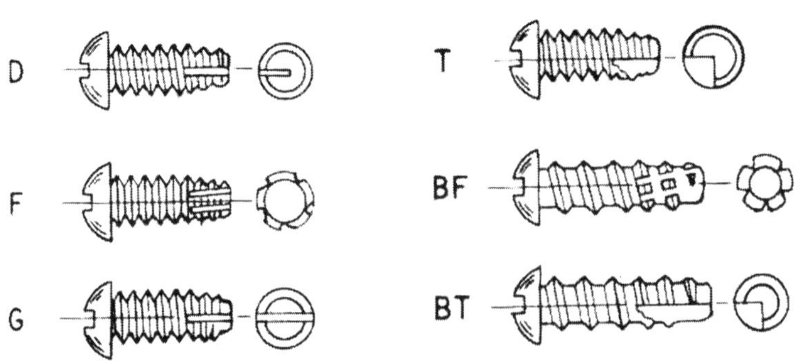

FIGURE 10.22 Thread-cutting screws most often used in thermosetting plastic materials.

10.8.1.3 Machine Screws

Parts molded of thermoplastic resin are sometimes assembled with machine screws or with bolts, nuts, and washers. These fastening methods are used especially if the plastic is very strong, such as in the case of engineering plastics or composites. Standard nuts, bolts, and machine screws made from both metals and plastics are used in common assembly methods. Coarse threads can be molded into most plastic materials. Threads of 32 or finer pitch should be avoided, along with tapered threads, because of excessive stress on the part. If the mating connector is metal, overtorquing will result in part failure. Threaded inserts can also be driven into or molded into plastic parts to allow for assembly with machine screws.

10.8.2 Inserts

Metal inserts are generally used when the plastic parts in an assembly must be disassembled repeatedly. Inserts that are used for plastic assembly consist of molded-in and postmolded inserts. In both cases the inserts are locked into the plastic part so that it is not removable. The inserts also can carry loads above the limits of threaded plastic. They can also be used as electrical contacts and other secondary functions to assembly.

Molded-in inserts are placed in the mold (generally by hand but also with automated equipment) before the plastic resin is injected. The insert is locked into the body of the plastic as the shaped plastic resin flows and hardens. Molded-in inserts provide very high strength and relatively low unit cost. However, part cycle time is generally affected adversely because of the time that it takes to place the insert into the mold. Thermal expansion differences between the insert and the plastic can also cause problems with molded-in inserts. Inserts must be preheated to avoid thermal shock and to help reduce the stresses that could occur because of differences in thermal expansion. Also, before inserts are placed in the mold they must be free of all foreign matter. Inserts with sharp points can produce poor impact values because of a notch effect.

Postmolded inserts are installed in a plastic part after it is molded. It is locked into the plastic part either by expansion of the insert, by threading, or by flow of the plastic material around the insert using several forms of heating. Generally, the outside of the insert is provided with projections of various configurations that penetrate the plastic and prevent movement under normal forces of assembly and disassembly. Postmolded inserts come in four general types: press-in, expansion, self-tapping, and thread forming.

An expansion insert design is illustrated in Figure 10.23. Expansion inserts are placed in preformed or drilled holes and expand against the inside walls of the holes when a screw is installed. Because of the allowable interference between the expanded insert and the hole, the practical use of these inserts is limited. Postmolded inserts can also be installed in thermoplastic parts quickly and effectively by means of ultrasonic, induction, or conduction heating. In these processes the thermoplastic resin is melted and the insert is forced into the hole, with molten plastic encapsulating the insert. Since the insert is anchored in place, it cannot be removed.

10.8.3 Rivets

Rivets are metal or plastic pins manufactured with a head on one side. They are inserted into a preformed hole through the materials being joined, and the straight end is flared to lock the rivet into place. Design components of a rivet assembly are shown in Figure 10.24. Rivets are inserted manually or by using automatic bench riveting

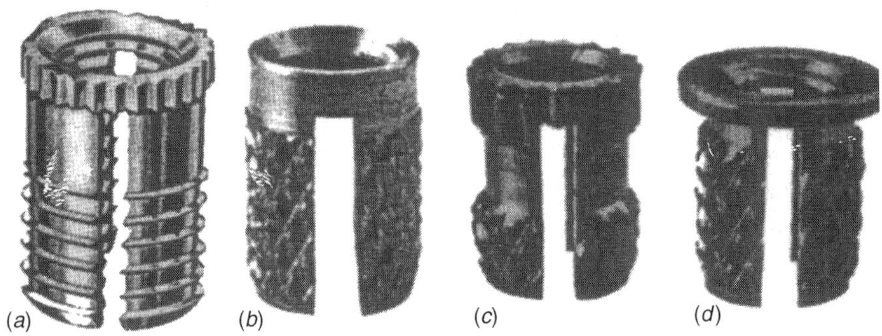

FIGURE 10.23 Metal expansion inserts: (*a*) vane design; (*b*) diamond knurl design; (*c*) straight-knurled design; (*d*) flange top. (From Ref. 18.)

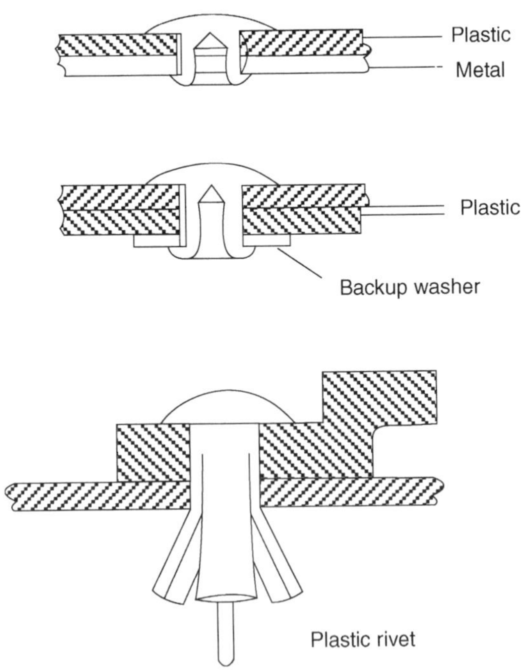

FIGURE 10.24 Riveted plastic assembly. (From Ref. 12.)

equipment. Rivets provide a permanent method of assembly for most plastics at low cost. The assembly cost, including the cost of the rivets, labor, and production time, are lower than for threaded fasteners. Large-volume high-speed assembly operations are easily achieved. Disadvantages of rivet assembly include lower tensile and fatigue strength compared to other methods of mechanical fastening, and rivets generally are not made with the precision of other mechanical fasteners.

Rivets can be used to fasten plastics to plastics and plastics to metals. The types of rivets used include solid shanks, semisolid shanks, and tubular rivets. When riveting plastics to metal, the head of the rivet is placed next to the plastic surface to avoid damage by the clinch end. Blind rivets made of metal or plastic are available for installation on only one side of the assembly. Clamp loads must be limited to low levels to prevent distortion of the part. To distribute the load, rivets with large heads should be used with washers under the flared end of the rivet. The heads should be about three times the shank diameter. Rivets should be used at least three times the shank diameter from the edges of the assembly.

Riveted joints are often used for high-strength plastics and composite parts, especially when they must be fastened to adjoining metal parts. Riveted composite joints should be designed to avoid loading the rivet in tension. Generally, a hole $\frac{1}{64}$ in. larger than the rivet shank is suggested for composite joints.

10.8.4 Staking

Staking is an assembly operation similar to rivets. However, in staking, a head is formed on a plastic stud by cold flow or melting of the plastic. The stud protrudes through a hole in the parts being joined, and forming of a mushroomed stud locks the two parts together mechanically, as shown in Figure 10.25. Staking can only be used to join thermoplastics. Staking can be produced by either cold or hot methods.

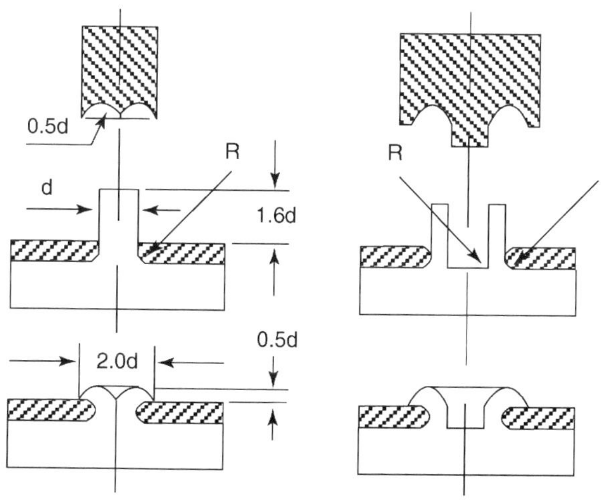

FIGURE 10.25 Staking process for plastic assembly. (From Ref. 12.)

In cold staking, high pressures are used to induce cold flow of the plastic material. Pressures of at least 6000 lb/in^2 are generally required. Only more malleable thermoplastic are suitable for cold staking. Soft or brittle materials are generally difficult to cold stake. Heat staking can be achieved by using a heated mold shaped so that the head can be formed. Ultrasonic horns have also been used for heat staking of certain plastics (see Section 10.6.2.1).

10.8.5 Springs and Clamps

Push-on spring steel fasteners can be used for holding light loads,as shown in Figure 10.26. Spring fasteners are used when low-cost rapid mechanical assembly is required. These are also often referred to as *speed nuts* and *speed clips*. Spring steel fasteners are simply pushed over a molded stud. The stud should have a minimum 0.38 mm (0.015 in.) radius at its base. Too large a radius could create a thick section, resulting in sinks or voids in the plastic molding.

10.9 DESIGN FOR SELF-ASSEMBLY

Self-assembly or *friction fitting* are terms commonly used to describe a number of pressure-tight joints of permanent or temporary assembly. These generally use an

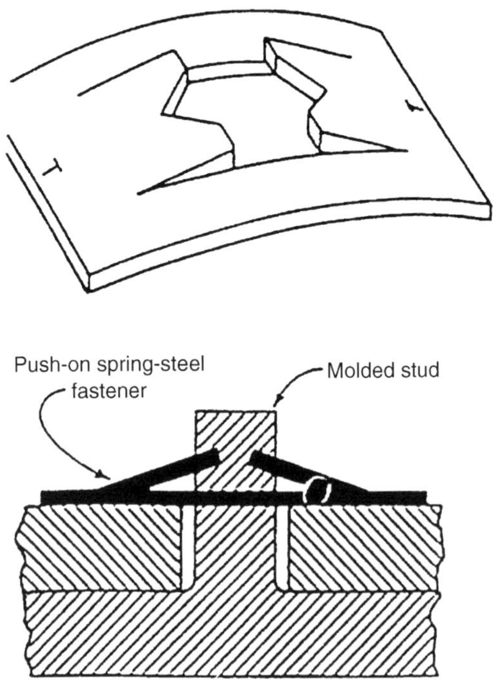

FIGURE 10.26 Typical spring fastener. (From Ref. 17.)

interference fit between the parts being assembled to lock the parts in place. The most common types of self-assembly are press fit, snap fit, and shrink fit. These techniques may join either like or unlike materials without the use of mechanical fasteners or adhesives. Whether these materials can be used will depend heavily on the nature of the plastic materials and the freedom that one has in part design.

Press fitting is a simple, fast assembly method that provides joints with strength and low cost. In press fitting, a shaft of one material is joined with the hub of another material by dimensional interference between the shaft's outside diameter and the hub's inside diameter. This process can be used to insert plastic or metallic parts into other plastic components. When two different materials are being assembled using press fitting, the harder material should be forced into the softer. For example, a metal shaft can be press-fitted into plastic hubs. Press-fitting joints can be made by simple application of force or by heating or cooling one part relative to the other part. Parts may be joined while the plastics are warm. When a shaft is press-fitted into a bearing or sleeve, the outside as well as the inside diameter may be expanded.

Press fitting produces very high stresses in the plastic parts. With brittle plastics such as thermosets, press-fit assembly may cause the plastic to crack if the conditions of assembly are not carefully controlled. For press-fitted joints, the effect of thermal cycling, stress relaxation, and environmental conditioning must be evaluated carefully. Testing of factory-assembled parts is obviously indicated under the expected service conditions, especially those that will cause changes to the dimensions or modulus of the plastic parts.

The following points need to be considered when examining the feasibility of using press-fit joints

- All parts must be clean and free of any foreign substance.
- Press fits of unlike materials should be avoided if they will be subjected to thermal cycling.
- Press fits should be avoided if the assembly is to be subjected to a harsh service environment.

Snap fitting is the assembly of parts by simply forcing a protruding part of one component to catch in a depression of the mating component (i.e., the parts are "snapped" into place). The protruding part can be a hook, stud, or bead. This method of assembly is uniquely suited to thermoplastic materials, due to their flexibility, high elongation, and ability to be molded into complex shapes.

Snap fits are an economical, rapid, and popular assembly method. Snap fits can be used to join dissimilar plastics or plastics to metals. Snap-fit joints are generally used with stiffer plastics such as nylon or acetal. They are not recommended for softer plastics such as polyethylene, polypropylene, fluorocarbons, and flexible polyvinyl chloride. Snap-fit joints are usually employed to attach lids or covers which are meant to be disassembled or which will be lightly loaded. Some examples of snap fitting are the simple locks or catches on plastic boxes or covers on many parts, such as pharmaceutical pill containers.

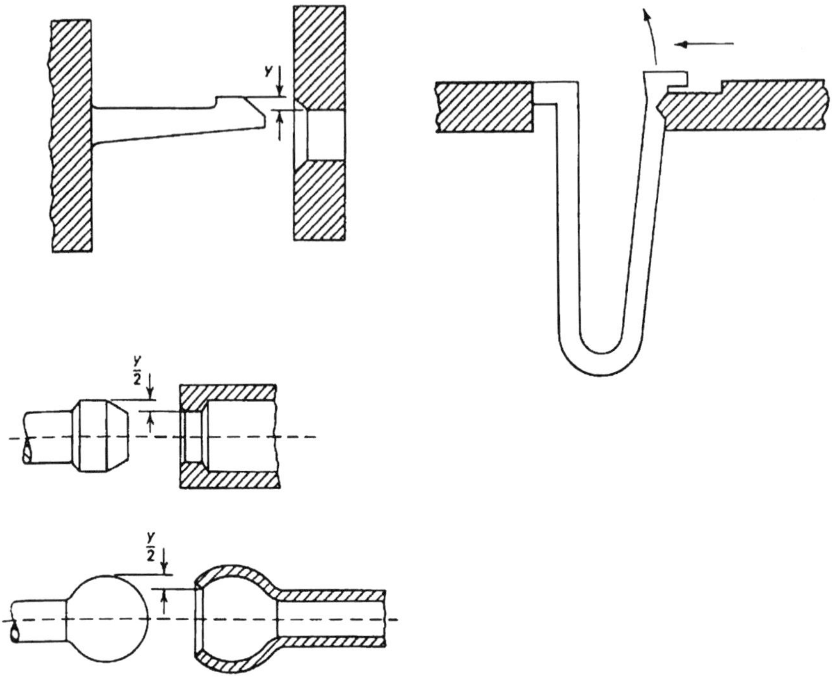

FIGURE 10.27 Various types of snap-fit joints. (From Ref. 17.)

The design should be such that after the assembly is made, the plastic parts will return to a stress-free condition. Snap-fit joints cannot carry loads greater than the force necessary to make or break the snap fit. Several types of snap fit joints are illustrated in Figure 10.27. Some design considerations for snap-fit joints include the forces required for assembly and disassembly, ease of molding, the material strain produced during assembly, and other requirements of the application. Based on a geometric model for the particular type of joint, stress analysis is generally performed to determine the characteristics of the joint.

Shrink fitting generally refers to the process of placing inserts in the plastic just after molding and allowing the plastics to cool. The difference in thermal expansion coefficients will cause the plastic material to contract around the insert and lock it in place. However, certain plastics (e.g., those that are stretched and irradiated) are known as shrink-fit plastics. These are generally available in tubular form and can be used to joint other tubes, electrical wire, and so on. The shrink-fit tube will contract when heated and contract around the part being joined.

REFERENCES

1. J. Chottiner, Dimensional stability of thermosetting plastics, *Materials Engineering,* February 1962.
2. A. V. Pocius et al., The use of adhesive in joining plastics, in *Treatise on Adhesion and Adhesives,* J. D. Minford, ed., Marcel Dekker, New York, 1991.
3. E. M. Petrie, *Handbook of Adhesives and Sealants,* McGraw-Hill, New York, 2000.
4. I. Skeist, ed., *Handbook of Adhesives,* 3rd ed., Van Nostrand Reinhold, New York, 1990.
5. J. Rotheiser, *Joining of Plastics,* Hanser Publishers, Munich, 1999.
6. *Handbook of Plastics Joining: A Practical Guide,* Plastics Design Library, Norwich, NY, 1997.
7. D. M. Brewis, Pretreatment of polymers, in *Handbook of Adhesion,* D. E. Backham, ed., Longman Scientific and Technical, Herlow, Essex, England, 1992.
8. J. G. Dillard, Adhesives and sealants, in *Engineering Materials Handbook,* Vol. 3, H. F. Brinson, ed., ASM International, Materials Park, OH, 1990.
9. R. L. Bersin, How to obtain strong adhesive bonds via plasma treatment, *Adhesive Age,* 1972, p. 37.
10. *Using Low Temperature Plasmas for Surface Treatment of Polymers,* Product Bulletin 2402, International Plasma Corporation, Hayward, CA.
11. E. M. Petrie, Trends in plastic surface modification to improve adhesion, SpecialChem4Adhesives.com, October 20, 2004.
12. *Assembly Guide,* GE Plastics, General Electric Company, released August 1998.
13. D. F. Gentle, Bonding systems for plastics, in *Aspects of Adhesion,* Vol. 5, D. J. Almer, ed., University of London Press, London, 1969.
14. H. F. Mark et al., *Encyclopedia of Polymer Science and Technology,* Vol. 1, Wiley, New York, 1964, p. 536.
15. EmaBond Corporation, Norwood, NJ.
16. S. S. Schwartz and S. H. Goodman, Fastening and joining techniques, Chapter 18 in *Plastics Materials and Processes*, Van Nostrand Reinhold, New York, 1982.
17. Engineering plastics, in *Engineered Materials Handbook,* Vol. 2, ASM International, Metals Park, OH, 1998.
18. *Handbook of Plastics Joining,* Plastics Design Library, Norwich, NY, 1997.

CHAPTER 11

Decorating and Finishing

EDWARD M. PETRIE
EMP Solutions, Cary, North Carolina

JOHN L. HULL
Consultant, Ivyland, Pennsylvania

11.1 INTRODUCTION

In this chapter we provide practical information and guidance on the processes commonly used to decorate plastics. Many of these processes are frequently used with metals, wood, and other nonplastic parts. However, because the part is polymeric, certain unique finishing opportunities are available to the design engineer. There are also many critical considerations that must be taken into account regarding the plastic substrate material.

Most plastic products experience some kind of decorating process, whether it is decorative coatings, attachment of a pricing label, electroplating, surface embossing, or surface printing of a trademark. Decorating processes for plastic parts are unfortunately not as well understood as the primary manufacturing processes such as injection molding. This can lead to expensive scrap and excessive production costs.

An understanding of the selection and correct application of decorating processes is extremely important because (1) the decorating process can affect the salability of the final product significantly, (2) the decorating process can be a large component of the finished product cost, and (3) misapplied decorating can result in complete loss of the finished product, including total part manufacturing costs as well as decorating costs. With plastic parts the decorating processes can be accomplished either during molding of the part, directly after molding, or any time before final assembly and packaging. The most inexpensive decorating process is to incorporate the decorative design into the mold or to apply the decoration as part of the molding operation.

Handbook of Plastic Processes Edited by Charles A. Harper

However, often this is not possible, and decoration must be accomplished as a secondary finishing process.

The most widely used decorating processes that are used in the plastic industry are coating, printing, and metallization. There are also a host of other decorating processes, such as hot stamping, application of labels and decals, and so on, that can be employed, depending on the type of plastic involved and the desired function of the finished part. In this chapter we discuss the most commonly used decorative processes used for plastic parts. Both in-mold and postmolding (secondary) processes are discussed. Decorating processes that are integral to the plastic material, such as pigmenting a resin for color, are not covered in this chapter, but the reader can find this information in another section of the book.

As with adhesive bonding (see Chapter 10), surface preparation and cleanliness are of primary importance with many decorating processes. All traces of surface contamination must be removed so that the decorative medium can adhere properly. With certain plastic substrates more aggressive chemical or physical surface modification may be necessary. Because of the importance of surface preparation, in the first section of this chapter we describe important pretreatment processes that need to be considered prior to all decorating processes.

11.2 PLASTIC PREPARATION PROCESSES PRIOR TO DECORATING

Prior to decorating plastic parts, the surface of the part generally must be cleaned or altered in some manner to improve the adhesion of the decorative application and for other purposes as well (aesthetics, reduction of stress crazing, specific surface texture, etc.). In this section we describe common plastic preparation processes that are used prior to decorating plastic parts. For a more detailed review of pretreatment processes specifically aimed at improving the bond strength between adhesives and plastics, the reader is referred to Chapter 10.

As a general rule, all substrates must be treated in some manner prior to processes such as decorating or adhesive bonding, if not to remove or prevent the formation of weak boundary layers, then to provide a consistent and repeatable surface. Surface preparation can range from simple solvent wiping or detergent cleaning to a combination of mechanical abrading, chemical cleaning, and acid etching. Sometimes, simple cleaning is the only surface preparation step required, depending on the type and condition of the substrate. When structural adhesive bonding is required or when significant stress is expected on the decorating medium, cleaning often is a process that comes before and then again after more aggressive processes that physically or chemically modify the surface.

The degree to which the plastic part must be prepared prior to decorating is related primarily to the type of plastic and to the service environment to which the part will be exposed. In many applications extensive surface preparation may be unnecessary. However, where permanence and reliability in aggressive environments are required, carefully controlled surface treating processes may be necessary. These processes could be expensive, require prolonged production time, and provide safety and environmental concerns. Thus, one should be careful not to o verspecify the surface treatment required. Only the minimal process necessary to accomplish the functional objectives of the application is required.

There are many processes that can be employed to clean or modify the surface of the part prior to secondary operations such as decorating. Surface treatments, in general, can be classified as follows:

- *Passive chemical treatments:* generally used when the requirements are minimal and service conditions are mild
- *Mechanical abrasion:* generally used when a structural bond is required to an adhesive or coating and service conditions are moderate
- *Active chemical or physical treatments:* generally used for maximum quality and when service conditions are relatively harsh
- *Surface primers:* generally employed when an intermediate surface is required between the part and the decorative medium; also employed when a prepared surface must be protected for a time prior to decoration

The various ways in which common treatments affect plastic surfaces to improve decorating characteristics are summarized in Table 11.1.

TABLE 11.1 Effect of Common Pretreatments on the Surface of Polymers

Pretreatment Type	Possible Effects of Pretreatment on the Surface of the Plastic Part
Solvent	Removal of organic contaminants and additives that have migrated to the surface. Roughening (e.g., trichloroethylene vapor with polypropylene). Weakening of surface regions if excessive attack by the solvent.
Detergent	Removal of nonorganic contaminants and particulate material on the surface.
Mechanical abrasion	Surface roughening to increase effective bond area and provide mechanical interlocking.
Oxidative (flame, chemical, etc.)	Introduction of functional groups. Change in topography (e.g., roughening of surface with chromic acid treatment of polyolefins).
Plasma	Removal of contaminants and cross-linking (if inert gas is used). Introduction of functional groups if active gases such as oxygen are used.Grafting of monomers to polymer surface after activation.
Surface primers	Provides an intermediate material between the plastic part and the decorating medium. Primers bond well to the plastic part and provide a surface for coatings/adhesives to bond. Often used to protect the surface of the part between molding and decorating.

Source: Ref. 1.

11.2.1 Passive Chemical Surface Treatments

Passive *chemical* cleaning processes merely remove obvious surface contamination by chemical means without altering the parent material. Contaminants removed by passive cleaning include dirt, oil, mold release, moisture, grease, fingerprints, and other foreign substances on the surface. Solvent washing, alkaline and detergent cleaning, and ultrasonic cleaning are typical examples of passive chemical cleaning processes.

The main purpose of cleaning is to remove loosely held contaminant from the substrate surface to either prepare the substrate for the attachment of a decorative medium or to prepare the substrate for further surface treatment. Sometimes, cleaning processes are used merely for aesthetic value, to prepare a product for packaging or shipment. Surface preparation chemicals must be safe to handle and should not be flammable or toxic. They should be inexpensive and provide fast processing times. The processes should be easy to monitor and control in a production situation. In addition, the process should not, in itself, leave contaminants or a weak boundary layer. If chemical solutions are used, they should rinse off easily.

All cleaning methods are improved by additional agitation that could take the form of scrubbing, aggressive stirring, or ultrasonic agitation. Table 11.2 provides an indication of the wide variation in surface cleanliness available as a function of cleaning medium and level of agitation. As with any process utilizing cleaning solutions, sudden or gradual contamination of the surface and/or the cleaning chemicals is always a possibility and must be considered in a quality control plan.

Most cleaning methods require the use of solvents or chemicals; thus, safety and consideration of the environment are of prime importance. Toxicity, flammability, materials incompatibility, and hazardous equipment are all important safety factors that must be considered in choosing the proper cleaning or surface treating system. Environmental factors that must be considered are volatile emissions and waste handling, storage, and disposal.

There are significant reforms occurring with cleaning and surface preparation processes, due to safety and health issues and the enforcement of new environmental regulations. Modern equipment and processes, such as solvent recovery systems, have been developed to eliminate or reduce harmful emissions, and new "safety solvents" have been developed to replace dangerous, environmentally unfriendly cleaning solvents and chemicals.

TABLE 11.2 Cleaning Efficiency as a Function of the Degreasing Process

Degreasing Method	Cleaning Efficiency (%)
Pressure washing with detergent solution	14
Mechanical agitation in solvent	30
Vapor degreasing in solvent	35
Wire brushing in detergent solution	92
Ultrasonic agitation in detergent solution	100

Source: Ref. 2.

11.2.1.1 Solvent Cleaning

Solvent cleaning is the process of removing soil and organic contaminants from a substrate surface with an organic solvent. Where loosely held dirt, grease, and oil are the only contaminants, simple solvent wiping alone with a brush or clean cloth will provide clean surfaces. Clean lint-free cloth or paper tissue is commonly used, and steps must be taken to assure that the cleaning materials do not become contaminated or a source of contamination. There are also more effective means of solvent cleaning than merely wiping, and these are described below.

Solvent cleaning is widely used and should precede (and generally follow) any chemical or abrasive surface preparation. The initial solvent cleaning is to remove any organics from the surface so that they will not be driven deeper into the part because of the mechanical abrasion operation. The final solvent cleaning step is to remove the residue from the mechanical abrasion. However, solvent cleaning, by itself, is the least effective substrate treatment in that it only cleans the surface of loosely held particulates and organic contamination.

The compatibility of cleaning solvents with plastic substrates is extremely important. Solvents can physically and chemically affect polymeric surfaces. The incorrect choice of solvent can provide unacceptable part appearance or even degrade mechanical properties of the part. Solvents that are recommended for the cleaning of plastics are shown in Table 11.3.

Since mold release agents are the most common form of contaminant on plastics parts, the suppliers of mold release agents are often the best source for information on solvents that will remove these contaminants. Certain mold release agents can also be

TABLE 11.3 Degreasing Solvents Recommended for Plastic Substrates

Plastic	Solvent
Acetal (copolymer)	Ketone
Acetal (homopolymer)	Ketone
Acrylonitrile–butadiene–styrene	Ketone
Cellulose, cellulose acetate, cellulose acetate butyrate, cellulose nitrate	Alcohol
Fluorocarbons	Chlorinated alcohol or ketone
Polyamide (nylon)	Ketone
Polycarbonate	Alcohol
Polyolefins	Ketone
Polyethylene terephthalate (Mylar)	Ketone
Polyimide	Ketone
Polymethyl methacrylate, methacrylate butadiene	Ketone or alcohol
Polyphenylene oxide	Alcohol
Polyphenylene sulfide	Ketone, chlorinated solvents
Polystyrene	Alcohol
Polyvinyl chloride, polyvinyl fluoride	Ketone, chlorinated solvents
Thermoplastic polyester	Ketone
Thermoset plastics	Ketone

Source: Ref. 3.

removed with detergent solutions. Silicone-based mold release agents provide several difficulties. Silicone has the tendency to travel easily in factory airstreams. Often, accidental contamination of parts occur due to the use of silicone mold release in operations that are nearby the decorating process. Silicone also provides a very low energy surface that will cause poor adhesion of coatings, adhesives, and decorative applications. In many plastic part manufacturing operations, silicone mold release agents have been replaced with fluorocarbon or another organic mold release.

A solvent should be selected that does not affect the plastic surface but is sufficiently strong to remove organic contamination. Certain solvents can attack plastic surfaces by causing softening and flow of the polymer or stress crazing. Stress crazing is the attack of solvent at highly stressed regions within a molded plastic part. *Stress crazing* by solvents or aggressive monomers is especially noticeable on acrylic or polycarbonate parts. When highly stressed parts are to be placed in contact with solvent, an annealing process is generally used first to reduce the level of internal stress. The annealing process requires heating of the part up to a temperature approximately 10°C below its heat distortion temperature.

Common solvents used to clean plastic surfaces are acetone, toluene, trichloroethylene, methyl ethyl ketone (MEK), low-boiling petroleum ether, and isopropanol. Volatile solvents are acceptable; however, the local and most recent safety and environmental restrictions must be consulted before selecting any solvent. Certain solvents are considered toxic and hazardous, and exposure limits as well as safety procedures must be maintained.

The solvent industry has made significant strides in developing newer grades and blends of solvents for a variety of applications that are either biodegradable and/or environmentally compatible. New low-volatile solvents are taking the place of the older, less environmentally safe solvents. There is an excellent Web site, SAGE (http://clean.rti.org), that provides a comprehensive guide to provide pollution prevention information on solvent and cleaning process alternatives. SAGE was developed by the U.S. Environmental Protection Agency Air Pollution Prevention and Control Division.

Several solvent cleaning processes are in common use: (1) solvent wiping, (2) immersion, (3) spraying, and (4) vapor degreasing. The choice is generally determined by the type of the plastic, the part size, and the number of parts that must be treated in a specified time. The various processes also differ by their potential for cross-contamination. Mechanical or ultrasonic scrubbing may also be used to facilitate these cleaning processes. After cleaning, the parts should be air-dried in a clean, dry environment before moving on to the decorating process.

The *solvent wiping process* is perhaps the most commonly used cleaning process for plastic parts. The parts are not exposed to the solvent for a long period of time, so the degradation of properties does not usually occur if the proper solvents are chosen. However, the cleanliness of the surface is difficult to measure, and special precautions are necessary to prevent the solvent from becoming contaminated. For example, the wiping cloth should never touch the solvent container, and new wiping cloths must be used often. A clean cloth should be saturated with the solvent and wiped across the area to be bonded until no signs of residue are evident on the cloth or substrate. With *solvent wiping*, the cleanliness of the surfaces tends to be dependent on the training and attention given by the operator. Automated spray or immersion processes are less dependent on the operator's skill.

The *solvent immersion method* is more suited for high-production volumes, and it is often sufficient to remove light contamination and soil. In this method the part is immersed in a container of solvent and mildly agitated by tumbling, solvent mixing, brushing, or wiping. After being soaked and scrubbed, the parts must be rinsed by a clean-flowing liquid or spray. A number of different solvents may be used in this process. It is important to note that the parts will be no cleaner than the final rinse solvent.

The *multiple-bath method* of solvent immersion (Figure 11.1) is the most common immersion method. The first tank is the wash tank, in which scrubbing may be performed. The second and third tanks are rinse tanks. With this method, one must prevent contamination of the cleaner solvents by changing the scrub and rinse solvents continually.

The *spray method* of solvent cleaning is very efficient, due to the scrubbing effect produced by the impingement of high-speed particles on the surfaces being cleaned. The spray causes flow and drainage on the surface of the substrate, which washes away loosened contamination.

Vapor degreasing is a form of solvent cleaning that is attractive when many parts must be prepared. This method is also more reproducible than solvent wiping. It will remove soluble soils and contaminants from a variety of metallic and nonmetallic parts. Vapor degreasing consists of suspending the adherends in a container of hot chlorinated solvent such as trichloroethylene (boiling point 122°C) for about 30 seconds. When the hot vapors come into contact with the relatively cool substrate, solvent condensation occurs, which dissolves the organic contaminants from the surface.

Vapor degreasing is preferred to solvent wiping because the surfaces are being washed continuously in distilled uncontaminated solvent. It is recommended that the vapor degreaser be cleaned and a fresh supply of solvent used when the contaminants dissolved in the old solvent lower the boiling point significantly. Extremely soiled parts may not be suitable for vapor degreasing unless initially cleaned by other methods. Vapor degreasing is often used on metal parts. The combination of solvent

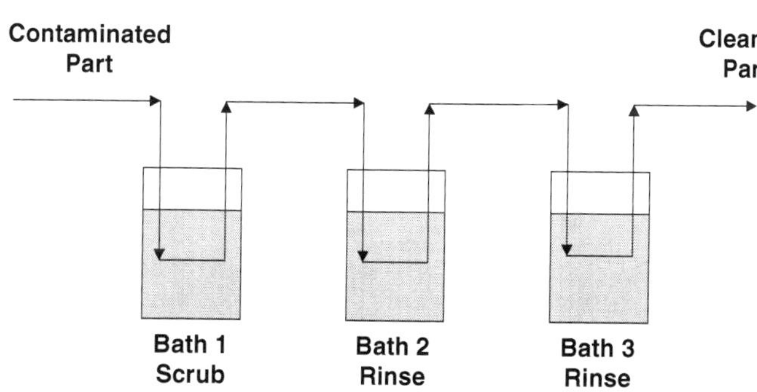

FIGURE 11.1 Three-bath method of solvent immersion cleaning.

contact and high temperatures, however, can degrade certain types of plastics. Thermoset, thermoplastic, and elastomeric materials should not be treated by the vapor-degreasing processes without first testing them in the process.

11.2.1.2 Chemical Cleaning

Strong detergent solutions are used to emulsify surface contaminants on both metallic and nonmetallic substrates. These methods are popular on polymeric surfaces, where solvent cleaning may degrade the part or on parts where the contamination is more easily removed by an aqueous cleaner (e.g., salt films, dust). Chemical cleaning is generally used in combination with other surface treatments. Chemical cleaning by itself will not remove heavy or strongly attached contaminants such as rust or scale. All of the chemical cleaning processes should be preceded and followed by solvent cleaning (when possible) or water rinsing.

Detergents, soaps, and other cleaning chemicals are the least expensive and easiest cleaning agents to handle. Two basic environmentally friendly cleaning chemistries are available: aqueous and nonchlorinated solvents. Aqueous cleaners are generally manufactured in concentrated forms and are diluted with water for use. They come in three major classes: acidic, neutral, and alkaline. The alkaline cleaners are generally used for cleaning metal surfaces prior to bonding. The following types of aqueous cleaners are available [4]. All of these products work well on plastic parts. Special concerns relative to metal parts are noted.

- *Caustics* (sodium or potassium hydroxide). Caustics work well on steel, removing scale, smut, light rust, and heavy oils, but cannot be used on aluminum. They are also dangerous if not handled carefully.
- *Silicates* (sodium metasilicate). Silicated cleaners work well on aluminum, but often are too aggressive on brass or copper.
- *Amines* (triethanolamine, monoethanolamine). Amines are good on all metals, but pose odor concerns.
- *Phosphates* (trisodium phosphate, tetrapotassium pyrophosphate). Phosphates are very safe for the user (9.5 to 10.5 pH), but often are not aggressive enough for heavy cleaning unless they are combined with large amounts of wetting agents and/or solvents.
- *Acids* (gluconic, phosphoric, hydrofluoric, citric, etc.). Acids are very effective on metals that may require oxide removal, but they often require strict safety precautions.
- *Chelates* (EDTA). In small doses, chelates can help extend the life of cleaning solutions. However, they often prompt wastewater concerns because of the dissolved metals in the spent waste.
- *Solvent additives*. In water-based formulations, when used in moderation (up to 15%), solvent additives can help many cleaning tasks become easier.

Many commercial detergents are capable of cleaning substrates prior to bonding. Generally, 1 to 5 oz of liquid detergent per gallon of tap water is used. These are often formulated with surfactants, pH buffers, inhibitors, emulsifiers, deflocculants, sequestering agents, chelating agents, and antifoaming agents.

A formulation and processing method for a mild alkaline cleaning process and a heavy-duty alkaline cleaning process are shown in Table 11.4. These processes can be used effectively on plastic as well as metal parts. The parts are immersed in the hot, agitated solution for 8 to 12 minutes. As dirt and other contaminants collect in the bath, more alkalinity must be added to restore the pH factor to a suitable level. The parts are finally rinsed with tap water followed by distilled water.

11.2.1.3 Other Cleaning Methods

A *wet abrasive* method of cleaning consists of scouring the surface of the plastic part with detergent. This process consists of scrubbing the substrate thoroughly with a clean cloth or nonmetallic bristle brush and a detergent solution maintained at 45 to 60°C. It is a common surface cleaning method used for many polymeric substrates. In cases where the substrate is so delicate that the usual abrasive treatments may be too rough, contaminants can be removed by *vapor honing*. This method is similar to grit blasting except that very fine abrasive particles are suspended in a high-velocity water or steam spray. Sometimes solvents are also used as the liquid medium in vapor-honing operations. Thorough rinsing after vapor honing is usually not required.

Ultrasonic cleaning employs a bath of cleaning liquid or solvent ultrasonically activated by a high-frequency transducer. The most often used frequency for ultrasonic cleaning is 20,000 to 50,000 Hz. The ultrasonic scrubbing action effectively removes contamination from small parts. The part to be cleaned is immersed in the liquid, which carries the sonic waves to the surface of the part. High-frequency vibrations then dislodge the contaminants. Ultrasonic cleaning is ideal for lightly soiled work with intricate shapes, surfaces, and cavities that may not be easily cleaned by spray or immersion techniques. Commercial ultrasonic cleaning units are available from a number of manufacturers.

Supercritical fluids, which result from subjecting substances to temperatures and pressures above their critical points, possess properties intermediate between liquid and gases. In this state the supercritical fluid can rapidly penetrate substrates and small spaces, dissolve the contaminants, and then be removed easily and completely

TABLE 11.4 Mild and Heavy-Duty Alkaline Cleaning Processes

	Mild Alkaline Cleaning Method	Heavy-Duty Alkaline Cleaning Method
Cleaning solution	Tetrasodium pyrophosphate, 15 pbw Sodium metasilicate, 80 pbw Nacconol 40F (a surfactant), 5 pbw	Sodium metasilicate, 46 pbw Trisodium phosphate, 23 pbw Sodium hydroxide, 23 pbw Nacconol 40F, 8 pbw
Mixing	Mix 6–8 oz of ingredients above into 1 gallon of water.	
Immersion temperature	70–85°C	
Immersion time	8–12min	

since the "fluid" lacks surface tension. This process is very good for cleaning complex parts. Carbon dioxide is the most commonly used supercritical fluid in cleaning applications. This process is relatively new and expensive.

11.2.2 Mechanical Abrasion

Abrasive treatments consist of scouring, machining, hand sanding, and dry and wet abrasive blasting. The abrasive medium can be fine sandpaper, carborundum or alumina abrasives, metal wools, or abrasive shot. The choice is generally determined by available production facilities and cost. Wet abrasive processes use a slurry of abrasive medium in water. Soft plastic surfaces can also be scoured with a mixture of tap water and scouring powder.

Abrasion is usually preferred for pretreating most plastic parts prior to application of a coating or adhesive. It is a cleaning process that removes surface contaminants, especially if it is preceded and followed by a solvent cleaning step. The abrasion processes improve adhesion through several mechanisms.

- Abrasion removes the surface layer, which may be incompatible with the decorating process, adhesive, or coating.
- Abrasion provides mechanical interlocking so that the decorative medium will adhere well to the substrate surface.
- Abrasion increases the area over which a coating or adhesive is in contact with the surface and thereby improves adhesion.

In addition to a mechanical effect, abrasion can also cause chemical effects at the surface. Chemical bonds are broken when the surface is abraded, and the free radicals that are created on the surface can react with the decorative medium or with a subsequent treating process. The topology of a plastic surface will also affect the bond strength of an adhesive or coating on the plastic. When a supposedly smooth solid surface is examined closely under a microscope, it is found to contain irregularities. It is not flat and smooth but contains many surface asperities, such as peaks and valleys, with a certain degree of roughness. The nature of the surface is most important with adhesive bonding and coating processes and less important for the other processes.

In most surface preparation processes, mechanical abrasion will provide improved adhesion. This operation provides mechanical interlocking and increased area over which the forces of adhesion can develop. For metals, ceramics, and other high-energy polymeric surfaces (Table 11.5), roughness provides greater bonding area. The greater effective surface area offers a larger area for the forces of adhesion to operate, thereby providing a stronger bond. As a result, abrasion and other surface roughening processes are often recommended as prebond treatments for high-energy plastic surfaces.

However, if the substrate has a lower surface energy than the decorating medium, adhesive, or coating, such as the case with many common plastics, surface roughening could actually inhibit the development of high bond strength. The combination of surface roughness and poor wetting could contribute to air pockets being trapped in the irregularities on the surface (Figure 11.2). This causes stress concentrations in the interface, which reduces its ultimate strength. When the voids are in the same plane along the joint interface, they can seriously weaken the bond strength.

As a result, prebond surface preparation processes for low-surface-energy materials generally involve changing the chemistry (to modify surface energy and wetting characteristics) rather than the physical topology of the surface. A complete explanation of surface energy and surface properties on adhesion can be found in Chapter 10 and in the sections below regarding the adhesion of coatings.

11.2.3 Conventional Surface Modification Processes

Both chemical and physical methods are employed to alter the surface of the plastic part to enhance adhesion of the decorative medium, adhesive, or coating. Generally, surface pretreating processes are employed only when the surface tension of the substrate is relatively low. This occurs in the case of low-surface-energy substrates such as polyolefins and fluorocarbons (see Table 11.5.) Many of these low-energy plastics are finding their way into applications such as automotive interiors, furniture, and sporting goods.

TABLE 11.5 High- and Low-Surface-Energy Plastics

High-Surface-Energy Plastics	Low-Surface-Energy Plastics
Acrylic	Fluorocarbons
Most thermosetting plastics (epoxy, phenolic, etc.)	Silicone
Polyphenylene sulfide	Polypropylene
Polyphenylene oxide	Polyethylene
Polystyrene	Thermoplastic polyolefin elastomers
Polycarbonate	
ABS	
Polyvinyl chloride	

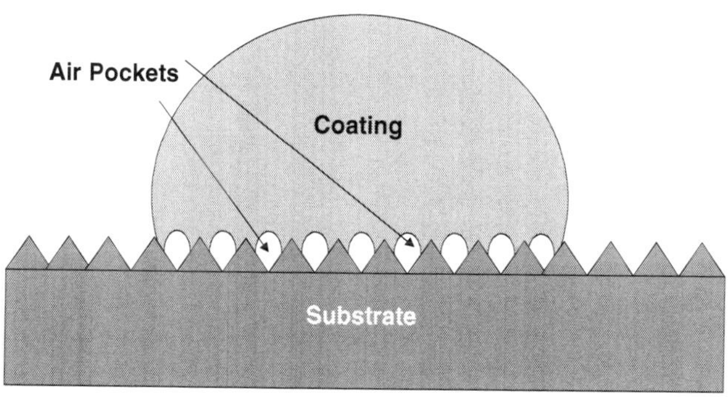

FIGURE 11.2 Surface roughness can cause voids to form at the interface if the decorating medium, adhesive, or coating does not effectively wet the surface of the plastic part.

Common surface treatments that are used prior to decorating low-surface-energy plastics are:

- Abrasion or surface roughening
- Primer
- Chemical etching
- Flame treatment
- Corona discharge
- Gas plasma

These processes either chemically or physically modify the surface of the plastic without affecting the bulk characteristics. The processes generally result in a higher energy surface that is more amenable to adhesion. For higher-surface-energy plastics, such as epoxies, acrylonitrile–butadiene–styrene (ABS), polycarbonate, polyphenylene sulfide, and so on, the only surface preparation required is generally cleaning with a solvent or detergent solution. These cleaning processes do not alter the chemical nature of the plastic, but they do remove unwanted contaminants such as oil, dirt, dust, and grease. The necessity for cleaning depends on how clean the parts are before the start of the process, how they are handled subsequently, and how critical the coating adhesion is to part performance.

Surface roughening by high-velocity-particle or other abrasive media will cause chemical bonds on the surface to break. Free radicals can be produced that can react with the surrounding media. If abraded in air, the surface may add oxygen, nitrogen, or water to form polar groups, which provide a higher surface energy and are more easily wetted by coatings.

Primers can also be coated onto plastic substrates to improve adhesion. Primers result in the deposition of a layer between the plastic and the decorative medium that does not itself modify the chemistry of the substrate. The primer will adhere well to the plastics and provides a surface conducive to adhesion. Common primers used for plastic substrates are:

- Polyethylene imide for polyolefin surfaces
- Resorcinol–formaldehyde resin for nylon, polyester, and rubber surfaces
- Polychlorinated polyolefins for polypropylene, thermoplastic polyolefin, EPDM, and so on.

Additional information regarding primers can be found in the following section.

Flame treatment consists of passing the part through a hot oxidizing flame of 1100 to 2800°C. The momentary exposure does not cause distortion of the plastic but oxidizes the surface, making it more receptive to decorating materials. Flame treating processes are generally used prior to coating or printing on plastics. In the flame treating process, a flame is moved across the part surface at an optimum distance of 1 to 5 cm. The flame tip should not touch the part or streaking could occur. Propane, butane, or natural gas can be used. For natural gas, the air-to-gas ratio can be critical (approximately a 1.7:1 ratio should be emplo yed), and in sophisticated flame treating equipment this is monitored. A typical flame-treating process for plastic bottles is shown in Figure 11.3.

Corona treatment is a process in which the surf ace of the plastic is oxidized by electron discharge (corona). The discharge is created by a frequency generator and a step-up transformer; the v oltage is generally in the range 15 to 30 kV. It is passed through an electrode system onto the polymer surf ace, as shown in Figure 11.4. The electrode is at a distance 1 to 2 mm from the surface, and the treatment speed is generally 1 to 2 m/s. The corona treatment process is v ery effective on continuous

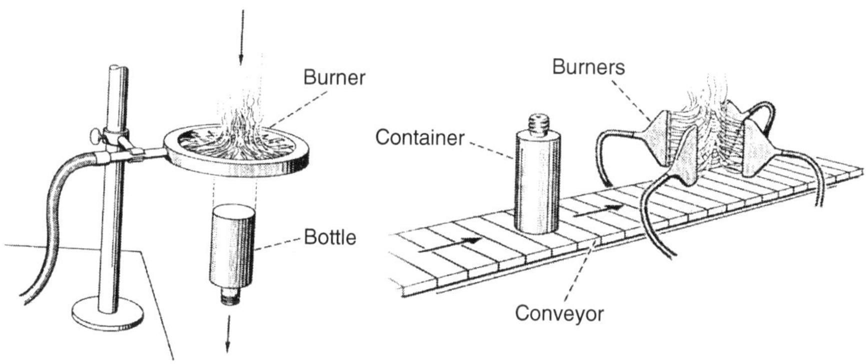

FIGURE 11.3 Flame treatment process for surface preparation of round plastic bottles prior to painting or printing. (From Ref. 5.)

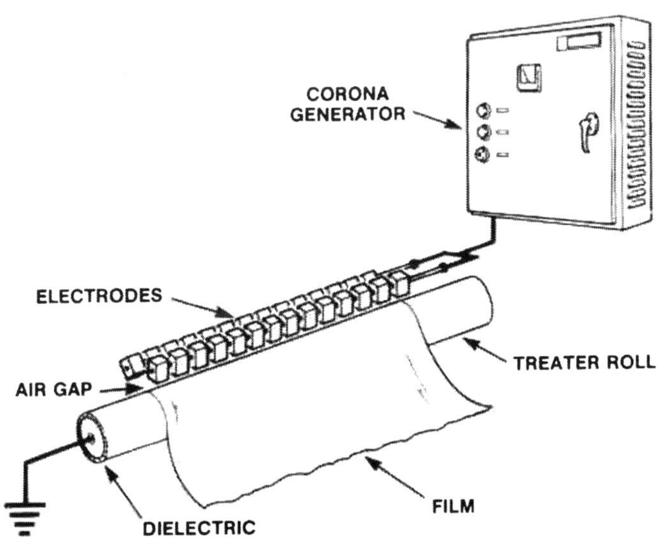

FIGURE 11.4 Corona treatment process. (From Ref. 6.)

films and thin webs. It enhances the adhesion of both low-energy (e.g., polyolefins, fluorocarbons, silicone) and high-energy (polyesters, polyvinyl chloride, nylon, polycarbonate) plastics. The plastic surface is activated through oxidation with ozone formation as a by-product.

Chemical treatment processes generally consist of submerging the surface of the part (the entire part can be immersed if resistant to the chemical solution) in an acid bath. The bath etches the surface, making it receptive to decorating. The solutions generally consist of oxidizing acids (chromic, sulfuric, nitric, and formic) and acid mixtures, depending on the type of plastic to be treated. Chemical treating processes can be both expensive and hazardous. An acid etching solution that is commonly used for polyolefins consists of a sodium dichromate–sulfuric acid mixture. A solution commonly used for fluorocarbons is a sodium naphthalene solution. Both of these are defined in ASTM D2093. The surfaces of ABS, polycarbonate, polyphenylene oxide, and acetal can also be modified using chemical treating. For information concerning the correct chemical treatment and the process parameters, one should contact the supplier of the base resin used in the plastic substrate.

Plasma treating processes are somewhat similar to corona treating processes except that the parts are subjected to electrical discharge while in a closed vacuum chamber. The type of gas used and the frequency range will depend on the type of plastic being treated. Oxygen and inert gases are often used, and the discharge is often generated at microwave frequencies (e.g., 2.45 GHz). When subjected to the activated gas that is generated in the plasma treater, atoms on the surface of the plastics are physically changed and rearranged. With many plastics, this increases the surface energy considerably. The main disadvantages with plasma treating are that it is a batch process and the equipment is relatively expensive. As a result, many academic papers are available regarding plasma treatment, but it is used relatively little in industry.

11.2.4 New Surface Modification Processes

Several new surface treatments and modifications of older, conventional surface treatments have been introduced over the last few years to provide alternatives to the common processes noted above. The driving factors for these developments have been related primarily to environment and safety. Harsh chemicals and elevated temperature processing associated with conventional chemical and flame treatment methods have inhibited many from using such processes.

In addition to providing safer and environmentally friendly processes, these newer surface treatments have also been shown to provide for easier and faster processing. They promise a potentially tremendous positive impact on both manufacturing cost and performance properties. The reduced cost impact can be in the form of equipment costs, implementation costs, operational costs, rework costs, and storage/waste removal costs.

The following new surface treatments for low-surface-energy plastics are in the late developmental stages or are being used commercially within selective industries. Generally, the developers are assessing the transfer of their technology to lager markets.

11.2.4.1 Atmospheric Plasma Treatment

Corona treatment has long been utilized as a surface treatment for plastics to promote the bonding of adhesives or coatings. The primary disadvantages of conventional corona treatment are well known:

- Treated substrates have a limited shelf life, and generally, the adhesive or coating must be applied as soon as possible after surface treatment.
- Thick substrates (generally thicker than 0.125 in.) are difficult to treat because of the electrical resistance provided by the substrate itself in the corona treater.
- The level of improvement in adhesion is measurable but not as great as with plasma treatment. Certain plastics, such as fluorocarbon-based materials, do not respond well to plasma treatment.

Although providing excellent surface adhesion on many hard-to-bond plastics and long-shelf-life treatments, conventional plasma treatment also has shortcomings. It requires a low pressure (partial vacuum), and thus the surface must be processed in a vacuum chamber. This generally necessitates a batch process. Continuous webs, films, and so on, are difficult to process with plasma.

A new process, atmospheric plasma treatment, has been developed that negates many of the disadvantages of both corona and plasma treatment. A significant benefit of atmospheric plasma over corona treatment is that the shelf life of the surface treatment is significantly longer. Table 11.6 illustrates the aging effect of standard corona treatment and atmospheric plasma on oriented polypropylene film. Atmospheric plasma treatment represents a new generation of surface treatment technology that allows the plasma to be sustained at atmospheric pressure in a way that permits the surface treatment of polymers and other substrates. It can be configured to work with a variety of feeding systems, including continuous web.

An atmospheric plasma source may be viewed as an electrical switch that turns chemical reactions on and off. When it's on, a high concentration of atoms and radicals

TABLE 11.6 Aging Effect of Corona and Atmospheric Plasma Surface Treatments on the Critical Surface Tension of Oriented Polypropylene Film

Hours Aging at Room Temperature After Surface Treatment	Critical Surface Tension (dyn/cm)		
		Atmospheric Plasma Treatment	
	Standard Corona Treatment	Proprietary Gas 1	Proprietary Gas 2
0	45	60	60
200	37	45	60
400	36	43	60
600	36	43	60

Source: Ref. 7.

are delivered to a surface to clean, remove, modify, or deposit materials of choice. The reaction chemistry depends on the gases fed to the plasma. These gases will depend on the substrate, and they can include oxygen, hydrogen, nitrogen, and carbon tetrafluoride. By using electrons instead of heat to turn reactions on and of f, one can independently control the temperature of the surf ace. Atmospheric plasma has been used effectively on films, papers, foams, nonwovens, wovens, fibers, metals, and powders. One company has used atmospheric plasma to surf ace-treat carbon-fiber-reinforced polyether ether ketone composites. After such treatment, lap shear samples exhibited substrate failure (lap shear strength greater than 6000 lb/in^2), whereas untreated specimens failed adhesively [7].

Atmospheric plasma treatment equipment has been de veloped for any extrusion, lamination, or production line width. Operational speeds v ary by substrate and can be up to 1000 ft/min. The process is claimed to have lower operational costs than conventional plasma treatment (due primarily to the lack of v acuum and high-energy requirements) with similar adhesion enhancement characteristics on difficult-to-treat materials.

11.2.4.2 Sicor Surface Treatment

Sicor is a ne w patented technology de veloped by Australia's Commonwealth Scientific and Industrial Research Organization (CSIRO) to engineer the surface of polymers to enable high-performance adhesion of surface coatings, including adhesives and paints [8].

Sicor is applied by (1) oxidizing the polymer surf ace using standard techniques such as flame or corona treatment, (2) depositing a graft chemical in water solution, and (3) drying excess moisture from the surface using air knives or other standard industrial drying systems. The chemical deposition consists of special types of chemicals (such as silanes and others), which form chemical connector molecules on the surface of the original substrate.

The improved performance of Sicor relative to corona and flame treatment is especially apparent on low-surface-energy plastics. The major advantages of this process are that it provides enhanced adhesion strength and durability, cost savings by allowing the use of cheaper materials and more efficient processes, elimination of the use of ozone-depleting substances, including solvents and chorine-based materials, and easy integration into existing manufacturing systems with treating speeds up to 300 m/min.

The process has demonstrated e xcellent results on normally dif ficult-to-bond homopolymers and blends based on polyethylene, polypropylene (PP), metallocene-based PP, polyvinylidene fluoride, polyacetal, and other plastics. Sicor is currently being used on the Australian-manufactured Toyota Camry to bond polyurethane foam to the polypropylene airbag cover. Target applications include the manufacture of interior components, such as instrument panel airbag covers, door trims, and consoles. It is also being used to develop special surfaces on plastic "corks" for bottling premium wines.

11.2.4.3 Chlorinated Surfaces

In a process somewhat similar to the Sicor process described abo ve, a group of scientists at Kettering University in Flint, Michigan, has developed a process that will allow adhesives and coatings to bond more efficiently to plastic parts [9]. This process is also expected to find major applications in the automotive industry. The Kettering University process chemically alters the surface structure of the polymer. It requires immersing or spraying the surface of a polymer with a water-based solution that includes a mild oxidizer (such as sodium hypochlorite) designed to react with a weak acid (dilute acetic acid or succinnic acid solution). This causes a controlled degradation of the oxidizer, allowing the addition of chlorine to the polymer's surface making the surface receptive to coatings or adhesives.

After extensive testing, the Kettering group feels that the process works by forming free radicals that extract hydrogen atoms from the polymer chains on the surface of the plastic. This enables the chlorine to react with the remaining carbon radicals, rendering the surface more compatible with adhesives and coatings. The process effects a change only at the surface; it does not change the structural integrity of the bulk material.

11.2.4.4 UV-Irradiated Fluorocarbons

Another Australian research organization (Johannes Kepler University) has developed a new method in which laser radiation is used to modify fluorocarbon surfaces so that adhesives and coatings will bond strongly [10]. The effect is achieved by irradiating the surface of fluorocarbons with ultraviolet (UV) light at extremely low wavelength (172 nm). The plastic substrate is modified only in those places where it will be bonded or coated. This is an advantage over other surface preparation processes where the entire part must be treated. The unmodified and bulk plastic retains its major properties. This is considered to be of special advantage in medical engineering, where the application of UV-irradiated fluorocarbons significantly facilitates the use of artificial blood vessels, cardiac valves, grafts, and artificial skin. The new technology ensures tissue adhesion and compatibility of the surface. The method is also being explored in the textile and the microelectronics industries, where surface modification of predetermined areas is desired.

11.2.4.5 Switchable Surfaces

Virginia Tech researchers have designed a new type of polymer surface material that could result in more universal adhesive because the material surface can be "switched" to interact with different kinds of molecules [11]. Block copolymers are made that can "tune" the adhesive properties on a surface so that the surface materials can interact with other molecules or a range of molecules.

11.2.5 Primers

Primers generally cannot be used as a substitute for chemical or physical surface preparation. However, there are several instances where primers have provided excellent adhesion without having to go through the process of surface preparation. This is a distinct advantage because surface treatment methods may be hazardous, inconvenient, time consuming, and often expensive. The use of a surface primer, although an extra step in the bonding process, is a more desirable alternative for use on the production line.

Primers function by providing an intermediate surface that will bond well to the plastic part and will also provide good adhesion for decorative medium, adhesives, or coatings. Primers are adhesion promoters that provide a stronger interphase region having improved adhesion and permanence. This is illustrated in Figure 11.5. Primers may also be used to protect treated plastic substrates. After surface treatment, the high-energy polymeric surface is an active surface that will readily adsorb atmospheric contamination. The primer protects the treated surface until the time when the adhesive or sealant is applied. Primers are especially useful in this way for the protection of polymeric parts that are treated by flame or corona discharge. Primers also find use on polymeric substrates in that their solvents will soften the surface and some of the primer resin will diffuse into the bulk of the substrate, thereby increasing adhesive strength by molecular diffusion.

Another production-related reason for priming is that with bonded assemblies having many subsections, the nature of the assembly may not be suitable for immersion in pretreating solution as a single structure. With primers, individual subsections can be treated, primed, and then fit into place before the bonding step without regard to time. This then allows the entire assembly to be bonded at one time.

Conventional primers for plastics substrates have been developed over the years. These have been included both generic and proprietary systems. The most notable are primers based on chlorinated polyolefins that are commonly used in the auto industry to improve paint adhesion to low-energy plastic parts. Chlorinated polyolefin primers have generally been available as either chlorinated polyethylene or polypropylene in solvent solutions. The chlorine atoms in the outer surface of the primed part are believed to increase polarity and enhance adhesion of paint and adhesives. Appropriate substrates for priming with chlorinated polyolefins include polyolefin plastics, TPO and EPDM. Several recent

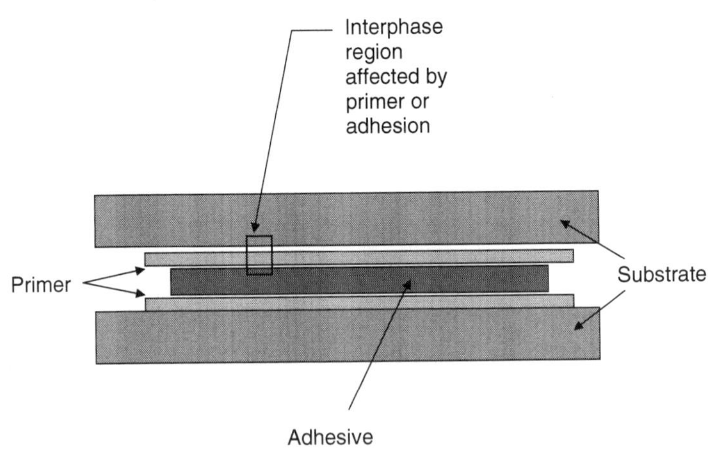

FIGURE 11.5 Primers and adhesion promoters provide a stronger interphase region having improved adhesion and permanence.

developments have been directed at reducing the amount of hazardous air pollutants in these primer systems.

The manufacturers of solvent-based chlorinated polyolefin primers have developed solvent-free waterborne chlorinated polyolefin primers based on emulsions and dispersions of chlorinated polyolefin. These primers provide an increase in bond strength and water resistance for polypropylene and other thermoplastic polyolefin joints that is similar to that of their solvent-borne counterparts. Because the waterborne chlorinated polyolefin primers are solvent-free systems, they are preferred from an environmental and safety standpoint in many applications.

High-solids-content (25 to 50% in xylene) chlorinated polyolefin primer solutions and 100% solids systems have also been developed for one-step coatings on polypropylene, TPO, or other polyolefin plastics. The chlorinated polyolefins can be added directly to the paint formulation. This gives coating manufacturers a way to reduce hazardous air pollutants, making it easier to remain within environmental limits or guidelines.

A new priming process has been developed to overcome the problem of poor adhesion of UV curable coatings to plastic surfaces [12]. In this process the plastic surface is activated by a plasma or corona treatment, and then a special primer composition is applied as a thin layer to form a covalently attached primer. The thermally stable primer layer contains latent groups, which can react with the UV-curable coating under UV light exposure. This novel process produces significantly enhanced adhesion to plastic substrates.

This technology is based on a sequential process, which results in a permanent modification of the polymer surface. In a first step, the surface is activated by a conventional plasma, flame, or corona treatment. Subsequently, an unsaturated photoinitiator is deposited on the activated polymer surface and grafted by the formation of covalent bonds. Due to the polar groups of the photoinitiator, the surface energy of the treated surface is enhanced and wetting is improved. In addition, the steric hindrance and the cross-linking of the photoinitiator molecules prevent diffusion of the polar groups into the polymer. By anchoring the photoinitiator at the surface, a practically unlimited storage stability of the surface is achieved as long as exposure to short-wavelength light is omitted.

11.3 COATINGS ON PLASTICS

Plastics are claiming significant strides in many markets due to property advantages such as reduced weight and lower system costs through design flexibility, increased process efficiency, and increased part performance. As a result, painting or coating on plastic materials has become essential in many applications, such as automotive components and consumer products, and this has posed new challenges for the coating industries.

The technology for coating on plastics is an extension of processes originally developed for metal and other nonpolymeric substrates. The equipment, techniques, terminology, and regulations evolved in the coatings industry apply equally well to the coating of plastics. However, the knowledge necessary to assure the quality and performance of painted plastic parts is different from that required with other substrates. This is due primarily to the low surface energy of plastic parts as compared to conventional metallic materials and the chemical similarity between the coating materials and the plastic substrate.

As a result, important considerations must be made regarding the nature of (1) the coating material, (2) the coating–substrate interface, and (3) the substrate itself. As shown in Figure 11.6, the completed paint system will only be as strong as any one of these links. In addition, as with any coating process, an understanding of the application processes and end-use environments is also required for successful implementation.

11.3.1 Function of the Coating Versus Alternative Processes

The main function of a coating on a plastic part is aesthetic. One generally wants to either alter the color of the part or change its "feel." For example, in the automobile

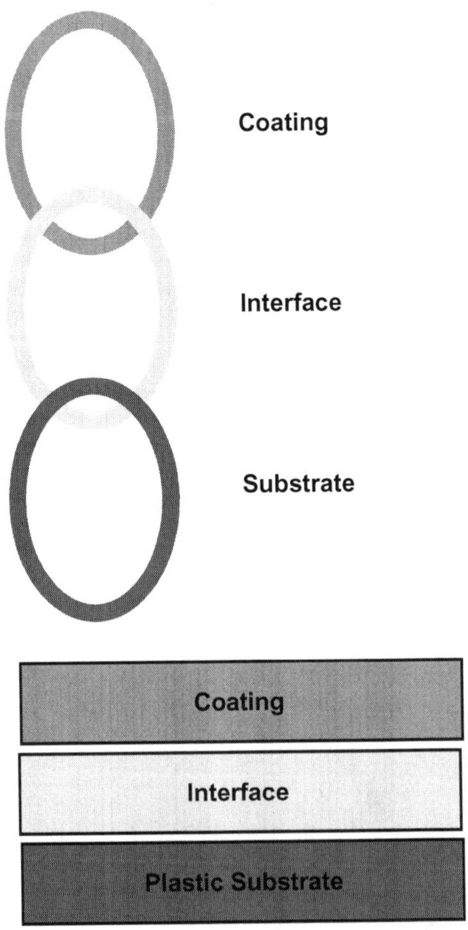

FIGURE 11.6 The coating system is only as strong as its weakest link.

industry, coatings are often applied to match the color of a secondary part or to provide a softer surface to the touch (e.g., interior dash components). Other functions of coatings on plastics may be to increase the abrasion resistance of the part or to provide for improved weathering and resistance to UV light. To accomplish this, the coating must adhere well to the plastic part for the life of the component. Thus, attention must be given to the most important performance criteria of painted plastic parts, and to their measurement. These criteria include adhesion, chemical resistance, impact performance, and scratch resistance.

There are several processes that can perform these functions. In addition to conventional coating or painting, they include internal coloration of the plastic with pigments or dyes; application of decals, labels, or tapes; and the application of dry films within the molding operation (in-mold decorating) or as a secondary process after molding. The advantages and disadvantages of the coating process for plastics are summarized in Table 11.7.

Although the advent of these alternative processes have reduced the growth rate of conventional coatings for plastics, many processors and designers consider coating or painting to be the only process that can assure a proper color match between adjacent plastic and nonplastic parts.

11.3.2 Nature of the Substrate

Although a significant amount has been written about plastics, it is worthwhile to review several factors about the nature of plastic substrates that can affect the performance of coatings. The nature of the substrate is important and can manifest itself in coating problems during three stages: (1) during molding of the part, (2) during application of the coating, and (3) during aging of the part in the service environment.

Processing conditions during part molding are extremely important for most types of plastics. As adhesion is determined primarily by the surface layer, factors such as mold temperature, injection speed, and melt temperature affect adhesion to a large extent. Mold design is important, as high-shear regions in thin-walled applications will also influence the adhesion in a negative manner. Chemicals that

TABLE 11.7 Advantages and Disadvantages of Painting Plastic Parts

Advantages	Disadvantages
Several inexpensive methods are possible.	Some plastics are solvent sensitive.
Pretreatment of most plastics (e.g., high-surface-energy plastics) is not necessary.	Hand methods have high labor costs.
Variation of methods and designs may hide imperfections.	Paint reduces cold impact resistance.
A significant number of coatings are available for various aesthetic and performance functions.	The plastic surface must be cleaned carefully. Poor adhesion and appearance defects (e.g., fisheyes) may occur due to contamination on the surface.
New solvent-free paint systems (waterborne, UV-cured, high solids content) have been developed.	Solvents may be a health and safety hazard.
	Oven drying may be a problem with some heat-sensitive thermoplastics.

are employed in paint formulation, such as solvents, monomers, and other additives, or solvents used in the cleaning process could attack the plastic part. Temperatures needed to dry or cure the coating may be dangerously close to the plastic's heat deflection temperature, resulting in part deformation and greater chemical attack. The impact resistance of the plastic part may also be lowered by stress crazing caused by solvents or monomers attacking molded-in stress within the plastic.

On aging during service, internal additives and low-molecular-weight constituents may migrate to the surface of the plastic part, creating a weak boundary layer at the coating–substrate interface, which results in poor adhesion. Dimensional changes in the plastic part as a result of thermal excursions or changing relative humidity could provide internal stress at the interface. To complicate this further, the modulus or stiffness of plastics vary greatly with both temperature and polymer type.

11.3.3 Nature of the Interface

The interface may be the most important link in the overall coating system since it is the interface that will ultimately control the adhesion of the coating to the plastic substrate. One must recognize that plastics materials and surface conditions vary from material to material and even vary within the same materials depending on the processing characteristics. There can be a different chemistry on the surface of the plastic than within the bulk. Since plastics are relatively dynamic materials, this surface can change with time and ambient conditions.

The performance of the coating part is generally determined by the adhesion of the coating to the part and the physical and chemical characteristics of the coating to the service environment. Adhesion is achieved by several processes occurring simultaneously. The adhesion forces are determined by (1) the surface tension characteristics of both the coating and the substrate, and (2) the interdiffusion of the coating into the substrate.

The surface tension of the coating system should be lower than the surface tension of the substrate to enable wetting of the coating on the surface of the substrate. Various degrees of wetting behavior are shown in Figure 11.7. Efficient wetting will maximize the bonding area and create stronger forces of adhesion. Surface roughness is also an important parameter in certain cases, as mechanical interlocking is an excellent way to improve adhesion.

Interdiffusion is the main adhesion mechanism for amorphous plastics, especially in the case of solvent-based coating systems. The solvent is used to diffuse the base coating resin into the plastic substrate to provide for molecular interlocking. However, one has to be careful in choosing a solvent. Chemical attack of the substrate by using too aggressive a solvent has to be avoided, as this will affect the mechanical properties of the painted part.

It is important that the surface of the plastic part be clean and free of any process oils or lubricants. The surface should also be consistent from part to part, and this may necessitate a surface preparation process to remove contaminants or surface layers that could occur due to variables in the molding process. For maximum adhesion, the critical surface tension of the plastic part should be higher than the surface

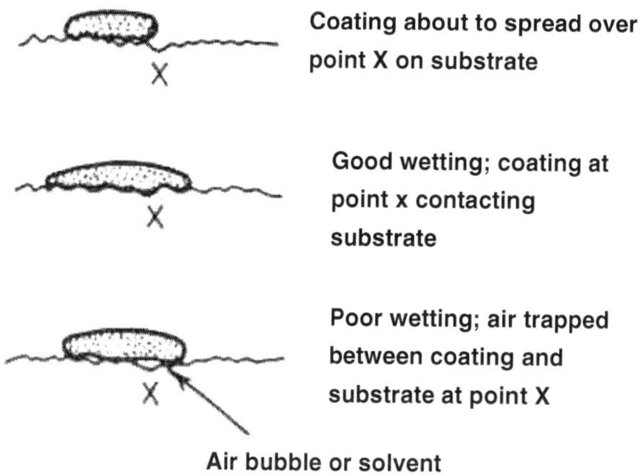

FIGURE 11.7 Good and poor wetting of a coating over a substrate.

tension of the coating. This may also necessitate pretreatment of the plastic part prior to coating. Various surface treatments for improved adhesion of coatings are described above and in Chapter 10.

11.3.4 Nature of the Coating

The majority of coatings used for plastics are, in fact, polymers themselves. Thermoset and thermoplastic paints and coatings are used in very specific applications. The advancements in polymer chemistry that yield stronger plastic materials also yield more durable paint systems. A few of the more common paint systems that are used to coat plastics are highlighted in Table 11.8.

The most important characteristics of the paint system are its wetting characteristics, its modulus or flexibility, and the solvent or monomers that are employed in the paint system to achieve interdiffusion. The effects of wetting and solvents on the coating performance properties are noted above. Coating systems commonly used for plastics are often formulated from flexible, tough polymers. The coating flexibility is required to resist possible dimensional changes between the substrate and the coating that could occur due to thermal expansion and contraction or due to moisture absorption in the plastic part. Thus, polyurethane, acrylic, and flexible epoxy coatings are commonly used.

Important performance properties for common plastic coatings are given in Table 11.9. Although aesthetics may often be the most important criteria in selecting a coating system, performance factors such as humidity resistance, corrosion resistance, exterior durability, chemical resistance, and mar or abrasion resistance can also be important in determining what coating system is to be used in a specific application.

Modern coating systems for plastics can be solvent-based: water-based: 100% solid liquids curable by chemical reaction, heat, or UV light: or solids that require heating for melting and wetting to occur. The choice will most often depend on

TABLE 11.8 Common Paint Systems Used to Coat Plastic Parts

Plastic	Urethane	Epoxy	Polyester	Acrylic	Waterborne
ABS	R	R	NR	R	R
Acrylic	NR	NR	NR	R	NR
PVC	NR	NR	NR	R	NR
Polystyrene	R	R	NR	R	R
PPO–PPE	R	R	R	R	R
Polycarbonate	R	R	R	R	R
Nylon	R	R	R	NR	NR
Polypropylene	R	R	R	NR	NR
Polyethylene	R	R	R	NR	NR
Polyester	R	R	R	NR	NR
RIM	R	NR	NR	R	R

[a]R, recommended; NR, not recommended.

production requirements and the nature of the substrate. Table 11.9 shows the various forms or physical states that are available for each common type of coating system. A significant trend in recent years has been the development of coating systems that are environmentally friendly and safe to use. These include waterborne, UV-curable, and powder-applied coatings.

11.3.4.1 Coating Materials

It is the base resin in the coating formulation that determines the adhesion and permanence properties of the coating. Thus, most coatings are classified by their base resin. The most widely used resins for use in coatings on plastic substrates are:

- Acrylics
- Alkyd
- Epoxy
- Polyester
- Polyurethane
- Vinyl

The important characteristics of coating formulations based on these resins are summarized below.

Acrylics are noted for color and gloss retention. They have excellent resistance to weathering and outdoor exposure. They can be supplied as either solvent- or water-based coating systems with high solids concentrations. They are also available as solid powder coatings, but these are not used on heat-sensitive plastics substrates. Solvent-based systems are relatively brittle and therefore have poor impact resistance. When acrylic resins are used to modify other resins, their properties are often imparted to the resulting formulation.

TABLE 11.9 Performance Properties and Physical Form of Common Coating Systems for Plastic Parts

	Property[a]				
Resin Type	Humidity Resistance	Corrosion Resistance	Exterior Durability	Chemical Resistance	Mar Resistance
Acrylic	E	E	E	G	E
Alkyd	F	F	P	G	G
Epoxy	E	E	G	E	E
Polyester	E	G	G	G	G
Polyurethane	E	G	E	G	E
Vinyl	E	G	G	G	G

	Form					
Resin Type	Conventional Solvent	Waterborne	High-Solids	Powder Coating	100% Solids	Two-component
Acrylic	×	×	×	×		
Alkyd	×	×	×			
Epoxy	×	×	×	×	×	×
Polyester	×		×	×	×	×
Polyurethane	×	×	×	×	×	×
Vinyl	×	×	×	×	×	

[a]E, excellent; G, good; F, fair; P, poor.

Alkyd resin–based coatings offer the advantage of good durability and relati vely low cost. They are used for finishing a wide variety of products either alone or modified with oils or other resins. The degree and type of modif ication will determine the final performance properties. Although alkyds are used in outdoor applications, they are not as durable in long-term exposure, and their color and gloss retention are inferior to those of acrylics.

Epoxy resins can be formulated with a wide range of properties. Their chemical and heat resistance are outstanding due to the thermosetting, cross-linked molecular structure *that results* when these coatings are fully cured. When epoxy top coats are used outdoors, they tend to chalk and discolor because of inherently poor UV resistance. Epoxy resins are generally brittle materials. Ho wever, with proper modification they can be made v ery flexible. The flexibility is especially important when bonding plastics that are also fle xible and have modulus and coefficient of thermal expansion that is often different than the coating material.

Polyesters are used alone or modif ied with other resins to formulate coatings ranging from clear finishes to highly pigmented industrial f inishes. They are available as one- and tw o-component systems. In both cases, the paint formulator can adjust properties to meet most e xposure conditions. Polyesters are also commonly used in powder coatings. Two-component polyesters are well known as gel coats for composites such as bathtubs, spas, boats, and automobile sections.

Polyurethane resin–based coatings are extremely versatile due to their formulating capability. They are available as one- or two-component systems. Certain one-component polyurethanes cure by reacting with moisture in the air or on the substrate. Polyurethanes can be formulated to be very tough and flexible. They have excellent abrasion and impact resistance, and they are often used for that function. Polyurethanes have moisture sensitivity and should be tested when the service environment is expected to provide high temperatures and high humidity.

Vinyl-based resin coatings (commonly made from polyvinyl chloride) are noted for their toughness, chemical resistance, and durability. They are available as solutions and water dispersions. Vinyl coatings have been used primarily in the food and beverage industry for applications such as can linings and packaging. However, they are also used in automobile interiors and office machine hardware because of their "soft" feel. Owing to their excellent chemical resistance, vinyl coatings are often used to coat tanks and mixing equipment used in the chemical industry.

11.3.4.2 Coating Processes

Coatings can be applied on plastic surfaces by a variety of methods. Most often the conventional coating processes used for metals and other nonpolymeric substrates are used. These include spray painting, dip painting, roller coating, screen painting, and electrostatic painting. In electrostatic painting the plastic surface must first be treated to take an electrical charge. This is usually done with a conductive primer.

Fill-in painting is a process, however, which is specific to molded plastic parts. In the fill-in painting process, letters, figures, or designs on plastic parts can be molded into the part, and the coating is applied by placing it into the depressions in the molded article. These recesses are filled by either spraying or wiping the paint into the depressions. The depression should be deep and narrow to accept the paint and provide good contrast. Excess paint is then removed by wiping or buffing operations. In a reverse of this process, coatings may also be applied to raised portions on molded surfaces. In this case, the paint is applied via a coating roller that only touches the raised surface. If edges and corners are sharp and highly railed, good coating detail can be obtained.

11.4 IN-MOLD DECORATING

The in-mold decorating process generally uses an overlay or carrier film called a *foil* rather than a liquid coating. The foil is placed into the mold before the part is manufactured and during molding becomes an integral part of the plastic part. Both thermosetting and thermoplastic materials are applicable to in-mold decorating processes. The decorative image and, if possible, the film carrier are made of the same materials as those used in the part to be molded.

The advantages and disadvantages to in-mold decorating over other methods of decoration are shown in Table 11.10. In addition to providing decoration and color, in-mold coatings have been used to eliminate porosity, provide hard, durable surfaces, and provide a primed surface for subsequent application of an adhesive. Mold

TABLE 11.10 Advantages and Disadvantages of In-Mold Decorating

Advantages	Disadvantages
Full color images, halftones, or combinations may be used.	Cost of foil, labor for hand loading, or capital cost of automated loading machines are high.
Strong bonding can be achieved.	Mold design must minimize washing and turbulence of resin in the mold relative to foil.
Designs and short runs are economical.	Mold design must cause foil to separate cleanly at the edges to minimize cleanup.
Very high efficiency due to little or no waste.	
Little space or capital equipment is required for secondary coating processes.	
Wraparound coatings are possible.	
Provides mass customization through assorted images, yet using a single molding tool.	

design becomes an important consideration before selecting in-mold decorating as the process of choice. The design of the mold should take into account placement of gates to prevent wrinkled or washed overlays. The overlay can be held in place in the mold by cutting so that it is tight fitting or by electrostatic attraction.

Blow-molded parts are often decorated by the in-mold processes. The ink or paint image is placed on a carrier film and placed in the mold. As the hot plastic expands, filling the mold cavity, the image is transferred from the carrier to the hot plastic surface. In-mold coatings are also becoming popular with sheet molding compounds and reaction injection-molded (RIM) materials. In-mold coatings are most commonly liquid or powdered materials that are carried on a foil. However, they can also be reactive liquids that are coated directly on the mold surface. Liquid in-mold coatings are primarily two-component urethane systems. The liquid coating is injected into the mold at high pressure. The coating will cure somewhat faster than the parent plastic. Electrostatic powder coatings can also be applied directly to the mold surface before injection of the plastic resin. Powder in-mold coatings generally are unsaturated polyesters. The primary advantage of the powder coating process is that uniform and thick coatings can be achieved easily and there are no hazards associated with use of a solvent.

With thermosetting production, the carrier film may be clear cellulose sheet covered with a partially cured coating. The overlay is placed in the mold cavity, while the thermosetting material is only partially cured. For thermoplastic, a polyester film is generally used as the carrier. The foil is generally placed in the mold before it closes prior to injection. Recently, polycarbonate films have been used for applications requiring high gloss, UV stability, and resistance to chemicals and scratches. Figure 11.8 is a simplified schematic of the in-mold decorating process using a polycarbonate film for decorating thermoplastics.

In-mold coatings can contain multiple layers. The bottom layer adheres to the parent plastic. The decoration is often protected by a transparent top coat or wear coat. This top coat will determine the abrasion and scratch resistance of the final surface. A heat-activated release coat causes separation of the decorating from the carrier film. In-mold decorating is well known for putting graphics, such as logos

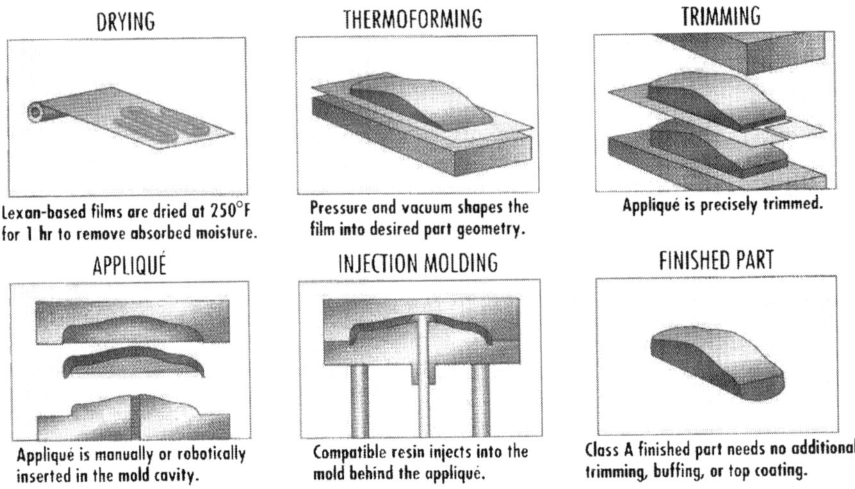

FIGURE 11.8 In-mold decorating process for thermoplastics. (From Ref. 13.)

and model numbers, directly on complex three-dimensional parts without secondary operations. Complex geometries ranging in size from small cell phones to large automobile parts are ideal candidates for in-mold coatings. The largest-volume application is automobile bodies. Here in-mold coating is advantageous because of its fast production speed, high degree of coating adhesion, coating permanence, and the elimination of hazards and volatile organic components (VOCs) associated with conventional coating proceses.

11.5 PRINTING

Printing on plastics serves both for decoration and for placing information, such as bar coding, logos, freshness dates, and so on, on a part surface. The conventional techniques for printing include stamp printing, silk screening, heat transfer, and non-contact methods. The surface condition of the part must be considered before any attempts are made at printing on plastic. Surface cleaning and treating methods described above for preparing the plastic surface prior to bonding or coating are also applicable for treating surfaces prior to printing.

11.5.1 Stamp Printing

Stamp printing produces code or lettering by a metal or rubber die exposed to an ink. Letterpress, letterflex, gravure, and flexography are types of stamp printing. They refer to methods of applying the ink using various types of dies. For layflat film products, gravure and flexographic printing are most commonly used.

Letterpress printing is a method where raised rigid printing plates are inked and pressed against the plastic part. The raised portion of the plate transfers the image. *Letterflex printing* is similar to letterpress except that flexible printing plates are used. Flexible plates may transfer their designs to irregular surfaces.

Gravure printing (also called *rotogravure*) transfers ink from cells etched or engraved on a copper and chrome-plated cylinder to the material being printed. The cylinder rotates in an ink fountain and the cells pick up the liquid ink. A doctor blade wipes excess ink from the roller, as shown in Figure 11.9, leaving ink only in the cells. The low-viscosity liquid inks used in this process contain volatile solvent.

The gravure processes are widely used for printing plastic films and applying various lightweight coatings and adhesives. The main advantage is the reliability of the coating deposit, which is independent of web tension and coating speed. Gravure is used to print directly on a product with a very high degree of color-to-color registration. The color quality of the gravure process is superior to other printing methods. It is used mainly when design requires good continuity of sharp pattern details and rich color effects. It is a high-speed process; however, it can be expensive for short runs because of the cost of the cylinders.

Pad transfer printing is a gravurelike process offering versatility for printing on a variety of shapes and surfaces. It is one of the most popular and flexible methods of plastics decorating. Pad transfer printing can be used to decorate a wide variety of part shapes and surfaces that are difficult, if not impossible, to decorate with other methods. In pad transfer printing, the image is transferred using a special silicone pad instead of the etched plate used in gravure. The flexible silicone rubber transfer pad is the key to this process. Pad printing can reproduce fine-line engraving, and it can apply wet ink on wet ink. Pad printing is generally most useful for printing small areas. Pad printers are generally more expensive than hot stamping or screen printing. The ink costs, however, are much less than the cost of hot stamping foil.

Flexographic printing is more widely used for short and medium-length runs. This method prints by transforming images from a flexible raised printing plate directly to the material. A uniform film of ink is applied to the raised portion of the

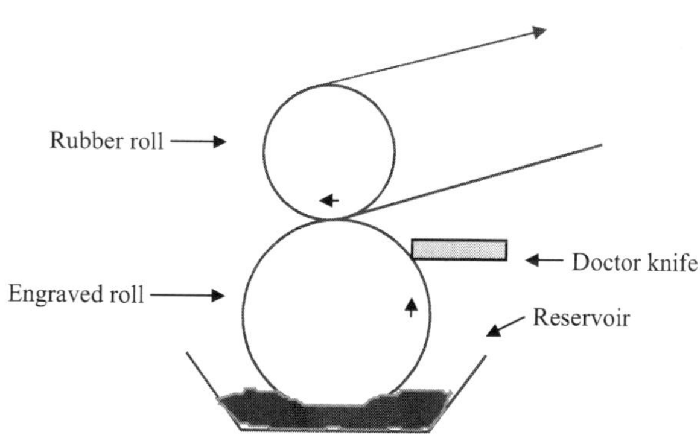

FIGURE 11.9 Direct gravure printing and coating process.

printing plate by bringing it in contact with a foam roller. The plate continues to rotate until it contacts the substrate and transfers the image to it. Individual colors are applied by a series of separate printing stations. Typical applications for flexographic printing include packaging films, cartons, labels, food containers, and plastic-based wall coverings.

11.5.2 Silk Screen Printing

Silk screening is the process where ink or paint is forced by a rubber squeegee through fine metallic or fabric screen onto the product. The screen is blank or blocked off in areas where no ink is wanted. The pattern is produced on the screen by a photo process that seals some of the holes in the fabric with an emulsion, leaving other mesh areas in the pattern open. Screen printing from flatbed or rotary systems makes it possible to decorate large or complex parts.

One of the advantages of the screen printing process is the laydown or thickness of the ink, which can be applied in one pass. However, only one color can be applied per pass. Multicolored parts require drying and curing between color applications. Image sharpness is of lower quality than with other printing processes. The major uses for screen printing have been short runs on molded products with complex contours, on flat substrates, or to get special color effects. General applications are for large, flat surface areas and cylindrical parts such as blow-molded bottles.

11.5.3 Heat Transfer Printing

Heat transfer printing is used as both a printing and a decorating process (see below). A carrier film is used to support a release layer and an ink image. The ink is made from a thermoplastic polymer so that it will soften when heated and transfer from the carrier film to the product. A heated rubber roll is used to supply the thermal energy and pressure required.

11.5.4 Noncontact Methods of Printing

Noncontact printing methods generally allow the designer to print on most plastics regardless of the shape of the surface. Common noncontact methods of printing are (1) electrostatic, (2) ink jet, and (3) laser etching. In electrostatic printing, dry ink is applied to areas by a difference in electrical potential. The information to be printed is supplied in digital form from a computer. It is imaged onto a continuously rotating drum. The drum then transfers the image to the passing substrate. Finally, the rotating drum is wiped clean to repeat the process for the next print. A block diagram of the electrostatic printing process is illustrated in Figure 11.10. Ink jet printing is a direct adaptation of the ink jet printer used to print on paper. Laser etching requires that a laser beam be aimed via a series of computer-controlled-mirrors. The laser burns the surface of the plastic, thus creating the printed image.

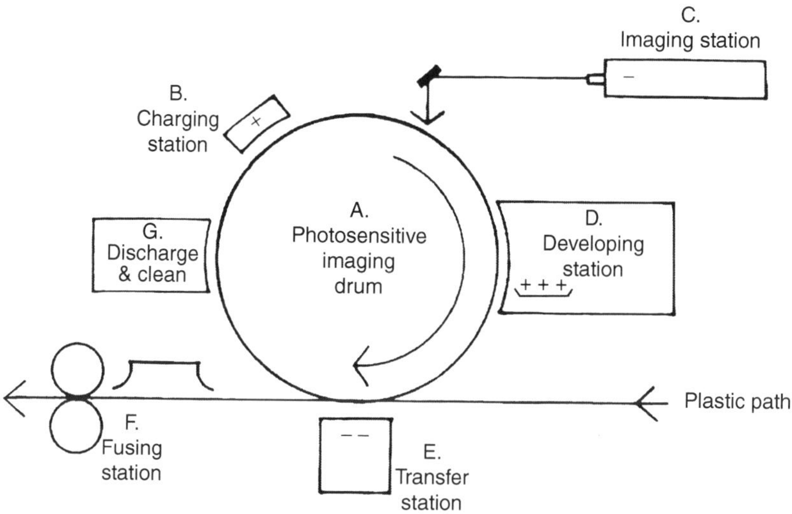

FIGURE 11.10 Electrostatic printing process. A, continuous rotating photosensitive imaging drum; B, charging station (makes entire passing surface positively charged); C, laser imaging station (puts a negative charge on drum where image is to appear); D, developing station (positively charged toner is attracted to negative charge on drum); E, transfer station (a stronger negative charge pulls toner off page onto passion plastic); F, fusing station (heat and pressure melt the toner onto the plastic); G, discharge and clean station (remaining toner is removed from the drum and it is discharged). (From Ref. 14.)

11.6 METALLIZATION PROCESSES

Plastic parts are metallized for aesthetics as well as performance. The mirrorlike finish on many consumer parts are examples of metallization serving an aesthetic function. Metallized electrically conductive surfaces for electronic items such as electronics and plating for corrosion resistance are examples of performance functions. Meal coatings can also provide a surface that is wear resistant or one that processes heat deflection characteristics.

Electroless, electrolytic plating, and vacuum metallizing are processes used to deposit metal surfaces on plastics materials. The construction of these various coatings are shown in Figure 11.11. However, metal surfaces can also be provided by adhesives or hot stamp methods. Some finished plastic parts must have shiny metallic surfaces. Besides providing a decorative finish, metal coatings may provide an electrical conducting surface, a wear- and corrosion-resistant surface, or added heat deflection.

Many metallized plastics have drastically altered physical properties compared to as-molded parts. This is due partly to the metal plate itself and partly to the metallizing process. For a part to be metallized, it should be smooth and blemish-free. There should be no weld lines or sink marks on the partAny surface irregularity will

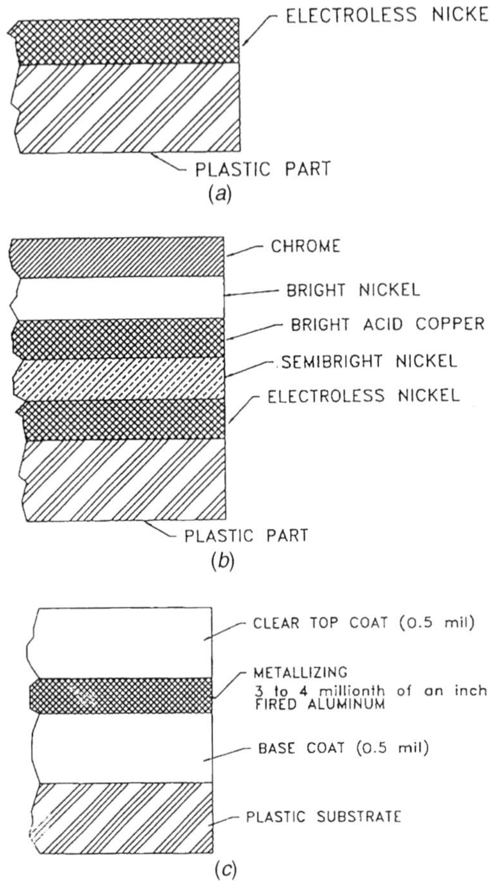

FIGURE 11.11 Layers of construction: (*a*) electroless plating; (*b*) electrolytic plating; (*c*) vacuum plating. (From Ref. 15.)

be magnified by the metallic coating. There should be very little surface stress, as this will affect adhesion of the coating to the part.

Several plastics can be plated easily by electroless or electrolytic processes on a commercial scale, as shown in Table 11.11. Some thermoplastics that can be readily etched and plated include acrylonitrile–butadiene–styrene (ABS), modified polyphenyl oxide, modified polypropylene, and polysulfone. Some grades of foamed plastics can also be plated. The greatest volume of plated plastic is represented by ABS and modified PPO. Most of the plated plastics market is for automotive parts. However, significant nonautomotive parts that are plated include marine hardware, plumbing fixtures, packaging, and appliance and furniture hardware. One of the most important applications for electroless plated plastics parts and all forms of metallized plastics is the radio-frequency interference (RFI) and electromagnetic interference (EMI) markets.

TABLE 11.11 Commercially Plated Plastics

Plastic	Note
ABS	Among the highest volume of plated plastics
Polycarbonate	Among the highest volume of plated plastics
ABS–polycarbonate	Among the highest volume of plated plastics
Polystyrene	
Phenolics	Typically plated for printed circuits or electronic applications
Epoxy–glass	Typically plated for printed circuits or electronic applications
Polyimide	Typically plated for printed circuits or electronic applications
Nylon	
Polyester	
Polyacetal	
Polypropylene	
Polysulfone	
Fluorocarbon–glass	Typically plated for printed circuits or electronic applications
Polyphenylene oxide	Among the highest volume of plated plastics

11.6.1 Electroless Plating

Electroless plating is a surface treatment process that does not require electric current as does electrolytic plating. Electroless plating deposits a dissolved metal, such as copper or nickel, on the surface of a plastic part through the use of a chemical solution. Also, any plastic part to be plated with the electrolytic process must first be electroless plated to create a conductive surface. Prior to electroless plating plastic parts, the surfaces have to be treated to ensure good adhesion. A flow diagram of the electroless plating process is shown in Figure 11.12*a*. The etching process usually involves the use of a chromic acid solution to provide a microscopically roughened surface to the plastic part. The catalytic process is sometimes referred to as *seeding*. Here very small particles of an inactive noble metal catalyst, normally palladium, are deposited into the microcracks created during the etching process. The palladium will act as active catalyst sites for chemical reduction of the electroless copper or electroless nickel.

The electroless metal layer formed is usually pure copper or a nickel–phosphorus alloy, depending on the corrosion requirements of the plated part. In an exterior environment electroless copper has been shown to have better corrosion resistance than electroless nickel. Usually, the thickness of the electroless metal is approximately 0.5 mil. Plastics that have been plated with electroless metal can then be given any electrolytic metal coating. This offers more possibilities to the design engineer regarding metal coating type and thicknesses.

11.6.2 Electrolytic Processes

Many plastic products require only electroless plating, but electrolytic plating can be added to the process to provide a metallized surface that is thicker and more like typical metal surfaces. The electrolytic plating process starts at the completion of the

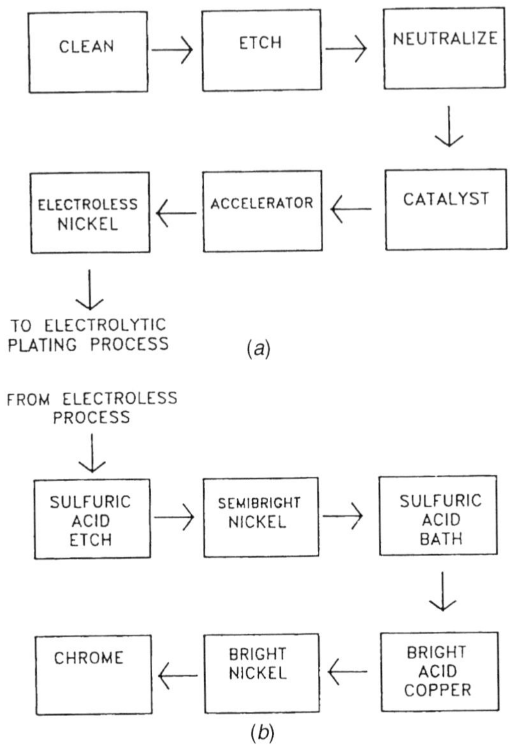

FIGURE 11.12 Flow diagrams for (*a*) electroless and (*b*) electrolytic plating processes. (From Ref. 15.)

electroless process. To deposit metal onto a plastic electrolytically, the plastic must be made electrically conductive and then grounded. The metal to be plated is positively charged, thus allowing the positively charged metal atoms to precipitate onto the negatively charged plastic.

Figure 11.12*b* shows a flow diagram for the electrolytic plating process. First the electroless plated plastic is etched once again, this time usually in sulfuric acid. The part then undergoes a semibright nickel deposition. This layer helps limit and control the amount of current that is exposed to the plated surface and thus protect it from burning off. A mild sulfuric acid bath follows. Bright acid copper is deposited electrolytically on the surface followed by bright nickel. The nickel layer is the immediate layer onto which the chrome will be deposited. The final electrolytically deposited layer is usually chrome, but many other finishes are possible, such as bright brass, satin nickel, silver, black chrome, and gold.

Electrolytic-plated parts are nearly indistinguishable from metal parts, and due to the plastic body, the plated plastic parts are lighter, less expensive, and can be used in a wider range of designs than their metal counterparts. To provide a standard rating for plated plastic, the American Society of Electroplated Plastics (ASEP) has developed a service condition code for common applications.

11.6.3 Vacuum Metallizing

Vacuum metallizing is a physical rather than a chemical process for depositing metal coatings on prepared surfaces. Vacuum metallizing, plastics, or films are thoroughly cleaned and/or etched and given a base coat of lacquer to eliminate surface defects. The parts are then placed in a vacuum chamber with small pieces of coating metal (e.g., chromium, gold, silver, zinc, or aluminum) placed on special heating filaments. Once sealed and under about 0.5 μm of vacuum, the heating filaments are energized until the metal melts and vaporizes. The vaporized metal coats everything it contacts in the chamber, condensing on the cooler surfaces. Parts must be rotated for full coverage, since the metal vapor travels in a line of sight from its origin. Once the plating is accomplished, the vacuum is released and the parts are removed and coated with lacquer to help protect the surface. The cross section of a typical vacuum-metallized part is shown in Figure 11.11*c*.

The metal thickness on a vacuum-metallized plastic part is generally 3 to 4 Å (1 to 4×10^{-6} in.), which is significantly less that the 1-mil thickness for conventional electroplating. Vacuum metallizing is a batch-type process and does not have the continuous flow attributes that electrolytic plating has. However, there are several cost savings associated with vacuum metallizing. Because vacuum metallizing inherently has a thinner coating, less metal is used and no toxic chemicals are involved.

The emphasis on metallizing parts has been to develop coatings with a mirror-like quality. However, gold, copper, silver, brass, and a wide variety of colored metallic finishes can be duplicated for furniture, hardware, appliance, and other industries where decoration is necessary. Virtually all plastic substrates can be metallized, except polyethylene and urethane. The most popular substrates are styrene, acrylonitrile–butadiene–styrene, acrylic, nylon, polypropylene, polycarbonate, and phenolics.

11.7 OTHER DECORATING PROCESSES

In addtion to those processes described above, several other decorating processes are commonly used: hot decorating (hot stamping and heat transfer) and application of labels, decals, and so on. These are used depending on the type of plastic that is being decorated, the special effect that is desired, and on the application equipment that is at hand.

11.7.1 Hot Decorating Processes

Several decorating processes use heat to transfer a decoration or a printed image to the plastic part surface. These techniques generally use a carrier element, the decorative media, and an adhesive. In certain cases, the adhesive is the molten surface of the part that the decoration comes into contact.

The primary processes for hot decorating are hot stamping, in-mold decorating, and heat transfer. Since these technologies employ no wet inks or paints, there is no offensive odor, environmental concerns over volatile emission, or storage problems. In addition, no ink mixing is necessary, and color or design changes involve merely changing a roll of dry printed foil, which minimizes setup time.

11.7.1.1 Hot Stamping

The hot stamping (as known as hot-leaf stamping) process is the most widely used in plastics decorating because of its convenience, versatility, and performance. The process of hot stamping consists of placing the part to be decorated under a hot stamping die. The hot die then strikes the surface of the part through a metallized or painted roll-leaf carrier. The paint carried by the roll is fused into the impression made by the stamp. The stamp is programmed to be pressed down on the thermoplastic surface at a controlled pressure and temperature and set to penetrate a controlled distance.

Hot stamping is best suited for flat surfaces, although contoured dies could provide hot stamping for curved surfaces. The fit between the plastic part and the contoured die face must be very close to get uniform hot stampings. For the most part, hot stamping is used to apply small decorations and explanatory lettering, logos, and so on, to plastic moldings. Hot stamping may be used to place gold, silver, or other metal foils (leaf), as well as paint pigments, onto plastic parts. Because these foils and pigments are dry, they are easy to handle and may be placed over painted surfaces. No masking is required and the process may be automatic or hand operated. Although hot stamping is used primarily on thermoplastics, the process is also applicable to thermosets.

Figure 11.13 shows a typical metallized hot stamping foil. The carrier film supports the decorative coatings until they are pressed on the plastics part. A lacquer coating is passed over the releasing layer to provide protection for the metal foil. If paints are used, the lacquer and paint pigment are combined in one layer. The bottom layer functions as a heat- and pressure-sensitive hot melt adhesive. Heat and pressure must have time to penetrate the various film coating and layers to bring the adhesive to a liquid state. Before the carrier film is stripped away, a short cooling time is required to ensure that the adhesive is solidified.

Typical hot stamping processes and the equipment employed are shown in Figure 11.14. The major components of a hot stamping installation include the equipment, tooling, and hot stamp foil. The hot stamp die is generally either metal or silicone rubber mounted to a heater hood. The die is heated to a temperature near the melting point of the plastic substrate. A typical application is an automotive taillight trim piece, where decorative bright silver graphics are stamped onto the exterior plastic surface.

11.7.1.2 Heat Transfer

In heat transfer decorating the image is transferred from a carrier film onto the plastic part. The structure of the heat transfer stock is shown in Figure 11.15. The preheated carrier stock is transferred to the product by means of a heated rubber roller. This process is commonly used for both decorating and printing on plastics.

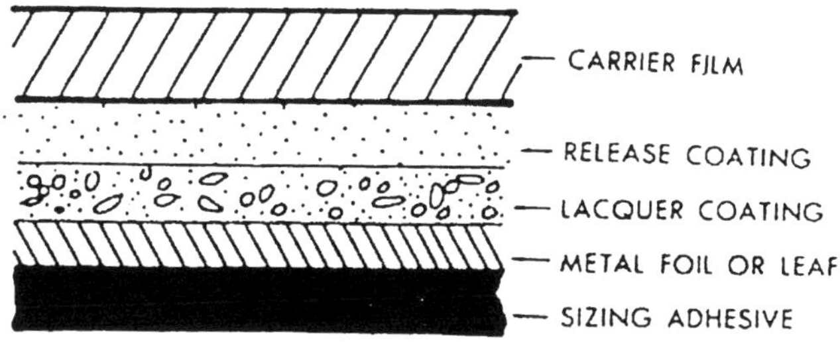

FIGURE 11.13 Typical hot stamping foil and structure. (From Ref. 16.)

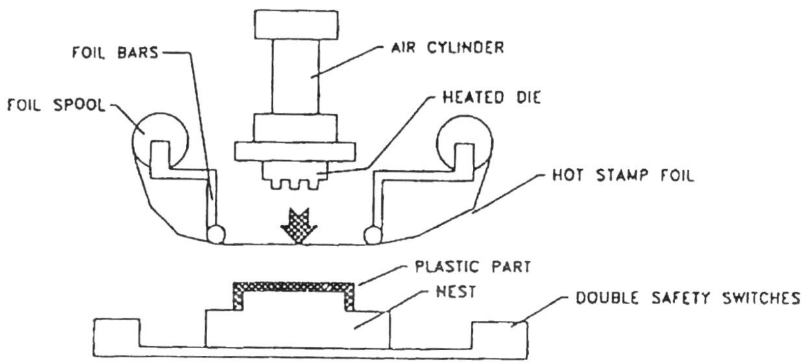

FIGURE 11.14 Hot stamping machine for flat plastics parts. (From Ref. 15.)

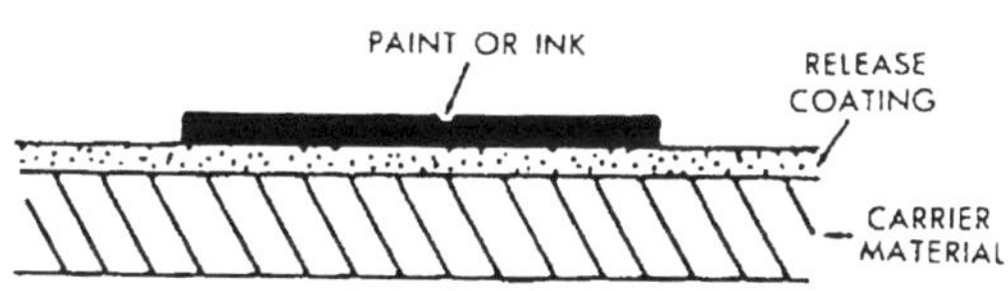

FIGURE 11.15 Heat transfer decorating stock. (From Ref. 14.)

11.7.2 Labels, Decals, and Surface Finishes

There are numerous miscellaneous decorating methods, including pressure-sensitive labels, decals, and flocking. Pressure-sensiti ve labels are easy to use and apply . The designs or printing is generally printed on an adhesive-backed foil or film, a pressure-sensitive adhesive is applied, and a release material covers the adhesive. To apply, the user must simply remo ve the release material and press the label to the part. Of course, the part must be clean and free of an y weak boundary layers so that adequate adhesion may develop. The stripping off of the *release* layer and application of the label can be accomplished either by hand or by automatic mechanical means.

Decals are also a means of transferring a design or printed message to a plastic part. With decals the decoration is placed on a decorative adhesive-backed film and then on a paper backing. The adhesive is such that it activates when in contact with water. To apply, the decal is moistened in water and the film is placed on the plastic surface. The process is not widely used because of the difficulty in placing the decal accurately in a specific location on the surface.

Flocking is a process of placing a velvetlike finish on virtually any surface. On plastics, flocking is applied generally by mechanical or electrostatic means. The process consists of coating the part with an adhesive and placing the flocking fibers on the adhesive areas before the adhesive sets.

Decorative wood-grained finishes are also possible by a number of processes. Some are accomplished by heat stamping or rolling an engraved plate over the surface of the part before applying a coating. Wood-grained decorative designs can be applied to hot stamp films. Various decorative wood-grained laminates and clad coatings are also applied with adhesives.

11.8 FINISHING OPERATIONS ON MOLDED OR CAST PLASTIC PARTS

Thanks to the versatility of most plastic materials to fill intricate cavities and to match the smooth (or intentionally mottled) surfaces and contours of a wide variety of cavities, the vast majority of molded parts need little or no secondary operations following removal from the mold.

11.8.1 Gate Scar Removal

Gate scars on injection-molded thermoplastic parts may simply be trimmed by snippers or razor-sharp knives. Similar scars on transfer and injection-molded thermoset parts can often be manually eliminated with a simple fine-cut file.

11.8.2 Parting Line Flash

Parting line flash, often about 0.005 in. thick, on well-maintained injection molds for thermoplastics, can be trimmed manually with razor-sharp knives. Such parting line flash on thermoset parts is generally brittle and can easily be removed by brushing it away with a gloved hand. For large quantities of smaller thermoset parts, deflashing is often accomplished by tumbling baskets, using a slow-speed motor for perhaps 5 to 10 minutes per load. If such parts have molded through-holes, with some flash partially or completely closing one end of such holes, adding a few six-pointed play jacks to the tumbling parts in the basket will often ensure removal of the unwanted flash across the hole.

11.8.3 Air Blast Deflashing

Air blast deflashers offer a more sophisticated removal of thin flash on thermoset parts. The process is similar to sandblasting, but the "sand" consists of less abrasive small particles or granules of a medium, possible nylon or acrylic (or other polymeric materials), or possibly crushed walnut shells or peach pits (or other organic media). Such a medium is propelled by compressed air blowing the medium through ceramic nozzles positioned close to the molded parts, and aimed to direct the high-speed medium where it will be most effective.

Air blast deflashers may aim the media blast at molded parts in a tumbling basket, to shorten the deflashing cycle. Other more elaborate blast-type deflashers use fixed-position nozzles to direct the fast-moving media jet against the molded parts, which are positioned on a continuous mesh belt conveyor which carries the parts through the blast chamber and then releases the parts into a container, allowing the continuous belt to travel to the entry end of the deflasher to pick up more parts as it feeds back into the blast chamber.

11.8.4 Cryogenic Deflashers

With thermoset parts, deflashing is generally carried out at room temperature. Cryogenic deflashers, performing similar functions, but with provision for chilling the media, are often used for deflashing thermoplastic and elastomeric molded parts. The chilling may be accomplished by propelling the medium with low-temperature refrigerated air or by adding particles of dry ice to the media. The chilling makes the flash brittle, and the blast easily breaks it away from the molded parts.

11.8.5 Wheel Deflashing

A variation on traditional air blasting deflashers is the wheel deflasher, which uses a high-speed rotating round plate with spiral or angled blades onto which media are fed. Propelled by centrifugal force, the media are "thrown" toward the molded parts at high velocity, causing the flash to be removed. Manufacturers of wheel deflashers claim that less energy is consumed than that of the air compressor energy costs of air blast deflashers.

11.8.6 Moisture Spray

Another feature, used on media deflashers for sensitive encapsulated electronic devices (integrated circuits, diodes, rectifiers, etc.), is a water spray chamber through which the deflashed parts are transported as a last step before the devices are carried out of the deflasher. The moisture spray removes any accumulated static charge that may have developed on the parts during bombardment by the media.

11.8.7 Machining

Machining operations such as grinding, drilling, tapping, reaming, sawing, and turning, may be performed on both thermoplastics and thermosetting parts, but the cutting tools and travel speeds are much more critical than those used for wood and metalworking.

Most machining generates heat. Plastics generally exhibit poor thermal conductivity. Localized heat generated by a grinding wheel or a cutting tool quickly reaches the level where the plastic, not only thermoplastic but also thermosetting plastics, becomes gummy or sticky, causing the cutting tool to bind and/or tear the plastic. Additionally, localized pressure at the cutting edge may distort the workpiece and often results in irregular tolerances on the machined part.

Machine tool manufacturers continue to develop cutting tools and techniques for machining plastics. But the cutting tools and speeds for, say, acrylics will not necessarily produce satisfactory results on polycarbonates. Similarly, cutting tools for thermoplastics may not be suited to thermosetting plastics, and vice versa.

While plastics molding enables relatively inexpensive manufacturing costs, secondary operations costs are often expensive. It is wise, therefore (and economically beneficial), to consult with the supplier of your plastic formulation, who may well have application engineering experience which can guide you to a workable solution for your machining needs. And if your material supplier cannot come up with a satisfactory solution, consider contracting with an experienced plastic part and mold designer to minimize (or even eliminate) the need for postmolding machining operations.

REFERENCES

1. D. M. Brewis, Pretreatment of polymers, in *Handbook of Adhesion*, D. E. Backham, ed., Longman Scientific and Technical, Harlow, Essex, England, 1992.
2. J. F. Watts, Degreasing, in *Handbook of Adhesion*, D. E. Backham, ed., Longman Scientific and Technical, Harlow, Essex, England, 1992.
3. J. G. Dillard, *Engineered Materials Handbook*, Vol. 3, *Adhesives and Sealants*, H. F. Brinson, ed., ASM International, Materials Park, OH, 1990.
4. R. Knipe, Green cleaning technologies, *Advanced Materials and Processes*, August 1997.
5. D. Satas, Surface modification, Chapter 3 in *Plastic Finishing and Decorating*, D. Satas, ed., Van Nostrand Reinhold, New York, 1986.
6. G. W. Schuelke, Modern trends in corona treating, in *Proceedings of the Polymers, Laminations and Coatings Conference TAPPI*, Boston, September 24–26, 1984, pp. 249–252.
7. Plasma3, Enercon Industries Corporation.
8. Sicor, CSIRO Manufacturing and Infrastructure Technology.
9. D. Hibbard, Sticky Science, *Adhesives Age*, November 2001.
10. Institute for Environmental Physics, Johannes Kepler University, Linz, Australia.
11. T. Long, et al., Central functionalized tri-block copolymer for surface modification with switchable surface properties, presented at the American Chemical Society's 220th National Meeting, Washington, DC, August 2000.
12. V. Sebastian et. al., *New Priming Technology for Top Adhesion Performances of UV Curable Coatings on Plastic Substrates*, Ciba Specialty Chemicals, Inc.

13. L. Saggese, Paintless parts, *Machine Design*, February 19, 2004, pp. 94–100.
14. R. K. Hughes, Electrostatic printing, Chapter 16 in *Plastic Finishing and Decorating*, D. Satas, ed., Van Nostrand Reinhold, New York, 1986.
15. E. A. Muccio, Finishing and decorating plastic parts, *Plastic Part Technology*, ASM International, Materials Park, OH, 1991.
16. T. A. Richardson, Decorating processes, in *Modern Industrial Plastics*, Howard W. Sams, New York, 1974.

CHAPTER 12

Polymer Nanocomposites in Processing

NANDIKA A. D'SOUZA and LAXMI K. SAHU
University of North Texas, Denton, Texas

AJIT RANADE
GE Advanced Materials, Mt. Vernon, Indiana

WILL STRAUSS
Raytheon Company, McKinney, Texas

ALEJANDRO HERNANDEZ-LUNA
Texas Instruments, Inc., Dallas, Texas

12.1 HETEROGENEOUS FILLED POLYMERS

At the macroscopic level, a *composite material* is generally described as a combination of two or more phases, which will result in a system with superior characteristics to those of the individual components themselves. All composites have two basic components, the matrix or host, and the reinforcement or filler. The *matrix* is an element that gives shape to a composite and acts as a load-transfer medium to the filler. The *filler* is designed to optimize selected mechanical or other properties, such as electrical or thermal conductivity, or EMI shielding, of the composite. Sandwich composites are an exception to this classification, since they could be constituted as individual layers independent of one another. In this case, a bonding phase is also a constituent that could affect the final performance of the entire system. *Fiber composites* are defined as composites where the fibers could be long or short, unidirectional, planar, or three-dimensional. In flake or platelet composites, the reinforcement architecture is such that one of the dimensions, generally the thickness, is much

Handbook of Plastic Processes Edited by Charles A. Harper

smaller than the other two. In particulate composites, the reinforcement is microscopic and all dimensions are similar. Different forms of particulate fillers can be used in a polymeric matrix:

- *Particles:* calcium carbonate, in natural or synthetic form
- *Fibers:* glass fibers, carbon fibers, and other processed minerals
- *Plate-shaped particles:* kaolin, mica, talcum, and aluminum hydroxide

Some of these materials have to be surface-modified by means of coupling agents to make the inorganic material compatible with the polymer. Laminar composites are generally formed with no matrix and are composed of several panels that have identical dimensions in a sandwich-like structure. Each panel could be a composite itself [1–4].

The rapidly growing area of nanocomposites has resulted in inclusion of the dimensions of the reinforcements in classifying them. When the scale of even one dimension of the filler is in the range 1 to 10 nm, the resulting composite is considered a *nanocomposite.* According to the size of the reinforcement, composites can be divided into:

- *Macroscopic composites:* when the phases can be easily distinguished from one another without the use of a magnifying instrument, as in glass fiber composites.
- *Microscopic composites:* composites that generally have a matrix with the reinforcement phase embedded in the matrix. Both phases are separate, but for clear observation, microscopic techniques are used. Cermets belong to this class of composite.
- *Nanocomposites:* systems where at least one of the dimensions of the reinforcement phase is nanometric.

The interface between the matrix and the reinforcement can by physical or chemical. In physically bonded composites there is no chemical bonding among phases and the load in the composite is transferred by the action of physical forces. In chemically bonded composites there is chemical interaction among the components. Each composite phase can be a composite itself. By combining one or more of the divisions defined above, new groups or divisions can be found or created, but all of them have an essential purpose: to optimize the performance of the resulting system.

In this chapter we focus on nanocomposites modified by layered silicates. In particular we utilize montmorillonite, which is the most common nanofiller based on the cost–benefit value of its use. To aid readers in gaining a general understanding of other nanofillers, we first review nanofillers being used, then introduce the general principles of nanocomposites. Applications to a representative set of polymers are then reviewed. Finally, application-driven material application is reviewed for coatings, films, and foams.

12.2 NANOFILLERS

12.2.1 Nanoclays

There are generally two types of clays: synthetically prepared and naturally occurring. Clay minerals belong to the phyllosilicates [5]. There are three principal groups of naturally occurring phyllosilicates: montmorillonite, illite, and kaolinite. Montmorillonite layered silicates (MLS) have a 2 : 1 layered structure, illite has a 2 : 1 layered structure, and kaolinite has a 1 : 1 layered structure. Phyllosilicates are two-dimensional arrays of silicon–oxygen tetrahedra and tw o-dimensional arrays of aluminum or magnesium–oxygen–hydroxyl octahedra. In silicon–oxygen sheets, silicon atoms are coordinated with four oxygen atoms located on the four corners of a re gular tetrahedron with the silicon atom in the center (Figure 12.1). In the sheet,three neighboring tetrahedra share three of the four oxygen atoms of each tetrahedron,and the fourth oxygen atom of each tetrahedron is pointed downward as shown in Figure 12.2. The silicon–oxygen sheet is called a *tetrahedral sheet* or *silica sheet*. In the case of Al– and Mg–O–OH sheets, the Al or Mg atoms are coordinated with six oxygen atoms or OH groups. A regular octahedron is formed with the corners occupied by oxygen or OH groups with Al or Mg at

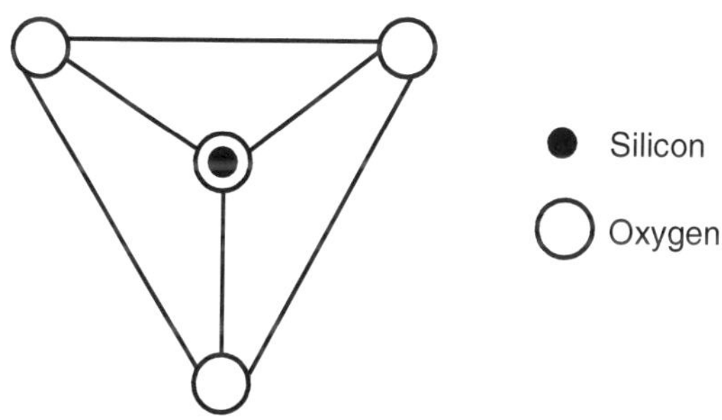

FIGURE 12.1 Tetrahedral arrangement of Si and O. (From Ref. 6.)

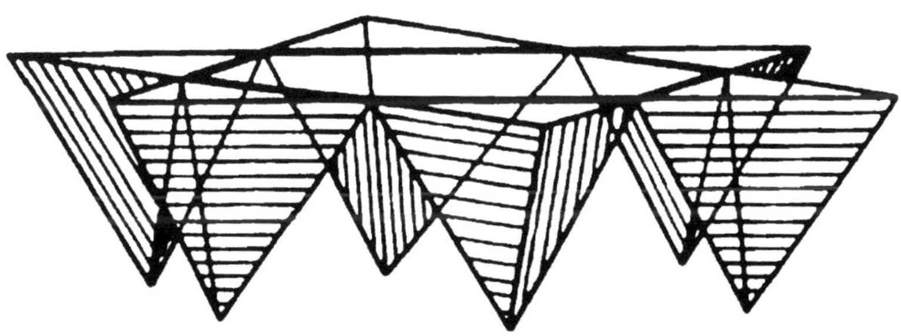

FIGURE 12.2 Tetrahedral linking. (From Ref. 6.)

the center (Figure 12.3). Sharing oxygen or OH groups with a neighboring octahedron results in the sheet structure shown in Figure 12.4, called an *octahedral sheet* or *alumina* or *magnesia sheet*.

Tetrahedral and octahedral sheets have similar symmetry and identical dimensions. This helps in the sharing of oxygen atoms between these sheets. Thus, an octahedral sheet shares the fourth protruding oxygen atom from a tetrahedral sheet. If sharing occurs between one silica and one alumina sheet, the result is a 1 : 1 layered mineral, and if one alumina sheet shares two oxygen atoms from two silica sheets, it is a 2 : 1 layered mineral. Within each layer there is a repetition of structure and it is therefore referred to as a *unit cell*. The distance between a certain plane in the layer and the corresponding plane in the next layer, called *basal* or *d spacing*, can be determined by x-ray defraction (XRD). The basal spacing is around 7.2 Å in 1 : 1 layer clays and 9.2 Å in 2 : 1 layer clays. The bonding between the layers is weak (van der Waals type), but a strong covalent bond exists between the atoms of the same layer.

The suggested crystallographic structure for mica-type layered silicate is shown in Figure 12.5. This structure is derived from pyrophyllite and talc by substitution of certain atoms for other atoms. In the tetrahedral sheet, Si^{4+} replaces Al^{3+}. In an octahedral sheet, Al^{3+} replaces Mg^{2+}. The substitution of higher valence atoms results in a deficit of positive charge or an excess of negative charge. The excess of negative charge within a layer is balanced by the adsorption of cations. These clays, which are hydrophilic in nature, have Na^{+}, K^{+}, or Ca^{2+} cations in the layer gallery. Mica-type

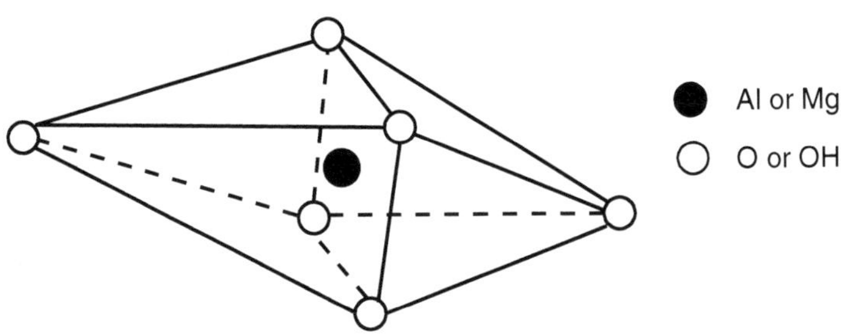

FIGURE 12.3 Octahedral arrangement of Al or Mg with O or OH.

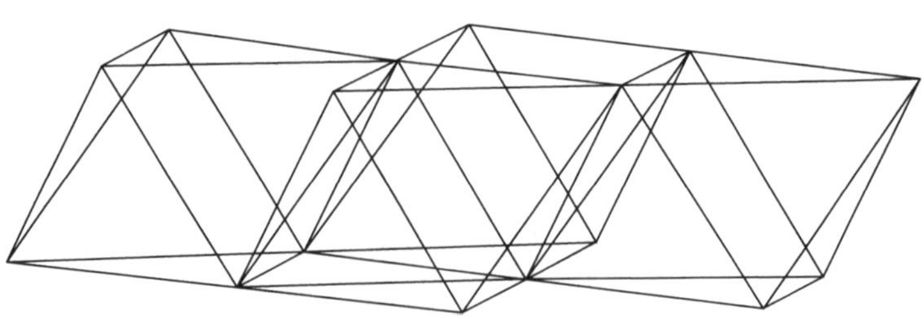

FIGURE 12.4 Octahedral linking.

layered silicates (MTSs) are distinguished by the location and type of cations in the layer. Table 12.1 summarizes the classification of MTSs. Due to a high concentration of negative charge, the surface oxygens of tetrahedrally substituted MTS are stronger electron donors than the surface oxygens of octahedrally substituted MTS.

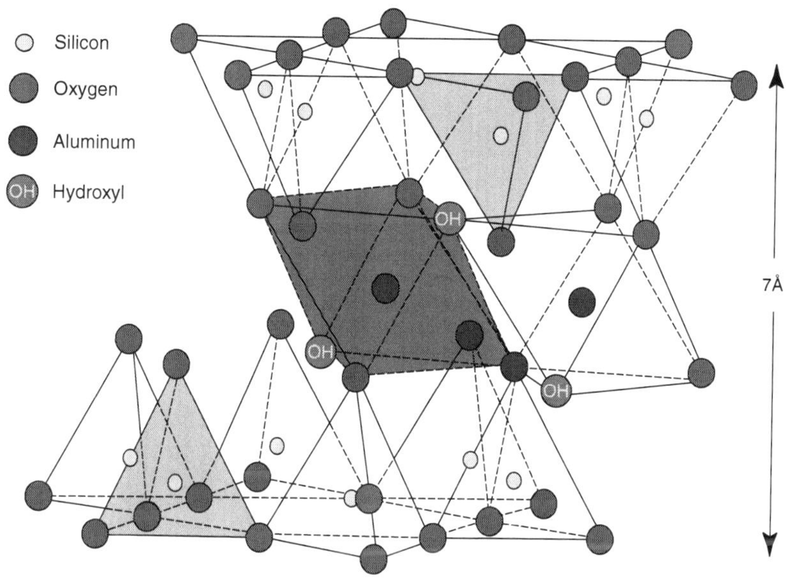

FIGURE 12.5 Suggested layered structure of mica-type layered silicates. (From Ref. 6.)

TABLE 12.1 Major Phyllosilicate Minerals in Soils

Layer Type	Group	Charge per Unit Formula	Common Minerals
1 : 1	Kaolinite–serpentine	~0	Kaolinite, halloysite, chyrsotile, lizardite, antigorite
2 : 1	Pyrophyllite–talc,	~0	Pyrophyllite, talc
	smectite, or montmorillonite	0.25–0.6	Montmorillonite (smectite), bedellite, notronite, saponite, hectorite, sauconite
	Mica	~1	Muscovite, paragonite, biotite, phlogopite
	Brittle mica	~2	Margarite, clintonite
	Illite	2	Illite
	Vermiculite	0.6–1.9	Vermiculite
2 : 1 : 1	Chlorite	Variable	Chlorite

Anionic clays such as layer double hydroxides (LDHs;Figure 12.6) have the general formula $M(II)_x\ M(III)(OH)_{2x+2}Y_n{\cdot}zH_2O$, where Y^{n-} is any of a large range of anions. Conventionally synthesized LDHs are strongly hydrophilic materials, either amorphous or microcrystalline with hexagonal shape, with the dominant faces developed parallel to the metal hydroxide layers. Adjacent layers are tightly bound to each other by their shared electrostatic attraction to the interlayer anion. Unlike smectite clays, such materials are not easily exfoliated, because of their high charge density and the fact that anions exposed by exfoliation have less exothermic solvation energy than do cations exposed by smectite exfoliation. In contrast, recent results from Braterman indicate that controlled shape, controlled charge density, and hydrophobicity can be obtained with his LDHs [7]. Additionally, Braterman has advanced the concept of shape and charge density control in LDH by synthesizing LDH as ribbons and needles [8]. Intercalation of polymeric anions into LDH gives a new class of polymer–clay nanocomposites [9–12], achieved in two ways. Oriakhi et al. [9] reported direct intercalation of polyacrylate, polyvinylsulfonate, and polyvinylstyrene into MgAl, CaAl, ZnAl, and CoAl LDHs during the hydroxide precipitation process [9]. Polyaspartate was inserted in the Mg_3Al LDH in the same way [10]. The other reported method is to introduce the monomer anion first and then to polymerize in situ. For example, Tanaka et al. obtained Mg_2Al-acrylate LDH by exchanging LDH chloride or nitrate with acrylate. The anion was then polymerized with thermal treatment at 80°C [11]. As a polymer filler, the LDH, existing as its own phase, has been found to stabilize the polymers to heat and light, especially for halogen polymers such as polyvinyl chloride [12–14].

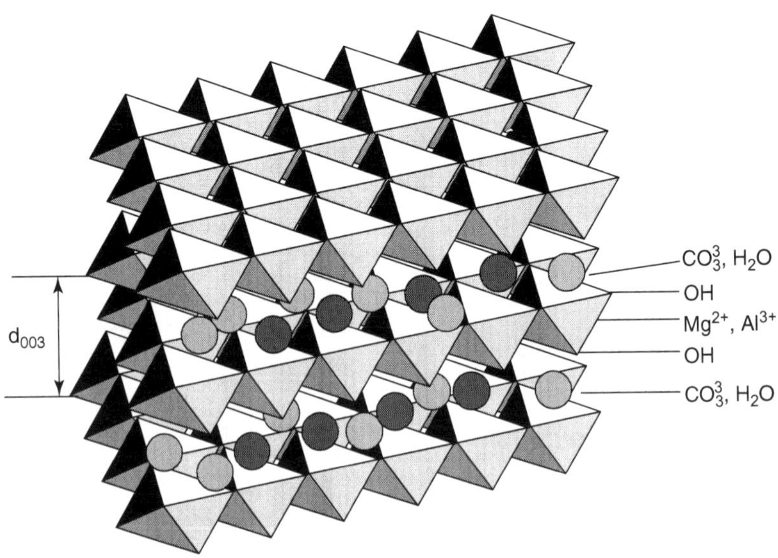

FIGURE 12.6 Layered double hydroxides or anionic-exchanged clays.

12.2.2 Carbon Nanofibers and Nanotubes

Since their discovery by Iijima in 1991 [15], carbon nanotubes have stimulated intensive research. Numerous investigators have reported unique mechanical and physical properties of carbon nanotubes due to their structure, size, and topology. A carbon nanotube (CNT) can be visualized as a single sheet of graphite rolled into a cylinder several micrometers in length and a few nanometers in diameter. The carbon atoms are arranged in a hexagonal array, and each carbon atom has three nearest neighbors.

Nanotubes can take two forms: single-walled (SWCNTs) and multiwalled (MWCNTs). Single-walled carbon nanotubes consist of one layer of carbon atoms through the thickness of the cylindrical wall and tend to assemble as a rope consisting of 10 to 100 nanotubes per bundle in ranom tangles. Multiwalled carbon nanotubes, the first to be discovered, consist of concentric cylinders around a common central hollow with constant separation between the layers held together by secondary, van der Waals bonding. Each cylinder ranges from 2 to 25 nm in diameter and is several micrometers in length [16]. Strengthening of polymers by carbon-based nanofillers has not been proven to be a reason for the incorporation of nanotubes. The costs are currently prohibitive. However, it has been shown experimentally that the introduction of nanotubes into a polymer matrix improves the electrical conductivity of the original polymer matrix. An ultralow electrical percolation threshold (0.0025 wt%) was reported for multiwalled carbon nanotube–epoxy composites [17]. Percolation occurs when a three-dimensional network of filler particle through the composites is formed, and the percolation threshold is characterized by a sharp drop in resistivity by several orders of magnitude. Nanotubes have been reported to be thermally stable up to 2800°C in vacuum and have a thermal conductivity twice as high as that of diamond. SWCNT–epoxy composites with only 1 wt% loading showed a 125% increase in thermal conductivity [18].

12.2.3 Polyhedral Oligomeric Silsesquioxane

Polyhedral oligomeric silsesquioxane (POSS) is a monodisperse nanoparticle containing a Si_8O_{12} core with an organic periphery. By functionalizing the Si on a POSS nanoparticle, reactive covalently bonded nanocomposites are possible. In some reports, seven of the eight Si atoms have an inert organic functionality such as isobutyl or cyclpentyl groups, and the eighth Si atom has a reactive group such as a styrenyl or norbonyl group. Copolymers with polyethylene, polypropylene, polybutadiene, polystyrene, and polynorbornene with POSS have been reported [19].

12.3 NANOCOMPOSITES

Most nanocomposites are based on the use of a low concentration of expandable smectite clays such as montmorillonite layered silicates (MLSs) in the matrix. Scientists have also tried to correlate and explain the behavior of nanocomposites through thermodynamic models to understand clay treatments and formation of nanocomposites [20–26]. Improved barrier properties [27–31], higher mechanical properties [32–36], improved flame retardancy [37–40], and increased dimensional stability [41–45] have been determined in polymer nanocomposites. All of these benefits are obtained without significant reduction in the density of the compound or light transmission [20]. The potential property enhancements have led to increased application in various fields, such as the automobile industry (exterior and interior body parts and fuel tanks), packaging industry (bottles, containers, plastic films), electronic industry (packaging material and exterior parts of electronic devices), coating industry (paints, wire enamel coatings) and aerospace industry (body parts of airplanes and exterior surface coatings) [46–62]. Multifunctional benefits such as mechanical performance, oxygen permeability resistance, and flame retardant characteristics are obtained simultaneously. In comparison to particulate-modified polymers, nanocomposites have a lower density than that of host polymers, due to perturbation of the polymer chains [63].

The addition of MLS in a polymeric matrix results in different clay distributions that do not always provide improved performance. When MLS is added to a polymer, it remains an immiscible system. This fact was known for years in the development of drilling fluid muds. However, engineering applications of MLS were recently found to be of interest when researchers at Toyota showed that surfactant-treated MLS is capable of hydrophobic interactions leading to partially or completely dispersed composites [53]. Prior to that, in 1961, Blumstein reported on polymerization of vinyl monomers intercalated into MLS [64].

Depending on the nature of the components used (layered silicate, organic cation, polymer matrix) and the preparation method, three principal types of nanocomposite structures are obtained: phase separated, intercalated, and exfoliated. Figure 12.7 shows the various types of structures observed in nanocomposites. A phase-separated microcomposite is obtained when a polymer is unable to intercalate between the silicate layers. Properties of polymer nanocomposites with the phase-separated structure

Phase Separated (*a*) — Intercalated (*b*) — Exfoliated (*c*)

FIGURE 12.7 Structures observed in polymer nanocomposites: (*a*) phase-separated; (*b*) intercalated; (*c*) exfoliated.

are in the same range as traditional microcomposites. Lack of miscibility between silicate and polymer is the primary reason for a phase-separated structure. An intercalated structure is formed when single or multiple polymer chains are intercalated between the silicate layers, resulting in a well-ordered multilayer morphology built up with alternating polymeric and inorganic layers. An intercalated structure results in finite expansion of intergallery spacing, but the long-range order of the layered silicate is retained. The degree of intercalation depends on the method of preparation, percentage of layered silicate, and miscibility between inorganic and organic phases. An exfoliated or delaminated structure is observed when there is a high degree of interaction between the charged silicates and the polymer. Nanoscale dispersion is observed and the structure is called *exfoliated* or *delaminated*. The silicate layers are well separated and there is no short-range order within the galleries. In general, exfoliated structures have resulted in better performance and are often sought as the desired structure in the final nanocomposite. This arises because the exfoliated dispersion gives rise to the largest interfacial area between the clay and the polymer.

A wide range of MLS concentrations have been sought in the quest for an ideal system that combines low concentration with good mechanical performance. The paradox that nanocomposites face mirrors the one that long-fiber-reinforced composite technology faced earlier: how to obtain both strength and toughness in the same system. The bulk of nanocomposite literature indicates that increased strength, or elastic modulus, considerably decreases strain to failure and ductility.

Nanocomposites have shown additional advantages over conventional macroscopic composites. Most of the emphasis has focused on the improvement in the mechanical properties of the various systems. These properties include an increase in mechanical performance where tensile strength has increased and the elastic modulus has almost doubled [65]. A decrease in the permeability of the nanocomposite has been a primary advantage of the technology [66]. The decreased permeability is attributed to increased path length for the permeating medium due to the MLS platelets in the path. The addition of MLS improved resistance to organic solvents such as toluene and chloroform, as well as increasing the heat distortion temperature, reducing the thermal expansion coefficients, and improving flame retardance and wear resistance [67,68]. Due to the light refraction nature of the fillers used, these composites are generally optically opaque, but due to the degree of molecular interaction it is possible to obtain nanocomposites with a good degree of transparency. Various polymers have been modified with MLS to obtain better performance in specific applications. Some of these systems are listed below.

12.4 PROCESSING POLYMER CLAY NANOCOMPOSITES

Techniques for polymerization include melt intercalation, solution polymerization, and in situ polymerization. Melt intercalation is primarily applicable to thermoplastics and is adaptable to conventional processing technologies such as injection and extrusion. Solution polymerization is followed when the polymer is mixed with an organic solvent, the clay swells, and the polymer chains diffuse between the clay platelets. When the solvent is evaporated, an intercalated nanocomoposite often results. Solvent polymerization is a useful route for coating applications.

12.4.1 Melt Intercalation

In 1993, Vaia and his colleagues were the first to report melt intercalation [62]. This method is used to make thermoplastic nanocomposites but the method is not applicable for thermosets because the process involves the melting of a polymer. The process involves blending a molten thermoplastic with organoclay and then annealing at a temperature above the glass transition temperature of the polymer. The mechanism is shown in Figures 12.8 and 12.9.

The observation that polymer chains can demonstrate a center of mass diffusion in essentially two dimensions is rather surprising because the unperturbed chain diameters (10 nm) are approximately an order of magnitude greater than the interlayer distance between the silicate sheets (1 nm). The proposed driving force for this mechanism is the important enthalpic contribution of polymer–clay interactions during the blending and annealing steps. Since the process involves the blending of polymer with clay, melt intercalation is one of the most popular methods to prepare nanocomposites in industry. Various thermoplastics (e.g., polypropylene, polyethylene, nylon, polystyrene) have been tried to make nanocomposites by this technique.

12.4.2 In Situ Polymerization

In situ polymerization was the first method used to synthesize polymer clay nanocomposites. The first polymer used was nylon (polyamide 6). At present this method is in common use for the preparation of thermoset nanocomposites.

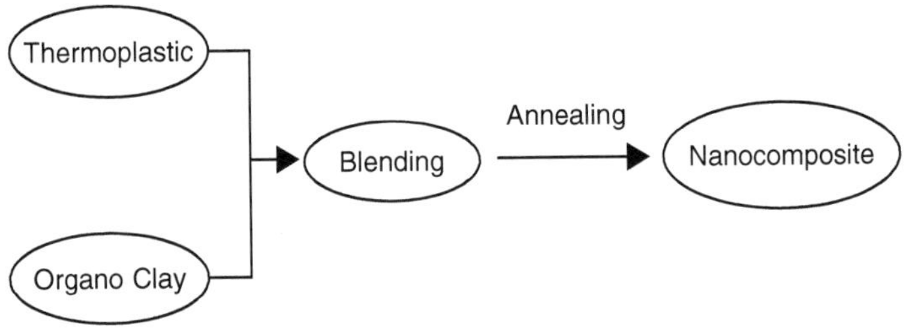

FIGURE 12.8 Melt intercalation approach.

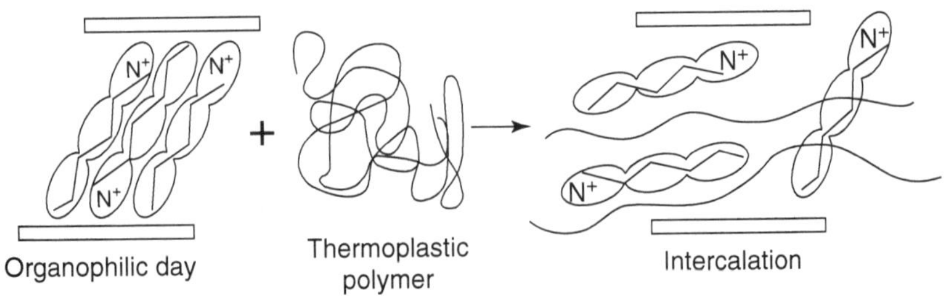

FIGURE 12.9 Chemical steps involved in the melt intercalation approach.

Figure 12.10 is a flowchart showing various steps involved in in situ polymerization. The first step is to achieve the swelling of organoclay in monomer. This is a time-consuming step since swelling depends on the polarity of monomer molecules, surface treatment of clay, and swelling temperature. Then depending on the type of polymer, the reaction is initiated. In thermoplastics, polymerization is achieved by addition of free radicals or by an increase in temperature. In case of thermosets, a curing agent is added to initiate polymerization.

The driving force for in situ polymerization depends on the polarity of monomer molecules. Due to the high surface energy of the clay, monomer molecules get diffused between clay layers. The diffusion is continued until equilibrium is reached. During polymerization, monomer starts to react with the curing agent or free radical. This reaction lowers the overall polarity of the intercalated molecules so that more polar molecules are driven between the clay layers. The overall reaction is shown in Figure 12.11. The most important thing is to control the polymerization occurring between layers (*intragallery polymerization*). If the cure kinetic is lower between the layers than outside the layers (*extragallery polymerization*), the delamination of clay is possible. Therefore, greater importance is given to intragallery polymerization than to extragallery polymerization.

This synthesis method allows control of the molecular weight of the polymer and the distribution of molecular weights. The polymer growth process, known as *living polymerization*, allows additional monomers to be added to grow block copolymers. But the method is limited to polymers that can be grown by living polymerization. Strategies with conventional epoxies are organized on two grounds. In one case, the epoxy monomer swells the clay, and subsequently the curing agent is added (two-stage). In the second case, epoxy and hardner are mixed and the clay added (one-stage). The two-stage method is generally followed.

12.5 FACTORS AFFECTING PROCESSING

Conditions favorable to the formation of a nanocomposite are the degree of compatibility or hydrophilicity between the clay and the host polymer. This is influenced by the length of the cation appended to the clay mineral. Determining the choice of clay that is suitable to a particular polymer when synthesis of clays in the laboratory is

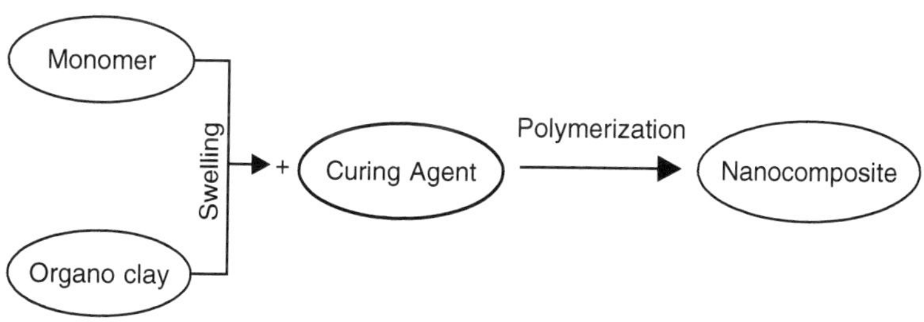

FIGURE 12.10 In situ polymerization approach.

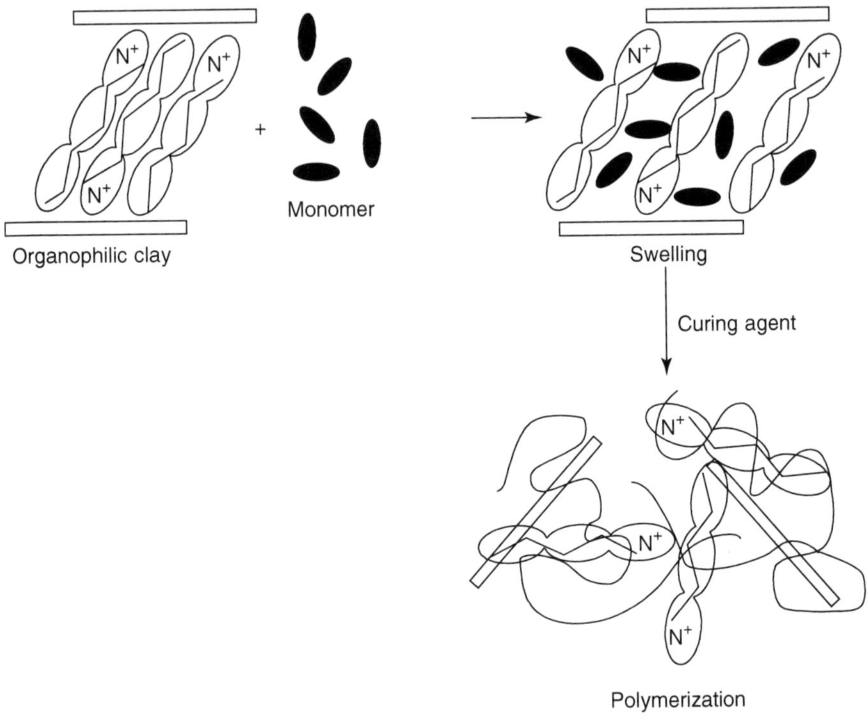

FIGURE 12.11 Chemical steps involved in the in situ polymerization approach.

not possible can be done by forming solutions of the polymer and determining the viscosity of the clay–polymer dispersion. By determing the type of clay that enables maximum viscosity, a suitable clay can be chosen for the melt compounding steps. The principle of this method is that when exfoliated, clay would have the maximum perturbation of the polymer solution and increase the viscosity via increased interfacial area. Surfactant treatment of clay is covered in Section 12.5.1.

Large-scale processing of nanocomposites depends on the thermal stability of the clay. Although the dimensions of the exfoliated platelet indicate that transparency and optical clarity are attainable, practical processing of nanocomposites has been limited by a yellowness, especially when temperatures exceed 220°C (common for most polymers). Factors affecting the formation of polymer organoclay nanocomposites by melt compounding inlcude molecular architechture of the modified, compatiblizer concentration and acid value, shear and processing temperatures, presence of additives, and polymer type and molecular weight [69].

12.5.1 Surface Treatment of Clay Minerals

Clays are hydrophilic by nature, which can affect the miscibility of clay with an organophilic polymer. Therefore, the surface of the clay is treated to make it hydrophobic and more organophilic. This is a cation-exchange reaction where inorganic cations such as Ca^{2+}, K^+, and Na^+ are replaced by organically modified cations such as alkyl ammonium ($R–NH_3^+$) groups, where R is a long aliphatic residue. The cationic head of the aliphatic group is found close to the surface layer, due to the presence of negative charge on the layer and the tail without charge is found away from the layer. Four types of structures are possible, depending on the packing density, temperature and chain length of the aliphatic groups. These are shown in Figure 12.12. When the alkyl ammonium chains lie parallel to the host layers, a lateral monolayer or lateral bilayer is formed. If alkyl ammonium chains radiate away from the host layers, an extended paraffin-type monolayer or paraffin-type bilayer structure is formed [70,71].

Hydrophilic clays are typically intercalated by alkyl ammonium ion (onium) using an ion-exchange reaction. Based on the alkyl chain length, x-ray diffraction and Fourier transform infrared spectroscopy (FTIR) are typically used to determine which structure is formed in the clay prior to introduction into the polymer. When alkyl ammonium ions with 12 to 18 carbons are ion-exchanged for the inorganic cations in smectic clays with a cation-exchange capacity (CEC) value of 100 to 120 mEq per 100 g, a lateral bilayer (18 Å) or an inclined paraffinlike structure (20 to 23 Å) results. At higher concentrations of onium ions, a lipidlike bilayer is also formed. For epoxies in particular, a detailed study was conducted to understand the influence of alkyl chain length on the basal spacing of the clay. Lan et al. [34] calculated the final structure by first assuming that the gallery cations would orient vertically from their initial monolayer or bilayer orientation. Basal spacings were expected to follow the relation $d_{001}=1.27(n-1)+d_A+r_M$, where $n-1$ is the number of methylene groups in the onium ion chain, d_A is the basal spacing for the NH_4^- montmorillonite (12.8 Å), and r_M is the van der Waals radius of the methyl end group (3 Å), and 1.17 Å is the contribution due to the CH_2 segments when the chain adopts an all-trans configuration. Table 12. 2 shows the results. The difference in the epoxy solvated value and the calculated value for $x = 1$ was determined to be caused by the onium ion taking an inclined orientation as opposed to a vertical one. The similar values between the measured and predicted values for $x > 8$ implied that for higher chain lengths, a vertical orientation was obtained and multiple epoxy monomers can be accommodated in the galleries. Vaia developed a schematic structure model for the organically modified fluorohectorite (FH–C_n)[6]. The schematic structure of a model is shown in Figure 12.13. Fourier transform infrared spectroscopy along with x-ray diffraction was used to investigate the model. By monitoring the frequency shifts of the asymmetric CH_2 stretching and bending vibrations, they found that the alkyl ammonium chains exist in states with varying degrees of order. Depending on the number of carbon atoms present in an alkyl ammonium chain, the model adopts a more ordered structure. For the least number of carbon atoms (i.e., when the chain is short), the molecules are isolated from each other. For intermediate chain lengths, a structure with a varying degree of in-plane disorder is formed. At higher chain lengths, an increase in interlayer order was determined. Thus as the interlayer packing density increases, the alkyl ammonium chains adopt a more disordered structure.

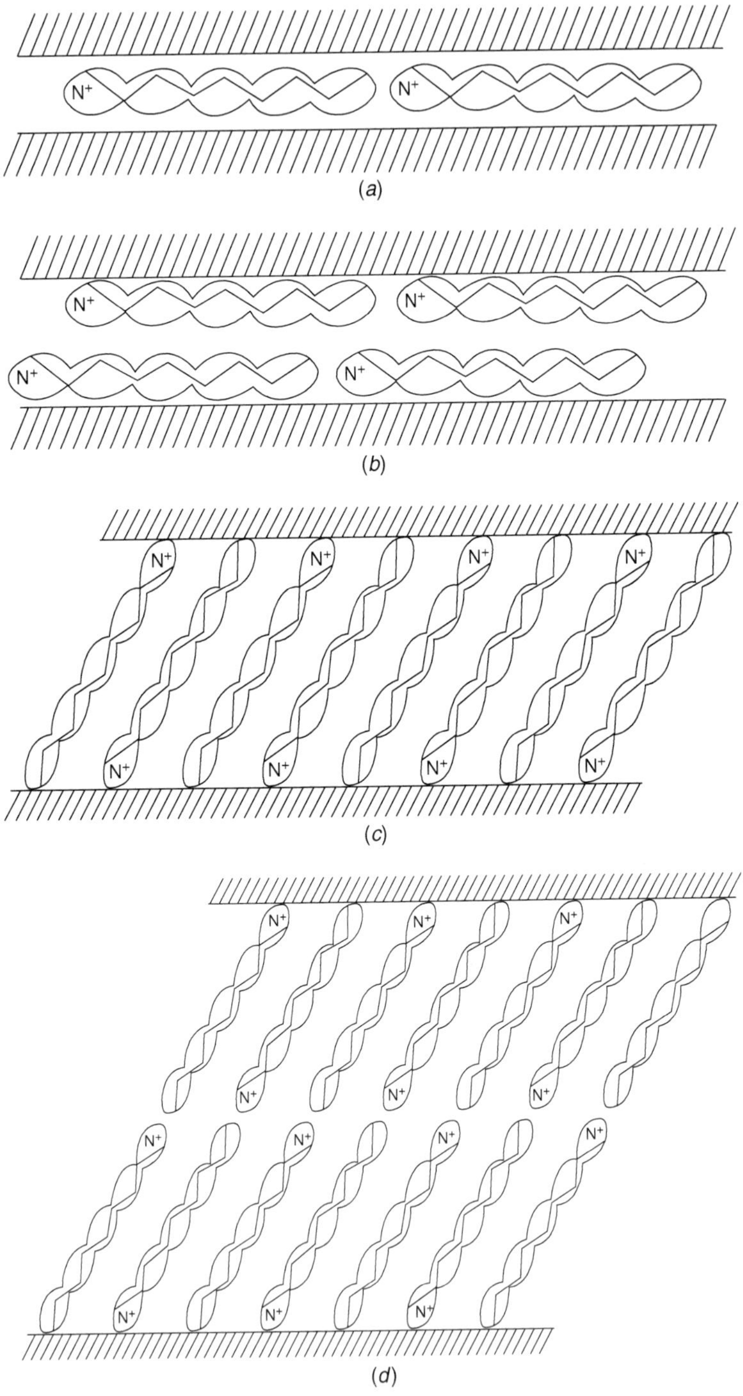

FIGURE 12.12 (*a*) Lateral monolayer; (*b*) lateral bilayer; (*c*) paraffin-type monolayer; (*d*) paraffin-type bilayer. (From Ref. 6.)

TABLE 12.2 Surfactant Length Effects on Clay Spacing (Å)

$CH_3(CH_2)_xNH_3^+$	Air Dried	Epoxy Solvated	Calculated Distance Between Plates	Initial Cation Orientation
3	13.5	16.5	19.6	Monolayer
7	13.8	27.2	24.7	Monolayer
9	13.8	30	27.2	Monolayer
11	15.6	31.9	29.8	Bilayer
15	17.6	34.1	34.9	Bilayer
17	18	36.7	37.4	Bilayer

The organic modifier chain length within the silicate galleries plays a crucial role in determining the dispersion beha vior in nanocomposites. Vaia et al. [20, 24–26] predicted the conformation of aliphatic chains between silicate galleries, depending on the number of carbon atoms n. Three types of conformations are possible, a lateral monolayer (up to six carbon atoms), a pseudo bilayer (9 to 12 carbon atoms), and finally, a pseudo trilayer (13 to 18 carbon atoms). The intercalation behavior depends on the number of carbon atoms. No intercalation occurs for a pseudo bilayer. For n greater than 12, intercalated hybrids are formed. The gallery height increases as the chain length increases, resulting in hybrid dispersions. Vaia and his colleagues also observed the relation between the type of intercalation and the number of carbon atoms. F or intercalated nanocomposites ($n>12$) the full width at half maximum of the basal reflections of the intercalated structure is comparable to that of the original nonintercalated silicate, indicating that polymer intercalation does not disrupt the stacking order of the silicate layers. F or a full pseudo bilayer ($n=12$), intercalated reflections are observed with lower intensities and greater breadth than the original silicate. The increasing n value is reflected by the enhanced ability to draw polymer chains between the silicate layers, to decrease the v an der Waals attractive forces, and to develop intercalated or exfoliated dispersions. Lan et al. [34] described how the structure of the organic modifier of the MLS, as well as the MLS itself, influenced exfoliation in polyimide nanocomposites. They found that use of surfactants with a longer chain length (4 to 18 carbon atoms) led to e xfoliation. The acidity of surfactants and charge density also played a crucial role in the dispersion. Lan et al. [35] have shown that fewer substituted ammonium ions (primary and secondary) leads to more e xfoliated epoxy-based nanocomposites rather than highly substituted ions. They attribute the difference in exfoliation behavior to the level of acidity of the ammonium ion and the catalytic ef fect this has on the polymerization of epoxy.

The polarity of the host matrix is also a central issue dictating the compatibility of a clay choice with a polymer. For instance, being nonpolar, polyolefins have a low interaction potential with the pendant cations and ha ve shown limited success as nanocomposites. Commonly, a compatibilizer such as maleated polypropylene (PP) is utilized to f acilitate the interaction. On the other end of the spectrum, the hydrophilic nature of n ylon enables exfoliated nanocomposites to be formed in in situ and melt-compounded nanocomposites. In forming polymethyl methacrylate

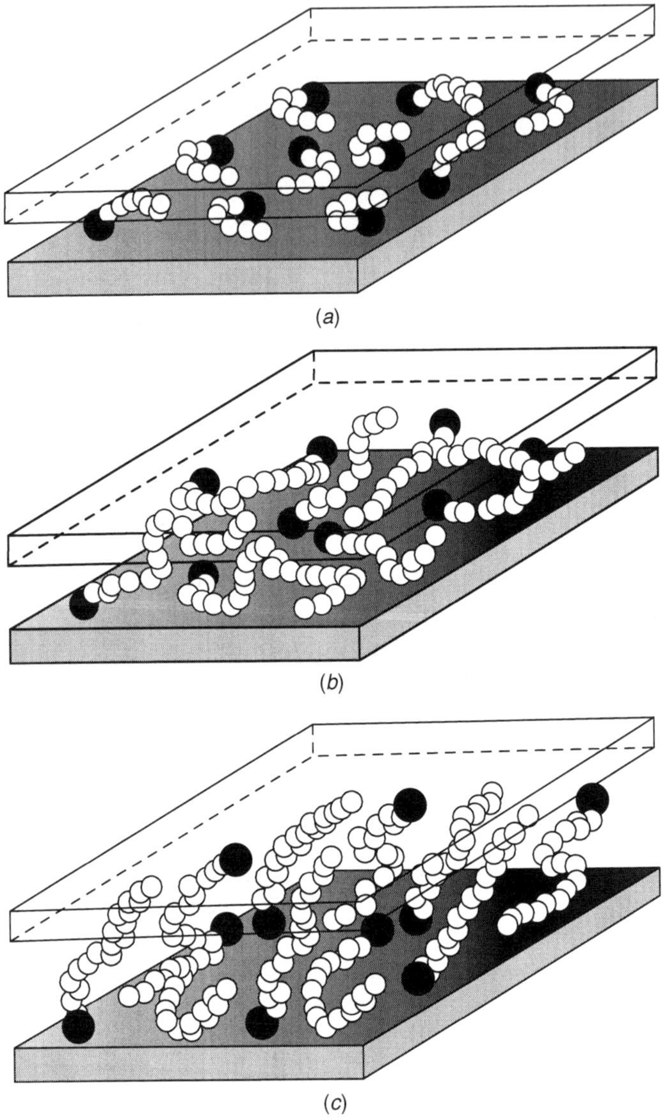

FIGURE 12.13 Alkyl chain aggregation model for (*a*) the shortest, (*b*) the intermediate, and (*c*) the longer lengths of carbon atoms. (Open circles represent CH_2 segments; filled circles represent cationic head groups.) (From Ref. 6.)

(PMMA) nanocomposites, Park and Jana [72] report on the use of epoxy in combination with PMMA and clay. Increases of 40% and 25% in tensile and impact strength at 2 wt% clay loading was obtained over the nanocomposites formed without epoxy as a compatibilizer. To form nanocomposites with epoxy, first epoxy, hardener, and clay intercalated mixtures were formed with up to 30% clay loading. Ground dried PMMA pellets were then mixed with the epoxy–clay mixtures to form a paste which was dried and then mixed in a Brabender.

Functionalization of the base sodium montmorillonite for compatibility with PP has been reported by Merinska et al. [73]. A blend of Na^+ MLS, octadecylamine (ODA), and stearic acid (STA) was stirred at 80°C, then held at 80° without stirring for 1 hour. The MLS was ground to a fine powder. PP and the modified clay materials were prepared using a one-screw kneader and a two-screw lab extruder. Optimal intercalation was achieved when 120 g of Na^+ MLS combined with 60 g of ODA and STA, Li reported on the use of MLS as a means to graft a maleated PP (mPP) nanocomposite [74]. A peroxide-treated MLS was first made. Three grams of stearyl ammonium–modified MLS was dispersed into acetone and stirred ultrasonically for 1 hour. One gram of dicumyl peroxide dissolved in acetone was added slowly. The entire mixture was stirred for 3 hours. The resulting MLS had a lower degree of chain scission than is generally obtained in nanocomposites, due to thermal instability of the reactive constituents. It is clear that chemical compatibility and shear history are critical in successful nanocomposite formation. Wang et al. [86,87] investigated the effect of shear and compatibilizer viscosity on dispersion. Comparing a highly maleated high-molecular-weight mPP with a low-molecular-weight low-MA PP, they found that effective compounding by maintainance of a steady torque required changes in the process temperatures. For low-MA systems 187°C was required, whereas 171°C was used for high-MA systems. Molecular weight was seen to have no impact on the degree of interacation, but maleic anhydride content had a higher impact. A parallel investigation was conducted by Hasegawa and Usuki [75]. Modified polyolefins grafted with a 0.09 to 4.5% MA group were investigated. The influence of shear and surfactant was investigated. The investigators found that the copolymers intercalate into the MLS galleries with and without shear However, when accompanied by shear, a higher degree of exfoliation occurs. In making polyethylene (PE) nanocomposites, the choice of extruder configuration is most important. Mehrabzadeh et al. [76] compared a corotating twin-screw five-zone extruder with intermeshing screws. A mean residence time of 75 seconds was maintained in the base configuration. The alternative configuration included more blister, mixing, and reverse elements, leading to more shear and residence times (95 seconds). Comparing the XRD results for both configurations for a 5% clay loading revealed a more exfoliated structure when intense shear and mixing were utilized.

Lew et al. [70] investigated layered silicate exfoliation in metallocene-catalyzed and conventional Ziegler–Natta-catalyzed LLDOE using melt extrusion. They found that metallocene-based LLDPEs had a significantly higher degree of exfoliation then did the Ziegler–Natta-based systems. The researchers attributed the differences to the smaller molecular-weight distribution in the metallocene-catalyzed LLDPEs, which resulted in higher extrusion shear stress. A combination of maleated PE with LLDPE prior to the addition of MLS was found to be optimal for exfoliated dispersions.

Bureau et al. [77] investigated the formulation impact on PP nanocomposites. They introduced a dry blend of PP and maleated PP into an extruder while the nanoclay was fed through a side feeder. The master batch was dry-blended with additional coupling agent and PP to give a range of compositions. In the absence of coupling agents the clay aggregates sheared and broke into large stacks. In the presence of the coupling agent, they reported that the diffusion process was accompanied by a shear effect. Polymer molecules diffused into clay galleries and then the layers peeled off by shear. The influence of the surfactant is determined by the surface tension changes in the MLS. In essence the problem with matrix polarity occurs because alkyl ammonium–modified clays have surface tensions in the range 280 to 30 mJ·m^2, whereas PP and mPP–modified PP has around 20.6 mJ/m^2 [78]. In choosing clays, the cation-exchange capacity is often considered. Typical surfactants and influence on clay basal spacing are shown in Table 12.3. Comparing clay MLS 30B to MLS 20A, the under- and over-exchanged MLS can be seen (the base MLS is 92.6 mEq/g).

12.6 THERMOPLASTIC NANOCOMPOSITES

The potential barrier property improvement of nanocomposites has resulted in an interest in polyolefins, nylon, ethylene vinyl alcohol, and polyethylene terephthalate. In this section representative polymers from thermoplastic nanocomposites that best represent other systems are reviewed. Since dispersion is of greatest importance, the challenges of making nonpolar polymers into nanocomposites are first reviewed. The hydrophilic systems and the relatively easier dispersion of MLS are reviewed in the nylon system. Finally, the challenges of materials that are yet to develop adequate wettability between the polymer melt and the clay are reviewed in PET.

12.6.1 Polyolefin Nanocomposites

PP has been used due to its low cost and good performance. With the addition of MLS, a cost-effective structural material can be obtained. PP nanocomposites have found a special place in the automotive industry, where its flexibility and low density have been used as a substitute for metallic parts. A tough nanocomposite will promote diversification of PP in automotive and other industries.

TABLE 12.3 Intergallery Spacing of MLS 125 mEq, 90 mEq, and 95 mEq

Type of MLS	Surfactant	CEC (mEq)	(001) Spacing (Å)
MLS	None	92.6	11.7
MLS 30B	Methyl, tallow, bis-2-hydroxyethyl, quaternary ammonium	90.0	18.5
MLS 20A	Dimethyl, dihydrogenated tallow, quaternary ammonium	95.0	24.0
MLS 15A	Dimethyl, dihydrogenated tallow, quaternary ammonium	125	31.5

PP nanocomposites have been prepared by in situ polymerization or melt processing. To obtain a nanocomposite with improved mechanical properties, a stacked layer structure of MLS separated into monolayers in a polymer matrix is achieved more easily. With this dispersion, an increment in the aspect ratio is obtained and reinforcement effects are enhanced. Bonding between the MLS and the matrix requires control of the interfacial affinity between the MLS particle surface and the matrix of the polymer. The influence of the MLS on the termination rates of PP polymerization has made the first option difficult. In the case of melt processing, the nonpolar nature of the propylene repeat unit renders intercalation of the chains between the silicate layers difficult (i.e., in situ polymerization). Blending PP with mPP or styrene–acrylonitrile (SAN) are two of the most often followed routes [45]. The poor mechanical properties in SAN-modified systems has led to an increased interest in mPP.

mPP has proven effective in PP+ glass fiber composites [79]. The maleic anhydride segment is compatible with the alkyl ammonium salt, while the PP is compatible with the PP host. Exfoliation of the smectite MLS is enabled through migration of the mPP between the MLS sheets. To compound these systems, two methods are followed. In one method, the mPP is blended with the smectic MLS to form an exfoliated system, which is then mixed into PP. In the second method, a three-component system is compounded together. The first method is followed more commonly, based on the premise that exfoliation of the MLS is highly likely in mPP+MLS mixtures. However, given that the surfactant degradation is around 220 to 280°C, single batches are also utilized.

Different techniques can be used to obtain a PP nanocomposite [49], such as mechanical shear, solvents, and monomer intercalation. An important reason for studying nanoparticle-filled nanocomposites is that the fracture mechanisms in these systems may be quite different from conventional composites. It has been speculated that the rigid particles will resist the propagation of cracks, causing them to bend between particles. However, when these rigid particle sizes are in the nanoscale region, this concept is questionable, requiring resolution.

Wang et al. [80] investigated the formation of nanocomposites of PP, mPP, and MLS. An intermeshing, corotating twin-screw extruder was used. The screw consisted of 10 segmented barrels with three kneading zones (Figure 12.14), the first of which began at the second barrel with high shear disk blocks and ended with neutral blocks. The second started at the fourth barrel with high shear elements and ended with reverse elements. The third one started at the seventh barrel with a wide-pitched low-shear element. In the first and second kneading zones, more severe shearing action was assumed due to the high-shear disk blocks. Increased residence time was ensured by the reverse blocks. A wide-pitched shear element ensured gentle shearing and homogenization of the polymer melt. The polymer–filler therefore experienced high-intensity dispersive mixing in the first and second kneading zones and distributive mixing in the third kneading zone. A wide range of systems were investigated with the variables being the molecular weight of the PP, maleic anhydride content of the the compatibilizer. The base PP resin had a melt index of 2. They found that the lower the molecular weight of the maleated PP and the higher the maleic anhydride ratio of the mPP the better the dispersion of the polymer. Mechanical properties were not investigated.

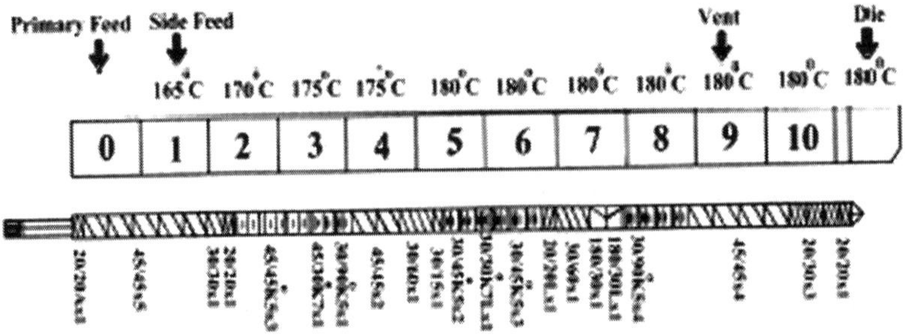

FIGURE 12.14 Schematics of twin screw configuration and barrel temperature profiles of PSM 30.

Some of the previous research in PP nanocomposites can be summarized using the work of Wang et al. observed in Table 12.4, where low concentrations of MLS (5%) but high concentrations of compatibilizer were used [43,47,81,82]. Different types of MLS and compatibilizer were used. Also included are results for PP and the compatibilizer by themselves. This work suggests that the compositions with the highest degree of improvement are composites 5 and 6, with high ultimate tensile strength. It can be observed from nanocomposite 7 that even when the tensile strength and the elastic modulus are not as high, impact strength is greatest. These results indicate that high compatibilizer concentration will not result in higher impact resistance, which could be extended to another property, fracture toughness. Thus, our own work has utilized low mPP concentrations.

12.6.1.1 Influence of Gating on Nanocomposite Formation

D'Souza's research group has investigated the influence of processing on the dispersion of montmorillonite in polypropylene nanocomposites. A single-gated injection-molded PP nanocomposite was analyzed (Figure 12.15). XRD was used to determine if the montmorillonite dispersion in PP was intercalated or exfoliated. The XRD scans of the sample surfaces are shown in Figure 12.16. First considering the interaction between the interlayer spacing of the MLS and the PP +mPP blend, one can see that there was retention of the peaks at low 2θ angles (<5°) in the samples of NC 1, NC 2, NC 3, and NC 5. There was also a slight shift to lower MLS interlayer d spacing or higher 2θ, which we attribute to the experimental error in obtaining MLS spectra from a different batch of material. Comparing the full width at half maximum for this peak at $2\theta=3°$ (Table 12. 5) for the various samples, it can be seen that the width of the peak increases gradually with increasing MLS concentration. This indicates decreased structural regularity of the MLS interlayers in PP as the MLS concentration is increased.

The transmission electron microscopy (TEM) images of nanocomposites with MLS concentration of 1, 3, and 5% are shown in Figure 12.17. A high degree of exfoliation is evident, as is platelet separation ranging from 30 to 300 nm, indicating a highly exfoliated structure. The TEM analysis of the specimens seemed to be

TABLE 12.4 Mechanical Properties of PP Nanocomposites

Material	Tensile Strength (MPA)	Elastic Modulus (MPA)	Impact Strength (Izod) (J/m)
PP	34.2 ± 0.2	1680 ± 10	19 ± 2
PB 3150	31.7 ± 0.2	1580 ± 60	
PP/20A (95/5)	35.7 ± 0.2	2060 ± 70	22 ± 2
PP/E43/20A (80/15/5)	36.6 ± 0.2	2400 ± 100	16 ± 2
PP/PB3200/20A (80/15/5)	38.2 ± 0.3	2500 ± 70	21 ± 2
PP/PB3150/20A (80/15/5)	39.0 ± 0.6	2440 ± 20	23 ± 2
PP/PB3150/20A (90/5/5)	36.3 ± 0.1	2220 ± 80	24.7 ± 1
PP/PB3150/I30 (80/15/5)	33.8 ± 0.2	1900 ± 100	14 ± 2
PP/PB3150/nylon 6/I30 (80/15/5)	30.0 ± 0.2	2000 ± 30	13 ± 2

Source: Data from Ref. 43.

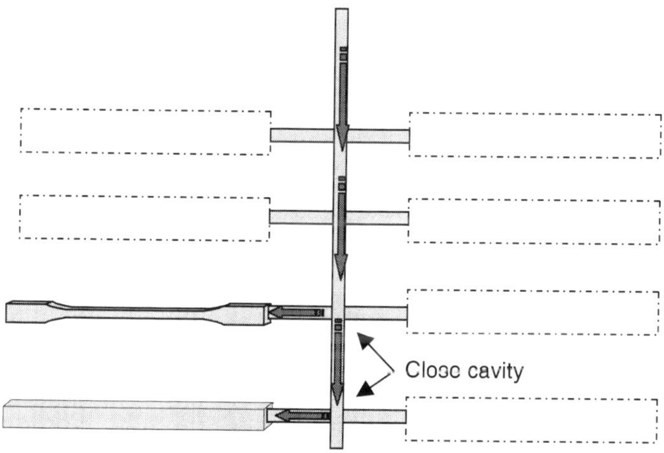

FIGURE 12.15 Injection flow inside a mold.

inconsistent with the x-ray diffraction results. One significant difference was that the x-ray analysis was done across the cross section of the surface, whereas the TEM samples were taken from the core of the specimens. Further analysis was therefore conducted after removing the surface layers of the sample by polishing at least 1 mm from the skin of the sample. We refer to the polished samples as the core. The x-ray spectra (Figure 12.18) indicated that all nanocomposites had no peak corresponding to the 001 interlayer basal spacing of the MLS, indicating an exfoliated structure within the core. Thus, we conclude that the PP nanocomposites show a skin–core structure of montmorillonite dispersion with an intercalated skin and exfoliated core.

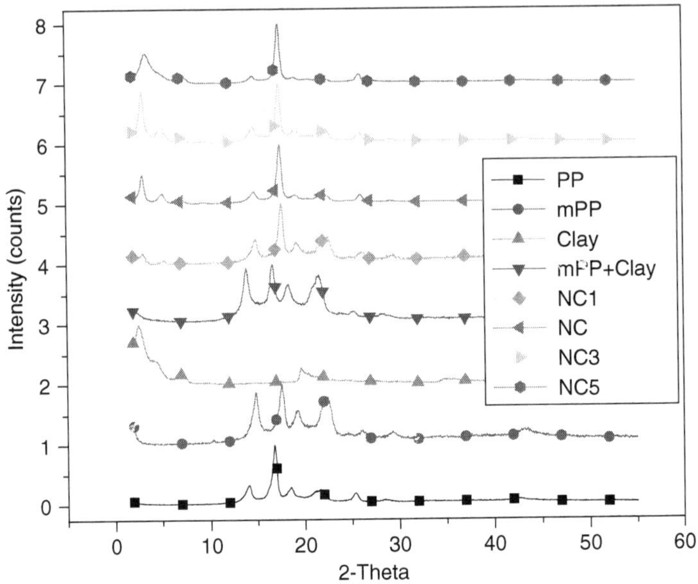

FIGURE 12.16 XRD of a PP injection-molded sample as processed.

TABLE 12.5 Full Width at Half Maximum of the XRD Peaks of a PP Nanocomposite

Material	2θ	FWHM
Clay	2.68	0.546
NC 1	3.195	0.553
	5.347	0.391
NC 2	3.101	0.43
	5.154	0.54
	7.253	0.482
NC 3	3.099	0.471
	5.198	0.564
	7.158	0.504
NC 5	3.548	0.776
	4.944	0.792
	7.351	0.85

12.6.1.2 Influence of Montmorillonite on γ-PP Crystallinity

The correlation between the presence of montmorillonite on PP crystallization w as investigated using XRD, DSC, and polarized OM. Structural changes in PP can be followed using x-ray dif fraction. Pure PP, with no filler, has five peaks in the 2 θ range between 10 and 30° which correspond to a monoclinic α phase (Table 12.6). The monoclinic α phase is the most studied phase, but other phases include hexagonal β and triclinic γ phases [83]. Analysis of the XRD and clear demarcation

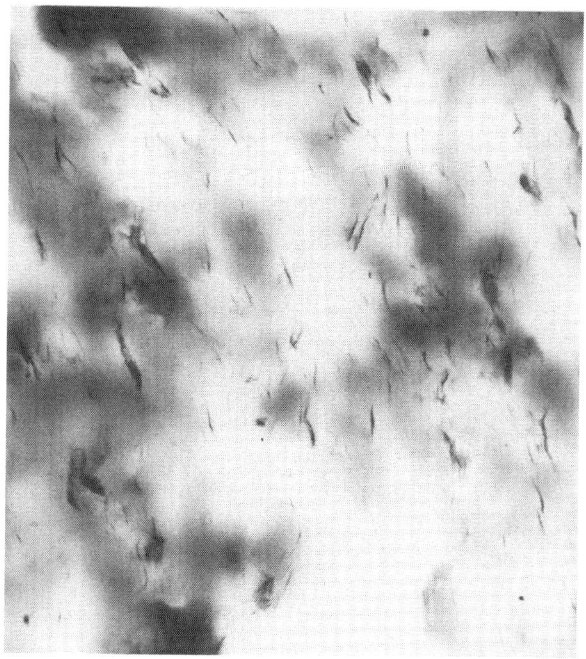

FIGURE 12.17 TEM of a PP nanocomposite.

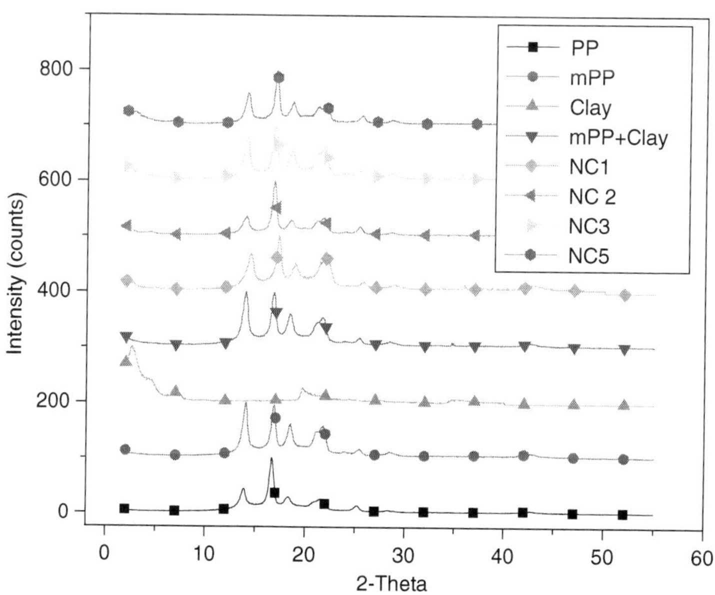

FIGURE 12.18 XRD of polished PP nanocomposite samples.

between the α and γ phase is complicated by the epitaxial relationship of the α phase to the γ phase such that each can grow on the lamella of the other [84,85]. A demarcation between the α and γ phases is made by comparing peak intensities at $2\theta = 14°$ to $2\theta = 17°$. The 14° peak is seen to be sum of the α peak at 14.08° and the γ peak at 13.84°, while the peak of 17° is the sum of the α peak at 16.95° and the γ peak at 16.72°. When the latter peak is higher, it implies a dominance of the γ phase [86] . The addition of filler can modify the relative intensity of these peaks, as in the case of nanocomposites, where the intensity of peak II is strongly af fected. As the filler size decreases, the intensity of peak II increases [86,87]. The development of γ–phase PP is influenced by crystallization under high pressure, use of nucleating agents, and ethylene comonomer units in the chain. The peak assignments of the α and γ forms relevant to this study and the results of our skin and core XRD are shown in Table 12.6. The ratio of intensities of the $2\theta = 16°$ peak to $2\theta = 14°$ peak are shown in Table 12.7. First analyzing the XRD spectra of the unpolished samples, it can be seen that the PP has a strong γ-phase dominance with the intensity ratio approaching 4. mPP, however, shows a α-phase dominance. The PP + mPP blend revealed that even a small fraction of the α mPP resulted in the dominant growth of the α phase in the blend. When the MLS was introduced into the system, there was a gradual increase in the γ-phase growth and NC5 indicates strong evidence of high γ-phase content. This is clear in the stack ed XRD spectra of the surf ace shown in

TABLE 12.6 XRD Peaks and Reference α- and γ-Phase Peaks

Unpolished						
PP	14.05	16.85	18.6	21.6		25.35
MPP	14.2	16.95	18.7	21.3	22	
PP+mPP	14.9	17.7	19.45	22.45		25.65
mPP–NC	14	16.75	18.45	21.65		25.45
NC1	15	17.75	19.4	22.6		26.2
NC3	14.85	17.5	19.35	22.45		26
NC5	14.8	17.55	Low res.	Low res.		26.15
Polished						
PP	14.1	16.7	18.45	21.65		25.4
MPP	14.2	16.95	18.7	21.3	22	
PP+mPP	14.1	16.9	18.6	21.2	21.8	25.6
mPP–NC	14.05	16.9	18.55	21.2	21.8	25.35
NC1	14.5	17.25	19	21.5	22.15	25.85
NC3	14.1	16.85	18.6	21.2	21.8	25.55
NC5	14.1	16.95	18.65	21.25	21.9	25.65
			Reference Values			
	I	II	III	IV	V	VI
	14.08	16.95	18.5	21.2	21.85	25.5
	16.1	16.6	21.3	24.7	28.07	
	13.84	15.05	16.72	20.07	21.2	21.88

TABLE 12.7 Intensity Ratio for the Surface and Inner Portions of the Specimens for Peaks at 2θ Values of 16° to 14°

	$I_{16°}/I_{14°}$	
	Surface	Inner
PP	3.80	2.26
mPP	0.67	0.67
PP+mPP	1.08	0.95
mPP+MLS	0.96	0.99
NC 1	2.28	1.47
NC 3	3.31	1.21
NC 5	7.64	1.72

Figure 12.16. Previous research on nylon has indicated that the presence of MLS in a α nylon transforms the α phase to a γ phase. Our research indicates that as the clay concentration increases, there is a significant increase in the γ-phase content.

The polished core sample XRD indicated that the phase beha vior of the PP and the mPP are similar in the core and b ulk, but the degree of γ phase is much lower than that of the skin. Figure 12.19 sho ws tensile stress–strain relationships for the materials investigated with the concentrations defined in Table 12.8. As can be seen, the yield strength for almost all nanocomposites are similar to that of PPwith a maximum value observed in the NC 5 nanocomposite. These values are tabulated in Table 12.9.

Important differences can be observed from this analysis. Taking as a reference the behavior of PP, the matrix, we observe that the mPP + MLS specimen has a marked inferior behavior in the elongation properties and is more brittle. The rest of the samples show plastic behavior to a varying degree. NC 1, NC 2, and NC 3 do not reach a level of elongation comparable to PP, and only NC 5 reaches a value close to it. However, two samples did not fail following the same cycle, the specimen of PP + mPP and NC 5. A strain-hardened necking of the gauge length w as observed. Considering the first case, PP + mPP, one can determine that this system is a very ductile material, showing a larger plastic deformation before failure. In the case of the sample with NC 5 of MLS, the situation is different. The values of yield strength, elastic modulus, and elongation to f ailure increased considerably . Actually, it can be observ ed that as the MLS content increases, the elongation to failure is also increasing.

The simultaneous increase in both elastic properties and ductility implies a change in base PP properties being influenced by the MLS and the mPP . This was resolved through polarized microscop y (Figure 12.20). PP by itself has lar ge spherulites along the entire surf ace, above 60 μm on a verage. When the mPP is added, this size decreases slightly and spherulites of different sizes can be observed at 40 μm. When MLS is added, the nucleation of spherulites increases, leading to

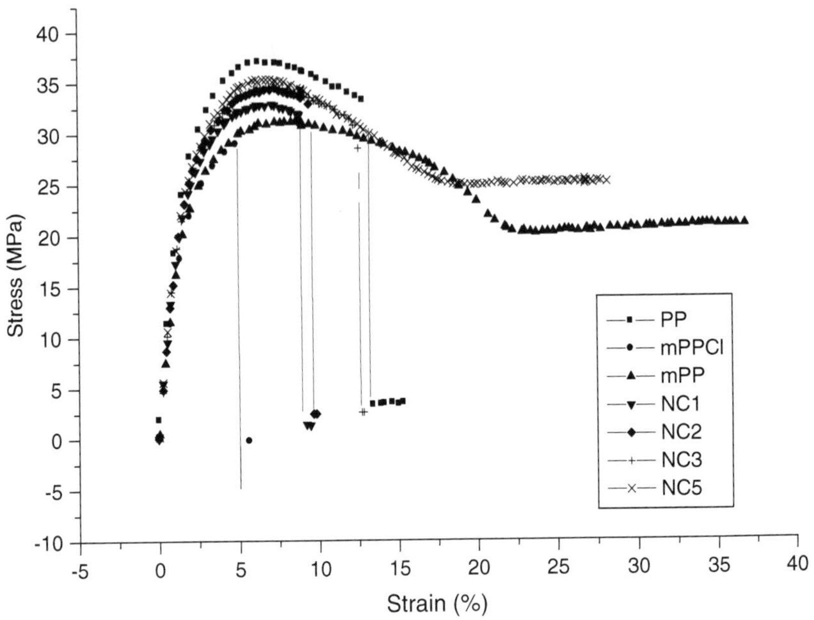

FIGURE 12.19 Tensile stress strain plots of PP nanocomposites showing that the PP and mPP work to synergistically form a PP+mPP blend. Nanocomposites show a gradual increase in strain to failure with concentration.

TABLE 12.8 Summary of the Mechanical Properties

Sample	*E* (GPa)	UTS (MPa)	Yield (MPa)
PP	1.90 ± 0.0266	37.24	19.93
mPP+MLS	1.49 ± 0.0313	29.16	16.20
PP+mPP	1.61 ± 0.0306	31.41	18.49
NC 1	1.62 ± 0.0194	32.89	18.31
NC 2	1.82 ± 0.0200	34.55	18.43
NC 3	1.70 ± 0.0306	34.78	18.73
NC 5	1.84 ± 0.0166	35.63	19.91

smaller sizes; spherulites $<20\,\mu m$ can be observed. For NC 5 the nucleation of spherulites generates distribution in the spherulitic size; both spherulite sizes as big as those found in PP and as small as in NC 1 are observed. Intermediate concentrations, as in NC 2 and NC 3, show a similar effect, but the spherulite sizes are not as large as in the PP specimen. The results in NC 5 seem to be the result of two effects: (1) the presence of the PP +mPP crystalline structure, where the small crystallites facilitate plastic deformation; and (2) the presence of another distribution of larger crystallites similar in dimension to NC 3 nanocomposites providing increased stiffness.

TABLE 12.9 Composition of the Specimens Used

Material	HPP	Polybond 3150	Cloisite
PP	100		
PP+mPP	98	2	
mPP+NC		98	2
NC 1	97	2	1
NC3	95	2	3
NC5	93	2	5

FIGURE 12.20 Optical microscopy of a PP +mPP nanocomposite showing bimodal crystallite distribution and maleated PP coating the spherulite surface.

Figure 12.21 show characteristic plots for each of the systems obtained from three-point bending testing. First, from the load–deformation curves we can see that for pure PP the fracture mechanism is the most brittle of all the samples and that the plastic zone is increasing as the concentration of MLS increases.

All the MLS-containing specimens showed higher fracture toughness than PP, with a maximum obtained at NC 5 content, where it is possible to observe that the K_q value is about 1.6 K_q of PP. This has a great impact, since no similar results have been obtained before in the literature. From these results we observe a trend similar to the one at room temperature. The larger values of K_q arise due to the higher load values, cause a deflection similar to that obtained at room temperature. This is significant from an application standpoint since automotive materials are specified to sustain stresses at temperatures as low as −30°C.

The use of maleated polyolefins has also been utilized in polyethylene [88,89]. Bae et al. [89] compared a 0.5 wt% maleic anhydride–grafted polyethyelene (PEMA) with a 1wt% maleic anhydride grafted polyethyelene. LLDPE, PEMA, and MLS were compounded using an internal mixer with a rotor speed of 80 rpm for 20

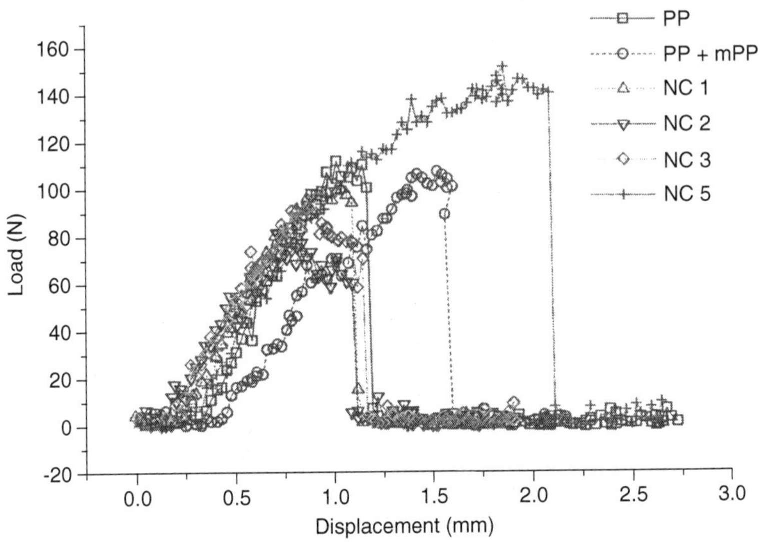

FIGURE 12.21 Characteristic three-point bending plots.

minutes and by a twin-screw extruder of 1-in. diameter and an *L/D* ratio of 30. They found that the twin-screw extruder produced an intercalated/exfoliated nanocomposite, while the internal mixer had a largely immiscible dispersion. In addition, as the concentration of PEMA increased, the degree of intercalation increased. The higher the concentration of maleic anhydride, the more the degree of dispersion. Mechanical properties of the twin-screw extruded sample were better than that of the internal mixer samples, but overall no improvement over the base LLDPE in mechanical properties was obtained. Barrier properties, however, showed substantial decrease (40% at 10% clay loading) over that of the base resin.

Bureau et al. [82] investigated the influence of shear history as opposed to compatibilizer polarity by comparing compression and injection-molded PP composites. They measured the essential work of fracture of the nanocomposites. They found that nanocomposites behave like microcomposites. By applying the Cox shear lag model and Halpin–Tsai equations, they found that the systems obey the microcomposite theory with an effective *l/d* ratio of 17 to 35. The influence of compatibilizer and shearing conditions are therefore related to the degree of dispersion or the size of the micro stacked layers.

12.6.2 Nylon Nanocomposites

Work on nylon nanocomposites includes an in situ polymerization method used by Kojima et al. to make nylon 6 nanocomposites [90]. Kojima found nylon 6 nanocomposites less crystalline compared to neat nylon 6. He also observed the presence of a γ-crystalline form in nylon 6 nanocomposites. Neat nylon 6 has only α-type crystalline structure, and formation of γ form was a direct evidence of change in crystallinity due to addition of MLS. Mathias used a solid-state NMR technique to study the crystallization behavior of nylon 6 nanocomposites [91]. He also concluded that clay addition favored γ-crystalline formation. Liu et al. also reported γ-crystalline formation [92]. Akkapeddi used a melt intercalation method where MLS and nylon 6 was mixed in a twin-screw extruder [93]. Surprisingly, nylon 6 nanocomposites showed higher crystallinity than did neat nylon 6. Akkapeddi related the crystallinity increase to orientation of the thin clay platelets, exerting high pressure on nylon 6 chains in a twin-screw extruder. Devaux observed the effect of clay addition on the crystalline melting temperature. The presence of clay in nylon 6 increases the polymer crystallization temperature and decreases the melting point [94]. Decreased melting is associated with smaller crystallites and more uniform thermal transfer, while increased nucleation is reflected in higher fusion or crystallization temperatures. The phenomenon hints at the role of clay as a nucleating agent. It must be recalled that nylon 6 is a semicrystalline and has two major crystalline phases: monoclinic α phase and hexagonal γ phase. Similar to polypropylene, pressure, nucleating agents, and processing conditions influence the crystalline phase development. The principal difference between the two phases is molecular packing. The α phase is the thermodynamically most stable phase where antiparallel chains are attached by hydrogen bonds. The γ phase consists of pleated parallel chains joined by hydrogen bonds. It is a less stable phase. Twin-screw compounded nylon 6 nanocomposites showed properties similar to those of in situ polymerized nanocomposites [95].

Ranade et al. investigated the mechanical properties and helium permeability of dispersed MLS in nylon 6 nanocomposite films [96]. A 20 wt% MLS in a nylon 6 master batch was prepared in a twin-screw extruder and then mixed with nylon to produce nylon+MLS nanocomposites with MLS content ranging from 1 to 5% for testing. The nylon nanocomposites were extruded into film. A highly exfoliated dispersion in different films was determined using x-ray diffraction. Nylon 6 nanocomposites showed formation of γ-crystalline form, which was absent in neat nylon 6. Differential scanning calorimetry (DSC) confirmed the crystallinity change and the role of MLS as an additional nucleation contributor. The glass transition and mechanical properties were related to the nucleation capabilities of MLS. Increase in degree and type of crystallinity plus tortuous path development due to exfoliated dispersion were primary reasons behind improvement in gas barrier properties of exfoliated nylon 6 nanocomposites.

Processing effects on nylon-6, formation of color and polymer degradation were reviewed thoroughly by Fornes et al. [97] investigated nylon-t. Processsing resulted in a significant decrease in molecular weight due to the reaction between the organoclay surfactant and the nylon 6 matrix. The level of degradation depended on the molecular weight of the nylon 6 used. The higher the molecular weight of the nylon, the more the degree of molecular weight reduction. The coloring of the polymer was related to the level of unsaturation on the surfactant. The more the double bonds, the more the discoloration of the polymer. The addition of an antioxidant was found to be promising in hindering the molecular weight reduction.

Chavarria and Paul [98] compared nylon 6 and nylon 6,6 nanocomposites. They found that the failure strain was significantly lower in the nylon 6,6 nanocomposites than in the nylon 6. The lower level of exfoliation in nylon 6,6 nanocomposites was attributed to lower interacation between the organoclay surfactant and the nylon 6,6 system. Borse et al. compared the crystallization of nylon 6 versus nylon 6,6. They found that the absence of a γ form in nylon 6,6 results in different growth parameters in the nanocomposites.

In situ polycondensation polymerization of nylon 6,6 nanocomposites with hectorite were reported by Kalkan and Goettler [99]. A remarkable result that they report is on the retention of molecular weight in the nanocomposite sample.

A takeup condition effect on the fracture of nylon nanocomposites was investigated by Avlar and Qiao [100]. They found that while nylon 6 fracture resistance is insensitive to cooling rate, nanocomposites are signficantly affected. Increasing cooling rate and addition of the MLS resulted in finer crystallites and higher fracture toughness. The result can be considered in the context of the predominance of the α and γ phases in nylon analyzed by Zhou et al. [101]. By annealing nylon at different temperatures and comparing the resulting crystal phase ratios, they found that lower temperatures of annealing (similar effects to high quench rates) result in a dominance of the α phase. Higher-temperature anneals (or slower cooling rates) result in a γ-phase dominance.

The influence of flow direction on the dispersion of MLS in nylon was investigated by Yalcin and Cakmak for an injection-molded nanocomposite [102]. They found that in fast crystallizable polymers such as polyolefins, the formation of spherulites through the thicknss of an injection-molded sample is possible. However, in nylon, crystallization is impeded against the mold wall, and only in the core are clearly formed spherulites visible in the nanocomposites. Nylon 6 fibers were melt spun by Razafimahefa et al. [103]. They found that the introduction of clay resulted in formation of the γ form of nylon increasing the crystallization temperature and decreasing the melting point. The lower enthalpy of melting is indicative of a larger amorphous moiety in the filled yarns. Dyeing the resulting polymer indicated that the greater amorphous region permitted increased permeation of the dye into the yarn.

12.6.3 Polyester Nanocomposites

The preparation of PET nanocomposites has not been as successful as that of other polymer nanocomposites. Matayabas and his colleagues prepared PET nanocomposites through melt compounding [104]. He observed a decrease in the inherent viscosity of PET nanocomposites, indicating degradation due to molecular weight loss. The degradation increased with increasing MLS concentration. He used higher molecular-weight PET to compensate for the molecular-weight degradation, but this did not alleviate the problem. Takekoshi and colleagues used the in situ polymerization method to make PET nanocomposites with quaternary ammonium salt–modified MLS [105]. Uniform nanoparticle dispersion and improved impact strength and modulus were observed in the samples. The melt mixing method was also tried for PET nanocomposites, but poor dispersion properties were observed. Matayabas et al. [104] related poor dispersion properties to the low decomposition temperature (240 °C) of organic modifier bound to MLS surface. Davis et al. [106] showed that better dispersion could be achieved by using an organic modifier which is thermally stable at PET processing conditions. Davis melt-blended PET with thermally more stable organically modified MLS nanocomposites. A higher level of dispersion and delamination was observed with MLS, which had a thermally stable modifier. Organically modified MLS with a decomposition temperature less than PET processing temperatures showed poor dispersion properties. This was due to early loss of modifiers, which affected the miscibility between organic polymer and inorganic MLS. Nanocomposites made with two MLSs cation-exchanged with dimethyl dihydrogenated tallow quaternary ammonium at two concentrations, 95 (MLS-95) and 125 mEq per 100 g (MLS-125), were investigated. The host MLS has a cation-exchange capacity (CEC) around 92 mEq per 100 g of clay. The effect of exchanging past the CEC is studied with relation to the dispersion, glass transition, and mechanical properties. Films were extruded on a single-screw extruder with a film die attached. XRD and optical transmission measurements indicated that increasing surfactant concentration changed the dispersion from intercalated to immiscible. The differences in the dispersion were related to the inherent degradation temperature of the MLS treated with the different surfactant concentrations relative to the processing temperature. The overexchanged MLS (MLS-125) had more surfactant degradation, leading to an immiscible dispersion, and narrower XRD peaks reflected a higher degree of ordering. In contrast, MLS-95 nanocomposites had increased breadth and decreased ordering. MLS orientation, however, did not influence birefringence measured by polarized infrared spectroscopy (FTIR). All nanocomposites did not show significant changes in the glass transition temperatures [107].

12.7 THERMOSET NANOCOMPOSITES

A variety of clays have shown a very high degree of reinforcement in epoxy systems. Alliphatic amine, aromatic amine, anhydride, and catalytic curing agents have been used to form epoxy matrices with broad glass transition temperatures. However, a more effective application of this nanolayer has been found in the mechanical properties when the polymer is in the rubbery state. The dimensional stability, thermal stability, and solvent resistance on the glassy matrix can be improved when the MLS nanolayers are present.

Since treatment of the clay is vital to generating nanocomposites and epoxy polymerization, a critical contributor to the generation of nanocomposites, it is important to consider if the alkyl ammonium surfactant itself contributes to the polymerization. Lan et al. [32,33] argue that the acidity of the alkyl ammonium ions may catalyze the homopolymerization of the digycidyl either of bisphenol A molecules inside the clay galleries. Kornmann et al. [42] investigated the effect of the cation-exchange capacity of the base clay on epoxy polymerization. If the CEC is high, more alkyl ammonium surfactant concentration would lie next to the clay, and therefore the effect of homopolymerizaiton would be clarified. First in treating the low-CEC clay, an increase in (001) spacing from 9.7 Å to 17.2 Å was obtained for the organoclay, whereas the low CEC clay swelled from 12.4 Å to 21.4 Å. When swollen in DGEBA resin at 75°C, the organoclays quickly increased in d spacing to 30 Å. This was associated with the development of a lateral bilayer of perpendicular orientation. As the time for swelling reached 24 hours, the interlamellar spacing of the higher CEC clay was constant around 34 Å, whereas the low CEC clay was completely exfoliated. An increase in viscosity of the low-CEC clay was also apparent, which the authors explained as resulting from self-polymerization. However, they also acknowledged that the exfoliated state with increased effective clay particle size would also lead to increased viscosity. Thus, Kornmann et al. [42] showed that a low CEC value permits higher diffusion of alkyl ammonium ions. Butzloff et al. [108] also investigated the competing homopolymerization versus epoxy + curing agent polymerization. For a given clay, they showed that the kinetics of the two-part system given by the epoxy + organoclay are substantially lower than that of epoxy + hardener; therefore, the homopolymerization of the epoxy and clay had little influence on the final nanocomposite. By increasing the swelling temperature to 80°C and decreasing the time for epoxy + clay swelling to 1 hour, they found that the DSC trace showed only a single hump as opposed to the two-staged hump shown by Lan et al. at 75°C. They investigated the influence of changing clay concentration and found that all the kinetics of the three-part (epoxy + clay + curing agent) system dominated the homopolymerization. However, there were definite signs that the alkyl ammonium ions were participatory in the reaction since changing the heating rate changed the area under the curve of the polymerization. One of the few attempts to study the high-T_g epoxy resins was recently made by Becker et al. [109]. They investigated the tri- and tetrafunctional epoxies. They found that whereas the DGEBA system had an exfoliated dispersion, the high cross-link density resins had intercalated dispersions. A general increase in modulus and fracture toughness was obtained. This was related to the intercalated dispersion. The comparative results of curing agent, epoxy, and clay on dispersion are shown in Table 12.10.

12.8 APPLICATION-DRIVEN NANOCOMPOSITES

There are a few significant areas where nanocomposites have the potential to be cost-effective. Coatings with montmorillonite have a higher hardness, films with nanocomposites have a higher barrier to oxygen and moisture, and foams with nanocomposites show a higher modulus and toughness than those of the unmodified systems. These three are investigated below.

12.8.1 Nanocomposite Coatings

Coatings are made using the solution method. It is important first to form a stable suspension. Often, real-world coatings are made up of a number of solvents.

TABLE 12.10 Resin Choice Effects on Dispersion

Epoxy Resin	Curing Agent	Clay Treatment	Intercalated or Exfoliated?
DGEBA	$NH_2CHCH_3CH_2(OCH_2CHCH_3)_xNH_2$, $x=2.6$; commercial name, Jeffamine 230	Clay treated with octadecylamine	Exfoliated, $d>88$ nm
	3,3-dimethylmethylenedi (cyclohexylamine) (3DCM)	Clay treated with octadecylamine	Intercalated, $d\sim 40$ Å
	Amicure bis-*para*-aminocyclohexyl-methane (PACM)	Clay treated with octadecylamine	Intercalated, $d=37$ nm
	DDM	Clay treated with octadecylamine	Exfoliated
	Jeffamine 230	I.30E (Nanocor)	Exfoliated
	Jeffamine 230	I.28E (Nanocor)	Intercalated–mixed at 65°C and degassed, cured at 75°C
	PDA	Clay treated with octadecylamine	Intercalated
	DDS	Clay treated with octadecylamine	Exfoliated
	Jeffamine D400 ($N=5.6$)	Bis (2-hydroxy-ethyl) methyl tallow ammonium MMT	Exfoliated
	Jeffamine D400 ($N=5.6$)	Dimethyl ditallow ammonium MMT	Intercalated
	Jeffamine D400 ($N=5.6$)	Dimethyl tallow benzylammonium MMT	Intercalated
	Jeffamine 230	Jeffamine 230	Exfoliated

One approach is to distribute the treated clay into the coating suspension and agitate until stability is achieved. Ranade et al. [110] followed a new approach, a clay suspension in a compatible solvent was formed, then introduced into the polymer suspension. PAI is formed when a triacid derivative is treated with a diamine. The resulting polymer contains both amide and imide links. PAI has the advantages of both polyamides and polyimides, such as excellent mechanical and thermal properties as well as solvent resistance. PAI is unaffected by aliphatic, aromatic, chlorinated, and fluorinated hydrocarbons, dilute acids, aldehydes, ketones, ethers, and esters. Like a thermoplastic, PAI can be processed by compression molding, injection molding, or by extrusion, but once cured it behaves like a thermoset by maintaining its mechanical properties around 200°C.

PAI has a glass transition temperature of around 250°C, making it thermally stable even at high operating temperatures. The combination of aromatic groups and imide linkages makes PAI mechanically strong. PAI has a found extensive use in the aerospace industry, affording weight reduction by replacing metals in such components as jet engine parts and electrical devices. PAI is used as a specialty coating applied to copper and aluminum conductors due to its high thermal endurance, high chemical resistance, good mechanical strength and sustained high-voltage resistance. PAI is used for coating applications because of its outstanding surface adhesion to different materials, including metals and Teflon. PAI also has a small coefficient of thermal expansion, making it ideal for coating applications. Suspensions of PAI are widely used in magnetic wire coatings in under-the-hood automotive applications and in coatings for high-voltage enamel wires. Increasing the glass transition temperature with retention of PAI thermooxidative stability and mechanical performance leads to increased reliability and service lifetimes.

PAI+MLS nanocomposites were prepared by solvent casting. Silicone molds of known size and shape were prepared. The clay was premixed with xylene for 1 hour to improve its swelling capacity. This premix was added very slowly into the PAI suspension with concurrent stirring. The suspension was stirred continuously with the help of a torque-controlled Calframo BDC 1850 digital mechanical stirrer with a stepwise increase in stirring speeds ranging from 50 to 1000 rpm. The schedule followed was 200 rpm for 5 seconds, 50 rpm for 20 seconds, 70 rpm for 90 seconds, and 1000 rpm for 10 seconds. This schedule was based on recording optimum conditions for obtaining nonaggregated films using a photoresist spinner. The fragility and brittle nature of the films made removal from the substrate difficult, so we switched to a controlled torque and casting in the silicon mold technique. The high rpm prevented the flocculation of clay particles. PAI suspension was a mixture of PAI oligomer plus four different aromatic solvents. A selective solvent evaporation technique was therefore used to decrease the effect of trapping the solvent and air. Samples were cured in a mechanical convection oven between 30 and 175°C. The temperature of the oven was raised by 10°C every 1½ hours until the final temperature was reached.

Five different compositions (1.0, 1.5, 2.0, 2.5, and 3.0%) of MMT by weight of PAI were prepared. The level of distribution of clay platelets in the PAI matrix was first determined using optical microscopy. Typically, the determination of an exfoliated nanocomposite has been based on XRD and TEM results. However, as we determined previously for epoxy–clay nanocomposites, scanning electron microscopy and optical microscopy has shown that at the *lower magnification* of SEM and OM scales, the exfoliated and intercalated nanocomposites might still show signs of aggregation similar to clay + water suspensions [111–114]. In aggregation, two or more particles clump together, touching only at certain points. Two states of aggregation are common in montmorillonite layered silicate (MLS)–water suspensions. Coagulation refers to dense aggregates produced by face–face-oriented particle associations. Flocculation is based on loose aggregates produced by among others, edge–face- or edge–edge-oriented particle association. Aggregated clays retain their identity but move kinetically as a single unit [115]. The state of aggregation is driven by heterogeneous charge distribution in natural occurring clays and leads to a number of particle interaction orientations in clay suspensions. Clay suspensions have shown many dispersions, including dispersed, face–face coagulated tactoids, edge–face flocculated, edge–edge flocculated, and coagulated particles. The state of aggregation hinders realization of uniform barrier properties. Optical microscopy images are shown in Figure 12.22 for the 1% MLS by weight sample and in Figure 12.23 for the MLS 3% by weight sample. With increasing clay concentration, the average clay agglomerate size was affected. The average agglomerate size were observed to be around 3.2, 5.0, 5.7, 7.4, and 8.1 μm for 1.0, 1.5, 2.0, 2.5, and 3.0% MLS by weight nanocomposites, respectively. As can be seen, the average platelet size ranged from 3 to 8 μm as the concentration of clay increased. As clay concentration increased, the average agglomerate separation of 12, 9, 8, 6, and 5 μm was observed for 1.0, 1.5, 2.0, 2.5, and 3.0% MLS by weight nanocomposites, respectively. As has been shown for clay+water mixtures, when clay is dispersed in an aqueous medium of near neutral to alkaline pH, the particles carry a net negative charge, due largely to isomorphic substitution of cations of lower charge for cations of higher charge (Al^{3+} for Si^{4+} in the tetrahedral sheet and Fe^{2+} or Mg^{2+} for Al^{3+} in the octahedral sheet). Meanwhile, the clay edges may carry a positive charge in nearly neutral to acid pH solution because of protonation of various atoms exposed at the edges. We surmise that in a solvent+polymer suspension, charge migration is facilitated through the aqueous medium. We have elsewhere determined increased ionic conductivity with increasing clay content in the solutions [6]. An exfoliated dispersion was observed at (1.0%) low clay concentration. Increase in clay concentration from 1.5

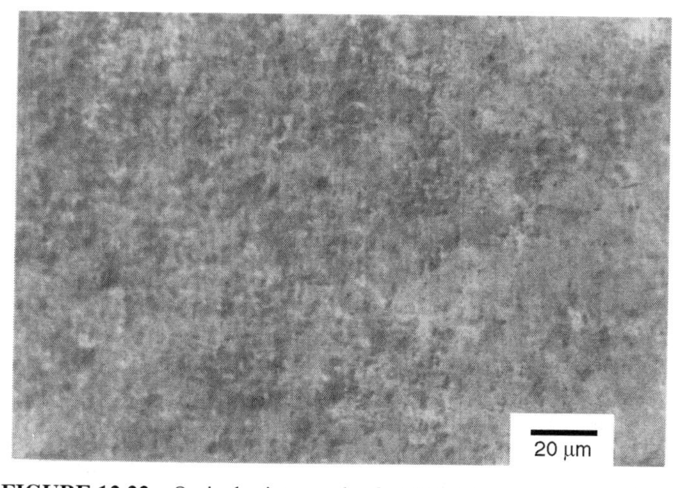

FIGURE 12.22 Optical micrograph of a 1% PAI nanocomposite film.

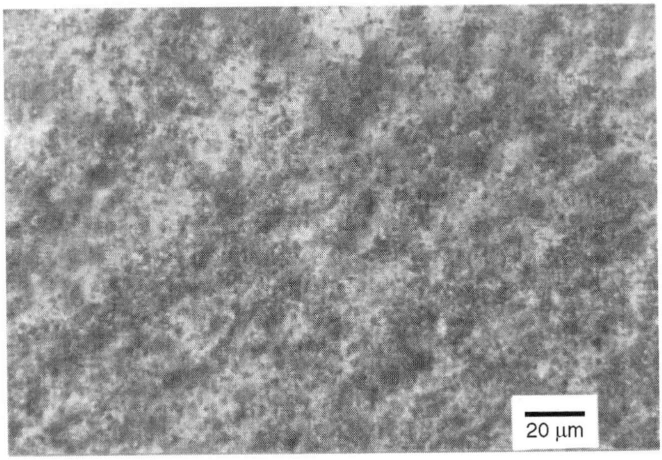

FIGURE 12.23 Optical micrograph of a 3% PAI nanocomposite film.

to 3.0% resulted in an intercalated dispersion. At low clay concentration, polymer clay interactions overcame the van der Waals forces between silicate galleries, resulting in complete disruption of clay structure. Due to an increase in clay concentration, van der Waals interaction dominated polymer clay interactions, resulting in a finite expansion of silicate galleries and retention of clay structure. We reached a critical concentration of 1.5 and 2.0% composite where high structural intercalation was observed; 2.5 and 3.0% showed partial intercalation and exfoliation, even though the *d* spacing increased to about 38 Å.

There has been much concern over the surfactant degradation temperature as being the upper limit of use of these materials. However, when FTIR (Figure 12.24) was conducted on samples before and after being subjected to 1 hour at 400°C, the results show no change in the base polymer.

Benefits of nanocomposite coatings are evident in the exemplary hardness obtained from a well-processed sample. Figure 12.25 shows the Vickers hardness of PAI and PAI+MLS nanocomposites of different MLS concentration. Addition of MLS showed a marked increase in hardness of around 32%. At 1.0%, which showed exfoliated dispersion, the highest hardness was observed. The uniformly distributed exfoliated platelets improved the stiffness of PAI. The presence of these stiff clay platelets and entanglement of polymer chains made the PAI nanocomposite harder. The intercalated dispersions showed almost similar hardness where polymer chains are penetrated into silicate galleries. Finite penetration of polymer chains did not improve the hardness of PAI over that of the 1.0% PAI nanocomposite, due mainly to retention of clay structure. The addition of MLS increased the hardness of PAI from 15 MPa to 20 MPa. Dispersion-dependent mechanical properties were obtained where exfoliated dispersion showed the highest hardness compared to the intercalated dispersion. Thus, the process must ensure dispersion, and it is not necessary to go higher in concentration to achieve excellent properties.

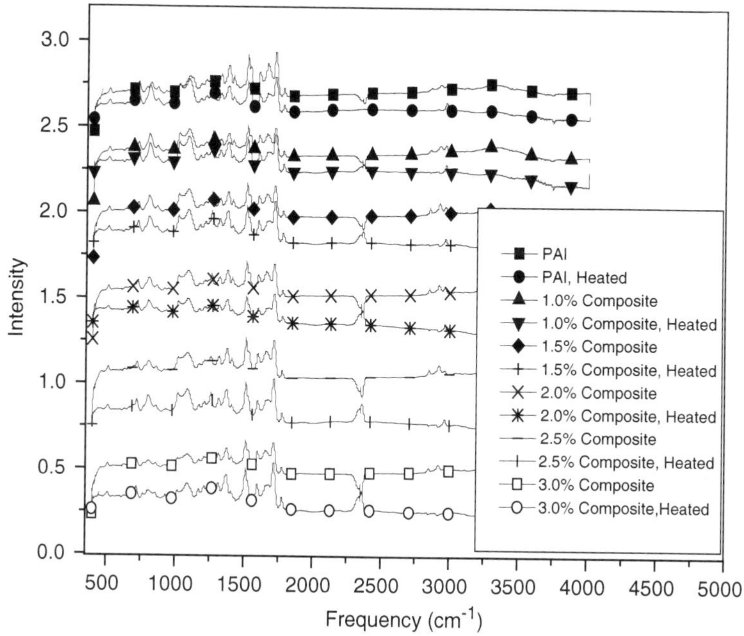

FIGURE 12.24 FTIR analysis of PAI nanocomposites before and after heat treatment.

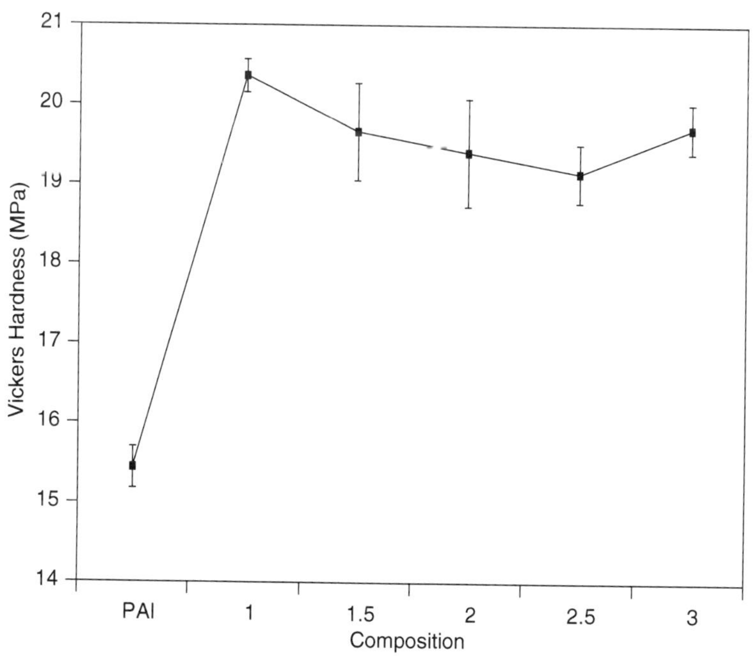

FIGURE 12.25 Mechanical behavior of PAI nanocomposites.

12.8.2 PET Nanocomposite Cast Films

Permeability of polymer films has been of significance to a number of packaging applications. Oxygen and water permeability of PET has been studied for the previous 10 years, due to its growing applications in areas such as food packaging and flexible displays [116,117]. Qureshi et al. [118] studied oxygen barrier properties of oriented and heat-set PET. The orientation was done at 65°C, and heat setting was done at a higher temperature (90 to 160°C), to improve crystallinity. Orientation decreased the permeability of PET by 33% and the phenomenon was related to decrease in diffusion coefficient and solubility coefficient. They proposed a two-phase model for oriented PET consisting of a permeable amorphous phase and an impermeable oriented phase. Heat-set PET samples showed increase in crystallinity and therefore decrease in permeability. McGonigle et al. [119] studied the permeability of nitrogen, argon, helium, oxygen, and carbon dioxide through biaxially oriented PET film. They observed that the diffusivity of a gas was more influenced by crystallinity and chain alignment than changes in void sizes during stretching. Seklik et al. [120] studied oxygen barrier properties of talc-filled PET with varying crystallinity. Talc flakes were added to make a controlled impermeable layered structure. They used Nielsen's model to describe the effect of crystallinity and talc on the permeability of PET. The model consists of initial non-steady-state diffusion (i.e., the process of penetration of oxygen into material). This part of the curve was determined mainly by the diffusion coefficient D. The steady-state flux J, was reached as the permeant concentration in the film reached a constant distribution. The permeability coefficient P is a ratio of thickness over permeant gas pressure multiplied by flux. Seklik et al. [120] observed a strong effect of impermeable phase such as crystallinity or talc on the steady and nonsteady states of the oxygen flux curve. As the crystallinity increased, the non-steady-state region broadened (slower diffusion) and the steady-state permeability decreased. Increase in crystallinity decreased the amount of amorphous material through which oxygen can diffuse. Amorphous phase affects the solubility of permeant gas, while the diffusion coefficient is affected by the tortuous crystalline paths. The reduction in permeability was due primarily to the decrease in the diffusion coefficient. The solubility coefficient remained constant, indicating that changes occurred in the amorphous regions due to crystallization. Seklik and his co-workers studied the effect of inorganic talc on oxygen permeability of PET using three different processing techniques. Microlayered, extruded, and compression-molded specimens with varying inorganic lading were prepared. The best orientation was achieved by a microlayered co-extrusion. Talc-filled PET showed a decrease in oxygen permeability, but the drop was independent of processing technique. Current theoretical approaches fail to explain the small or negligible effect of particle orientation in permeation through filled media. Microlayered structure showed the best mechanical properties, and the improvement was due to better orientation of talc flakes.

Eichler and Miltz [121] predicted the oxygen permeability of nylon 6 from the glass transition temperature and observed a good agreement between predicted and experimental values. Lasoski and Cobbs [122] studied the effect of crystallinity and orientation on the moisture permeability of nylon 6. The moisture permeability dropped with increased crystallinity and orientation. They also compared the effects due to crystallinity and orientation and found that improving crystallinity is the more effective way to decrease permeability. Tak et al. [123] studied the transport of water vapor in nylon 6,6 at varying relative humidities. The water transmission rate increased with increased humidity. Moisture from humidity acted as a plasticizer, thereby affecting the crystallinity of polymer. Nobile et al. [124] studied the barrier properties of a series of polyamides used in food-packaging applications. He also observed the dependence of permeability on relative humidity. Baiano and Nobile [125] studied the shelf life of almond paste pastries wrapped in nylon films. The loss of moisture severely affected the freshness of pastries.

Cast films of PET nanocomposites were formed. PET was dried overnight in a vacuum oven at 65°C. A 10% by weight master batch of MLS and PET was prepared on a twin-screw extruder followed by individual (1, 2, 3, and 5% by weight) compositions. The viability of the films as flexible substrates in the end product requires examination of the influence of fatigue on the permeability. An MTS 810 hydraulic system was used for the samples to undergo the fatigue of 50 and 10,000 cycles at the rate of 10 mm/s; 50 and 10,000 cycles are chosen to consider both extreme ends that the system will undergo during its use. Permeability of these samples is then measured immediately after fatigue cycling, to avoid aging of the samples. Table 12.11 shows that the permeability of nylon fatigued samples is less than that of the as-processed samples. For 10,000 fatigue cycles, pure nylon and 3% MLS films permeability value drops by 19 % and 18%, respectively, whereas for pure PET it drops by 6%. For PET + 3% MLS film, an increase in permeability is evident. This increase in permeability could be reflective of defects generated during fatigue cycling in the materials. A 50-fatigue-cycle sample shows an increase in permeability except in the case of pure nylon. It could be due to the mobility of the chains during fatigue in such a way that it changes the free volume for the permeants to diffuse through. The difference in the magnitude of change in permeability during fatigue differs in different polymers, depending on the structure of the polymer and on the processing history. So the free volume is the most important factor in determining the permeability of fatigued samples.

For the better dispersion of MLS in the matrix and polymer chain to be oriented, films are biaxially stretched. The permeability of these biaxial stretched samples

TABLE 12.11 Permeabilities of Various PET and Nylon Samples (cm^3/m-s-torr)

Sample	As-Processed Permeability	Fatigued (50 cycles) Permeability	Fatigued (10,000 cycles) Permeability	Stretched Permeability
PET (helium)	3.70E-09	6.92E-09	3.47E-09	4.68E-09
PET+3% MLS	1.35E-09	3.44E-09	3.60E-09	3.65E-09
Nylon (helium)	7.89E-10	7.67E-10	6.36E-10	6.90E-10
Nylon+3% MLS	5.48E-10	6.58E-10	4.50E-10	4.67E-10

was then measured. The permeability of biaxially stretched samples shows less permeability than do unstretched films in nylon. It is due to the increased tortuous path due to better MLS dispersion during stretching. The maximum drop in permeability value observed for stretched film is nylon + 3% MLS film. This may be reflective of the increase in tortuous path by a higher degree of MLS alignment parallel to the surface. We did not observe any decrease of permeability for helium gas in PET, but we observed a decrease in the trend for oxygen gas (the data are not reported here). During biaxial stretch, because of the chain mobility, microholes might get formed in the film that would assist helium gas atoms to permeate because of their small size. The rate-limiting step for the penetrating permeability could be the creation of a transient gap in the polymer matrix via the movement of a polymer segment in the local scale involving several polymer chains. This free volume is redistributed continuously as a result of the random thermally stimulated molecular motion of the polymer segment, which affects permeability.

12.8.3 Biodegradable Blown Film Nanocomposites

A number of biodegradable polymers are considered for nanocomposite applications. Poly(L-Lactic) acid (PLA), for instance, has ester bonds pronte to chemical and enxyamtic hydrolysis. Nanocomposites of PLA have shown increased propensity for biodegradability and compositing. This was associated with the terminal hydroxylated edge groups of the silicate. PLA,which is typically formed through the fermentation process of corn and corn-based products. The hydrolysis of this polymer back into its monomer state provides a major advantage over many petroleum-based polymers and polymer resins. PLA's biodegradable nature has made the polymer popular in biomedical applications such as sutures, bone screws, and scaffolding for tissue engineering [126].

Ratto's group has investigated the barrier properties of PLA; nanocomposites of PLA–MLS blown films were investigated in this work. The study offers an opportunity to assess the influence of process methodologies on nanocomposite formation. Prior studies had shown that solvent casting [127] resulted only in microcomposites and a generally immiscible mixture. Ratto's group compounded the PLA–MLS resin, and the pelletized PLA resin was again dried for 4 hours at 80°C to remove moisture absorbed during the pelletizing process. The resin was allowed to cool to room temperature for 1 hour. Once the material had cooled to room temperature, 10 wt% plasticizer was added to the pellets and mixed mechanically for 20 minutes. Once the mixing was finished, the pellets and plasticizer were allowed to sit for 2 hours so that the pellets could absorb the plasticizer. The polymer pellets with plasticizer were loaded into the extruder hopper of the 16-mm corotating twin-screw extruder (ThermoPrism) and fed by way of the primary volumetric feeder. A spiral flow blown-film die with a 25.4-mm diameter was used to produce the film. The blow-up ratio (BUR) of the bubble was 2 : 1. This setting produced a bubble with an average film thickness of 0.076 mm (about 0.003 in.). Films were cooled using a single-lip air ring.

To examine how processing parameters affected the film properties, the screw speed and feeding rate were varied during the blown-film process. The temperature of zones 1, 2, 3, and 4 were 150, 155, 160, and 165°C, respectively while the die temperature was varied from 165 to 175°C. The screw speed was varied from 80, 110, and 130 rpm to produce neat PLA blown films as well as nanocomposite blown films. All films did had an intercalated–exfoliated morphology.

12.8.4 Supercritically Foamed Polymer Nanocomposites

Nanocomposite thermoplastic foams are a newly developed group of polymeric materials that have rapidly become a vital focus of materials research. The primary benefits of these foams lie in their unique combination of high strength-to-weight ratio, thermal stability, and tailorable electrical and chemical properties. They hold great potential for use in structural applications such as foam core panels, self-skinning molded parts, and impact-tolerant structures for the aerospace industry, as well as in applications that take advantage of other properties, such as low vapor permeability, controlled chemical release, and thermal insulation. Later uses might include insulated food packaging, biochemical hazard containment, and self-medicated bone implants, to name a few.

Facilitating the development of nanocomposite foams is the technology of foam generation with supercritical CO_2 ($scCO_2$), which has become a popular topic of research over the past decade and has been shown to have many attractive advantages over other physical foaming methods [128–131]. Because $scCO_2$ methods permit foaming of solid samples without requiring that they be melted or dissolved completely, carefully prepared nanocomposite thermoplastics may now be foamed without destroying their finely mixed microstructure. Also, the $scCO_2$ foaming process is environmentally benign and involves no toxic chemicals.

While the potential for nanocomposite thermoplastic foams is seemingly endless and ever more appealing, their technical development is rigorous and multidisciplinary. The optimization of nanocomposite foams requires a well-rounded knowledge of thermodynamics, mass transfer, mechanics of materials, foam nucleation and growth, polymer processing, rheology, nanocomposite chemistry, and supercritical fluids. This thesis attempts to address these disciplines by first discussing individually the subjects of nanocomposite polymers, supercritical fluids, and thermoplastic foams, and subsequently integrating these three into the discussion of a new and original engineered material. At issue in the current work are the specific mechanical and thermal benefits of nanocomposite foam as well as the benefits of a simple isochoric $scCO_2$ process for batch production of these foams.

Strauss and D'Souza investigated polystyrene nanocomposite foams [132]. Supercritical foaming was performed in a Polaron critical point drier (Model E3100) (Figure 12.26). For each experiment, a 40-mm disk was cut from one of the nanocomposite laminates. The disk was rested on a wire grate and placed inside the drier's reactor chamber. The chamber was cooled to 10°C and pressurized with liquid CO_2 to approximately 5 to 6 MPa. The temperature was then raised at 1°C/min to the process temperature. During this time, the pressure in the chamber increased according to the ideal gas law and was regulated using a vent valve to maintain process pressure. Process temperatures used were 60, 75, and 85°C. Process pressures, except as noted, were 9.5 to 10.5 MPa. After reaching the desired temperature

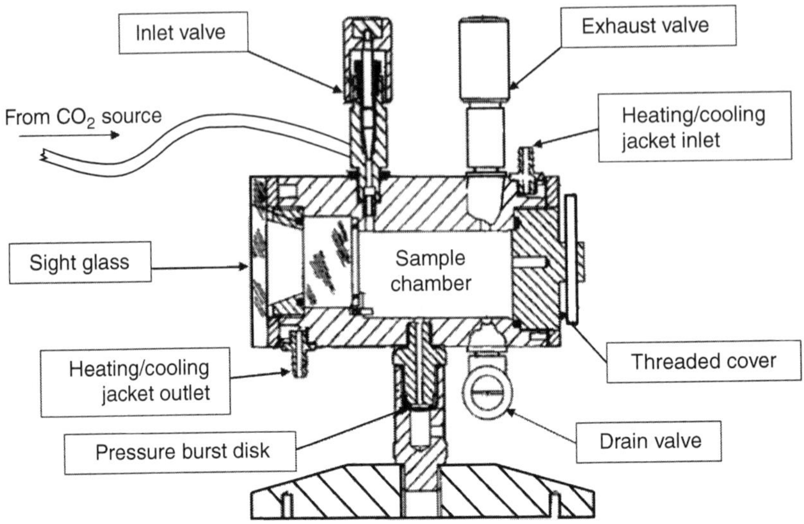

FIGURE 12.26 Supercritical reactor used to generate foam samples.

and pressure, the sample was allowed to soak for a set time and then was rapidly vented to the atmosphere, or quenched, in < 5 seconds. The sample was then cooled at a rate of 1°C/min to room temperature before removal from the reactor.

Figure 12.27 shows a reactor setup. The reactor was designed to be mounted vertically in a benchtop, with acme-threaded lids on each end for simplicity of accessing and servicing the pressure vessel. Dovetail O-ring grooves are provided on each lid for a face seal with the body when tightened. The design of the vessel body includes a welded jacket for circulation of heat transfer fluid from a thermal control unit (TCU). CO_2 is supplied simply from a cylinder or pump to a manifold on the upper lid.

Foaming experiments were successful in creating macrocellular closed-cell foams in both pure polystyrene and PS + MLS laminates. For complete foaming of the 1.1-mm-thick pure PS laminates, we found that a minimum 1-hour soak was required for diffusion of supercritical CO_2 throughout the sample prior to pressure quench. This parameter is consistent with the findings of similar work, in which soak times for SCF foaming of small samples has been published at 3 to 24 hours for process temperatures of 80°C and lower [128]. In our experiments, uniform polystyrene foams were obtained at temperatures at or above 60°C, pressures at or above 9 MPa, and soak times at or above 4 hours; 60°C was very near the effective glass transition temperature of the samples when they were saturated with SCF CO_2 in the reactor vessel. Because of this proximity to the T_g, variations in temperature had a large effect on the morphology of the foams. At temperatures below the effective T_g, foaming did not occur at all; instead, when the reactor pressure was released to generate foaming, expansion of the absorbed CO_2 in the laminate caused crazing and cracking in the laminate. At the lowest successful foaming temperatures, cell size was smallest, often less than 10 μm, cell shape was ellipsoidal, and cell walls were

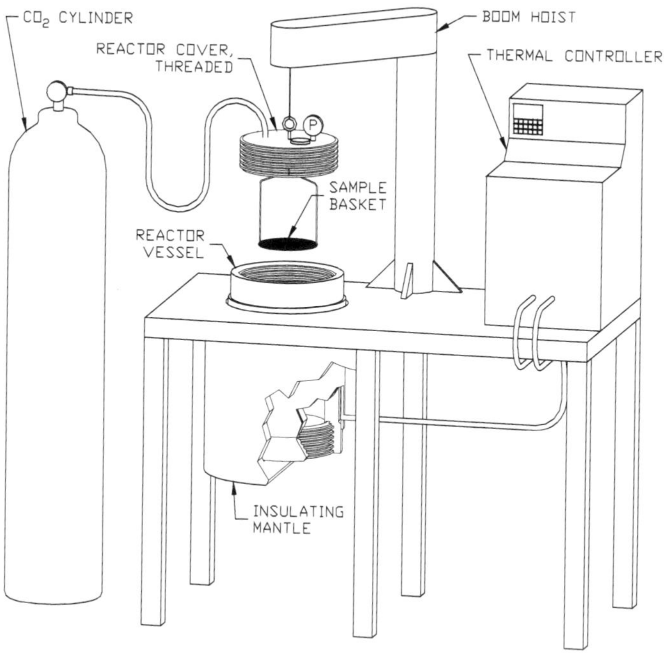

FIGURE 12.27 Preliminary sketch of a scaled-up $ScCO_2$ reactor.

thick. When higher process temperatures were used, cells became larger, and interaction between cells caused the cell walls to become thin and flat; if the temperatures used were too high, these walls would rupture, creating voids and/or local areas of open-celled foam structure. The effects of soak temperature on foam development are due to more than one simultaneous influence. First, temperature has an inverse relationship to the viscosity of the polystyrene, thereby enhancing both nucleation and growth of cells. As viscosity decreases, so does the critical radius of void nucleation; and this effect tends to increase cell density (cells/cm ^{3}). At the same time, it also decreases the energy required for cell growth, which permits greater volumetric expansion of the foam.

A second effect of temperature, however, lies in its relationship to diffusion rates. As temperature increases, so does the rate of desorption of the foaming agent from the bulk polymer, both into the growing foam cells and into the atmosphere surrounding the sample in the reactor chamber. This desorption leaves localized depleted regions in the foam where there is insufficient dissolved CO_2 to initiate void nucleation, thus decreasing the total density of cells while increasing the sizes of the extant cells. Depletion of CO_2 near the exposed surfaces of the foam also causes the *skinning effect*, in which cell size and density gradually decreases with distance from the surface, leaving a nonporous polymer "skin" over the outside of the sample. In Figure 12.28 the skinning effect can be seen toward the bottom of the cross-sectional micrograph. The end result of the process temperature effect on the foam product was that cell density was seen initially to decrease as temperature increased, in

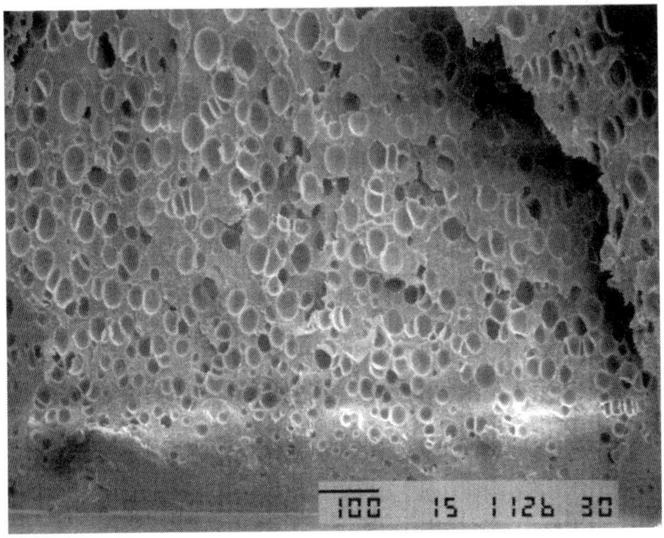

FIGURE 12.28 Micrograph of PS foam, showing skinning effect at the bottom edge.

agreement with previous findings [128,131]. At 85°C, however, cell density showed a moderate increase. This may be attributed to both the lowered critical radius of void nucleation and to the limitation of cell geometry that occurs as a finite volume of CO_2 desorbs into them. Figure 12.29 shows the effect of processing temperature on foam morphology as observed by scanning electron microscopy.

Pressure has been shown to have an effect on foam morphology in several key ways. First, increased pressure in the reactor during soak increases the total mass uptake of CO_2 by the sample, which in turn increases potential volumetric expansion of the foam product while decreasing the effective T_g of the sample. As stated previously, the T_g of the PS laminates in our experiments was decreased to around 60°C at 10 MPa; had we doubled our pressures to 20 MPa, the T_g would have been lowered to approximately 40°C. Another variable influenced by process pressure is the total pressure drop during quench. Because the pressure differential between the CO_2 inside the sample and the atmosphere outside the sample provides the energy for void nucleation and foam expansion, higher pressures obviously present the theoretical potential for increasing the total number of homogeneous nucleation sites as well as the volume of the foam generated. In our experiments, however, the highest pressures attainable were around 12 MPa, and changes in soak pressure were not shown to have as significant an impact on cell morphology as other factors. To fully experiment with the effects of higher pressure on foam morphology, one would have to perform experiments in the regime of 20 to 25 MPa, where homogeneous nucleation is a dominant mechanism in foam growth; at those pressures and above, changes in process pressure have been shown to be the most critical parameter to cell density and size [128]. Below those pressures, heterogeneous nucleation is dominant, and changes in pressure do not have as substantial an effect on the population of nucleated cells.

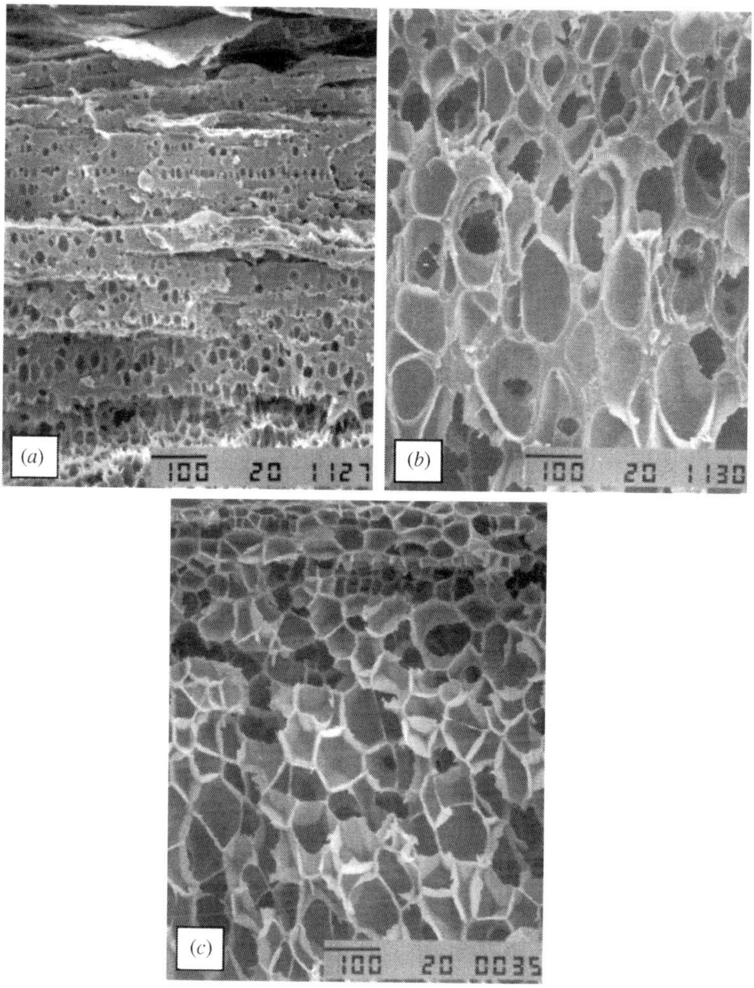

FIGURE 12.29 Micrographs of PS showing the effect of temperature on a foam structure, processing at *(a)* 60°C (N_f=2.4E7); *(b)* 75°C (N_f=7.6E5); *(c)* 85°C (N_f=3.0E6)

Of much greater impact to our experiments was the rate of pressure quench. If the pressure was released from the reactor too slowly, mass transfer of CO_2 within the sample prevented the generation of uniform foams. Immediately upon decrease of reactor pressure, supercritical CO_2 begins a phase transformation into gaseous CO_2, which has a much higher diffusion rate even than SCF. As the gas desorbs into nucleated cells and out of the polymer , the T_g of the composite begins to rises. It is this effect that "freezes" the foam cells after expansion. By slowing the quench rate, more CO_2 is permitted to escape during the process; T_g increases and the viscosity rises. Simultaneously, there is less pressure differential in the material to drive void nucleation. The result is similar to that of lowering both processing temperature and pressure. Foams formed with slow quench rates exhibited thick, crustlike skins, shear zones in which expanding foam cells were ruptured as they moved across hardened polymer, and large irregularities in cell size (Figure 12.30). To create uniform foams, quench times were maintained below 5 seconds.

During foaming, the highly strained material in the foam cell walls may have inherited some localized alignment of the polymer chains (Figure 12.31). After annealing in the melt state for 30 minutes at 170°C, however, the samples were cycled a second time in the DSC, and all of them had returned to their prefoamed endothermic behavior. Scanning electron microscopy was used to examine physical features of the foamed samples. Scaled micrographs were scanned and measured to determine foam density, cell size, and so on. At low cell densities and sizes, where cell shapes were uniformly ellipsoidal and cell walls thick, the ratio of average cell width to height was measured at approximately 1:1.5, with the longest dimension of each cell being in the through-thickness direction of the laminate (Figure 12.32).The same ratio was consistent through all the samples, although more irregularity was found in those samples in which the cells had grown to the point of impinging on each other.

Cell density was calculated in a method similar to that used by Kumar and Suh [133], adjusted for asymmetrical cell growth. In this model, number of cells/cm^3, N_f, may be expressed as

$$N_f = \left(\frac{nM^2}{A}\right)^{3/2} (r_c)^{1/2} \tag{12.1}$$

where n is the number of cells seen in a micrograph, M is the magnification factor, A is the area of the micrograph, and r_c is the ratio of cell width over height.

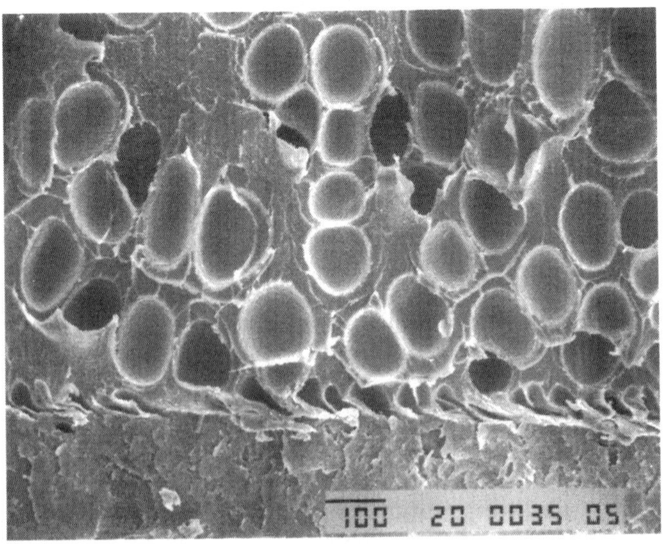

FIGURE 12.30 Micrograph showing the negative effects of a slow pressure quench rate.

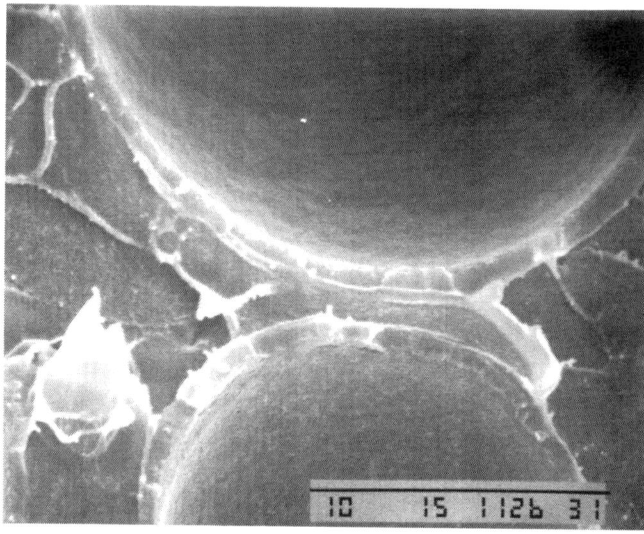

FIGURE 12.31 Micrograph of PS foam showing strain-induced polymer alignment around the cells.

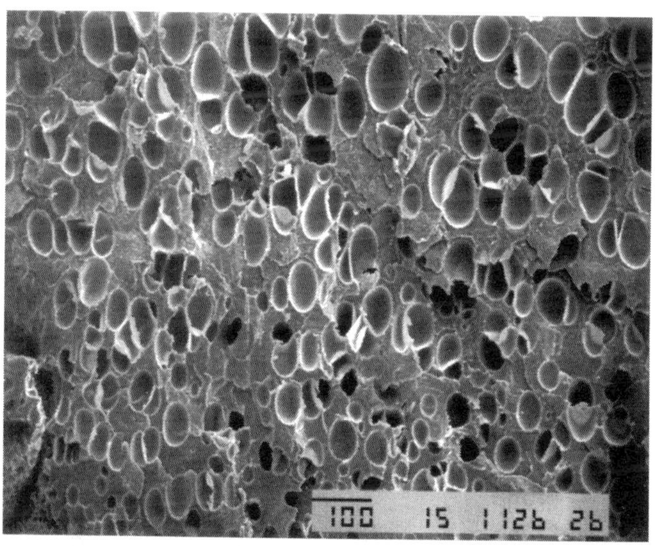

FIGURE 12.32 Micrograph of PS foam showing typical cell elongation in the z-axis.

The addition of the last term was required, due to the preferential elongation of foam cells in the through-thickness direction that was observed.

The cell densities and sizes varied widely based on processing and material parameters; densities achieved spanned the range from 10^4 to 10^8 cells/cm^3. Cell size in the bulk foam also varied accordingly, with cell diameters most typically in the range 10 to 100 μm. In regions near the surface of the foam, however, where the skinning effect was apparent, cell diameters gradually decreased to as low as 1 μm or less. In many micrographs, a distinct "eggshell" phenomenon was displayed, in which a smooth thin layer of material was made visible in the fracture surface surrounding each bubble of foam (see Figure 12.32). This "shell" appears to have been generated by the straining of the CO_2-depleted polymer around each cell during cell growth. The resulting layer evidently contains residual stresses, and its highly strained nature may contribute to the toughness and/or other mechanical properties of the foam as a whole. It may also be responsible for the enlarged endothermic peaks in the DSC curves, discussed earlier.

MLS concentration in the nanocomposite influenced foam morphology strongly, as might be expected. Higher concentrations of MLS created a higher density of nucleation sites for heterogeneous nucleation, thus increasing the cell density. Cell size is shown to decrease as cell density increases, preserving the net volume expansion of the foam as a function of temperature, pressure, and CO_2 saturation (Figure 12.33). The distribution of MLS in the polymer also had an effect on cell nucleation, especially at low temperatures and high viscosities. In the compression molding of the samples, a preferential orientation was introduced that aligned the silicate into layers parallel to the laminate major axis. As evidence of this and of heterogeneous nucleation, PS + MLS foams generated at low temperatures exhibited void nucleation solely within those segregated layers (Figure 12.34).

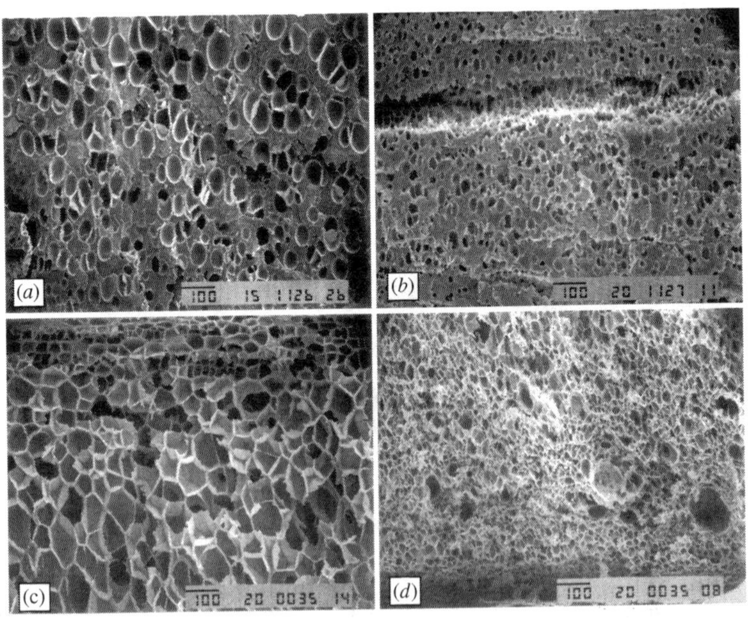

FIGURE 12.33 SEM micrographs showing increased cell density (N_f) with MLS concentration (100 × magnification): *(a)* pure PS, 60°C; *(b)* 1% MLS, 60°C; *(c)* 1% MLS, 85°C; *(d)* 3% MLS, 85°C.

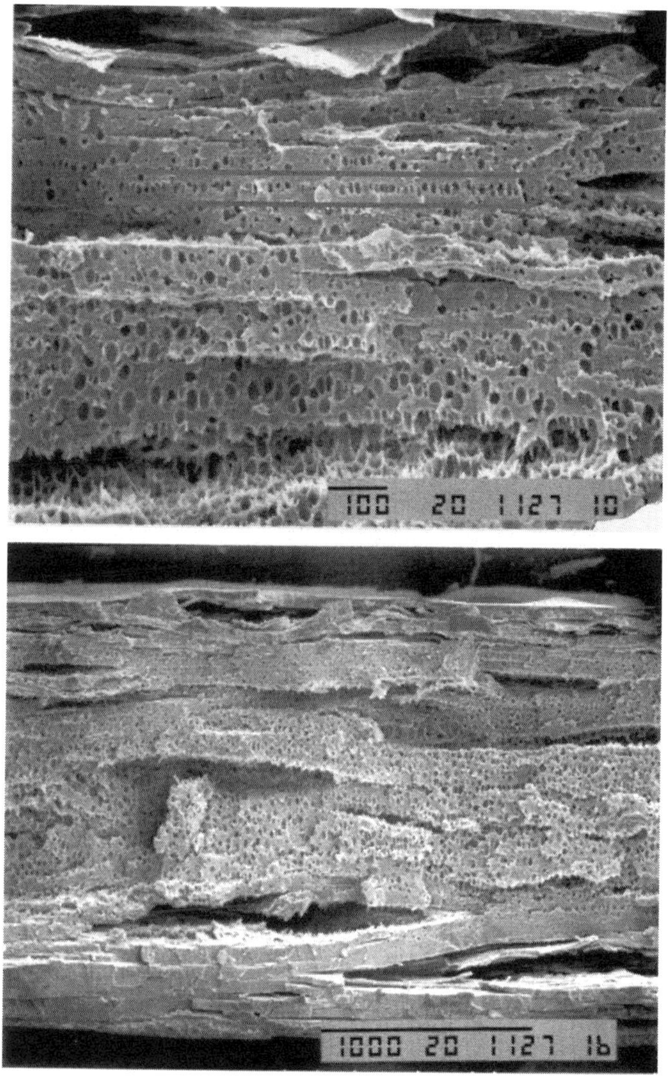

FIGURE 12.34 SEM micrographs showing pastrylike layered void formation corresponding to MLS-rich planes.

Foam formation at higher temperatures was able to disturb those layers and cause the silicate to reorient around the cells, and the resulting foam materials showed preferential orientation of grainlike nanocomposite structures parallel to the cell walls (Figure 12.35). In general, the temperatures required to generate uniform nanocomposite foams were higher than those required to generate pure PS foams, even though the T_g of the nanocomposite laminates was not higher (and in the case of the 1% composite it was lower) than that of the PS laminates. The reason that higher temperatures were required

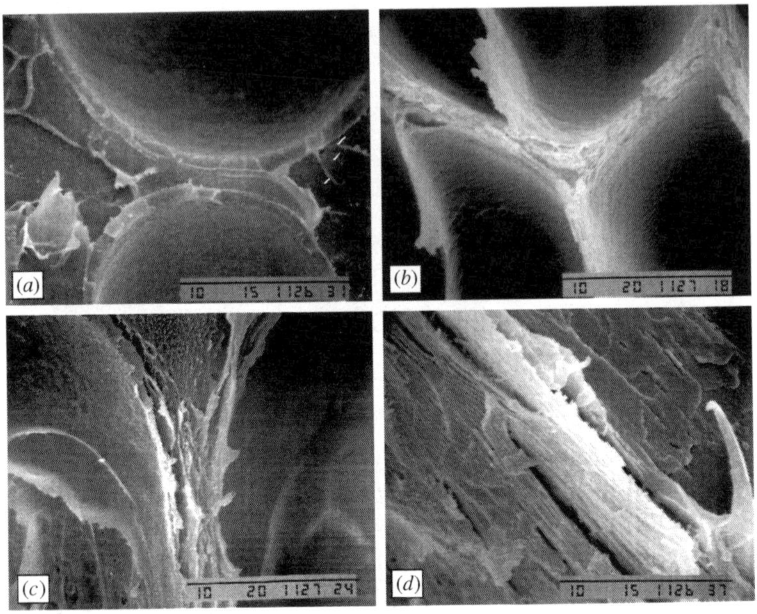

FIGURE 12.35 Micrographs showing aligned grain structure in the cell walls of a nanocomposite (5000 × magnification): (*a*) pure PS (eggshell effect), 60°C; (*b*) 1% MLS, 60°C; (*c*) 1% MLS, 85°C; (*d*) 1% MLS, 75°C.

is that the presence of the silicate in the polymer both increased its viscosity above the glass transition and provided localized anisotropic deformation mechanisms (such as delamination) in place of normal cell nucleation and growth. As described above, low process temperatures for the PS + MLS laminates resulted in segregated layers of bubbles. These bubbles often coalesced into large sheetlike pockets that delaminated the sample into a flaky, pastry-like material.

Another, unexpected result of the presence of clay in the sample was a highly accelerated absorption rate for the supercritical CO_2. Unlike the pure PS samples, which required hours of soak time in the supercritical CO_2 chamber (Figure 12.36 depicts the effects of incomplete diffusion), PS + MLS laminates created uniform foams after as little as only 5 minutes in the presence of the SCF. This implies that there is a secondary mechanism of mass transfer of CO_2 in the nanocomposite foams, one that is significantly faster than the linear diffusion model predicts. We suspect that diffusion at the polymer–MLS interfaces may be accelerated by the differences in CO_2 solvency of the two materials.

12.9 DEGRADATION OF POLYMERS

There are many issues influencing the degradation of nanocomposites. First, during processing the surfactant degradation temperature (above 220°C and below 300°C) makes nanocomposite formation in the melt problematic for most polymers. To

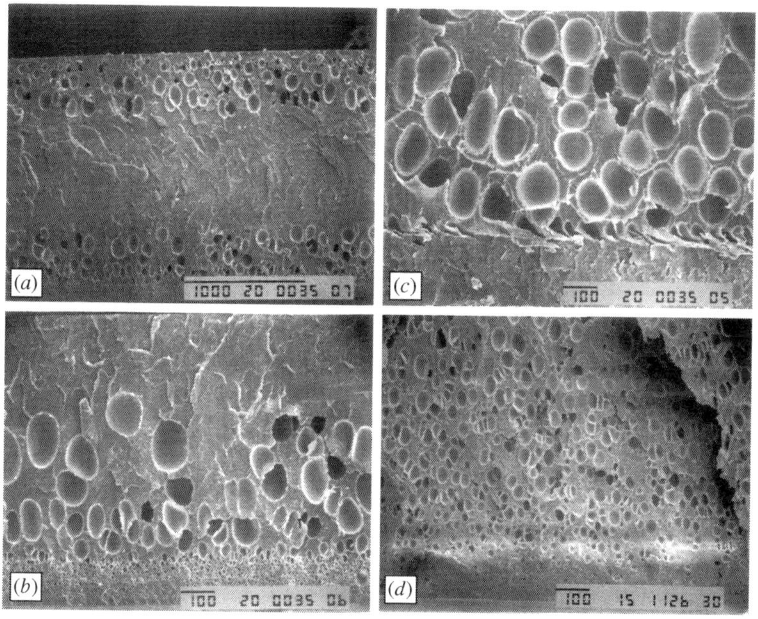

FIGURE 12.36 SEM micrographs showing the effect of saturation time on CO_2 diffusion through pure PS at 10 MPa: (*a*) 75°C, 15-min soak (35×) (incomplete diffusion); (*b*) detail of same (100×); (*c*) 75°C, 1-hr soak (100×); (*d*) 60°C, 5-hr soak (100×).

ensure an exfoliated nanocomposite, the polymer must wet the clay surface. However, premature surfactant degradation results in gallery collapse and inadequate dispersion. Further, if the degradation products react with the polymer, chain scission or polymer discoloration results. As we have indicated in the coatings section, this is not necessarily predetermined. In PAI, for instance, no interaction between the polymer and the degraded surfactant ensured retention of the host polymer structure and properties. Given the most likely occurrence of surfactant degradation in processing engineering polymers, we have found that an under-exchanged clay or a clay with a low cation-exchange capacity performs better than an over- or equi-exchanged clay. This was described in the PET nanocomposite section above. Although one might consider more surfactant concentration good for the clay dispersion, the more the surfactant, the more likely the group is available to degradation.

Finally, in the application area, a cautious note in the use of polyethylene nanocomposites was sounded by Huali et al. [134]. They compared PE nanocomposites and found that degradation under UV lamps at elevated temperatures was substantially higher in the nanocomposites than in the base resin. Degradation of the polymer nanocomposite is polymer dependent. In polycarbonate Sloan et al. [135] determined that UV-accelerated weathering showed lower degradation in the PC nanocomposite compared to the base PC resin. Degradation of nylon during processing has been attributed to the water content in the polymer, which is considered to cause hydrolytic cleavage [136].

REFERENCES

1. W. F. Smith, *Principles of Materials and Engineering,* 2nd ed., McGraw-Hill, New York, 1990 (Chapters 6, 7, and 13).
2. C. T. Heracovich, *Mechanics of Fibrous Composites,* Wiley, New York, 1998 (Chapter 1).
3. M. M. Schw artz, *Composites Materials Handbook,* McGraw-Hill, New York, 1984 (Chapters 1 and 2).
4. J. M. Shackelford, *Introduction to Materials Science for Engineers,* 4th ed., Prentice Hall, London, 1996 (Chapter 10).
5. H. van Olphen, *An Introduction to Clay Colloid Chemistry,* Wiley, New York, 1977, pp. 57–76.
6. A. Ranade, Polyamide-imide nanocomposites, M.S. thesis, University of North Texas, 2001, http://www.library.unt.edu/ theses/open/20012/ranade_ajit/thesis.pdf.
7. P. S. Braterman and Y. Wang (Sandia National Laboratories), U.S. patent application, November 2001.
8. P. S. Braterman and N. A. D'Souza, unpublished results.
9. C. O. Oriakhi, I. V. Farr, and M. M. Lerner, *J. Mater. Chem.* 1996; 6: 103.
10. N. T. Whilton, P. J. Vickers, and S. Mann, *J. Mater. Chem.* 1997; 7: 1623.
11. M. Tanaka, I. Y. Parky, K. Kuroda, and C. Kato, *Bull. Chem. Soc. Jpn.* 1989; 62: 3442.
12. S. Miyata, U.S. patent 4,299,759, 1981.
13. H. R. Fischer, and L. H. Gielgens, U.S. patents 6,372,837 and 6,365,661, 2002.
14. *Kyowa Chem. Ind.,* EP-B 0,189,899, 1986; EP-B 0,256,872, 1987; EP-B 0,052,331, 1981.
15. S. Iijima, *Nature* 1991; 354–356.
16. W. Tang, M. H. Santare, and S. G. Advani, *Carbon* 2003; 41: 2779–2785.
17. J. K. W. Sandler, J. E. Kirk, I. A. Kinloch, M. S. P. Shaffer, and A. H. Windle, *Polymer* 2003; 44, 5893-5899.
18. M. J. Biercuk, M. C. Llaguno, M. Radosavljevic, J. K. Hyun, A. T. Johnson, and J. E. Fischer, *Appl. Phys. Lett.* 2002; 80 (15): 2767–2769.
19. G. Li, L.Wang, H. L. Ni, and C. U. J. Pittman, *Inorg. Organomet. Polym.* 2001; 11: 123.
20. R. A. Vaia, Ph.D. dissertation, Cornell University, May 1995.
21. R. Krishnamoorti, R. Vaia, and E. P. Giannelis, *Chem. Mater.* 1996; 8: 1728.
22. R. A. Vaia, B. B. Sauer, O. K. Tse, and E. P. Giannelis, *J. Polym. Sci.* 1997; 35: 59.
23. R. A. Vaia, S. Vasudevan, W. Krawiec, L. G. Scanlan, and E. P. Giannelis, *Adv. Mater.* 1995; 7(2): 154.
24. R. A. Vaia, K. D. Jandt, E. J. Kramer, and E. P. Giannelis, *Chem. Mater.* 1996; 8: 2628.
25. R. A. Vaia and E. P. Giannelis, *Macromolecules* 1997; 30: 8000.
26. R. A. Vaia and E. P. Giannelis, *Macromolecules* 1997; 30: 7990.
27. Z. Wang and T. J. Pinnavaia, *Chem. Mater.* 1998; 10: 1820.
28. M. S. Wang and T. J. Pinnavaia, *Chem. Mater.* 1994; 6: 468.
29. T. Lan, P. D. Kaviratna, and T. J. Pinnavaia, *J. Phys. Chem. Solids* 1996; 57: 1005.
30. P. D. Kaviratna, T. J. Pinnavaia, and P. A. Schroeder, *J. Phys. Chem. Solids* 1996; 57 (12): 1897.
31. P. B. Messersmith and E. P. Giannelis, *J. Polym Sci.* 1995; 33: 1047.

32. T. Lan and T. J. Pinnavaia, *Chem. Mater.* 1994; 6: 2216.
33. Z. Wang and T. J. Pinnavaia, *Chem. Mater.* 1998; 10: 3769.
34. T. Lan, D. Padmananda, P. D. Kaviratna, and T. J. Pinnavaia, *Chem. Mater.* 1994; 6: 573.
35. T. Lan, D. Padmananda, P. D. Kaviratna, and T. J. Pinnavaia, *Chem. Mater.* 1995; 7: 2144.
36. E. P. Giannelis, *J. Miner.* 1992; 44:3: 28.
37. P. B. Messersmith and E. P. Giannelis, *Chem. Mater.* 1994; 6: 1719.
38. R. Dagani, *Chem. Eng. News* 1999; 77:23: 25.
39. B. Miller, *Plast. Formulat. Compound.,* May/June 1997; 30.
40. L. M. Sherman, *Plast. Technol.* June 1999; 52.
41. Z. Shen, Ph.D. dissertation, Monash University, Australia, October 2000.
42. X. Kornmann, H. Lindberg, and L. A. Berglund, *Polymer* 2001; 42: 1303.
43. S. D. Burnside, H. C. Wang, and E. P. Giannelis, *Chem. Mater.* 1999; 11: 1055.
44. E. P. Giannelis, *Chem. Mater.* 1990; 2: 627.
45. P. B. Messersmith and E. P. Giannelis, *Chem. Mater.* 1993; 5: 1064.
46. D. Hull and T. W. Clyne, *An Introduction to Composite Materials,* 2nd ed., Cambridge University Press, Cambridge, 1996 (Chapter 1).
47. R. B. Se ymour and C. E. Carraher , *Structure Property Relationships in Polymers*, Plenum Press, New York, 1984 (Chapters 3, 5, and 12).
48. N. M. Bikales, *Mechanical Properties of Polymers*, Wiley, New York, 1971 (Chapters 1 and 3).
49. S. L. Rosen, *Fundamental Principles of Polymeric Materials,* 2nd ed., Wiley, New York, 1993 (Chapters 17 and 18).
50. F. W. Billmeyer, *Textbook of Polymer Science*, 2nd ed., Wiley, New York, 1971 (Chapters 4, 6, and 7).
51. T. J. Pinnavaia, *Science* 1983; 220 (4595): 365–371.
52. M. Alexandre and P. Dubois, *Mater. Sci. Eng. R.* 2000; 28: 1–63
53. T. J. Pinnavaia and G. W. Beall, *Polymer Clay Nanocomposites*, Wiley, New York, 2000 (Chapters 7 and 8).
54. S. C. Tjong, Y. Z. Meng, and A. S. Hay, *Chem. Mater.* 2002; 14: 44–51.
55. A. Okada, M. Kawasumi, A. Usuki, Y. Kojima, T. Marauchi, and O. Kamigaito, *Mater. Res. Soc. Proc.* 1990; 171: 45–50.
56. G. Lagaly, *Appl. Clay Sci.* 1999; 15: 1–9.
57. H. Fong, R. A. Vaia, J. H. Sanders, D. Lincoln, A. J. Vreugdenhil, W. Liu, J. Bultman, and C. Chen, *Chem. Mater.* 2001; 13: 4123–4129.
58. E. Manias, A. Touny, L. Wu, K. Strawhecker, B. Lu, and T. C. Chung, *Chem. Mater.* 2001; 13: 3516–3523.
59. A. Tabtiang and R. A. Venables, *Polymer* 2002; 43 (17): 4791–4801.
60. P. C. LeBaron, Z. Wang, and T. Pinavaia, *Appl. Clay Sci.* 1999; 15: 11–29.
61. A. Gonzalez-Montiel, H. Keskkula, and D. R. Paul, *J. Polym. Sci.* 1995; 33: 1751–1767.
62. R. A.Vaia, R. K. Teukolsky, and E. P. Giannelis, *Chem. Mater.* 1994; 6 (7): 1017–1022.
63. T. S. Ellis and J. S. D'Angelo, *J. Appl. Polym. Sci.* 2003; 90: 1639.
64. A. Blumstein, *Bull. Chim. Soc.* 1961; 899.
65. A. Okada, M. Kawasumi, A. Usuki, Y. Kojima, T. Marauchi, and O. Kamigaito, *Mater. Res. Soc. Proc.* 1990; 171: 45–50.

66. H. Shi, T. Lan, and T. J. Pinnavia, *Chem. Mater.* 1996; 8: 1584–1587.
67. H. R. Dennis, D. L. Hunter, D. Chang, S. Kim, J. L. White, J. W. Cho, and D. R. Paul, *Polymer* 2001; 42: 9513–9522.
68. Y. Kurokawa, H. Yasuda, M. Kashiwagi, and A. Oyo, *J. Mater. Sci. Lett.* 1997; 16: 1670–1672.
69. C. Y. Lew, W. R. Murphy, and G. M. McNally, *Polym. Eng. Sci.* 2004; 44: 1027.
70. G. Lagaly, *Solid State Ion.* 1986; 22: 43.
71. E. Hackett, E. Manias, and E. P. Giannelis, *J. Chem. Phys.* 1998; 108: 7410.
72. J. Park and S. C. Jana, *Antec 2003,* p. 1443.
73. D. Merinska, J. Vaculik, A. Kalendova, M. Kristkova, and J. Simonik, *Antec 2003,* p. 2744.
74. R. Li and D. Shi, *Antec 2004,* p. 1810.
75. N. Hasegawa and A. Usuki, *J. Appl. Polym. Sci.* 2004; 93: 464.
76. M. Mehrabzadeh, M. R. Kamal, and V. Mollet, *Antec 2003,* p. 2260.
77. M. N. Bureau, M. T. Ton-that, and F. Perrin-Sarazin, *Antec 2004*, p. 1518.
78. A. Leueritz, D. Pospiech, B. Kretzschamar, M. Wileke, D. Jehnichen, U. Jentzsch, K. Grundke, and A. Janke, *Adv. Eng. Mater.* 2003; 5: 678.
79. C. S. Own, D. Seader, N. A. D'Souza, and W. Brostow, *Polym. Compos.* 1998; 19: 107–115.
80. Y. Wang, F. B. Chen, Y. Li, and K. Wu, *Antec 2003,* p. 3670.
81. H. Wang, M. Elkovitch, L. J. Lee, and K. W. Koelling, *Antec 2001*, Dallas, TX.
82. H. Wang, C. Zeng, P. Svoboda, and J. L. Lee. *Antec 2000*, Orlando, FL.
83. E. Assouline, S. Pohl, R. Fulchiron, J. F. Gerard, A. Lustiger, H. D. Wagner, and G. Marom, *Polymer* 2000; 41: 7843–7854.
84. B. Lotz, S. Graff, and J. C. Wittman, *J. Polym. Sci. Polym. Phys. Ed.* 1986; 24: 2017.
85. B. Lotz, S. Graff, C. Straupe, and J. C. Wittman, *Polymer* 1991; 32: 2902.
86. T. Foresta, S. Piccarolo, and G. Goldbeck-Wood, *Polymer* 2001; 42: 1167–1176.
87. C. Saujanya and S. Radhakrishnan, *Polymer* 2001; 42: 6723–6731.
88. A. Ranade, D. Fairbrother, K. Nayak, and N. A. D'Souza, *Polymer* 2005; accepted.
89. J. H. Bae, S. H. Ryu, and Y. W. Chang, *Antec 2004*, p. 2196.
90. Y. Kojima, T. Matsuoka, H. Takahashi, and T. Kurauchi, *J. Appl. Polym. Sci.* 1995; 51: 683.
91. L. J. Mathais, R. D. Davis, and W. L. Jarrett, *Macromolecules* 1999; 32: 7958.
92. L. Wu, Z. Qi, and X. Zhu, *J. Appl. Polym. Sci.* 1999; 71: 1133.
93. M. K. Akkapeddi, *Antec 1999*, p. 1619.
94. E. Devaux, S. Bourbigot, and A. Achari, *J. Appl. Polym. Sci.* 2002; 86: 2416.
95. J. W. Cho and D. R. Paul, *Polymer* 2001; 42: 1083.
96. A. Ranade, N. A. D'Souza, B. Gnade, and A. Dharia, *J. Plast. Films Sheet.* 2003; 19: 271.
97. T. D. Fornes, P. J. Yoon, and D. R. Paul, *Polymer,* 2003; 7545.
98. F. Chavarria and D. R. Paul, Comparison of nanocomposites based on nylon 6 and nylon 6, 6.

99. Z. Kalkan and L. A. Goettler, *Antec 2004,* p. 2902.
100. S. Avlar and Y. Qiao, *Composites A,* 2005; 36: 624.
101. Z. Zhao, W. Yu, Y. Liu, J. Zhang, and Z. Shao, *Mater. Lett.* 2004; 58: 802.
102. B. Yalcin and M. Cakmak, *Polymer* 2004; 45: 2691.
103. L. Razafimahefa, S. Chlebicki, I. Vroman, and E. Devaux, *Dyes Pigments* 2005; 66: 55.
104. J. W. Matayabas, S. R. Turner, B. J. Sublett, G. W. Conell, and R. B. Barbee, *PCT Int. Appl.* 1998; WO98: 29499.
105. T. Takekoshi, F. F. Khouri, J. R. Campbell, T. C. Jordan, and K. H. Dai (General Electric Co.), U.S. patent 5,530,052 June 25, 1996.
106. C. H. Da vis, L. J. Mathias, J. W. Gilman, D. A. Schiraldi, R. Shields, P. Trulove, T. E. Sutto, and H. C. Delong, *J. Polym. Sci. B Polym. Phys.* 2002; 40: 2661.
107. A. Rande, N. A. D'Souza, C. Thellen, and J. A. Ratto, *Polymer International,* 2004; 54: 875.
108. P. Butzloff, N. A. D'Souza, D. Garrett, and T. D. Golden, *Polym. Eng. Sci.* 2001; 41: 1794–1801.
109. O. Becker, R. Varley, and G. Simon, *Polymer* 2002; 43: 4365.
110. A. Ranade, N. A. D'Souza, and B. Gnade, Polyamide-imide nanocomposites for wire coatings in polyimides and other high temperature polymers: synthesis and characterization, VSP (Boston), 2005, p. 283.
111. P. Butzloff, Master's thesis, University of North Texas, December 2000.
112. P. Butzloff and N. A. D'Souza, *Antec 2000,* p. 1527.
113. P. Butzloff and N. A. D'Souza, *Antec 2000,* p. 1531.
114. P. Butzloff, N. A. D'Souza, T. D. Golden, and D. Garrett, *Polym. Eng. Sci.,* in press.
115. L. L. Schramm, ed., *Suspensions: Fundamentals and Applications in the Petroleum Industry,* American Chemical Society, Washington DC, 1996.
116. D. H. Weinkauf and D. R. Paul, *J. Polym. Sci. B Polym. Phys.* 1992; 30: 837.
117. A. J. Hill, S. Weinhold, G. M. Stack, and M. Tant, *Eur. Polym.* 1996; 32: 843.
118. N. Qureshi, E. V. Stepanov, D. Schiraldi, A. Hiltner, and E. Baer, *J. Polym. Sci. B Polym. Phys.* 2000; 38: 1679.
119. E. A. McGonigle, J. J. Liggat, R. A. Pethrick, S. D. Jenkins, J. H. Daly, and D. Hayward, *Polymer* 2001; 42: 2413.
120. D. J. Seklik, E. V. Stepanov, S. Nazarekno, D. Schiraldi, and E. Baer, *J. Polym. Sci. B Polym. Phys.,* 1999; 37: 847.
121. S. Eichler and J. Miltz, *J. Appl. Polym. Sci.* 1993; 50: 2095.
122. S. W. Lasoski and W. H. Cobbs, *J. Polym. Sci.* 1959; 36: 21.
123. L. Tak, J. Britt, and M. Tung, *J. Appl. Polym. Sci.* 1999; 71: 197.
124. M. A. Nobile, G. G. Buonocore, L. Palmieri, A. Aldi, and D. Acierno, *J. Food Eng.* 2002; 53: 287.
125. A. Baiano and M. A. Nobile, *J. Food Eng.* 2004; 30: 30.
126. M. Vert and F. Chabot, *Makromol. Chem. Supplement,* 1981; 5:30
127. N. Ogata, G. Jimenez, H. Kawai, and T. Ogihara, *J. Polym. Sci. Part B Poly. Phys.* 1997; 35: 389.
128. S. K. Goel and E. J. Beckman, *Polym. Eng. Sci.* 1994; 34 (14): 1137.

129. S. K. Goel and E. J. Beckman, *Polym. Eng. Sci.* 1994; 34 (14): 1148.
130. S. G. Kazarian, *Polym. Sci. Ser. C.* 2000; 42(1): 78.
131. M.-T. Liang and C.-M. Wang, *Ind. Eng. Chem. Res.* 2000; 39(12): 4622.
132. W. Strauss and N. A. D'Souza, *J. Cell. Plast.,* 2004; 40: 229.
133. V. Kumar and N. P. Suh, *Polym. Eng. Sci.* 1990; 30: 1323.
134. Q. Huali, Z. Chungui, Z. Shimin, C. Guangming, and Y. Mingshu, *Polym. Degrad. Stabil*, 2003; 81: 497.
135. J. M. Sloan, P. Patterson, and Hsieh, *Polym. Mater. Sci. Eng.* 2003; 88: 354.
136. R. D. David, J. W. Gilman, and D. L. VanderHart, *Polym. Degrad. Stabil.* 2003; 79: 111.

INDEX

Handbook of Plastic Processes Edited by Charles A. Harper

Printed in the USA/Agawam, MA
December 21, 2015

627448.001